# Universitext



[illegible]

[illegible]

[illegible]

# Universitext



*Universitext* is a series of textbooks that presents material from a wide variety of mathematical disciplines at master's level and beyond. The books, often well class-tested by their author, may have an informal, personal, even experimental approach to their subject matter. Some of the most successful and established books in the series have evolved through several editions, always following the evolution of teaching curricula, into very polished texts.

Thus as research topics trickle down into graduate-level teaching, first textbooks written for new, cutting-edge courses may make their way into *Universitext*.

More information about this series at http://www.springer.com/series/223

Roger Godement

# Analysis IV

## Integration and Spectral Theory, Harmonic Analysis, the Garden of Modular Delights

Translated by Urmie Ray

Roger Godement
Paris, France

Translator: Urmie Ray

Translation from the French language edition:
Analyse mathematique IV by Roger Godement

Springer-Verlag GmbH Berlin Heidelberg is part of Springer Science+Business Media


ISSN 0172-5939 ISSN 2191-6675 (electronic)
Universitext
ISBN 978-3-319-16906-4 ISBN 978-3-319-16907-1 (eBook)
DOI 10.1007/978-3-319-16907-1

Library of Congress Control Number: 2015935238

Springer Cham Heidelberg New York Dordrecht London


Printed on acid-free paper

Springer International Publishing AG Switzerland is part of Springer Science+Business Media (www.springer.com)

# Table of Contents of Volume IV


**XI – Integration and Fourier Transform** .......... 1
§ 1. *The Upper Integral of a Positive Function* .......... 5
1 – Integral of an lsc Function .......... 5
*(i)* Positive measures .......... 5
*(ii)* Dini's Theorem .......... 6
*(iii)* Integral of an lsc function .......... 8
2 – Upper Integral of a Positive Function. Null Sets, Reasonable Sets .......... 10
*(i)* Upper integrals .......... 10
*(ii)* Null sets .......... 13
*(iii)* Reasonable sets and functions .......... 14
3 – $F^p$ Spaces .......... 15
*(i)* Definition of $F^p$ spaces .......... 15
*(ii)* Convergence in mean and almost everywhere .......... 17
§ 2. *$L^p$ Spaces* .......... 20
4 – Integrable Functions, $L^p$ Spaces .......... 20
*(i)* Integral of an integrable function .......... 20
*(ii) $L^p$ spaces*; the Riesz-Fischer theorem .......... 23
*(iii)* The case of lsc or usc functions .......... 26
5 – Lebesgue's Theorems .......... 27
*(i)* The dominated convergence theorem .......... 27
*(ii)* Relation between $L^p$ and $L^1$ ; Hölder's inequality .......... 30
*(iii)* Applications to Fourier transforms on $\mathbb{R}$ .......... 34
§ 3. *Measurable Sets and Functions* .......... 37
6 – Measurable and Integrable Sets .......... 37
*(i)* Properties of integrable sets .......... 37
*(ii)* Measurable sets .......... 39
7 – Measurable Functions .......... 41
*(i)* Separable spaces .......... 41
*(ii)* Measurable maps .......... 42
8 – Measurability and Continuity .......... 45
*(i)* Egorov's and Lusin's theorems .......... 45
*(ii)* Lusin-measurable functions .......... 49

9 – Measurability and Integrability .......... 50
§ 4. *Lebesgue-Fubini's Way* .......... 53
10 – The Lebesgue-Fubini Theorem (LF) .......... 53
*(i)* Product of measures .......... 53
*(ii)* The Lebesgue-Fubini theorem .......... 54
*(iii)* Additions to the LF theorem .......... 57
*(iv)* The Fourier inversion formula .......... 61
11 – A Topological Interlude: Polish Spaces .......... 61
*(i)* Polish spaces .......... 61
*(ii)* Lsc functions on a locally compact Polish space .......... 67
*(iii)* Borel sets in a Polish space .......... 68
12 – Continuous Sums of Measures: Examples .......... 70
*(i)* Product measures .......... 70
*(ii)* Measures induced by locally integrable densities .......... 70
*(iii)* Image of a measure under a map .......... 71
*(iv)* Quotient of an invariant measure .......... 72
13 – Integrable Functions with respect to a Continuous Sum .......... 74
*(i)* The case of lsc functions .......... 74
*(ii)* The generalized Lebesgue-Fubini theorem .......... 76
14 – Integrable Functions with respect to the Image of a Measure. 79
15 – Invariant Measures under Group Actions .......... 84
*(i)* Invariant measures on a group .......... 84
*(ii)* Continuous linear representations .......... 86
*(iii)* Quotient of a space by a group .......... 89
*(iv)* Quotient of an invariant measure .......... 91
*(v)* An example: the orthogonal group on $\mathbb{R}^n$ .......... 94
*(vi)* The case of homogeneous spaces .......... 96
*(vii)* The case of discrete groups .......... 98
§ 5. *The Lebesgue-Nikodym Theorem* .......... 103
16 – Measures With Respect To a Base Measure $\lambda$: Integrable Functions .......... 103
17 – The Lebesgue-Nikodym Theorem (LN) .......... 107
*(i)* Characterization of absolutely continuous measures .......... 107
*(ii)* Application to complex measures .......... 108
*(iii)* Lebesgue decomposition .......... 112
18 – Continuous Linear Functionals on $L^p$. The $L^\infty$ Space .......... 114
§ 6. *Spectral Decomposition on a Hilbert Space* .......... 118
19 – Operators on a Hilbert Space .......... 118
*(i)* Definitions, continuous linear functionals .......... 118
*(ii)* Orthonormal bases .......... 121
*(iii)* Adjoints, Hermitian operators .......... 122
*(iv)* Spectrum of a Hermitian operator .......... 124
*(v)* Weak topology .......... 126
*(vi)* Hilbert-Schmidt operators .......... 128
*(vii)* Von Neumann algebras .......... 130

20 – Gelfand's Theorems on Normed Algebras .......... 134
21 – A Characterization of Algebras of Continuous Functions .... 141
22 – Spectral Decompositions .......... 143
*(i)* The GN algebra of a normal operator .......... 143
*(ii)* Spectral measure of an operator algebra .......... 145
*(iii)* Integration with respect to a spectral measure .......... 146
*(iv)* Spectral decomposition of a normal operator .......... 151
23 – Self-Adjoint Operators .......... 153
*(i)* Inverse of an injective Hermitian operator .......... 153
*(ii)* Canonical extension of a positive symmetric operator ..... 156
24 – Continuous Sum Decomposition .......... 162
*(i)* Virtual eigenvectors .......... 162
*(ii)* Continuous sums of Hilbert spaces .......... 167
*(iii)* The $L^2$ space of the integral of a measure .......... 170
§ 7. *The Commutative Fourier Transform* .......... 173
25 – Convolution Product on a Lcg .......... 173
*(i)* Convolutions and representations .......... 173
*(ii)* Convolution of two measures .......... 175
*(iii)* Convolution of a measure and a function .......... 178
*(iv)* Convolution of two functions .......... 181
*(v)* Dirac sequences .......... 183
26 – Fourier Transform on $L^1(G)$ .......... 185
*(i)* Characters of a commutative lcg .......... 185
*(ii)* The topology on the dual group .......... 188
*(iii)* The canonical homomorphism $G \longrightarrow \widehat{\widehat{G}}$ .......... 191
27 – Fourier Transform on $L^2(G)$ .......... 192
*(i)* The algebra $\mathbf{A}(G)$ and its characters .......... 192
*(ii)* Spectral decomposition of the regular representation ...... 195
*(iii)* The invariant measure on the dual .......... 196
*(iv)* Fourier inversion formula and biduality .......... 201
§ 8. *Unitary Representations of Locally Compact Groups* .......... 204
28 – Further Representation Theory .......... 204
29 – Fourier Transform on a Compact Group .......... 207
*(i)* Irreducible representations of central groups .......... 207
*(ii)* Central functions on a compact group .......... 209
*(iii)* Spectral decomposition of $\mathbf{Z}(G)$ .......... 214
*(iv)* Characters of $\mathbf{Z}(G)$ and irreducible representations ...... 217
*(v)* Easy generalizations .......... 220
30 – Measures and Functions of Positive Type .......... 224
*(i)* Measures of positive type .......... 224
*(ii)* Case of a commutative group .......... 226
*(iii)* Functions of positive type .......... 227
31 – Quasi-Regular Representations of a Unimodular Group ..... 233
*(i)* Central measures of positive type .......... 233
*(ii)* The commutation theorem .......... 235

*(iii)* Traces on a Hilbert algebra ... 241
*(iv)* Case of a commutative group ... 246
*(v)* Characters of a locally compact group ... 247
*(vi)* Characters of type (I) ... 249
32 – Discrete Components of the Regular Representation ... 251

**XII –The Garden of Modular Delights or The Opium of Mathematicians** ... 261
§ 1. *Infinite Series and Products in Number Theory* ... 261
1 – The Mellin Transform of a Fourier Transform ... 261
2 – The Functional Equation of the $\zeta$ Function ... 266
3 – Weil's method for the Function $\eta(z)$ ... 273
§ 2. *The series* $\sum 1/\cos\pi nz$ *and* $\sum \exp\left(\pi i n^2 z\right)$ ... 281
4 – The series $\sum 1/\cos\pi nz$ ... 281
5 – The Identity $\sum 1/\cos\pi nz = \theta(z)^2$ ... 283
*(i)* The fundamental domain of $\Gamma(\theta)$ ... 283
*(ii)* A general method ... 285
*(iii)* The identity $f(z)/\theta(z)^2 = 1$ ... 286
6 – The Infinite Product of the Function $\theta(u,z)$ ... 288
7 – The Reciprocity Law for Gauss Sums ... 291
*(i)* Cauchy's method ... 291
*(ii)* The Dirichlet method ... 294
*(iii)* The quadratic reciprocity law ... 295
§ 3. *The Dirichlet Series* $L(s;\chi)$ ... 298
8 – The Functional Equation of $\eta(z)$: bis ... 298
9 – Arithmetic Interlude ... 300
*(i)* Quotient rings ... 300
*(ii)* The groups $G(m)$ ; characters $\bmod m$ ... 302
*(iii)* Orthogonality relations ... 304
*(iv)* Gauss sums ... 306
*(v)* Case of the unit character ... 308
10 – The Series $\theta_f(x;\chi)$ and $L(s;\chi)$ ... 311
*(i)* Functional equation of $\theta_f(x;\chi)$ ... 311
*(ii)* The series $L(s,\chi)$ ... 312
§ 4. *Elliptic Functions* ... 316
11 – Liouville's Theorems ... 316
12 – Elliptic Functions and Theta Series ... 318
*(i)* Abel's theorem ... 318
*(ii)* General theta functions ... 321
*(iii)* Metamorphoses of the Jacobi series ... 323
13 – Eisenstein and Weierstrass's Point of View ... 328
*(i)* Convergence of Eisenstein series ... 328
*(ii)* The Weierstrass $\wp$-function ... 329
*(iii)* The series $\sum \pi^2/\sin^2\pi(u+nz)$ and $G_2(z)$ ... 333

*(iv)* Relation between $\wp$ and $\theta_1$ functions ... 335
*(v)* Elliptic functions with given simple poles ... 338
*(vi)* The functions $\zeta_L$ and $\sigma_L$ ... 340
14 –Elliptic Integrals ... 341
*(i)* The field of elliptic functions ... 341
*(ii)* The Riemann surface of the field of elliptic functions ... 342
*(iii)* Addition formula ... 345
§ 5. *$SL_2(\mathbb{R})$ as a Locally Compact Group* ... 349
15 – Subgroups, Invariant Measure ... 349
*(i)* Actions of $SL_2(\mathbb{R})$ on the half-plane ... 349
*(ii)* Automorphic forms as functions on $G$ ... 349
*(iii)* Subgroups of $SL_2$ ... 353
*(iv)* Fixed points and eigenvalues ... 355
*(v)* Invariant measure ... 356
*(vi)* The point of view of the unit disc ... 357
16 – The Discrete Series of Representations of $SL_2(\mathbb{R})$ ... 359
*(i)* Integrable holomorphic functions on the half-plane ... 360
*(ii)* The spaces $\mathcal{H}_r^p$ of the unit disc ... 364
*(iii)* A theorem of Paley-Wiener type for $\mathcal{H}_r^2(P)$ ... 366
*(iv)* The kernel function of $\mathcal{H}_r^2(P)$ ... 368
*(v)* The holomorphic discrete series of irreducible representations of $SL_2(\mathbb{R})$ ... 370
*(vi)* Solutions of the equation $f * \omega_r = f$ ... 372
§ 6. *Modular Functions: The Classical Theory* ... 376
17 – Fundamental Domain, Modular Forms ... 376
*(i)* Generators of the modular group ... 376
*(ii)* Fundamental domain ... 376
*(iii)* The classical definition of modular forms ... 378
*(iv)* Eisenstein and Poincaré series ... 380
18 – Analogues of Liouville's Two Theorems ... 385
*(i)* The Riemann surface of $SL_2(\mathbb{Z})$ ... 385
*(ii)* Zeros and poles ... 388
*(iii)* Construction of modular forms using $\Delta(z)$ and Eisenstein series ... 390
*(iv)* Application to elliptic functions ... 391
§ 7. *Fuchsian Groups* ... 394
19 – Generalities on Automorphic Forms ... 394
*(i)* Two lemmas on discrete subgroups ... 394
*(ii)* Generalities on automorphic forms ... 397
*(iii)* The topology of horocycles. The Riemann surface of $\Gamma$ ... 401
*(iv)* Fuchsian groups ... 404
20 – Parabolic Forms and Representations of $G$ ... 408
21 – Poincaré, Eisenstein and Maaß-Selberg Series ... 413
*(i)* Poincaré series ... 413
*(ii)* Poincaré-Eisenstein series ... 419

*(iii)* Eisenstein series .... 422
22 – Fourier Series Expansions .... 427
*(i)* General method .... 427
*(ii)* The case of Poincaré-Eisenstein series .... 430
*(iii)* The case of Maaß-Selberg forms ; analytic extensions .... 434
§ 8. *Hecke Theory* .... 441
23 – Modular Forms and Dirichlet Series .... 441
*(i)* Hecke series .... 441
*(ii)* Weil series .... 442
*(iii)* Generalization to non-holomorphic forms .... 447
24 – Hecke Operators .... 450
*(i)* Operators $T(x)$ for an abstract group .... 450
*(ii)* Operators $T(x)$ for a locally compact group .... 453
*(iii)* Operators $T(x)$ for the modular group .... 455
*(iv)* Operators $T(p)$: the case of functions on $\Gamma\backslash G$ .... 456
*(v)* Eigenfunctions of Hecke operators .... 458
*(vi)* Applications to modular forms .... 460
§ 9. $SL_2(\mathbb{R})$ *as a Lie Group* .... 463
25 – Lie Groups .... 463
*(i)* Definition and examples .... 463
*(ii)* Operations on tangent vectors .... 464
*(iii)* Differentiation and invariant vector fields .... 465
*(iv)* Canonical coordinates .... 467
*(v)* The Lie algebra of a group .... 469
*(vi)* Lie algebras of classical groups .... 471
*(vii)* Invariant distributions and differential operators .... 471
26 – Lie Algebras in Infinite-Dimension .... 476
*(i)* The subspace $\mathcal{H}^\infty$ .... 476
*(ii)* Weak differentiability and strong differentiability .... 478
*(iii)* Convolution operators on $\mathcal{H}^\infty$ .... 480
*(iv)* The Dixmier-Malliavin theorem .... 481
*(v)* Analytic vectors .... 484
*(vi)* The case of unitary representations .... 485
27 – Differential Operators on $SL_2(\mathbb{R})$ .... 488
*(i)* The Lie algebra of $SL_2(\mathbb{R})$ .... 488
*(ii)* Differential operators on the half-plane .... 490
28 – The Representation of $\mathfrak{g}$ Associated to a Holomorphic Function .... 494
*(i)* The $\mathfrak{g}$-module $HC_r(f) = HC(f_r)$ .... 494
*(ii)* The case $r = -p \leq 0$ .... 495
*(iii)* Finite-dimensional simple $\mathfrak{g}$-modules .... 496
*(iv)* Condition for $\dim HC_r(f) < +\infty$ .... 496
*(v)* A theorem of Maaß .... 497
29 – Irreducible Representations of $\mathfrak{g}$ .... 498
*(i)* Classification .... 498
*(ii)* Function space models for representations of $\mathfrak{g}$ .... 500

30 – Irreducible Representations of $G$ .......................... 502
*(i)* The multiplicity one theorem .......................... 502
*(ii)* Function space models for $G$: the discrete series .......... 504
*(iii)* Function space models for $G$: the even principal series .... 505

**Index** .......................................................... 511

**Table of Contents of Volume I** ................................ 517

**Table of Contents of Volume II** ............................... 521

**Table of Contents of Volume III** .............................. 525

# XI – Integration and Fourier Transform

*§ 1. The Upper Integral of a Positive Function – § 2. $L^p$ Spaces – § 3. Measurable Sets and Functions – § 4. Lebesgue-Fubini's Way – § 5. The Lebesgue-Nikodym Theorem – § 6. Spectral Decompositions in a Hilbert Space – § 7. The Commutative Fourier Transform – § 8. Unitary Representations of Locally Compact Groups*

Bourbaki has sometimes been blamed for giving priority to function integrals instead of first defining measures on sets like everyone and like Borel and Lebesgue. From a historical viewpoint, it should be borne in mind that since no later than Leibniz, an integral is, strictly speaking, not the measure of a surface area: an area is always positive, whereas in all classical definitions of an integral, areas above the $x$-axis are counted positively while those below the $x$-axis are counted negatively. Why? Naturally, in order to ensure that the *linearity* formula

$$\int \alpha f + \beta g = \alpha \int f + \beta \int g$$

holds, a formula made obvious by Leibniz's notation and without which the whole machinery of *Calculus* would fall apart. In fact, from Newton to Borel and Lebesgue, everyone, even Cauchy and Dirichlet, integrated functions assuming this does not present any problems, and from Riemann to our times, endeavouring to extend the definition to more and more general functions. Before Cauchy, integration was considered either as the inverse operation of differentiation, especially as this is our only chance of computing an integral explicitly, or in Euler and Fourier's work, as a method for proving countless identities, for example by integrating series of functions term by term – a central problem in Lebesgue theory. By the time we get to Riemann, the area alluded to had virtually disappeared from the definition of the integral, becoming merely an intuitive geometric interpretation of the latter.

Soon afterwards, others – Cantor, Peano, Jordan – attempted to define the measure $m(A)$ of a subset $A$ of the line or the plane by approximating it by finite unions of intervals or slabs. They thus obtained a category of sets for which the measure of a finite union of pairwise disjoint sets is the sum of the measures of the individual sets. Definitions of multiple integrals extended to simple sets could thereby be made more rigorous, but over the line, stay within the framework established by Riemann. The first decisive step is due to Borel who, in 1898, defined the measure of an open subset of the line to be the sum of the series of the lengths of its constituent intervals (connected components), introduced "Borel" sets and assigned them a measure (the infimum of the measures over all open sets containing them). With this definition, not only the measure of a finite union but of a *countable* union of pairwise disjoint sets is the sum of the measures of the individual sets. In his thesis in 1902, Lebesgue, who was primarily trying to extend the classical relation between integration and differentiation to very general functions, expanded the definition to *measurable* sets, i.e. to unions of a Borel set and of a set of measure zero. This decisive step enabled him to obtain limit theorems, now named after him, which he presented in a famous book in 1905.

In 1894, Stieltjes had defined the notion of a far more general integral over the line – a notion, however, not any "deeper" than Riemann's (Chap. V, n° 32) as it applies to the same functions – from which the notion of surface area had totally disappeared. It was founded on an easy generalization of the measure of an *interval*; any physicist familiar with mass distribution could have invented it. In 1909, Frédéric Riesz proved that any continuous linear functional on the space of continuous functions on a compact interval $K$ is the same as a Stieltjes integral; this time, the *function* integral is at the centre of the construction. Riesz's theorem together with the equally fundamental Riesz-Fischer theorem (1907) established the theory in the almost algebraic framework of linear functional analysis; it is even one of its main starting points. The rest of the story is simply a matter of generalizations to the space $\mathbb{R}^n$ (Radon, 1913) or to "abstract" measures, i.e. defined on sets without a topology (Fréchet 1915 and especially Caratheodory in a 1918 book). Around the same period, W. H. Young (1911) and P. J. Daniell (1918) showed how to define, in the framework of the usual measure, Lebesgue integrable functions from the integral of continuous functions *without* resorting to measures on sets. Their method did not meet with success before being systematically developed by N. Bourbaki in his book on this topic.

When he came up with his theory, Lebesgue had no idea of what a functional space or a vector space was. These concepts progressively appeared during the following years, when the theory of integral equations and Hilbert spaces, on the one hand, and "abstract" and "modern" algebra, on the other, were developed. Lebesgue was too focused on "fine" analysis arguments about functions of real variables and trigonometric series to convert to the new fields that were being developed before his very eyes. This is also true of many of those who presented the subject before Bourbaki, a polycephalous author familiar with as much algebra as analysis.

Therefore, those who blamed Bourbaki for having discarded the point of view of Borel, Lebesgue and others just because it is more natural (why?) to try and measure sets than to integrate functions, are mistaken, all the more so as in general, with Bourbaki, theorems related to the measure of sets, that were obviously not forgotten, follow directly from properties of integrals. Besides, all that is needed is opening randomly some analysis books and articles to see that, with few exceptions, they contain countless integrals and very little measures of sets. There are no sacred texts in mathematics and every day theories are developed using methods radically different from those of their creators without anyone protesting.

The most serious objections to the point of view presented here came from probabilists who, following Kolmogorov (1933), have interpreted a probability as a measure on a set without a topology. This point of view, developed in the first drafts of N. Bourbaki's book, had led to a presentation, the *Diplodocus*, whose complexity and ugliness ended up disgusting all members of the group. I joined it right when the decision was taken to stick to the view subsequently published. If I remember correctly I was instructed to prepare the first detailed drafts of chapters III and V and the two first ones of chapter IV in this new perspective. So I am in a fairly good position to comparer both methods. Moreover in general, almost all authors taking the "abstract" view refrain – prudently? – from dealing with questions studied in detail by N. Bourbaki: measures defined by a density, image of a measure, integration of measures, quotient measures, etc.

In fact, Radon measures are enough for almost all purposes in Analysis and in most cases, are better adapted than "abstract" measures. As used to say not so long ago a specialist of partial differential equations, « *all I ask of integration theory is to provide me with Banach spaces* » . The definition of a measure as a linear functional on continuous functions is perfectly adapted to this point of view since Bourbaki's goal is the construction of $L^p$ Banach spaces. By contrast, Rudin's book, *Real and Complex Analysis* (1966), which does not refer to those of N. Bourbaki though they were published ten years earlier, is convincing in this respect: after the "compulsory" chapter on abstract measures, its author is forced to make the connection with the topological viewpoint and prove F. Riesz's theorem. Thus he is forced to start by defining the measure of an open *set* using a method which, in N. Bourbaki, defines more easily (because it is linear) the integral of any lower semi-continuous *function*; after which abstract measures disappear when he applies the theory to analytic functions and Fourier transforms. The book by P. Malliavin and H. Airault, *Intégration, analyse de Fourier, probabilités, analyse gaussienne* (Masson, 1994) or *Integration and Probability* (Springer, 1995), also starts with a presentation about abstract measures, followed by an equally long chapter on Radon measures, and then by an excellent chapter on Fourier transforms where abstract measures are not involved. The latter only occur (almost always together with Radon measures) in the two chapters on probability theory. Not only can their usefulness be seen in this framework, but also the conceptual difficulties and the complicated notations resulting from the simultaneous consideration of measures defined on varying "tribes" – all this in a textbook meant for students who probably have not yet mastered the fundamental classical theorems.

Finally, from a pedagogical viewpoint, it is sufficiently clear that someone who has taken in N. Bourbaki's viewpoint will not have any serious difficulty in understanding, if necessary, abstract measures and more recent developments beyond the framework of locally compact spaces.

Therefore, like Dieudonné's presentation in his *Eléments d'analyse*, my presentation closely follows that of N. Bourbaki, probably for the same reasons as him: we have both spent quite some time developing it, and have not yet come across anything better or more complete if we keep to the framework of locally compact spaces, and at the very least, this framework covers 95% of the needs of analysts.

We warn the reader learning about this topic that this chapter includes several parts of unequal importance. Lebesgue's fundamental theorems, without which nothing can be accomplished, are the subject matter of §§ 1 to 3 and of the beginning (Lebesgue-Fubini) of § 4. The rest of § 4 shows how using Lebesgue-Fubini's arguments, generalized to "continuous sums" of measures, it is possible to unify apparently dissimilar developments: integration with respect to the image of a measure, or with respect to a measure defined by a density, group invariant measures and quotient measures, etc. § 5 characterizes integrable functions with respect to a measure defined by a density, a result that could have been inserted in § 4, and proves the Lebesgue-Nikodym theorem. Although §§ 4 and 5 contain important results, mainly theorem 25 about measures defined by a density (constantly used, usually without realizing it as it seems so natural...) and the Lebesgue-Nikodym theorem, these §§ are less universally useful than §§ 1 to 3. Like the countless exercises found in Dieudonné and N. Bourbaki, they will help the reader to understand these fundamental theorems since it is in general hard to master them without using them again and again: as Niels Bohr once said about quantum physics, in this field, « *the experts are those who have already made all possible mistakes* ». He should have added: once and only once, but in his case this was self-evident...

The same can be said about §§ 6 and 7 on spectral theory and Fourier transforms on commutative groups: these topics are perfect illustrations of the previous §§ (and as far as I am concerned, mutually influenced each other at the end of the 1940s). § 7 is not a mere gratuitous generalization of classical Fourier transforms: John Tate's thesis in the 1950s on zeta functions of algebraic number fields, where these results are first used, also gives the first proof of possible simplifications and clarifications of classical results introduced by general Fourier transforms and the "adelic" viewpoint. The latter was later extended to noncommutative groups in the theory of automorphic functions. Finally, § 8 is an introduction to "integral" methods applicable to general locally compact groups.

# §1. The Upper Integral of a Positive Function

## 1 – Integral of an lsc Function

Let $X$ be a locally compact space, i.e. a topological space satisfying the Hausdorff separation axiom and in which all points have a compact neighbourhood. $L(X)$ will denote the vector space of continuous functions $f : X \longrightarrow \mathbb{C}$ with compact support.[1] It is a union of subspaces $L(X, K)$ where, for any compact set $K \subset X$, $L(X, K)$ denotes the set of $f \in L(X)$ vanishing outside $K$. Let $L_{\mathbb{R}}(X)$ (resp. $L_{\mathbb{R}}(X, K)$) be the set of real-valued functions on these spaces and, likewise, $L_+(X)$ (resp. $L_+(X, K)$) be the set of positive-valued functions. The subspaces $L(X, K)$ are complete with respect to the norm

$$\|f\|_X = \sup |f(x)|$$

of uniform convergence on $X$, which we shall denote by $\|f\|$ when no ambiguity is possible. However, $L(X)$ itself may not be so if $X$ is not compact.

As shown for metrizable spaces in Chap. IX, n° 17 – but this assumption only helps to simplify the proof[2] –, there are enough functions in $L(X)$ for the construction of *partitions of unity*: if $A \subset X$ is compact and $(U_i)_{1 \leq i \leq n}$ a finite open cover of $A$, then there are $f_i \in L_+(X)$ such that $\sum f_i(x) = 1$ on $A$ and whose supports are contained in the sets $U_i$.

(i) *Positive measures.* A *measure* on $X$ is a linear functional

$$\mu : L(X) \longrightarrow \mathbb{C}$$

whose restriction to each $L(X, K)$ is continuous with respect to the topology of uniform convergence; this amounts to attributing to every compact set $K \subset X$ a constant $M(K)$ such that

$$f \in L(X, K) \Longrightarrow |\mu(f)| \leq M(K)\|f\| \,.$$

$\mu$ will be said to be real if $\mu(f)$ is real for $f$ real, and to be *positive* if $\mu(f) \geq 0$ for all $f \geq 0$. Examples of such measures were given in Chap. V, §9. If $X$ is an oriented differential manifold and $\omega$ a differential form of maximum degree

[1] The support of a function is the smallest closed set outside which the function is zero.

[2] See N. Bourbaki, *Topologie générale*, Chap. X, in particular §4 on normal spaces, i.e separated and satisfying the following condition: if $A$ and $B$ are two disjoint closed sets, there exist open disjoint sets $U \supset A$ and $V \supset B$. As shown by Urysohn, this condition means that there is a continuous map from $X$ to $[0, 1]$ equal to 1 on $A$ and 0 on $B$, or that any continuous map from a closed set $F \subset X$ to $[0, 1]$ extends to $X$. A locally compact space is not necessarily normal, but every compact subspace of $X$ is. Thus, in this case, Urysohn's theorem continues to hold if $A$ is supposed to be compact and $B$ to be the complement of a compact neighbourhood of $A$.

on $X$, the support of the form $f\omega$ is compact for all $f \in L(X)$. Thus, setting $\mu(f) = \int_X f\omega$ (Chap. IX, n° 17), we get a measure on $X$.

As we will see, Leibniz's notation

$$\mu(f) = \int_X f(x)d\mu(x)\,,$$

essential in applications, is hardly needed in most of the theory.

For positive measures, continuity follows in fact from positivity thanks to the next lemmas:

**Lemma 1.** *Let $\mu$ be a positive linear functional on $L(X)$. Then*

$$|\mu(f)| \leq \mu(|f|) \quad \text{for all } f \in L(X)\,. \tag{1.1}$$

This is obvious if $f$ is real since $f = f^+ - f^-$, $|f| = f^+ + f^-$. If $f = g + ih$ is complex-valued, multiplying $f$ by a complex factor with absolute value 1, $\mu(f)$ may be assumed to be real and $> 0$. Then $\mu(h) = 0$ and $|\mu(f)| = \mu(f) = \mu(g) \leq \mu(|f|)$ since $g \leq |g| \leq |f|$.

**Lemma 2.** *For any compact set $K \subset X$, there is a constant $M_K(\mu)$ such that*

$$|\mu(f)| \leq M_K(\mu)\|f\| \quad \text{for all } f \in L(X, K)\,. \tag{1.2}$$

Indeed, there is a positive-valued continuous function $p$ on $X$ equal to 1 on $K$ and zero outside a compact set $K' \supset K$. Then $|f(x)| \leq \|f\|p(x)$ for all $f \in L(X, K)$, and so

$$|\mu(f)| \leq \mu\,(|f|) \leq \mu\,(\|f\|p) = \|f\|\mu(p)\,,$$

qed.

Every measure is obviously of the form $\mu' + i\mu''$ where $\mu'$ and $\mu''$ are real. Though elementary, it is slightly less obvious that every real measure is the difference of positive measures. We will prove this in n° 17 (theorem 29) when we will need it: integration theory is mostly about positive measures, and in what follows, we will almost always refrain from mentioning that the measures considered are.

(ii) *Dini's Theorem.* Let us start with a locally compact space $X$ and a (positive!) measure $\mu$ on $X$. As already stated in Chap. V, the first step towards Lebesgue theory is to define $\mu(\varphi)$ for every lower semi-continuous function $\varphi$ with values in $[0, +\infty]$.

We start by recalling (Chap. V, § 2, n° 10) that an *increasing philtre* is a non-empty set $\Phi$ of functions defined on $X$, with values in $]-\infty, +\infty]$ and satisfying the following condition: For all $f, g \in \Phi$, there exists $h \in \Phi$ such that $\sup(f, g) \leq h$. If there are finitely many functions $f_i \in \Phi$, then there exists $g \in \Phi$ such that $g \geq f_i$ for all $i$.

Increasing philtres generalize increasing sequences of functions and enables us to generalize the notion of limit. Given a function $I : \Phi \longrightarrow \mathbb{C}$ and some $a \in \mathbb{C}$, we will write

$$\lim_{\Phi} I(f) = a$$

if, for any *number* $r > 0$, there is a *function* $r' \in \Phi$ such that

$$f \in \Phi \quad \& \quad f \geq r' \Longrightarrow |I(f) - a| < r\,.$$

The reader will easily be able to extend the elementary properties of limits to this general framework: the function $f$ plays the role of the traditional index $n \in \mathbb{N}$ and $I(f)$ replaces $u_n$; the fact that neither $f \leq g$ nor $g \leq f$ in general ( i.e. no "total" order), contrary to what happens in $\mathbb{N}$, makes no difference.

In practice, we mostly need to consider functions $I$ with values in $]-\infty, +\infty]$ and *increasing*, i.e. such that $f \leq g$ implies $I(f) \leq I(g)$. Then

$$\lim_{\Phi} I(f) = \sup_{f \in \Phi} I(f)\,,$$

which is readily explained by the case of an increasing sequence, and by convention, the left hand side is defined using this relation when the right hand side is equal to $+\infty$. If it is written $a$, then the following properties hold:

(i) $I(f) \leq a$ for all $f \in \Phi$,

(ii) for all $M < a$, there exists $r' \in \Phi$ such that $I(r') > M$, in which case

$$f \geq r' \text{ implies } M \leq I(f) \leq a\,.$$

Choosing $I(f) = f(x)$ for a given $x \in X$, this give the definition of a function

$$x \longmapsto \sup_{f \in \Phi} f(x) \leq +\infty\,,$$

which is just the *upper envelope* of the $f \in \Phi$, written $\sup(\Phi)$.

**Lemma 3 (Dini's Theorem).** *Let $\Phi \subset L_{\mathbb{R}}(X)$ be an increasing philtre $\varphi = \sup(\Phi)$. Suppose that $\varphi \in L(X)$. Then*

$$\lim_{\Phi} \|\varphi - f\|_X = 0 \quad \text{and} \quad \mu(\varphi) = \lim_{\Phi} \mu(f) = \sup_{\Phi} \mu(f)\,. \tag{1.3}$$

This result says that $\Phi$ converges uniformly on $X$ if its limit is continuous and with compact support.

The proof is the same as in Chap. V, n° 10. We first suppose that $\Phi \subset L_+(X)$. Let $r$ be a number $> 0$ and $K$ the support of $\varphi$. As $\varphi(x) = \sup f(x)$ is everywhere $< +\infty$, for all $a \in K$, there exists $f \in \Phi$ such that $f(a) > \varphi(a) - r$. This relation continues to hold in an open neighbourhood $V(a)$ of $a$ since $f$

and $\varphi$ are continuous. As $K$ is compact, there are finitely many points $a_i \in K$ such that the set $V(a_i)$ cover $K$, as well as functions $f_i \in \Phi$ that are upper bounds for $\varphi - r$ in $V(a_i)$. There is a function $r' > \sup(f_i)$ in $\Phi$ since there are finitely many $f_i$, and so $\varphi(x) - r < r'(x) \leq \varphi(x)$ for all $x \in K$ and perforce for all $f \in \Phi$ such that $r' \leq f$. As these functions $f$ are positive-valued and bounded above by $\varphi$, they all vanish outside $K$. As a result,

$$r' \leq f \Longrightarrow \|\varphi - f\|_X \leq r\,,$$

which proves the first relation (3). Moreover,

$$\mu(\varphi) - \mu(f') = \mu(\varphi - f') \leq M_K(\mu)\|\varphi - f'\|_K \leq M_K(\mu)r\,,$$

proving the second one.

If the functions $f \in \Phi$ are not all positive, replace them with $f - f_0$, where $f_0 \in \Phi$ is chosen once and for all, and only consider functions $f \geq f_0$. The $f - f_0$ clearly form an increasing philtre converging to $\varphi - f_0$. The results concerning $\Phi$ readily follow.

(iii) *Integral of an lsc function.* Given a function $\varphi : X \longrightarrow ]-\infty, +\infty]$, we shall denote by $L_{\text{inf}}(\varphi)$ the set of $f \in L_{\mathbb{R}}(X)$ such that $f \leq \varphi$. It is necessary to assume that $\varphi(x) \geq 0$ outside a compact subset of $X$. Otherwise $L_{\text{inf}}(\varphi)$ would be empty. This is an increasing philtre since $f', f'' \in L_{\text{inf}}(\varphi) \Longrightarrow \sup(f', f'') \in L_{\text{inf}}(\varphi)$. The functions $\varphi : X \longrightarrow ]-\infty, +\infty]$ for which

$$\varphi(x) = \sup_{L_{\text{inf}}(\varphi)} f(x) \tag{1.4}$$

for all $x \in X$ are characterized by the condition of being positive outside a compact set and *lower semi-continuous* (lsc): for all $a \in X$ and all $M < \varphi(a)$, $\varphi(x) > M$ in the neighbourhood of $a$. The condition is necessary since, by definition of a supremum, there exists $f \in L_{\text{inf}}(\varphi)$ such that $f(x) > M$ for $x = a$, hence by continuity, also in the neighbourhood of $a$. So $\varphi(x) > M$ in this neighbourhood. Conversely, suppose that $\varphi$ lsc and positive outside a compact set $K$. As shown in Chap. V, n° 10, (vi) by generalizing the classic argument applied to continuous functions, $\varphi$ is bounded below over $K$; if $m$ is its minimum over $K$, then $\varphi(x) \geq m$ everywhere. Then there exists $g \in L_{\mathbb{R}}(X)$ such that $g \leq \varphi$ (take $g \leq 0$ everywhere and $\leq m$ over $K$), and so $\varphi = g + \varphi'$ where $\varphi'$ is lsc and positive. As $L_{\text{inf}}(\varphi)$ is the set of $g + f'$ where $f' \in L_{\text{inf}}(\varphi')$, the proof reduces to showing that $\varphi'$ is the upper envelope of functions $f' \in L_{\text{inf}}(\varphi')$. Suppose that $\varphi'(x) > M$ in an open neighbourhood $V$ of a given point $a$. Clearly, there is some $f \in L_+(X)$ zero outside $V$, bounded above by $M$ in $V$ and equal to $M$ at $a$. Then $f \in L_{\text{inf}}(\varphi')$, and the result follows.[3]

[3] In practice, an lsc function positive outside a compact set is the upper envelope of a *countable* family of continuous functions with compact support. This result for the moment unnecessary will be proved later [n° 11, (ii)].

$\mathcal{I}$ (resp. $\mathcal{I}_+$) will denote the set of lsc functions on $X$ with values in $]-\infty, +\infty]$ and positive outside a compact set (resp. everywhere). The upper envelope of a finite or infinite family of functions $f_i \in \mathcal{I}$ is also in $\mathcal{I}$. This is also the case of the lower envelope or of the sum of a *finite* family. The unordered sum (Chap. II, n° 15) of a series of positive lsc functions is also lsc since it is the upper envelope of its partial sums, all of which are lsc.

For all $\varphi \in \mathcal{I}$, define the *upper integral* of $\varphi$ by

$$\mu^*(\varphi) = \sup_{L_{\inf}(\varphi)} \mu(f) \leq +\infty\,. \tag{1.5}$$

Clearly, $\mu^*(f) = \mu(f)$ for all $f \in L_{\mathbb{R}}(X)$ and

$$\varphi \leq \psi \Longrightarrow \mu^*(\varphi) \leq \mu^*(\psi)\,.$$

**Lemma 4.** *Let $\Phi$ be an increasing philtre of continuous functions with compact support and $\varphi$ its upper envelope. Then*

$$\mu^*(\varphi) = \sup_{\Phi} \mu(f) = \lim_{\Phi} \mu(f)\,. \tag{1.6}$$

The first expression is obviously greater than the second one. So, it suffices to prove the reverse inequality. For any $g \in L_{\inf}(X)$, let $\Phi_g \subset L_{\mathbb{R}}(X)$ be the set of functions of the form $\inf(f, g)$ where $f \in \Phi$. Since, generally speaking,

$$f \geq f' \Longrightarrow \inf(f, g) \geq \inf(f', g)\,,$$

$\Phi_g$ is an increasing philtre; it converges simply to $g$. Indeed, for all $x \in X$ and $M < \varphi(x)$, there exists $f \in \Phi$ such that $f(x) > M$. As $g(x) \leq \varphi(x)$, we can take $M = g(x) - r$, where $r > 0$ is given. For $f' = \inf(f, g) \in \Phi_g$, $f'(x) \geq g(x) - r$. The result follows.

Lemma 3 applied to $\Phi_g$ then shows that

$$\mu(g) = \sup_{f \in \Phi_g} \mu\left[\inf(f, g)\right] \leq \sup_{\Phi} \mu(f)$$

since $\inf(f, g) \leq f$. As the right hand side does not dependent on the function $g \in L_{\inf}(\varphi)$, the supremum of all $\mu(g)$, namely $\mu^*(\varphi)$, is inferior to it, proving the lemma.

**Theorem 1.** *The function $\varphi \mapsto \mu^*(\varphi)$, $\varphi \in \mathcal{I}$, has the following properties:*

(i) *Additivity:*

$$\mu^*(\varphi + \psi) = \mu^*(\varphi) + \mu^*(\psi)\,;$$

(ii) *Passage to the limit for increasing sequences:*

$$\mu^*\left(\sup \varphi_n\right) = \sup \mu^*\left(\varphi_n\right)\,,$$

*and more generally*

$$\mu^*\left(\sup_\Phi \varphi\right) = \sup_\Phi \mu^*(\varphi)$$

*for any increasing filtering set $\Phi \subset \mathcal{I}$;*

(iii) *Term by term integration of all positive function series:*

$$\mu^*\left(\sum \varphi_n\right) = \sum \mu^*(\varphi_n) \leq +\infty\,.$$

The proofs are the same as in Chap. V, n° 10 and 11. To prove (i), choose two increasing philtres $\Phi$ and $\Psi$ in $L_+(X)$ converging to $\varphi$ and $\psi$ and note that the functions $f+g$, with $f \in \Phi$ and $g \in \Psi$, form an increasing philtre converging to $\varphi+\psi$. It then remains to apply lemma 4. To prove (ii), consider the set $\Psi = \bigcup L_{\text{inf}}(\varphi)$ of functions bounded above by at least one of the $\varphi \in \Phi$. That it is an increasing philtre is immediate. It converges to $\psi = \sup \varphi$ since, for all $x \in X$ and $M < \psi(x)$, there exists $\varphi \in \Phi$ such that $\varphi(x) > M$, hence also some $f \in L_{\text{inf}(\varphi)} \subset \Psi$ such that $f(x) > M$. This done, lemma 4 shows that

$$\mu^*(\psi) = \sup_\Psi \mu(f) = \sup_\Phi \sup_{L_{\text{inf}}(\varphi)} \mu(f) = \sup_\Phi \mu^*(\varphi)$$

[associativity of suprema: Chap. II, (9.9)]. Finally, (iii) is obtained by applying (ii) to the partial sums of the series.

Clearly, $\mu^*(t\varphi) = t\mu^*(\varphi)$ for all scalars $t \geq 0$, agreeing, as will always be the case, that

$$\alpha. + \infty = +\infty \quad \text{if } 0 < \alpha \leq +\infty\,, \qquad 0. + \infty = 0\,.$$

Like in Chap. V, n° 10 and 11, an entirely similar theory could be constructed for upper semi-continuous functions admitting functions of $L_\mathbb{R}(X)$ as upper bounds. It reduces to the previous case since a function $f$ is usc if and only if $-f$ is lsc.

## 2 – Upper Integral of a Positive Function. Null Sets, Reasonable Sets

(i) *Upper integrals.* Let us begin with a function $f$ on $X$ taking values in $[0, +\infty]$. There are functions $\varphi \in \mathcal{I}$ such that $\varphi \geq f$, for example the function which is everywhere equal to $+\infty$. Then define the *upper integral* of $f$ by setting

$$\mu^*(f) = \inf_{\substack{\varphi \geq f \\ \varphi \text{ lsc}}} \mu^*(\varphi)\,, \tag{2.1}$$

so that $\mu^*(f) \in [0, +\infty]$.

Despite the adopted notation – I will also sometimes write

$$\mu^*(f) = \int f(x)d\mu^*(x)$$

–, this definition is different[4] from the one given in Chap. V, n° 1, which at best has only the pedagogical merit of being simple.

If $\varphi$ is lsc, definition (1) clearly provides the same value as (1.5). Another trivial but constantly used point is that

$$f \leq g \Longrightarrow \mu^*(f) \leq \mu^*(g)\,. \tag{2.2}$$

Also,

$$\mu^*(tf) = t\mu^*(f) \quad \text{for } t \geq 0$$

provided that here too, the calculation rules for the symbol $+\infty$ set out above are respected.

In particular, we define the *outer measure* of a set $A \subset X$ by setting

$$\mu^*(A) = \mu^*(\chi_A)\,, \tag{2.3}$$

which is the upper integral of the characteristic function of $A$. Then

$$\mu^*(A) = \inf \mu^*(U)\,, \tag{2.3'}$$

where $U$ varies in the set of all open subsets containing $A$. As the functions $\chi_U$ are lsc and upper bounds of $\chi_A$, the left hand side is less than the right hand one. For the reverse inequality, there is nothing to show if $\mu^*(A) = +\infty$; otherwise, for all $r > 0$, there is a function lsc $\varphi \geq \chi_A$ such that $\mu^*(\varphi) \leq \mu^*(A) + r$; for the open set $U = \{\varphi(x) > 1\}$, $\chi_U \leq \varphi$ and so $\mu^*(U) \leq \mu^*(A) + r$. (3') follows since $A \subset U$.

The outer measure of every compact set $K \subset X$ is clearly finite: indeed, there is some $f \in L_+(X)$ which is $\geq 1$ on $K$, and so $\mu^*(K) \leq \mu(f)$ by (2).

As already seen in Chap. V, n° 11 in the case of open sets, most statements related to the measure of sets are obtained more or less trivially by applying the statements valid for arbitrary functions to their characteristic functions.

Non-trivial properties of $\mu^*(f)$ proved for lsc functions (theorem 1) can be partly extended to the general case:

---

[4] If $X$ is a compact interval in $\mathbb{R}$, then as was seen in n° 2 of Chap. V, in Riemann theory, the upper integral of the function equal to 1 on $X \cap \mathbb{Q}$ and 0 everywhere else is equal to 1. In Lebesgue theory, it is equal to 0, as we will see later for any countable set.

**Theorem 2.** *The function* $\mu^*$ *has the following properties:*
(i) *Countable convexity:*

$$\mu^*\left(\sum f_n\right) \leq \sum \mu^*(f_n) \leq +\infty \tag{2.4}$$

*for all series of functions with values in* $[0, +\infty]$ *;*
(ii) *Passage to the limit for increasing sequences:*

$$\mu^*(\lim f_n) = \lim \mu^*(f_n) \leq +\infty\,. \tag{2.5}$$

To prove (4), it suffices to show that

$$\mu^*(f + g) \leq \mu^*(f) + \mu^*(g) \tag{2.6}$$

for all $f$ and $g$, and then to apply (5) to the partial sums of the series $f_n$.
(6) is obvious if the right hand side is infinite. Otherwise, for all $r > 0$, there are lsc functions $\varphi \geq f$ and $\psi \geq g$ such that $\mu^*(\varphi) \leq \mu^*(f) + r$ and $\mu^*(\psi) \leq \mu^*(g) + r$. Thus

$$\mu^*(f + g) \leq \mu^*(\varphi + \psi) = \mu^*(\varphi) + \mu^*(\psi) \leq \mu^*(f) + \mu^*(g) + 2r\,,$$

which leads to (6).
To prove (ii), first note that, by (2), the right hand side of (5) is less than the left hand one. Hence it suffices to show the reverse inequality. This requires a proof only if the right hand side is finite.
Let us take a sequence of numbers $r_n > 0$ and, for all $n$, consider a lsc function $\psi_n$ satisfying

$$\psi_n \geq f_n \quad \text{and} \quad \mu^*(\psi_n) \leq \mu^*(f_n) + r_n\,.$$

Then setting

$$\varphi_n = \sup(\psi_1, \ldots, \psi_n)\,, \quad \text{whence } \varphi_n \geq \sup(f_1, \ldots, f_n) = f_n\,,$$

$\varphi_{n+1} = \sup(\varphi_n, \psi_{n+1})$. So

$$\varphi_{n+1} + \inf(\varphi_n, \psi_{n+1}) = \varphi_n + \psi_{n+1}\,.$$

As a result (theorem 1),

$$\mu^*(\varphi_{n+1}) + \mu^*[\inf(\varphi_n, \psi_{n+1})] = \mu^*(\varphi_n) + \mu^*(\psi_{n+1})\,.$$

Since $\inf(\varphi_n, \psi_{n+1}) \geq \inf(f_n, f_{n+1}) = f_n$, the left hand side is greater than $\mu^*(\varphi_{n+1}) + \mu^*(f_n)$, and so

$$\begin{aligned}\mu^*(\varphi_{n+1}) &\leq \mu^*(\varphi_n) - \mu^*(f_n) + \mu^*(\psi_{n+1}) \\ &\leq \mu^*(\varphi_n) - \mu^*(f_n) + \mu^*(f_{n+1}) + r_{n+1}\,.\end{aligned}$$

Adding the first $n-1$ relations sidewise, it follows that

$$\mu^*(\varphi_n) - \mu^*(f_n) \leq r_1 + \ldots + r_n$$

since $\varphi_1 = \psi_1$. Choosing $r_n = r/2^n$ for a given $r > 0$ thus gives an *increasing* sequence of lsc functions $\varphi_n \geq f_n$ satisfying

$$\mu^*(\varphi_n) \leq \mu^*(f_n) + r$$

for all $n$. $f = \sup(f_n)$ is bounded above by the upper envelope $\varphi$ of all $\varphi_n$ and (theorem 1)

$$\mu^*(f) \leq \mu^*(\varphi) = \sup \mu^*(\varphi_n) \leq \sup \mu^*(f_n) + r\,,$$

qed.

**Corollary.** *For*[5] *any countable family of sets* $A_n \subset X$,

$$\mu^*\left(\bigcup A_n\right) \leq \sum \mu^*(A_n) \tag{2.4'}$$

*and for any increasing sequence,*

$$\mu^*\left(\bigcup A_n\right) = \lim \mu^*(A_n)\,. \tag{2.5'}$$

These results must not be applied to arbitrary increasing philtres of positive functions; countability is essential here.

(ii) *Null sets.* Let us now introduce the basic notion of a *measure-zero-set* or *null set*, i.e. such that

$$\mu^*(N) = \mu^*(\chi_N) = 0\,. \tag{2.7}$$

By (3'), this is equivalent to the requirement that, for all $r > 0$, there exists $U$ such that

$$N \subset U \quad \& \quad \mu(U) \leq r\,. \tag{2.7'}$$

**Theorem 3.** *All subsets of a null set are null; the* finite or countable *union of a family of null sets is null.*

Obvious.

With respect to the usual Lebesgue measure on $\mathbb{R}$, a set reduced to a unique point is obviously null. Hence this is also the case of all countable sets, for instance $\mathbb{Q}$. The converse is, however, false. The most well-known

[5] Do not assume that if the sets $A_n$ were pairwise disjoint, we would get an equality: as we will see further down, for this the sets $A_n$ need to be assumed to be measurable.

counterexample is the *Cantor triadic set* of numbers $x \in K = [0,1]$ which do not contain the digit 1 in their base-3 expansion. It can be easily shown (n° 6) to have measure zero (fanciful proof: the probability that the expansion of a randomly chosen number does not contain the digit 1 is zero). It is equipotent to $\mathbb{R}$: the map $f$ from the Cantor set to $[0,1]$ associating the binary number $f(x) = 0, x'_1 x'_2 \ldots$, with $x'_i = x_i$ if $x_i = 0$ and $x'_i = 1$ if $x_i = 2$, to every $x = 0, x_1, x_2, \ldots$ whose base-3 expansion contains only 0s and 2s is surjective, which suffices.

When a relation $P\{x\}$ depends on one variable $x \in X$, $P\{x\}$ is said to be *true almost everywhere* (ae.) if the set of points $x$ for which $P\{x\}$ is not true has measure zero: with respect to the Lebesgue measure, almost every real number is irrational and even transcendental because the set of algebraic equations with rational coefficients is countable, and hence so is the set of their roots. If, taken separately, each one of a finite or countable family of propositions $P_n\{x\}$ is true almost everywhere then, by theorem 3, they are all simultaneously true almost everywhere. For example, the sum of a series of functions zero almost everywhere is zero almost everywhere. The set of complex-valued *null functions*, i.e. zero almost everywhere, is clearly a vector space $\mathcal{N}(X;\mu) = \mathcal{N}$. Despite the terminology, supposedly null sets and functions are far from being negligible. They actually serve to cover up horrors that are useless to consider when calculating integrals because they do not occur in the calculations.

(iii) *Reasonable sets and functions.* Let us start by prove the following result:

**Theorem 4.** *Let $f$ be a function with values in $[0, +\infty]$; then*

$$\mu^*(f) = 0 \Longleftrightarrow f(x) = 0 \text{ ae.}, \tag{2.8}$$

$$\mu^*(f) < +\infty \Longrightarrow f(x) < +\infty \text{ ae.} \tag{2.9}$$

*Any function such that $\mu^*(|f|) < +\infty$ vanishes outside a countable union of sets of finite outer measure.*

Supposing that $f \geq 0$ and $\mu^*(f) = 0$, let us consider the set $N_p$ of $x$ where $f(x) > 1/p$ and its characteristic function $\chi_p$. Then $f \geq \chi_p/p$, whence

$$\mu^*(N_p) = \mu^*(\chi_p) \leq p\mu^*(f) = 0\,.$$

Being the union of the sets $N_p$, the set $N = \{f(x) > 0\}$ has measure zero. To prove the converse, choose $N_p = \{p \leq f(x) < p+1\}$. This time $f\chi_p \leq (p+1)\chi_p$. So $\mu^*(f\chi_p) \leq (p+1)\mu^*(\chi_p) = 0$. As $f = \sum_{p \geq 0} f\chi_p$, $\mu^*(|f|) = 0$ by theorem 2, leading to (8).

To prove (9), we set $A_p = \{f(x) > p\}$ and again let $\chi_p$ be the characteristic function of $A_p$. Then $f > p\chi_p$, whence $\mu^*(\chi_p) \leq \mu^*(f)/p$. As the set $N = \{f(x) = +\infty\}$ in the intersection of the sets $A_p$, $\mu^*(N) \leq \mu^*(f)/p$ for all $p$, giving (9).

Finally, if $\mu^*(|f|) < +\infty$, there is a lsc function $\varphi \geq |f|$ such that $\mu^*(\varphi) < +\infty$. So it suffices to prove the third proposition of the statement for $\varphi$. Now the set $\{\varphi \neq 0\}$ is the union of the open sets $U_n = \{\varphi(x) > 1/n\}$ and each one of them has finite outer measure since its characteristic function is bounded by $\varphi$, up to a factor $n$, qed.

To simplify the language and at the risk of introducing unorthodox terminology, I will say that a set $E \subset X$ is *reasonable* if it is contained in the union of a countable family of sets of finite outer measure, and that a function $f$ is reasonable if so is the set $\{f \neq 0\}$. By the previous theorem, that is the case if $\mu^*(|f|) < +\infty$. Clearly, every subset of a reasonable set is reasonable, and so is the countable union of reasonable sets. Finally, the measure $\mu$ itself will be said to be reasonable if $X$ is the union of a countable family of sets of finite outer measure, i.e. is reasonable, in which case all sets and functions are reasonable with respect to $\mu$.

Since the outer measure of every compact set is finite, any set $E \subset X$ contained in the union of a countable family of compact sets and of a measure-zero-set is reasonable. The converse will be proved later. If the space $X$ is *countable at infinity*, i.e. the union of a countable family of compact sets, then all subsets of $X$ as well as all measures on $X$ are reasonable. Due to this remark, these notions lose much of their importance since in practice, spaces over which integration is done are almost always countable at infinity. But introducing assumptions of countability at infinity for $X$ or the measure $\mu$ would not affect the validity of the fundamental theorems of § 3 nor simplify their proofs. These assumptions will be useful, even essential, § 4 onwards.

*Exercise.* Let $X$ be a discrete set: all subsets of $X$ are open, the compact subsets are the finite subsets of $X$, all functions are continuous. Choose the measure $\mu(f) = \sum f(x)$. Show that a subset of $X$ has finite outer measure if and only if it is finite, and is reasonable if and only if it is countable.

### 3 – $F^p$ Spaces

(i) *Definition of $F^p$ spaces* . For a complex-valued function $f$, we set

$$N_1(f) = \mu^*(|f|) \leq +\infty \tag{3.1}$$

and more generally

$$N_p(f) = \mu^*(|f|^p)^{1/p} \tag{3.1'}$$

for any real number $p \geq 1$. By far the two most important cases are $p = 1$ and $p = 2$, but to deal with them simultaneously as will be done here, one might as well consider the general case: it is not harder and avoids unnecessary repetitions imposed on their readers by the vast majority of authors. Moreover, general $L^p$ spaces occur in many applications of the theory. Like N. Bourbaki in *Intégration*, Chap. III, the case of functions with values in a Banach space $\mathcal{H}$ could be dealt with at the same time without changing notations: it would suffice to denote the norm of a vector $\mathbf{a}$ by $|\mathbf{a}|$, which is not forbidden by higher authorities, and the function $x \mapsto |f(x)|$ by $|f|$, and to agree that what we denote by $L(X)$ is the set of continuous functions with compact support and values in $\mathcal{H}$; at least up to the Lebesgue-Fubini theorem, results and proofs will be exactly the same as those for complex-valued functions; this the reader can check step by step.[6] The only additional difficulty lies in the definition of the integral of a vector-valued function; we will do this in the next n°.

For $p \geq 1$, the only case considered, the function $N_p$ has quite simple properties:

(3.2) $$N_p(\alpha f) = |\alpha| N_p(f)$$

for all scalars $\alpha$. It is equally clear that

(3.3) $$|f| \leq |g| \Longrightarrow N_p(f) \leq N_p(g)\,.$$

Last but not least, *Minkowski's inequality*

(3.4) $$N_p(f+g) \leq N_p(f) + N_p(g)$$

and *Hölder's inequality*

(3.4') $$N_r(fg) \leq N_p(f)N_q(g) \quad \text{if } 1/p + 1/q = 1/r$$

of Chap. V, n° 14 hold in this general context because they are only based on the formal properties (2.3), (2.4) and (2.5) of the function $\mu^*(f)$ for $f \geq 0$. Inequality (4) is trivial for $p = 1$ – it suffices to write that $|f+g| \leq |f| + |g|$ – and (4') is mostly useful for $r = 1$ in the form

(3.4") $$|\mu^*(fg)| \leq N_p(f)N_q(g) \quad \text{if } 1/p + 1/q = 1\,.$$

This assumes $p > 1$ and $q > 1$ since $p$ and $q$ must be finite.

The set $\mathcal{F}^p(X;\mu) = \mathcal{F}^p$ of complex functions such that $N_p(f) < +\infty$ is, therefore, a vector space on which the function $N_p$ is a norm, except for one detail: by (2.8),

(3.5) $$N_p(f) = 0 \quad \text{is equivalent to} \quad f(x) = 0 \text{ ae.}$$

and not to $f = 0$. If two functions $f$ and $g$ are equal almost everywhere, then, by (4) and (5), $N_p(f) \leq N_p(g) + N_p(f-g) = N_p(g)$. Thus symmetry implies

(3.6) $$f = g \text{ ae.} \Longrightarrow N_p(f) = N_p(g)\,.$$

[6] The case of functions with values in a Banach space is not an unwarranted generalization of the classical theory. It is an essential part of the representation theory of locally compact groups (n° 19) and of many other questions. This does not force the reader to be concerned with it before needing to do so.

Hence the number $N_p(f)$ does not depend on the class of $f$ modulo the subspace $\mathcal{N}$ of null functions; denoting – very provisionally – this class by $\dot{f}$, we set

$$\|\dot{f}\|_p = N_p(f)\,. \tag{3.7}$$

Relations (2) and (4) hold for these classes, and as all has been done so that $\|\dot{f}\|_p = 0$ implies $\dot{f} = 0$, the quotient space

$$F^p(X;\mu) = F^p = \mathcal{F}^p(X;\mu)/\mathcal{N}(X;\mu)$$

becomes a genuine normed vector space. Convergence in this space, defined by

$$\lim N_p(f - f_n) = 0\,,$$

is called *convergence in mean* of order $p$; instead of writing

$$\lim \int |f(x) - f_n(x)|^p \, d\mu^*(x) = 0\,,$$

which for explicitly given functions can be quite cumbersome, it is sometimes convenient to use the expression

$$f(x) = \text{l.i.m.}^p\, f_n(x)$$

(*limit in mean*).

(ii) *Convergence in mean and almost everywhere.* Proving relations between convergence in mean and simple convergence or, more generally, almost everywhere, is one of the basic features of the theory, as the next result already shows.

**Theorem 5**. *Let $\sum f_n(x)$ be a series of complex-valued functions such that $\sum N_p(f_n) < +\infty$ for given $p \geq 1$. Then*

$$\sum |f_n(x)| < +\infty \text{ ae.} \tag{3.8}$$

*and any function $f$ such that*

$$f(x) = \sum f_n(x) \text{ ae.} \tag{3.9}$$

*satisfies*

$$N_p(f) \leq \sum N_p(f_n)\,, \tag{3.10}$$

$$f(x) = \text{l.i.m.}^p\, f_1(x) + \ldots + f_n(x)\,. \tag{3.11}$$

*The normed vector space $F^p$ is* complete, *and any sequence of functions $f_n \in \mathcal{F}^p$ converging in mean to a limit $f$ contains a subsequence converging almost everywhere to $f$.*

Let us start with the case of a series $\sum f_n$ whose terms are $\geq 0$ and denote the partial sums by $s_n$. The increasing sequence of functions $s_n^p$ converges to $f^p$, where for all $x$, we set $f(x) = \sum f_n(x) \leq +\infty$. Then, by theorem 2, $\mu^*(f^p) = \lim \mu^*(s_n^p)$ and so, raising to the power of $1/p$,

$$N_p(f) = \lim N_p(s_n) \leq \lim \left[N_p(f_1) + \ldots + N_p(f_n)\right] \leq \sum N_p(f_n) ,$$

whence (10). As this implies $N_p(f) < +\infty$, $f(x) < +\infty$ ae., and (8) follows.

In the general case, apply the results obtained above to the functions $|f_n|$ : Since $N_p(f_n) = N_p(|f_n|)$ (8) follows, and since (9) shows that $|f| \leq \sum |f_n|$ ae., relation (10) for the functions $|f_n|$ implies (10) for functions $f_n$. Relation (11) is obtained by removing the first $n$ terms from the given series, which subtracts the partial sum $s_n$ of the series from $f$. Then (10) shows that

$$N_p(f - s_n) \leq \sum_{m>n} N_p(f_m), \tag{3.12}$$

which leads to the result since the series $\sum N_p(f_n)$ converges.

To show that $F^p$ is complete, let $(f_n)$ be a sequence of functions in $\mathcal{F}^p$ and suppose that, for all $r > 0$,

$$N_p(f_j - f_i) \leq r \quad \text{for large } i \text{ and } j .$$

It is a matter of proving the existence of a function $f \in \mathcal{F}^p$ such that $\lim N_p(f - f_n) = 0$. Like in any metric space, this means that it suffices to show that a subsequence converging in $F^p$ can be extracted from any given Cauchy sequence.

By successively taking $r = 1/2, 1/2^2, \ldots$, a subsequence such that

$$N_p\left(f_{n_{k+1}} - f_{n_k}\right) \leq 1/2^k \tag{3.13}$$

can readily be extracted from the given sequence. Since the differences

$$g_k = f_{n_{k+1}} - f_{n_k} \tag{3.14}$$

satisfy $\sum N_p(g_k) < +\infty$, the series $\sum g_k$ converges absolutely almost everywhere *and in mean* to a function $g$ which, from the first part of the proof, is also the limit (almost everywhere and in mean) of the partial sums of the series. But (14) shows that

$$g_1(x) + \ldots + g_k(x) = f_{n_{k+1}}(x) - f_{n_1}(x) . \tag{3.15}$$

As the left hand side tends to $g(x)$ ae., one can deduce that

$$\lim f_{n_k}(x) = g(x) + f_{n_1}(x) \text{ ae.}$$

Denoting by $f(x)$ the limit function – regardless of its value at points where it is defined –, by (15),

$$g - g_1 - \ldots - g_k = f - f_{n_{k+1}} \text{ ae..}$$

So

$$\lim N_p\left(f - f_{n_{k+1}}\right) = 0$$

since the series $\sum g_k$ converges *in mean* to $g$. This leads to a subsequence converging to $f$ both in $F^p$ and almost everywhere, qed.

*Exercise.* Let $(a_n)$ be a sequence of real numbers (for example the sequence of rational numbers). Show that the series $\sum 1/|n^2x - a_n|^{1/2}$ converges almost everywhere in $\mathbb{R}$.

The previous theorem shows that $F^p$ is a *Banach space* (Appendix to Chap. III, n° 5). In practice, it is not very useful because the functions used to define it are far too general for other non-trivial properties apart from the previous theorems to be proved. Proper integration theory will be obtained by replacing $F^p$ with a far more practical closed space, namely the space of limits in mean of *continuous* functions *with compact support.*

# § 2. $L^p$ Spaces

## 4 – Integrable Functions, $L^p$ Spaces

(i) *Integral of an integrable function.* A complex-valued function $f$ is said to be *integrable* if, for all $r > 0$, there is a continuous function $g$ with compact support such that

$$N_1(f-g) = \mu^*(|f-g|) \leq r\,. \tag{4.1}$$

To require the existence of a sequence of functions $f_n \in L(X)$ such that

$$\lim N_1(f-f_n) = 0 \tag{4.1'}$$

would be equivalent. The integral of $f$ is then defined by writing

$$\mu(f) = \lim \mu(f_n)\,. \tag{4.2}$$

This limit exists and does not depend on the chosen sequence $(f_n)$. Indeed,

$$\begin{aligned}|\mu(f_p) - \mu(f_q)| = |\mu(f_p - f_q)| &\leq \mu(|f_p - f_q|) = N_1(f_p - f_q)\\ &\leq N_1(f_p - f) + N_1(f - f_q)\,,\end{aligned}$$

an arbitrarily small result when $p$ and $q$ are sufficiently large. Therefore, the sequence $\mu(f_n)$ satisfies the usual Cauchy criterion and converges. If another sequence $(g_n)$ of continuous functions satisfies (1'), then similarly

$$|\mu(f_n) - \mu(g_n)| \leq \mu(|f_n - g_n|) = N_1(f_n - g_n) \leq N_1(f_n - f) + N_1(f - g_n)\,,$$

whence $\lim \mu(f_n) - \mu(g_n) = 0$, qed. There is also a general theorem about the extension of a continuous linear functions. Its proof is similar...

Any continuous function $f \in L(X)$ is integrable [take $g = f$ in (1)] and its Lebesgue integral equals $\mu(f)$. As the relation

$$|\mu(f)| \leq N_1(f) \tag{4.3}$$

holds for all continuous functions, by passing to the limit, it can be generalized to integrable functions. If $f$ is integrable, so is[7] $|f|$ because of inequality $\big||f| - |g|\big| \leq |f-g|$, and

$$N_1(f) = \mu(|f|) \quad \text{for } f \text{ integrable}\,. \tag{4.3'}$$

Indeed, the same inequality and (1') show that $|f_n|$ converges in mean to $|f|$, whence

$$\mu(|f|) = \lim \mu(|f_n|) = \lim N_1(|f_n|) = N_1(|f|) = N_1(f)\,.$$

[7] Thus Lebesgue theory generalizes the *absolutely* convergent integrals of Chap. V, § 7.

On the other hand, it is clear that all functions equal ae. to an integrable function are integrable and that the integrals of these functions are equal.

**Lemma 1.** *If $f$ and $g$ are integrable, so is $\alpha f + \beta g$ for all $\alpha, \beta \in \mathbb{C}$, and*

$$\mu(\alpha f + \beta g) = \alpha\mu(f) + \beta\mu(g) .$$

This follows from the inequality

$$N_1\left[(f+g) - (f_n + g_n)\right] \leq N_1\left(f - f_n\right) + N_1\left(f - f_n\right) \tag{4.4}$$

and from the linearity of the integral of continuous functions.

It was stated above that this theory can be generalized word for word to functions with values in a Banach space $\mathcal{H}$. Nonetheless, the integral $\mu(\mathbf{f}) \in \mathcal{H}$ has to be defined for every integrable function $\mathbf{f}$ with values in $\mathcal{H}$ and, to begin with, for every continuous function with compact support. Though elementary, the construction requires several stages.

(a) We first suppose that

$$\mathbf{f}(x) = \sum f_i(x)\mathbf{a}_i \tag{*}$$

with finitely many $f_i \in L(X)$ and $\mathbf{a}_i \in \mathcal{H}$. We then naturally set

$$\mu(\mathbf{f}) = \sum \mu(f_i)\mathbf{a}_i . \tag{**}$$

If $\mathbf{f}(x) = \sum g_j(x)\mathbf{b}_j$ is another similar representation of $\mathbf{f}$, then there are finitely many linearly independent elements $\mathbf{c}_k$ satisfying relations[8] $\mathbf{a}_i = \sum \alpha_{ik}\mathbf{c}_k, \mathbf{b}_j = \sum \beta_{jk}\mathbf{c}_k$, whence

$$\sum f_i(x)\alpha_{ik} = \sum g_j(x)\beta_{jk} \quad \text{for all } k .$$

Integrating, one can deduce that $\sum \mu(f_i)\alpha_{ik} = \sum \mu(g_j)\beta_{jk}$ and so, multiplying by $\mathbf{c}_k$ and summing, that

$$\sum \mu(f_i)\mathbf{a}_i = \sum \mu(g_j)\mathbf{b}_j .$$

This shows that $\mu(\mathbf{f})$ only depends on $\mathbf{f}$, justifying definition (**).

(b) As assumptions on $\mathbf{f}$ remain the same, inequality

$$|\mu(\mathbf{f})| \leq \mu\left(|\mathbf{f}|\right) \tag{***}$$

needs to be generalized to this case. Recall that absolute values denote the norm in $\mathcal{H}$ and $|\mathbf{f}|$ is the function $x \mapsto |\mathbf{f}(x)|$. The quickest proof is

[8] Choose the elements $\mathbf{c}_k$ so that they form a basis for the (finite-dimensional) vector subspace generated by the $\mathbf{a}_i$ and the $\mathbf{b}_j$.

based on the following consequence of the Hahn-Banach theorem, which I will admit:[9] for all $\mathbf{a} \in \mathcal{H}$,

$$(****) \qquad |\mathbf{a}| = \sup |\varphi(\mathbf{a})| \,,$$

where $\varphi$ varies in the set of *continuous linear functionals* having norm $\leq 1$ on $\mathcal{H}$. Now, for such a form, (**) shows that

$$\varphi[\mu(\mathbf{f})] = \sum \mu(f_i)\varphi(\mathbf{a}_i) = \mu\left[\sum f_i\varphi(\mathbf{a}_i)\right] = \mu(\varphi \circ \mathbf{f}) \,,$$

where $\varphi \circ \mathbf{f}$ is the numerical function

$$x \longmapsto \varphi[\mathbf{f}(x)] = \sum f_i(x)\varphi(\mathbf{a}_i) \,.$$

Since by assumption $|\varphi| \leq 1$, $|\varphi \circ \mathbf{f}(x)| \leq |\mathbf{f}(x)|$, and so

$$|\varphi[\mu(\mathbf{f})]| \leq \mu(|\varphi \circ \mathbf{f}|) \leq \mu(|\mathbf{f}|) \,.$$

As this holds for every $\varphi$ having norm $\leq 1$, inequality (****) for $\mathbf{a} = \mu(\mathbf{f})$ proves (***).

(c) Now let $\mathbf{f}$ be a continuous function with values in $\mathcal{H}$ and vanishing outside a compact set $K \subset X$. For all $r > 0$, there is (BW) a finite open cover $(U_i)$ of $K$ such that $\mathbf{f}$ is constant on each $U_i$, up to $r$ (uniform continuity...); and there exist $f_i \in L_+(X)$ zero outside $U_i$ and such that $\sum f_i(x) = 1$ on $K$. So $\mathbf{f}(x) = \sum f_i(x)\mathbf{f}(x)$ for all $x \in X$. Let us choose some $x_i \in U_i$ and set $\mathbf{a}_i = \mathbf{f}(x_i)$, $\mathbf{g}(x) = \sum f_i(x)\mathbf{a}_i$, whence

$$|\mathbf{f}(x) - \mathbf{g}(x)| = \left|\sum f_i(x)[\mathbf{f}(x) - \mathbf{a}_i]\right| \leq \sum f_i(x)\,|\mathbf{f}(x) - \mathbf{a}_i| \,.$$

For any given $x \in X$, the only terms that matter correspond to the indices $i$ for which $f_i(x) \neq 0$. But this implies $x \in U_i$ and so $|\mathbf{f}(x) - \mathbf{a}_i| \leq r$. Hence $|\varphi(x) - \mathbf{g}(x)| \leq r\sum f_i(x) = r$. This shows that $\mathbf{f}$ is the uniform limit of functions of type (*) vanishing outside a fixed compact set.

---

[9] The framework for the present case is actually a finite-dimensional subspace of $\mathcal{H}$, so that the Hahn-Banach theorem in finite dimension is sufficient. It is obvious (Cauchy-Schwarz) for a Hilbert norm, but less so in the general case. Rudin, *Real and Complex Analysis*, Chap. 5, gives a proof of a special version of the HB theorem. The proper statement, found in N. Bourbaki, *Espaces vectoriels topologiques*, is as follows: let $C$ be a convex and closes set in a topological vector space $\mathcal{H}$ whose topology is defined by seminorms (i.e. locally convex). $C$ is then defined by relations $\operatorname{Re}[\mathbf{f}_i(x)] \leq 1$, where the $\mathbf{f}_i$ are continuous linear functionals on $\mathcal{H}$. In other words, $C$ is an intersection of "closed real half-spaces". In exercises 3 and 4 of his Chap. XII.15, Dieudonné also proves some particular cases, since he obviously does not wish to have to invoke transfinite induction, which in some form or other is essential in the general case (including for Banach spaces). The theory was finalized during the war by Dieudonné and independently by George W. Mackey.

(d) $\mu(\mathbf{f})$ can now be defined by approximating it with functions $\mathbf{f}_n$ of type (*) zero outside a fixed compact set: inequality (***) shows that $(\mu(\mathbf{f}_n))$ is a Cauchy sequence, and so converges to a limit $\mu(\mathbf{f})$, which is obviously the same for all sequences $(\mathbf{f}_n)$. Inequality (***) readily generalizes to this case.

(e) Finally, let $\mathbf{f}$ be an integrable function, i.e. such that there exist continuous functions $\mathbf{f}_n$ with compact support satisfying

$$\lim N_1(\mathbf{f}-\mathbf{f}_n) = \lim \mu^*\left(|\mathbf{f}-\mathbf{f}_n|\right) = 0\,.$$

As $(\mathbf{f}_n)$ is a Cauchy sequence with respect to the norm $N_1$, (***) again shows that their integrals converge to a limit depending only on $\mathbf{f}$. This gives the definition of $\mu(\mathbf{f})$ in the general case, inequality (***) remaining valid.

Finally note the following linearity property: let $\mathbf{f}$ be an integrable function with values in a Banach space $\mathcal{H}$ and $T$ a continuous linear map from $\mathcal{H}$ to a Banach space $\mathcal{H}'$. The function $T\circ\mathbf{f}$ is then integrable and

$$\mu(T\circ\mathbf{f}) = T\left[\mu(\mathbf{f})\right]$$

or, in more standard notation,

$$\text{(*****)}\qquad \int T\mathbf{f}(x).d\mu(x) = T\int \mathbf{f}(x)d\mu(x)\,.$$

This is obvious for functions (*), and, given the continuity of $T$, the general case follows by passing to the limit. The previous relation is frequently used when $T: F\mapsto\mathbb{C}$ is a continuous linear functional on $F$.

(ii) *$L^p$ spaces; the Riesz-Fischer theorem.* More generally, let us now define $p^{th}$ *power integrable functions* for all $p\geq 1$ by requiring that for each such function $f$, there be functions $f_n\in L(X)$ such that

$$\text{(4.5)}\qquad \lim N_p(f-f_n) = 0\,.$$

The set $\mathcal{L}^p(X;\mu)=\mathcal{L}^p$ of these functions is clearly a vector subspace of $\mathcal{F}^p$. Since $N_p(f)$ remains invariant if we modify $f$ on a null set, only the class of $f$ mod $\mathcal{N}$ occurs in $f\in\mathcal{L}^p$.

The quotient space

$$L^p(X;\mu) = \mathcal{L}^p(X;\mu)/\mathcal{N}(X;\mu)\,,$$

denoted simply by $L^p$ when no confusion is possible, is a normed vector space; it is the closure in $F^p$ of the set of classes mod $\mathcal{N}$ of continuous functions with compact support. Since $F^p$ is complete, so is $L^p$, which is, therefore, a Banach space (and historically, probably the first of its kind apart from Hilbert spaces). Almost always, by abuse of language, one writes $f\in L^p$ instead of $f\in\mathcal{L}^p$ and $\|f\|_p$ instead of $N_p(f)$. Another convenient convention consists in extending the previous definitions to functions only defined outside a null set or taking the value $+\infty$ or $-\infty$ on such a set. Assigning to them arbitrary finite values on the null set in question reduces the definition to the previous one. The norm $N_p$ of the function thus modified and its class modulo $\mathcal{N}$ only depend on the initial function. Hence hereinafter, an expression of type « let $f$ be a function defined on $X$ » may also mean a function defined on the complement of a null set. This convention may seem somewhat strange when encountered for the first time, but one quickly gets used to it because of its convenience, provided some minimal caution is exercised.

**Lemma 2.** *Let $(f_n)$ be a sequence of functions in $L^p$ and $f$ a function such that* $\lim N_p(f - f_n) = 0$. *Then $f$ is in $L^p$. If $p = 1$, then*

$$\mu(f) = \lim \mu(f_n)\,. \tag{4.6}$$

Obvious.

**Theorem 6.** *Let $(f_n)$be a sequence of functions in $L^p$ such that $\sum N_p(f_n) < +\infty$. Then $\sum |f_n(x)| < +\infty$ ae. Any function $f$ satisfying $f(x) = \sum f_n(x)$ ae. is in $L^p$ and satisfies*

$$\lim N_p\,(f - f_1 - \ldots - f_n) = 0\,, \tag{4.7}$$

$$\mu(f) = \sum \mu(f_n) \quad \textit{if } p = 1\,. \tag{4.8}$$

This is theorem 5 applied to functions in $L^p$. By lemma 2, the (class of the) limit function $f$ is still in $L^p$ and (8) follows from (4) and (7).

**Corollary.** *Let $\sum f_n(x)$ be a sequence of* positive *integrable functions. The sum of the series is integrable if and only if $\sum \int f_n(x)d\mu(x) < +\infty$. The series then converges almost everywhere and in the space $L^1$, and*

$$\int d\mu(x).\sum f_n(x) = \sum \int f_n(x)d\mu(x)\,.$$

It suffices to observe that for a positive integrable function $f$, $N_1(f) = \mu(|f|) = \mu(f)$.

Compare with theorems 20 and 21 of Chap. V, n° 23: uniform or normal convergence has disappeared from assumptions and restricting oneself to regulated functions is no longer necessary.

**Theorem 7 (Riesz-Fischer).** *The normed vector space $L^p$ is complete. A sequence of functions $f_n(x)$ in $L^p$ converging* in mean *to a limit $f$ contains a* subsequence *converging* almost everywhere *to $f$. If, moreover, the sequence $f_n(x)$ converges almost everywhere to a limit $g(x)$, then $f(x) = g(x)$ almost everywhere.*

Being a closed subspace of a complete space, $L^p$ is complete. The second proposition of the statement is here too theorem 5 in the special case of $L^p$.

The appearance of "exceptional" null sets in these statements cannot be avoided: the behaviour *everywhere* of a sequence or of a series cannot be expected to follow from integral computations. Similarly, statistics provides no information on an individual case (the reason why Condorcet had already realized around 1750 that statistics are only useful for political leaders). It is already noticeable in Riemann theory, but in Lebesgue theory, it can be done only if one specifically tolerates null sets about which nothing can be said.

In some very particular cases, classical analysis can reinforce integration theory.

If for example $X$ is an open subset in $\mathbb{C}$, and if $\mu$ is the usual measure $dxdy$, then a sequence of *holomorphic* functions $f_n \in L^p(X, \mu)$ converging in $L^p$ converges *uniformly in every compact set* to a holomorphic function [Chap. VIII, n° 4, (iv)], which is necessarily their limit in $L^p$. This proves that in the space $L^p$ considered, the set of holomorphic functions is a possibly trivial *closed* vector subspace. But it was obviously not in order to integrate such harmless functions that Lebesgue and his successors worked for half a century. And even in this idyllic case, compact convergence does not imply convergence in mean.

*Exercise 1.* Suppose that $X = [0, 1]$, $d\mu(x) = dx$ and set

$$f_p(x) = \cos^{2n} 2\pi nx \qquad \text{where } n = 2^p\,.$$

Calculate $\mu(f_p)$ [observe that $f_p$ is a trigonometric polynomial whose constant term can be calculated by using Euler's relation and the binomial formula]. Using Stirling's formula show that $f_p$ converges in mean to 0. For which $x \in X$ does $\lim f_p(x) = 0$?

As any function $f \in L^p$ is by definition the limit in mean of continuous functions, theorem 7 shows the existence of a sequence of continuous functions with compact support that converges almost everywhere to $f(x)$. The proof of theorem 5 even leads to a somewhat more precise though obvious result:

**Corollary 1.** *For any $f \in L^p$, there is a series of continuous functions with compact support such that*

$$\sum N_p(f_n) < +\infty\,, \quad f(x) = \sum f_n(x) \text{ ae.}\,,$$
$$\lim N_p\,(f - f_1 - \ldots - f_n) = 0\,.$$

If $f$ is real-valued, the $f_n$ may be assumed to be real. Writing $\varphi'(x) = \sum f_n^+(x)$, $\varphi''(x) = \sum f_n^-(x)$ defined lsc functions with values in $[0, +\infty]$, but finite almost everywhere, since $N_p(\varphi') \leq \sum N_p(f_n^+) \leq \sum N_p(f_n) < +\infty$, and we get

$$f(x) = \varphi'(x) - \varphi''(x) \text{ ae.}\,,$$

as well as $\mu(f) = \mu(\varphi') - \mu(\varphi'')$ if $p = 1$.

This result shows that, for real-valued functions, Lebesgue theory would not go beyond the framework of lsc functions *if* null functions did not exists.

**Corollary 2**. *Every function belonging to the space $L^p$ ($p < +\infty$) is reasonable.*

It suffices to consider the supports $K_n$ of the functions $f_n$ of corollary 1 and to note that $f$ vanishes outside the union of these $K_n$ and a null set. This result is of little interest when $X$ is countable at infinity...

**Corollary 3**. *For all $f \in L^p$, there is a sequence of functions $f_n \in L(X)$ and a function $F \in L^p$ such that*

$$f(x) = \lim f_n(x) \text{ ae.}, \qquad |f_n(x)| \leq |F(x)| \text{ ae.}$$

*If $|f(x)| \leq M$ almost everywhere, we may assume that $|f_n(x)| \leq M$ for all $x$ and $n$.*

To prove the first point, it suffices to consider the partial sums $s_n(x)$ of the series whose existence is ensured by corollary 1 and to choose $F(x) = \sum |f_n(x)|$. By theorem 6 applied to the $|f_n|$, $F \in L^p$. To prove the second point, consider the contraction $p$ of the complex plane onto the disc $|z| \leq M$ given by

$$p(z) = Mz/\sup(|z|, M) .$$

The functions $g_n = p \circ f_n$ are still in $L(X)$, $|g_n(x)| \leq |F(x)|$ still holds ae., and finally $\lim f_n(x) = \lim g_n(x)$ since $f(x) \leq M$, qed.

When, for a sequence of *a priori* arbitrary functions $f_n$, there is a function $F$ such that

$$N_p(F) < +\infty \quad \& \quad |f_n(x)| \leq |F(x)| \text{ ae.},$$

the sequence $(f_n)$ is said to be *dominated* in $L^p$. This notion, almost always applied to sequences $f_n \in L^p$, plays a fundamental role in Lebesgue's convergence theorems. In particular, we will see (theorem 9) that, under the assumptions of the corollary, the functions $f_n$ converge to $f$ in $L^p$ and not only almost everywhere.

(iii) *The case of* lsc *or* usc *functions.*

**Lemma 3**. *A* lsc *or* usc *function $\varphi$ is integrable if and only if $N_1(\varphi) < +\infty$.*

As $-\varphi$ is lsc if $\varphi$ is usc, we need only consider the case when $\varphi$ is lsc and show that condition $N_1(\varphi) < +\infty$ is sufficient. As $\varphi$ is lsc, $\varphi^+$ is lsc, $\varphi^-$ is usc and these two functions satisfy the condition of the lemma.

Let us first consider $\varphi^+$. For all $r > 0$, there is a function $f \leq \varphi^+$ in $L_{\mathbb{R}}(X)$ such that

$$\mu(f) \leq \mu^*(\varphi^+) = N_1(\varphi^+) \leq \mu(f) + r .$$

As $\varphi^+ = f + (\varphi^+ - f)$, and as $f$ and $\varphi^+ - f$ are lsc since $f$ is continuous, by property (i) of theorem 1,

$$N_1(\varphi - f) = \mu^*(|\varphi - f|) = \mu^*(\varphi - f) = \mu^*(\varphi) - \mu(f) \leq r\,,$$

whence $\varphi^+ \in L^1$.

Next consider $\varphi^-$. Since $\mu^*(\varphi^-) = N_1(\varphi^-) < +\infty$, there exists a lsc function $\psi$ such that $\psi \geq \varphi^-$ and $\mu^*(\psi) < +\infty$. The function $\psi - \varphi^-$ being lsc,[10] positive and bounded above by $\psi$, $\mu^*(\psi - \varphi^-) < +\infty$, so that, by the first part of the proof, $\psi - \varphi^-$ and $\psi$ are integrable. As a result, $\varphi^- = \psi - (\psi - \varphi^-)$ is integrable, qed.

**Lemma 4.** *A function $f$ with values in $[-\infty, +\infty]$ is integrable if and only if, for all $r > 0$, there exist* integrable *functions $\varphi$ and $\psi$,* lsc *and* usc *respectively, such that*

$$\psi \leq f \leq \varphi \quad \& \quad \mu^*(\varphi - \psi) \leq r\,.$$

If $f$ is integrable, there is a function $u \in L_{\mathbb{R}}(X)$ such that $\mu^*(|f - u|) < r$, hence a lsc function $\theta$ such that

$$|f - u| \leq \theta\,, \qquad \mu^*(\theta) \leq r\,.$$

As $-\theta \leq f - u \leq \theta$, $u - \theta \leq f \leq u + \theta$ since $u$ is finite-valued. Since $u$ is continuous, $u + \theta = \varphi$ is lsc, $u - \theta = \psi$ is usc, and like $u$ and $\theta$ (lemma 3) these two functions are integrable. Seeing that

$$\mu^*(\varphi - \psi) = \mu^*(2\theta) \leq 2r$$

the condition is necessary.

Conversely, if it holds, the function $\varphi - \psi$, which is lsc and positive, is integrable (lemma 3). As $N_1(\varphi - f) \leq \mu^*(\varphi - \psi) \leq r$, the function $f$ is the limit in mean of usc and integrable functions (put $r = 1/n$), hence is integrable, qed.

### 5 – Lebesgue's Theorems

(i) *The dominated convergence theorem.*

**Lemma 1.** *The upper (resp. lower) envelope of a finite family of functions of $L^p$ is in $L^p$.*

For real functions, inequality $|f^+ - g^+| \leq |f - g|$ shows that $N_p(f^+ - g^+) \leq N_p(f - g)$. If $f$ is in $\mathcal{L}^p$, so are $f^+$ and $f^-$. We conclude the proof using relations

[10] As $\varphi(x) > -\infty$ everywhere, the function $-\varphi^-$ does not take the value $-\infty$, and hence is in $\mathcal{I}$, and consequently so is $\psi - \varphi^-$.

$$\sup(f,g) = f + (g-f)^+ , \quad \inf(f,g) = g - (g-f)^+ .$$

Likewise, an analogous argument using inequality $||f| - |g|| \leq |f - g|$ shows that

$$f \in L^p \Longrightarrow |f| \in L^p .$$

The converse is *almost* true (theorem 10).

**Theorem 8.** *Let* $(f_n)$ *be an* increasing *sequence of* positive *functions in* $L^p$. $f = \sup f_n = \lim f_n$ *is in* $L^p$ *if and only if* $\sup N_p(f_n) < +\infty$. *Then* $\lim N_p(f - f_n) = 0$.

The condition is clearly necessary since $0 \leq f_n \leq f$ for all $n$. To obtain the converse, it suffices to show that the sequence $(f_n)$ satisfies Cauchy's criterion in $L^p$.

This is easy if $p = 1$. Indeed, since $f_j - f_i$ is integrable and positive for $i \leq j$,

$$N_1(f_j - f_i) = \mu(f_j) - \mu(f_i) .$$

As the sequence $\mu(f_n)$ is increasing and bounded above, Cauchy's criterion follows readily. One can also apply the corollary of theorem 6 to the series $\sum(f_{n+1} - f_n)$.

In the general case, we first note that, if two *positive* functions $f$ and $g$ are continuous and with compact support, then

$$N_p(f)^p + N_p(g)^p \leq N_p(f+g)^p . \tag{5.1}$$

To see this, it suffices to apply $\mu$ to the inequality $f^p + g^p \leq (f+g)^p$. Now, any positive function $f \in L^p$ is the limit in mean of positive functions $f_n \in L(X)$ since, if a sequence $f_n \in L(X)$ converges to $f$ in $L^p$, so must functions $\mathrm{Re}(f_n)^+$. Passing to the limit, we conclude that (1) holds for positive $f, g \in L^p$.

We then consider an increasing sequence $(f_n)$ of positive functions in $L^p$. For $i \leq j$, (1) can be applied to the positive functions $f_i$ and $f_j - f_i$, whence

$$N_p(f_j - f_i)^p \leq N_p(f_j)^p - N_p(f_i)^p .$$

If $\sup N_p(f_n) < +\infty$, the increasing sequence of numbers $N_p(f_n)^p$ converges to a finite limit, and so Cauchy's criterion holds in $L^p$ for the sequence $(f_n)$, qed. As mentioned in the case $p = 1$, one could also apply theorem 6 to the series $(f_{n+1} - f_n)$.

**Lemma 2.** *Any decreasing sequence* $(f_n)$ *of* positive *functions in* $L^p$ *converge in mean.*

Apply theorem 8 to functions $f_1 - f_n$.

**Lemma 3.** *The lower envelope of a countable family* $(f_n)$ *of* positive *functions in* $L^p$ *is in* $L^p$.

Apply the previous result to the functions $\inf(f_1, \dots, f_n)$; we already know (lemma 1) they are in $L^p$.

**Lemma 4.** *The upper envelope of a countable family of* positive *functions* $f_n \in L^p$ *is in* $L^p$ *if and only if there is function* $F \geq 0$ *such that*[11]

$$N_p(F) < +\infty \quad \& \quad f_n(x) \leq F(x) \text{ ae. } \quad \text{for all } n .$$

The condition is necessary: for $F$ take the upper envelope of the functions $f_n$. If it holds, the functions $g_n = \sup(f_1, \dots, f_n) \in L^p$ form an increasing sequence and satisfy $g_n \leq F$ ae., whence $N_p(g_n) \leq N_p(F) < +\infty$. Apply theorem 8 to functions $g_n$ completes the proof.

For real-valued functions $f_n$, it is better to assume that $|f_n(x)| \leq F(x)$ ae.. The upper and lower envelopes of the $f_n$ are then clearly in $L^p$. In this case:

**Theorem 9 (dominated convergence[12]).** *Let* $(f_n)$ *be a sequence of functions in*$L^p$ *converging ae. to a function* $f$. *Suppose that there is a function* $F \geq 0$ *such that*

$$N_p(F) < +\infty \quad \& \quad |f_n(x)| \leq F(x) \text{ ae. for all } n .$$

*Then* $f$ *is in* $L^p$ *and*

$$\lim N_p(f - f_n) = 0 , \quad \lim \mu(f_n) = \mu(f) \quad \text{if } p = 1 .$$

It suffices to show that $(f_n)$ is a Cauchy sequence with respect to convergence in mean.

As $|f_i - f_j| \leq 2F$ ae., lemma 4 shows that the functions

$$g_n(x) = \sup_{i,j \geq n} |f_i(x) - f_j(x)|$$

are in $L^p$, and lemma 2 that this *decreasing* sequence of positive functions converges in mean in $L^p$. But since the given sequence converges outside a null set $N$, the usual Cauchy criterion tells us that $\lim g_n(x) = 0$ for $x \notin N$.[13]

---

[11] The relation $f_n(x) \leq f(x)$ almost everywhere for all $n$ means that, for all $n$, there is null set $N_n$, possibly dependent on $n$, such that $f_n(x) \leq f(x)$ for $x \notin N_n$, hence for $x \notin N = \bigcup N_n$.

[12] Throughout this theory, a function $f$ or a sequence of functions $f_n$ is said to be *dominated* by a function $g$ if $|f_n(x)| \leq |g(x)|$ ae. for all $n$. As it is constantly used, theorem 8 is often called Lebesgue's theorem.

[13] For a sequence of complex numbers, the relation

$$i, j \geq n \Longrightarrow |u_i - u_j| \leq r$$

is equivalent to

$$v_n = \sup_{i,j \geq n} |u_i - u_j| \leq r ,$$

so that Cauchy's criterion means that $\lim v_n = 0$.

Theorem 7 then shows that $\lim N_p(g_n) = 0$. However, $|f_i(x) - f_j(x)| \leq g_n$ and so

$$N_p(f_i - f_j) \leq N_p(g_n) \quad \text{for } i, j \geq n\,.$$

This gives Cauchy's criteria and convergence in $L^p$, qed.

Compare with theorem 19 of Chap. V, n° 23.

**Corollary**. *Let $p$ and $q$ be two real numbers such that $1 \leq p,\ q < +\infty$. For all functions $f \in L^p \cap L^q$, there exists a sequence of functions $f_n \in L(X)$ converging to $f$ almost everywhere, as well as in $L^p$ and in $L^q$.*

Indeed, it is possible to extract from $L(X)$ two sequences $(g_n)$ and $(h_n)$ that are convergent to $f$ ae. and such that $f$ is the limit in $L^p$ (resp. $L^q$) of the sequence $(g_n)$ (resp. $(h_n)$). They can even be assumed to be dominated by two functions $G \in L^p$ and $H \in L^q$ (corollary 3 of Riesz-Fischer). The function $\omega_n(x) = g_n(x)/|g_n(x)|$ is defined and continuous on the open set $U_n = \{g_n \neq 0\}$. Let us define $f_n$ by

$$f_n(x) = \begin{array}{l} \inf(|g_n(x)|, |h_n(x)|)\, \omega_n(x) \quad \text{if } g_n(x) \neq 0\,, \\ 0 \quad \text{if } g_n(x) = 0\,. \end{array}$$

The function $f_n$ is continuous on $U_n$, and as $|f_n(x)| \leq |g_n(x)|$ everywhere, it is also continuous at every $x \in X - U_n$. As a result, $f_n \in L(X)$. $f_n(x)$ obviously converges ae. to $f(x)$, and as $f_n$ is dominated by $G$ *and* by $H$, convergence in $L^p$ *and* in $L^q$ follow, qed.

(ii) *Relation between $L^p$ and $L^1$; Hölder's inequality.* Lebesgue's theorem is going to enable us to explain the expression "$p$-th power integrable functions":

**Theorem 10**. *A function $f$ is in $L^p$ if and only if the function $|f|^{p-1}f$ is in $L^1$. Then so is the function $|f|^p$ and*

$$N_p(f) = \mu\left(|f|^p\right)^{1/p}\,. \tag{5.2}$$

We obviously assume that $p > 1$ and confine ourselves to finite-valued functions as others reduce to them thanks to null sets.

Let us suppose that $f \in L^p$, choose (corollary 1 of theorem 7) a series of functions $f_n \in L(X)$ such that

$$\sum N_p(f_n) < +\infty\,, \quad f(x) = \sum f_n(x) \text{ ae.}$$

and set

$$s_n(x) = f_1(x) + \ldots + f_n(x)\,, \quad S(x) = \sum |f_n(x)|\,.$$

The functions $g_n = |s_n|^{p-1}s_n$ are still in $L(X)$ since $p > 1$. On the other hand, $|g_n| = |s_n|^p \leq S^p = G$ with a function $G$ having values in $[0, +\infty]$ satisfying

$$N_1(G) = N_1(S^p) = N_p\left(\sum |f_k|\right)^p \leq \left(\sum N_p(f_k)\right)^p < +\infty .$$

The dominated convergence theorem can therefore be applied in $L^1$ to the sequence $(g_n)$, and as it converges ae. to the function $g = |f|^{p-1}f$, it is integrable, and hence so is its absolute value $|f|^p$. The same argument applies to functions $|g_n|$, which converge ae. to $|f|^p$, whence

$$\mu(|f|^p) = \lim \mu(|g_n|) = \lim \mu(|s_n|^p) = \lim N_p(s_n)^p$$

since the $s_n \in L(X)$ are integrable. Since the series $\sum f_n$ converges to $f$ in $L^p$, it finally follows that

$$\mu(|f|^p) = N_p(f)^p ,$$

whence (2).

Conversely, suppose that $g = |f|^{p-1}f \in L^p$. The theorem being trivial for $p = 1$, one may suppose that $p > 1$. The map $z \mapsto |z|^{p-1}z$ from $\mathbb{C}$ to $\mathbb{C}$ is a homeomorphism. Its inverse is $z \mapsto |z|^{-1/q}z$, with $1/p + 1/q = 1$ and so $0 < 1/q < 1$. To ensure continuity at 0, one needs to assume $|0|^{-1/q}.0 = 0$. As a result,

$$f(x) = |g(x)|^{-1/q}\, g(x) . \tag{5.3}$$

Let $g_n \in L(X)$ be a sequence of functions such that

$$\sum N_1(g_n) < +\infty , \quad g(x) = \sum g_n(x) \text{ ae.} \tag{5.4}$$

and we set

$$f_n = |g_1 + \ldots + g_n|^{-1/q} (g_1 + \ldots + g_n) .$$

The map $z \mapsto |z|^{-1/q}z$ from $\mathbb{C}$ to $\mathbb{C}$ being continuous and zero at $z = 0$, like $g_k$ for all $k$, $f_n$ is in $L(X)$ for all $n$. Then

$$|f_n| = |g_1 + \ldots + g_n|^{1/p} \leq \left(\sum |g_k|\right)^{1/p} = h .$$

The first relation (4) implies that $N_1(\sum |g_k|) < +\infty$, and as a result, $N_p(h)^p = N_1(|h|^p) = N_1(\sum |g_k|) < +\infty$. Therefore, the dominated convergence theorem in $L^p$ applies to the sequence $(f_n)$. However, (4) shows that it converges ae. to $|g(x)|^{-1/q}g(x) = f(x)$, whence $f \in L^p$, qed.

**Theorem 11 (Hölder).** *Let $p, q > 1$ and $r \geq 1$ be real numbers such that $1/p + 1/q = 1/r$. Then*

$$f \in L^p \quad \& \quad g \in L^q \Longrightarrow fg \in L^r \tag{5.5}$$

*and*

$$N_r(fg) \leq N_p(f)N_q(g)\,. \tag{5.6}$$

Relation (6) is a particular case of (3.4'), which holds without any integrability assumptions. It, therefore, suffices to prove (5). To this end, we choose sequences $(f_n)$ and $(g_n)$ in $L(X)$ such that

$$\lim N_p(f - f_n) = 0 \quad \& \quad \lim N_q(g - g_n) = 0\,.$$

The proof reduces to showing that $\lim N_r(fg - f_n g_n) = 0$. Now, $(f_n)$ may be assumed to converge almost everywhere to $f$ and to be dominated by a function $F \geq 0$ such that $N_p(F) < +\infty$; even $(g_n)$ may be assumed to converge ae. to $g$ and to be dominated by some $G \geq 0$ such that $N_q(G) < +\infty$. Then the product $f_n g_n$, being dominated by the function $FG$ whose norm $N_r$ is finite by (6), converges ae. to $fg$. Applying Lebesgue's theorem completes the proof.

**Corollary 1.** *If $f, g \in L^2$, then $fg$ is integrable and*

$$|\mu(fg)| \leq N_2(f)N_2(g)\,. \tag{5.7}$$

This corollary shows that a *Hilbert inner product*

$$(f|g) = \mu(f\overline{g}) = \int f(x)\overline{g(x)}d\mu(x) \tag{5.8}$$

can be defined on $L^2$. It is obviously linear[14] in $f$, satisfies the condition $(f|g) = \overline{(g|f)}$, and finally, by theorem 9,

$$(f|f) = \mu\left(|f|^2\right) = N_2(f)^2 \geq 0\,,$$

which reduces (7) to the Cauchy-Schwarz inequality. This shows that $(f|f) = 0$ implies $f = 0$ ae., or $f = 0$ if one argues, as one should, in terms of classes mod $\mathcal{N}$. As, moreover, $L^2$ is complete, we conclude that $L^2$ is a genuine *Hilbert space.* This result explains its importance in applications since the theory of Hilbert spaces is far simpler and more complete than that of other classes of topological vector spaces. We will again come across the $L^2$ spaces in relation to Fourier transforms, but they have many other applications.

[14] One can formulate the same definition on $F^2$, but the result would not be an inner product: the function $\mu^*$ is not linear on $F^1$.

Relation (6) is mainly of interest for $1/p+1/q=1$, from which $r=1$ and the same conclusion as above follow: for $f \in L^p$ and $g \in L^q$, the function $fg$ is integrable and

$$|\mu(fg)| \leq N_1(fg) \leq N_p(f)N_q(g) \,. \tag{5.9}$$

As a result, for given $g \in L^q$, the map

$$f \longmapsto \mu(fg)$$

is a *continuous linear functional* on $L^p$. It will be shown later (n° 18, theorem 30) that it is the only one. As $q$ must be finite, this supposes that $p>1$. For $p=1$, a space $L^\infty$, which we shall also define later, is needed.

**Corollary 2.** *If $f \in L^p$, then $f\chi_K \in L^1$ for every compact set $K \subset X$ and $fg \in L^1$ for all $g \in L(X)$.*

Indeed, the function $\chi_K$ is in $L^1$ (n° 4, lemma 7) and hence is in $L^q$ for all $q$ (theorem 10), whence $f\chi_K \in L^1$. Use the same argument for the product $fg$.

In fact, both propositions of corollary 2 are equivalent:

**Lemma 5.** *The following two properties are equivalent for all functions* $\mathbf{j}$*:*
*(LI 1) the function $\mathbf{j}(x)\chi_K(x)$ is integrable for all compact sets $K$;*
*(LI 2) the function $\mathbf{j}(x)f(x)$ is integrable for all $f \in L(X)$.*
*The map $f \mapsto \mu(f\mathbf{j})$ is then a complex measure on $X$.*

(LI 1) $\Longrightarrow$ (LI 2): if $f$ vanishes outside a compact set $K$, then $f\mathbf{j} = f.\mathbf{j}\chi_K$. Hence, corollary 2 applied to $f\chi_K$ gives the result.

(LI 2) $\Longrightarrow$ (LI 1): if $f\mathbf{j} \in L^1$ for all $f \in L(X)$, $f$ can be chosen to be equal to 1 on $K$, so that, by corollary 2, $\mathbf{j}\chi_K = f\mathbf{j}.\chi_K$ is integrable.

The map $f \mapsto \mu(f\mathbf{j})$ is obviously a linear functional on $L(X)$. If $f \in L(X,K)$, where $K$ is a compact set, then

$$|f(x)\mathbf{j}(x)| = |f(x)\chi_K(x)\mathbf{j}(x)| \leq \|f\| . \,|\mathbf{j}(x)\chi_K(x)|$$

and so

$$|\mu(f\mathbf{j})| \leq \|\mathbf{j}\chi_K\|_1 \cdot \|f\| \,,$$

whence the continuity of $f \mapsto \mu(f\mathbf{j})$ on $L(X,K)$, qed.

A function $\mathbf{j}$ is said to be *locally integrable* with respect to $\mu$ if it satisfies the equivalent conditions of lemma 5. We then set[15]

$$\int_K \mathbf{j}(x)d\mu(x) = \int \mathbf{j}(x)\chi_K(x)d\mu(x) \tag{5.10}$$

[15] One can define the integral over any measurable set $A$ such that $\mathbf{j}(x)\chi_A(x)$ is integrable in a similar way, for example if $A$ is contained in a compact set.

for every compact set $K \subset X$, and we say that $f \mapsto \mu(f\mathbf{j}) = \nu(f)$ is the *measure with density* $\mathbf{j}$ (for "Jacobian") *with respect* to $\mu$. This is symbolically written as

$$d\nu(x) = \mathbf{j}(x)d\mu(x).$$

Hence, by definition,

$$\int f(x).\mathbf{j}(x)d\mu(x) = \int f(x)\mathbf{j}(x).d\mu(x)$$

for all $f \in L(X)$. A punctuation mark has been inserted in every integral so as to separate the function being integrated from the measure with respect to which it is integrated. We will see in n° 16 that the previous relation generalizes to integrable functions $f$ with respect to the measure $\mathbf{j}d\mu$, but this result, "obvious from a physics viewpoint" is far from being so from a mathematical one...

For example, $d^*x = dx/|x|$ is a measure on $\mathbb{R}^*$ (but not on $\mathbb{R}$).

(iii) *Applications to Fourier transforms on* $\mathbb{R}$. The definition of $L^p$ spaces adopted above can be directly applied. Let us show for example that *there exists a unique isomorphism from* $L^2(\mathbb{R})$ *onto* $L^2(\mathbb{R})$ *which, on the Schwartz space* $\mathcal{S}$ [16]*, reduces to the Fourier transform.* Indeed, we know that the map $f \mapsto \widehat{f}$ from $\mathcal{S}$ to $\mathcal{S}$ is bijective and preserves the $L^2$ norm (Chap. VII, n° 31, theorem 28). On the other hand, $\mathcal{S}$ is everywhere dense in $L^2(\mathbb{R})$ because any function $f \in L(\mathbb{R})$ can be approximated using $C^\infty$ functions zero outside a fixed compact set (Chap. V, n° 27, theorem 26), which implies convergence in all $L^p$ spaces. Thus the Fourier transform extends to $L^2(\mathbb{R})$ and the extension is unique. The same would be the case if $\mathbb{R}$ were replaced with $\mathbb{T}$: for every $f \in L^2(\mathbb{T})$, there is a Fourier series $\sum \widehat{f}(n)u^n$ for which

$$\sum \left|\widehat{f}(n)\right|^2 = \|f\|_2^2,$$

and conversely. The convergence of the series is not an obvious consequence. However, it is possible to show that the symmetric partial sums of the series $\sum \widehat{f}(n)e_n(u)$ converge almost everywhere to $f(u)$, a difficult result to prove – and false for $L^1(\mathbb{T})$.

One can also define the Fourier transform in $L^1(\mathbb{R})$: since $L(\mathbb{R})$ is everywhere dense in $L^1(\mathbb{R})$ and since

$$\|\widehat{f}\| \leq \|f\|_1$$

holds trivially for all $f \in L(\mathbb{R})$, the left hand side being as usual the uniform norm on $\mathbb{R}$, the map $f \mapsto \widehat{f}$ can be extended by continuity to a map from

[16] Recall that, it is the space of $C^\infty$ functions $\varphi$ on $\mathbb{R}$ for which all functions $x^p\varphi^{(q)}(x)$ are bounded.

$L^1(\mathbb{R})$ to the space of continuous functions on $\mathbb{R}$, which is obviously given by the formula

$$\widehat{f}(y) = \int f(x)\overline{\mathbf{e}(xy)}dx \text{ ae.} \tag{5.11}$$

where, like in Chap. VII, we set $\mathbf{e}(x)$ for $\exp(2\pi ix)$. As we know that $\widehat{f}$ (Chap. VII, n° 27, theorem 23) tends to 0 at infinity for $f \in L(\mathbb{R})$, the Fourier transform maps $L^1(\mathbb{R})$ to the space $L_\infty(\mathbb{R})$ of continuous functions tending to 0 at infinity.

It should be noted that for $f \in L^1(\mathbb{R}) \cap L^2(\mathbb{R})$, we thus have at our disposal two possible definitions of the Fourier transform. Let us show that they are compatible. Indeed, we know (corollary of theorem 9) that there is a sequence of functions $f_n \in L(\mathbb{R})$ converging to $f$ both in $L^1$ and in $L^2$. Then $(\widehat{f_n})$, given by (11), converges in $L_\infty(\mathbb{R})$ (uniform convergence) to function (11) for $f$, and in $L^2(\mathbb{R})$ to the Fourier transform of $f$ in the $L^2$ sense. Hence by the Riesz-Fischer theorem for $L^2$, it is the (class of) function (11).

This result enables us to calculate $\widehat{f}$ using a formula analogous to (11) when$f \in L^2(\mathbb{R})$. Indeed, let $f_n$ denote the function equal to $f$ for $|x| \leq n$ and 0 elsewhere. By corollary 2 above, it is in $L^1$. It is also in $L^2$ since, being the product of $|f|f \in L^1$ and the characteristic function of $[-n, n]$, $|f_n|f_n$ is in $L^1$. Its Fourier transform can therefore be calculated using (11). On the other hand, $f_n$ clearly converges (dominated convergence) to $f$ in $L^2$. As a result,

$$\widehat{f}(y) = \text{l.i.m.}^2 \int f_n(x)\overline{\mathbf{e}(xy)}dx$$

or using notation (10),

$$\widehat{f}(y) = \text{l.i.m.}^2 \int_{-n}^{+n} f(x)\overline{\mathbf{e}(xy)}dx\,. \tag{5.12}$$

If, for example $f$ is a regulated function approaching 0 monotonously at infinity, the (ordinary) limit of the integral exists for all $y \neq 0$ (Chap. V, n° 24, theorem 23). So, if we know beforehand that $f \in L^2$, in other words (exercise!) that the *Riemann* integral $\int |f(x)|^2 dx$ converges, we can replace the symbol l.i.m.$^2$ with an ordinary limit for $y \neq 0$, which by the way shows that it is necessary to include functions that are only defined almost everywhere.

We will see in n° 10, (iv) how to generalize the Fourier inversion formula (Chap. VII, n° 30, theorem 26) to functions $f \in L^1(\mathbb{R})$, namely that, if $\widehat{f} \in L^1(\mathbb{R})$ as well, then

$$f(x) = \int \widehat{f}(y)\mathbf{e}(xy)dy\,. \tag{5.13}$$

Finally note that all this falls in the framework of Fourier transforms of tempered distributions [Chap. VII, n° 32, formula (32.11)]. It should first be observed that every $f \in L^p(\mathbb{R})$ defines a distribution $T_f$ given by

$$T_f(\varphi) = \int \varphi(x) f(x) dx \quad \text{for } \varphi \in \mathcal{S}\,.$$

The integral is well-defined since for all $q$, $\varphi \in L^q$, and Hölder's inequality readily shows (exercise!) that $T_f$ is "tempered". By definition, the Fourier transform of $T_f$ is the distribution

$$T_{\widehat{f}}(\varphi) = T_f(\widehat{\varphi}) = \int \widehat{\varphi}(y) f(y) dy\,.$$

For $f \in L^2$, this integral is the inner product of $\widehat{\varphi}$ and $\overline{f}$, and so is also equal to the inner product of $\widehat{\widehat{\varphi}}$ and the Fourier transform of $\overline{f}$, which is the conjugate of $\widehat{f}(-x)$. As $\widehat{\widehat{\varphi}}(x) = \varphi(-x)$ for $\varphi \in \mathcal{S}$,

$$T_{\widehat{f}}(\varphi) = \int \varphi(x) \widehat{f}(x) dx$$

readily follows. As a result, *the Fourier transform of the distribution defined by $f$ is the distribution defined by $\widehat{f}$.*

If it is possible to identify a function $f \in L^p$ with the distribution $T_f$, then the latter can be assigned a Fourier transform, but it is not always defined by a *function.*

# § 3. Measurable Sets and Functions

The definition of integrable functions as limits in mean of continuous functions with compact support cannot always be used. This § will show that a function $f$ is in $L^p$ if and only if it is not too complicated and, of course, that it satisfies $N_p(f) < +\infty$.

## 6 – Measurable and Integrable Sets

(i) *Properties of integrable sets.* In line with the general principle stated in n° 2, a set $A \subset X$ is *integrable* if so is its characteristic function $\chi_A$; the *measure* of A is then the number

$$\mu(A) = \mu(\chi_A) = \mu^*(A). \tag{6.1}$$

The following properties are immediate restatements of results from the previous n°. An open set $U$ is integrable if and only if $\mu^*(U) < +\infty$, and then $\mu(U) = \mu^*(U)$ (n° 4, lemma 3). The intersection of a finite or countable family of integrable sets is integrable (n° 5, lemma 3). In addition, for a decreasing sequence of integrable sets $(A_n)$,

$$\mu\left(\bigcap A_n\right) = \inf \mu(A_n) = \lim \mu(A_n). \tag{6.2}$$

The union of a finite family of integrable sets is integrable (n° 5, lemma 1). The same is true for the union of a countable family $(A_n)$ provided its outer measure is finite (n° 5, lemma 4) or, equivalently, provided there is a set of finite outer measure containing all the sets $A_n$; then

$$\mu\left(\bigcup A_n\right) \leq \sum \mu(A_n) \tag{6.3}$$

since the characteristic function of $A$ is bounded above by the sum of those of the sets $A_n$. Furthermore, the relation $\sum \mu(A_n) < +\infty$ suffices to ensure the integrability of $\bigcup A_n$, and if the sets $A_n$ are pairwise disjoint, inequality (3) becomes an equality (theorem 6). If, moreover, the sequence $A_n$ is increasing, then

$$\mu\left(\bigcup A_n\right) = \sup \mu(A_n) = \lim \mu(A_n) \tag{6.4}$$

(theorem 8 of n° 5).

Finally, it is clear that if $A$ and $B$ are integrable and if $A \subset B$, then $B - A$ is integrable and

$$\mu(B - A) = \mu(B) - \mu(A). \tag{6.5}$$

Any compact set $K$ is integrable since its characteristic function is usc and it upper integral is finite.

Let us, for example, show that the Cantor set $C$ has measure zero. This follows by removing from $K = [0,1]$ its median interval $]1/3, 2/3[$, then by repeating this process for the two remaining intervals, then for each of the four remaining ones, and so on. The sum of the measures of the strictly open intervals equals

$$1/3 + 2/3^2 + 2^2/3^3 + \ldots = 1\,,$$

and as they are pairwise disjoint, $\mu(K - C) = 1$, and so $\mu(C) = 0$.

**Lemma 1.** *A set $A \subset X$ is integrable if and only if, for all $r > 0$, there is an open set $U$ and a compact set $K$ such that*

$$K \subset A \subset U \quad \& \quad \mu(U - K) < r\,.$$

*Every integrable set is the union of a null set and a countable family of compact sets.*

By lemma 4 and n° 4, the condition is sufficient since the characteristic functions of $K$ and $U$ are respectively usc and lsc. To show that it is necessary, we choose integrable functions $\varphi$ and $\psi$ respectively lsc and usc such that

$$\psi \leq \chi_A \leq \varphi \quad \text{and} \quad \mu(\varphi - \psi) \leq \epsilon\,, \tag{6.6}$$

where $\epsilon > 0$ is given. They can be assumed to be positive, if need replacing them with $\varphi^+$ and $\psi^+$.

For any integer $n \geq 1$, the relation $\varphi(x) > 1 - 1/n$ defines an open set $U \supset A$, and $(1 - 1/n)\mu(U) \leq \mu(\varphi) \leq \mu(A) + \epsilon$. So $\mu(U) \leq \mu(A) + r$ if $\epsilon$ is sufficiently small and $n$ sufficiently large.

On the other hand, the function $\psi$ being usc and positive, and so bounded above by a continuous, positive function with compact support, the relation $\psi(x) \geq 1/2$ defines a compact set $K \subset A$. As $\psi(x) < 1/2$ on $A - K$ and $\psi(x) \leq 1$ on $K$, $\psi(x) \leq \chi_{A-K}(x)/2 + \chi_K(x)$ for all $x \in A$, and in fact for all $x \in X$. As a result, $\mu(\psi) \leq \mu(A - K)/2 + \mu(K)$. But (6) shows that

$$\mu(\chi_A - \psi) = \mu(A) - \mu(\psi) \leq \mu(\varphi - \psi) \leq \epsilon\,.$$

Hence, it follows that

$$\mu(A) - \epsilon \leq \mu(\psi) \leq \mu(A - K)/2 + \mu(K)\,.$$

So $\mu(A - K) \leq \epsilon + \mu(A - K)/2$ and finally $\mu(A - K) \leq 2\epsilon$.

At last, for compact sets $K_n \subset A$ and open sets $U_n \supset A$ such that $\mu(U_n - K_n) < 1/n$, the set $N = A - \bigcup K_n$, contained in $U_n - K_n$ for all $n$, has measure zero, qed.

A consequence of the previous lemma is that *any reasonable set $E \subset X$ is contained in the union of a null set and a countable family of compact sets.*

The converse is obvious. Indeed, by definition, $E \subset \bigcup E_n$, where $\mu^*(E_n) < +\infty$ for all $n$. By (2.3'), each $E_n$ is contained in an open set $U_n$ such that $\mu^*(U_n) < +\infty$, and so is integrable. As a result, $U_n = N_n \cup \bigcup K_{np}$, where the sets $K_{np}$ are compact and $N_n$ null sets. $E$ is, therefore, contained in the union of a null set $N = \bigcup N_n$ and a family of sets $K_{np}$, qed.

The entire space $X$ may be integrable, in particular if $X$ is compact. The measure $\mu$ is then said to be *bounded.* The whole of $X$ is then clearly reasonable, and hence so are all subsets of $X$. As, on the other hand, $|f| \le \|f\|\chi_X$ for any function $f \in L(X)$, it follows that

$$|\mu(f)| \le M\|f\| \qquad \text{where } M = \mu(X)\,.$$

Conversely, every measure satisfying an inequality of this type is bounded since $\mu^*(X)$, a supremum of $\mu(f)$ for $f \in L(X)$ such that $0 \le f \le 1$, is then $\le M$. Hence if we equip $L(X)$ with the norm $\|f\| = \|f\|_X$ of uniform convergence on $X$, bounded measures are just the *continuous* linear functionals on $L(X)$ that are positive for $f \ge 0$. If for example $\mathbf{j}$ is a positive integrable function, then the function $f\mathbf{j}$ is integrable for all $f \in L(\mathbb{R})$ (corollary 2 of theorem 11) and the map

$$f \longmapsto \mu(f\mathbf{j}) = \int f(x)\mathbf{j}(x)d\mu(x)$$

is a bounded measure since $|f\mathbf{j}| \le \|f\|\mathbf{j}$ and so

$$|\mu(f\mathbf{j})| \le N_1(\mathbf{j})\|f\|\,.$$

We will show later [n° 17, (ii)] that this property *characterizes* integrable functions.

(ii) *Measurable sets.* The notion of a measurable set is at least as useful as that of an integrable one. That is the name given to any set $A \subset X$ such that $A \cap K$ is *integrable* for all compact sets $K \subset X$. This is the case of all integrable, closed or open sets. Properties of integrable sets regarding unions and intersections readily show that

(i) *any finite or countable intersection of measurable sets is measurable,*

(ii) *the complement of a measurable set is measurable*, and hence

(iii) *any finite or countable union of measurable sets is measurable.*

In addition:

(iv) *a set $A$ is integrable if and only it is measurable and of finite outer measure.*

Indeed, if $A$ is integrable, then so is $A \cap K$ for any compact set $K$. Thus $A$ is measurable. Conversely, suppose that $A$ is measurable and that $\mu^*(A) < +\infty$. There is a lsc function $\varphi \ge \chi_A$ such that $\mu^*(\varphi) < +\infty$. It is integrable (n° 4, lemma 3), and so vanishes outside a set of the form $N \cup \bigcup K_n$, where $N$ is null and the $K_n$ are compact (n° 4, corollary 2 of theorem 7). The same is true for $\chi_A$. Thus

$$A = (A \cap N) \cup \bigcup (A \cap K_n) ,$$

which is the countable union of integrable sets since $A$ is measurable. As $\mu^*(A) < +\infty$, this union is also integrable, qed.

One could add that

(v) *a measurable set $A$ is reasonable if and only if it is the union of a null set and a countable family of compact sets.*

We saw above that, if $A$ is reasonable, then it is contained in a set of the form $N \cup \bigcup K_n$, where $N$ is null and the $K_n$ are compact. $A \cap N$ is null and $A \cap K_n$ integrable, and so of the form $N_n \cup \bigcup K_{np}$, qed.

Properties (i) and (ii) are sometimes expressed as saying that the set of measurable subsets of $X$ is a *tribe* or a *$\sigma$-algebra*, a barbarous expression[17] used by abstract measure specialists. These results enable us to construct more and more complicated measurable sets from open or closed sets: countable intersections of countable unions of countable intersections of... As no one has ever constructed a non- measure set without resorting to the axiom of choice in some form or other, it may be thought that none are encountered in classical mathematics. In fact, as the Lebesgue-Fubini theorem will show, there are fields where caution is warranted.

A tribe is a subset of the set $\mathcal{P}(X)$ of subsets of $X$ (Chap. I, n° 4), and it readily follows from the definition that any intersection of tribes is again a tribe. Hence, one may talk of a *smallest* tribe containing the open sets. By definition, its elements are the *Borel* sets. They are measurable with respect to all measures $\mu$ on $X$, which is why they are of interest. The next simplest ones are the countable intersections of open sets (the $G_\delta$ sets in usual notation, a category containing the closed subsets if $X$ is metrizable[18]), then the countable unions of $G_\delta$ sets (the $G_{\delta\sigma}$ sets), then the countable intersections of $G_{\delta\sigma}$ sets (the $G_{\delta\sigma\delta}$ sets), etc. From these classes of sets, in general pairwise distinct and each larger than the preceding one, the reader may get the impression that all Borel sets can be obtained by this rather simple mechanical process. False. Choosing in each of these classes a set not belonging to any of the previous ones gives the union of a countable family of Borel sets that does not in general belong to any of these classes. The construction would need to be continued beyond countability.

---

[17] In the special issue of *Pour la Science* dedicated to N. Bourbaki, it is said that the terms "clan, phratry, tribe" used by this author before he decided to reject abstract measures show his lack of taste for the topic. It was more a matter of wishing to express himself in French. The English translation of N. Bourbaki's book came back to *$\sigma$-algebras*, probably because of what the English and Americans themselves call the *NIH syndrome (not invented here)*...

[18] If $X$ is a metric space, any closed subset $F$ is a $G_\delta$ set since it is the intersection of the open sets defined by the inequalities $d(x, F) < 1/n$. I take it that the letters $\delta$ and $G$ comes from the German for intersection, *Durchschnitt* and open *geffnet*.

The notion of a tribe explains the theory of abstract measures.[19] Take a set $E$ without topology, a tribe of subsets of $E$ and a function $\mu$ on it with values in $[0, +\infty]$ having the same formal properties (countable additivity) as a Radon measure. This leads to the definition of integrable functions, $L^p$ spaces and to Lebesgue's theorems. For example, if there is a Radon measure $\lambda$ on a locally compact space $X$ and if $E$ is taken to be a *measurable* subset of $X$, the measurable sets contained in $E$ form a tribe on which the function $\mu(A) = \lambda(A)$ is a measure on $E$ in the previous sense. If $E$ is locally compact (i.e. the intersection of an open and a closed set), $\mu$ is the Radon measure on $E$ which can be directly defined by putting

$$\mu(f) = \lambda(f')$$

where, for all $f \in L(E)$, $f'$ denotes the function equal to $f$ on $E$ and 0 elsewhere. As a compact subset of $E$ is also a compact subset of $X$, $f' \in L(X)$. $\mu$ is said to be the *measure induced* by $\lambda$ on $E$. For example, this enables us to talk of the Lebesgue measure on a locally compact subset of a space $\mathbb{R}^n$. If $\mu$ is defined by the previous formula, we immediately encounter the problem of having to prove the following result, "obvious from a physics viewpoint": a function $f$ on $E$ is integrable with respect to $\mu$ if and only if so is its extension $f'$ with respect to $\lambda$, in which case the previous formula continues to hold. This in fact require a non-trivial proof.[20]

## 7 – Measurable Functions

(i) *Separable spaces.* Let us consider a locally compact space $X$, a positive measure $\mu$ on $X$ and a sequence of maps $f_n$ from $X$ to a *metric* space $P$. Suppose that the sequence $(f_n)$ converges everywhere to a limit function $f(x)$. For a given $a \in P$ and a given number $R > 0$, let $A \subset X$ be the set of $x$ where

$$d\,[f(x), a] < R\,.$$

We intend to compute it using similar sets with respect to the functions $f_n$.

The relation $x \in A$ means that there are integers $m$ and $k$ (dependent on $x$) such that

$$d\,[f_n(x), a] < R - 1/m \quad \text{for all } n \geq k\,. \tag{7.1}$$

We set

$$A_{m,n} = \{d\,[f_n(x), a] < R - 1/m\}\,.$$

For given $m$ and $k$, (1) clearly means that

$$x \in \bigcap_{n \geq k} A_{m,n}\,.$$

[19] See for example Chap. 1 in Walter Rudin's *Real and Complex Analysis*.

[20] See N. Bourbaki, *Integration*, Chap. V.

Hence

(7.2) $$A = \bigcup_{m,k>0} \bigcap_{n \geq k} A_{m,n} \,.$$

If, instead of converging everywhere, the sequence $f_n$ converged almost everywhere to $f$, the result would obviously remain the same up to a null set. This shows that if the inverse images of the open balls of $P$ under the functions $f_n$ are measurable, then $f$ also has the same property.

This argument, however, relies on using a metric $d$ defining the topology on $P$. To obtain a result involving only the topology on $P$, we would need to show that it is preserved if $d$ is replaced by another metric compatible with the topology on $P$. Now, any open $d'$-ball (and more generally any open subset of $P$) is then a union of open $d$-balls. Hence there is no problem if every open $d'$-ball is known to be the *countable* union of open $d$-balls and conversely. It is simplest to suppose that all open subsets $U$ of $P$ is the countable union of open balls, a property independent of the chosen metric since a family $(B_{mn})$ dependent on two integer indices is still countable.

This is for example the case if $P = \mathbb{R}^n$: every open subset of $P$ are the unions of open balls of rational radii and centred at points with rational coordinates. Let us suppose that in the general case there is a *countable and everywhere dense* set $D$ in the metric space $P$. $P$ is then said to be *separable.* If $U$ is open in $P$ and if $a \in U$, $U$ contains an open ball $B(a,r)$ with $r \in \mathbb{Q}$, and the ball $B(a,r/2)$ contains some $d \in D$. The ball $B(d,r/2)$ is then contained in $U$ and contains $a$. Since $a \in U$ is arbitrary, $U$ is the union of open balls centred at points of $D$ and of rational radii, whence the result.

The existence of $D$ is equivalent to the next property:

(TD) There is a countable family of open set $(U_n)$ such that any open subset of $P$ is the union of those $U_n$ contained in it.

We saw above that (TD) is clearly necessary. Conversely, if (TD) is satisfied and if, for all $n$, some $a_n \in U_n$ is chosen arbitrarily, the set $D$ of the points $a_n$ meets every open subset of $P$, and so is everywhere dense. For metrizable spaces, (TD) is thus equivalent to separability.

A separated topological space (Hausdorff axiom) satisfying (TD) is said to be *of countable type*; any family $(U_n)$ satisfying (TD) is called a *basis* for the topology of $X$. We will return in much greater detail to all this in n° 11.

(ii) *Measurable maps.* After these preliminaries, let us return to $X$ and to the measure $\mu$. A map $f$ from $X$ to a *metrizable and separable space* $P$ is said to be *measurable*[21] when it satisfies the condition

(FM): the inverse image under $f$ of every open subset of $P$ is measurable.

[21] A definition applicable to all topological spaces will be given later.

Every continuous map is obviously measurable. The arguments that have led to relation (2) then prove the next result:

**Theorem 12.** *If a sequence of measurable functions with values in a metrizable and separable space converges almost everywhere, then the limit function is measurable.*

An immediate corollary:

**Theorem 13.** *All functions belonging to a $L^p$ space are measurable.*

Because it is the limit almost everywhere of a sequence of continuous functions with values in $\mathbb{C}$, the archetype of a metrizable and separable space.

For a function $f$ with values in $\mathbb{R}$ or, more generally, in $[-\infty, +\infty]$, measurability means that $f^{-1}(I)$ is measurable for every interval $I$, including for example if $I = \{+\infty\}$, or $\{-\infty\}$, or $[3, +\infty[$, etc.. It would even suffice to require this for intervals of the form $I = [a, +\infty]$ where $a$ is finite since every other interval can be obtained from this type of intervals using a countably infinite number of standard operations. For example, $[-\infty, a[$ is the complement of $[a, +\infty]$, and $[0, a[ = [0, +\infty] \cap [-\infty, a[$. So

$$[0, 1] = [0, +\infty] \cap \bigcap [-\infty, 1 + 1/n[ .$$

Theorem 13 continues to hold for these functions since they are anyhow finite almost everywhere. We will see later that, modulo the condition $N_p(f) < +\infty$, this theorem *characterizes* the functions of $L^p$.

A measurable function remains so if it is modified on a null set. Hence the definition can be applied to functions defined almost everywhere.

Property (FM) can be generalized:

**Theorem 14.** *Let $P$ be a metrizable and separable set. The inverse image of every Borel set $B \subset P$ under a measurable map $f : X \longrightarrow P$ is measurable.*

Let $\mathcal{T}$ be the set of subsets $E$ of $P$ such that $f^{-1}(E)$ is measurable. Elementary formulae about the inverse image of an intersection or a complement show that, like the family of measurable subsets of $X$, $\mathcal{T}$ is a tribe. However, by definition, $\mathcal{T}$ contains all the open subsets of $P$, and hence also all its Borel subsets.

**Lemma 1.** *Let $f_1, \ldots, f_n$ be maps from $X$ to metrizable and separable sets $P_1, \ldots, P_n$. Then the map*

$$f : x \mapsto (f_1(x), \ldots, f_n(x))$$

*from $X$ to the product space $P = P_1 \times \ldots \times P_n$ is measurable if and only if so are the maps $f_i$.*

First, the product space is metrizable and separable since if, for each $i$, $D_i$ is everywhere dense in $P_i$, then the product of the sets $D_i$ is everywhere dense in $P$. If $f$ is measurable, then, for any Borel set $B \subset P_1$, the set

$$f^{-1}(B \times P_2 \times \ldots \times P_n) = f_1^{-1}(B)$$

is measurable. Hence the condition is necessary.

To show that it is sufficient, note that, if $U_i$ is an open subset of $P_i$, the product of the sets $U_i$ is open in $P$,. Since

$$f^{-1}(U_1 \times \ldots \times U_n) = \bigcap f_i^{-1}(U_i),$$

the left hand side is measurable. Hence so is $f^{-1}(U)$ for any open set $U \subset P$ if $U$ is shown to be the countable union of a product of open sets. To see this, we equip $P$ with the distance

$$d(a,b) = \sup d_i(a_i, b_i),$$

where $d_i$ is the distance on $P_i$. It defines the topology of the product. The open balls of the product are then products of open balls in the sets $P_i$, which are unions of open balls of rational radii centred at points of countable and everywhere dense sets $D_i \subset P_i$, qed.

In the next statement, if $P$ and $Q$ are separable metric spaces, $\varphi : P \longrightarrow Q$ will be said to be a *Borel* map if, for any open ball $V \subset Q$, $\varphi^{-1}(V)$ is a Borel subset of $P$. Then, more generally, the same holds for all Borel sets $V$ since the set of subsets $V$ for which $f^{-1}(V)$ is a Borel set is a tribe. A Borel map from $X$ to a separable and metrizable space is, therefore, measurable. The argument used to prove theorem 12 shows that *the limit $f$ of a sequence of Borel maps $f_n$ converging EVERYWHERE is a Borel map,*[22] since the sets $A_{m,n}$ of relation (2) and so $A$ are then Borel sets. For example, any regulated function on $\mathbb{R}$ is a Borel map.

**Lemma 2.** *Let $P$ and $Q$ be two separable and metrizable spaces, $\varphi$ a Borel map from $P$ to $Q$ and $f$ a measurable map from $X$ to $P$. Then $g = \varphi \circ f : X \longrightarrow Q$ is measurable.*

It suffices to write that $g^{-1}(B) = f^{-1}(B')$, where $B' = \varphi^{-1}(B)$.

**Theorem 15.** *Let $P_1, \ldots, P_n$ and $Q$ be separable and metrizable spaces, $\varphi$ a* Borel *map from $P_1 \times \ldots \times P_n$ to $Q$ and $f_i : X \longrightarrow P_i$ measurable maps. Then the map*

$$x \longmapsto \varphi[f_1(x), \ldots, f_n(x)]$$

*is measurable.*

This is a direct consequence of the previous two lemma.

Theorems 12 and 15 enable us to check that *all usual operations (quasi-algebraic formulae, passages to the limit) that can be performed on measurable*

[22] Were convergence ae. sufficient, then (among others) any integrable function would be a Borel map, any null set would be a Borel set, and hence so would be any subset of such a set, etc.

*functions lead to measurable functions.* In fact, far more ingenuity would be needed to invent, using explicit formulae, non-measurable maps, possibly even non-Borel maps. However, see end of n° 11: the image of a measurable (resp. Borel) set under a *continuous* map may not be a measurable (resp. Borel) map, including in the case of the projection $(x, y) \mapsto x$ from $\mathbb{R}^2$ onto $\mathbb{R}$.

The set of measurable functions with values in $\mathbb{C}$ (or in a separable Banach space) is a vector space since $(u, v) \mapsto \alpha u + \beta v$ from $\mathbb{C} \times \mathbb{C}$ to $\mathbb{C}$ is a Borel map. Together with theorem 12, this result shows that if a series $\sum f_n(x)$ of measurable functions converges ae., its sum, i.e. the limit of its partial sums, is measurable. The same holds for functions $f_n$ with values in $[-\infty, +\infty]$, provided the expression $(-\infty) + (+\infty)$ is given a meaning. This can be done (in integration theory...) by assigning once and for all an arbitrary value to it since, if $(f_n)$ is a countable family with values in $[-\infty, +\infty]$, the set of $x$ where $-\infty$ *and* $+\infty$ appear in the sequence $(f_n(x))$ is measurable as it is the intersection of the measurable sets

$$f_n^{-1}(\{-\infty\}) \cap f_n^{-1}(\{+\infty\}) .$$

The product of two complex-valued measurable functions is measurable. This result holds for any continuous bilinear map. For example, if $f$ and $g$ take values in a separable Hilbert space, the function $x \mapsto (f(x)|g(x))$ is measurable. If $f$ with values in a Banach space $F$ is measurable, so is the function $x \mapsto \|f(x)\|$ since $u \mapsto \|u\|$ is a continuous map from $F$ to $\mathbb{R}$. If $f$ and $g$, with values in a separable and metrizable space $P$, are measurable, so is the function $x \mapsto d[f(x), g(x)]$ with respect to any distance compatible with the topology of $P$, since the map $(u, v) \mapsto d(u, v)$ from $P \times P$ to $\mathbb{R}$ is continuous. If $f$ and $g$, with values in $[-\infty, +\infty]$, are measurable, so are $\sup(f, g)$ and $\inf(f, g)$. Thanks to theorem 12, we can deduce that the upper and lower envelopes of a countable family of real measure functions are measurable. Etc.

*Exercise.* Define the measurable functions with values in the Riemann sphere $\widehat{\mathbb{C}}$ [Chap. VIII, n° 5, (vi)] and show that if $f$ is measurable, then so is $1/f$. Define the map $x \mapsto 1/x$ from $[-\infty, +\infty]$ to itself by agreeing, either that $1/0 = +\infty$, or $-\infty$, or any other value. Show that if $f : X \longrightarrow [-\infty, +\infty]$ is measurable, then so is $1/f$ in all three cases.

## 8 – Measurability and Continuity

(i) *Egorov's and Lusin's theorems.* We saw above that any limit ae. of measurable functions is measurable. There is actually a more precise result whose proof, like the calculations of n° 7, (i), is a set theory exercise:

**Theorem 16 (Egorov).** *Let $(f_n)$ be a sequence of measurable functions with values in a separable metric space $P$ and converging ae. to a function $f : X \longrightarrow P$. For all $\epsilon > 0$ and all integrable sets $A \subset X$, there exists an integrable set $B \subset A$ such that $\mu(A - B) < \epsilon$ and in which the sequence $(f_n(x))$ converges* uniformly *to $f(x)$.*

Removing a null set from $A$, we may suppose that $\lim f_n(x) = f(x)$ for all $x \in A$ without exception. Henceforth, all arguments apply exclusively to $A$.

For integers $m, p \geq 1$, consider the measurable set

$$A_m(p) = \{d[f(x), f_n(x)] < 1/m \text{ for all } n \geq p\}.$$

Obviously

$$A_m(p) \subset A_m(p+1). \tag{8.1}$$

By definition of *simple* convergence,

$$A = \bigcup_{p \geq 1} A_m(p) \quad \text{for all } m. \tag{8.2}$$

Convergence is uniform in a subset $B$ of $A$ if and only if, for all $m$, there exists $p$ such that $B \subset A_m(p)$, in other words, if and only if there exists a sequence $(p_m)$ such that $B \subset \bigcap A_m(p_m)$. Therefore, the proof reduces to showing that there exist integers $p_m$ such that the set

$$B = \bigcap A_m(p_m) \tag{8.3}$$

satisfies $\mu(B) \geq \mu(A) - \epsilon$. As $A - B = \bigcup [A - A_m(p_m)]$, for this to be the case it suffices that

$$\mu(A) - \mu[A_m(p_m)] \leq \epsilon/2^m \tag{8.4}$$

for all $m$. Since for given $m$, by (1), the sequence $(A_m(p))$ is increasing, for each $m$, the existence of an integer $p_m$ satisfying (4) follows from (2), qed.

Using Egorov's theorem, measurability can be related to continuity. Indeed, the function $f$ is the limit ae. of a sequence of functions continuous everywhere – for example if $f$ belongs to a $L^p$ space –, and if the result is applied to a compact set $K \subset X$, then $f$ has the following property:

(LUS) For all compact sets $K \subset X$ and all $\epsilon > 0$, there exists a compact set $K' \subset K$ such that $\mu(K - K') < \epsilon$ and such that the restriction of $f$ to $K'$ is continuous.[23]

More generally:

**Theorem 17 (Lusin).** *A function $f$ with values in a metrizable and separable space $P$ is measurable if and only if $f$ satisfies condition* (LUS).

[23] In the following, by abuse of language, we will often say that "$f$ is continuous on $A$" meaning that the *restriction* of $f$ to $A$ is continuous. This does not mean that $f$ is continuous at every point $x \in A$ as a function on $X$ (example: the characteristic function of $A$ is continuous on $A$ in the previous sense, but discontinuous at every boundary point of $A$ as a function on $X$).

We break up the proof of this result into several parts.

(a) Let us first suppose that there is a countable partition of $X$ into measurable sets $E_n$ such that $f$ is constant on each of them. We will then say that $f$ is a *step* function. This is a less trivial analogue of the step functions of Chap. V, n° 1. All functions of this type are measurable. Indeed, if $f(x) = u_n$ on $E_n$ and if $U$ is a subset of $P$, the set $f^{-1}(U)$ is the union of the sets $E_n$ for $n$ such that $u_n \in U$. Hence it is measurable.

(b) Let us now show that all step functions $f$ have the stated property. Indeed, let $K$ be a compact set. $K$ is the disjoint union of integrable sets $K \cap E_n$. Hence, for $\epsilon > 0$ and all $n$, there is a compact set $K_n \subset K \cap E_n$ such that $\mu(K \cap E_n - K_n) < \epsilon_n = \epsilon/2^n$. Since $f$ is constant on each $E_n$, it is continuous on each $K_n$, hence also on

$$K'_n = K_1 \cup \ldots \cup K_n$$

for all $n$ since the $K_i$ are closed and pairwise disjoint. But

$$\mu(K - K'_n) = \sum_{p \leq n} \mu(K \cap E_p - K_p) + \sum_{p > n} \mu(K \cap E_p) .$$

The second sum is arbitrarily small for large $n$ since

$$\sum_{p \geq 1} \mu(K \cap E_p) = \mu(K) < +\infty$$

and the first one is bounded above by $\epsilon_1 + \ldots + \epsilon_n < \epsilon$, qed.

(c) If there is sequence $(f_n)$ of measurable functions, then for all $\epsilon > 0$, there is a compact set $K' \subset K$ such that $\mu(K - K') \leq \epsilon$ and on which *all* functions $f_n$ are continuous. Indeed, for each $n$, there are sets $K_n \subset K$ on which $f_n$ is continuous and such that $\mu(K - K_n) \leq \epsilon/2^n$. Then $\mu\left(K - \bigcap K_n\right) \leq \epsilon$ and the compact set $K' = \bigcap K_n$ answers the question.

(d) Let $f : X \longrightarrow P$ be an arbitrary measurable map. To show that it has Lusin's property, choose a distance $d$ on $P$ and a number $r > 0$. As $P$ is separable, it is the union of a sequence of open balls $A_n$ of radius $r$. Replacing the $A_n$ by the sets

$$B_1 = A_1, \ldots, B_n = A_n - A_n \cap (A_1 \cup \ldots \cup A_{n-1}) ,$$

we get a partition of $P$ into Borel sets of diameter $\leq 2r$. Since $f$ is measurable, the sets $E_n = f^{-1}(B_n)$ form a partition of $X$ into measurable sets on which the function $f$ is constant up to $2r$. For each $n$, we choose some $x_n \in E_n$ and we define a step map $f$ from $X$ to $P$ by the condition $g(x) = f(x_n)$ on $E_n$. Then the uniform distance $d(f, g) = \sup d[f(x), g(x)]$ is clearly $\leq 2r$.

Giving values of type $1/n$ to $r$, we thus see that $f$ is the *uniform* limit in $X$ of a sequence of measurable step functions $g_n$ with values in $P$. Now, section (c) of the proof shows that there is a compact set $K' \subset K$ satisfying $\mu(K - K') \leq \epsilon$ and on which the functions $g_n$ are continuous. Hence so is $f$, and property (LUS) follows for $f$.

(e) It remains to prove that conversely, any function $f$ having this property is measurable, i.e. that the set $f^{-1}(U) \cap K$ is integrable for any open set $U \subset P$ and any compact set $K \subset X$. By assumption, $K$ is the union of a sequence of compact sets $K_n$ on each of which $f$ is continuous and a null set $N$. Therefore, $f^{-1}(U) \cap K_n$ is open in $K_n$, and so is a Borel set, and thus is measurable. Hence, so is $f^{-1}(U) \cap K$, the union of the$f^{-1}(U) \cap K_n$ and a null set. As $f^{-1}(U) \cap K$ has finite outer measure, it is integrable [n° 6, property (iv)], qed.

The proof highlights the analogy between continuity and measurability: $f$ is continuous (resp. measurable) if and only if there is a countable cover of $X$ by open (resp. measurable) sets on which $f$ is constant up to $\epsilon$. This obviously supposes that $P$ is separable.

Here is a useful application of Lusin's theorem:

**Corollary**. *Let $X$ be a locally compact space, $\mu$ a positive measure on $X$ and $\mathbf{j}$ a locally integrable function with respect to $\mu$.*

$$\int f(x)\mathbf{j}(x)d\mu(x) = 0 \quad \text{for all } f \in L(X)$$

*if and only if* $\mathbf{j}(x) = 0$ *almost everywhere on every compact set*[24] $K \subset X$ [hence $\mathbf{j}(x) = 0$ almost everywhere if $X$ is countable at infinity or if $\mu$ is reasonable].

The condition is clearly sufficient. To show that it is necessary, we argue by contradiction. As $\mathbf{j}$ is measurable, there is a compact set $K$ and a measurable set $A \subset K$ with non-zero measure such that $\mathbf{j}(x) \neq 0$ for all $x \in A$. Lusin's theorem enables us to assume that $A$ is compact and that $\mathbf{j}$ continuous on $A$. The function $\overline{\mathbf{j}(x)}\chi_A(x)$ being integrable by assumption or definition, there is a sequence of functions $f_n \in L(X)$ converging almost everywhere to $\bar{\mathbf{j}}\chi_A$. As $\bar{\mathbf{j}}\chi_A$ vanishes outside the compact set $A$, the $f_n$ may be assumed to be zero outside a compact neighbourhood $B$ of $A$. As $\mathbf{j}\chi_A$ is bounded, the functions $|f_n|$ can also be assumed to be bounded above by a constant $M$ independent of $n$ (corollary 3 of theorem 7). Then the functions $f_n(x)\mathbf{j}(x)$ converge almost everywhere to $|\mathbf{j}(x)|^2\chi_A(x)$ and are dominated by the function $M\mathbf{j}(x)\chi_A(x)$, which are integrable by assumption. Hence

$$\int |\mathbf{j}(x)|^2 \chi_A(x)d\mu(x) = \lim \int \mathbf{j}(x)f_n(x)d\mu(x) = 0\,,$$

and thus $\mathbf{j}(x) = 0$ almost everywhere on $A$, a contradiction.

It should be noted that the condition in the statement says that the *measure* $\mathbf{j}(x)d\mu(x)$ with density $\mathbf{j}$ [n° 5, (ii)] is zero.

*Exercise 1.* Let $K$ be a compact set and $f$ a function vanishing outside $K$ and lsc on $K$. Show that $f$ is measurable. Deduce that for real functions

[24] This means that the set $N = \{\mathbf{j}(X) \neq 0\}$ satisfies $\mu(N \cap K) = 0$ for any compact set. Such a set is said to be *locally null*.

with values $\leq +\infty$, "continuous" can be replaced by "lsc" in the (LUS) statement.

(ii) *Lusin-measurable functions.* Apart from the importance of Lusin's theorem for its own sake, it can be used to construct a theory of measurable functions with values in a general (separated) topological space: by definition, these are maps $f$ having property (LUS) of Lusin's theorem. This point of view, taken by N. Bourbaki, often enables us to free ourselves from separability assumptions. Though they hold in most applications, as we will see later in relation to the vague topology on measure spaces, in functional analysis, we also need to for example consider functions with values in non-separable Banach spaces and even in non-metrizable topological vector spaces. In the exercises below, (LUS) is taken as the definition of measurability.

*Exercise 2.* A function equal almost everywhere to a measurable function is measurable.

*Exercise 3.* If the restrictions of a function $f$ to a sequence of measurable sets $A_n$ are continuous and if $X = N \cup \bigcup A_n$, where $\mu(N) = 0$, then $f$ is measurable.

*Exercise 4.* Let $f$ be a measurable map from $X$ to a topological space $P$. Show that $f^{-1}(B)$ is measurable for every Borel set $B \subset P$ (consider first the case of open sets). The converse may be false if $X$ is not separable.

*Exercise 5.* Let $f_i$ be finitely many maps from $X$ to topological spaces $P_i$ and $g$ a continuous map from the Cartesian product of the $P_i$ to a topological space $Q$. Show that the map

$$x \longmapsto g\left[f_1(x), \ldots, f_n(x)\right]$$

from $X$ to $Q$ is measurable.

*Exercise 6.* Let us say that a family $(f_i)$ of measurable maps from $X$ to a topological space $P$ is *equimeasurable* if for all compact sets $K$ and all $\epsilon > 0$, there exists a compact set $K' \subset K$ such that $\mu(K - K') < \epsilon$ and on which the maps $f_i$ are continuous. Show that this is always the case for a countable family of measurable functions.

*Exercise 7.* Show that Egorov's theorem continues to hold for functions with values in an arbitrary metrizable space (use exercises 4 and 6).

*Exercise 8.* (a) Let $A$ be a compact subset of a Banach space $\mathcal{H}$. Show that the closed vector subspace of $\mathcal{H}$ generated by $A$ is separable. (b) Let $X$ be a locally compact space, $\mu$ a measure on $X$ and $f$ a map from $X$ to $\mathcal{H}$. Suppose that $f$ is measurable and reasonable. Show that there exists a separable closed subspace $\mathcal{H}'$ of $\mathcal{H}$ such that $f(x) \in \mathcal{H}'$ almost everywhere.

The importance of equimeasurable families is due to the next result, which will prove useful later:

**Lemma 1.** *Let $(f_i)_{i \in I}$ be an equimeasurable and increasing filtering family of functions with values in $[0, +\infty]$. The function $\sup f_i(x) = f(x)$ is measurable and*

(8.5) $$\mu^*(\chi_K f) = \sup \mu^*(\chi_K f_i)$$

*for all compact sets $K$. If $f$ is reasonable, then*

(8.6) $$\mu^*(f) = \sup \mu^*(f_i) .$$

Let us first suppose that $f$ and hence that the functions $f_i$ are bounded. If $K$ is a compact set on which all the functions $f_i$ are continuous, then the function $f$ is lsc on $K$, so that $\chi_K f$ is measurable (exercise 1), and hence is continuous on a compact set $K' \subset K$ such that $\mu(K - K')$ is arbitrarily small. Thus $f$ satisfies condition (LUS), and so is measurable. As $f$ is continuous on $K'$, the functions $f_i$ converge uniformly to $f$ on $K$ (Dini's theorem), which proves (5) for $K'$ since

$$|\mu(\chi_{K'} f_i - \chi_{K'} f)| \leq \mu(K') \|f_i - f\|_{K'} .$$

The case of an arbitrary compact set $K$ is obtained by passing to the limit using an increasing sequence of compact sets $K_n \subset K$ whose union is $K$, up to a null set. If the set $\{f(x) \neq 0\}$, which is measurable, is reasonable [use property (v) of n° 6], then (6) is proved in a similar manner.

If $f$ is not bounded, consider the family of functions $\inf(f_i, n) = f_{i,n}$ with $i \in I$ and $n \in \mathbb{N}$. Like the given family, it is equimeasurable. It also an increasing filtering family for fixed $i$ and varying $n$, or for fixed $n$ and varying $i$, or for varying $i$ and $n$. For fixed $i$, its upper envelope is $f_i$; for fixed $n$, it is $f_n = \inf(f, n)$, a function which the first part of the proof tells us is measurable; hence so is $f = \sup f_n$. Therefore (associativity!), supposing first that $f$ vanishes outside a compact set,

$$\begin{aligned} \mu^*(f) &= \sup_n \mu^*(f_n) = \sup_n \sup_i \mu^*(f_{i,n}) \\ &= \sup_i \sup_n \mu^*(f_{i,n}) = \sup_i \mu^*(f_i) ; \end{aligned}$$

this proves (5) in the general case, and (6) as above if $f$ is reasonable.

## 9 – Measurability and Integrability

The fact that a measurable set of finite outer measure is integrable (n° 6) is merely a particular case of the next result:

**Theorem 18.** *A function $f$ with values in*[25] *$\mathbb{C}$ or $[-\infty, +\infty]$ is in $L^p$ if and only if it is measurable and $N_p(f) < +\infty$.*

(a) As $|f(x)| < +\infty$ ae., we need only consider the case of finite-valued function. By Lusin, there is a sequence of compact sets $K_n$, which may be assumed to be increasing, such that $f$ is continuous on each $K_n$ and zero

[25] or in a Banach space $\mathcal{H}$, provided Lusin's definition is used.

ae. outside the union of these $K_n$ because $N_p(f) < +\infty$. Let $f_n$ denote the function equal to $f$ on $K_n$ and 0 elsewhere; then $f(x) = \lim f_n(x)$ ae. and $|f_n(x)| \leq |f(x)|$ ae. So the proof reduces (dominated convergence) to showing that $f_n \in L^p$ for all $n$. Thus, it reduces to the case where $f$ is zero outside a compact set $K$ and is continuous on $K$, which we will now suppose.

(b) For all $n \geq 1$, $K$ may be covered by finitely many open subsets $U_{k,n}$ of $X$ such that $f$ is constant on $K \cap U_{k,n}$, up to $1/n$. As in point (d) in the proof of Lusin's theorem, for given $n$, the Borel sets $K \cap U_{k,n}$ can be replaced by pairwise disjoint Borel sets $A_{k,n}$. Let $u_{k,n} = f(a_{k,n})$, where $a_{k,n} \in A_{k,n}$ is chosen arbitrarily, and consider a step function $f_n$ equal to $u_{k,n}$ in $A_{k,n}$ and 0 outside $K$. Denoting the characteristic function of $A_{k,n}$ by $\chi_{k,n}$,

$$f_n(x) = \sum_K \chi_{k,n}(x) u_{k,n} \quad \text{everywhere}.$$

We already know that the characteristic function of a measurable set of finite outer measure is integrable; hence (n° 5, theorem 10) is in $L^p$ for all $p$, so are the functions $f_n$. In $A_{k,n}$,

$$|f(x) - f_n(x)| = |f(x) - u_{k,n}| \leq 1/n,$$

and so

$$|f_n(x)| \leq |f(x)| + \chi_K(x).$$

As everything vanishes outside the union $K$ of the sets $A_{k,n}$, one can conclude that the sequence $f_n$ converges everywhere to $f$ while remaining dominated by the function $g = |f| + \chi_K$. However, its norm $N_p$ is finite like those of $f$ and $\chi_K$. The dominated convergence theorem then shows that the limit $f$ of the functions $f_n$ is in $L^p$, proving the theorem.

Its usefulness is due to the already mentioned fact that in practice, and to a large extent also in theory, all functions that one encounters are measurable. It then suffices to check that $N_p(f)$ is finite, which in general only requires finding easy upper bounds.

**Corollary 1.** *The product of a function of $L^p$ and a bounded measurable function is in $L^p$.*

Obvious.

**Corollary 2.** *Let $f$ be a* reasonable *and measurable function. $f \in L^p$ if and only if*

$$\sup_{K \subset X} \int_K |f(x)|^p \, d\mu(x) < +\infty.$$

The condition is obviously necessary. As $f$ vanishes outside the union of a null set and an increasing sequence of compact sets $K_n$, by theorem 2,

$$\mu^*\left(|f|^p\right) = \lim \mu^*\left(|f|^p \chi_{K_n}\right) ;$$

hence the result. It can be applied for any measure on $\mathbb{R}$ by taking $K_n = [-n, n]$.

**Corollary 3**. *Let $f$ be a* reasonable *and measurable function with values in $[0, +\infty]$. Then there is an increasing sequence $(f_n)$ of* bounded *positive integrable functions such that*

$$f(x) = \lim f_n(x) \quad \text{for } all \ x \in X .$$

Indeed, let $A_n$ be an increasing sequence of integrable sets whose union is the measurable set $\{f(x) \neq 0\}$. We set

$$f_n(x) = \begin{cases} \inf\,[f(x), n] & \text{if } x \in A_n , \\ 0 & \text{otherwise} . \end{cases}$$

$f_n$ is the product of the measurable function $\inf(f, n)$ and the characteristic function of $A_n$, and so is measurable and even integrable since $\mu^*(|f_n|) \leq n\mu(A_n) < +\infty$. These functions answer the question.

# § 4. Lebesgue-Fubini's Way

## 10 – The Lebesgue-Fubini Theorem (LF)

(i) *Product of measures.* Let $X$ and $Y$ be two locally compact spaces, $\lambda$ and $\mu$ two positive measures on $X$ and $Y$. Like in Chap. V, the *product measure* $\nu$ on $X \times Y$ is given by the formula

$$\iint_{X\times Y} f(x,y)d\nu(x,y) = \int_X d\lambda(x) \int_Y f(x,y)d\mu(y) \tag{10.1}$$

for all functions $f \in L(X \times Y)$. Checking that (1) is well-defined is the only issue since if that is the case then obviously, the result depends linearly on $f$ and is $\geq 0$ if $f \geq 0$.

For given $x$, integration over $Y$ does not raise any problems since the function $f_x : y \mapsto f(x,y)$ is in $L(Y)$. On the other hand, $f$ vanishes outside a product $K \times H$, where $K \subset X$ and $H \subset Y$ are compact.[26] The integral in $y$ is thus zero for $x \notin K$. It remains to show that the function $x \mapsto \int f(x,y)d\mu(y)$ is continuous, and hence that, as $x$ tends to some $a \in X$, the function $f_x(y) = f(x,y)$ converges uniformly to $f_a(y)$ on $Y$, in other words that, for all $r > 0$, there is a neighbourhood $U$ of $a$ in $X$ such that

$$|f(x,y) - f(a,y)| \leq r \text{ for all } x \in U \text{ and all } y \in H\,.$$

There are general theorems on the topic (uniform continuity in the classical case), but we may as well argue directly. As $f$ is continuous, for all $y$, there is a neighbourhood of the point $(a,y)$ in $X \times Y$ on which $f$ is constant, up to $r$. This neighbourhood may be supposed to be the product of a neighbourhood $U(a,y)$ of $a$ in $X$ and a neighbourhood $V(y,a)$ of $y$ in $Y$. As $H$ is compact, it can be covered by finitely many sets $V(y_i,a)$. Then the set $U = \bigcap U(a,y_i)$ clearly answers the question.

It goes without saying that the formula

$$\int_X d\lambda(x) \int_Y f(x,y)d\mu(y) = \int_Y d\mu(y) \int_X f(x,y)d\lambda(x)\,,$$

proved in Chap. V for particular cases, can be extended to the general case. One shows that functions $f \in L(X \times Y)$ vanishing outside a compact set $K \times H$ are uniform limits of finite sums $\sum k_p(x)h_q(y)$, with functions $k_p \in L(X)$ and $h_q \in L(Y)$ that may be assumed to vanish outside fixed compact neighbourhoods $K'$ and $H'$ of $K$ and $H$: uniform continuity together with the existence of partitions of unity [Chap. IX, n° 17, (ii), lemma 4]. One may also apply the Stone-Weierstrass to the sums of functions $k(x)h(y)$. We leave it to the reader to fill in the gaps in this "proof".

[26] For example for $K$ and $H$ take projections of a compact set $A \subset X \times Y$ outside which $f$ vanishes.

This done, the LF theorem aims, if not to characterize in a precise manner integrable functions $f$ with respect to the product measure – it can be done if $f$ is known in advance to be measurable with respect to $\nu$ –, at least to generalize formula (1) to these function, as was done in Chap. V in very particular cases.

(ii) *The Lebesgue-Fubini theorem.*

**Theorem 19.** *Let $X$ and $Y$ be two locally compact spaces, $\lambda$ and $\mu$ measures on $X$ and $Y$, $\nu = \lambda \times \mu$ the product measure and $f \in L^1(X \times Y, \nu)$ an integrable function with respect to $\nu$. Then:*

*(i) the function $y \mapsto f(x,y)$ is $\mu$-integrable for almost all $x \in X$,*

*(ii) the function $x \mapsto \int f(x,y)d\mu(y)$, defined almost everywhere, is $\lambda$-integrable,*

*(iii) the next equality holds:*

$$\nu(f) = \iint_{X \times Y} f(x,y)d\lambda(x)d\mu(y) = \int_X d\lambda(x) \int_Y f(x,y)d\mu(y)\,. \tag{10.2}$$

*(iv) For any reasonable and measurable function $f$ with respect to $\nu$,*[27]

$$\nu^*(|f|) = \int d\lambda^*(x) \int |f(x,y)|\, d\mu^*(y)\,; \tag{10.2'}$$

*$f$ is integrable with respect to $\nu$ if and only if*

$$\int d\lambda^*(x) \int |f(x,y)|\, d\mu^*(y) < +\infty\,. \tag{10.2''}$$

Obviously, results permuting the order of integration should also be mentioned, in particular one should take not of the constantly used formula

$$\int d\lambda(x) \int f(x,y)d\mu(y) = \int d\mu(y) \int f(x,y)d\lambda(x)\,,$$

valid for all integrable functions with respect to the product measure. If $f$ is measurable, it holds whenever the right hand side of (2') – or else the similar expression obtained by switching the roles played by $\lambda$ and $\mu$ – is finite. Finally, similar results clearly hold for all functions $f$ belonging to an $L^p$ space: it suffices to apply the theorem to the function $|f|^{p-1}f$.

*Exercise 1.* Interpret the theorem when one or both spaces are discrete.

The proof of the LF theorem comprises several parts. In what follows, the letters $f$ and $\varphi$ will denote functions defined on $Z$ and we shall write $f_x(y)$ for $f(x,y)$.

[27] Recall that, generally speaking, one writes $\int f(x)d\lambda^*(x) = \lambda^*(f)$ for all measures and positive functions.

For $f \in L(Z)$, the definition of $\nu$ can then be written as

$$\nu(f) = \int \mu(f_x) d\lambda(x) .$$

(a) Let us start with a positive lsc function $\varphi$ on $Z = X \times Y$ and the set $L_{\mathrm{inf}}(\varphi)$ of $f \in L_+(Z)$ such that $f \leq \varphi$. By definition,

$$\nu^*(\varphi) = \sup_{L_{\mathrm{inf}}(\varphi)} \nu(f) = \sup \int \mu(f_x) d\lambda(x) .$$

The functions $f_x$ are continuous, their set is an increasing philtre since $L_{\mathrm{inf}}(\varphi)$, and their upper envelope is $\varphi_x$. Thus (theorem 1)

$$\sup \mu(f_x) = \mu^*(\varphi_x) ,$$

which shows that the function $x \mapsto \mu^*(\varphi_x)$ is lsc. As the functions $x \mapsto \mu(f_x)$ are also continuous and form an increasing philtre, theorem 1 shows that

$$\sup \int \mu(f_x) d\lambda(x) = \int \mu^*(\varphi_x) d\lambda^*(x) .$$

So the relation

$$\nu^*(\varphi) = \int \mu^*(\varphi_x) d\lambda^*(x) \tag{10.3}$$

finally follows for all positive lsc functions on $Z$.

(b) We next consider an arbitrary function $f$ with values in $[0, +\infty]$. By definition, $\nu^*(f)$ is the infimum of the numbers $\nu^*(\varphi)$ for lsc functions $\varphi \geq f$. Since this inequality implies $\mu^*(\varphi_x) \geq \mu^*(f_x)$ for all $x$, and hence by (3) that

$$\nu^*(\varphi) \geq \int \mu^*(f_x) d\lambda^*(x) ,$$

it follows that

$$\nu^*(f) \geq \int \mu^*(f_x) d\lambda^*(x) \quad \text{for all } f \geq 0 \text{ on } Z . \tag{10.4}$$

(c) Let us in particular suppose that $f$ is the characteristic function of a null set $N \subset Z$. The function $f_x$ is then the characteristic function of the "vertical cut" $N_x$ of $N$, i.e. the set of $y$ such that $(x, y) \in N$. The right hand side of (4) is zero, and so (n° 2, (iii), theorem 4)

$$\mu^*(N_x) = 0 \quad \text{for almost all } x \in X . \tag{10.5}$$

(d) Now let $f$ be an integrable function with values in $\mathbb{C}$ or $[-\infty, +\infty]$ (or a Banach space). Denoting by $N_1$ the norm $L^1$ with respect to the measure

$\nu$, there is (n° 4, (ii), theorem 7) a null set $N \subset Z$ with respect to $\nu$ and a series of functions $f_n \in L(Z)$ such that

$$\sum N_1(f_n) < +\infty \quad \& \quad f(z) = \sum f_n(z) \quad \text{for all } z \in Z - N\,.$$

The series $\sum f_n$ then converges in mean to $f$ in $L^1(Z,\nu)$ and

$$\nu(f) = \sum \nu(f_n) = \sum \int \mu\left(f_{n,x}\right) d\lambda(x)\,. \tag{10.6}$$

However,

$$\sum \int \mu\left(|f_{n,x}|\right) d\lambda(x) = \sum \nu\left(|f_n|\right) = \sum N_1\left(f_n\right) < +\infty\,. \tag{10.7}$$

Hence, by theorem 4, $\sum \mu(|f_{n,x}|) < +\infty$ outside a $\lambda$-null set $M' \subset X$, and the series $f_{n,x}$ converges in $L^1(Y,\mu)$ for all $x \in X - M'$.

But $\sum f_n(z) = f(z)$ for all $z \in Z - N$ and, by (5), we know that $\mu^*(N_x) = 0$ for all $x \notin M''$, where $M''$ is a null set with respect to $\lambda$. If $x \notin M' \cup M'' = M$, then the series $\sum f_{n,x}(y)$ converges both $\mu$-almost everywhere to $f_x(y)$ and in $L^1(Y,\mu)$. So, for all $x \notin M$, (theorem 7)

$$f_x \in L^1(Y,\mu) \quad \& \quad \mu(f_x) = \sum \mu\left(f_{n,x}\right)\,. \tag{10.8}$$

Finally, as $\sum |\mu(f_{n,x})| \leq \sum \mu(|f_{n,x}|)$, theorem 6 on series shows that the function $\sum \mu(f_{n,x})$ is $\lambda$-integrable and that

$$\int \sum \mu\left(f_{n,x}\right) d\lambda(x) = \sum \int \mu\left(f_{n,x}\right) d\lambda(x) = \nu(f)$$

by (8). As $\sum \mu(f_{n,x}) = \mu(f_x)$ for $x \notin M$ by (8), the function $x \mapsto \mu(f_x)$, defined almost everywhere with respect to $\lambda$, is $\lambda$-integrable, with in addition

$$\nu(f) = \int \mu(f_x) d\lambda(x)\,, \tag{10.9}$$

which proves propositions (ii) and (iii) of the statement.

(e) To prove (iv), let $f$ be a reasonable function with values in $[0,+\infty]$ and $\nu$-measurable. It is the simple limit of an increasing sequence of $\nu$-integrable functions $f_n \geq 0$ converging *everywhere* (n° 9, corollary 3 of theorem 18). Theorem 2 of n° 2 then shows that

$$\nu^*(f) = \lim \nu^*(f_n) = \lim \nu(f_n)$$

and so, by proposition (iii) of the theorem, that

$$\nu^*(f) = \lim \int \mu^*\left(f_{n,x}\right) d\lambda(x)\,.$$

However, the functions $x \mapsto \mu^*(f_{n,x})$, with values in $[0, +\infty]$, form an increasing sequence like $(f_n)$. Applying theorem 2 to $\lambda$ then leads to

$$\nu^*(f) = \int d\lambda^*(x). \lim \mu^* (f_{n,x}) \, .$$

Since the same theorem 2, now applied to $\mu$, shows that

$$\lim \mu (f_{n,x}) = \mu^* (f_x) \, ,$$

$$\nu^*(f) = \int \mu^* (f_x) \, d\lambda^*(x) \, .$$

Integrability condition (2") readily follows, qed.

*Exercise 2.* Suppose that $X$ is the set $\mathbb{R}$ equipped with the discrete topology, that $Y = \mathbb{R}$ is equipped with the usual topology, that $\lambda(\{x\}) = 1$ for all $x \in X$ and that $\mu$ is the Lebesgue measure. Show that

$$\nu^*(\varphi) = \sum_x \int \varphi(x, y) dy$$

for any positive lsc function $\varphi$.[28] Let $f(x, y)$ be the (unreasonable) function equal to 1 if $y \in \mathbb{Q}$ and 0 otherwise. Show that $f$ is measurable with respect to $\nu$ and that for this function, the left hand side of (2') is infinite and the right hand side zero.

(iii) *Additions to the LF theorem.* Let us now present some consequences of the LF theorem and additions to it.

Applying it to a characteristic function shows that:

(C1) *Let $E \subset X \times Y$ be an integrable set; then the set $E_x$ of $y \in Y$ such that $(x, y) \in E$ is $\mu$-integrable for almost all $x$, the function $x \mapsto \mu(E_x)$ is $\lambda$-integrable and*

$$\nu(E) = \int_X \mu (E_x) \, d\lambda(x) \, , \tag{10.10}$$

a result already obtained in Chap. 5, n° 33 by assuming $E$ to be compact. In particular,

$$\nu(E) = 0 \Longleftrightarrow \mu(E_x) = 0 \quad \text{for almost all } x \, . \tag{10.11}$$

By (2'), this equivalence continues to hold for all *$\nu$-measurable* sets $E \subset X \times Y$.

[28] The sum of the right hand side series is, by definition, the supremum of its partial sums (unconditional convergence for a series extended to an uncountable set).

(C2) *If $U \subset X$ and $V \subset Y$ are open, then*

$$\nu^*(U \times V) = \lambda^*(U)\mu^*(V)\,.$$

Obvious by (3).

(C3) *If $K \subset X$ and $H \subset Y$ are compact, then*

$$\nu(K \times H) = \lambda(K)\mu(H)\,.$$

Apply (2) to the characteristic function of the compact set $K \times H$.

(C4) *If $A \subset X$ and $B \subset Y$ are integrable, $A \times B$ is integrable and*

$$\nu(A \times B) = \lambda(A)\mu(B)\,. \tag{10.12}$$

*If a product $A \times B$ is integrable and if neither $A$ nor $B$ is a null set, then both $A$ and $B$ are integrable.*

Indeed, for all $\epsilon > 0$, there are compact and open sets such that

$$\begin{aligned} K \subset A \subset U\,, \qquad & \lambda(U - K) < \epsilon\,, \\ H \subset B \subset V\,, \qquad & \mu(V - H) < \epsilon\,. \end{aligned}$$

As $U \times V - K \times H \subset U \times (V - H) \cup (U - K) \times V$,

$$\nu(U \times V) - \nu(K \times H) \leq \lambda(U)\mu(H - U) + \lambda(U - K)\mu(V)\,,$$

a quantity $< r$ for sufficiently small $\epsilon$. Hence the first proposition.

Conversely, let us suppose $A \times B = E$ to be integrable. Then, for all $x \in X$, $E_x = B$ or $\varnothing$ depending on whether $x$ belongs to $A$ or not. However, by (C1), the set of $x$ for which $E_x$ is not integrable has measure zero. Hence, if $A$ is not a null set, then $B$ is integrable, qed.

(C5) *If $N \subset X$ is a null set and if $B \subset Y$ is reasonable, then $N \times B$ has measure zero with respect to $\nu$.*

This is obvious by (C4) since $B = N \cup \bigcup K_n$. The assumption on $B$ always holds if $\mu$ is countable at infinity; otherwise, $N \times Y$ may well not be a null set.

Let us for example suppose that $Y$ is an *un*countable discrete space and that $\mu(f) = \sum f(y)$ for any function $f \in L(Y)$, i.e. with finite support, whence obviously $\mu^*(f) = \sum f(y)$ for any function $f \geq 0$. If the product $E = N \times Y$ is a null set, there is a lsc function $\varphi$ on $X \times Y$ which is everywhere positive, $\geq 1$ on $N \times Y$ and such that $\nu^*(\varphi) < +\infty$. But, by (3),

$$\nu^*(\varphi) = \sum \int \varphi(x, y) d\lambda^*(x)\,,$$

and this sum is finite only if the set of $y$ such that $\int \varphi(x, y) d\lambda^*(x) \neq 0$ is countable. Thus the set of $x$ such that $\varphi(x, y) > 0$ must have measure zero for uncountably many values of $y$. However, for all $y$, this set contains $N$ and is open since $\varphi$ est lsc. We, therefore, infer that $N \times Y$ may be a null set only if $N$ is contained in an *open* null set, a condition which, by (C2), is anyhow sufficient. Hence if $X = \mathbb{R}$ is chosen to have the Lebesgue measure and $Y$ to be as indicated, then $N \times Y$ is a null set only if $N$ is empty. Measures of this type are not encountered in Nature, but that of mathematicians is not that of physicists.

(C6) *If $f \in L^p(X, \lambda)$ and $g \in L^p(Y, \mu)$, then the function*

$$f \times g : (x, y) \longmapsto f(x)g(y)$$

*is in $L^p(X \times Y, \nu)$ and*

$$\|f \times g\|_p = \|f\|_p \|g\|_p \,, \tag{10.13}$$

$$\nu(f \times g) = \lambda(f)\mu(g) \quad \text{for } p = 1 \,. \tag{10.13'}$$

*Finite sums of such functions are everywhere dense in $L^p(X \times Y)$.*

To show this, choose $f_n \in L(X)$ and $g_n \in L(Y)$ such that

$$\sum N_p(f_n) < +\infty \,, \quad \sum f_n(x) = f(x) \quad \text{for } x \in X - M \,,$$
$$\sum N_p(g_n) < +\infty \,, \quad \sum g_n(y) = g(y) \quad \text{for } y \in Y - N \,,$$

where $M$ and $N$ are null sets. Then

$$N_p(f_m \times g_n) = N_p(f_m) N_p(g_n) \,,$$

and so $\sum N_p(f_m \times g_n) < +\infty$.

On the other hand,

$$\sum f_m(x) g_n(y) = f(x)g(y) \tag{10.14}$$

unless $x \in M$ or $y \in N$, i.e. outside $(M \times Y) \cup (X \times N)$. As $f$ (resp. $g$) is in $L^p$, there is a reasonable set $A$ (resp. $B$) outside which the functions $f$ and $f_m$ (resp. $g$ and $g_n$) are all zero (all unions of countable reasonable sets are reasonable), so that (14) holds outside $(A \times Y) \cup (X \times B)$. We may suppose that $M \subset A$ and $N \subset B$. Hence relation (13) in fact holds outside the set $(M \times B) \cup (A \times N)$, so by property (C5) above, almost everywhere. Therefore the series $\sum f_m \times g_n$ converges in mean to $f \times g$ (n° 4, theorem 6), giving the first result.

To show that any $h \in L^p(X \times Y)$ can be approximated by functions such as $\sum f_i \times g_i$, with $f_i$ and $g_i$ in the $L^p$ spaces with respect to $\lambda$ and $\mu$, $h$ may be assumed to be continuous and zero outside $K \times H$, where $K \subset X$ and $H \subset Y$ are compact. The Stone-Weierstrass theorem then shows that $h$ is the uniform limit on $K \times H$ of functions $f_i \times g_i$, where all $f_i$ are defined and continuous on $K$ and all $g_i$ on $H$.

Extending these function by 0 outside $K$ or $H$ gives integrable functions on $X$ or $Y$, and

$$\|h - \sum f_i \times g_j\|_p \leq \nu(K \times H)^{1/p} \|f - \sum f_i \times g_j\|,$$

where, the norm on the right hand side is the uniform norm on $K \times H$, which may be made arbitrarily small, qed. (As in Chap. V, n° 30, theorem 30, one can also use partitions of unity).

(C7) *If $f : X \longrightarrow P$ and $g : Y \longrightarrow Q$ are measurable, then the function $(f, g) : (x, y) \mapsto (f(x), g(y))$ is measurable.*

Let $K \subset X$ and $H \subset Y$ be compact sets. Then there are decompositions

$$K = M \cup \bigcup K_n, \qquad H = N \cup \bigcup H_n,$$

where $K_n$ and $H_n$ are compact, $M$ and $N$ null sets, so that the restriction of $f$ (resp. $g$) is continuous on each $K_n$ (resp. $H_n$). Then, $(f, g)$ is clearly continuous on the sets $K_p \times H_q$, whose union, thanks to (C5), is the compact set $K \times H$, up to a null set. As all compact subsets of $X \times Y$ are contained in such a product, the function $(f, g)$ satisfies Lusin's theorem, qed.

Corollary. If $f$ and $g$ are complex-valued, the function $f(x)g(y)$ is measurable [and we recover (C6) by taking (2') into account].

(C8) *Let us suppose $\lambda$ and $\mu$ to be reasonable and let $f$ be a measurable map from $X \times Y$ to a topological space $P$. Then the function*

$$x \longmapsto f(x, y) \quad (\text{resp.} \quad y \longmapsto f(x, y))$$

*is measurable for almost all $y$ (resp. $x$).*

Since $\lambda$ and $\mu$ are reasonable, by statements (C3) and (C4) above, so is the product measure $\nu$. As was seen at the end of n° 10, then $Z = N \cup \bigcup A_n$, where $N$ is a null set and for each $n$, $A_n$ is an integrable set on which the restriction of $f$ is continuous. For all $x \in X$, let $A_{n,x}$ be the set of $y$ such that $(x, y) \in A_n$ and let $N_x$ be the similar set relatively to $N$. The restriction of the function $y \mapsto f(x, y)$ to $A_{n,x}$ is clearly continuous for all $x$. However, by (C1), $A_{n,x}$ is integrable for almost all $x$, and $\mu(N_x) = 0$ for almost all $x$. The exceptional sets occurring here depend on $n$, but there is a countably infinite number of them, so that their union $M \subset X$ is a null set. As the relation $Z = N \cup \bigcup A_n$ implies $Y = N_x \cup \bigcup A_{n,x}$ for all $x$, for all $x \notin M$, $Y$ is the union of a null set (namely $N_x$) and a sequence of integrable sets (namely $A_{n,x}$) on which the restrictions of $f_x$ are continuous. The function $f_x$ is, therefore, measurable for $x \notin M$, qed.

When $P$ is a Polish space (n° 11), for the function $f_x : y \mapsto f(x, y)$ to be measurable, $f_x^{-1}(U)$ needs to be measurable for all open sets $U \subset P$, which leads to another proof in this case. Exercise!

Applying (C8) to the characteristic function of a measurable set $A \subset X \times Y$ shows that $A_x$ is measurable for almost all $x$.

(iv) *The Fourier inversion formula.* Let us consider two functions $f, g \in L^1(\mathbb{R})$. Since their Fourier transforms are continuous and bounded, the functions $f(x)\widehat{g}(x)$ and $\widehat{f}(x)g(x)$ are integrable, and

$$\int f(x)\widehat{g}(x)dx = \int f(x)dx \int g(y)\overline{\mathbf{e}(xy)}dy\,.$$

The exponential being continuous and bounded, the function $f(x)g(y)\overline{\mathbf{e}(xy)}$ is integrable over $\mathbb{R} \times \mathbb{R}$ by (C6), and permuting the integrations gives

$$\int f(x)\widehat{g}(x)dx = \int \widehat{f}(y)g(y)dy\,. \tag{10.15}$$

As the Fourier transform maps the Schwartz space $\mathcal{S}$ to itself, $\widehat{g}$ may be replaced with an arbitrary function $h \in \mathcal{S}$. By the Fourier inversion formula (Chap. VII, n° 30, theorem 26) in $\mathcal{S}$, this substitution replaces $g(y)$ with $\widehat{h}(-y)$ ; and so

$$\int f(x)h(x)dx = \int \widehat{f}(-y)\widehat{h}(y)dy\,.$$

Let us also suppose that $\widehat{f}$ is integrable. (15) can then be applied to the function $\widehat{f}(-y)$, whose Fourier transform is

$$\int \widehat{f}(y)e(xy)dy = \widehat{\widehat{f}}(-x)\,.$$

So

$$\int f(x)h(x)dx = \int \widehat{\widehat{f}}(-x)h(x)dx$$

for all $h \in \mathcal{S}$. The function $f(x) - \widehat{\widehat{f}}(x)$, which is the difference of an integrable function and a continuous one, is, therefore "orthogonal" to $\mathcal{S}$, and hence to $L(\mathbb{R})$ since every $h \in L(\mathbb{R})$ is the uniform limit of a sequence of functions $h_n \in \mathcal{S}$ vanishing outside a fixed compact. The *Fourier inversion formula*, namely

$$f(x) = \int \widehat{f}(y)\mathbf{e}(xy)dy \text{ ae. if } f \text{ and } \widehat{f} \in L^1(\mathbb{R})\,, \tag{10.16}$$

now follows from the corollary of Lusin's theorem [n° 8, (i)].

*Exercise 3.* Prove (16) assuming that $f \in L^2(\mathbb{R})$ and $\widehat{f} \in L^1(\mathbb{R})$.

## 11 – A Topological Interlude[29]: Polish Spaces

(i) *Polish spaces.* Metrizable and separable spaces have frequently occurred in previous sections. Clearly, the completion of such a space $P$ with respect

[29] The results of this n° will not be used before n° 13.

to a metric compatible with its topology, is again metrizable and separable (a sequence everywhere dense in $P$ is everywhere dense in its completion) and, a function with values in $P$ is measurable if and only if it is measurable as a function with values in the completion of $P$.

From a classical viewpoint where measurability is defined using condition (FM) of n° 7, one may confine oneself to functions with values in spaces satisfying the following two conditions, involving in fact only the topology of $P$:

(EP 1) the topology of $P$ can be defined by a distance with respect to which $P$ is *complete*,

(EP 2) $P$ is*separable.*

As was seen in n° 7, the second condition is equivalent to saying that $P$ contains either an everywhere dense countable set, or else a countable family of open sets $U_n$ such that any open subset of $P$ is the union of the sets $U_n$ contained in it.

These are the *Polish spaces*[30] of N. Bourbaki, *Topologie générale*, Chap. IX, § 6. Their importance also stems from the fact that they are the only spaces in which non-trivial theorems about Borel sets can be proved. Apart from some exceptions, most results presented in this n° are not really essential for the rest of this chapter, but it is useful to know how far one can go without encountering unannounced pitfalls.

**Lemma 1.** *Eery metric space is Polish.*

First, by Cauchy's criterion together with BW, it is complete with respect to a metric compatible with its topology. On the other hand, for all $n$, it can be covered by a finite number of open balls of radii $1/n$. The centres of the balls thus obtained are everywhere dense in $X$, qed.

Conversely, every compact space of countable type can be shown to be metrizable, and hence Polish.

**Lemma 2.** *A locally compact metric space $X$ is separable if and only if it is countable at infinity.*

Indeed if $X$ is the countable union of compact sets $K_n$, each $K_n$ contains an everywhere dense countable subset $D_n$. This gives the result for $X$ by

[30] An abbreviation that I had jokingly suggested to N. Bourbaki in 1949 after having learnt the subject from Casimir Kuratowski's *Topology* (Warsaw Acad. of Sciences, 1933, in French, re-published in two volumes in the 1950s) and realized the contribution made by the Polish. I could have also called them " Polono-Russian". The joke, in fact not completely a joke, was taken seriously and since then all experts have adopted this strange terminology, generally without mentioning N. Bourbaki except Kuratowski himself who, in a small book published in 1974 on the history of Polish mathematics, justifiably saw in it a tribute to the latter. Just like we make a distinction between metric and metrizable space, so should we make a distinction between Polish spaces (in which a complete metric is given) and *Polishable* ones (in which such a metric exists).

considering the union of all $D_n$. Conversely, if $(U_n)$ is a countable basis for the topology on $X$, for all $a \in X$, let us choose a compact neighbourhood $V(a)$. Its interior being the union of sets $U_n$, one of them contains $a$. Its closure, contained in $V(a)$ is compact. As $a$ is arbitrary, $X$ is the union of the sets $U_n$ whose closure is compact. This gives countability at infinity.

All locally compact separable and metrizable sets $X$ will later be shown to be Polish.

Many spaces encountered in classical analysis are Polish, but familiar spaces that are not saw can be found without venturing too far. For example, the space of continuous and bounded numerical functions on $\mathbb{R}$ equipped with the norm of uniform convergence, though being a complete metric space, is not separable. One would obtain a separable and hence a Polish space by confining ourselves to functions tending to finite limits as $x$ tends to $+\infty$ or $-\infty$. Indeed, these functions are just the continuous functions on the compact metrizable space[31] $[-\infty, +\infty]$, the *finished line* of N. Bourbaki. Hence the result using the next exercise.

*Exercise 1.* (a) Let $X$ be a compact metric space, $d$ a distance on $X$, $(a_n)$ a sequence everywhere dense in $X$ and we set $f_n(x) = d(x, a_n)$. Let $\mathcal{A}$ be the set in $L(X)$ of polynomials in a finite number of functions $f_n$ whose coefficients are complex. Show that $\mathcal{A}$ is everywhere dense in the Banach space $L(X)$ (use the Stone-Weierstrass theorem). Replace $\mathcal{A}$ by the set $\mathcal{A}_0$ of polynomials in $f_n$ whose coefficients are rational. Show that $\mathcal{A}_0$ is countable and everywhere dense in $\mathcal{A}$, hence in $L(X)$. (b) Let $X$ be a locally compact metrizable space and countable at infinity (i.e. Polish, see further down). Show that there is a countable subset $D$ of $L(X)$ with the following property: for every compact set $K \subset X$, there is a compact neighbourhood $H$ of $K$ such that any $f \in L(X, K)$ is the uniform limit of functions belonging to $D$ and vanishing outside $H$ [first show that $L(X, K)$ is separable for all $K$]. Deduce that *the $L^p(X; \mu)$ spaces of measures on $X$ are separable for $p < +\infty$.*

If a metric space $X$ is separable, so is also every subspace $Y$ of $X$ since if $(U_n)$ is a basis for the topology on $X$, then the sets $U_n \cap Y$ play the same role for the topology of $Y$. It already follows that every *closed* subset $Y$ of a Polish space $X$ is Polish since it is separable and complete with respect to every metric of $X$ with respect to which $X$ is Polish. As will be seen so are all $G_\delta$ sets in $X$. Let us start with the simplest case:

**Lemma 3**. *Every open subspace $P$ of a Polish space $X$ is Polish.*

Let $P = X - F$, where $F$ is closed, and we set

$$f(x) = d(x, F)$$

assuming that $X$ is complete with respect to $d$. The relation $f(x) \neq 0$ is equivalent to $x \in P$. In the Cartesian product $\mathbb{R} \times X$, which is obviously

[31] To transform $[-\infty, +\infty]$ into a Polish space, choose a strictly increasing continuous function $f(x)$ on $\mathbb{R}$ converging to $-1$ (resp. $+1$) as $x$ tends to $-\infty$ (resp. $+\infty$), write $f(-\infty) = -1$, $f(+\infty) = +1$, and use the metric $d(x, y) = |f(x) - f(y)|$.

Polish, the set $Z$ of ordered pairs $(u, x)$ such $u.f(x) = 1$ is closed, and hence Polish. However, $(u, x) \mapsto x$ is a continuous bijection from $Z$ onto $P$; its inverse $x \mapsto (1/f(x), x)$ is continuous since $1/f$ is continuous on $P$. As a result, $P$ is homeomorphic to a Polish space, qed.

The proof equally shows that every open subspace of a complete metric space is a complete metric space – obviously not with respect to the metric of $X$, but that of $Z$ in $\mathbb{R} \times X$, namely

$$d'(x, y) = d(x, y) + |1/d(x, F) - 1/d(y, F)| .$$

If $X_1, \ldots, X_n$ are complete metric spaces, their Cartesian product, equipped for example with the distance

$$d[(x_i), (y_i)] = \sum d_i(x_i, y_i) ,$$

is a complete metric space. If the spaces $X_i$ are Polish, then so is their product.

More generally, let $(X_n)$, $n \in \mathbb{N}$, be a *countable* family of complete metric spaces and, for each $n$, choose a distance $d_n \leq 1$ in $X_n$ with respect to which $X_n$ is also complete.[32] We denote a generic element $u = u(n)$ of $X = \prod X_n$ as a function defined on $\mathbb{N}$ with values in the spaces $X_n$. The formula

$$d(u, v) = \sum 2^{-n} d_n [u(n), v(n)] \tag{11.1}$$

defines a metric on $X$. As series (1) converges normally in $X \times X$, the relation $\lim u_p = u$ with respect to the metric $d$ is equivalent to

$$\lim u_p(n) = u(n) \quad \text{for all } n \in \mathbb{N} \tag{11.2}$$

(Chap. III, n° 13, theorem 17). Hence, if the spaces $X_n$ are complete, then so is $X$.

If the spaces $X_n$ are separable, then so is $X$. Indeed, the definition of the distance shows that every ball centered at $a \in X$ contains an open set of the form $\prod E_n$, where, for a finite set $F$ of values for $n$, $E_n$ is a ball centered at $a(n)$ and where $E_n = X_n$ for $n \notin F$. Choosing for all $n$, a basis $(U_{np})$ for the topology of $X_n$, we get a basis for the topology of $X$ by considering the products $\prod E_n$, where $E_n$ is one of the $U_{np}$ for $n \in F$ and where $E_n = X_n$ otherwise. Now, the results of Chap. I, n° 7 tell us that the set of these products is countable. A countable product of Polish spaces is, therefore, Polish.

This is in particular the case of the set $\mathbb{R}^{\mathbb{N}}$ of functions or sequences $u : \mathbb{N} \longrightarrow \mathbb{R}$ equipped with distance (1). It is necessary to choose a distance $\leq 1$ on each component $\mathbb{R}$, for example

---

[32] If $d$ is a metric on a set $E$, the function $d'(x, y) = d(x, y)/[1 + d(x, y)]$ is a metric $\leq 1$ defining the same topology and having the same Cauchy sequences as $d$. If $E$ is complete with respect to $d$, it is equally so with respect to $d'$. Hence the existence of the distances $d_n$.

$$d_n(u, v) = |u - v| / (1 + |u - v|) .$$

As in the general case, the topology of $\mathbb{R}^{\mathbb{N}}$ is just that of simple convergence: a sequence $u_n \in \mathbb{R}^{\mathbb{N}}$ converges to some $u \in \mathbb{R}^{\mathbb{N}}$ if, for all $p$, $u_n(p)$ converges to $u(p)$.

These constructions enable us to justify what had been announced before lemma 3:

**Theorem 20.** *A subspace $P$ of a Polish space $X$ is Polish if and only if $P$ is the intersection of a countable family of open subsets of $X$.*

Since a subspace of a separable space $X$ is itself separable, we might as well prove a more general result:

**Theorem 20 bis.** *The topology of a subspace $P$ of a complete metric space $X$ is induced by a metric with respect to which $P$ is complete if and only if $P$ is a $G_\delta$ set in $X$.*

The proof is similar to that of lemma 3. Since a closed subspace of a metric space $X$ is complete and is a $G_\delta$ set in $X$, we may assume $P$ to be everywhere dense in $X$: replace $X$ by the closure of $P$.

(a) Let us suppose that $P$ is complete with respect to a distance $d$. For all $x \in X$, let $\omega(x)$ be the infimum of diameters[33] of the (by assumption non-empty) sets $V \cap P$, where $V$ is an arbitrary open neighbourhood of $x$ in $X$. For all $r > 0$, $V$ may be chosen so that the diameter of $P \cap V$ is $< \omega(x) + r$. As $V$ is also an open neighbourhood of each of its points, $\omega(y) < \omega(x) + r$ for all $y \in V$. Therefore, the set $G_n$ of $x$ such that $\omega(x) < 1/n$ is open in $X$. For $x \in P$, the sets $V \cap P$ are just the open neighbourhoods of $x$ in $P$, which include all open balls. Hence $\omega(x) = 0$ and as a result, $P \subset \bigcap G_n$. We show that $P = \bigcap G_n$.

To do this, we consider some $a \in \bigcap G_n$. There are neighbourhoods $V_n$ of $a$ in $X$ such that the diameters of the sets $V_n \cap P$ are $\leq 1/n$. As any smaller neighbourhood is perforce suitable, $X$ being metrizable, the sets $V_n$ can be assumed to be closed, decreasing and with intersection $\bigcap V_n = \{a\}$. Choosing some $u_n \in V_n \cap P$ for all $n$, for $n \geq p, q$, we see that the set $V_n$ contains $V_p$ and $V_q$, so that $u_p$ and $u_q$ are in $V_n \cap P$. Hence $d(u_p, u_q) \leq 1/n$ for $p, q \geq n$. As $P$ is complete with respect to $d$, $(u_n)$ converges to some $u \in P$. But for each $p \geq n$, $u_p$ is in the closed set $V_n$. So $u \in V_n$ for all $n$. Thus $u \in \bigcap V_n = \{a\}$ and $a \in P$, which proves that $P = \bigcap G_n$ as expected.

(b) Conversely, let us suppose that $P = \bigcap G_n$ where the sets $G_n = X - F_n$ are open in $X$. Let $d$ now denote a distance with respect to which $X$ is complete. The relation $x \in P$ is then equivalent to

$$d(x, F_n) = f_n(x) \neq 0 \quad \text{for all } n .$$

[33] The diameter of a non-empty set $E \subset X$ is the supremum of numbers $d(x, y)$ where $x, y$ vary in $E$. Attention should be paid to the fact that the metric $d$ of $P$ is not that of $X$ .

For all $x \in X$, let $f(x) \in \mathbb{R}^{\mathbb{N}}$ denote the sequence $n \mapsto f_n(x)$. Like the maps $f_n$, $f$ is continuous since the relation $\lim x_p = x$ implies that $\lim f_n(x_p) = f_n(x)$ for all $n$. Let $Z$ be the subset of the Cartesian product $\mathbb{R}^{\mathbb{N}} \times X$, a complete metric space like $\mathbb{R}^{\mathbb{N}}$ and $X$, of ordered pairs $(u, x)$ such that

$$u(n)f_n(x) = 1 \quad \text{for all } n\,. \tag{11.3}$$

It is obviously closed, and hence a complete, metric space (Polish if so is $X$). The projection $(u, x) \mapsto x$ is clearly a bijection from $Z$ onto $P$. Its inverse, which maps each $x \in P$ to the point $(u, x)$ for which $u(n) = 1/f_n(x)$ is continuous for $n$, since if the sequence of points $x_p \in P$ converges to some $x \in P$, then the sequences $(f_n(x_p))$ converge to *non-zero* limits $f_n(x)$. Finally, $P$ is homeomorphic to $Z$, qed.

**Corollary 1.** *Let $P$ a metrizable space. Suppose that $P$ is complete with respect to* some *distance compatible with its topology. Then $P$ is a $G_\delta$ set in its completion with respect to* any *distance compatible with its topology.*

Obvious.

The proof does not use separability. This result enables one to show a bit:

**Corollary 2.** *Every Polish space is homeomorphic to a $G_\delta$ set in a* compact *Polish space.*

To prove this, let us consider the infinite-dimensional cube $I^{\mathbb{N}}$, where $I = [0, 1]$. It is the set of maps or sequences $u : \mathbb{N} \longrightarrow I$ equipped with the distance

$$d(u, v) = \sum 2^{-n} |u_n - v_n|$$

which, as was seen above for the product $\mathbb{R}^{\mathbb{N}}$, induces the topology of simple convergence. We show it is compact by using Bolzano-Weierstrass together with Cantor's "diagonal argument":[34] if there is a sequence $u(p) = (u_n(p))$ in $I^{\mathbb{N}}$, the sequence $u_1(p)$ contains (BW for $I$) a subsequence $u_1(p_{m,1})$ converging to a limit $u_1$, the sequence $u_2(p_{m,1})$ a subsequence $u_2(p_{m,2})$ converging to a limit $u_2$, etc. The diagonal sequence $(u(p_{n,n}))$ then converges to $u = (u_n)$. To see this it suffices to show that

$$\lim_m u_n(p_{m,m}) = u_n = \lim_m u_n(p_{m,n})$$

for all $n$, which follows from the fact that, for given $n$, the sequence $(p_{m,m})$ for $m > n$ is is a subsequence of the sequence $(p_{m,n})$.

This done, to construct a set $Z$ in $I^{\mathbb{N}}$ *homeomorphic* to a given Polish space $P$, one chooses a sequence $(a_n)$ everywhere dense in $P$ and a distance

[34] The following argument applies to all products of a countable family of compact metric spaces (see for example Dieudonné, (12.5.9)). A result without any countability or metrizability assumptions using a far more sophisticated version of BW ("ultrafilters") can be found in N. Bourbaki's *Topologie générale*.

$d \leq 1$ with respect to which $P$ est complete. Then the image $Z$ of $X$ under the map $x \mapsto (d(a_n, x))$ from $X$ to $I^{\mathbb{N}}$ answers the question. To see this, it suffices to show that in $X$, the relation $\lim x_p = x$ is equivalent to

$$\lim d\,(a_n, x_p) = d\,(a_n, x) \quad \text{for all } n\,.$$

If this holds, then indeed, for all $r > 0$, there exists $n$ such that $d(a_n, x) < r$. As $d(a_n, x_p) < r$ for large $p$, we deduce that $d(x_p, x) < 2r$, whence $\lim x_p = x$. As $P$ is a complete metric space, its image in $I^{\mathbb{N}}$ is necessarily a $G_\delta$ set, qed.

If in particular a locally compact space $P$ is metrizable and separable (i.e. countable at infinity: lemma 2), then its completion with respect to a metric compatible with its topology, being Polish, is homeomorphic to a subset of $I^{\mathbb{N}}$, hence so is $P$ itself. As $P$ and $I^{\mathbb{N}}$ are locally compact, $P$ is the intersection of an open and of a closed subset of $I^{\mathbb{N}}$. But a closed subset in a metrizable space is a $G_\delta$ set. Hence $P$ is a $G_\delta$ set in $I^{\mathbb{N}}$, and so:

**Corollary 3**. *Every locally compact metrizable space countable at infinity is Polish.*

(ii) *Lsc functions on a locally compact Polish space.* A good way to understand the quasi-necessity of introducing Polish spaces consists in analyzing the proof of the next result:

**Theorem 21**. *Let $X$ be a locally compact* Polish *space. Every positive* lsc *function on $X$ is the limit of an increasing* sequence *of continuous functions with compact support.*

It suffices to show that every positive lsc function $\varphi$ is the upper envelope of a countable family of functions $f_n \in L(X)$ because the functions $\sup(f_1, \ldots, f_n)$ converge to $\varphi$.

Let us consider the open sets

$$U_{pn} = \{\varphi(x) > p/n\} \tag{11.4}$$

for $p \geq 0, n \geq 1$. Denoting by $\chi_{pn}$ the characteristic of the function of $U_{pn}$, the functions $\chi_{pn}$, the functions

$$\varphi_n(x) = \frac{1}{n} \sum_{p \geq 1} \chi_{pn}(x) \tag{11.5}$$

are lsc, and $\varphi = \sup(\varphi_n)$. Indeed, let $x$ be a point of $X$. If $\varphi(x) \leq 1/n$, then $\chi_{pn}(x) = 0$ for all $p \geq 1$. Hence $\varphi_n(x) = 0$ (if $\varphi(x) = 0$) or 1 (if $\varphi(x) > 0$), and so

$$\varphi_n(x) \leq \varphi(x) \leq \varphi_n(x) + 1/n \tag{11.6}$$

in this case. If $q/n < \varphi(x) \leq (q+1)/n$ for some $q \geq 1$, the point $x$ belongs to all the sets $U_{pn}$ such that $1 \leq p \leq q$ and to none of the following ones, so that

$\varphi_n(x) = q/n$, leading again to (6). Finally, if $\varphi(x) = +\infty$, then $\chi_{pn}(x) = 1$ for all $p$. Hence $\varphi_n(x) = +\infty = \varphi(x)$. Therefore, relation (6) holds for all $x \in X$ and $n \geq 1$, whence $\varphi = \sup(\varphi_n)$.

On the other hand, if the theorem holds for a sequence of lsc functions, it clearly holds for their upper envelope and for their sum. Hence the proof reduces to showing this for the functions $\chi_{pn}$, i.e. for the characteristic function of an open set $U \subset X$.

Now, if an increasing sequence of functions $f_n \in L_+(X)$ converges to $\chi_U$, clearly $f_n(x) \geq 1 - 1/p$ for large $n$, for all $x \in U$ and $p \geq 1$. As the functions $f_n$ vanish outside $U$, $U$ is the union of the compact sets $K_{p,n} = \{f_n(x) \geq 1 - 1/p\}$. Hence to go on with the proof, one needs to show that *all open subsets of $X$ are countable at infinity.*[35] This is indeed the case since $X$ being locally compact and Polish, so is $U$ (lemma 1), which is, therefore, countable at infinity (lemma 2).

The open set $U$ is then the union of an increasing sequence of compact sets $K_n$. As any compact set has a compact *neighbourhood* in $U$,[36] one may even assume that each $K_{n+1}$ contains an open set $U_{n+1}$ containing $K_n$. Then, for each $n$, there exists a continuous function on $U$ with values in $[0, 1]$, equal to 1 on $K_n$ and 0 on $X - U_{n+1}$ (Chap. IX, n° 17, lemma 1), hence on $U - K_{n+1}$. If $f_n(x)$ is replaced by 0 for all $x \notin U$, the functions $f_n$ are in $L(X)$ and their upper envelope is, as desired, the function $\chi_U$, qed.

*Exercise 2.* Let $X$ be a metric space and $U = X - F$ an open subset of $X$. Set $f_n(x) = n.\inf[d(x, F), 1/n]$. Show that $\chi_U(x) = \sup f_n(x)$ for all $x$. Deduce that *every positive* lsc *function on a metric space is the limit of an increasing sequence of continuous functions on $X$*. Does this lead to an alternative proof of theorem 21?

(iii) *Borel sets in a Polish space.* Polish spaces have far stranger properties than the above ones. We are going to state some of them without proof in this n°, not because they will be immediately needed, but in order to show how quite simple theoretical constructions in set theory and general topology can lead to non-measurable sets, even in familiar spaces such as $\mathbb{R}$.

*Exercise 3.* Let $B$ be the set in $\mathbb{R}^{\mathbb{N}}$ consisting of $u = (u_n)$ for which $\lim u_n$ exists. Show that $B$ is a Borel set.

*Exercise 4.* Let $X$ be the set of real-valued continuous functions on $I = [0, 1]$. Equipped with the distance associated to uniform convergence, it is Polish (exercise 1). Let $S$ be the unit ball of $X$ and $S^{\mathbb{N}}$ the set of sequences $f = (f_n)$ with $f_n \in S$, equipped with the distance

$$d(f, g) = \sum \|f_n - g_n\| / 2^n .$$

[35] $X$ being countable at infinity is not sufficient for this. Any locally compact space, even uncountable at infinity, is an open subset of a compact set, for example, its Alexandrov compactification $\widehat{X} = X \cup \{\infty\}$.

[36] Recall that a neighbourhood of a set $E$ must contain an open set containing $E$.

Show that $S^{\mathbb{N}}$ is Polish. Let $B$ be the set of $f \in S^{\mathbb{N}}$ for which $\lim f_n(x)$ exists for all $x \in I \cap \mathbb{Q}$. Show that $B$ is a Borel set. Answer the same question for the set of $f$ for which the sequence of functions $f_n$ is increasing.

The simplest example of an apparently strange Polish space is the set

$$P = \mathbb{R} - \mathbb{Q} = \bigcap_{\xi \in \mathbb{Q}} \mathbb{R} - \{\xi\}$$

of irrational numbers. It is a $G_\delta$ set in $\mathbb{R}$ and even in the compact space $[-\infty, +\infty]$. Equipped with the topology of $\mathbb{R}$, $P$ is, therefore, Polish. Here the sets $F_n$ in the proof of theorem 20 bis are the sets $\{a\}$, $a \in \mathbb{Q}$. The space $P$ is *totally discontinuous*: for all $x \in P$, any of its neighbourhoods contains a neighbourhood which is both open and closed in $P$, namely the intersection of $P$ and an interval centered at $x$ and with rational endpoints. It is then obvious that $P$ cannot be locally compact.

This space has a literally miraculous "universal" property concerning Borel subsets: an uncountable subset $B$ of a Polish set $X$ is a Borel set if and only if it is the image of $\mathbb{R} - \mathbb{Q}$ under a *continuous and injective* map. In particular, this property holds for $X$ itself. Hence, if $X$ and $Y$ are Polish, *any continous and injective map from* $X$ *to* $Y$ *transforms Borel subsets of* $X$ *into Borel subsets of* $Y$.

The theory of Borel sets is connected to the more general theory of Suslin's[37] *analytic* sets. These are images of Polish spaces (or, more simply of $\mathbb{R} - \mathbb{Q}$) under continuous maps that are not necessarily injective. Any union or intersection of countably analytic sets is analytic, but in a Polish space, the complement of analytic space $A$ is analytic if and only if $A$ is a Borel set. It can be shown that *all analytic sets are measurable* with respect to any Radon measure in a locally compact Polish set.

The image of an analytic set under a continuous map is clearly also analytic, but the image of the *complement* of an analytic non-Borel set under such a map may well be neither analytic, nor the complement of an analytic set. Thus, the class of analytic sets leads to the definition of more and more

[37] Yakovlevich Suslin (1894–1920), son of "poor peasants" who, as stated by one of his biographers, brilliantly completed his secondary school studies thanks to the help of "rich people" from his village, was admitted to Moscow university in 1913 where he worked, with some other future brilliant mathematicians, under the supervision of N. Lusin. Reading an article of 1905 where Lebesgue "proves" that, in $\mathbb{R}^2$, the projection of a Borel set is again a Borel set, Suslin noticed that the proof was wrong and published a counterexample in 1917 in a note to the *Comptes-rendus* of the Paris Académie des sciences (Lebesgue reacted with delight at his mistake: it made Science progress...). This is practically Suslin's only publication: lack of funds, health problems (seemingly tuberculosis) and hunger led him to return to his village, where he died of typhus. Lusin, who took over, remained for a long time one of the bosses of Soviet mathematics. In Bourbaki, any metrizable space which is the image of a Polish space under a surjective continuous map (resp. bijective and continuous) is called a Suslin (resp. Lusin) space. This terminology seems to have been adopted by all specialists.

complicated sets: complements of analytic sets, continuous images of these complements, complements of these continuous images, continuous images of complements of continuous images of complements of analytic sets, etc. In this way, we get the *projective* sets. In a locally compact space equipped with a measure they may be *non-measurable* contrary to analytic sets.

These results as well as many others were obtained using classical techniques of set theory and general topology with much ingenuity and imagination. The topic continues to be the subject of active research that often causes serious problems of mathematical logic.[38]

## 12 – Continuous Sums of Measures: Examples

(i) *Product measures.* As in n° 10, let $X$ and $Y$ be two locally compact spaces, $\lambda$ and $\mu$ positive measures on $X$ and $Y$ and $\nu$ the product measure on $Z = X \times Y$. For all $x \in X$, the map

$$\mu_x : f \longmapsto \int f(x,y)d\mu(y) \tag{12.1}$$

is a measure $\mu_x$ on $Z$, and the function $x \mapsto \mu_x(f)$ is continuous with compact support, hence $\lambda$-integrable, for all $f \in L(Z)$. The definition of the product measure can then be put in the form

$$\nu(f) = \int \mu_x(f)d\lambda(x)\,, \quad f \in L(Z)\,. \tag{12.2}$$

If the space $X$ is finite (resp. discrete), a sum (resp. series) of measures is thereby obtained. In the general case, the "discrete" sum is replaced by a "continuous" sum of measures. As will be seen, this situation occurs elsewhere.

(ii) *Measures induced by locally integrable densities.* The most elementary analysis provides examples of measures induced by a "density" from the Lebesgue measure on $\mathbb{R}$, for example the map

$$f \longmapsto \int f(x)x^2dx$$

where $f \in L(\mathbb{R})$, or

$$f \longmapsto \int_0^{+\infty} f(x)x^{-1}dx$$

[38] See Alexander S. Kechris, *Classical Descriptive Set Theory* (Springer, 1995) and his bibliography. For an introduction to the topic, see Chap. IX, § 6, of N. Bourbaki, *Topologie générale.* The book of the Polish mathematician W. Sierpinski, *Les ensembles projectifs et analytiques* (Gauthier-Villars, 1952), as well as *Topologie* by Kuratowski are other references.

in the case of $\mathbb{R}_+^*$. In the general case of a measure $\lambda$ on a locally compact space $X$, a function $\mathbf{j}$ with values in $\mathbb{C}$ or in $[-\infty, +\infty]$ will be said to [n° 5, (ii)] be *locally integrable* if $f\mathbf{j}$ is integrable for all $f \in L(X)$. This is for example the case if $\mathbf{j} \in L^p$ for an exponent $p \geq 1$ (corollary 2 of theorem 11: Hölder). The map $f \mapsto \mu(f\mathbf{j})$ is then a complex measure $\nu$ (n° 5, lemma 5), the measure $\mathbf{j}(x)d\mu(x)$ with density $\mathbf{j}$ with respect to $\mu$, defined by

$$\int f(x)d\nu(x) = \int f(x)\mathbf{j}(x).d\lambda(x)\,. \tag{12.3}$$

A measure of this type is also called a *measure with base* $\mu$, a N. Bourbaki terminology, or, in a more traditional terminology, an *absolutely continuous measure with respect to* $\mu$. We will see later how to characterize them (Lebesgue-Nikodym theorem).

These measures can be written in similar way to (2). If $d\nu = \mathbf{j}d\lambda$ and if the measure

$$\mu_x : f \longmapsto \mathbf{j}(x)f(x) \tag{12.4}$$

proportional to the Dirac measure at the point $x$ is associated to every $x \in X$, then clearly

$$\nu(f) = \int \mu_x(f)d\lambda(x) \quad \text{for all } f \in L(X)\,; \tag{12.4'}$$

In this case, functions $x \mapsto \mu_x(f)$ are integrable with respect to $\lambda$ but are not continuous if neither is $\mathbf{j}$.

(iii) *Image of a measure under a map.* Let us consider two locally compact spaces $X$ and $Z$, a positive measure $\lambda$ on $X$ and a Lusin-measurable map $p : X \longrightarrow Z$. For all $f \in L(Z)$, the composite function $f \circ p$ is then measurable and bounded. Let us suppose it is integrable. This is for example the case if $\lambda$ is bounded. Then the formula

$$\nu(f) = \int f\,[p(x)]\,d\lambda(x) \tag{12.5}$$

defines a positive measure on $Z$, the *image of* $\lambda$ *under* $p$. Setting

$$\mu_x(f) = f\,[p(x)] \tag{12.6}$$

to be the Dirac measure at $p(x)$, one recovers formula (2).

A particularly simple case is that of a continuous and *proper* map $p$, i.e. such that $p^{-1}(K)$ is compact for any compact set $K \subset Z$. The function $f \circ p$ is then in $L(X)$ for all $f \in L(Z)$. For $X$, one can for example take in $\mathbb{R}^2$, for $Z$, the space $\mathbb{R}$, and for $p$, the projection $(x, y) \mapsto x$. The latter is proper if and only if, for every compact interval $K \subset \mathbb{R}$, the set $X \cap (K \times \mathbb{R})$ is compact. If $\lambda$ is the Lebesgue measure on $X$, the measure $\nu$ is given by

$$\int_{\mathbb{R}} f(x)d\nu(x) = \iint_X f(x)dxdy\,.$$

Hence, if $m(x)$ denotes the (linear) Lebesgue measure of the compact set consisting of $y \in \mathbb{R}$ such that $(x, y) \in X$, then

$$\int f(x)d\nu(x) = \int f(x)m(x)dx\,,$$

where integration is over $\mathbb{R}$.

Another important example: we consider a compact group $G$ and the map $(x, y) \mapsto xy$ from $G \times G$ to $G$. If $\lambda$ and $\mu$ are two positive measures on $G$, the image of the product measure is the *convolution product* $\lambda * \mu = \nu$, defined by

$$\nu(f) = \iint f(xy)d\lambda(x)d\mu(y)$$

for all $f \in L(G)$. If $\lambda = \epsilon_a$ and $\mu = \epsilon_b$ are Dirac measures, then obviously

$$\epsilon_a * \epsilon_b = \epsilon_{a*b}\,.$$

The generalization to the case of a *locally compact group* (lcg) will be given in n° 25 . This how a group $G$ equipped with a locally compact topology with respect to which the maps $(x, y) \mapsto xy$ and $x \mapsto x^{-1}$ are continuous is called. These groups occur everywhere: discrete groups, $\mathbb{R}^n$, $\mathbb{T}^n$, closed subgroups of $GL_n(k)$ where $k$ is a locally compact field such as $\mathbb{R}$, $\mathbb{C}$, $p$-adic fields, etc.

(iv) *Quotient of an invariant measure.*[39] To give another example of relation (2), we consider a locally compact space $Z$ and a lcg $G$ acting "on the left" on $Z$ by a continuous map $(g, x) \mapsto gx$ satisfying obvious algebraic conditions. One can then define a quotient set $G\backslash Z$ whose elements are the orbits $Gx$ of $G$, as well as a map $p : Z \mapsto G\backslash Z$ taking each $x$ to its class $Gx$. As in every quotient of a topological space by an equivalence relation, there is a natural topology on $G\backslash Z$: a set $U \subset G\backslash Z$ is open if and only if $p^{-1}(U)$ is open in $Z$. As will be seen in n° 15, in "good cases", the quotient space is locally compact.

As $G$ is locally compact, there is a measure on $G$ which is invariant under *right* translations $g \mapsto ga$, the (*Haar measure*[40]), i.e. such that

[39] Do not look for proofs in this section, which will be detailed in n° 15.

[40] Such a measure always exists and is unique up to a constant factor, its image under $g \mapsto g^{-1}$ also being an invariant measure under left translations $g \mapsto ag$. A proof of this result can be found in Dieudonné, *Eléments d'analyse*, XIV.1, but as the proof, however ingenious, does not teach anything more than the theorem, it can be admitted without any inconvenience. To know the properties of Haar measures is far more important, especially everything that concerns convolution products (n° 15 and 25).

$$\int \varphi(ga)d\mu(g) = \int \varphi(g)d\mu(g) \tag{12.7}$$

for all $\varphi \in L(G)$ and $a \in G$. For $f \in L(Z)$, the function

$$z \longmapsto \int_G f(gz)d\mu(g)$$

on $Z$ may then be considered. The function being integrated is in $L(G)$ *if* the set of $g$ such that $gK \cap H \neq \emptyset$ is *supposed to be* compact for every compact sets $K, H \subset Z$. Replacing $z$ by $az$ for some $a \in G$ replaces $g$ by $ga$, hence does not change the integral. So function (7) is in fact a function on the space $X = G\backslash Z$, given by

$$f^G(x) = \int f(gz)d\mu(g) \quad \text{if } x = p(z)\,. \tag{12.8}$$

Since, for given $x$, the right hand side is a positive linear functional on $L(Z)$, one may write

$$f^G(x) = \mu_x(f)\,, \tag{12.9}$$

where $\mu_x$ is a measure $> 0$ on $Z$, namely the image of $\mu$ under $g \mapsto gz$.

As will be seen in n° 15, in "good cases", $f \mapsto f^G$ is a surjective map from $L(Z)$ to $L(X)$ and even from $L_+(Z)$ onto $L_+(X)$.

Then suppose we are given a measure $\nu > 0$ on $Z$, *G-invariant*, i.e. such that

$$\int f(gz)d\nu(z) = \int f(z)d\nu(z)$$

for all $g \in G$. If $d\mu(g) = d\mu(g^{-1})$, we show that

$$f^G = 0 \Longrightarrow \nu(f) = 0\,.$$

Hence the integral $\nu(f)$ only depends on $f^G$, and depends on it linearly, and as the map $L_+(Z) \longrightarrow L_+(X)$ is surjective, it follows there is a measure $\lambda \geq 0$ on $X$ such that[41]

$$\nu(f) = \lambda\left(f^G\right) \tag{12.10}$$

for all $f \in L(Z)$. Taking account of (9), it is written

$$\nu(f) = \int \mu_x(f)d\lambda(x)\,, \tag{12.11}$$

which again gives a continuous sum of measures.

[41] We have here a particularly clear example of the usefulness of *defining* a measure as a linear functional on functions. Adopting the traditional viewpoint of tribes or abstract measures would give rise to serious complications.

Choosing $Z = \mathbb{R}$, $G = \mathbb{Z}$ acting on $\mathbb{R}$ by $z \mapsto z+n$ and for $\nu$ the Lebesgue measure, the quotient $X = \mathbb{R}/\mathbb{Z}$ is just the group $\mathbb{T}$ of the theory of Fourier series, $\lambda$ is the measure $dm(u)$ on $\mathbb{T}$ (Chap. VII, § 1) and (11) is just the relation

$$\int_{\mathbb{R}} f(t)dt = \int_{\mathbb{T}} dm(u) \sum_{\mathbb{Z}} f(t+n) \qquad (u = e^{2\pi it})$$

which is the cornerstone of the Poisson summation formula.

## 13 – Integrable Functions with respect to a Continuous Sum

(i) *The case of* lsc *functions.* As in the case of a product measure, the examples in the previous n° give rise to the problem of characterizing if possible measurable and integrable functions with respect to the measure

$$\nu(f) = \int \mu_x(f)d\lambda(x)\,. \tag{13.1}$$

Hence let us consider two locally compact spaces $X$ and $Z$, a measure $\lambda > 0$ on $X$ and a map $x \mapsto \mu_x$ from $X$ to the set $\mathcal{M}_+(Z)$ of positive measure on $Z$. (1) defines a measure if and only if – linearity and positivity – the following condition holds:

(INT) for all $f \in L(Z)$, the function $F_f(x) = \mu_x(f)$ is integrable with respect to $\lambda$.

The first problem is then to compute the upper integral $\nu^*(\varphi)$ of positive lsc function over $Z$. By definition,

$$\nu^*(\varphi) = \sup \nu(f) = \sup \int \mu_x(f)d\lambda(x) = \sup \lambda\left(F_f\right) \tag{13.2}$$

and

$$\mu_x^*(\varphi) = \sup \mu_x(f) = \sup F_f(x)\,, \tag{13.3}$$

where $f$ varies in the set $L_{\text{inf}}(\varphi)$ of $f \in L_+(Z)$ dominated by $\varphi$. Now, the corresponding functions $F_\varphi$ are positive and their set is an increasing philtre since, if $h$ is an upper bound for $f$ and $g$, $F_h$ is obviously an upper bound for $F_f$ and $F_g$. By (3), the upper envelope of this philtre is the function $F_f$. But the functions $F_f$ have no reason to be continuous or even lsc. Hence without any additional assumptions, it is not possible to readily deduce that

$$\sup \lambda\left(F_f\right) = \lambda^*\left(F_\varphi\right)$$

as we did when proving LF. We can only write that

$$\sup \lambda\left(F_f\right) \leq \lambda^*\left(F_\varphi\right)\,.$$

Nonetheless, it is possible to obtain the equality in some simple cases.

($\alpha$) Suppose first that the function $x \mapsto \mu_x(f)$ is lsc for all $f \in L_+(Z)$. This is the case in example (i) of the previous n° – they are even continuous –, in example (ii) if the density $\mathbf{j}(x)$ is lsc, and in example (iii) if the map $p$ is continuous. Then, by (3), the function $x \mapsto \mu_x^*(\varphi)$ is lsc, and (2) shows (theorem 1)[42] that

$$\nu^*(\varphi) = \sup \int \mu_x(f) d\lambda(x) = \int \mu_x^*(\varphi) d\lambda^*(x) . \tag{13.4}$$

($\beta$) If $Z$ is *Polish*, we know (theorem 21) that $\varphi$ is the simple limit of an increasing *sequence* of functions $f_n \in L_+(Z)$. Theorem 2 of n° 2, applied first to $\nu$ then to the measures $\mu_x$, first shows that

$$\nu^*(\varphi) = \sup \nu(f_n) = \sup \int \mu_x(f_n) d\lambda(x) ,$$

then that

$$\mu_x^*(\varphi) = \sup \mu_x(f_n) .$$

Since, by condition (INT), the functions $x \mapsto \mu_x(f_n)$ are integrable with respect to $\lambda$, the same theorem 2, now applied to $\lambda$, enables us to pass to the limit under the $\int$ sign and again obtain relation (4). In practice, it is by far the most important case.

($\gamma$) In the general case, let us impose an additional condition to the measures $\mu_x$

(MVM) The family of functions

$$F_f(x) = \mu_x(f) , \qquad f \in L(Z) ,$$

is *equimeasurable* with respect to $\lambda$.

This means (n° 8, exercise 6) that for every compact set $K \subset X$ and for all $r > 0$, there is a compact set $K' \subset K$ such that $\lambda(K - K') < r$ and on which *all* functions $F_f$ are continuous. Under this assumption, the lemma of n° 8 justifies relation (4) if the set $\{\varphi \neq 0\}$ is reasonable. This restriction cannot be removed from the general theorem (including the classical version of LF), but the lemma of n° 8 shows that, for want of anything better,

$$\nu^*(\chi_K \varphi) = \sup \int \mu_x(\chi_K f) d\lambda(x) = \int \mu_x^*(\chi_K \varphi) d\lambda^*(x) \tag{13.4'}$$

always holds for every compact set $K \subset X$ and even, passing to the limit, for every reasonable set.

Condition (MVM) trivially holds in case (i) of the previous n°. In case (ii), $F_f(x) = \mathbf{j}(x)f(x)$, where $\mathbf{j}$ is locally integrable and hence measurable. If

[42] Recall that, for us, the notation $d\lambda^*(x)$ indicates an upper integral with respect to $\lambda$.

$\mathbf{j}$ is continuous on a compact set, so are the functions $F_f$. In case (iii) of the image of a measure under a map $p$, $\mu_x(f) = f[p(x)]$, i.e. $F_f = f \circ p$, so that these functions are continuous on all sets on which $p$ is continuous. Therefore, if $p$ satisfies (LUS), condition (MVM) is then satisfied.

In what follows, we will express condition (MVM) by saying that $x \mapsto \mu_x$ is *vaguely measurable.*

To justify this expression, we need to equip the space $\mathcal{M}(Z)$ of (complex) measures on $Z$ with a topology such that condition (MVM) rephrases property (LUS). *Vague topology* is the one that is suitable: by identifying every measure $\mu$ on $Z$ to the map $f \mapsto \mu(f)$ from $L(Z)$ to $\mathbb{C}$, the vague topology is that of simple convergence. In other words, a variable measure $\mu$ converges vaguely to a limit $\mu_0$ if and only if $\mu(f)$ tends to $\mu_0(f)$ for all $f \in L(Z)$. There is an obvious analogy with the notion of convergence for Laurent Schwartz's distributions[43] (Chap. V, § 10), which incidentally is influenced by the vague topology used at the time by Henri Cartan in potential theory. Saying that $x \mapsto \mu_x$ is continuous on a subset $A$ of $X$ then amounts to saying that, for *all* $f \in L(Z)$, the function $x \mapsto \mu_x(f)$ is continuous on $A$: express that $\mu_x$ converges vaguely to $\mu_a$ if $x \in A$ converges in $A$ to $a \in A$. Choosing a compact set for $A$ and applying Lusin's definition, one sees that condition (MVM) means precisely that $x \mapsto \mu_x$ is measurable with respect to the vague topology. By the way, note that it is not in general metrizable, even when $Z$ is as simple a space as $[0, 1]$.

Condition (MVM) holds if $x \mapsto \mu_x$ is *vaguely continuous*, i.e. if $\mu_x(f)$ is continuous for all $f \in L(Z)$. This is the case in example (i) of the previous n°, in example (ii) if the function $\mathbf{j}$ is continuous, and in example (iii) if $p$ is continuous. Assumption ($\alpha$) is then satisfied, and hence (4) holds without any restrictions on $\varphi$.

(ii) *The generalized Lebesgue-Fubini theorem.* Though the results need to be rounded off in each particular case, the next theorem makes it possible to unify the situations described in the previous n°.

**Theorem 22**. *Let $X$ and $Z$ be locally compact spaces, $\mu_x$ a family of positive measures on $Z$ depending on a parameter $x \in X$, and $\lambda$ a positive measure on $X$. Suppose that condition* (INT) *holds and that this is also the case of one of the following conditions:*

($\alpha$) *the function $x \mapsto \mu_x(f)$ is* lsc *for all positive $f \in L(X)$ positive;*
($\beta$) *$Z$ is Polish;*

[43] As in this case, the open subsets of $\mathcal{M}(Z)$ can be easily defined. For given $\mu_0 \in \mathcal{M}(Z)$ and $f_1, \ldots, f_n \in L(Z)$, let $W(\mu_0; f_1, \ldots, f_n)$ be the set of $\mu \in \mathcal{M}(Z)$ such that $|\mu(f_i) - \mu_0(f_i)| < 1$ for $1 \leq i \leq n$. A set $U \subset \mathcal{M}(Z)$ is then open if and only if it is the union of sets of this type. See some additional fact in Dieudonné, XIII.4.

($\gamma$) *the map $x \mapsto \mu_x$ is vaguely measurable with respect to $\lambda$.*

*The following results then hold.*

(i) *The function*

$$\nu(f) = \int \mu_x(f) d\lambda(x), \qquad f \in L(Z), \tag{13.5}$$

*is a positive measure on $Z$.*

(ii) *If $f \in L^1(Z;\nu)$, then $f$ is $\mu_x$-integrable for almost all $x \in X$, the function $x \mapsto \mu_x(f)$, defined for almost all $x$, is $\lambda$-integrable and*

$$\int_Z f(z) d\nu(z) = \int_X d\lambda(x) \int_Z f(z) d\mu_x(z). \tag{13.5'}$$

(iii) *If $f$ is measurable with respect to $\nu$ and vanishes outside a reasonable set, then*

$$\nu^*(|f|) = \int \mu_x^*(|f|)\, d\lambda^*(x). \tag{13.5''}$$

The following arguments bear a strong resemblance, and with good reason, to those used to prove the Lebesgue-Fubini theorem.

Proposition (i) in the statement is obvious.[44]

To prove (ii), we start with the equality

$$\nu^*(\varphi) = \int \mu_x^*(\varphi) d\lambda^*(x),$$

proved above for all positive lsc functions on $Z$ in cases ($\alpha$), ($\beta$) and, if $\nu^*(\varphi) < +\infty$, in case ($\gamma$); we then imitate the proof of the classical LF theorem. Its part (b) and relation (4) lead to the inequality

$$\nu^*(f) \geq \int \mu_x^*(f) d\lambda^*(x) \tag{13.6}$$

which holds for every function $f \geq 0$ on $Z$, including in case ($\gamma$): there is nothing to prove if $\nu^*(f) = +\infty$, and otherwise it suffices to calculate $\nu^*(f)$ using lsc functions $\varphi$ dominating $f$ and for which $\nu^*(\varphi) < +\infty$; in that case, (4) can be applied to them. For a characteristic function, (6) shows that

$$\{\nu^*(N) = 0\} \Longrightarrow \{\mu_x^*(N) = 0 \text{ for almost all } x\}. \tag{13.7}$$

Now, let $f$ be a function on $Z$ with values in $\mathbb{C}$ or $[-\infty, +\infty]$, and suppose it is integrable with respect to $\nu$. Denoting by $N_1$ the $L^1$ norm with respect

[44] Because measures have been *defined* as positive linear functionals on $L(X)$. If the other possible definition had to be used (countably additive functions on Borel sets, for example), we would come across additional difficulties...

to the measure $\nu$, there is (n° 4, corollary 1 of theorem 7) then a null set $N \subset Z$ with respect to $\nu$ and a series of functions $f_n \in L(Z)$ such that

$$\sum N_1(f_n) < +\infty \quad \& \quad f(z) = \sum f_n(z) \text{ for all } z \in Z - N.$$

The series $\sum f_n$ converges in mean to $f$ in $L^1(Z, \nu)$ and

$$\nu(f) = \sum \nu(f_n) = \sum \int \mu_x(f_n)\, d\lambda(x). \tag{13.8}$$

However,

$$\sum \int \mu_x(|f_n|)\, d\lambda(x) = \sum \nu(|f_n|) = \sum N_1(f_n) < +\infty. \tag{13.9}$$

By theorem 6, $\sum \mu_x(|f_n|) < +\infty$ outside a $\lambda$-null set $M' \subset X$, the series $f_n$, moreover, converging in $L^1(Z, \mu_x)$ for all $x \in X - M'$.

But $\sum f_n(z) = f(z)$ for all $z \in Z - N$ and (7) tells us that $\mu_x^*(N) = 0$ for all $x \notin M''$, where $M''$ is a $\lambda$-null set. Hence, if $x \notin M' \cup M'' = M$, a $\lambda$-null set, then the series $f_n(z)$ converges to $f(z)$ both $\mu_x$-almost everywhere and in $L^1(Z, \mu_x)$. Thus (theorem 7) $f \in L^1(Z, \mu_x)$ and

$$\mu_x(f) = \sum \mu_x(f_n) \quad \text{for } x \notin M. \tag{13.10}$$

As $\sum |\mu_x(f_n)| \leq \sum \mu_x(|f_n|)$, relation (9) shows that the function $\mu_x(f_n)$ is $\lambda$-integrable and that

$$\sum \int \mu_x(f_n) d\lambda(x) = \int d\lambda(x). \sum \mu_x(f_n) = \nu(f).$$

But $\sum \mu_x(f_n) = \mu_x(f)$ for $x \notin M$ by (10). The function $x \mapsto \mu_x(f)$, defined almost everywhere with respect to $\lambda$, is, therefore, $\lambda$-integrable with, furthermore,

$$\nu(f) = \int \mu_x(f) d\lambda(x), \tag{13.11}$$

which proves point (ii) of the statement (including for functions with values in a Banach space).

To prove (iii), let $f$ be a reasonable and measurable function on $Z$ with respect to $\nu$, with values in $[0, +\infty]$. Then $f$ is the simple limit of an increasing sequence of $\nu$-integrable functions $f_n \geq 0$ converging *everywhere* (n° 9, corollary 3 of theorem 18). Theorem 2 shows that

$$\nu^*(f) = \lim \nu^*(f_n) = \lim \nu(f_n)$$

and hence, by proposition (ii) proved above, that

$$\nu^*(f) = \lim \int \mu_x^*(f_n) d\lambda^*(x) .$$

The same theorem 2 applied to a $\mu_x$ shows that

$$\lim \mu_x^*(f_n) = \mu_x^*(f) \quad \text{for } all\ x \in X .$$

The sequence of functions $x \mapsto \mu_x^*(f_n)$ being increasing, applying theorem 2 once again this time to $\lambda$, shows that

$$\lim \int \mu_x^*(f_n) d\lambda(x) = \int \mu_x^*(f) d\lambda^*(x) ,$$

whence (5"), which completes the proof.

The necessary and sufficient condition for integrability should also be added, namely

$$\int \mu_x^*(|f|)\, d\lambda^*(x) < +\infty$$

if $f$ is measurable and reasonable with respect to $\nu$.

Finally, a $p^{th}$ power $\nu$-integrable function also has this property with respect to $\mu_x$ for almost all $x$. Denoting by $f$ and $f_x$ its classes in $L^p(Z;\nu)$ and $L^p(Z;\mu_x)$,

$$\|f\|_p^p = \int \|f_x\|_p^p \, d\lambda(x) . \tag{13.12}$$

For $p = 2$, we compute the inner product in $L^2(Z;\nu)$ by

$$(f|g) = \int (f_x|g_x)\, d\lambda(x) , \tag{13.13}$$

a formula analogous to the one defining inner products on direct products of Hilbert spaces. But the analogy is misleading; it is already so in the case of a space $X$ with two elements: the $L^2$ space of a sum $\mu_1 + \mu_2$ is not in general the direct sum of the $L^2$ spaces of the given measures (trivial counterexample: take $\mu_1 = \mu_2$). We will come back to this point in the context of the Lebesgue-Nikodym theorem.

## 14 – Integrable Functions with respect to the Image of a Measure

Theorem 19, i.e. the classical formulation of LF, is not exactly a simple rewording of theorem 22. The latter asserts that, for almost all $x \in X$, the function $f$ is $\mu_x$-integrable, whereas theorem 19 asserts that the function $f_x(y) = f(x,y)$ is $\mu$-integrable. To obtain proposition (i) – and hence (ii) and (iii) – of theorem 19, it is thus necessary to show that

$$f \in L^1(\mu_x, Z) \Longleftrightarrow f_x \in L^1(Y,\mu) \tag{14.1}$$

and that, in this case, $\mu_x(f) = \mu(f_x)$. Similarly, (10.5') reduces to formula (13.2') of theorem 22 if

$$\mu_x^*(f) = \mu^*(f_x) \tag{14.2}$$

for all reasonable functions $f$. All this may seem obvious since $\mu_x$ is obtained by transposing to the vertical

$$Y(x) = \{x\} \times Y$$

of $x$ the measure $\mu$ given on $Y$ or, equivalently, since $\mu_x$ is the image of $\mu$ under the map $y \mapsto (x, y)$ from $Y$ to $Z$. It still needs to be proved.

As an exercise, the reader will easily find a direct ad hoc proof. But a general result holding in the framework of example (iii) of n° 12 can also be proved.[45] It will come in useful later in the very particular case where $p : X \mapsto Z$ is continuous and proper. While somewhat simplifying the proof, this assumption does get rid of all difficulties.

**Theorem 23.** *Let $X$ and $Z$ be two locally compact spaces, $\lambda$ a positive measure on $X$ and $p$ a measurable map from $X$ to $Z$. Suppose that, for all $f \in L(Z)$, the function $f \circ p$ is $\lambda$-integrable. Let*

$$\nu(f) = \lambda(f \circ p), \qquad f \in L(Z),$$

*be the image measure of $\lambda$ under $p$.*

(i) *A* reasonable *function $f$ on $Z$ is $\nu$-integrable if and only if $f \circ p$ is $\lambda$-integrable; then*

$$\nu(f) = \lambda(f \circ p). \tag{14.3}$$

(ii) *A function $f$ on $Z$ is $\nu$-measurable if and only if $f \circ p$ is $\lambda$-measurable.*
(iii) *the equality*

$$\nu^*(f) = \lambda^*(f \circ p) \tag{14.3'}$$

*holds for any reasonable* (but not necessarily measurable) *function $f \geq 0$, and for all $f \geq 0$ if the map $p$ is* continuous and proper.

Let us first state two results which we will will need:

**Lemma 1.** *Let $X$ and $Z$ be two locally compact spaces and $p$ a continuous* proper *map from $X$ to $Z$. The image $p(F)$ of every closed set $F \subset X$ is closed in $Z$.*

The restriction of $p$ to $F$ being proper, we may assume that $X = F$. Let $z \in Z$ be a closure point of $p(X)$. Since every compact neighbourhood $W$ of $z$ has non-trivial intersection with $p(X)$, the compact sets $p^{-1}(W)$ are

[45] See N. Bourbaki, *Intégration*, Chap. IV, §4, from which I got rid of the "essential upper integrals" (limits of integrals over compact sets).

non-empty. The intersection of two sets of this type being again of the same time, they have a common point $x$ (Chap. V, § 2, n° 6, corollary 1 of the Borel-Lebesgue theorem, which holds in all generality). As $p(x)$ belongs to all compact neighbourhoods $W$ of $z$, $z = p(x) \in p(X)$, qed.

**Lemma 2.** *Let $X$ and $Z$ be two locally compact spaces, $p$ a continuous* proper *surjective map from $X$ to $Z$ and $f$ a map from $Z$ to a topological space. $f$ is continuous if* (and only if) *$f \circ p$ is continuous.*

By lemma 1, $p$ maps every closed subset of $X$ onto a closed subset of $Z$. Hence for a set $A \subset Z$ to be closed it is sufficient (and necessary since $p$ is continuous) for $p^{-1}(A)$ to be closed. Hence $V \subset Z$ is open if and only if $p^{-1}(V)$ is open in $X$. By the way, this shows that $Z$ *is the quotient space of $X$ by the equivalence relation $p(x) = p(y)$*. If $f$ is a map from $Z$ to a topological space $E$ and if $U \subset X$ and $V \subset Z$ are the inverse images of an open set $W$ of $E$ under $f \circ p$ and $f$, then $U = p^{-1}(V)$ and $V = p(U)$. As a result, $U$ is open if and only if so is $V$, qed.

Let us come back to the family of measures

$$\mu_x(f) = f\left[p(x)\right] .$$

As was seen before the statement of theorem 22, it satisfies condition $(\gamma)$ of theorem 22, and even condition $(\alpha)$ if $p$ is continuous. In all cases, $\nu = \int \mu_x d\lambda(x)$.

*Proof of* (i)*: necessity of the condition.* If $f$ is $\nu$-integrable, then theorem 22 shows that $f$ is $\mu_x$-integrable for almost all $x$ – not very surprising[46] –, that the function $\mu_x(f) = f[p(x)]$, defined for almost all $x$, is $\lambda$-integrable, and finally that $\nu(f) = \int f[p(x)]d\lambda(x)$, whence (3).

If in particular $f$ is the characteristic function of a $\nu$-integrable set $E \subset Z$, then $f \circ p$ is that of $p^{-1}(E)$. If $E \subset Z$ is $\nu$-integrable, then one deduces that $p^{-1}(E)$ is $\lambda$-integrable, and one gets

$$\nu(E) = \lambda\left[p^{-1}(E)\right] . \tag{14.4}$$

If $E$ is a null set, then so is $p^{-1}(E)$.

*Proof of* (ii). Suppose that $f \circ p$ is $\lambda$-measurable and let us show that so is $f$ with respect to $\nu$. Let $H \subset Z$ be compact. As stated above, $A = p^{-1}(H)$ is integrable. Since $p$ and $f \circ p$ are measurable, there is a sequence of compact sets $K_n \subset A$ on which $p$ *and* $f \circ p$ are continuous and such that $M = A - \bigcup K_n$ is a null set. We consider the compact sets $H_n = p(K_n)$. As $p$ is a continuous and surjective map from $K_n$ to $H_n$ and as $f \circ p$ is continuous on $K_n$, lemma 2 shows that $f$ is continuous on $H_n$. It remains to show that $N = H - \bigcup H_n$ is a null set with respect to $\nu$. Now,

$$p^{-1}(N) = A - \bigcup p^{-1}(H_n) \subset A - \bigcup K_n = M ,$$

[46] but not totally trivial: it means that $|f(x)| < +\infty$ for almost all $x$.

which is a null set. As $N$ is a Borel set contained in $H$, it is $\nu$-integrable, and so by (4), $\nu(N) = \lambda[p^{-1}(N)] \leq \lambda(M) = 0$. Thus $f$ is measurable.

Conversely let us suppose that $f$ is $\nu$-measurable and that $K \subset X$ is compact. $K$ is the union of compact sets $K_n$ on which $p$ is continuous, up to a null set. To show that $f \circ p$ satisfies condition (LUS) on $K$, it suffices to show it for each $K_n$. In other words, $p$ may be assumed to be continuous on $K$. But then $p(K) = H$ is compact, and so $H = N \cup \bigcup H_n$, where $N = H - \bigcup H_n$ is a null set and the sets $H_n$ are compact ones on which $f$ is continuous. The function $f \circ p$ is continuous on $K_n = K \cap p^{-1}(H_n)$. These are compact sets since they are closed in $K$. On the other hand, (4) applied to $N$ shows that $K \cap p^{-1}(N) = M$ is a null set. As

$$K \cap p^{-1}(H) = p^{-1}(N) \cap \bigcap p^{-1}(H_n),$$

$K = M \cup \bigcup K_n$. Hence $f \circ p$ is measurable even if $f$ is a function with values in a topological space.

*Proof of* (i): *sufficiency of the condition.* We suppose that $f \circ p$ is integrable, and hence measurable. By (ii), $f$ is measurable. If $f$ is reasonable, proposition (iii) of theorem 22 enables us to apply (3'). Hence, if $f \circ p$ is integrable, then $\nu^*(f) < +\infty$ and $f$ is integrable.

In particular, this shows that a reasonable set $E \subset Z$ is integrable (resp. null) if and only if so is $p^{-1}(E)$, in which case (4) applies.

*Proof of* (iii): *case where $p$ is continuous and proper.* Let $\varphi$ be a lsc function on $X$ and $\psi$ the function on $Z$ given by

$$\psi(z) = \inf_{p(x)=z} \varphi(x). \tag{14.5}$$

We show that $\psi$ is lsc. Since $p^{-1}(\{z\})$ is compact for all $z$, there is a point in $p^{-1}(\{z\})$ where $\varphi$ reaches its *minimum* over this compact set [Chap. V, n° 10, (vi)]. Hence for given $m \in \mathbb{R}$, $\psi(z) \leq m$ if and only if there exists $x$ such that $p(x) = z$ and $\varphi(x) \leq m$. So putting

$$F = \{\varphi(x) \leq m\}, \qquad G = \{\psi(z) \leq m\},$$

we get $G = p(F)$. Since $p$ is continuous and proper, $p(F)$ is closed (lemma 1), whence the result.

This being so, let $f$ be a positive function on $Z$. By (13.6), $\nu^*(f) \geq \lambda^*(f \circ p)$, so that it suffices to prove that $\nu^*(f) \leq \lambda^*(\varphi)$ for any lsc function $\varphi \geq f \circ p$. But for the corresponding function (5), obviously $\psi \circ p \leq \varphi$. As $f \circ p \leq \varphi$ and so$f(z) \leq \varphi(x)$ if $p(x) = z$, we also get $f \leq \psi$. Hence both

$$\nu^*(f) \leq \nu^*(\psi) \quad \text{and} \quad \lambda^*(\psi \circ p) \leq \lambda^*(\varphi)$$

hold.

The map $p$ being continuous, assumption ($\alpha$) of theorem 22 holds. As $\psi$ is lsc, by (13.4), $\nu^*(\psi) = \lambda^*(\psi \circ p)$, which gives (iii) in this particular case. It will come in useful for the completion of the proof in the general case. The idea is the same as in the previous proof, but contains additional technical details.

*Proof of* (iii)*: general case.* We first note that (3') is proposition (iii) of theorem 22 if $f$ is measurable. But (3') only assumes $f$ to be reasonable.

By (13.6), $\nu^*(f) \geq \lambda^*(f \circ p)$, which, as in the preceding case, reduces the proof to showing that

$$(14.6) \qquad \nu^*(f) \leq \lambda^*(\varphi) \quad \text{for any lsc function } \varphi \geq f \circ p\,.$$

As $f$ is reasonable, it suffices (theorem 2) to do this when $f$ vanishes outside a compact set $H \subset Z$; $f \circ p$ is then zero outside the set $A = p^{-1}(H)$, which is integrable by the already proved proposition (ii). Since $p^{-1}[p(A)] = A$ is measurable, (ii) proves that $p(A) = B$ is measurable, and hence integrable since it is contained in $H$.

Let us choose an increasing sequence of compact sets $K_n \subset A$ on which $p$ is continuous and such that $M = A - \bigcup K_n$ are a null sets, and consider the compact sets $H_n = p(K_n)$ and the measurable set $N = B - \bigcup H_n$. Since $p^{-1}(N) \subset M$, $N$ is a null set by (4). Then $B = N \cup \bigcup H_n$.

Let $\varphi$ be a positive lsc function on $X$ dominating $f \circ p$. Setting

$$(14.7) \qquad \psi_n(z) = \inf_{\substack{p(x)=z \\ x \in K_n}} \varphi(x) \quad \text{for } z \in H_n\,,$$

$$(14.7') \qquad \psi_n(z) = +\infty \quad \text{for } z \in B - H_n\,,$$

$$(14.7'') \qquad \psi_n(z) = 0 \quad \text{for } z \in Z - B\,,$$

let us show that

$$(14.8) \qquad \psi_n(z) \geq f(z) \quad \text{for all } z \in Z\,.$$

If $z \in H_n$, then $\varphi(x) \geq f[p(x)] = f(z)$ for all $x \in K_n$ such that $p(x) = z$, whence $\psi_n(z) \geq f(z)$ by (7). If $z \in B - H_n$, the same holds by (7'). Lastly, if $z \in Z - B$, then both sides are zero by (7").

On the other hand, clearly,

$$(14.9) \qquad \varphi(x) \geq \psi_n \circ p(x) \quad \text{on } K_n\,.$$

Finally, the sequence $\psi_n$ is decreasing since $K_n \subset K_{n+1}$ for all $n$.

Setting $\psi = \inf(\psi_n)$, $\psi(z) \geq f(z)$ by (8). Thus

$$(14.10) \qquad \nu^*(\psi) \geq \nu^*(f)\,.$$

Let us show that $\varphi(x) \geq \psi \circ p(x)$ ae. For $x \in K_n = A - M$, this follows from (9). If $x \in M$, it does not matter if it is false since $M$ is a null set. Finally, if $x \in X - A$, then $p(x) \in Z - B$, whence $\psi \circ p(x) = 0$ by (7").

This implies that

$$\lambda^*(\varphi) \geq \lambda^*(\psi \circ p)\,. \tag{14.11}$$

By (10) and (11), the proof of relation (6) and hence of proposition (iii) will follow if we show that

$$\nu^*(\psi) = \lambda^*(\psi \circ p)\,. \tag{14.12}$$

As $\psi$ is reasonable by (7"), for this purpose, it suffices to check that $\psi$ is *measurable* [theorem 22, (13.5")] or, since $\psi = \inf(\psi_n)$, that so are $\psi_n$.

The function $\psi_n$ being constant on the measurable sets $B - H_n$ and $Z - B$, the proof reduces to verifying it on the compact set $H_n = p(K_n)$. Now, $p$ is continuous (and, trivially, proper) on $K_n$, and the restriction of $\varphi$ to $K_n$ is lsc. Replacing $X$ and $Z$ by $K_n$ and $H_n$, we are back in the particular case proved above: the restriction of the function $\psi_n$ to $H_n$ is lsc, qed.

## 15 – Invariant Measures under Group Actions

We will now consider case (iv) of n° 12, but the topics covered in this n° go far beyond the framework of a simple application of the generalized LF theorem. It is rather a matter of presenting the abc of what André Weil called *Integration in topological groups and its applications* in a famous book which was like the Bible to me in the 1940s (Paris, Hermann, 1941).

(i) *Invariant measures on a group.* On every lcg $G$, there is measure $d\mu_r(g)$ invariant under right translations $g \mapsto ga$, as well as a measure $d\mu_l(g)$ invariant under left translations $g \mapsto ag$. Though the existence theorem of these *Haar measures* is essential for the general case, as mentioned above, it does not teach us anything beyond its simple statement. In practice, the problem is the explicit calculation of these measures; this requires quite different methods, in particular for Lie groups often needing the Chap. IX results on differential forms.[47]

*Exercise 1.* Let $G = GL_n(\mathbb{R})$ be the group of $n \times n$ invertible real matrices, open in $M_n(\mathbb{R})$. Show that the measure $\det(g)^{-n}dm(g)$, where $m$ is the Lebesgue measure on $M_n(\mathbb{R})$, is right and left invariant on $G$. Is it well-defined as a measure on $M_n(\mathbb{R})$?

In fact, the most useful result is that these invariant measures are unique up to constant factors. The method presented in Chap. IX, n° 10, lemma a

[47] A Lie group is equipped with the structure of a $C^\infty$ (and even analytic) manifold such that the map $(x, y) \mapsto xy^{-1}$ is $C^\infty$. For all $g$, the map $x \mapsto gx$ is a diffeomorphism whose tangent map at $e$ is an isomorphism from $G'(e)$ onto $G'(g)$. Choosing an alternating multilinear form of maximum degree $\omega \neq 0$ on $G'(e)$ and denoting by $\omega(g)$ the form on $G'(g)$ it induces by $x \mapsto gx$, we get a differential form (also denoted $\omega$) of maximum degree and everywhere $\neq 0$ on $G$. This shows that the manifold $G$ is orientable and enables us to define a positive measure $\mu$ on $G$ by $\mu(f) = \int f\omega$. It is the left invariant measure.

shows this as easily as for $\mathbb{R}^n$. On the other hand, the map $g \mapsto g^{-1}$ obviously transforms every left invariant measure into a right invariant one.

It is often troublesome to use the same letter $\mu$ to denote the Haar measures of several different groups, and hardly more convenient to denote them by $\lambda$, $\mu$, $\nu$, etc. Just like $dx$ denotes the Haar (i.e. Lebesgue) measure on $\mathbb{R}^n$, in his book, André Weil only used $dg$ (or $dh$, $dk$, etc. if the group is written $H$, $K$, etc.) for the *left* invariant measure of $G$. He never used the right invariant measure since, to integrate a function $f$ with respect to it, one computes $\int f(g^{-1})dg$. This convention simplifies notations when only general groups are considered. Other authors prefer to distinguish between $d_l g$ and $d_r g$ by setting $d_r g = d_l(g^{-1})$. In his Chap. VII et VIII on the Haar measure and the convolution product, N. Bourbaki denotes by $\beta$ – why not? – the left invariant measure. Hereinafter, I will follow Weil's convention, even if it means coming back to more precise notations where required. The notation $L^p(G)$ will refer to the measure $dg$, and so will any mention of integrable and measurable functions unless otherwise stated.

The uniqueness of the measure $dg$ up to a factor has consequences. If, for a given $a \in G$, we consider the image of the *left* invariant measure $dx$ under the *right* translation $x \mapsto xa$, we again get a left invariant measure because $b(xa) = (bx)a$. Hence there is a number $\Delta_G(a) = \Delta(a) > 0$ such that[48]

$$\int f(xa^{-1})dx = \Delta(a)\int f(x)dx \tag{15.1}$$

for all $f \in L(G)$ or more generally integrable. Then $\Delta(xy) = \Delta(x)\Delta(y)$ and the function $\Delta$ is clearly continuous. In many cases, $\Delta(x) = 1$, especially if $G$ is *compact* (Chap. IX, n° 10, lemma e). $G$ is then said to be *unimodular.*

The measure $\Delta(x)^{-1}dx$ is right invariant since, in short, $\Delta(xa)^{-1}d(xa) = \Delta(xa)^{-1}\Delta(a)dx = \Delta(x)^{-1}dx$. The measure $d(x^{-1})$ is also right invariant since $(xa)^{-1} = a^{-1}x^{-1}$. Hence $d(x^{-1}) = c\Delta(x)^{-1}dx$ with a constant $c > 0$, and so, $x \mapsto x^{-1}$ gives $dx = c\Delta(x).d(x^{-1})$ and thus $c = 1$. As a result,

$$d(xa) = \Delta(a)dx\,, \qquad dx^{-1} = \Delta(x)^{-1}dx = d_r x\,. \tag{15.2}$$

Since the second relation can be set out more explicitly as

$$\int f\left(x^{-1}\right)dx = \int f(x)\Delta(x)^{-1}dx\,,$$

it, in particular, shows that *if $G$ is unimodular, then the Haar measure is invariant under $x \mapsto x^{-1}$*.

[48] Applying (1) to the characteristic function of a measure set $X \subset G$ gives the $m(Xa) = \Delta(a)m(X)$, where $dm(x) = dx$. If, like Leibniz, we consider that the symbol $dx$ denotes an "infinitesimal volume element" at point $a$, then $dx.a = \Delta(a).dx$, which is more conveniently written as $d(xa) = \Delta(a)dx$. Then Formula (1) is obtained by the variable change $x \mapsto xa$ in the first integral since it transforms $f(xa^{-1})$ into $f(x)$ and $dx$ into $d(xa)$. The notation $d(x^{-1})$ used in (2) can be interpreted in a similar way. Compare with the change of variable formulas for multiple integrals (Chap. IX, n° 10).

*Exercise 2.* Let $G$ be the group of matrices[49] $(x\ y|0\ 1)$ with $x \in \mathbb{R}^*$, $y \in \mathbb{R}$. It acts on $\mathbb{R}$ by $t \mapsto xt + y$. Show that its invariant measures are of the form $\mathrm{j}(x,y)d^*xdy$ and calculate the $\Delta$ function of the group.

(ii) *Continuous linear representations.* In all cases, the group $G$ defines continuous linear operators on $L^p(G)$ spaces, whose study, for $p = 1$ and mostly for $p = 2$, is the foundation of harmonic analysis on $G$. For example, for all $a \in G$, denoting by $L(a)$ the operator transforming every $f \in L^p(G)$ into the function

$$L(a)f : x \longmapsto f\left(a^{-1}x\right), \tag{15.3}$$

which is also in $L^p(G)$, defines an *isometric* operator (i.e. preserves the norm) on this Banach space, and for $p = 2$, even a *unitary* one,[50] and $L(xy) = L(x)L(y)$ for all $x, y$. An operator $R(a)$ in $L^p$ defined by

$$R(a)f : x \longmapsto f(xa) \tag{15.3'}$$

can be associated to all $a \in G$. Then $R(xy) = R(x)R(y)$ and

$$L(x)R(y) = R(y)L(x),$$

but the operators $R(y)$ are isometric only if $G$ is unimodular. The maps $x \mapsto L(x)$ and $x \mapsto R(x)$, called left and right *regular representations* from $G$ to the $L^p$ spaces, are not continuous as maps on the Banach space of bounded operators on $L^p(G)$, but the maps $(x, f) \mapsto L(x)f$ and $(x, f) \mapsto R(x)f$ from $G \times L^p$ to $L^p$ are. To see this, first notice that

$$\|L(x)f - L(x_0)f_0\|_p \leq \|L(x)\| \cdot \|f - f_0\|_p + \|L(x)f_0 - L(x_0)f_0\|_p .$$

As $(x, f)$ converges to $(x_0, f_0)$, the first term on the left hand side tends to 0 since $\|L(x)\| = 1$. It, therefore, suffices, to analyze the second term, namely to check that the map $x \mapsto L(x)f$ from $G$ to $L^p$ is continuous for all $f \in L^p$. Checking it for everywhere dense functions $f$ is in fact sufficient because, if $f_n$ converges to $f$, $L(x)f_n$ converges uniformly to $L(x)f$ on $G$. However, if $f \in L(G)$, then this is immediate since if $x$ tends to $x_0$, thus stays in a compact neighbourhood $V$ of $x_0$, the function $L(x)f$ vanishes outside the compact set $VK$, where $K$ is the support of $f$ and, $f$ being continuous, converges uniformly to $L(x_0)f$ while remaining zero outside $VK$. This implies convergence in $L^p$. Same arguments apply to $R(x)$.

[49] To simplify typing, it is convenient to set

$$(a\ b|c\ d) = \begin{pmatrix} a & b \\ c & d \end{pmatrix}.$$

[50] i.e. bijective and isometric. See n° 19.

This situation can be generalized using the notion of a *linear representation* (implied to be continuous) of $G$: It consists of an ordered pair[51] $(\mathcal{H}, U)$ where $\mathcal{H}$ is a Banach space and $U$ a map $x \mapsto U(x)$ to the group of continuous invertible operators of $\mathcal{H}$, satisfying the following two conditions:

(RL 1) $U(xy) = U(x)U(y)$, where $U(e) = 1$ and $U(x^{-1}) = U(x)^{-1}$,
(RL 2) the map $x \mapsto U(x)\mathbf{a}$ from $G$ to $\mathcal{H}$ is continuous for all $\mathbf{a} \in \mathcal{H}$.

Condition (RL 2) actually implies that $(x, \mathbf{a}) \mapsto U(x)\mathbf{a}$ is continuous. This can be shown by writing as above that

$$\|U(x)\mathbf{a} - U(x_0)\mathbf{a}_0\| \leq \|U(x)\| \,.\, \|\mathbf{a} - \mathbf{a}_0\| + \|U(x)\mathbf{a}_0 - U(x_0)\mathbf{a}_0\|$$

and by taking into account the fact that [52]

$$\sup_{x \in K} \|U(x)\| < +\infty \tag{15.4}$$

for any compact set $K \subset G$. This enables one to show that the first term of the right hand side tends to 0 (for $K$ take a neighbourhood of $x_0$), the same holding for the second term by (RU 2). It is actually enough to check (RU 2) for vectors everywhere dense in $\mathcal{H}$.

If $\mathcal{H}$ is a Hilbert space[53] and if all $U(x)$ are unitary, $(\mathcal{H}, U)$ is said to be a *unitary representation* of $G$. In dimension 1, such a representation is just a continuous solution with absolute value 1 of the functional equation

[51] What I write as $U(g)$ is usually written $\pi(g)$, following Harish-Chandra. In algebra, which (theoretically) does not involve topology, a linear representation of a group $G$ is a homomorphism $g \mapsto U(g)$ from $G$ to the group $GL(E)$ of in general a finite-dimensional vector space $E$ over a commutative field.

[52] Since continuous functions $x \mapsto \|U(x)\mathbf{a}\|$ are bounded on every compact set for all $\mathbf{a} \in \mathcal{H}$, it suffices to use a Banach theorem, a quick proof for which follows. We start by proving the famous *Baire's theorem*: let $\mathcal{H}$ be a complete metric space and $(F_p)$ a sequence of closed sets such that $\mathcal{H} = \bigcup F_p$. Then at least one of the sets $F_p$ contains an open subset of $\mathcal{H}$. Otherwise, if $B$ is a closed ball, the interior of $B$ is not contained in any $F_p$, and so for each $p$, contains a closed ball which does not intersect $F_p$. Hence, starting from an arbitrarily chosen closed ball $B_0$ of radius 1, we can construct a decreasing sequence of closed balls $B_p$, of radius $\leq 1/p$, such that $B_p \cap F_p = \varnothing$ for all $p$. As $\mathcal{H}$ is complete, there exists $x \in \bigcap B_n$, a contradiction since $x \in F_p$ for at least some $p$. (Exercise: the theorem continues to hold if $\mathcal{H}$ is replaced by some $G_\delta$ set in $\mathcal{H}$ and the sets $F_n$ by their intersections with it). *Banach-Steinhaus theorem*: Let $E$ be a set of continuous linear maps from a Banach space $\mathcal{H}$ to another. Suppose that $\sup \|T\mathbf{a}\| = p(\mathbf{a}) < +\infty$ for all $\mathbf{a} \in \mathcal{H}$. Then $\sup \|T\| < +\infty$. Otherwise, there exist $T_n \in E$ such that $\sup \|T_n\| = +\infty$. For all $p$, the set $F_p$ of $x \in \mathcal{H}$ satisfying $\|T_n x\| \leq p\|x\|$ for all $n$ is closed, and as $\sup \|T_n x\| < +\infty$ for all $x$,$\mathcal{H} = \bigcup F_p$. Hence, one of the sets $F_p$ contains a ball $B(\mathbf{a}, r)$. Then, for $\|x - \mathbf{a}\| \leq r$,

$$\|T_n(x - \mathbf{a})\| \leq \|T_n x\| + \|T_n \mathbf{a}\| \leq p\|x\| + p(\mathbf{a}) = c\,,$$

a constant independent of $n$; whence $\sup \|T_n\| \leq c/r$, a contradiction. We also state the *closed graph theorem*: a linear map $T : \mathcal{H}_1 \longrightarrow \mathcal{H}_2$ is continuous if and only if its graph $G \subset \mathcal{H}_1 \times \mathcal{H}_2$ is closed.

[53] For elementary properties of Hilbert spaces see n° 19.

$$\chi(xy) = \chi(x)\chi(y) .$$

This is called a *character* of the group $G$. If $G$ is commutative, there are enough of them so that, like on $\mathbb{T}$ or $\mathbb{R}$, one can construct a theory of Fourier transforms on $G$ without any explicit calculations (§ 7). But if $G$ is not commutative, the function 1 may well be the only solution, for example as in the case of $SL_2(\mathbb{R})$ which does admit any non-trivial finite-dimensional *unitary* representations. Characters are then replaced by *irreducible unitary* representations, i.e for which the only *closed* subsets of $\mathcal{H}$ stable under the operators $U(x)$ are $\{0\}$ and $\mathcal{H}$. However, since

$$\left(U(x)\mathbf{a}\middle|U(x)\mathbf{b}\right) = (\mathbf{a}|\mathbf{b}) ,$$

*if a closed vector subspace of* $\mathcal{H}$ *is invariant under operators* $U(x)$*, then so its orthogonal subspace* and conversely. Among other things, this property explain the importance of unitary representations. If $\mathcal{H}$ is finite-dimensional, in which case it contains a non-zero invariant subspace of minimal dimension (possibly $\mathcal{H}$ itself), it then follows that $\mathcal{H}$ is a *direct sum of minimal and pairwise orthogonal invariant subspaces,* i.e. on which the representation is irreducible. As it is based on a simple dimension argument, the latter no longer holds in infinite dimension, though there are cases where it does (compact groups), but one can prove a similar but imperfect result by broadening the notion of a direct sum. In the case of a general lcg, one first shows (Gelfand and Raïkov, 1943), using quite ingenious arguments from functional analysis (see end of n° 30), that there are many irreducible unitary representations,[54] then that any unitary representation is a "continuous sum" or a "direct integral" of irreducible representations in the sense of n° 24. Chap. XII will give concrete non-trivial examples of infinite-dimensional irreducible representations. If $G$ is *compact*, the irreducible representations are finite-dimensional and for any unitary representation $(\mathcal{H}, U)$ of $G$, there is a decomposition of $\mathcal{H}$ into a Hilbert ("discrete") direct sum of minimal invariant subspaces, in which the $U(x)$ define irreducible representations (end of n° 29, (iii). This is not hard to prove for the regular representation of a *linear* group since in this case the existence of several minimal finite-dimensional invariant subspaces is obvious:

*Exercise 3.* Let $G$ be a compact subgroup of $GL_n(\mathbb{R})$. (a) Show that polynomial functions in coefficients of $g$ are everywhere dense[55] in $L^2(G)$ (use Stone-Weierstrass). (b) Let $\mathcal{H}_d$ be the subspace of $\mathcal{H} = L^2(G)$ consisting of polynomials of total degree $\leq d$ in coefficients of $G$. Show that it

[54] This means that for all $x \neq e$, there is an irreducible representation such that $U(x) \neq 1$. Gelfand and Raïkov's article being in Russian, it is probably better to look up Henri Cartan and Roger Godement, *Analyse harmonique et théorie de la dualité dans les groupes abéliens localement compacts* (Ann. Sc. de l'Ecole normale supérieure, 1947).

[55] Polynomial functions on a non-compact closed subgroup $G$ of $GL_n(\mathbb{R})$ can never be in $L^2(G)$. In this case, the exercise shows that there are finite-dimensional linear representations of $G$, but nothing more.

is finite-dimensional and invariant under operators $L(x)$ and $R(x)$ defined in (3) and (3') or as the saying goes, bi-invariant. (c) Show that every left invariant (resp. bi-invariant) finite-dimensional subspace $\mathcal{M} \neq \{0\}$ of $L^2(G)$ contains a *minimal* subspace of the same type and that $\mathcal{M}$ is the direct sum of pairwise orthogonal minimal left invariant (resp. bi-invariant) subspaces. (d) Show that $L^2(G)$ is a Hilbert direct sum of minimal left invariant (resp. bi-invariant) subspaces. (e) For $G = \mathbb{T}$, rewrite the result in the language of the theory of Fourier series.

Unlike the above, the next exercise shows an intractable situation.

*Exercise 4* (von Neumann, 1943). Let $G$ be a discrete group, so that the measures or Dirac functions

$$\epsilon_x(y) = 1 \quad \text{if } y = x\,, \quad = 0 \quad \text{if } y \neq x$$

form an orthonormal basis for $L^2(G)$. Show that $L(x)\epsilon_y = \epsilon_{xy}$. Let $A$ be a continuous operator in $L^2(G)$ such that $AL(x) = L(x)A$ for all $x$. Show that

$$Af(x) = \sum a\left(xy^{-1}\right) f(y)$$

for all $f \in L^2(G)$ where $a = A\epsilon_e \in L^2(G)$. Show that $AL(x) = L(x)A$ for all $x$ if and only if the function $a$ satisfies $a(xyx^{-1}) = a(y)$ for all $x$ and $y$. Suppose that, for all $y \neq e$, the set of conjugates $xyx^{-1}$ of $y$ is infinite. Show that $A$ is a scalar and deduce that in this case, $L^2(G)$ does not contain any non-trivial closed bi-invariant subspace (observe that the orthogonal projection operator onto such a subspace commutes with all $L$ and $R$. This also mean that the unitary representation $(x, y) \mapsto L(x)R(y)$ from $G \times G$ to $L^2(G)$ is irreducible. [in this case, the quotient of the group $SL_2(\mathbb{Z})$ by its centre $\{1, -1\}$ is, like all "arithmetically defined" groups such as $SL_n(\mathbb{Z})$, the group of matrices with integer entries preserving an (indefinite) quadratic form with integer coefficients, etc.]

(iii) *Quotient of a space by a group.* Let $Z$ be a locally compact space and $G$ a lcg acting on the left (or the right) on $Z$ by a continuous map $(g, x) \mapsto gx$ (or $xg$) satisfying obvious algebraic conditions. For $A \subset G$ and $B \subset Z$, let $AB$ be the set of products $gz$, where $g \in A$ and $z \in B$. For $B = \{z\}$, the set $GB$, written simply $Gz$, is the *orbit* of $z$. $gB$ is defined likewise by taking $A = \{g\}$.

One can then define a quotient set $X = G\backslash Z$ (or $Z/G$ if $G$ acts on the right), a map $p : Z \mapsto G\backslash Z$ taking each $z$ to its orbit $Gz$, and a topology on $X$ with respect to which the set $U \subset G\backslash Z$ is open if and only if $p^{-1}(U)$ is open in $Z$. For simplicity's sake, it is often convenient to use the notation

$$p(z) = \dot{z}\,.$$

The continuous map $p$ is *open*, i.e. transforms every open subset $W$ of $Z$ into an open subset $U$ of $X$ since $p^{-1}[p(W)] = \bigcup gW$ is a union of open sets.

Hence, if $X$ is separated, $p$ maps every compact neighbourhood of $z \in Z$ onto a compact neighbourhood of $p(x)$ in $X$. As a result, $X$ is locally compact *if it is separated.*

**Lemma 1**. *Let $Z$ be a separated topological space and $R \subset Z \times Z$ the graph*[56] *of an equivalence relation on $Z$ such that the map $p : Z \mapsto Z/R$ is* open. *$Z/R$ is separated if and only if $R$ is closed in $Z \times Z$.*

Let us set $Z' = Z/R$. The map $(x, y) \mapsto (p(x), p(y))$ from $Z \times Z$ to $Z' \times Z'$ is continuous. However, the set $R$ of ordered pairs such that $p(x) = p(y)$ is the inverse image of the diagonal of $Z' \times Z'$, which is closed if and only if $Z$ is separated (exercise!). Hence the necessity of the condition follows without any assumption on $p$. Conversely, if $R$ is closed and if $p(x) \neq p(y)$, the ordered pair $(x, y)$ is exterior to $R$, hence there are neighbourhoods $U$ and $V$ of $x$ and $y$ in $Z$ such that $U \times V \cap R = \varnothing$. Then the images $p(U)$ and $p(V)$ are disjoint open *neighbourhoods* of $p(x)$ and $p(y)$ in $Z'$, qed.

**Lemma 2**. *Let $G$ be a group acting on a separated topological space $Z$. The quotient space $G\backslash Z$ is separated if and only if the following condition holds: for every pair of points $c, c' \in Z$ such that $Gc \neq Gc'$, there are neighbourhoods $U$ and $U'$ of $c$ and $c'$ in $Z$ such that*

$$gU \cap U' = \varnothing \quad \text{for all } g \in G. \tag{15.5}$$

*For any* compact *set $K \subset Z$, the set $GK$ is then closed. If $Z$ is Polish, then $GB$ is a Borel set for any open or closed set $B \subset Z$.*

We denote by $R$ the equivalence relation defined by $G$. As was seen above, the map $p$ is open. Let $(c, c')$ be a closure point of $R$. Any neighbourhood of $(c, c')$ contains a set $U \times U'$, where $U$ and $U'$ are neighbourhoods of $c$ and $c'$ in $Z$. Saying that $U \times U'$ intersects $R$ non-trivially means that there is some $z \in U$, $z' \in U'$ and $g \in G$ such that $z' = gz$, i.e. some $g$ such that[57] $gU \# U'$. The statement $R$ is closed is expressed by saying that if this condition holds for all $U$ and $U'$, then $(c, c') \in R$, i.e. $Gc = Gc'$. This gives condition (10).

To prove the second proposition of the lemma, suppose first that $B$ is compact. Then $GB = p^{-1}[p(B)]$. As $p(B)$ is compact and hence closed $G\backslash Z$ being separated, $GB$ is closed, whether or not $Z$ is Polish. If it is, then it is metrizable and countable at infinity, and as was seen in n° 10, so are all its locally compact subspaces. Hence, if $B$ is the countable union of compact sets, for example if $B$ is closed or open, then $GB$ is the countable union of closed sets, and so is a Borel set,[58] qed.

---

[56] It is the set of ordered pairs $(x, y)$ satisfying the given relation.

[57] Generally speaking, one writes $E \# F$ instead of $E \cap F \neq \varnothing$.

[58] One may wonder whether the result continues to hold for all Borel sets $B$. In fact, it is only possible to sat that $p(B)$ is *analytic* (n° 10), and hence that so is $p^{-1}[p(B)] = GB$, so that $p(B)$ is *measurable* with respect to any measure on $Z$.

Lemma 2 does not assume that $G$ is equipped with a topology. Suppose that that $G$ is locally compact and that the map $(g, z) \mapsto gz$ is continuous. $R$ is the image of $X = Z \times G$ under the continuous map

$$\pi : (z, g) \longmapsto (z, gz)$$

to $Y = Z \times Z$. If $G$ is locally compact, so are $X$ and $Y$. On the other hand, the graph of the equivalence relation defined by $G$ being $\pi(Z \times G)$, it is closed if $\pi$ is *proper* (n° 14, lemma 1), i.e. if the inverse image of any compact set $K \subset Z \times Z$ is compact. As it is closed, it suffices to express this for a product $B \times A$, where $A$ and $B$ are compact. Then $\pi^{-1}(K) = B \times C$, where $C = C(A, B)$ is the set of $g$ such that $gA\#B$. As a result:

**Lemma 3.** *Let $G$ be a locally compact set acting on a locally compact space $Z$. The quotient space $G\backslash Z$ is locally compact* if *the following condition is satisfied:*

> (GOP)*: for all compact sets $A, B \subset Z$, the set $C(A, B)$ of $g \in G$ such that $gA\#B$ is compact.*

$G$ is then said to act *properly* on $Z$. All *closed* subgroups of $G$ act properly on $Z$.

*Exercise 5.* Suppose that $G$ and $Z$ are metrizable so as to be able to use BW. Show by a direct argument that condition (GOP) implies (5) for sufficiently small $U$ and $U'$ (consider sequences $g_n$, $z_n$ and $z'_n$ such that $z'_n = g_n z_n$, $\lim z_n = c$, $\lim z'_n = c'$).

(iv) *Quotient of an invariant measure.* This preliminary stage completed, let $G$ be a lcg acting properly (on the left) on a locally compact space $Z$. For $f \in L(Z, K)$ and $z \in Z$, the function $g \longmapsto f(gz)$ vanishes outside the set of $g$ such that $gz \in K$. As it is compact by (GOP), like in n° 12, (iv) one can define a function

$$z \longmapsto \int f(gz) d_r g = \int f\left(g^{-1} z\right) dg \tag{15.6}$$

using Weil's notation for the *left* invariant measure $dg$. This function is $G$-invariant since replacing $z$ by $az$ amounts to replacing $g$ by $ga$. This leads to a function on $X = G\backslash Z$ given by

$$f^G(\dot{z}) = \int f(gz) d_r g\,, \tag{15.7}$$

where recall that $\dot{z} = p(z)$. If $K$ is the support of $f$, the integral vanishes outside $GK$, so that $f^G$ vanishes outside the compact set $p(K)$. Finally, it is more or less obvious that function (7) is continuous, and thus

$$f \in L(Z) \Longrightarrow f^G \in L(X)\,. \tag{15.8}$$

**Lemma 4.** *The map $f \longmapsto f^G$ from $L_+(Z)$ to $L_+(X)$ is surjective.*

First note that *any compact set $K \subset X$ is the image under $p$ of a compact subset of $Z$*. For all $x = p(z)$ and all compact neighbourhoods $W$ of $z$, the image $p(W)$ is indeed a compact neighbourhood of $x$. By BL, $K$ can be covered by finitely many such sets $p(W_i)$. As $H = \bigcup W_i$ satisfies $p(H) \supset K$, the compact set $H \cap p^{-1}(K)$ answers the question.

This being so, first note[59] that if some $f \in L_+(Z)$ is $> 0$ on a compact set $H \subset Z$, then $f^G > 0$ on $p(H)$. Hence, for all $\varphi \in L_+(X)$ with support $p(H)$, there exists $\psi \in L_+(X)$ such that $\varphi = f^G\psi$. The function $h(z) = f(z)\psi[p(z)] = f(z)\psi(\dot{z})$ also being in $L_+(Z)$, it is possible to compute

$$h^G(\dot{z}) = \int f(gz)\psi\,[p(gz)]\,d_rg = \int f(gz)\psi(\dot{z})d_rg = \psi(\dot{z})\int f(gz)d_rg =$$
$$= \psi(\dot{z})f^G(\tilde{z}) = \varphi(\dot{z})\,,$$

whence the lemma. It obviously also applies to $L(Z)$ and $L(X)$.

We now suppose that there is a $G$-invariant measure on $Z$, i.e. such that

$$\int f(gz)d\nu(z) = \int f(z)d\nu(z)\,. \tag{15.9}$$

We will also need functions

$$f_G(z) = \int f(gz)d_lg = \int f(gz)\Delta_G(g)d_rg\,. \tag{15.7'}$$

For all $a \in G$, one readily gets

$$f_G\left(a^{-1}z\right) = \Delta_G(a)f_G(z)\,. \tag{15.10}$$

One can show as above that the map $f \longmapsto f_G$ from $L_+(Z)$ to the set $L_+(X;\Delta_G)$ of continuous, positive solutions with compact support mod $G$ of (10) is surjective. The distinction between $f^G$ and $f_G$ is of interest only if $G$ is not unimodular, but that it is precisely what often happens in practice.

**Lemma 5.** *For $f \in L(Z)$, the relation $f_G = 0$ implies $\nu(f) = 0$.*

Let $A$ be the support of $f$ and $B$ that of a function $h \in L(Z)$. The relation $h(z)f(gz) \neq 0$ requires $z \in B$ and $gz \in A$, hence $g \in C(B, A)$, a compact set by (GOP). The function $(z, g) \longmapsto h(z)f(gz)$ is, therefore, in $L(Z \times G)$. The most elementary version of LF then shows that

[59] Every non-empty open set $U$ has $> 0$ measure with respect to the left invariant measure of $G$. Otherwise, all the cosets $aU$ would be null sets, and hence compact subsets of $G$ (apply BW).

$$\begin{aligned}\int f_G(z)h(z)d\nu(z) &= \int d_l g \int h(z)f(gz)d\nu(z) = \\ &= \int d_l g \int h\left(g^{-1}z\right) f(z)d\nu(z) = \\ &= \int f(z)d\nu(z) \int h\left(g^{-1}z\right) d_l g = \\ &= \int f(z)d\nu(z) \int h(gz)d_r g = \int f(z)h^G(z)d\nu(z)\,.\end{aligned}$$

Choosing $h$ in such a way that the function $h^G$ equals 1 on the support $A$ of $f$, we find $\nu(f) = \nu(f_G h)$, qed.

Since lemma 5 shows that the integral $\nu(f)$ only depends on the function $f_G$, the analogue of lemma 4 for the map $f \longmapsto f_G$ proves that

$$\nu(f) = \lambda\left(f_G\right), \tag{15.11}$$

where $\lambda$ is a well-defined positive linear functional on $L(X; \Delta_G)$. It is the *quotient pseudo-measure* of $\nu$ by $G$. So, using the integral notation for $\lambda$,

$$\int f(z)d\nu(z) = \int d\lambda(z) \int f(gz)d_l g\,. \tag{15.11'}$$

In what precedes, starting from an invariant measure $\nu$ on $Z$, we inferred a pseudo-measure $\lambda$ on $X$. We could as well have started from $\lambda$ and defined $\nu$ by (10) since $f_G \in L(X; \Delta_G)$. Clearly this would give an invariant measure on $Z$. Invariant measures on $Z$ thus correspond bijectively to genuine measures on $X$ *if $G$ is unimodular.*

To apply theorem 22 to the measure $\nu$ *in this case*, we set

$$\mu_x(f) = \int f(gz)dg \quad \text{if } x = p(z)\,. \tag{15.12}$$

This gives for each $x \in X = G\backslash Z$, a positive measure $\mu_x$ on $Z$. It is the image of $dg$ under the (continuous and proper) map $g \longmapsto gz$. Relation (11) then becomes

$$\nu(f) = \int \mu_x(f)d\lambda(x)\,. \tag{15.13}$$

Here the situation is particularly simple since (12) shows that $x \longmapsto \mu_x(f)$ is continuous for all $f \in L(Z)$. Assumption (a) of theorem 22, therefore, holds. It follows that if a function $f$ on $Z$ is $\nu$-integrable, then $f$ is $\mu_x$-integrable for almost all $x \in X$, the function $x \longmapsto \mu_x(f)$ is $\lambda$-integrable and the equivalent formulas (12) and (13) continue to hold. Moreover, theorem 23, (i), shows that $f \in L^1(\mu_x; Z)$ if and only if the function $g \longmapsto f(gz)$ is $dg$-integrable, in which case (12) continues to hold. Hence finally, relation (13) can be written in form (11) in all cases. This assumes that $Z$ is countable at infinity and that $G$ is *unimodular.* If $G$ is not unimodular, there is an analogous result, but before that integration theory needs to be rewritten for linear functionals on $L(X; \Delta_G)$ – an excellent way to spend rainy days.

*Exercise 5.* Suppose that $Z$ is a unimodular lcg, that $d\nu(z) = dz$ and that $G$ is a closed subgroup of $Z$ acting by $(g, z) \longmapsto gz$, which gives a positive linear functional $\lambda$ on the space $L(G\backslash Z, \Delta_G)$ of continuous solutions with compact support of $f(g^{-1}z) = \Delta_G(g)f(z)$. Suppose there is a closed subset $H \subset Z$ such that $(g, h) \longmapsto gh$ is a homeomorphism on $Z$, which enables us to identify $L(G\backslash Z, \Delta_G)$ with $L(H)$ by restricting ourselves to $H$. Show that

$$\int f(z)dz = \iint f(gh)d_lgd_rh \quad \text{for all } f \in L(Z).$$

(v) *An example: the orthogonal group on* $\mathbb{R}^n$. For $Z$ take the space $\mathbb{R}^n$, for $G$ the compact group $O_n(\mathbb{R})$ of orthogonal matrices ($g'g = 1$) acting in an obvious way on $\mathbb{R}^n$ and for $\nu$ the Lebesgue measure $dz$ on $\mathbb{R}^n$, which is clearly $G$-invariant [Chap. IX, n° 10, (i)]. For $z', z'' \in \mathbb{R}^n$, there exists $g \in G$ such that $z'' = gz'$ if and only if $\|z'\| = \|z''\|$, where this is the standard Euclidean norm. Here the quotient $G\backslash Z$ is $\mathbb{R}^*_+$, the map $p$ being just $p(z) = \|z\|$. Hence, letting $dg$ denote the invariant measure of $G$ normalized by $\mu(G) = 1$, there is a unique measure $\lambda$ on $\mathbb{R}^*_+$ such that

$$\int_{\mathbb{R}^n} f(z)dz = \int_0^{+\infty} d\lambda\left(\dot{z}\right) \int_G f(gz)dg\,.$$

Denoting by $e_1$ the first vector of the canonical basis of $\mathbb{R}^n$, any point $z$ can be written $z = t.ue_1$ for some $u \in G$ and $t = \|z\| = \dot{z}$, and so $f(gz) = f(t.gue_1)$. As $\mu$ is invariant, by theorem 22, the previous formula becomes

$$\int_{\mathbb{R}^n} f(z)dz = \int_0^{+\infty} d\lambda(t) \int_G f\left(t.ge_1\right) dg \tag{15.14}$$

for all $f \in L^1(\mathbb{R}^n)$. This being so, let us apply (14) to the function $f(\alpha z)$,where $\alpha \in \mathbb{R}^*_+$. The left hand side is multiplied by $\alpha^{-n}$. On the right, $t$ is replaced by $\alpha t$. Therefore the measure $\lambda$ must satisfy

$$\alpha^{-n} \int f^G(t)d\lambda(t) = \int f^G(\alpha t)d\lambda(t)\,,$$

whence

$$d\lambda(t) = c_n t^n d^*t\,,$$

with a constant $c_n > 0$. To determine the latter, we may choose the function

$$f(z) = \exp\left[-\pi\|z\|^2\right].$$

The left hand side of (14) can be calculated by LF and equals $1^n = 1$ (Chap. VII, n° 28 for $n = 1$). On the right, $f(t.ge_1) = f(te_1)$, so that the integral in $g$ equals $f(te_1) = \exp(-\pi t^2)$. Thus the right hand side equals

$$c_n \int_0^{+\infty} t^n \exp\left(-\pi t^2\right) d^*t = \frac{1}{2}\pi^{-n/2}\Gamma(n/2)c_n \,,$$

and so $c_n = 2\pi^{n/2}/\Gamma(n/2)$. Note that t $\Gamma(n/2)$ is readily computed if $n$ is even and reduces to $\Gamma(1/2) = \pi^{1/2}$ if $n$ is odd. Finally, relation (13) becomes

$$\int_{\mathbb{R}^n} f(z)dz = \frac{2\pi^{n/2}}{\Gamma(n/2)} \int_0^{+\infty} t^n d^*t \int_G f\left(t.ge_1\right) dg \,. \tag{15.15}$$

For $n = 1$, $\Gamma(n/2) = \pi^{1/2}$, whence $c_n = 2$. Here, the group $G$ reduces to two maps $x \mapsto x$ and $x \mapsto -x$, the measure $\mu$ being induced by assigning weight 1/2 to each of these two elements of $G$. Hence, the value of the integral over $G$ is $1/2[f(t) + f(-t)]$, $t^n d^*t = dt$ and a known formula is recovered...

For $n = 2$, in which case $\mathbb{R}^n$ can be identified to $\mathbb{C}$, $c_2 = 2\pi$, the group $G$ is simply $\mathbb{T}$ and, keeping the notation of Chap. VII, § 1,

$$\int_{\mathbb{R}^2} f(x)dx = 2\pi \int_0^{+\infty} tdt \int_{\mathbb{T}} f\left[t.\mathbf{e}(u)\right] dm(u) \,.$$

Replacing $\mathbf{e}(u) = \exp(2\pi iu)$ by $\exp(iu)$, replaces integration over $\mathbb{T}$ by integration over $[0, 1]$ and the invariant measure $dm(u)$ by $du/2\pi$. The previous formula reduces to one enabling us to compute the integrals in polar coordinates. For $n = 3$, $\Gamma(3/2) = \Gamma(1/2 + 1) = \sqrt{\pi}/2$, whence $c_3 = 4\pi$ and

$$\int_{\mathbb{R}^3} f(x)dx = 4\pi \int_0^{+\infty} t^2 dt \int_G f\left(t.ge_1\right) dg \,.$$

As will be seen a bit later, (15) is often written in a slightly different and more familiar form.

All this is very simple because the group $G$ is compact. The question would become far more complicated if, instead of the subgroup of $GL_n(\mathbb{R})$ leaving the quadratic form $\sum z_i^2$ invariant, where the $z_i$ are the canonical coordinates of some $z \in \mathbb{R}^n$, one considered the orthogonal group of a indefinite quadratic form, i.e. reducible to the form $q(z) = \sum \epsilon_i z_i^2$ with positive *and* negative signs $\epsilon_i$. In $\mathbb{R}^4$, the quadratic form $x^2 + y^2 + z^2 - c^2t^2$ of physicists falls in this case, $G$ being the Lorentz group. One can then show that the orbits (Witt's theorem[60]) are the hyperboloids $q(z) = Cte$, but the group no longer acts properly [even outside the null set of "isotropic" vectors, i.e. such that $q(z) = 0$]. To start with, the stabilizer $G_z$ of a point $z$ is not even always compact.[61] Hence, less brutal

[60] See for example my *Cours d'Algèbre*, exercises 19 and 20, pp. 645–646.

[61] If $a$ is a non-isotropic vector, $\mathbb{R}^n$ is the direct sum of $\mathbb{R}a$ and the subspace $E(a)$ of orthogonal vectors to $a$ (with respect to $q$). The stabilizer of $a$ in $G$ is thus isomorphic to the orthogonal group with respect to $q$ in $E(a)$. As a result, it is compact only if the restriction of $q$ to $E(a)$ is positive or negative definite. For the Lorentz form, any non-isotropic vector $x$ is of the form $Ge_1$ if $q(x) > 0$ and $Ge_4$ if $q(x) < 0$, up to a factor $> 0$, where $(e_i)$ is the canonical basis of $\mathbb{R}^4$. The stabilizer if $x$ is compact in the second case, but not in the first. In fact, $G$ acts properly on the open set $q(x) < 0$, but not on the open set $q(x) > 0$.

methods than formulas (6) and (11), which are no longer well-defined, need to be used. So, the analytic theory of quadratic forms with integer coefficients, which aims, as a first approximation, to find the number of integer solutions of $q(x) = r$ for a given integer $r$, become much more complicated.

(vi) *The case of homogeneous spaces.* Let us consider the extremely important case of a lcg $G$ and a closed subgroup $K$ of $G$ acting on $G$ either on the left by $(k, g) \mapsto kg$, or on the right by $(k, g) \mapsto gk$. The corresponding quotient spaces, written $K\backslash G$ and $G/K$, are by definition *homogeneous spaces* . In the former case, their elements are the cosets $Kg$, in the latter one $gK$. In this case, condition (GOP) always hold. Indeed, as $kg = g'$, i.e. $k = g'g^{-1}$, $C(A, B) = K \cap BA^{-1}$, where $BA^{-1}$, the image of $A \times B$ under the continuous map $(g, g') \mapsto g'g^{-1}$, is compact. Hence so is $C(A, B)$ because of the essential assumption that $K$ is closed. As a result, *every homogeneous space is locally compact.*

On the other hand, any closed subgroup $\Gamma$ of $G$, including $G$ itself, acts on the left on $G/K$ by the formula $\gamma(gK) = (\gamma g)K$. This enables us to defined new quotient spaces $\Gamma\backslash G/K$; their elements are the *double cosets* $\Gamma gK \subset G$, i.e. the sets of products $\gamma gk$, where $\gamma \in \Gamma$ and $k \in K$. These are also the orbits of $\Gamma \times K$ acting on $G$ by $(\gamma, k)g = \gamma gk^{-1}$. To express (GOP), note that here $C(A, B)$ is the set of $(\gamma, k)$ such that $\gamma Ak^{-1}\#B$, i.e. such that $\gamma A\#Bk$. By far the most important case is that of a compact subgroup $K$. Since the previous relation implies that $\gamma$ belongs to the compact set $BKA^{-1}$, the set of $\gamma$ which satisfy it is then contained in a compact subset $M$ of $\Gamma$, and so are the corresponding $k$. Thus $C(A, B)$ is contained in a compact subset of $\Gamma \times K$, and the closure of $C(A, B)$ remains to be shown. But if some $(\gamma_n, k_n) \in C(A, B)$ converge to a limit $(\gamma, k)$, and if one chooses $a_n \in A$ and $b_n \in B$ such that $\gamma_n a_n = b_n k_n$, taking a subsequence if necessary, $(a_n)$ may be assumed to converge to some $a \in A$ and the $b_n$ to some $b \in B$. Then $\gamma a = bk$, whence $(\gamma, k) \in C(A, B)$ and the result. In conclusion:

**Theorem 24**. *Let $G$ be a locally compact group, $K$ a* compact *subgroup of $G$ and $\Gamma$ a closed subgroup of $G$. Then $\Gamma$ acts properly on the homogeneous space $G/K$ and the quotient space $\Gamma\backslash G/K$ is locally compact.*

The general formula (11) applies to any homogeneous space $G/K$ provided $K$ is unimodular. As we make $K$ act on the right on $G$, we choose $\nu$ to be the measure $d_r g$ of $G$, invariant under $g \mapsto gk$. The measure $\lambda$ is determined by

$$\int_G f(g)d_r g = \int_{G/K} d\lambda\,(\dot{g}) \int_K f(gk)dk\,, \tag{15.16}$$

where $\dot{g} = gK$. The effects of the operations of $G$ on $\lambda$ are easily calculated. Denoting by $\mu_r$ the right invariant measure of $G$, (16) becomes

$$\mu_r(f) = \int f^K(z)d\lambda(z) .$$

Replacing $f(g)$ by $f(a^{-1}g)$ with $a \in G$ replaces $f^K(z)$ by $f^K(a^{-1}z)$ and the right hand side of (16) by $\int f^K(a^{-1}z)d\lambda(z)$. On the other hand, the left hand side of (16) is replaced by

$$\int f\left(a^{-1}g\right) d_r g = \Delta_G(a) \int f(g)d_r g = \Delta_G(a) \int f^K(z)d\lambda(z) ,$$

where $\Delta_G$ is with respect to $G$. As $f^K$ is arbitrary in $L(Z)$,

$$\int f\left(a^{-1}z\right) d\lambda(z) = \Delta_G(a) \int f(z)d\lambda(z) \tag{15.17}$$

for all $f \in L(Z)$. Therefore, the transformation formula is the same as that for the measure $d_r g$ (case where $K = \{e\}$) and could be written

$$d\lambda(az) = \Delta_G(a)d\lambda(z) .$$

In particular, *if the group* $G$ *is unimodular, the measure* $\lambda$ *on* $G/K$ *is* $G$-*invariant* for all closed subgroup $K$ of $G$.

With different assumptions on $G$ and $K$, we get some other analogous results, and even a general result without any assumptions whatsoever. See N. Bourbaki's chapter on the Haar measure.

*Exercise 6.* Let $G$ be a unimodular lcg, $K$ and $H$ two closed subgroups of $G$ such that the map $(h, k) \mapsto hk$ from $H \times K$ to $G$ is a homeomorphism.[62] Show that

$$\int_G f(g)dg = \iint_{H\times K} f(hk)d_l h dk .$$

(Identify $G/K$ with $H$, set out explicitly the action of $H$ on $G/K$ and use the invariance of the quotient measure).

These calculations enable us to put (15) in its classical form. Points of type $ge_1$, where $g \in O_n(\mathbb{R}) = G_n$ are just those of the unit sphere $S_{n-1}$ of $\mathbb{R}^n$. Since the stabilizer of the vector $e_1$ in $G_n$ is obviously the orthogonal group $O_{n-1}(\mathbb{R}) = G_{n-1}$ and since the map $g \mapsto ge_1$ is constant on the cosets $gG_{n-1}$, the homogeneous space $G_n/G_{n-1}$ can be identified with the sphere. Thus the invariant measure of $G_n$ defines a measure $\sigma$ on it which is invariant under the groups $G_n$ and *of total mass* 1 [apply (16) by choosing the normalized invariant measures of $G_n$ and $G_{n-1}$]. Then relation (15) becomes

$$\int_{\mathbb{R}^n} f(x)dx = \frac{2\pi^{n/2}}{\Gamma(n/2)} \int_0^{+\infty} t^n d^*t \int_{S_{n-1}} f(t.u)d\sigma(u) . \tag{15.18}$$

[62] Example: $G = GL_n(\mathbb{R})$, $K = O_n(\mathbb{R})$, and for $H$ take the subgroup of triangular matrices whose diagonal entries are $> 0$.

But users do not work with measures $\sigma$ of total mass 1, but with "the surface element" $d\Sigma(u)$ of the unit sphere to obtain the simpler formula[63]

$$\int_{\mathbb{R}^n} f(x)dx = \int_0^{+\infty} t^n d^*t \int_{S_{n-1}} f(t.u)d\Sigma(u)\,.$$

As the measure $\Sigma$ is invariant under rotations, and hence proportional to $\sigma$, in effect we are multiplying $\sigma$ by $2\pi^{n/2}/\Gamma(n/2)$. We thus discovered that in $\mathbb{R}^2$ (resp. $\mathbb{R}^3$), the length (resp. surface) in the usual sense of the unit sphere is equal to $2\pi$ (resp. $4\pi$). One could also calculate the volume of the ball $\|x\| \leq R$. It suffices to apply the previous relation to its characteristic function, which reduces the calculation to integrating $t^{n-1}dt$ from 0 to $R$. One would thus recover $\pi R^2$, $4\pi R^3/3$, etc.

(vii) *The case of discrete groups.* The case of a *discrete* group is particularly simple and, moreover, important .

**Lemma 6.** *Let $\Gamma$ be a discrete group acting on a locally compact space $Z$. The quotient space $\Gamma\backslash Z$ is locally compact,* if *the following condition holds:*

> (GPD) *for all compact sets $A, B \subset Z$, the set of $\gamma \in \Gamma$ such that $\gamma A \# B$ is finite.*

*If $z, z'$ are two points of $Z$ and $U$, $U'$ are sufficiently small neighbourhoods of $z$ and $z'$, then*

$$\gamma U \# U' \Longleftrightarrow \gamma z = z'\,. \tag{15.19}$$

The first proposition is clear. If $\Gamma z \neq \Gamma z'$, $\gamma U \cap U'$ is empty for all $\gamma$ for sufficiently small $U$ and $U'$ (lemma 2). So (19) follows in this case. If $\Gamma z = \Gamma z'$, let $V$ and $V'$ be compact neighbourhoods of $z$ and $z'$ . The set $C$ of $\gamma$ such that $\gamma V \# V'$ is finite by (GPD). Let $\gamma_1, \ldots, \gamma_p$ be the elements of $C$ such that $\gamma_i z \neq z'$. As $Z$ is separated, there are neighbourhoods $W_i \subset V$ and $W_i' \subset V'$ of $z$ and $z'$ such that $\gamma_i W_i \cap W_i' = \varnothing$. We set $U = \bigcap W_i$ and $U' = \bigcap W_i'$. If $\gamma U \# U'$, then $\gamma V \# V'$, and so $\gamma \in C$. Thus either $\gamma z = z'$ or else $\gamma = \gamma_i$ for some $i$. The second case does not occur since $\gamma U \cap U' \subset \gamma_i W_i \cap W_i' = \varnothing$, qed.

Under the assumptions of lemma 6, $\Gamma$ is said to act *properly discontinuously* on $Z$, a terminology inherited from the theory of automorphic functions, as well as the notation $\Gamma$, the letter $G$ being set aside for "continuous" groups like $SL_2(\mathbb{R})$.

Some simple consequences follow from condition (GPD). First of all, *the orbits $\Gamma z$ are discrete and closed in $Z$.* The second claim is clear even if $\Gamma$

[63] Arguments following Leibniz consist in calculating the volume of the portion of space comprised between the spheres of radius $t$ and $t + dt$ and in the interior of the cone with base the infinitesimal surface element $d\Sigma(u)$ of the unit sphere. "Obviously", $t^{n-1}dtd\Sigma(u) = t^n d^*td\Sigma(u)$. The formula thus follows by integration.

is not discrete: $\Gamma A$ is closed for any compact set $A \subset Z$ (lemma 2). On the other hand, (19) for $z = z'$ shows that $\Gamma z \cap U = \{z\}$ for all sufficiently small neighbourhood of $z$, hence the first claim.

By (GPD), for all $z \in Z$, the set of $\gamma$ such that $\gamma z = z$ is clearly a *finite* subgroup $\Gamma_z$ of $\Gamma$. It is the *stabilizer* of $z$ in $\Gamma$. Hence, for any open neighbourhood $V$ of $z$, the set

$$U = \bigcap_{\gamma \in \Gamma_z} \gamma V$$

is an open neighbourhood of $z$, obviously stable under the $\gamma \in \Gamma_z$. For this neighbourhood, (19) becomes

$$\text{(15.20)} \qquad \gamma U \# U \Longleftrightarrow \gamma z = z \Longleftrightarrow \gamma U = U\,.$$

To express (11) in the case of a discrete group $\Gamma$, one may suppose that

$$\int h(\gamma) d\gamma = \sum h(\gamma)$$

for all functions with compact (i.e. finite) support in $\Gamma$. Then

$$h^{\Gamma}(\dot{z}) = \sum h(\gamma z)\,,$$

and (11) becomes

$$\text{(15.21)} \qquad \int_Z f(z) d\nu(z) = \int_{\Gamma \backslash Z} d\lambda(\dot{z}) \sum_{\gamma} f(\gamma z)\,.$$

For $Z = \mathbb{R}$ and $G = \mathbb{Z}$ acting by $(z, n) \mapsto z+n$, this relation is the cornerstone of the Poisson summation formula. Theorem 22 shows that *if $f$ is $\nu$-integrable, then the series $\sum f(\gamma z)$ converges unconditionally for almost all $\dot{z} \in \Gamma \backslash Z$, its sum is $\lambda$-integrable and relation (21) remains valid.*

The general notion of a quotient measure does not seem to have appeared in the literature before d'André Weil's book, which only dealt with homogeneous spaces. He already defined the quotient measure in the way we have done, thus giving an example of the simplifications due to the definition of measures as linear functionals. He was the first to use it in this context at a time when his partisans were very rare. The general case, which is an easy extension of the case studied by Weil, is hardly mentioned before the N. Bourbaki's chapter dedicated to the Haar measure, but it was obviously known, at least implicitly, by some other others, especially Gelfand and his school even if, before 1950, they confined themselves to particular cases such as $SL_n(\mathbb{C})$. In fact, they were the first to publish the earliest good methods to calculate Haar measures by exploiting the existence of subgroups (see exercise 6). At the time, some authors were also studying quotient measures for one-parameter groups ($G = \mathbb{R}$) acting improperly on compact spaces, in the framework of ergodic theory – a much more difficult question. Chap. V of Bourbaki defines such measures for general "measurable" equivalence relations.

Around 1880, the notion of a quotient space by a properly discontinuous discrete group appears, very vaguely of course, in the work of Henri Poincaré, Felix Klein and others, in the framework of the classical theory of automorphic forms since it is essential in the definition of the corresponding Riemann surfaces (Chap. XII). But these authors and their successors, including Carl Ludwig Siegel much later than 1940, never wrote relation (21) in the "natural" form that we are now familiar with: they replaced the quotient space by a *fundamental domain* of $\Gamma$.

This is how any *measurable* set $F \subset Z$ such that, for any function $f \in L^1(\Gamma\backslash Z; \lambda)$,

$$\int_{\Gamma\backslash Z} f(x)d\lambda(x) = \int_F f\,[p(z)]\,d\nu(z) \tag{15.22}$$

is called (we still hope for something much better). With this definition, (21) becomes

$$\int_Z f(z)d\nu(z) = \sum_\gamma \int_F f(\gamma z)d\nu(z) = \sum_\gamma \int_{\gamma F} f(z)d\nu(z) \tag{15.23}$$

for any function $f \in L^1(Z; \nu)$. The most obvious case is obtained by making $\mathbb{Z}$ act on $\mathbb{R}$. One can take $F = [0, 1]$ and thus recover the equality

$$\int_{-\infty}^{+\infty} f(z)dz = \sum \int_0^1 f(z+n)dz\,.$$

Relation (23) holds if

$$Z = \bigcup \gamma F \tag{15.24}$$

and if, moreover, the sets $\gamma F$ are pairwise disjoint. But this at the very least supposes that $\Gamma$ acts without *fixed points*, i.e. that $\Gamma_z = \{e\}$ for all $z$. This is the case if one makes $\Gamma$ act on $G$ since then $\gamma g = g$ implies $\gamma = e$. It is rarely the case in quotients $G/K$ as we will see in the context of modular functions.[64] In fact, (23) remains valid if one only assumes that

[64] The general framework of the theory of automorphic forms consists in considering a space $Z = G/K$, where $K$ is compact and $G$ unimodular, and a discrete subgroup $\Gamma$ of $G$. One then studies functions $f(z)$ $\Gamma$-invariant or functions satisfying a slightly less simple condition. All cases reduce to vector functions on $G$ satisfying $f(\gamma gk) = p(k)^{-1}f(g)$, where $p$ is a finite-dimensional unitary representation of $K$. By (9) and (21), integration over $\Gamma\backslash Z$ reduces to integration over $\Gamma\backslash G$. Then, if we so wish, we can use a fundamental domain of $\Gamma$ in $G$. See Chap. XII for the case $G = SL_2(\mathbb{R})$.

(15.25) $$\nu(F \cap \gamma F) = 0 \quad \text{for all } \gamma \neq e.$$

If this condition holds, $\gamma F \cap \gamma' F = \gamma(F \cap \gamma^{-1}\gamma' F)$ is a null set whenever $\gamma \neq \gamma'$ because of the invariance of the measure. Since in practice a discrete group is always countable,[65] (23) then follows from the general lemma:

**Lemma 7.** *Let $X$ be a locally compact space, $\mu$ a positive measure on $X$ and $(F_n)$ a countable family of measurable sets in $X$. Assume that $X$ is the union of sets $F_n$, up to a null set, and that $F_p \cap F_q$ is a null set whenever $p \neq q$. A measurable function $f$ is integrable if and only if*

$$\sum \int_{F_n} |f(x)|\, d\mu(x) < +\infty .$$

*Then*

(15.26) $$\int_X f(x) d\mu(x) = \sum \int_{F_n} f(x) d\mu(x) .$$

Indeed, let $\chi_n$ be the characteristic function of $F_n$. The value taken by the function

$$\chi(x) = \sum \chi_n(x)$$

at $x \in F_p$ is the number of indices $q$ such that $x \in F_q$, and so equals 1 in

$$F_p - \bigcup_{p \neq q} F_p \cap F_q .$$

Since the union of $F_p \cap F_q$ is a null set, $\chi(x) = 1$ almost everywhere on $F_p$, hence almost everywhere on $X$. Hence, for given $f$, one sets $f_n = f\chi_n$, $|f| = \sum |f_n|$ ae., and as $\mu(f_n)$ is the integral of $f$ over $F_n$, the lemma follows from theorem 6.

The existence of such sets $F$ remains to be shown when $\Gamma$ is *discrete*, an essential assumption to ensure that $\Gamma$ is countable and apply lemma 6. Apart from purely measure theoretic arguments, Poincaré uses a method applicable to all cases encountered in the theory of automorphic functions of one or many variables. $G/K$ is then a $C^\infty$ manifold, the group $G$ acts through diffeomorphisms and in $G/K$ there is a $G$- invariant Riemannian $ds^2$. This gives an indefinitely differentiable distance function $d(z, z')$ such that $d(gz, gz') = d(z, z')$ and with respect to which the balls $d(a, z) \leq r$ are compact submanifolds. The invariant measure $\nu$ is defined by a $G$-invariant differentiable form of maximal degree, so that (n° 16, exercise 2) any subvariety of $G/K$ has measure zero as in $\mathbb{R}^n$. For all $\gamma \neq e$, the relation $\gamma z = z$ defines a subvariety of $G/K$, hence a null set if, as is natural, ones assumes

[65] Every discrete subgroup of a locally compact group countable at infinity is countable since its intersection with any compact set is finite.

that the relation $gz = z$ implies $g = e$ for all $z$. Choosing once and for all some $\omega \in Z$ outside the union of this countable family of null sets, we see that, for given $z$, there are only finitely many points $z' \in \Gamma z$ in a ball $d(\omega, z') \leq r$ since the orbit of $z$ is discrete. Hence there is at least some $z' \in \Gamma z$ for which $d(\omega, z')$ is minimal. So, denoting by $F$ the set of $z$ such that

$$d(\omega, z) \leq d(\omega, \gamma z) \quad \text{for all } \gamma \in \Gamma\,,$$

$Z = \bigcup \gamma F$. The set $F$ is closed since, for given $\gamma$, the previous relation defines a closed set. To be able to apply lemma 7, we still need to know that $\nu(F \cap \gamma F) = 0$ for all $\gamma \neq e$. But for all $z \in F \cap \gamma F$, $\gamma^{-1} z \in F$, and so $d(\omega, z) = d(\omega, \gamma^{-1} z) = d(\gamma\omega, z)$. In the cases considered, any equation $d(a, z) = d(b, z)$, where $a \neq b$, defines a submanifold of measure zero; whence the result since $\omega$ was chosen in such a way that $\gamma\omega \neq \omega$ for all $\gamma \neq e$. Poincaré in fact proved much more. Dieudonné transformed all this into exercises (Chap. XXII, 3, exercise 15), without as usual giving any references.[66]

[66] Perhaps C.L. Siegel, *Topics in Complex Function Theory*, Vol. II (Wiley, 1971).

# § 5. The Lebesgue-Nikodym Theorem

## 16 – Measures With Respect To a Base Measure $\lambda$: Integrable Functions

Let $X$ be a locally compact Polish space, $\lambda$ a positive measure on $X$ and $\mathbf{j}$ a locally integrable function with respect to $\lambda$ [n° 5, (ii)]. By definition, the function $\mathbf{j}f$ is integrable for $f \in L(X)$, whence a measure

$$(16.1) \qquad \nu(f) = \int f(x)\mathbf{j}(x)d\lambda(x)$$

on $X$. When $\mathbf{j} \geq 0$, the measure $d\nu = \mathbf{j}d\mu$ is positive, which gives rise to the problem of characterizing integrable functions with respect to it. Formula (1) irresistibly suggests the answer: a function $f$ is integrable with respect to $\mathbf{j}d\lambda$ if and only if so is the function $f\mathbf{j}$ with respect to $\lambda$. Then,

$$(16.2) \qquad \int f(x).\mathbf{j}(x)d\lambda(x) = \int f(x)\mathbf{j}(x).d\lambda(x)\,.$$

The similarity between (2) and the associativity formula of multiplication does not, however, make the result obvious, except perhaps for a physicist identifying $d\lambda(x)$ to an infinitesimal volume element and the measure $\mathbf{j}(x)d\lambda(x)$ to a mass or electric charge distribution with density $\mathbf{j}(x)$ and calculating in Leibniz's manner. One could also invoke the latter's theory of pre-established harmony [Chap. IX, n° 12, (ii)]. For mathematicians, results explain the notation adopted.

**Theorem 25.** *Let $X$ be a locally compact Polish space, $\lambda$ a positive measure on $X$ and $\mathbf{j} \geq 0$ a locally $\lambda$-integrable function. A function $f$ with values in $\mathbb{C}$ or $[-\infty, +\infty]$ is integrable with respect to the measure $d\nu = \mathbf{j}d\lambda$ if and only if $f\mathbf{j}$ is $\lambda$-integrable. Relation (2) then holds.*

*A complex-valued function $\mathbf{j}'$ is locally $\nu$-integrable if and only if $\mathbf{j}'\mathbf{j}$ is $\lambda$-integrable. The measure with density $\mathbf{j}'$ with respect to $\nu$ is then identical to the measure with density $\mathbf{j}'\mathbf{j}$ with respect to $\lambda$:*

$$(16.3) \qquad \int f(x).\mathbf{j}'(x)d\nu(x) = \int f(x)\mathbf{j}'(x)\mathbf{j}(x).d\lambda(x)$$

*for all $f \in L(X)$.*

The proof is rather long and comprises various parts.

(a) As observed at the end of section (ii) of n° 12, it suffices to set $Z = X$ and

$$(16.4) \qquad \mu_x(f) = \mathbf{j}(x)f(x) \quad \text{for all } f \in L(X)$$

in order to put $\nu$ into the form

$$\nu(f) = \int \mu_x(f) d\lambda(x). \tag{16.5}$$

Then theorem 22 shows that, if a complex-valued functions $f$ is $\nu$-integrable, then $f$ is $\mu_x$-integrable (i.e. finite at $x$) for almost all $x$, the function $x \mapsto \mu_x(f) = f(x)$ is $\lambda$-integrable, and finally formula (5), i.e. (2), remains valid.

(b) But theorem 25 also affirms that it *suffices* that $f\mathbf{j} \in L^1(X, \lambda)$ for $f$ to be in $L^1(X, \nu)$. *If* the function $f$ is $\nu$-measurable, then[67]

$$\nu^*(|f|) = \int |f(x)| \mathbf{j}(x).d\lambda^*(x) \tag{16.6}$$

by (13.5"). Hence in this case, the condition is sufficient. Thus the proof reduces to showing that if $f\mathbf{j} \in L^1(X, \lambda)$, then $f$ is $\nu$-measurable. We might as well prove a general relation between $\nu$-measurable and $\lambda$-measurable functions:

**Theorem 26.** *Let $X$ be a locally compact Polish space, $\lambda$ a positive measure on $X$, $\mathbf{j}$ a locally $\lambda$-integrable positive function and $d\nu = \mathbf{j}d\lambda$ the measure with density $\mathbf{j}$ with respect to $\lambda$. Let $f$ be a map from $X$ to a topological space $Y$, $S$ the set of $x$ for which $\mathbf{j}(x) \neq 0$, and $f_S$ a map with values in $Y$, equal to $f$ on $S$ and constant on $X - S$. $f$ is $\nu$-measurable if and only if $f_S$ is $\lambda$-measurable.*

**Lemma 1.** *For all $N \subset X$,*

$$\nu^*(N) = 0 \Longleftrightarrow \lambda^*(N \cap S) = 0. \tag{16.7}$$

Indeed if the first relation holds, then the function $f = \chi_N$ is $\nu$-integrable and formula (5) applies. Thus $f(x)\mathbf{j}(x) = 0$ outside a $\lambda$-null set and relation (7) holds.

Conversely let us suppose that it holds. For any integer $p \geq 1$, let $\chi_p$ be the characteristic function of the set $S_p$ of $x \in X$ where $\mathbf{j}(x) \geq 1/p$. Since $\mathbf{j}(x) \geq p^{-1}\chi_p(x)$ for $x$,

$$\nu(f) = \lambda(f\mathbf{j}) \geq p^{-1}\lambda(f\chi_p) \tag{16.8}$$

for all positive $f \in L(X)$. Taking the upper limit in (8) of an increasing *sequence* of functions, one gets the same result for all lsc functions (n° 11, (ii), theorem 21), then, taking infima, one sees that

$$\nu^*(F) \geq p^{-1}\lambda^*(F\chi_p) \quad \text{for all } F \geq 0. \tag{16.9}$$

[67] We will see at the end of this n° that in fact relation (6) holds for all $f$, but this does not follow directly from theorem 22.

If $F = \chi_N$ for some $N \subset X$, it follows that the relation $\nu^*(N) = 0$ implies $\lambda(N \cap S_p) = 0$ for all $p$, which proves the lemma since $S = \bigcup S_p$.

In particular

$$(16.10) \qquad \nu^*(X - S) = 0 .$$

Thus the $\nu$-measurability of a function $f$ only depends on its restriction to $S$. This is a very natural result... Also

$$(16.11) \qquad \lambda^*(N) = 0 \Longrightarrow \nu^*(N) = 0$$

since the first relation trivially implies $\lambda^*(N \cap S) = 0$.

Having settled this point, let us return to theorem 26. Two functions equal almost everywhere being simultaneously measurable or non-measurable, by (10) $f$ is $\nu$-measurable if and only if so is $f_S$. On the other hand, as $S$ is $\lambda$-measurable, if $f$ is $\lambda$-measurable, then clearly so is $f_S$ as well (the converse is obvious false if $X - S$ is not $\lambda$-null).

This being so, suppose that $f_S$ is $\lambda$-measurable. For every compact set $K \subset X$, (Lusin) $K = N \cup \bigcup K_n$ with $\lambda(N) = 0$ and compact sets $K_n$ on which the restrictions of $f$ are continuous. Since $\nu(N) = 0$ by (11), $f_S$ and hence $f$ are $\nu$-measurable.

To prove the converse, one can once again confine oneself to the case where $f$ is constant outside $S$. As $f$ is $\nu$-measurable, for every compact set $K \subset X$, there is a sequence of compact sets $K_n$ such that $N = K - \bigcup K_n$ is $\nu$-null and $f$ is continuous on $K_n$. Writing $K = (N \cap S) \cup (N - N \cap S) \cup K_n$, one sees that $\lambda(N \cap S) = 0$ since $\mu(N) = 0$, and that $f$ is continuous on $K_n$ and on $N - N \cap S$ (as it is constant). Thus $f$ is $\lambda$-measurable, qed.

An immediate consequence of the previous theorem is that *all $\lambda$-measurable* functions are $\nu$-measurable, for if $f$ is $\lambda$-measurable, so is $f_S$ since $S$ is $\lambda$-measurable.

(c) We can now return to the characterization of $\nu$-integrable functions $f$. As was seen by using (6), the proof reduces to showing that if $f\mathbf{j}$ is $\lambda$-measurable, then $f$ $\nu$-measurable. Given theorem 26, proving the next result is sufficient:

**Lemma 2.** *Let $f$ be a function with values in $\mathbb{C}$ or $[-\infty, +\infty]$. Then $f$ is measurable with respect to $d\nu = \mathbf{j}d\lambda$ if and only if $f\mathbf{j}$ is $\lambda$-measurable.*

The function $\mathbf{j}'(x)$ equal to $1/\mathbf{j}(x)$ on $S$ and 0 on $X - S$, being $\lambda$-measurable like $\mathbf{j}$ and constant outside $S$, is $\nu$-measurable by theorem 26. So, if $f\mathbf{j}$ is $\lambda$-measurable, and hence $\nu$-measurable, then so is $f\mathbf{j}\mathbf{j}'$, a function equal to $f$ on $S$ and 0 elsewhere. Thus $f$ is $\nu$-measurable by the same theorem. Conversely, if $f$ is $\nu$-measurable, then the function $f\mathbf{j}\mathbf{j}'$, equal to $f$ on $S$ and 0 elsewhere, is $\lambda$-measurable. Hence so is also $f\mathbf{j}\mathbf{j}'.\mathbf{j} = f\mathbf{j}$, proving the lemma.

(d) The the proof of theorem 25 will be complete once we have characterized the locally integrable functions $\mathbf{j}'(x)$ with respect to the measure $d\nu = \mathbf{j}d\lambda$. This is readily done since we need to show that, for all $f \in L(X)$, the function $f\mathbf{j}'$ is $\nu$-integrable, i.e. that the function $f\mathbf{j}'\mathbf{j}$ is $\lambda$-integrable, etc.

A convenient form of the previous result is as follows:

$$\mathbf{j}'.\mathbf{j}d\lambda = \mathbf{j}'\mathbf{j}.d\lambda\,, \tag{16.12}$$

which makes it obvious, but this is misleading.

Relation (6) was obtained using theorem 22 and assuming $f$ to be $\nu$-measurable. Actually, it holds without this assumption. In what follows, we will suppose that $f$ takes its values in $[0, +\infty]$ and we will not forget conventions about $+\infty$:

$$\alpha. + \infty = +\infty \quad \text{if } \alpha < 0 \leq +\infty\,, \qquad 0. + \infty = 0\,. \tag{16.13}$$

The general relation (13.6) first shows that inequality

$$\nu^*(f) \geq \lambda^*(f\mathbf{j})$$

holds in all cases. Thus it suffices to prove the opposite inequality, i.e. by definition of the right hand side, that

$$\nu^*(f) \leq \lambda^*(\varphi) \quad \text{for any lsc function } \varphi \geq f\mathbf{j}\,. \tag{16.14}$$

To prove this we consider the function

$$g(x) = \varphi(x)/\mathbf{j}(x) \quad \text{if } \mathbf{j}(x) \neq 0\,, \quad g(x) = +\infty \quad \text{otherwise}\,.$$

It is $\lambda$-measurable like $\varphi$ and $\mathbf{j}$. $g\mathbf{j} = \varphi \geq f\mathbf{j}$ and hence $g \geq f$ on $S$,. This is also the case on $X - S$ since then $g(x) = +\infty$. Hence

$$\nu^*(f) \leq \nu^*(g)\,.$$

But as $g$ is $\lambda$-measurable, (6) shows that

$$\nu^*(g) = \lambda^*(g\mathbf{j})\,.$$

However, $g\mathbf{j}$ equals $\varphi$ on $S$ and 0 on $X - S$ since $\mathbf{j}(x) = 0$. So $g\mathbf{j} \leq \varphi$ *everywhere.* As a result,

$$\lambda^*(g\mathbf{j}) \leq \lambda^*(\varphi)\,,$$

which leads to the expected result.

Functions $\mathbf{j} \geq 0$ for which $\mathbf{j}d\mu$ is a *bounded* measure can be characterized. This means that the function 1 must be integrable and hence, by theorem 22, that $\mathbf{j} \in L^1(X;\mu)$. (This result will be later generalized to complex-valued functions $\mathbf{j}$). In the case of a general locally compact space, this condition should be replaced by

$$\sup_K \int_K \mathbf{j}(x)d\mu(x) < +\infty\,,$$

where $K$ varies in the set of compact subsets of $X$.

*Exercise 1.* Assuming that the functions $f$ considered are "reasonable", show that theorems 25 and 26 hold in every locally compact space, Polish or not [note that assumption (c) of theorem 22 holds].

*Exercise 2.* (a) Let $X$ be a $n$-dimensional oriented manifold and $\omega$ a differential form of degree $n$ on $X$. Suppose that, in every local chart $(U, \varphi)$ compatible with the orientation of $X$, there is a relation of the form

$$\omega = \mathbf{j}(\xi) d\xi^1 \wedge \ldots \wedge d\xi^n ,$$

where $\xi^i = \varphi(x)^i$ and $\mathbf{j}(\xi) \geq 0$. Show that the map

$$f \longmapsto \int f\omega$$

is a positive measure on $X$. (b) Show that every submanifold of $X$ of dimension $< n$ has measure zero with respect to this measure.

## 17 – The Lebesgue-Nikodym Theorem (LN)

(i) *Characterization of absolutely continuous measures.*

**Theorem 27.** *Let $\lambda$ and $\nu$ be two positive measures on a locally compact Polish space $X$.*[68] *There is a locally $\lambda$-integrable function $\mathbf{j}$ such that $d\nu(x) = \mathbf{j}(x)d\lambda(x)$ if and only if every $\lambda$-null set is $\nu$-null.*

Relation (16.2) shows that the condition is necessary. The converse requires some developments.

(a) Let us first suppose that $\nu \leq \lambda$ and that there is a compact set $K$ such that $\nu(X - K) = 0$. The first assumption shows that $\nu^*(f) \leq \lambda^*(f)$ for any function $f \geq 0$. Then, for all $f \in L(X)$,

$$|\nu(f)| = |\nu(f\chi_K)| \leq |\lambda(f\chi_K)| \leq N_2(\chi_K) N_2(f),$$

where $N_2$ is a $\lambda$-norm. The map $f \mapsto \nu(f)$ thus extends to a continuous linear functional on the Hilbert space $L^2(X; \lambda)$.

However, we know (n° 19, theorem 31) that on a Hilbert space $\mathcal{H}$, the only continuous linear functionals are the inner products $f \mapsto (f|g)$ with $g \in \mathcal{H}$. Hence there is a function $\mathbf{j} \in L^2(X; \lambda)$ such that

$$\nu(f) = \lambda(f\bar{\mathbf{j}}) \quad \text{for all } f \in L(X).$$

This proves that $d\nu = \bar{\mathbf{j}} d\lambda$. We leave it to the reader to show that $\mathbf{j}(x) \geq 0$ $\lambda$-almost everywhere and that $\mathbf{j}(x) = 0$ ae. outside $K$. This argument simplifies

---

[68] Superfluous assumption provided "null" is replaced by "locally null" in the statement: $\lambda(N \cap K) = 0$ for any compact set $K$ implies $\nu(N \cap K) = 0$. The proof of the general case, which was unknown before N. Bourbaki, uses transfinite induction.

Lebesgue's proofs for the measure named after him and those of Nikodym (1930) for general abstract measures. It is due to von Neumann (1940).

(b) We assume that $\nu \leq \lambda$ still holds, but we drop the assumption that the support of $\nu$ is compact. For every compact set $K$, von N's arguments can be applied to the measure $d\nu_K(x) = \chi_K(x)d\nu(x)$. Hence there is a function $\mathbf{j}_K \in L^2(X;\lambda)$ such that $d\nu_K = \mathbf{j}_K d\lambda$. One can obviously suppose that $\mathbf{j}_K$ vanishes outside $K$.

If $K$ and $H$ are two compact sets (theorem 25), then

$$\chi_H d\nu_K = \chi_H \chi_K .d\nu(x) = \chi_K d\nu_H = \chi_K \chi_H d\nu$$

and hence $\chi_H \mathbf{j}_K d\lambda = \chi_K \mathbf{j}_H d\lambda$. As a result (exercise!),

$$\chi_H \mathbf{j}_K = \chi_K \mathbf{j}_H = \chi_{K\cap H}\mathbf{j} \quad \lambda-\text{almost everywhere}.$$

Let $K_n$ be an increasing sequence of compact sets such that $X = \bigcup K_n$, up to a $\lambda$-null set. Let $\chi_n$ (resp. $\mathbf{j}_n$) denote the function $\chi_K$ (resp. $\mathbf{j}_K$) corresponding to $K_n$. These functions vanish outside $K_n$. Then the previous relation shows that, for all $n$, there is a $\lambda$-null set $N_n \subset K_n$ such that

$$\mathbf{j}_{n+1}(x) = \mathbf{j}_n(x) \quad \text{for all } x \in K_n - N_n .$$

If $N = \bigcup N_n$, then there is a function $\mathbf{j}$ on $X$ such that $\mathbf{j} = \mathbf{j}_n$ on $K_n - K_n \cap N$ for all $n$. Its value on the $\lambda$-null set $X - \bigcup K_n$ is immaterial. If $K$ is a compact set, then $\mathbf{j} = \mathbf{j}_K = \mathbf{j}_n$ almost everywhere on $K \cap K_n$, hence almost everywhere on $K$. Thus $d\nu = \mathbf{j}d\lambda$.

(c) In the general case, $\lambda$ and $\nu$ are bounded above by the measure $\rho = \lambda + \nu$. Hence

$$d\lambda = pd\rho\,, \qquad d\nu = qd\rho$$

with locally $\rho$-integrable functions $p$ and $q$. Let us consider the sets $S = \{p(x) \neq 0\}$ and $T = \{q(x) \neq 0\}$. We know from lemma 1 of the previous n° that $X - S$ (resp. $X - T$) is a $\lambda$- (resp. $\nu$-) null set. As $\lambda^*(N) = 0$ is assumed to imply $\nu^*(N) = 0$, one can deduce that $\nu^*(X - S) = 0$ and hence (n° 16, lemma 1) that $\rho^*[(X - S) \cap T] = 0$. Replacing $q(x)$by 0 at each point $x \in (X - S) \cap T$, which preserves the relation $d\nu = qd\rho$, one may, therefore, assume that $q = 0$ in $X - S$, in other words that $T \subset S$.

We then consider the function $\mathbf{j}(x)$ equal to $q(x)/p(x)$ on $S$ and 0 elsewhere, whence $q = \mathbf{j}p$ everywhere. Since $\mathbf{j}p$ is locally $\rho$-integrable, $\mathbf{j}$ is locally integrable with respect to $pd\rho = d\lambda$ and

$$\mathbf{j}d\lambda = \mathbf{j}.pd\rho = \mathbf{j}p.d\rho = qd\lambda = d\nu$$

[theorem 25, (16.3)], proving the Lebesgue-Nikodym theorem.

(ii) *Application to complex measures.* When $0 \leq \lambda \leq \mu$, $\lambda^*(f) \leq \mu^*(f)$ clearly holds for any function $f$ with values in $[0, +\infty]$ and so $\lambda^*(A) \leq \mu^*(A)$

for all $A \subset X$. In conclusion, $d\lambda(x) = \mathbf{j}(x)d\mu(x)$ with a function $\mathbf{j}$ obviously satisfying $0 \leq \mathbf{j}(x) \leq 1$. The next result will follow from this:

**Theorem 28**. *For every complex measure $\lambda$, there is a positive measure $\mu$ and a function $\mathbf{j}$ locally $\mu$-integrable such that*

$$d\lambda(x) = \mathbf{j}(x)d\mu(x)\,.$$

Let us set $\lambda = \lambda_1 + i\lambda_2$, where $\lambda_1$ and $\lambda_2$ are real and suppose there are positive measures such that

$$\lambda_1 = \lambda_1' - \lambda_1''\,, \qquad \lambda_2 = \lambda_2' - \lambda_2''\,.$$

The measure

$$\mu = \lambda_1' + \lambda_1'' + \lambda_2' + \lambda_2''$$

is positive and is an upper bounded for the four measures on the right hand side. Hence they are all of the form $\mathbf{j}(x)d\mu(x)$. Thus so is $\lambda$. A far more elementary and frequently proclaimed result remains to be proved:

**Theorem 29**. *Every real Radon measure $\nu$ on a locally compact $X$ is the difference of two positive measures.*

It suffices to prove that a positive measure $\lambda \geq \nu$ exists. For such a measure, the relation $0 \leq h \leq f$ implies $\nu(h) \leq \lambda(h) \leq \lambda(f)$. For all $f \in L_+(X)$, this leads one to write

$$\nu^+(f) = \sup_{\substack{0 \leq h \leq f \\ h \in L(X)}} \nu(h)\,. \tag{17.1}$$

Thus $\nu(f) \leq \nu^+(f) \leq \lambda(f)$ for all $f \geq 0$ and every positive measure $\lambda \geq \nu$. If $\nu^+$ is shown to be the restriction of a positive measure to $L_+(X)$, not only will the theorem have been proved but also that $\nu^+$ is the *smallest* positive measure bounding $\nu$ above.

First, $0 \leq \nu^+(f) < +\infty$ since functions $h$ such that $0 \leq h \leq f$ vanish outside a fixed compact set and satisfy $\|h\| \leq \|f\|$. Obviously, $\nu^+(\alpha f) = \alpha\nu^+(f)$ for $0 \leq \alpha < +\infty$. So is

$$\nu^+(f+g) \geq \nu^+(f) + \nu^+(g)$$

for $f, g \in L_+(X)$ since $h' + h'' \leq f + g$ if $h' \leq f$ and $h'' \leq g$.

To prove the opposite inequality, let us consider some $h \in L_\mathbb{R}(X)$ such that $0 \leq h \leq f + g$ and set $h' = hf/(f+g)$, $h'' = hg/(f+g)$, by prescribing that $h'(x) = h''(x) = 0$ at all points where $f + g$ vanishes. Let us show that $h'$ and $h''$ have continuous compact supports. The first claim is clear. So is continuity at all points where $f(x) + g(x) > 0$. If $f$ and $g$ are zero at $x$, hence also $h$, continuity follows from the fact that $h'$ and $h''$ are everywhere contained between 0 and $h$, the latter being a continuous function vanishing at $x$.

This being so, $h = h' + h''$ with $0 \leq h' \leq f$ and $0 \leq h'' \leq g$, whence

$$\nu(h) = \nu(h') + \nu(h'') \leq \nu^+(f) + \nu^+(g).$$

Taking the supremum of the left and side for the functions $h$ considered gives $\nu^+(f+g) \leq \nu^+(f) + \nu^+(g)$, which leads to the expected equality:

$$\nu^+(f+g) = \nu^+(f) + \nu^+(g). \tag{17.2}$$

We next define $\nu^+$ in $L_{\mathbb{R}}(X)$ by setting

$$\nu^+(f) = \nu^+(f') - \nu^+(f'') \tag{17.3}$$

if $f = f' - f''$ with positive $f'$ and $f''$. As $f' - f'' = g' - g''$ implies $f' + g'' = g' + f''$ and hence, by (2), the same relation by applying $\nu^+$ to the functions considered, (3) defines $\nu^+(f)$ without any ambiguities and reduces to (1) for $f \geq 0$. As a result, $\nu^+$ is a positive result. As (1) shows that $\nu(f) \leq \nu^+(f)$ for all $f \geq 0$, the difference $\nu^+ - \nu$ is also a positive measure, qed.

Note that this proof applies to other cases. For example, $L(X)$ can be replaced by the Banach space $E$ of bounded continuous real functions on an arbitrary topological space and $\nu$ can be chosen to be a linear functional on $E$.

To justify notations, it is necessary to show that

$$\nu = \nu^+ - \nu^- \quad \text{where } \nu^- = (-\nu)^+.$$

Now, by definition,

$$\begin{aligned}(-\nu)^+(f) &= \sup_{0\leq h\leq f} -\nu(h) = \sup_{0\leq h\leq f} -[\nu(f) - \nu(f-h)] \\ &= -\nu(f) + \sup_{0\leq h\leq f} \nu(f-h) = -\nu(f) + \nu^+(f)\end{aligned}$$

since $h \mapsto f - h$ permutes the functions $h$ considered, whence the result.

Additional discussions on upper envelopes of a family of measures will be found in N. Bourbaki.

Theorem 28 enables us to define the *absolute value* of a complex measure $\lambda$: we choose any measure $\mu \geq 0$ such that $d\lambda(x) = \mathbf{j}(x)d\mu(x)$ and set $d|\lambda|(x) = |\mathbf{j}(x)|d\mu(x)$. The result is independent of the choice of $\mu$. Indeed, if $d\lambda = \mathbf{j}'d\mu' = \mathbf{j}''d\mu''$, writing $\mu = \mu' + \mu''$ gives $d\mu' = \varphi'd\mu$, $d\mu'' = \varphi''d\mu$ with $\varphi', \varphi'' \geq 0$. Then theorem 25 shows that

$$d\lambda = \mathbf{j}'.\varphi'd\mu = \mathbf{j}'\varphi'.d\mu = \mathbf{j}''\varphi''.d\mu.$$

Thus $\mathbf{j}'\varphi' = \mathbf{j}''\varphi''$ $\mu$-almost everywhere, and so also $|\mathbf{j}'|\varphi' = |\mathbf{j}''|\varphi''$. Then

$$|\mathbf{j}'|\, d\mu' = |\mathbf{j}'|\, \varphi' d\mu = |\mathbf{j}''|\, \varphi'' d\mu = |\mathbf{j}''|\, d\mu''$$

as announced.

Then for any function $f \in L(X)$,

$$|\lambda(f)| = |\mu\,(\mathbf{j}f)| \leq \mu\,(|\mathbf{j}f|) = |\lambda|\,(|f|)\,, \tag{17.4}$$

or, using more traditional notation,

$$\left|\int f(x) d\lambda(x)\right| \leq \int |f(x)| d|\lambda|\,(x)\,,$$

a result that may seem as obvious as the triangle inequality! More generally, a function $f$ will be said to be $\lambda$-integrable if it is $|\lambda|$-integrable, i.e. if $\mathbf{j}f$ is $\mu$-integrable, in which case, we obviously set $\lambda(f) = \mu(\mathbf{j}f)$.

Choosing as above a measure $\mu \geq 0$ such that $d\lambda = \mathbf{j}.d\mu$, whence $d|\lambda| = |\mathbf{j}|d\mu$, the function

$$\omega(x) = \begin{matrix} \mathbf{j}(x)/\,|\mathbf{j}(x)| & \text{if} & \mathbf{j}(x) \neq 0\,, \\ 1 & \text{if} & \mathbf{j}(x) = 0 \end{matrix}$$

is $\mu$-measurable, and so locally integrable since it is bounded. As $\mathbf{j}(x) = \omega(x)|\mathbf{j}(x)|$ for all $x$ and as the right hand side is locally integrable, (16.3) applies and shows that

$$d\lambda(x) = \omega(x) d|\lambda|(x)\,. \tag{17.5}$$

Conversely if, for a measure $\mu \geq 0$, $d\lambda = \omega d\mu$ with $|\omega(x)| = 1$ almost everywhere, then clearly $\mu = |\lambda|$.

Finally, we state a formula, analogous to (1). It will come in useful in the next n° and will enable us to define $|\lambda|$ bypassing the LN theorem: for all $f \in L_+(X)$,

$$|\lambda|(f) = \sup_{\substack{|g|\leq f \\ g\in L(X)}} |\lambda(g)|\,. \tag{17.6}$$

Setting $|\lambda| = \mu$ and $d\lambda = \omega d\mu$, the proof reduces to showing that

$$\mu(f) = \sup_{\substack{|g|\leq f \\ g\in L(X)}} |\mu(\omega g)|\,. \tag{17.6'}$$

First of all, $|\mu(\omega g)| \leq \mu(|\omega g|) \leq \mu(f)$ for all functions $g$ considered. The left hand side of (6) is thus greater than the right hand side. To prove the opposite inequality, it suffices to show that there are functions $g_n$ such that

$\lim \mu(\omega g_n) = \mu(f)$. However, there is an increasing sequence of compact sets $K_n$ contained in the support $K$ of $f$, such that $K - \bigcup K_n$ is a null set, and on each of them the function $\omega$ is continuous and with absolute value 1. For all $n$, there is a continuous function $h_n$ on $X$[69] equal to $\overline{\omega}$ on $K_n$. Taking its composition with a continuous map from $\mathbb{C}$ to the disc $D = \{|z| \leq 1\}$ equal to $z \mapsto z$ on $D$ and in particular on its boundary, one may suppose that $|h_n(x)| \leq 1$ everywhere. The function $g_n = h_n f$, which is in $L(X)$, then satisfies $|g_n| \leq f$ everywhere and $\omega g_n = f$ on $K_n$. Thus, the decomposition of integrals on $K$ into integrals on $K_n$ and $K - K_n$ gives

$$|\mu(\omega g_n) - \mu(f)| \leq \int_{K-K_n} |\omega(x)g_n(x) - f(x)|\, d\mu(x)\,.$$

As $|\omega g_n - f| \leq 2f$ everywhere, the right hand side tends to 0, proving (6).

*Exercise.* (a) Show that $\lambda$ is bounded if and only if so is $|\lambda|$. (b) Suppose that $d\lambda = \mathbf{j}d\mu$ with respect to a reasonable measure $\mu \geq 0$. Show that $\lambda$ is bounded if and only if $\mathbf{j} \in L^1(X;\mu)$, and then that

$$\|\lambda\| = \int |\mathbf{j}(x)|\, d\mu(x)\,.$$

(iii) *Lebesgue decomposition.* Let $\mu$ and $\mu'$ be two positive measures on $X$ and let us consider the measure $\nu = \mu + \mu'$. The LN theorem enables us to write $d\mu = pd\nu$, $d\mu' = p'd\nu$ for some positive locally $\nu$-integrable functions $p$ and $p'$ such that $p + p' = 1$ ae. We set

$$q(x) = p'(x)/p(x) \quad \text{if } p(x) \neq 0\,, \quad q(x) = 0 \quad \text{if } p(x) = 0\,,$$
$$q'(x) = 0 \quad \text{if } p(x) \neq 0\,, \quad q'(x) = p'(x) \quad \text{if } p(x) = 0\,.$$

These functions are $\nu$-measurable. As $p' = pq + q'$,

$$d\mu' = pq.d\nu + q'd\nu = q.pd\nu + q'd\nu = qd\mu + q'd\nu\,.$$

[69] This follows from Urysohn's theorem: if $X$ is a normal topological space [i.e. such that, for any closed disjoint sets $A, B \subset X$, there exist open disjoint sets $U \supset A$ and $V \supset B$], then any function with values in $\mathbb{R}$ (or $\mathbb{R}^n$) defined, continuous and bounded on a closed set $A$ has a continuous extension to $X$ (if $X$ is locally compact without being normal, $A$ must be assumed to be compact, which is the case here); N. Bourbaki, *Topologie générale*, Chap. IX, §4.1. The case of metrizable spaces is dealt with in Dieudonné, IV.5: if $f$ is defined on a closed subset $A$ of $X$, the function $g$ equal to $f$ on $A$ and defined outside $A$ by

$$d(x, A)g(x) = \inf_{y \in A} d(x, y)f(y)$$

is continuous and answers the question. The proof reduces to showing that $g$ is continuous at all boundary points of $A$.

Setting $d\lambda = q'd\nu$,

$$d\mu' = qd\mu + d\lambda \tag{17.7}$$

follows. The sets $S = \{p(x) \neq 0\}$, $S' = \{q'(x) \neq 0\}$ are *disjoint* by definition of $q'$. Since $d\mu = pd\nu$, $\mu(X-S) = 0$ by lemma 1 of n° 16, hence also $\mu(S') = 0$; similarly $\lambda(X - S') = 0$ since $d\lambda = q'd\nu$.

Note that $X$ being countable at infinity, any $\nu$-measurable function is equal ae. to a Borel function (observe that a continuous function on a compact subset and vanishing elsewhere is Borel, then use Lusin's theorem). Hence one may assume $p$ and $p'$ to be Borel, and thus also $q$ and $q'$, in which case $S'$ is a *Borel* set.

Relation (7) therefore gives a decomposition of $\mu'$ into the sum of two measures, one absolutely continuous with respect to $\mu$ and the other concentrated (obvious definition) on a *Borel and $\mu$-null* set $S'$. This is the Lebesgue decomposition of $\mu'$ with respect to $\mu$. It is unique (exercise!).

Two positive measures $\mu$ and $\mu'$ are said to be *disjoint* when they are concentrated on disjoint sets. Setting $\nu = \mu + \mu'$ and $d\mu = pd\nu$, $d\mu' = p'd\nu$ as above, this obviously means that

$$\inf [p(x), p'(x)] = 0 \quad \text{almost everywhere with respect to } \nu\,.$$

As any measure $\lambda \geq 0$ bounded above by $\mu$ and $\mu'$ is of the form $qd\nu$ with $q \leq p, p'$, this means that 0 *is the only measure bounded above by $\mu$ and $\mu'$.*

The notion of disjoint measures can be interpreted in a completely different manner. Let us consider finitely-many $n$ positive measures $\mu_i$ and their sum $\nu$. Thus $d\mu_i = p_i d\nu$ with functions satisfying $\sum p_i(x) = 1, 0 \leq p_i(x) \leq 1$ almost everywhere with respect to $\nu$. We consider the $L^2$ spaces of these measures. By theorem 25, which is nothing special in this case, any square integrable (resp. null) function $f$ with respect to $\nu$ is also of this type with respect to any of the measures $\mu_i$ and conversely. Hence one can associate *classes* $f_i \in L^2(X;\mu_i)$ to each *class* $f \in L^2(X;\nu)$. This gives an injective linear map

$$L^2(X;\nu) \longrightarrow L^2(X;\mu_1) \times \ldots \times L^2(X;\mu_n) \tag{17.8}$$

compatible with the Hilbert structures of these spaces since

$$(f|g) = \int f(x)\overline{g(x)}d\nu(x) = \sum \int f(x)\overline{g(x)}d\mu_i(x) = \sum (f_i|g_i)$$

for all $f, g \in L^2(X;\nu)$. This said, let us show that *map (8) is bijective if and only if the measures $\mu_i$ are pairwise disjoint,* i.e. concentrated on pairwise disjoint sets $S_i$.

If this condition holds, taking for all $i$, a square integrable function $f_i$ with respect to $\mu_i$, it may be assumed to vanish outside $S_i$. The function

$f = \sum f_i$ is equal $\mu_i$-ae. to $f_i$ for all $i$ since $\mu_i(S_j) = 0$ for $j \neq i$. Thus (8) is surjective. Conversely, if (8) is surjective, then, for all $f_i \in L^2(X; \mu_i)$, there exists $f \in L^2(X; \mu)$ such that

$$f(x) = f_i(x) \quad \mu_i \ - \text{ almost everywhere}.$$

For a given $i$, let us choose $f_j = 0$ for all $j \neq i$ and $f_i = \chi_K$, where $K$ is a compact set. Setting $N = \{f(x) \neq 0\}$, $\mu_j(N) = 0$ and so $\mu_j(K \cap N) = 0$ for all $j \neq i$. Since $f(x) = 1$ almost everywhere on $K$ with respect to $\mu_i$, $\mu_i(K - K \cap N) = 0$. The application of this construction to an increasing sequence of compact sets whose union is $X$ shows that the union $N$ of the $N_p$ corresponding to the $K_p$ satisfies

$$\mu_j(N) = 0 \quad \text{for all } j \neq i\,, \mu_i(X - N) = 0\,.$$

Hence $\mu_i$ and $\mu_j$ are disjoint for $i \neq j$, qed.

This result can be partly generalized to "continuous sums"

$$\nu = \int \mu_x d\lambda(x)$$

of n° 12 and 13, by associating to each class $f \in L^2(Z; \nu) = \mathcal{H}$ the corresponding classes $f_x \in L^2(Z; \mu_x) = \mathcal{H}_x$. Theorem 22, (13.13), shows that

$$(17.9) \qquad (f|g) = \int (f_x|g_x)\, d\lambda(x)$$

for all $f, g \in \mathcal{H}$, which suggests a continuous analogue of a Hilbert direct sum. n° 23, (v) will clarify this vague notion encountered in many fields, mostly under a form related to Leibniz 's idea of an integral. But von Neumann has passed this way, even if many people seem or prefer to ignore it.

## 18 – Continuous Linear Functionals on $L^p$. The $L^\infty$ Space

Let $X$ be a locally compact space and $\mu$ a positive measure on $X$. For all $p > 1$, Hölder's inequality shows that, if $\mathbf{j} \in L^q$, the map

$$f \longmapsto \mu(f\mathbf{j})$$

is a continuous linear functional on the $L^p$ Banach space . The same is true for $p = 1$ if $\mathbf{j}$ is chosen to be a bounded measurable function. We intend to show that all continuous linear functionals on $L^p$ are obtained in this way.

So let $\nu$ be a continuous linear functional on $L^p$ or, equivalently, since $L(X)$ is everywhere dense in $L^p(X; \mu)$, on the space $L(X)$ equipped with the $L^p$ norm.

Since continuity reduces to the existence of a constant $M$ such that

$$(18.1) \qquad |\nu(f)| \leq M\|f\|_p = M\left[\int |f(x)|^p\, d\mu(x)\right]^{1/p}$$

for all $f \in L(X)$, for any compact set $K \subset X$,

$$|\nu(f)| \leq M\mu(K)^{1/p}\|f\| \quad \text{if } f \in L(X, K),$$

where $\|f\|$ is the norm of uniform convergence. As a result, $\nu$ is a complex *measure* on $X$. Assuming $M = 1$ for simplicity's sake and writing (1) as $|\nu(f)|^p \leq \mu(|f|^p)$, formula (17.6) defining $|\nu|$ shows that

$$|\nu|(f)^p = \sup |\nu(g)|^p \leq \sup \mu\left(|g|^p\right),$$

whence

$$|\nu|(f) \leq \mu\left(f^p\right)^{1/p} = \|f\|_p \tag{18.2}$$

for all $f \in L_+(X)$. Hence $\nu$ may be assumed to be positive.

Then let $\varphi$ be a positive lsc function. If $\Phi$ is the increasing philtre of $f \in L_+(X)$ bounded above by $\varphi$, the set of $f^p$ where $f \in \Phi$ is an increasing philtre whose upper envelope is $\varphi^p$. Taking suprema in (2), (2) follows for $\varphi$. Hence, if $f$ is now an arbitrary function with values in $[0, +\infty]$, then

$$\nu^*(f)^p = \inf_{\varphi \geq f} \nu(\varphi)^p \leq \inf_{\varphi \geq f} \mu\left(\varphi^p\right).$$

However, the strictly increasing functions $t \mapsto t^p$ and $t \mapsto t^{1/p}$ transform every lsc function into an lsc function. The functions $\varphi^p$ are, therefore, all lsc and upper bounds for $f^p$. In conclusion,

$$\nu^*(f)^p \leq \mu^*\left(f^p\right)$$

for any function $f \geq 0$. In particular any $\mu$-null set is $\nu$-null. Thus $d\nu = \mathbf{j}d\mu$ for a locally $\mu$-integrable function $\mathbf{j} \geq 0$. So $\nu(f) = \mu(\mathbf{j}f)$ for all $f \in L_+(X)$ and $\nu^*(f) = \mu^*(\mathbf{j}f)$ for all $f \geq 0$ by theorem 25.

But if $\mathbf{j}$ satisfies

$$\mu^*(\mathbf{j}f)^p \leq \mu^*\left(f^p\right) \quad \text{for all } f \geq 0, \tag{18.3}$$

so does every positive, locally integrable function and $\leq \mathbf{j}$, in particular $\mathbf{j}_n = \inf(\mathbf{j}, n)$ for all $n \in \mathbb{N}$ and so perforce $\mathbf{j}_{n,K} = \mathbf{j}_n\chi_K = \mathbf{j}'$ for any compact set $K$. We next need to distinguish between two cases.

If $p > 1$, there exists $q > 1$ such that $1/p + 1/q = 1$, which can also be written $p(q-1) = q$, whence $(\mathbf{j}'^{q-1})^p = \mathbf{j}'^q$. This function is $\mu$-integrable since it is measurable, bounded and zero outside a compact set (n° 9, theorem 18). Replacing $\mathbf{j}$ by $\mathbf{j}'$ and $f$ by $\mathbf{j}'^{q-1}$ in (3) gives

$$\mu^*\left(\mathbf{j}'^q\right)^p \leq \mu^*\left(\mathbf{j}'^q\right)$$

and as a consequence, $\mu^*(\mathbf{j}'^q) \leq 1$. As this holds for all $n$, (theorem 8) $\mathbf{j}\chi_K = \sup_n \mathbf{j}_{n,K}$ is in $L^q(X;\mu)$ and $\|\mathbf{j}\chi_K\|_q \leq 1$ for all $K$. Hence, if $\mu$ is reasonable, then $\mathbf{j}$ is in $L^q(X;\mu)$ and satisfies $\|\mathbf{j}\|_q \leq 1$.

If $p = 1$, relation (3) means that the measure $\mu - \mathbf{j}d\mu$ is positive, and so $\mathbf{j}(x) \leq 1$ almost everywhere.

Hence, coming back to a complex measure $\nu$, relation (1) shows that $d\nu = \mathbf{j}d\mu$ where the locally integrable (complex) function $\mathbf{j}$ satisfies $\|\mathbf{j}\|_q \leq M$ if $p > 1$, and $|\mathbf{j}(x)| \leq M$ almost everywhere $p = 1$. By Hölder's inequality,the expression $\mu(f\mathbf{j})$ is then well-defined for all $f \in L^p$ and is a continuous linear functional on $L^p$. Since it coincides with the given linear function on the everywhere dense subspace $L(X)$, relation $\nu(f) = \mu(f\mathbf{j})$ is true for all $f \in L^p$. In conclusion, and assuming $\mu$ to be reasonable:

**Theorem 30**. *For every continuous linear functional $\nu$ on the Banach space $L^p(X;\mu)$ with $1 < p < +\infty$ (*resp. $p = 1$*), there is a function $\mathbf{j} \in L^q(X;\mu)$ (*resp. *bounded) such that*

$$\nu(f) = \mu(f\mathbf{j}) \tag{18.4}$$

*for all $f \in L^p(X;\mu)$. $|\nu(f)| \leq M.\|f\|_p$ for all $f \in L^p(X;\mu)$ if and only if $\|\mathbf{j}\|_q \leq M$ (*resp. *$|\mathbf{j}(x)| \leq M$ almost everywhere).*

This shows that for $p > 1$, the norm of the given linear functional $\nu$ on $L^p$ is

$$\|\mathbf{j}\|_q = \sup |\nu(f)| / \|f\|_p . \tag{18.5}$$

For $p = 1$, $L^q$ needs to be replaced by the space $L^\infty(X;\mu)$ of classes of *bounded measurable* functions with respect to $\mu$. To get a relation similar to (5) in this case, denote by $\|\mathbf{j}\|_\infty$ the smallest number $M$ such that $|\mathbf{j}(x)| \leq M$ ae. Its existence is immediate since, if $M$ is the infimum of the set of $M'$ such that $|\mathbf{j}(x)| \leq M'$ ae., then $|\mathbf{j}(x)| \leq M + 1/n$ ae. for all $n$ and so $|\mathbf{j}(x)| \leq M$ ae.

Theorem 30 says that the dual[70] of $L^p$ is $L^q$ for $1 \leq p < +\infty$. But *it should not be thought that the dual of $L^\infty$ is $L^1$*. The arguments used to prove theorem 29 fall apart in this case for the simple raison that $L(X)$ *is not everywhere dense in $L^\infty$*. Hence by the general Hahn-Banach theorem,[71] there are continuous linear functionals $\neq 0$ on $L^\infty$ that vanish on $L(X)$. Such a form is obviously not defined by a measure on $X$. This is already the case if $\mu$ is the measure $\mu(f) = \sum f(x)$ on $\mathbb{N}$ though, in this apparently trivial case, it is impossible to explicitly construct a continuous linear functional on $L^\infty$ (the space of bounded sequences) not defined by some $\mathbf{j} \in L^1$. For this,

[70] The topological dual $E'$ (not to be confused with the algebraic dual $E^*$) of a Banach space $E$ is the space of continuous linear functionals on $E$.

[71] In other words: if $\mathcal{H}_0$ is a closed vector subspace of a Banach space $\mathcal{H}$, there is a family $(u_i)$ of continuous linear functionals on $\mathcal{H}$ such that

$$x \in \mathcal{H}_0 \Longleftrightarrow u_i(x) = 0 \quad \text{for all } i .$$

Dieudonné (XII.15, exercises 3 and 4) proves a more general result by assuming the space is separable, which is the case of $L^\infty$ only if $\mu$ reduces to a finite sum of Dirac measures.

"ultrafilters" from N. Bourbaki's *Topologie générale* or similar procedures founded on transfinite induction need to be used. This does not prevent the space $L^\infty$, a space that should be handled prudently, from playing as important a role as $L^1$ and $L^2$ in analysis.

**Corollary.** *A locally $\mu$-integrable function $\mathbf{j}$ is in $L^p(X;\mu)$ for some $p \in [1,+\infty]$ if and only if there is a constant $M$ such that*

$$|\mu(\mathbf{j}f)| \le M.N_q(f) \quad \text{for all } f \in L(X)\,.$$

*Then $\|\mathbf{j}\|_p \le M$.*

Indeed, the map $f \mapsto \mu(\mathbf{j}f)$ is a linear functional on $L(G)$ which, being continuous with respect to the norm $N_q$, can be extended to $L^q$ if $q < +\infty$. This gives a $\mathbf{j}' \in L^p$ such that $\mu(\mathbf{j}f) = \mu(\mathbf{j}'f)$ for all $f \in L(G)$, proving the corollary in this case. If $q = +\infty$, i.e. if $p = 1$, in which case $N_q(f) = \|f\|_\infty \le \|f\|$, where this is the uniform norm of $f$, then the measure $f \mapsto \mu(\mathbf{j}f)$ is bounded and the norm $\le M$. Thus $N_1(\mathbf{j}) \le M$, qed.

*Exercise.* For any function $\varphi \in L^\infty(X;\mu)$, set $M(\varphi)f = \varphi f$ for all $f \in L^p(p < +\infty)$. Show that all continuous operators on $L^p$ commuting with the $M(\varphi)$ are of the same type.

# § 6. Spectral Decomposition on a Hilbert Space

## 19 – Operators on a Hilbert Space

(i) *Definitions, continuous linear functionals.* Recall that a Hilbert space $\mathcal{H}$ is a complex vector space equipped with an "inner product" $(x|y)$ satisfying the following conditions:

(H 1) the function $x \mapsto (x|y)$ is linear for all $y$,

(H 2) $(x|y) = \overline{(y|x)}$ for all $x$ and $y$,

(H 3') $(x|x)$ is always $\geq 0$,

(H 3") $(x|x) = 0$ implies $x = 0$,

(H 4) $\mathcal{H}$ is complete with respect to the distance
$$d(x,y) = (x-y|x-y)^{1/2} = \|x-y\| .$$

In practice, a complex vector space $\mathcal{H}$ is often equipped with a *positive Hermitian form,* i.e. a map $B : \mathcal{H} \times \mathcal{H} \longrightarrow \mathbb{C}$ satisfying conditions (H 1), (H 2) and (H 3'), but not the following ones. The Cauchy-Schwarz inequality

$$|B(x,y)|^2 \leq B(x,x)B(y,y)$$

then shows that the set $\mathcal{N}$ of $x$ such that $B(x,x) = 0$ is a vector subspace of $\mathcal{H}$ and that $B(x,y)$ only depends on the classes of $x$ and $y \bmod \mathcal{N}$. Hence denoting by $x \mapsto x_B$ the map associating to each $x$ its class $x_B \in \mathcal{H}/\mathcal{N}$ provides the quotient space with a positive Hermitian form satisfying (H 3") as well as with a distance. To get a proper Hilbert space, it remains to replace $\mathcal{H}/\mathcal{N}$ by its completion with respect to this distance. By abuse of language it will be called the *completion* of $\mathcal{H}$ with respect to $B$. The spaces $L^2(X;\mu)$ of integration theory are obtained in this way: one endows $L(X)$ with the Hermitian form

$$B(f,g) = \int f(x)\overline{g(x)}d\mu(x)$$

and then one proceeds to its completion. Here, proceeding to the quotient corresponds to going from *functions* to *classes* of functions defined up to null sets.

In a complex vector space, every Hermitian form $B(x,y)$ satisfies the identity

$$\begin{aligned}(19.1)\qquad 4B(x,y) = {} & B(x+y,x+y) - B(x-y,x-y) \\ & + B(x+iy,x+iy) - B(x-iy,x-iy)\end{aligned}$$

following from axioms (AH 1) and (AH 2). It is frequently used. For a Hilbert inner product, it is written

$$(19.1')\qquad 4\,(x|y) = \|x+y\|^2 - \|x-y\|^2 + \|x+iy\|^2 - \|x-iy\|^2 .$$

The Cauchy-Schwarz inequality shows that, for all $a \in \mathcal{H}$, the map $x \mapsto (x|a)$ is a continuous linear functional on $\mathcal{H}$. Conversely:

**Theorem 31 (M. Fréchet, 1907).** *For any continuous linear functional $f$ on $\mathcal{H}$, there is a unique $a \in \mathcal{H}$ such that $f(x) = (x|a)$ for all $x \in H$.*

One may assume that $f \neq 0$. The subspace $\mathcal{E}$ defined by $f(x) = 0$ is then closed and distinct from $\mathcal{H}$. Let us suppose that there exists $b \neq 0$ orthogonal to $\mathcal{E}$; the relation $f(x) = 0$ then implies $(x|b) = 0$, whence $f(b) \neq 0$. So one may assume that $f(b) = 1$. Then

$$f(x) = f\left[f(x)b\right]$$

and thus $x - f(x)b \in \mathcal{E}$. So $(x|b) = (f(x)b|b) = f(x)(b|b)$. The vector $a = b/(b|b)$ answers the question.

Hence the proof reduces to showing the existence of $b$. This is a consequence of lemma 2 below.

**Lemma 1.** *Let $C$ be a closed convex set in $\mathcal{H}$. For all $a \in \mathcal{H}$, there is a unique $a' \in C$ such that*

$$\|x - a\| < \|a' - a\| \tag{19.2}$$

*for all $x \in C$ other than $a'$.*

Indeed, let $m$ be the distance from $a$ to $C$ and let us choose elements $a_n \in C$ such that $\lim \|a_n - a\| = m$. Applying to $a - x$ and $a - y$, the identity

$$\|x + y\|^2 + \|x - y\|^2 = 2\left(\|x\|^2 + \|y\|^2\right), \tag{19.3}$$

which follows from formal properties of the inner product, shows that

$$\|x - y\|^2/4 = \frac{1}{2}\left(\|a - x\|^2 + \|a - y\|^2\right) - \|a - \frac{1}{2}(x + y)\|^2$$

for all $x$ and $y$. For $x = a_p$ and $y = a_q$, $\frac{1}{2}(x + y) \in C$, so that the last term is $\geq m$. Thus

$$\|a_p - a_q\|^2/4 \leq \frac{1}{2}\left(\|a - a_p\|^2 + \|a - a_q\|^2\right) - m,$$

an arbitrarily small result for large $p$ and $q$. Therefore, $(a_n)$ converges to some $a' \in C$ obviously satisfying $\|a - a'\| = m$. If $a'' \in C$ is , like $a'$, the minimum distance from $a$, then by (3)

$$\|a' - a''\|^2/4 = \left(\|a - a'\|^2 + \|a - a''\|^2\right) - \|a - \frac{1}{2}(a' + a'')\|^2.$$

The first term on the right hand side is equal to $m$ and the second term is $\geq m$. The left hand side is thus $\leq 0$, qed.

**Lemma 2**. *Let $\mathcal{E}$ be a closed vector subspace of $\mathcal{H}$. Then $\mathcal{H}$ is the direct sum of $\mathcal{E}$ and of the orthogonal complement of $\mathcal{E}$.*

For all $a \in \mathcal{H}$, let $a'$ be the unique point of $\mathcal{E}$ at a minimum distance from $a$. For all $x \in \mathcal{E}$ and all $\lambda \in \mathbb{C}$,

$$\|a - (a' + \lambda x)\|^2 = \|a - a'\|^2 + 2\,\mathrm{Re}\,[\lambda\,(a - a'|x)] + |\lambda|^2\|x\|^2 \geq \|a - a'\|^2$$

and so

$$|\lambda|^2\|x\|^2 + 2\,\mathrm{Re}\,[\lambda\,(a - a'|x)] \geq 0$$

for all $\lambda \in \mathbb{C}$. Thus $(a - a'|x) = 0$ for all $x \in \mathcal{E}$, which proves that $a - a'$ is orthogonal to $\mathcal{E}$. The sum of $\mathcal{E}$ and its orthogonal complement is direct since $(x|x) = 0$ for all common vectors, qed.

Denoting by $\mathcal{E}^\perp$ the orthogonal complement of $\mathcal{E}$, a corollary of lemma 2 says that conversely

$$\mathcal{E} = \left(\mathcal{E}^\perp\right)^\perp$$

since $\mathcal{H} = \mathcal{E} \oplus \mathcal{E}^\perp = \mathcal{E}^\perp \oplus (\mathcal{E}^\perp)^\perp$; however, $\mathcal{E} \subset (\mathcal{E}^\perp)^\perp$.

Theorem 31 also has a very important corollary:

**Corollary**. *Let $\mathcal{H}$ and $\mathcal{H}'$ be two Hilbert spaces, $\mathcal{D}$ and $\mathcal{D}'$ subspaces every dense in $\mathcal{H}$ and $\mathcal{H}'$, and $B : \mathcal{D} \times \mathcal{D}' \longrightarrow \mathbb{C}$ a sesquilinear form*[72]. *There is a continuous linear map $A : \mathcal{H} \longrightarrow \mathcal{H}'$ such that*

$$B(x, y) = (Ax|y) \quad \text{for } x \in \mathcal{D}\,, \quad y \in \mathcal{D}'$$

*if and only there is constant $M$ such that*

$$|B(x, y)| \leq M\|x\|.\|y\| \quad \text{for } x \in \mathcal{D}\,, \quad y \in \mathcal{D}'\,.$$

Necessity is obvious. Conversely, the stated condition shows that, for all $x \in \mathcal{D}$, the linear functional $y \mapsto \overline{B(x,y)}$ is continuous on $\mathcal{D}'$, and so can be extended to $\mathcal{H}'$. Thus $B(x|y) = (x'|y)$ for some unique $x' \in \mathcal{H}'$ obviously depending linearly on $x$. Setting $x' = Ax$,

$$\|Ax\| = \sup |(x'|y)| \,/\, \|y\| = \sup |B(x, y)| \,/\, \|y\| \leq M\|x\|\,,$$

qed.

Throughout this chapter, the notation $\mathcal{L}(\mathcal{H})$ will denote the set of continuous linear maps ("operators") from $\mathcal{H}$ to $\mathcal{H}$. It is a Banach space with respect to the norm[73]

---

[72] This means that the partial maps $x \mapsto B(x, y)$ and $y \mapsto \overline{B(x, y)}$ are linear.
[73] It should be pointed out that the vector $x = 0$ is excluded from the relation. I prefer relying on the reader's common sense.

$$\|A\| = \sup \|Ax\|/\|x\| .$$

(ii) *Orthonormal bases.* A family $(e_i)$ of vectors in a Hilbert space $\mathcal{H}$ is said to be *orthonormal* if $(e_i|e_j) = 0$ or 1 according to whether $i$ and $j$ are distinct or equal. Every Hilbert space has *orthonormal bases.* This is how one calls an orthonormal family $(e_i)_{i\in I}$ such that 0 is the only vector orthogonal to all $e_i$; in other words, any *maximal* orthonormal family. Lemma 2 shows that the closed subspace generated by the vectors $e_i$ is $\mathcal{H}$. For a finite subset $F$ of $I$, letting $\mathcal{H}_F$ denote the subspace generated by the vectors $e_i$ $(i \in F)$, and for all $x \in \mathcal{H}$, setting $\xi_i = (x|e_i)$, the orthogonal projection $x_F$ of $x$ onto $\mathcal{H}_F$ is clearly the partial sum of the series $\sum \xi_i e_i$ extended to $F$. Thus

$$\sum_{i\in F} |\xi_i|^2 = \|x\|^2 - \|x_F\|^2 \leq \|x\|^2 .$$

The set of $i$ such that $\xi_i \neq 0$ is, therefore, countable. Moreover, as the finite linear combinations of the vectors $e_i$ are every dense in $\mathcal{H}$, enabling us to approximate $x$ by the points $x_F$ (points in $\mathcal{H}_F$ at a minimum distance from $x$), taking the limit,

$$\|x\|^2 = \sum |\xi_i|^2 = \sum |(x|e_i)|^2$$

and

$$x = \lim_{F\subset I} \sum_{i\in F} \xi_i e_i ,$$

where the limit is extended to all finite subsets $F$ of $I$, as in the case of unconditional convergence. A learned way of putting this would be to say that $\mathcal{H}$ is isomorphic (obvious definition) to the space $L^2(I;\mu)$ where $I$ is equipped with the discrete topology and where the measure $\mu$ assigns the weight 1 to each point of $I$.

The previous formula does not mean that the series $\sum \xi_i e_i$ converges absolutely to $x$: for this to be the case, it is necessary that $\sum |\xi_i| < +\infty$.

*Exercise 1.* Suppose $\mathcal{H} = L^2(X;\mu)$ where $X$ and $\mu$ are arbitrary. Let $K(x,y)$ be a square integrable function with respect to the product measure. (a) Show that there is continuous linear operator $A_K$ on $\mathcal{H}$ such that

$$(A_K f|g) = \iint K(x,y)f(x)\overline{g(y)}d\mu(x)d\mu(y)$$

for all $f, g \in \mathcal{H}$ and that

$$A_K f(x) = \int K(x,y)f(y)d\mu(y)$$

for all $f \in \mathcal{H}$, the integral being convergent for almost all $x$. (b) Let $(e_i)$ be an orthonormal basis for $\mathcal{H}$. Show that the functions $e_i(x)e_j(y)$ form an orthonormal basis for $L^2(X \times X; \mu \times \mu)$. (c) Set

$$A_K e_i = \sum a_{ij} e_j$$

with $a_{ij} \in \mathbb{C}$. Show that $\sum_{i,j} |a_{ij}|^2 < +\infty$. Converse?

In the general case, the existence of orthonormal bases can only be proved by using transfinite inductions (Zorn's theorem). But when $\mathcal{H}$ is *separable*, some may be constructed by applying the *Schmidt orthonormalization process*: choose linearly independent vectors $a_n$ whose linear combinations are everywhere dense in $\mathcal{H}$ and change them as follows: replace $a_1$ by $a_1/\|a_1\|$, subtract from $a_2$ its projection on the subspace generated by $a_1$ and divide the result by its norm. subtract from $a_3$ its projection on the subspace generated by $a_1$ and $a_2$ and divide the result by its norm, etc. On could write down explicit formulae; they can be useful at times:

*Exercise 2.* Since any continuous function on $X = [0, 1]$ is the uniform limit of polynomials, these are everywhere dense in the space $\mathcal{H} = L^2(X)$ of Lebesgue measure. Hence an orthonormal basis for $\mathcal{H}$ can be obtained by orthonormalizing the sequence of functions $1$, $t$, $t^2$,... Calculate the functions obtained.

*Exercise 3* (von Neumann). Let $X$ be a locally compact space, $\mu$ a positive measure on $X$, $\mathcal{H}$ a separable Hilbert space and $a_n(x)$ a sequence of measurable maps from $X$ to $\mathcal{H}$. Suppose that, for almost all $x$, the closed subspace of $\mathcal{H}$ generated by the maps $a_n(x)$ is the all of $\mathcal{H}$. Show that there are *measurable* maps $e_n(x)$ from $X$ to $\mathcal{H}$ such that, for almost all $x$, the family of maps $e_n(x)$ is an orthonormal basis of $\mathcal{H}$.

(iii) *Adjoints, Hermitian operators.* Now consider a continuous linear operator $A$ in $\mathcal{H}$,[74] or more generally a continuous linear map from $\mathcal{H}$ to another space $\mathcal{H}'$. The form $B(x, y) = (x|Ay)$ is accountable for the corollary of theorem 31, with $M = \|A\|$. This gives a continuous operator $A^* : \mathcal{H}' \longrightarrow \mathcal{H}$, the *adjoint* of $A$, characterized by the identity

$$(Ax|y) = (x|A^*y) \; . \tag{19.4}$$

We then deduce that

$$(A + B)^* = A^* + B^* \, , \quad (AB)^* = B^*A^* \, , \quad (A^*)^* = A \tag{19.5}$$

as in finite dimension. Since it has been proved above that $\|A^*\| \leq \|A\|$, the last equality shows that actually

$$\|A^*\| = \|A\| \, . \tag{19.6}$$

The sequence also satisfies the essential property

$$\|A^*A\| = \|A\|^2 \, . \tag{19.7}$$

[74] The terms "operator" and "map" are synonymous. The former, in use since long before the latter and still very widespread, is almost exclusively employed to denote a *linear* map from a Banach space $\mathcal{H}$ (or a subspace of it) to itself.

To see this, it suffices to show that the left hand side is greater than the right hand one. Now, for all $x \in \mathcal{H}$,

$$(A^*Ax|x) = (Ax|Ax) = \|Ax\|^2 ,$$

whence

$$\|A\|^2 = \sup_x \|Ax\|^2/\|x\|^2 = \sup (A^*Ax|x) / \|x\|^2 \leq \|A^*A\| ,$$

qed.

In this chapter, we will repeatedly make use of *self-adjoint* operator *algebras.* These are sets $\mathbf{A}$ of operators that are vector subspaces and subrings of $\mathcal{L}(\mathcal{H})$ and such that

$$T \in \mathbf{A} \Longrightarrow T^* \in \mathbf{A} .$$

**Lemma 3.** *Let $\mathcal{E}$ be a closed subspace of $\mathcal{H}$ and $\mathbf{A}$ a self-adjoint operator algebra on $\mathcal{H}$. The following properties are then equivalent:*

(i) *$\mathcal{E}$ is $\mathbf{A}$-invariant;*[75]

(ii) *the operator $P$ of orthogonal projection onto $\mathcal{E}$ commutes with all $T \in \mathbf{A}$.*

*The orthogonal complement of $\mathcal{E}$ is then $\mathbf{A}$-invariant.*

(i) $\Longrightarrow$ (ii): if $x$ is orthogonal to $\mathcal{E}$, then $0 = (x|Ty) = (T^*x|y)$ for all $y \in \mathcal{E}$ and all $T \in \mathbf{A}$, so that the orthogonal complement of $\mathcal{E}$ is $\mathbf{A}$-invariant.

(ii) $\Longrightarrow$ (i): the elements $x \in \mathcal{E}$ are characterized by the relation $Px = x$, which implies that $Tx = TPx = PTx$. Hence $Tx \in \mathcal{E}$ for all $T \in \mathbf{A}$, qed.

(Continuous) operators $H$ such that $H^* = H$ are said to be *Hermitian*. They are characterized by the fact that

$$(Hx|x) \in \mathbb{R} \quad \text{for all } x \in \mathcal{H} \tag{19.8}$$

or that $(Hx|y)$ is a Hermitian form on $\mathcal{H}$. If a closed vector subspace $\mathcal{E}$ of $\mathcal{H}$ is $H$-invariant, so is its orthogonal complement (lemma 3). Since $(Hx|y) = (x|Hy)$, *the orthogonal complement of* $\text{Im}(H)$ *is* $\text{Ker}(H)$ and that of $\text{Ker}(H)$ is the closure (which is not generally closed) of the subspace $\text{Im}(H)$.

*Positive Hermitian* are defined by the condition that

$$(Hx|x) \geq 0 \quad \text{for all } x \in \mathcal{H} . \tag{19.9}$$

This is the case of $H = A^*A$ for all $A \in \mathcal{L}(\mathcal{H})$. We write $H \leq H'$ when $H' - H$ is positive, i.e. if $(Hx|x) \leq (H'x|x)$ for all $x$.

[75] This means that every $T \in \mathbf{A}$ maps $\mathcal{E}$ to $\mathcal{E}$.

If $H \geq 0$, $(Hx|y)$ is a positive Hermitian form satisfying the Schwarz inequality

$$(19.10) \qquad |(Hx|y)|^2 \leq (Hx|x)\,(Hy|y)\,.$$

It follows that

$$(19.10') \qquad \|H\| = \sup |(Hx|x)| \,/\, (x|x)\,.$$

Hence $\|H\| \leq \|H'\|$ if $0 \leq H \leq H'$.

Relation (10) also shows that

$$(19.10") \qquad Hx = 0 \Longleftrightarrow (Hx|x) = 0\,.$$

It follows that $\mathrm{Ker}(H'+H'') = \mathrm{Ker}(H') \cap \mathrm{Ker}(H'')$ if $H'$ and $H''$ are positive.

If $\mathcal{E}$ is a closed vector subspace, the operator $P_{\mathcal{E}}$, which associates to all $x \in \mathcal{H}$ its orthogonal projection on $\mathcal{E}$, is Hermitian positive since, $x - P_{\mathcal{E}}x$ being orthogonal to $P_{\mathcal{E}}x$,

$$(P_{\mathcal{E}}x|x) = (P_{\mathcal{E}}x|P_{\mathcal{E}}x) \geq 0\,.$$

The relation

$$(19.11) \qquad P^* = P^2 = P$$

is easily seen to characterize these operators, the corresponding subspace $\mathcal{E}$ being the image of $\mathcal{H}$ under $P$.

(iv) *Spectrum of a Hermitian operator.* In finite dimension, hermitian operators $H$ are known to be diagonalizable. There is even an orthonormal basis whose elements are the eigenvectors of $H$, the corresponding eigenvalues being real. In infinite dimension, the situation is not so simple, but results heading in that direction can be easily obtained.

**Lemma 4.** *Let $H$ be a Hermitian operator on a Hilbert space $\mathcal{H}$. Then $H-\zeta$ is invertible for all $\zeta \notin \mathbb{R}$.*

If $\zeta = \alpha + i\beta$ with $\beta \neq 0$, then $H - \zeta = \beta[(H-\alpha)/\beta - i]$. As $(H-\alpha)/\beta$ is again Hermitian, it suffices to prove the lemma for $\zeta = i$. First of all, $H - i$ is injective since the imaginary part of

$$((H-i)x|x) = (Hx|x) - i\|x\|^2$$

is $\|x\|^2$. To show that $H - i$ is surjective, start with the relation

$$\|(H-i)x\|^2 = \|Hx\|^2 + \|x\|^2 \geq \|x\|^2\,.$$

If, for a sequence $(x_n)$, the sequence with general term $y_n = (H-i)x_n$ converges, and so satisfies Cauchy's criterion, so does $(x_n)$. It follows that

$H - i$ maps $\mathcal{H}$ onto a *closed* subspace $\mathcal{H}'$ of $\mathcal{H}$. Any $y \in \mathcal{H}$ orthogonal to $\mathcal{H}'$ must satisfy $((H-i)x|y) = 0$ for all $x$, and so $(Hx|y) = i(x|y)$. Hence $(Hy|y) = i(y|y)$, which gives $y = 0$ since the left hand side is real. Finally, the inequality used above tells us that $H - i$ is bijective and its inverse is continuous, qed.

Lemma 4 enables us to associate the *Cayley transform*

$$U = (H-i)(H+i)^{-1}$$

to $H$. Then $UH = HU$ and, by (5),

$$U^*U = UU^* = 1\,.$$

This relation, which defines *unitary* operators and means that $x \mapsto Ux$, is an isomorphism from $\mathcal{H}$ onto $\mathcal{H}$: $U$ is bijective and preserves the inner product. The definition of $U$ from $H$ shows that $H(1-U) = i(1+U)$. To deduce that

$$H = i(1+U)(1-U)^{-1}\,, \tag{19.12}$$

one needs to show that $1-U$ is invertible. Now, $(1-U)x = 0$ implies that $(1+U)x = 0$. Thus $Ux = 0$ and so $x = 0$. On the other hand, the equation $a = (1-U)x$, where $a$ is given, is equivalent to

$$a = x - (H-i)(H+i)^{-1}x\,,$$

i.e. to $(H+i)a = 2ix$ and so always has a solution. As a result, $1-U$ is bijective, which justifies the formula. As

$$(1-U)^{-1} - 2i(H+i)^{-1}\,,$$

the inverse of $1-U$ is continuous.

Conversely, (12) defines a Hermitian operator for every unitary operator $U$ *such that* $1-U$ *is invertible*. See n° 23, (ii), exercise 4.

For a continuous operator $A$ on $\mathcal{H}$, the set of $\zeta \in \mathbb{C}$ for which $A - \zeta$ is not invertible is called the *spectrum* of $A$, and written $Sp(A)$. We will see in the next n° that it is a *non-empty compact* set. For a Hermitian operator $H$, the infimum and supremum can be determined by introducing the numbers

$$m_H = \inf (Hx|x)\,/\,(x|x)\,, \quad M_H = \sup (Hx|x)\,/\,(x|x) \tag{19.13}$$

contained between $-\|H\|$ and $+\|H\|$. So saying that $H$ is positive mean that $m_H \geq 0$.

**Lemma 5**. *The spectrum of a Hermitian operator $H$ is contained in the interval $[m_H, M_H]$ and contains its endpoints.*

Let us suppose that for example $\zeta < m_H$ and replace $H$ by $H - \zeta$, which subtracts $\zeta$ from $m_H$ and $M_H$. For the new operator, $m_H > 0$ and the proof reduces to deducing that $H$ is invertible. As $(Hx|x) \geq m_H(x|x)$, $H$ is injective and $m_H\|x\| \leq \|Hx\|$, so that the image of $\mathcal{H}$ under $H$ is closed. A vector $y$ orthogonal to it must satisfy $(Hx|y) = 0$ for all $x$, and so $(x|Hy) = 0$, whence $Hy = 0$. Thus $y = 0$ since $H$ is injective. The image of $H$ is, therefore, (lemma 2) equal to $\mathcal{H}$. For $\zeta > M_H$, apply arguments to the operator $\zeta - H$.

It remains to show that $H - \zeta$ is not invertible for $\zeta = m_H$ or else that if a positive Hermitian operator $H$ is invertible, then $m_H > 0$. Setting $m = \|H^{-1}\|$, by (10),

$$m = \inf \|Hx\|/\|x\| = \inf |(Hx|y)| / \|x\|.\|y\|,$$

whence

$$\begin{aligned} m^2 &= \inf |(Hx|y)|^2 / \|x\|^2.\|y\|^2 \\ &\leq \inf (Hx|x)(Hy|y) / |\|x\|^2 \|y\|^2 \\ &= \inf (Hx|x) / \|x\|^2. \inf (Hy|y) / \|y\|^2 = m_H^2 . \end{aligned}$$

Thus $m_H \geq m > 0$ (and in fact equality holds), qed.

We will show later that, short of being diagonalizable as in finite dimension, any Hermitian operator $H$ has the following property: there are closed $H$-invariant subspaces $\mathcal{H}'$ other than 0 and $\mathcal{H}$. For all $r > 0$, the spectral theory of n° 22 even enables us to decompose $\mathcal{H}$ into a direct sum of finitely many pairwise orthogonal, closed $H$-invariant subspaces $\mathcal{H}_i$, and such that for each of them there is a scalar $\lambda_i \in \mathbb{R}$ for which

$$\|Hx - \lambda_i x\| \leq r\|x\| \quad \text{for all } x \in \mathcal{H}_i .$$

Thus the operator $H$ satisfies $\|H - \lambda_i\| \leq r$ on each $\mathcal{H}_i$. This is an approximative form of diagonalization, precisions about which we will given in n° 24 by constructing "virtual eigenspaces" that are not contained in $\mathcal{H}$.

(v) *Weak topology.* Recall (Chap. III, Appendix) that the *seminorm* on a complex vector space $\mathcal{H}$ is a function $p(x) \geq 0$ satisfying identities

$$p(ax) = |a|p(x), \quad p(x+y) \leq p(x) + p(y).$$

Any family $(p_i)_i \in I$ of seminorms enables us to define a topology on $\mathcal{H}$ which is compatible with its vector structure and with respect to which the functions $p_i$ become continuous: any set defined by *finitely* many inequalities

$$|p_i(x) - \alpha_i| < r_i , \tag{19.14}$$

where $r_i > 0$ and $\alpha_i \in \mathbb{C}$ are given, being necessarily open, it suffices to define the open subsets of $\mathcal{H}$ to be the finite and infinite unions of sets (14). That the axioms hold is readily verified. In fact one only does the bare minimum to make the functions $p_i$ continuous. The Hausdorff axiom holds if and only if 0 is the only vector for which $p_i(x) = 0$ for all $i$.

The topology on a Banach space is obtained through this process by using the seminorm $p(x) = \|x\|$. But there is another way of obtaining it, consisting of choosing functions

$$x \longmapsto |f(x)| ,$$

where $f$ varies in the set $\mathcal{H}'$ (topological dual of $\mathcal{H}$) of continuous linear functionals on $\mathcal{H}$. This gives the *weak topology* on $\mathcal{H}$, defined by the norm which most authors call *strong topology* . Since $|f(x)| \leq M\|x\|$ for all $f \in \mathcal{H}'$, all open (resp. closed) subsets $E$ of $\mathcal{H}$ with respect to the weak topology are clearly of the same type with respect to the strong topology. The converse holds if $E$ is a vector subspace for if $E$ is (strongly) closed, it is defined by a family (in general infinite) of equations $f_i(x) = 0$, where the $f_i$ are continuous linear functionals (a consequence of the Hahn-Banach theorem). It is, therefore, also weakly closed. In the case of a Hilbert space, lemma 2 enables one to avoid (or prove) Hahn-Banach.

In this case, $\mathcal{H}' = \mathcal{H}$ by theorem 31. Thus the weak topology is obtained by doing the bare minimum to make the function

$$x \longmapsto (x|u)$$

continuous for all $u \in \mathcal{H}$. Hence saying (for example) that a sequence $x_n \in \mathcal{H}$ converges weakly to some $x \in \mathcal{H}$ means that $\lim(x_n|u) = (x|u)$ for all $u$.

Choosing an orthonormal basis $(e_n)$, $\sum |(e_n|x)|^2 < +\infty$ and so $\lim(e_n|x) = 0$. Thus $e_n$ converges weakly to 0, but not strongly since $\|e_n\| = 1$ for all $n$. The identity

$$\|x_n - x\|^2 = \|x_n\|^2 - 2\,\mathrm{Re}\,(x_n|x) + \|x\|^2 ,$$

shows that *a sequence $x_n$ converging weakly to a limit $x$ converges strongly to $x$ if and only if* $\lim \|x_n\| = \|x\|$.

There are some useful consequences of Baire's theorem [n° 15, (ii), note 53] in this field. These are only particular cases of results holding for far more general topological vector spaces, the Fréchet spaces, but which can essentially be proved in the same way; see N. Bourbaki's book on the topic. In all cases, countability and Baire's theorem play a crucial role.

**Lemma 6**. *Let $(a_n)$ be a* sequence *of vectors such that* $\lim(a_n|y)$ *exists for all $y$ . Then* $\sup \|a_n\| < +\infty$ *and the sequence $(a_n)$ converges weakly.*

Setting $f_n(x) = (a_n|x)$ and $f(x) = \lim f_n(x)$, the proof reduces to showing that the linear functional $f$ is continuous, and to obtain this, that the norms of $f_n$ (i.e. of $a_n$) are bounded. For all $N$, let us consider the set $F_N$ of $x$ for which $|f_p(x) - f_q(x)| \leq 1$ for all $p, q \geq N$. Cauchy's criterion shows that $\mathcal{H} = \bigcup F_N$. However, the sets $F_N$ are closed since the $f_p - f_q$ are continuous. Thus one of the $F_N$ (Baire's theorem) has an interior point $a$. So there exists $r > 0$ such that

$$\|x\| \leq r \Longrightarrow |f_p(x+a) - f_q(x+a)| \leq 1$$

for all $p, q \geq N$. As $|f_p(a) - f_q(a)| \leq 1$, it follows that $\|f_p - f_q\| \leq 2/r$ for all $p, q \geq N$, whence $\sup \|f_n\| < +\infty$. Therefore, the linear functional $f$ is continuous, and so there exists $a \in \mathcal{H}$ such that $\lim(a_n|x) = (a|x)$, qed.

**Lemma 7**. *Let $(A_n)$ be a sequence of continuous linear operators such that $\lim(A_n x|y)$ exists for all $x$, $y$. Then $\sup \|A_n\| < +\infty$ and there is a continuous linear operator $A$ such that*

$$\lim (A_n x|y) = (Ax|y) .$$

The corollary of theorem 31 tells us that it is sufficient to prove the first proposition. By the previous lemma, we already know that $\sup \|A_n x\| < +\infty$ for all $x \in \mathcal{H}$, which is sufficient to prove the lemma [n° 15, (ii), note 53].

**Lemma 8**. *Let $H_1 \leq H_2 \leq \ldots$ be an increasing sequence of Hermitian operators. Suppose that $\sup(H_n x|x) < +\infty$ for all $x \in \mathcal{H}$. Then there is a continuous linear operator $H$ such that $\lim \|Hx - H_n x\| = 0$ for all $x \in \mathcal{H}$.*

As $H_{pq} = H_q - H_p \geq 0$ for $q \geq p$,

$$|(H_{pq}x|y)|^2 \leq (H_{pq}x|x)\,(H_{pq}y|y) .$$

Hence like the sequences $(H_n x|x)$ and $(H_n y|y)$, $(H_n x|y)$ satisfies Cauchy' criterion, which shows the existence of some continuous $H$ such that $(Hx|y) = \lim(H_n x|y)$. Obviously, $H \geq H_n$ for all $n$, whence

$$\begin{aligned} |(Hx - H_n x|y)|^2 &\leq (Hx - H_n x|x)\,(Hy - H_n y|y) \\ &\leq \|H - H_n\| \,.\|y\|.\, (Hx - H_n x|x) \end{aligned}$$

et $\|Hx - H_n x\|^2 \leq \|H - H_n\|(Hx - H_n x|x)$. However, $\sup \|H - H_n\| < +\infty$ by lemma 7.

(vi) *Hilbert-Schmidt operators.* In finite dimensional, any operator $A$ has a *trace* $Tr(A)$: this is the sum of its diagonal entries with respect to an arbitrary basis of the space $\mathcal{H}$ considered. The positive Hermitian form

$$Tr(AB^*) = \sum a_{ij}\overline{b_{ij}}$$

(obvious notation) transforms $\mathcal{L}(\mathcal{H})$ into a new Hilbert space. To make the symbol $Tr(A^*A)$ well-defined when $A$ is a continuous operator on an infinite-dimensional Hilbert space $\mathcal{H}$, the most naive idea is to choose an orthonormal basis $(\mathbf{e}_i)$ of $\mathcal{H}$ and to set, as in finite dimension,

$$Tr(A^*A) = \sum (A^*A\mathbf{e}_i|\mathbf{e}_i) = \sum \|A\mathbf{e}_i\|^2 .$$

At the very least, this supposes that

$$(19.15) \qquad \sum \|A\mathbf{e}_i\|^2 < +\infty$$

and that the the sum is independent of the chosen basis. However, for any orthonormal basis $(\mathbf{e}_i')$,

$$\begin{aligned}\sum_i \|A\mathbf{e}_i\|^2 &= \sum_{i,j} \left|(A\mathbf{e}_i|\mathbf{e}_j')\right|^2 = \sum_{i,j} \left|(\mathbf{e}_i|A^*\mathbf{e}_j')\right|^2 \\ &= \sum_j \|A^*\mathbf{e}_j'\|^2 .\end{aligned}$$

Hence if (15) holds for a particular basis, then $\sum \|A^*\mathbf{e}_i'\|^2 < +\infty$ in all bases. So applying arguments to $A^*$, we see that $\sum \|A\mathbf{e}_i''\|^2 < +\infty$ in all bases, qed.

Condition (15) defines the *Hilbert-Schmidt operators* (HS), a very important class of operators in most analysis problems and which includes all finite rank operators, i.e. for which the dimension of $\mathrm{Im}(A)$ is finite. Exercise 1 of section (ii) explains their historical origin.

The sum of two HS operators is HS thanks to the Cauchy-Schwarz inequality applied to the "square integrable" sequences $\|A\mathbf{e}_i\|$ and $\|B\mathbf{e}_i\|$ (Chap. II, n° 15, corollary of theorem 7). The set $\mathcal{L}_2(\mathcal{H})$ of these operators is, therefore, a vector subspace of $\mathcal{L}(\mathcal{H})$. If $A$ is HS, clearly so is $A^*$ as well. For all $P \in \mathcal{L}(\mathcal{H})$, $\|PAe_i\| \leq \|P\|.\|Ae_i\|$, so that $PA$ is a HS operator, and hence so is $A^*P^*$ as well. Replacing $A$ by $A^*$, in conclusion, if $A$ is HS, then $PAQ$ is HS for all $P$ and $Q$. In other words, $\mathcal{L}_2(\mathcal{H})$ is a two-sided ideal of $\mathcal{L}(\mathcal{H})$.

It is useful to observe that, for every family of vectors $(a_i)$ such that $\sum \|a_i\|^2 < +\infty$, there is a unique operator $A \in \mathcal{L}_2(\mathcal{H})$ such that $a_i = Ae_i$ for all $i$: it suffices to set

$$Ax = \sum_{i \in I} \xi_i a_i \quad \text{for } x = \sum_{i \in I} \xi_i e_i .$$

Then

$$\sum |\xi_i| . \|a_i\| \leq \left(\sum |\xi_i|^2\right)^{1/2} \left(\sum \|a_i\|^2\right)^{1/2} ,$$

so that the series defining $Ax$ converges absolutely. Moreover,

$$\|Ax\| \leq \sum \|\xi_i a_i\| \leq M\|x\|$$

where $M = (\sum \|a_i\|^2)^{1/2}$. $A$ is clearly a HS operator.

If $A$ and $B$ are HS, then $Tr(B^*A) = \sum (Ae_i|Be_i)$ is well-defined thanks to Cauchy-Schwarz. It does not depend on the choice of the basis: this was shown for $B = A$, and the general case can be deduced by using identity (1). It remains invariant if $A$ and $B$ are replaced by $U^{-1}AU$ and $U^{-1}BU$, where $U$ is unitary: this amounts to changing orthonormal bases. The same calculations using the matrices of $A$ and $B$ with $A$ and $B$ with respect to an orthonormal basis show that

$$Tr(AB) = Tr(BA)$$

like in finite dimension. $Tr(B^*A)$ is an inner product on the space $\mathcal{L}_2(\mathcal{H})$. The corresponding norm is written

$$\|A\|_2 = Tr(A^*A)^{1/2} = \left(\sum \|Ae_i\|^2\right)^{1/2}.$$

$\mathcal{L}_2(\mathcal{H})$ is *complete* with respect to this norm, and so is a Hilbert space. Indeed, associating the matrix $a_{ij} = (Ae_i|e_j)$ to each $A \in \mathcal{L}(\mathcal{H})$ gives $\|Ae_i\|^2 = \sum |a_{ij}|^2$. The space of Hilbert-Schmidt operators equipped with its inner product is, therefore, isomorphic to the space of all families $(a_{ij})$ such that $\sum |a_{ij}|^2 < +\infty$. Like any $L^2(X;\mu)$ space, it is complete, even and especially if $X = I \times I$ is discrete... If $E_{ij}$ denotes the operator mapping $e_i$ onto $e_j$ and and other basis vectors to 0, this clearly gives an orthonormal basis for $\mathcal{L}_2(\mathcal{H})$, the $a_{ij}$ being the coordinates of $A$ with respect to this basis.

As any norm 1 vector $x \in \mathcal{H}$ belongs to an orthonormal basis, $\|Ax\|^2 \leq Tr(A^*A)$ and so

$$\|A\| \leq Tr(A^*A)^{1/2} = \|A\|_2 \quad \text{if } A \text{ is HS}.$$

Also

$$\|PAQ\|_2 \leq \|P\|.\|A\|_2.\|Q\|$$

for all $P, Q \in \mathcal{L}(\mathcal{H})$.

*Exercise 4.* Let $\mu$ be a positive measure on a space $X$ and $K(x,y)$, $H(x,y)$ two square integrable functions with respect to $\mu \times \mu$, whence (exercise 1) HS operators

$$A_K f(x) = \int K(x,y) f(y) d\mu(y)$$

and $A_H$ on $\mathcal{H} = L^2(X;\mu)$. Show that

$$Tr(A_H^* A_K) = \iint K(x,y)\overline{H(x,y)} d\mu(x) d\mu(y);$$

The Hilbert space $\mathcal{L}_2(\mathcal{H})$ is, therefore, canonically isomorphic to the $L^2$ space of $\mu \times \mu$.

(vii) *Von Neumann algebras.*[76] Von Neumann began his work on operator algebras with F. J. Murray by introducing several useful topologies, all of them defined by families of seminorms, on the space $\mathcal{L}(\mathcal{H})$ of continuous

[76] The content of this section will be rarely used before n° 31 of this chapter.

operators on $\mathcal{H}$. They are the following ones, which impose more and more restrictive conditions on the notion of convergence.

(a) The *weak topology*[77] (or of simple weak convergence) is obtained by making the maps $T \mapsto Tx$ continuous with respect to the weak topology of $\mathcal{H}$, in other words by making all functions $T \mapsto (Tx|y)$ in $\mathcal{L}(\mathcal{H})$ continuous in $\mathcal{L}(\mathcal{H})$. It is, therefore, defined by the seminorms $T \mapsto |(Tx|y)|$. Hence every *finite* system of inequalities

$$|(Tx_i - Ax_i|y_i)| < r_i$$

is an open neighbourhood of $A$ with respect to the weak topology, and every open subset is the (finite or infinite) union of such open subsets.

(b) The *strong topology* (or of simple strong convergence) is obtained by doing the bare minimum to make the map $T \mapsto Tx$ continuous for all $x \in \mathcal{H}$. It is defined by the seminorms $T \mapsto \|Tx\|$. A subset $E$ of $\mathcal{L}(\mathcal{H})$ is a strong neighbourhood of a given $A$ if and only if there are finitely many $x_i \in \mathcal{H}$ and $r_i > 0$ such that $E$ contains all operators $T$ satisfying inequalities

$$\|Tx_i - Ax_i\| < r_i\,.$$

(c) The *ultrastrong topology* is given by the seminorms

$$p_A(T) = \|TA\|_2$$

for all Hilbert-Schmidt operators $A$. This time, the bare minimum is done to ensure that all maps $T \mapsto TA$ from $\mathcal{L}(\mathcal{H})$ to $\mathcal{L}_2(\mathcal{H})$ are continuous. It can also be defined differently: any family $(a_i)$ of vectors such that $\sum \|a_i\|^2 < +\infty$ defines a seminorm

$$T \longmapsto \left(\sum \|Ta_i\|^2\right)^{1/2},$$

on $\mathcal{L}(\mathcal{H})$ and these are the seminorms which give rise to the ultrastrong topology.

(d) The last and most obvious one is defined by the norm $\|T\|$, and is the one that we will most frequently use. Some authors call it the *uniform topology* (understood: of uniform convergence on the unit sphere of $\mathcal{H}$), but I will often say convergence in norm.

Studying the properties of algebraic operations in relation to the above topologies is important. Results are readily obtained, but risks of confusion cannot be ignored. Statement ($\delta$) is the most useful and simplest.

($\alpha$) The maps $(A, B) \mapsto AB$ and $A \mapsto A^*$ are continuous with respect to the*uniform* topology (obvious).

[77] Not to be confused with the weak topology on the Banach space $\mathcal{L}(\mathcal{H})$: there are many other types of continuous linear functionals on this space apart from the functions considered here. See J. Dixmier, *Les fonctionnelles linéaires sur l'ensemble des opérateurs bornés d'un espace de Hilbert* (Annals of Math., 51, 1950)

($\beta$) The map $(S,T) \mapsto ST$ is unrestrictedly continuous only with respect to the convergence in norm. For example, considering the unitary operators $U$ and $V$ in $L^2(\mathbb{Z})$ given by,

$$Uf(n) = f(n+1)\,, \quad Vf(n) = f(n-1)\,,$$

it follows that $U^p f$ and $V^p f$, converge weakly to 0 for all $f$ as $p$ increases indefinitely (exercise!), but $U^p V^p = 1$ does not converge weakly to 0.

($\gamma$) The inequality

$$\|STx - S_0T_0x\| \leq \|S\|.\,\|Tx - T_0x\| + \|ST_0x - S_0T_0x\|$$

shows that the map $(S,T) \mapsto ST$ is strongly continuous if $S$ (or $T$) is forced to stay in a *bounded* subset of $\mathcal{L}(\mathcal{H})$. In particular, the map $T \mapsto ATB$ is strongly continuous for given $A$ and $B$.

($\delta$) The map $T \mapsto ATB$ is *weakly* continuous for all $A$ and $B$ since

$$(ATBx|y) = (TBx|A^*y)\ .$$

It is also *strongly* continuous since, for all $x \in \mathcal{H}$,

$$\|AT'Bx - AT''Bx\| \leq \|A\|.\,\|(T' - T'')\,Bx\|\ ,$$

and a similar calculation shows that it is continuous with respect to the ultrastrong topology. The map $T \mapsto T^*$ is *weakly* continuous since $|(T^*x|y)| = |(Ty|x)|$.

These topologies are mainly useful in the theory of self-adjoint operator algebras. If $\mathbf{A}$ is such an algebra, the set $\mathbf{A}'$ of operators commuting with all the $T \in \mathbf{A}$ is obviously also a self-adjoint algebra, containing the unit operator, *closed with respect to the weak topology* [because of property ($\delta$)] and perforce with respect to the others. It is called the *commutator algebra* of $\mathbf{A}$. $\mathbf{A}'$ induces a new self-adjoint algebra $\mathbf{A}'' = (\mathbf{A}')'$ containing $\mathbf{A}$, its *bicommutator*, then $(\mathbf{A}'')' = (\mathbf{A}')''$ and so on. Fortunately, the sequence is periodic since it is always the case that $(\mathbf{A}')'' = \mathbf{A}'$.

**Von Neumann density theorem.** *Let $\mathbf{A} \subset \mathcal{L}(\mathcal{H})$ be a self-adjoint algebra. Then $(\mathbf{A}')'' = \mathbf{A}'$. If $\mathbf{A}$ contains the unit operator, $\mathbf{A}''$ is the closure of $\mathbf{A}$ with respect to each topology, weak, strong, and ultrastrong. If $\mathbf{A}$ is closed with respect to one of these topologies, then $\mathbf{A}'' = \mathbf{A}$.*

(i) The first step is to show that, for all $T \in \mathbf{A}''$, all $a \in \mathcal{H}$ and all $r > 0$, there exists $S \in \mathbf{A}$ such that $\|Ta - Sa\| < r$. Now, the closure $\mathcal{E}$ of the set of $Sa$ is a $\mathbf{A}$-invariant vector subspace. As a result, the projection onto $\mathcal{E}$ is in $\mathbf{A}'$ (lemma 3), hence commutes with the elements of $\mathbf{A}''$, so that $\mathcal{E}$ is $\mathbf{A}''$-invariant. However, $\mathcal{E}$ contains $a$ since $1 \in \mathbf{A}$. So $Ta \in \mathcal{E}$, and the result follows.

(ii) Let us now show that for all $r > 0$, $T \in \mathbf{A}''$ and $a_i \in \mathcal{H}$ such that $\sum \|a_i\|^2 < +\infty$, there exists $S \in \mathbf{A}$ such that

$$\sum \|Ta_i - Sa_i\|^2 < r^2, \tag{19.16}$$

which will prove that $\mathbf{A}$ is dense in $\mathbf{A}''$ with respect to the ultrastrong topology. To do this, we consider the space $\mathcal{H}'$ of $\mathbf{x} = (x_i)$ such that $\sum \|x_i\|^2 < +\infty$, equipped with the inner product $(\mathbf{x}|\mathbf{y}) = \sum (x_i|y_i)$. It is complete since it is actually the space of Hilbert-Schmidt operators. For all $A \in \mathcal{L}(\mathcal{H}')$, there is a matrix $(A_{ij})$ with entries in $\mathcal{L}(\mathcal{H})$ such that

$$\mathbf{y} = A\mathbf{x} \Longleftrightarrow y_i = \sum_j A_{ij} x_j .$$

The usual multiplication rule holds for these matrices if the order of the factors is respected. The matrices $A_{ij}$ cannot obviously be chosen arbitrarily unless, for example, they are all zero up to finitely many exceptions.

Let us associate to each $S \in \mathcal{L}(\mathcal{H})$ the "diagonal" operator $\pi(S)$ in $\mathcal{L}(\mathcal{H}')$ which multiplies each component of $\mathbf{x}$ by $S$. An operator $A$ on $\mathcal{H}'$ clearly commutes with these $\pi(S)$ if and only the matrices $A_{ij}$ commute with all $S$, i.e. are in $\mathbf{A}'$. An operator $B$ commutes with all these operators $A$ if and only if

$$\sum_k B_{ik} A_{kj} = \sum_k A_{ik} B_{kj}$$

for all $i, j$ and $A_{ij} \in \mathbf{A}'$. Supposing all except one of these $A_{ij}$ to be zero, one sees that $B = \pi(T)$ with $T \in \mathbf{A}''$, and conversely. In other words,

$$\pi(\mathbf{A})'' = \pi(\mathbf{A}'')$$

Applying the result obtained in (i) to $\pi(A)$ and setting $\mathbf{a} = (a_i)$ the conclusion follows that, for all $T \in \mathbf{A}''$, there exists $S \in \mathbf{A}$ such that

$$\|\pi(T)\mathbf{a} - \pi(S)\mathbf{a}\| < r,$$

which leads to (16) and to the fact that $\mathbf{A}''$ is the *ultrastrong* closure of $\mathbf{A}$.

(iii) Let $\mathbf{B} \subset \mathbf{A}''$ be the *weak* closure of $\mathbf{A}$. It is a self-adjoint algebra containing 1, which is weakly and thus ultrastrongly closed. As a consequence, by the result just obtained, $\mathbf{B}'' = \mathbf{B}$. However, anything commuting with $\mathbf{A}$ commutes with $\mathbf{B}$ because of property $(\delta)$ above, and conversely since $\mathbf{B} \supset \mathbf{A}$. Hence $\mathbf{B}' = \mathbf{A}'$, and so $\mathbf{B} = \mathbf{B}'' = \mathbf{A}''$.

(iv) If $1 \notin \mathbf{A}$, then $1 \in \mathbf{A}'$, and so $\mathbf{A}' = (\mathbf{A}')''$, qed.

In practice, one frequently needs to apply the previous theorem to self-adjoint algebras $\mathbf{A}$ not containing the unit operator. One can then apply it to operators $T + \lambda 1$, where $T \in \mathbf{A}$ and $\lambda \in \mathbb{C}$, but it would be good to be able to avoid additional scalars. The next lemma enables us to choose:

**Lemma 9.** *Let* $\mathbf{A}$ *be a self-adjoint operator algebra on a Hilbert space* $\mathcal{H}$. $\mathbf{A}$ *is everywhere dense in* $\mathbf{A}''$ *with respect to the ultrastrong topology if and only if*

$$(19.17) \qquad Ax = 0 \quad \text{for all } A \in \mathbf{A} \Longrightarrow x = 0\,.$$

(17) is obviously necessary. So suppose that (17) holds.

The previous theorem tells us operators $T + \lambda 1$, where $T \in \mathbf{A}$, are ultrastrongly dense in $\mathbf{A}''$. If one shows that the operator 1 is in the ultrastrong closure of $\mathbf{A}$, the result will follow.

Coming back to the space $\mathcal{H}'$ in the proof of the density theorem, the proof reduces to showing that every $\mathbf{a} \in \mathcal{H}'$ belongs to the closure $\mathcal{E}$ of the subspace of the set of vectors $\pi(T)\mathbf{a}$. To this end, let us write $\mathbf{a} = \mathbf{a}' + \mathbf{a}''$ with $\mathbf{a}' \in \mathcal{E}$ and $\mathbf{a}''$ orthogonal $\mathcal{E}$. Since operators $\pi(T)$ map $mcalE$ to $\mathcal{E}$, they transform every vector orthogonal to $\mathcal{E}$ into a similar vector (lemma 3). The vector $\pi(T)\mathbf{a}''$ is, therefore, orthogonal to $\mathcal{E}$. But it is equal to $T\mathbf{a} - T\mathbf{a}'$, which is in $\mathcal{E}$ like $T\mathbf{a}$ and $\mathbf{a}'$. Hence $\pi(T)\mathbf{a}'' = 0$ for all $T$, so that $T\mathbf{a}_i'' = 0$ for all components of $\mathbf{a}''$. Thus, if condition (17) holds, then $\mathbf{a}'' = 0$ and $\mathbf{a} \in \mathcal{E}$, qed.

Self-adjoint algebras containing 1 and weakly (or strongly, or ultrastrongly) closed, i.e. von N's *rings of operators*, are not now called *von Neumann algebras*.[78] They are characterized by the relation $\mathbf{A}'' = \mathbf{A}$. $\mathbf{A}' \cap \mathbf{A}'' = \mathbf{A}' \cap \mathbf{A}$ is clearly the *centre* of $\mathbf{A}$, i.e. the set of $S \in \mathbf{A}$ such that $ST = TS$ for all $T \in \mathbf{A}$. Von N's entire theory is about algebras for which $\mathbf{A}' \cap \mathbf{A}$ reduces to scalars, i.e. algebras he used to call *factors* and which he classified, a classification that has since been much improved.

## 20 – Gelfand's Theorems on Normed Algebras

The set $\mathcal{L}(\mathcal{H})$ of continuous operators on a Hilbert space, or more generally a Banach space, is an example of a *complete normed algebra*, a notion introduced by Gelfand in his famous 1941 article. Like everyone, I will follow it closely since attempts at simplification are bound to fail. This theory gave rise to much hope in the 1940s. It also showed its limits fairly quickly: one cannot prove everything without any explicit calculations or only using *abstract nonsense*, as Serge Lang calls it.

This is the name given to any algebra[79] $\mathbf{A}$ over $\mathbb{C}$, with a unit element, equipped with a norm satisfying

[78] See Jacques Dixmier, *Les algèbres d'opérateurs dans l'espace hilbertien* (Gauthier-Villars, 1957) and*Les $C^*$-algèbres et leurs représentations* (Gauthier-Villars, 1964) and, more recent and less concise, Richard Kadison and John R. Ringrose, *Fundamentals of the Theory of Operator Algebras* (AMS, 4 volumes, two of text and two of solved exercises, 1999–2001).

[79] An algebra $A$ over a commutative field $k$ is a ring with unit element equipped with the structure of a vector space over $k$ such that $\lambda 1.x = x.\lambda 1 = \lambda x$ for all

(20.1) $$\|xy\| \leq \|x\|.\|y\|$$

and with respect to which **A** is a complete space, and hence a Banach one. Other examples occur readily: the algebra of bounded continuous functions on a topological space $X$, equipped with the usual multiplication and the norm of uniform convergence; the algebra of bounded measures on some lcg $G$, equipped with the convolution product and the usual norm. And any closed subalgebra containing 1 of a complete normed algebra is also a complete normed algebra. If for example $H$ is a Hermitian operator on a Hilbert space $\mathcal{H}$ then considering the set **A** of continuous operators that are limits, with respect to the usual norm, of polynomials $\sum a_n H^n$ in $H$, we get a complete normed algebra which, as we will see, is isomorphic to the algebra of continuous functions on the spectrum of $H$. The spectral decomposition of $H$ will follow easily from this result.

The *spectrum* of $x \in \mathbf{A}$, defined to be the set $Sp(x)$ of $\zeta \in \mathbb{C}$ for which $x - \zeta$ is not invertible[80] in **A** generalizes the notion of the spectrum of an operator. If $P$ is a single-variable polynomial with complex coefficients, the relation

$$P(x) - P(\zeta) = (x - \zeta)Q(x) = Q(x)(x - \zeta),$$

where $Q$ is also a polynomial, shows that

$$\zeta \in Sp(x) \Longrightarrow P(\zeta) \in Sp\,[P(x)]\,.$$

Indeed, if $P(x) - P(\zeta)$ has an inverse $y$, then

$$yQ(x)(x - \zeta) = (x - \zeta)Q(x)y = 1\,,$$

so that $x - \zeta$ is invertible.

**Lemma 1.** *Let* **A** *be a complete normed algebra. For all* $x \in \mathbf{A}$*, the spectrum of* $x$ *is a non-empty compact set contained in the disc* $|\zeta| \leq \|x\|$.

To prove this, one sets that the geometric series $\sum x^n$ converges for $\|x\| <$ 1. Since calculations already encountered in Chap. II show that

$$(1 - x)(1 + x + \ldots) = 1\,,$$

it follows that $1 - x$ is invertible for $\|x\| < 1$ and that

$$(1 - x)^{-1} = 1 + x + x^2 + \ldots.$$

---

$\lambda \in k$ and all $x \in A$. $\lambda 1$ is usually shortened to $\lambda$. Examples: the algebra $M_n(k)$ of $k$-valued functions on a given set, etc. Finite-dimensional algebras over $\mathbb{Q}$ have long been studied using methods totally unrelated to those we are interested in here, and that are far more difficult; see André Weil, *Basic Number Theory* (Springer, 1967).

[80] An element $x$ of a ring is invertible if $yx = 1$ and $xz = 1$ have solutions. Then $y = z = x^{-1}$.

For arbitrary $x$ and $\zeta \neq 0$,

$$\zeta - x = \zeta(1 - x/\zeta),$$

so that left hand side is invertible for $|\zeta| > \|x\|$, which shows that the spectrum of $x$ is contained in the disc of radius $\|x\|$. If $\zeta - x = y$ is invertible and if $h \in \mathbb{C}$, then

$$y + h = y\left(1 + hy^{-1}\right)$$

is invertible for sufficiently small $|h|$, which shows that $\zeta' - x$ is invertible for $\zeta'$ sufficiently near $\zeta$. The complement of the spectrum is thus open, and the spectrum is compact.

The function $f(\zeta) = (\zeta - x)^{-1}$ is defined on the open set $\mathbb{C} - Sp(x)$. It is analytic on it since standard calculations show that, like in $\mathbb{C}$, $f(\zeta + h)$ is a power series with coefficients in $\mathbf{A}$ for the variable $h \in \mathbb{C}$, when $|h|$ is sufficiently small. For $|\zeta| > \|x\|$,

$$(\zeta - x)^{-1} = \zeta^{-1}(1 - x/\zeta)^{-1} = \zeta^{-1} \sum x^n/\zeta^n, \tag{20.2}$$

from which it follows that $f(\zeta)$ tends to 0 at infinity.

If the spectrum of $x$ was empty, then $f$ would be an analytic function on all of $\mathbb{C}$, tending to 0 at infinity. Liouville's theorem would then show that $f = 0$, which is absurd, qed.

This supposes a generalization of Liouville's theorem to functions with values in a Banach space $\mathcal{H}$. In fact, the entire theory of analytic or holomorphic functions (and much more: derivatives and differentials of functions of real variables, integrals, Fourier series and integrals, etc.) is valid in this framework, the proofs being almost the same as in the standard case. For example, an analytic function is holomorphic, i.e. has has a derivative

$$f'(\zeta) = \lim \left[f(\zeta + h) - f(\zeta)\right]/h,$$

and actually $f(\zeta + h) = \sum f_n(\zeta)h^{[n]}$ implies that $f'(\zeta) = \sum f_n(\zeta)h^{[n-1]}$, with coefficients $f_n(\zeta) \in \mathcal{H}$. The path integral of a holomorphic function with values in a Banach space can be defined as in Chap. VIII because, since n° 4 of the present chapter, we know how to integrate vector-valued continuous functions. Invariance under homotopy is not a problem, neither is Cauchy's integral formula for a circle [Chap. VIII, n° 4, (iii)] or a more complicated contour. It follows that a holomorphic function $f$ is represented in a disc centered at some $a$ where it is defined by its power series at $a \in \mathbb{C}$, hence in all of $\mathbb{C}$ if $f$ is entire. Then Liouville's theorem follows as in Chap. VII, n° 18 by taking upper bounds of the coefficients of the power series using Cauchy's formula for a circle.

**Lemma 2.** *If every non-trivial element of a complete normed algebra* $\mathbf{A}$ *is invertible, then* $\mathbf{A}$ *has dimension* 1.

Indeed, for all $x \in \mathbf{A}$, there exists $\zeta \in \mathbb{C}$ such that $x - \zeta$ is not invertible, whence $x - \zeta = 0$.

In a ring $\mathbf{A}$ with unit element, a *left ideal* is an additive subgroup $I$ of $\mathbf{A}$ for which

$$x \in I \quad \& \quad y \in \mathbf{A} \Longrightarrow yx \in \mathbf{A}\,.$$

Right and two-sided ideals can be similarly defined. If $\mathbf{A}$ is commutative, one talks of ideals. For example, in $\mathbb{Z}$, any ideal is the set of multiples $n\mathbb{Z}$ of a given integer $n$, namely the smallest $n > 0$ such that $n \in I$. In $\mathcal{L}(\mathcal{H})$, where $\mathcal{H}$ is a vector space, the operators with a given $x \in \mathcal{H}$ as a zero form a left ideal. In the ring of continuous functions on a space $X$, the functions with a given point as a zero form an ideal. Etc.

In a commutative ring $\mathbf{A}$, the set $\mathbf{A}x$ of multiples of a given $x \in \mathbf{A}$ is an ideal. It is equal to $\mathbf{A}$ if and only if $x$ is invertible. A *commutative* ring $\mathbf{A}$ is, therefore, a field if and only if its only ideals are $\{0\}$ and $\mathbf{A}$.

A left ideal $I \subset \mathbf{A}$ is said to be *maximal* if $I \neq \mathbf{A}$ and $I$ is not strictly contained in any other left ideal apart from $\mathbf{A}$. In $\mathbb{Z}$, this means that $I = p\mathbb{Z}$ for some *prime* $p$.

**Lemma 3 (Gelfand-Mazur).** *Let $I$ be a maximal ideal of a* commutative *complete normed space* $\mathbf{A}$. *Then* $\dim(\mathbf{A}/I) = 1$.

Indeed, the quotient of a commutative ring by an ideal $I$ can be considered to be a commutative ring by observing that, for $x, y \in \mathbf{A}$, the coset of $xy \bmod I$ only depends on those of $I$: if $a, b \in I$, then indeed

$$(x + a)(y + b) = xy + ay + xb + ab \in xy + I$$

because of commutativity. The inverse images under the ring homomorphism $p : \mathbf{A} \longrightarrow \mathbf{A}/I$ of the ideals of $\mathbf{A}/I$ are obviously the ideals of $\mathbf{A}$ containing $I$. As a result, an ideal $I \neq \mathbf{A}$ is maximal if and only if $\mathbf{A}/I$ is a *field.*

But if $\mathbf{A}$ is a complete normed space, the elements $x$ such that $\|x - 1\| < 1$ are invertible. Hence any non-trivial ideal has trivial intersection with this open ball, so that the closure of a non-trivial ideal is again a non-trivial ideal. Any maximal ideal is, therefore, closed.

But then the quotient $\mathbf{A}/I$ is also a complete normed space with respect to the norm

$$\|p(x)\| = \inf_{p(y)=p(x)} \|y\| \tag{20.3}$$

which can be more generally defined on the quotient of Banach spaces by closed vector subspaces. The outcome is again a Banach space since any Cauchy sequence in the quotient is the image under $p$ of a Cauchy sequence in the original space. Relation (1) is also be readily verified to hold in $\mathbf{A}/I$, which thus becomes a commutative complete normed algebra without any non-trivial ideals. All its non-zero elements are, therefore, invertible. The result now follows from lemma 2 applied to this algebra.

The question of the existence of maximal ideals can be radically solved:

**Krull's Theorem**. *Every left ideal $I \neq \mathbf{A}$ in a ring $\mathbf{A}$ with unit element is contained in at least one maximal left ideal.*

This is natural more difficult than proving the existence of prime divisors in $\mathbb{Z}$. If $\mathbf{A}$ is a finite-dimensional algebra over a field, the result follows immediately from a dimension argument. This is also the case if $\mathbf{A}$ is a Noetherian commutative ring (i.e. in which every increasing chain of ideals is stationary), but such objects are not encountered in functional analysis. In the general case, one is forced to use methods based on transfinite induction.[81] This is also the case when, for example, proving the Hahn-Banach theorem. Krull's theorem is extraordinarily simple, as general as can be and above all is useful outside functional analysis. It is therefore better to admit it rather than inventing ad hoc arguments for separable normed algebras as Dieudonné does (15.3.4.1).

**Theorem 32**. *Let $I$ be a maximal ideal of a commutative complete normed space $\mathbf{A}$. Then there is a linear functional $\chi$ on $\mathbf{A}$ such that $\chi(1) = 1$,*

$$\chi(xy) = \chi(x)\chi(y) \quad \text{for } x, y \in \mathbf{A} \tag{20.4}$$

*and for which $I$ is defined by the relation $\chi(x) = 0$. For every solution of (4),*

$$|\chi(x)| \leq \|x\| \quad \text{for all } x \in \mathbf{A}\,. \tag{20.5}$$

Since $\mathbf{A}/I$ has dimension 1, for all $x \in \mathbf{A}$ there is a unique scalar $\chi(x)$ such that

$$x - \chi(x).1 \in I\,.$$

The multiplicativity formula follows trivially. So does the characterization of $x \in I$. Conversely, any non-trivial solution of (3) clearly defines an ideal $I \neq \mathbf{A}$. As $x - \chi(x).1 \in I$ cannot be invertible, $\chi(x)$ belongs to the spectrum of $x$, which gives $|\chi(x)| \leq \|x\|$ and the continuity of $\chi$.

---

[81] The universal tool to prove this type of result is Zorn's theorem, which we now state. Consider a set $E$ equipped with an order relation $x \leq y$. Suppose that every *totally* ordered subset $F$ (i.e. $x \leq y$ or $y \leq x$ always holds in it) of $E$ has a supremum in $E$: there exists $M \in E$ such that, for $x \in E$,

$$x \geq y \quad \text{for all } y \in F \Leftrightarrow x \geq M\,.$$

Then $E$ has at least one maximal element $x$, i.e. such that $y \geq x$ implies that $y = x$. Krull's theorem follows by taking $E$ to be the set of left ideals $J \supset I$ distinct from $\mathbf{A}$, ordered by inclusion. For a *totally* ordered subset $F$ of $E$, the union of $J \in F$, distinct from $\mathbf{A}$ as they do not contain 1, is again an ideal and is obviously the supremum of $F$ in $E$.

Non-trivial solutions of (4) are called the *characters* of $\mathbf{A}$. The *spectrum* of $\mathbf{A}$, i.e their set, will be written $\widehat{\mathbf{A}}$. To each, $x \in \mathbf{A}$ associate the function $\widehat{x}$ on $\widehat{\mathbf{A}}$, the *Gelfand transform* of $x$, given by

$$\widehat{x}(\chi) = \chi(x)\,. \tag{20.6}$$

The map $x \mapsto \widehat{x}$ is, therefore, linear and satisfies

$$\widehat{(xy)} = \widehat{x}\widehat{y}\,, \quad \|\widehat{x}\| \leq \|x\|\,,$$

where $\|\widehat{x}\|$ is the uniform norm on $\widehat{\mathbf{A}}$. Furthermore, *the spectrum of $x$ is the set of values of the function $\widehat{x}$.* Indeed, $x \in \mathbf{A}$ is not invertible if and only if the ideal $\mathbf{A}x$ of multiples of $x$ is distinct from $\mathbf{A}$, hence is contained in a maximal ideal. So $\zeta$ belongs to the spectrum of $x$ if and only if there exists $\chi$ such that $\chi(x) = \zeta$, proving the result. In conclusion, *the function $f(\zeta) = (\zeta - x)^{-1}$ is defined and holomorphic on $|\zeta| > \|\widehat{x}\|$.*

The inequality $\|\widehat{x}\| \leq \|x\|$ can be made more precise:

**Lemma 4.** *For all $x \in \mathbf{A}$,*

$$\|\widehat{x}\| = \lim \|x^n\|^{1/n}\,. \tag{20.7}$$

For all $n$ and character $\chi$, $|\chi(x)^n| = |\chi(x^n)| \leq \|x^n\|$. Thus $|\chi(x)| \leq \|x^n\|^{1/n}$ for all $\chi$ and so

$$\|x^n\|^{1/n} \geq \|\widehat{x}\| \quad \text{for all } n\,. \tag{20.8}$$

On the other hand, the function $f(\zeta) = \zeta(\zeta - x)^{-1} = (1 - x/\zeta)^{-1}$ is defined and holomorphic for $|\zeta| \geq \|\widehat{x}\|$, hence has a Laurent series expansion in this open subset. However, $f(\zeta) = \sum x^n/\zeta^n$ for $|\zeta| > \|x\|$. This series, therefore, converges for $|\zeta| > \|\widehat{x}\|$ (uniqueness of the Laurent series). For $|\zeta| = \|\widehat{x}\| + r$ with $r > 0$, its terms tend to 0, and so $\|x^n\| \leq |\zeta|^n$. Hence $\|x^n\|^{1/n} \leq \|\widehat{x}\| + r$ for large $n$. This together with (8) proves (7), including the existence of the limit.

There is a natural topology on $\widehat{A}$, which is obtained by doing the bare minimum to make the functions $\widehat{x}$ continuous. This means that (a) for all $\chi_0 \in \widehat{A}$, $x \in A$ and $r > 0$, the inequality

$$|\widehat{x}(\chi) - \widehat{x}(\chi_0)| < r\,,$$

i.e.

$$|\chi(x) - \chi_0(x)| < r\,, \tag{20.9}$$

must define an *open* subset of $\widehat{\mathbf{A}}$, (b) a subset of $\widehat{\mathbf{A}}$ is open if and only if it is the union of sets defined by finitely many inequalities of the form

$$|\chi\,(x_i) - \chi_i\,(x_i)| < r_i\,.$$

As stated above, this is the *weak topology*, which can be defined on the dual of any topological vector space.[82]

**Lemma 5**. *Let $\mathcal{H}$ be a Banach space, $\mathcal{H}'$ the space of continuous linear functionals on $\mathcal{H}$ and $B'$ the unit ball $\|f\| \leq 1$ of $\mathcal{H}'$. Then $B'$ is compact with respect to the weak topology.*

The proof is easy provided any (finite or infinite) Cartesian product of compact spaces is admitted to be compact, which is another application of Zorn's theorem mentioned above in a footnote. This being admitted, let $D$ be the disc $|z| \leq 1$ of $\mathbb{C}$ and $B$ the unit ball $\|x\| \leq 1$ of $\mathcal{H}$. To each $x \in B$, let us associate a copy $D_x$ of $D$ and let $\Omega$ the Cartesian product of the sets $D_x$: it is therefore the set of families $\zeta = (\zeta_x)$ of elements of $D$, equipped with the minimum topology making the maps $\zeta \mapsto \zeta_x$ from $\Omega$ to $D$ continuous. As $D$ is compact, so is $\Omega$.[83] This done, let us associate to each $f \in B'$ the element $\zeta = j(f)$ of $\Omega$ given by $\zeta_x = f(x)$ for all $x \in B$. The map $j$ from $B'$ to $\Omega$ is obviously injective, and is a homeomorphism from $B$ onto its image since in both cases, we did the bare minimum to make the maps $f \mapsto f(x)$ and $\zeta \mapsto \zeta_x$ continuous. Hence the proof reduces to showing that $j(B')$ is closed in $\Omega$, i.e. that if a linear functional $f \in B'$ varies in such a way that $\lim f(x) = g(x)$ exists for all $x \in B$, then $g$ is the restriction of a continuous linear functional $g \in B'$ to $B$. This is clear since the limit of a sum is the sum of the limits and since taking limits preserves the inequality $|f(x)| \leq \|x\|$.

If we come back to the normed algebra $\mathbf{A}$, the space $\widehat{\mathbf{A}}$ is obviously contained in the unity ball of the topological dual $\mathbf{A}'$ of $\mathbf{A}$. It is closed with respect to the weak topology since, by definition of the weak topology, for given $x$ and $y$, both sides of the equality $\chi(xy) = \chi(x)\chi(y)$ are continuous functions of $\chi$. Taking limits also preserves equality $\chi(1) = 1$. In conclusion, *the space $\widehat{\mathbf{A}}$ is compact.*

[82] Even more generally, taking a set $X$, a family $(F_i)$ of topological spaces and a family $(f_i)$ of maps from $X$ to $F_i$, there are topologies on $X$ with respect to which the maps $f_i$ are continuous. With respect to such a topology, the sets $f_i^{-1}(U_i)$, as well the intersection of *finitely* many such sets, must be open for all $i$ and all open subsets $U_i$ of $F_i$. The simplest is then to stipulate that a subset of $X$ is open if and only it is the union of such intersections. Checking the axioms reduces to the calculation rule on sets of Chap. I. With respect to this topology, a varying $x \in X$ converges to a limit $a \in X$ if and only if $\lim f_i(x) = f_i(a)$ for all $i$. For example, if $X$ is a set of linear functionals on a real or complex vector space $\mathcal{H}$, the weak topology on $X$ is obtained by applying the previous procedure to the functions $f \mapsto f(x)$, $x \in \mathcal{H}$. If $X$ is the Cartesian product of a family of topological spaces $X_i$, the product topology on $X$ is obtained by choosing the projections $x \mapsto x_i$, onto the factors $X_i$.

[83] If $\mathcal{H}$ is separable and if $(x_n)$ is a sequence everywhere dense in the unit ball $B$ of $\mathcal{H}$, then the weak topology on $B'$ (but not on $\mathcal{H}'$) can be defined using the functions $f \mapsto f(x_n)$, with values in $D$. This defines a map $f \mapsto (f(x_n))$ from $B'$ to the Cartesian product $D^{\mathbb{N}}$, whose compactness has been proved without applying Zorn's theorem in n° 11, (i) about corollary 2.

## 21 – A Characterization of Algebras of Continuous Functions

Let $X$ be a compact space and $\mathbf{A} = L(X)$ the normed algebra whose elements are the continuous functions on $X$, equipped with the norm

$$\|f\| = \sup |f(x)| \tag{21.1}$$

of uniform convergence. If

$$f^*(x) = \overline{f(x)}\,, \tag{21.2}$$

denotes the conjugate function of a function $f \in \mathbf{A}$, relations (19.5), (19.6) and (19.7) obviously hold. I will use the terminology *GN algebra* (following Gelfand and Neumark, 1943) rather than "$C^*$-algebra", like many authors do, for a complete normed algebra $\mathbf{A}$ equipped with an *involution* $x \mapsto x^*$ satisfying relations:

$$(\lambda x)^* = \overline{\lambda} x^*\,, \quad (xy)^* = y^* x^*\,, \quad (x^*)^* = x\,, \tag{21.3}$$

$$\|x^*\| = \|x\|\,, \|x^* x\| = \|x\|^2\,. \tag{21.3'}$$

The Gelfand-Naimark theorem affirms that, for a commutative GN algebra, the map $x \mapsto \widehat{x}$ from $\mathbf{A}$ to the algebra $L(\widehat{\mathbf{A}})$ of continuous functions on the spectrum of $\mathbf{A}$, in other words the "Gelfand transform" (20.6), *is an isomorphism no matter how you look at it.* This means that it is a ring homomorphism – obvious –, that it transforms the given norm $\|x\|$ into the norm (1) of the function $\widehat{x}$, that it transforms the given involution $x \mapsto x^*$ into the involution (2), and finally that it is bijective.

The proof is simple but very ingenious. Its authors were quite obviously inspired by what was already known about operators on a Hilbert space, and in particular by lemmas 4 and 5 of n° 19. In what follows, we shall set $X = \widehat{\mathbf{A}}$.

We start by observing that, if $x \in \mathbf{A}$ is invertible, so is $x^*$. Since $(\zeta - x)^* = \overline{\zeta} - x^*$, *the spectrum of $x^*$ is the image of the spectrum of $x$ under $\zeta \mapsto \overline{\zeta}$.*

Secondly, *the spectrum of every $x$ such that $x = x^*$ is real.* As in the case of a Hermitian operator, it suffices to show that $x - i$ is invertible. Suppose this is not true. As $x - i = (x - \zeta i) - i(\zeta - 1)$, $i(\zeta - 1) \in Sp(x - \zeta i)$ for all $\zeta \in \mathbb{C}$. Hence $|\zeta - 1| \leq \|x - \zeta i\|$, and so, for *any* real $\zeta$,

$$(\zeta - 1)^2 \leq \|x - \zeta i\|^2 = \|(x - \zeta i)^*(x - \zeta i)\| = \left\|x^2 + \zeta^2\right\| \leq \|x\|^2 + \zeta^2\,,$$

which contradicts the first order binomial theory.

Since any $x \in \mathbf{A}$ can be written $x = x' + ix''$ with Hermitian $x' = \frac{1}{2}(x + x^*)$ and $x'' = (x - x^*)/2i$, and so $x^* = x' - ix''$, more generally,

$$\chi(x^*) = \overline{\chi(x)} \tag{21.4}$$

for any character of $\mathbf{A}$.

The image of $\mathbf{A}$ in $L(X)$ under $x \mapsto \widehat{x}$ is, therefore, stable under conjugation. This image is a subalgebra of $L(X)$. It separates the points of $X$ since the relation $\chi(x) = \chi'(x)$ is satisfied by all $x$ only if $\chi = \chi'$. The Stone-Weierstrass theorem then shows that $x \mapsto \widehat{x}$ maps $\mathbf{A}$ onto an *everywhere dense* subspace of $L(X)$.

To show that the image of $\mathbf{A}$ is equal to $L(X)$, it then suffices to show that it is closed, and for this purpose to use the last relation

$$\|x\| = \|\widehat{x}\|$$

which needs to be proved. Since, by assumption $\|x^*x\| = \|x\|^2$ and, by (4), $|\chi(x^*x)| = |\chi(x)|^2$, it suffices to prove it for $x^*x$, i.e. for Hermitian $x$. However, we know (n° 21, lemma 4) that $\|\widehat{x}\| = \lim \|x^n\|^{1/n}$. Since $x^n$ is Hermitian for all $n$, by (4),

$$\left\|x^2\right\| = \|x\|^2, \quad \left\|x^4\right\| = \left\|x^2\right\|^2 = \|x\|^4$$

etc. So for $n = 2^p$, $\|x^n\|^{1/n} = \|x\|$, whence the theorem:

**Theorem 33.** *Let $\mathbf{A}$ be a commutative GN algebra. Then the Gelfand transform $x \mapsto \widehat{x}$, defined by*

$$\widehat{x}(\chi) = \chi(x),$$

*is an isomorphism from $\mathbf{A}$ onto the GN algebra of continuous functions on the spectrum $\widehat{\mathbf{A}}$ of $\mathbf{A}$.*

A corollary of this result is that, if $f(\zeta)$ is a complex-valued function defined and *continuous* (but not necessarily analytic!) on the spectrum of some $x \in \mathbf{A}$, then there is a unique $y \in \mathbf{A}$ such that

$$\widehat{y}(\chi) = f\left[\widehat{x}(\chi)\right] \quad \text{for all } \chi :$$

it suffices to check that the right hand side is a continuous functions on $X = \widehat{\mathbf{A}}$. We set $y = f(x)$. The spectrum of $y$ is the image under $f$ of the spectrum of $x$, and $\|y\| = \|f\|$, where this is the uniform norm on $Sp(x)$. If $f = g + h$ or $gh$, then $f(x) = g(x) + g(h)$ or $g(x)h(x)$. Composing $f$ with a continuous function $g$ on the image of $Sp(x)$ under $f$ gives $g \circ f(x) = g[f(x)]$, etc.

Let us for example consider the algebra $\mathbf{A}$ of bounded continuous complex functions on a topological space $E$, with the obvious norm and involution. Every $x \in E$ defines a character $\chi_x : f \mapsto f(x)$ of $\mathbf{A}$ which, by definition, is a continuous function of $x$ with respect to the topology of $\widehat{\mathbf{A}}$. This gives a continuous map from $E$ to the *Čech compactification* $\widehat{\mathbf{A}}$ of $E$. It is a compact space which in general is far too gigantic to be of any interest other than theoretical. In particular, it is not metrizable, even for $E = \mathbb{N}$.

But if $E$ is *compact,* in which case $B(E) = L(E)$, the embedding of $E$ into $X = \widehat{\mathbf{A}}$ is a *homeomorphism.* As it is continuous, it suffices to show that it is bijective. Injectivity is equivalent to saying that, for given $x \neq y$, there exists $f \in L(E)$ such that $f(x) \neq f(y)$. This is obvious. Surjectivity is equivalent to saying that, for every character $\chi$ of $L(E)$, there exists $x$ such that $\chi(f) = f(x)$ for all $f \in L(E)$. However, let $I$ be the ideal $\chi(f) = 0$. If there exists $x$ such that $f \in I$, the linear functionals $f \mapsto \chi(f)$ and $f \mapsto f(x)$ have the same kernel, hence are proportional and in fact identical since these are characters, proving the result. So it suffices to show that all the functions $f \in I$ vanish at some point of $X$. Otherwise, for all $x \in X$, there exists $f \in I$ which does not vanish at $x$ and hence in the neighbourhood of $x$. By BL, we get finitely many $f_i \in I$ such that the open sets $\{f_i(x) \neq 0\}$ cover $E$. The function $g = \sum f_i^* f_i = \sum |f_i|^2$ is again in $I$, it is $> 0$ everywhere, and so invertible in $\mathbf{A}$, a contradiction since $I \neq \mathbf{A}$.

*Exercise 1.* Let $E$ be a locally compact space and $\mathbf{A}$ the algebra of continuous functions on $E$ which tend to a finite limit at infinity. Equip $\mathbf{A}$ with the obvious norm and involution. Using the same method, show that $\widehat{\mathbf{A}}$ is the Alexandrov compactification $E \cup \{\infty\}$ of $E$.

## 22 – Spectral Decompositions

(i) *The GN algebra of a normal operator.* Let $\mathcal{H}$ be a Hilbert space and $\mathbf{A} \subset \mathcal{L}(\mathcal{H})$ a GN algebra, i.e. satisfying the following conditions: $\mathbf{A}$ is self-adjoint, i.e. $T \in \mathbf{A}$ implies $T^* \in \mathbf{A}$, $\mathbf{A}$ is closed with respect to the topology defined by the norm of the operators. Let us suppose that $\mathbf{A}$ is commutative. The elements of $\mathbf{A}$ then satisfy $N^*N = NN^*$. This property characterizes *normal operators*, which were studied by von Neumann around 1927 in the context of Hilbert spaces. In finite dimensional, they are diagonalizable, for if $\zeta$ is an eigenvalue of $N$, the corresponding eigenspace $\mathcal{H}(\zeta)$ is invariant under $N$ and $N^*$ since $N$ and $N^*$ commute, hence so is its orthogonal complement (n° 19, lemma 3), and diagonalization follows by induction on the dimension of $\mathcal{H}$.

Starting from a normal operator $N$ and considering operators that are limits (in norm) of polynomials

$$P(N, N^*) = \sum a_{pq} N^p N^{*q}, \tag{22.1}$$

one gets an algebra $\mathbf{A}(N) = \mathbf{A}$ of the former type, hence isomorphic to the algebra of continuous functions on the spectrum of $\mathbf{A}$. It can be easily determined:

**Lemma.** *Let $N$ be a normal operator on a Hilbert space $\mathcal{H}$. Then the map $\chi \mapsto \chi(N)$ is a homeomorphism from the spectrum of $\mathbf{A}(N)$ onto the spectrum of $N$ in $\mathcal{L}(\mathcal{H})$.*

For a character $\chi$ of $\mathbf{A} = \mathbf{A}(N)$, clearly

$$\chi\left[P(N, N^*)\right] = P\left[\chi(N), \overline{\chi(N)}\right] \tag{22.2}$$

since $\chi(T^*) = \overline{\chi(T)}$ for all $T \in \mathbf{A}$. As the linear functional $\chi$ is continuous, it is fully determined by the number $\zeta = \chi(N)$. This gives a continuous injective map (hence a homeomorphism) $\chi \mapsto \chi(N)$ from the compact space $\widehat{\mathbf{A}}$ onto a compact subspace of $\mathbb{C}$, namely the spectrum $S$ of $N$ in $\mathbf{A}$ as was seen in n° 20 for every normed algebra. Thus $\mathbf{A}(N)$ is isomorphic to the algebra of continuous functions on $S$ by the GN theorem. Let $\sigma$ be the spectrum of $N$ in $\mathcal{L}(\mathcal{H})$. Clearly $\sigma \subset S$. To prove the inverse inclusion, it suffices to show that if $N - \zeta$ is invertible as an operator on $\mathcal{H}$, then its inverse is in $\mathbf{A}$. Replacing $N$ by $N - \zeta$, the proof then reduces to showing that *if $N$ is invertible in $\mathcal{L}(\mathcal{H})$, then $N^{-1}$ is the limit in norm of polynomials of $N$ and $N^*$*. This property is special to normal operators.

To prove this, we consider the set $\mathbf{B} = \mathbf{A}(N, N^{-1})$ of limits in norm of polynomials of $N, N^{-1}$ and of their adjoints. It is again a GN algebra. As above, the map $\chi \mapsto \chi(N)$ from the spectrum $\widehat{\mathbf{B}}$ of $\mathbf{B}$ to $\mathbb{C}$ is clearly injective and continuous, and so is a homeomorphism from $\widehat{\mathbf{B}}$ onto a compact subset $K$ of $\mathbb{C}^*$, and the map $f \mapsto f(N)$ is an isomorphism from the algebra $L(K)$ onto $\mathbf{B}$. The elements $N, N^{-1}$ and $N^*$ of $\mathbf{B}$ correspond to the functions $\zeta, 1/\zeta$ and $\overline{\zeta}$ on $K$. However, the function $1/\zeta$ is the uniform limit on $K$ of polynomials in $\zeta$and $\overline{\zeta}$ (Stone-Weierstrass). As $f \mapsto f(N)$ preserves norms, $N^{-1}$ is the limit of polynomials of $N$ and $N^*$, qed.

In fact this proves that $\mathbf{A}(N, N^{-1}) = \mathbf{A}(N)$ if $N$ is an invertible normal operator in $L(\mathcal{H})$. An immediate consequence of the previous lemma is that if $N$ belongs to a closed self-adjoint subalgebra $\mathbf{A}$ of $\mathcal{L}(\mathcal{H})$, commutative or not, then the spectrum of $N$ in $\mathbf{A}$ is the same as its spectrum in $\mathcal{L}(\mathcal{H})$, in other words does not depend on $\mathbf{A}$, since if, for example, $N$ is invertible as an operator, then $N^{-1} \in \mathbf{A}(N) \subset \mathbf{A}$. Hence it is possible to talk of the spectrum of a normal operator without mentioning the GN algebra considered.

We saw in n° 19, lemma 5 that, for a Hermitian operator $H$, the smallest compact interval of $\mathbb{R}$ containing the spectrum of $H$ is $[m_H, M_H]$. If $H$ is positive, i.e. if $m_H \geq 0$, the function $f(\zeta) = \zeta^{1/2}$ is defined and continuous on this interval. Setting $H^{1/2} = f(H)$ gives a Hermitian operator such that

$$(22.3) \qquad (H^{1/2})^2 = H\,, \quad H^{1/2} \geq 0\,.$$

These properties fully determine $H^{1/2}$ among all Hermitian operators in $\mathcal{H}$. Indeed, let $H'$ be a positive Hermitian operator such that $(H')^2 = H$. The algebra $\mathbf{A}(H')$ is isomorphic to $L(S')$, where the spectrum $S'$ of $H'$ is contained in $\mathbb{R}_+$ (n° 19, lemma 5). $\mathbf{A}(H')$ contains $H$, hence also $H^{1/2} \in \mathbf{A}(H) \subset \mathbf{A}(H')$. The continuous functions corresponding to $H$ and $H'$ in $L(S')$ are $\zeta \mapsto \zeta^2$ and $\zeta \mapsto \zeta$. The one corresponding to $H^{1/2}$ is positive-valued since its values are the elements of the spectrum of $H^{1/2}$. It is, therefore, *the* positive square root of the function $\zeta \mapsto \zeta^2$ on $S'$, hence the function $\zeta \mapsto \zeta$; thus $H^{1/2} = H'$, qed.

*Exercise 1.* Let $H$ and $H'$ be two commuting positive Hermitian operators. Show that their square roots commute. Deduce that $HH'$ is positive Hermitian.

*Exercise 2.* Let $B$ be the set of Hermitian operators $H$ such that $0 \leq H \leq 1$. (a) Show that, for all $r > 0$, there is a real polynomial $p(\zeta)$ such that $\|H^{1/2} - p(H)\| < r$ for all $H \in B$. (b) Let $H_n \in B$ be a sequence of operators strongly converging to some $H \in B$ [for example, an increasing sequence: n° 19, (v), lemma 8]. Show that $H_n^{1/2}$ strongly converges to $H^{1/2}$ [first show that, for any polynomial $p$, the map $H \mapsto p(H)$ is strongly continuous on $B$].

*Exercise 3.* For a positive Hermitian operator $H$, define a positive Hermitian operator $H^s$, where $s \in \mathbb{R}$ and show that $H^s H^t = H^{s+t}$. Show that the function $s \mapsto H^s$ extends to a holomorphic function on $\mathbb{C} - \mathbb{R}_-$ and that the set of $H^{it}$, $t \in \mathbb{R}$, is a group of unitary operators.

(ii) *Spectral measure of an operator algebra.* Let us now consider the general case of a commutative GM algebra $\mathbf{A}$ in $\mathcal{L}(\mathcal{H})$ and denote its spectrum by $X = \widehat{\mathbf{A}}$; it is a compact space. Conversely, the Gelfand transform $T \mapsto \widehat{T}$ being bijective, one can associated the operator $T = M(f) \in \mathbf{A}$ defined by the condition that

$$\widehat{T} = f$$

to each $f \in L(X)$. Relations

$$M(f+g) = M(f) + M(g)\,, \quad M(\overline{f}) = M(f)^*\,, \qquad M(fg) = M(f)M(g) \tag{22.4}$$

as well as

$$\|M(f)\| = \|f\|$$

trivially hold. The map $f \mapsto M(f)$ from $L(X)$ to $\mathbf{A}$ will be called a *spectral measure* of $\mathbf{A}$. It is linear and transforms every real (resp. positive) function into a Hermitian (resp. positive Hermitian) operator. The characteristic of these "measures" is the relation $M(fg) = M(f)M(g)$: the only usual numerically-valued measures satisfying it are the Dirac measures.

The isomorphism $f \mapsto M(f)$ transforms every continuous linear functional on $\mathbf{A}$ into a continuous linear functional on $L(X)$, i.e. into a measure on $X$. This the case of the map

$$T \longmapsto (Tx|y)$$

for all $x, y \in \mathcal{H}$. This gives a unique measure $\mu_{x,y}$ on $X$ such that

$$(M(f)x|y) = \mu_{x,y}(f) = \int f(\chi) d\mu_{x,y}(\chi) \tag{22.5}$$

for all $f \in L(X)$ or, using a different notation,

$$(Tx|y) = \int \widehat{T}(\chi) d\mu_{x,y}(\chi)$$

for all $T \in \mathbf{A}$ and $x, y \in \mathcal{H}$. These measures satisfy fairly obvious properties:

(MS 1) the map $x \mapsto \mu_{x,y}$ is linear;
(MS 2) the measures $\mu_{y,x}$ are $\mu_{x,y}$ are imaginary and conjugate;
(MS 3) the measures $\mu_{x,x}$ are positive and of total weight $\|x\|^2$;
(MS 4) $d\mu_{Tx,y}(\chi) = \widehat{T}(\chi)d\mu_{x,y}(\chi)$ for all $T \in \mathbf{A}$;
(MS 5) An operator $T$ commutes with the elements of $\mathbf{A}$ if and only if

$$\mu_{Tx,y} = \mu_{x,T^*y} \quad \text{for all } x, y \in \mathcal{H}. \tag{22.6}$$

Property (MS 2) follows from the observation that, for real $f \in L(X)$, the operator $M(f)$ is Hermitian, so that $(M(f)x|y) = \overline{(M(f)y|x)}$, which can also be written $\mu_{x,y}(f) = \overline{\mu_{y,x}(f)}$. Hence the result. (MS 3) says that the operators $M(f)$ are positive Hermitian for $f$ positive. So $(M(f)x|x) \geq 0$. As

$$|\mu_{x,y}(f)| = |(M(f)x|y)| \leq \|M(f)\| \,.\|x\|.\|y\|$$

and $\|M(f)\| = \|f\|$, in fact

$$\|\mu_{x,y}\| \leq \|x\|.\|y\|\,.$$

(MS 4) follows from the fact that, for $f, g \in L(X)$,

$$\mu_{M(f)x,y}(g) = (M(g)M(f)x|y) = (M(fg)x|y) = \mu_{x,y}(fg)\,.$$

(MS 5) follows from relations

$$\mu_{Tx,y}(f) = (M(f)Tx|y)\ , \quad (TM(f)x|y) = \big(M(f)x\big|T^*y\big) = \mu_{x,T^*y}(f)\,.$$

(iii) *Integration with respect to a spectral measure.* Instead of starting from a spectral measure associated to a GN algebra, more generally, it is possible to define a spectral measure for any locally compact space $X$ as a map $M : f \mapsto M(f)$ from $L(X)$ to the algebra $\mathcal{L}(\mathcal{H})$ of a Hilbert space $\mathcal{H}$, which must satisfy above conditions (4). By (5), one then gets measures $\mu_{x,y}$, and they clearly satisfy (MS 1) and (MS 2). They also satisfy (MS 4) under the form

$$d\mu_{M(f)x,y}(t) = f(t)d\mu_{x,y}(t)$$

for $f \in L(X)$, and (MS 5) for operators $T$ commuting with all $M(f)$. The fact that $\mu_{x,x} \geq 0$ is obvious. $\|M(f)\| \leq \|f\|$ always holds since, for $|z| > \|f\|$, there exists $g \in L(X)$ such that $(f + z)(g + z^{-1}) = 1$. This relation implies $[M(f)+z][M(g)+z^{-1}] = 1$ and shows that the spectrum of $M(f)$ is contained in the disc of radius $\|f\|$, whence the result. One then readily sees that the measures $\mu_{x,x}$ are bounded and that

$$\|\mu_{x,x}\| = \sup_{\substack{|f|\leq 1 \\ f\in L(X)}} \mu_{x,x}\left(|f|^2\right) = \sup \|M(f)x\|^2 \leq \|x\|^2\,.$$

Relation $\mu_{x,x}(X) = \|x\|^2$, which implies $\mu_{x,y}(X) = (x|y)$, nonetheless requires an additional assumption when $X$ is not compact; it will be given later. The Cauchy-Schwarz inequality

$$|\mu_{x,y}(f)|^2 \leq \mu_{x,x}(f)\mu_{y,y}(f) \tag{22.7}$$

shows that $\mu_{x,y}$ is bounded for all $x$ and $y$ and that

$$\|\mu_{x,y}\| \leq \|x\|.\|y\| .$$

When there is measure, even if it is with values in $\mathcal{L}(\mathcal{H})$, continuous functions are not the only ones that are integrable. A function $\varphi$ on $X$ will be said to be $M$-measurable (resp. $M$-integrable) if it is measurable (resp. integrable) with respect to *all* measures $\mu_{x,x}$, which is the case of Borel functions, if need be bounded to ensure integrability. A set $N \subset X$ will even be said to be $M$-null if $\mu_{x,x}(N) = 0$ for all $x$, and so $\mu_{x,y}(N) = 0$. The map $(x, y) \mapsto \mu_{x,y}$ having all the formal properties of a Hermitian form – except that its values are measures –, the relation

$$4\,(x|y) = (x + y|x + y) - (x - y|x - y) + \ldots$$

indeed shows that the $\mu_{x,y}$ are linear combinations of positive measures $\mu_{z,z}$.

For a $M$-integrable function $\varphi$, the number $\mu_{x,y}(\varphi)$ depends linearly on $x$ by (MS 1) and semi-linearly on $y$ by (MS 2). Moreover, as

$$|\mu_{x,y}(\varphi)| \leq \|\mu_{x,y}\| \cdot \|\varphi\| \leq \|\varphi\|.\|x\|.\|y\| ,$$

there is a unique continuous operator $M(\varphi)$ on $\mathcal{H}$ such that

$$(M(\varphi)x|y) = \mu_{x,y}(\varphi)$$

for all $x$, $y$ (n° 19, corollary of theorem 33). The map $\varphi \mapsto M(\varphi)$ also satisfies relations (4) as we are now going to see.

The first two are obvious by (MS 1) and (MS 2). To prove the third one, for which theorem 25 on the definition of measures by a density is essential, let us begin with an "arbitrary" function $\varphi$ and a continuous function $g$. Applying (M 5),

$$\begin{aligned}(M(\varphi g)x|y) &= \int \varphi(t)g(t).d\mu_{x,y}(t) = \int \varphi(t).g(t)d\mu_{x,y}(t) \\ &= \int \varphi(t)d\mu_{M(g)x,y}(t) = (M(\varphi)M(g)x|y) ,\end{aligned}$$

whence $M(\varphi g) = M(\varphi)M(g)$. Taking adjoints, $M(\varphi)M(g) = M(g)M(\varphi)$ follows since

$$\begin{aligned}\int g d\mu_{M(\varphi)x,y} &= (M(g)M(\varphi)x|y) = (M(\varphi)M(g)x|y) \\ &= \int \varphi d\mu_{M(g)x,y} = \int \varphi.g d\mu_{x,y} \quad \text{by (MS 4)} \\ &= \int \varphi g.d\mu_{x,y} = \int g.\varphi d\mu_{x,y} .\end{aligned}$$

Thus

$$(22.8) \qquad d\mu_{M(\varphi)x,y}(t) = \varphi(t) d\mu_{x,y}(t)\,,$$

which is the analogue of (MS 4) for $\varphi$. Now if $\psi$ is an "arbitrary" function on $X$, then theorem 25 shows that

$$(M(\psi)M(\varphi)x|y) = \mu_{M(\varphi)x,y}(\psi) = \int \psi.\varphi d\mu_{x,y}$$

$$\int \psi\varphi.d\mu_{x,y} = (M(\psi\varphi)x|y)\,.$$

So we finally get $M(\varphi\psi) = M(\varphi)M(\psi)$.

The latter shows that $M(\varphi) = M(1)M(\varphi)$ for all $\varphi$, and (4) that $M(1) = M(1)^* = M(1)^2$. As a consequence, $M(1) = E$ is the projection onto a closed $M(\varphi)$-invariant subspace $\mathcal{H}'$ of $\mathcal{H}$ mapping the orthogonal complement $\mathcal{H}'$ onto 0. Obviously, $\mathcal{H}' = \mathcal{H}$ if $X$ is compact. This is the case of the spectral measure of a GN algebra. $M$ is then said to be *non-degenerate.* As $\mu_{x,x}(X) = \mu_{x,x}(1) = \|M(1)x\|^2$, condition

$$\mu_{x,x}(X) = \|x\|^2$$

says that $M$ is non-degenerate. Otherwise, $\mathcal{H}$ may as well be replaced by $\mathcal{H}'$. Hence in what follows, $M$ will be assumed to be non-degenerate.

Likewise, a trivial calculation using (MS 5) shows that $M(\varphi)$ *commutes with every operator commuting with* $M(f)$, for all $f \in L(X)$, in other words, belongs to the von Neumann algebra generated by all $M(f)$ [n° 19, (vii), density theorem].

Lebesgue's theorems lead to limit properties for operators $M(\varphi)$. First of all,

$$(22.9) \qquad \|M(\varphi)x\|^2 = (M(\varphi)x|M(\varphi)x) = (M(\varphi)^*M(\varphi)x|x) = \\ = \int |\varphi(t)|^2\, d\mu_{x,x}(t)\,.$$

So if a sequence $(\varphi_n)$ converges $M$-almost everywhere to a function $\varphi$, relation

$$\|M(\varphi)x - M(\varphi_n)\,x\|^2 = \int |\varphi(t) - \varphi_n(t)|^2\, d\mu_{x,x}(t)$$

shows that

$$(22.10) \qquad \lim M(\varphi_n)\,x = M(\varphi)x \quad \text{for all } x \in \mathcal{H}$$

provided $(\varphi_n)$ converges in mean to $\varphi$ in all $L^2(\mu_{x,x})$ spaces, for example if the functions $\varphi_n$ are dominated by a bounded Borel function. Theorems on increasing series and sequences can be similarly interpreted.

We next consider a $M$-measurable set $\omega \subset X$ and let $f$ be its characteristic function. Then $f = \overline{f} = f^2$, and so

$$M(f) = M(f)^* = M(f)^2 .$$

As a result, $M(f) = M(\omega)$, traditionally written $E(\omega)$, is an orthogonal projection operator onto a closed subspace $\mathcal{H}(\omega)$ of $\mathcal{H}$. By definition,

$$(22.11) \qquad (M(\omega)x|y) = \int_\omega d\mu_{x,y}(t) = \mu_{x,y}(\omega)$$

and in particular, $M(\omega)$ being a projection,

$$(22.12) \qquad \|M(\omega)x\|^2 = (M(\omega)x|x) = \mu_{x,x}(\omega) .$$

Therefore, as $x \in \mathcal{H}(\omega)$ is equivalent to $\|M(\omega)x\| = \|x\|$, hence to

$$\|x\|^2 = \mu_{x,x}(\omega) ,$$

and as $\mu_{x,x}(X) = \|x\|^2$, one sees that

$$(22.13) \qquad x \in \mathcal{H}(\omega) \Longleftrightarrow \mu_{x,x}(X - \omega) = 0 .$$

The $\mathcal{H}(\omega)$ are the *spectral manifolds* of the measure $M$ (or of the algebra $\mathbf{A}$ when $M$ is associated to $\mathbf{A}$).

Let us for example suppose that $\omega = \{\chi_0\}$ for some $\chi_0 \in X$. Elements $x \in \mathcal{H}(\omega)$ are then characterized by the fact that

$$(22.14) \qquad Tx = \chi_0(T)x \quad \text{for all } T \in \mathbf{A} .$$

Indeed, these $x$ are characterized by $\mu_{x,x}(X - \{\chi_0\}) = 0$ or, if $\varphi$ denotes the characteristic function of $\omega$, by the fact that $\widehat{T}(\chi) - \chi_0(T)\varphi(\chi) = \psi(\chi)$ is zero almost everywhere with respect to $\mu_{x,x}$. As

$$M(\psi) = T - \chi_0(T)M(\varphi) = T - \chi_0(T)M(\omega)$$

and as $\mu_{x,x}(\psi) = \int |\psi(\chi)|^2 d\mu_{x,x}(\chi) = \|M(\psi)x\|^2$, the result follows.

The map $\omega \mapsto M(\omega)$ has properties analogous to those of a measure and others without any parallel. First of all, for any finite or countable family $(\omega_n)$, relation (13) shows that

$$(22.15) \qquad \mathcal{H}\left(\bigcap \omega_n\right) = \bigcap \mathcal{H}(\omega_n) .$$

If the sets $\omega_n$ are pairwise disjoint, the corresponding projections cancel pairwise since $M(\varphi\psi) = M(\varphi)M(\psi)$. So the subspaces $\mathcal{H}(\omega_n)$ are pairwise orthogonal. Moreover, $\mathcal{H}\left(\bigcup \omega_n\right)$ is the Hilbert direct sum of the spaces $\mathcal{H}(\omega_n)$. Indeed, let $\varphi_n$ be the characteristic function of $\omega_n$ and $\varphi = \sum \varphi_n$ that of $\omega = \bigcup \omega_n$. For all $x \in \mathcal{H}$,

$$M(\omega)x = \lim \left[M(\omega_1)x + \ldots + M(\omega_n)x\right]$$

by (10) applied to the sequence $\varphi_1 + \ldots + \varphi_n$. The vectors $x_n = M(\omega_n)x$ are pairwise orthogonal. If $x \in \mathcal{H}(\omega)$, then $M(\omega)x = x$, and so

$$x = \lim (x_1 + \ldots + x_n)$$

and $\|x\|^2 = \sum \|x_n\|^2$. Conversely, taking $x_n \in \mathcal{H}(\omega_n)$ such that $\sum \|x_n\|^2 < +\infty$, the limit $x$ of the partial sums exists and $M(\omega)x$ is the limit of the partial sums of the series $\sum M(\omega)x_n$. Now, $M(\omega)x_n = x_n$ since

$$x_n = M(\varphi_n)\, x_n = M(\varphi\varphi_n)\, x_n = M(\varphi)M(\varphi_n)\, x_n = M(\omega)x_n\,.$$

The projections $M(\omega)$ can be zero; but, for a GN algebra, $M(\omega) \neq 0$ if $\omega$ is a non-empty *open* set. Indeed, there is then a *continuous* function $f$ on $\widehat{\mathbf{A}}$ which is not identically zero, but vanishes outside $\omega$. As it is equal to its product with the characteristic function of $\omega$, $M(f)M(\omega) = M(f)$. Since $f$ is continuous, the operator $M(f) = A$ is in $\mathbf{A}$ and like $f$, is not zero. Hence $M(\omega) \neq 0$.

This shows that, *if* $\mathbf{A}$ *does not only consist of scalar operators,*, i.e. if the spectrum $\widehat{\mathbf{A}}$ of $\mathbf{A}$ has at least two points, then *there exist spectral manifolds other than* $\{0\}$ *and* $\mathcal{H}$: indeed, there are two non-empty disjoint open subsets $\omega$ and $\omega'$ in $\widehat{\mathbf{A}}$, hence the that result since then $\mathcal{H}(\omega)$ and $\mathcal{H}(\omega')$ are non-zero and orthogonal.

An integral may be calculated approximatively using Lebesgue sums. Similarly, let us consider for an arbitrary spectral measure $M$, a $M$-integrable function $\varphi$ and, for given $r > 0$, let us partition $X$ into finitely or countably many $M$-measurable sets $\omega_n$ on each of which $\varphi$ is constant, up to $r$. Let $\varphi_n$ be their characteristic functions. Choosing $t_n \in \omega_n$,

$$\left|\varphi(t) - \sum \varphi(t_n)\, \varphi_n(t)\right| \leq r \quad \text{for all } t$$

and so

$$\left\|M(\varphi) - \sum \varphi(t_n)\, M(\omega_n)\right\| \leq r\,. \tag{22.16}$$

This result justifies or explains the *spectral decomposition formula*

$$M(\varphi) = \int \varphi(t) dM(t)\,, \tag{22.17}$$

which is merely notation. For an algebra $\mathbf{A}$ and $\varphi \in L(\widehat{\mathbf{A}})$, the operator $M(\varphi) = A$ is in $\mathbf{A}$ and $\varphi(\chi) = \widehat{A}(\chi)$. So (17) shows that in particular

$$A = \int \widehat{A}(\chi) dM(\chi) \quad \text{for all } A \in \mathbf{A}\,. \tag{22.17'}$$

The fact that each $A \in \mathbf{A}$ can be approximated by linear combinations of projections $M(\omega)$ shows that *a continuous operator* $T$ *on* $\mathcal{H}$ *commutes with all* $A \in \mathbf{A}$ *if and only if it commutes with the projections* $M(\omega)$, in which case it commutes with $M(\varphi)$ for any $M$-measurable bounded function $\varphi$ on $\widehat{\mathbf{A}}$.

This result has an easy but important consequence, another version of which will be given in n° 23.

**Schur's Lemma**[84] **I.** *Let* $\mathbf{A}$ *be a self-adjoint subalgebra of* $\mathcal{L}(\mathcal{H})$ *containing the unit operator. The following properties are equivalent:*

(i) *the only* $\mathbf{A}$*-invariant* closed *subspaces of* $\mathcal{H}$ *are* $\{0\}$ *and* $\mathcal{H}$*;*

(ii) *the only Hermitian operators commuting with all* $A \in \mathbf{A}$ *are scalars.*

$\mathbf{A}$ *is then everywhere dense in* $\mathcal{L}(\mathcal{H})$ *with respect to the ultrastrong topology. If* $\mathcal{H}$ *is finite-dimensional, then* $\mathbf{A} = \mathcal{L}(\mathcal{H})$.

Since $\mathbf{A}$ is self-adjoint, the orthogonal complement of a closed invariant subspace is again invariant (n° 19, lemma 3), so that the corresponding projection commutes with all $A \in \mathbf{A}$. This shows that (ii) $\Longrightarrow$ (i).

If (i) holds – in which case the algebra $\mathbf{A}$ is said to be *irreducible* –, let $H$ be a Hermitian operator commuting with $\mathbf{A}$. As was seen above, so do its spectral projections [apply arguments to $\mathbf{A}(H)$], the corresponding subspaces are $\mathbf{A}$-invariant, and hence are trivial.

Finally, under the assumption of the Lemma, the von Neumann algebra $\mathbf{A}'$ of operators commuting with $\mathbf{A}$ consisting only of scalars, $\mathbf{A}'' = \mathcal{L}(\mathcal{H})$. Von Neumann's density theorem [n° 19, (vii)] then shows that $\mathbf{A}''$ is the closure of $\mathbf{A}$ with respect to the ultrastrong topology (but obviously not with respect to the topology defined by the norm if $\mathcal{H}$ is infinite-dimensional). If $\mathcal{H}$ is finite-dimensional, so is $\mathcal{L}(\mathcal{H})$, all topologies coincide, and $\mathbf{A} = \mathcal{L}(\mathcal{H})$ since in finite dimensional, every subspace is closed, qed.

Above all, this result is relevant for *unitary representations* of a locally compact group $G$. Such a representation is [n° 15, (ii)] an ordered pair $(\mathcal{H}, U)$ consisting of a Hilbert space $\mathcal{H}$ and of a linear representation $x \mapsto U(x)$ by unitary operators. As $U(x)^* = U(x)^{-1} = U(x^{-1})$, the set $\mathbf{A}$ of linear combinations of $U(x)$ is a self-adjoint algebra, and the $\mathbf{A}$-invariant subspaces (closed or not...) are invariant under operators $U(x)$ and conversely. If the representation $(\mathcal{H}, U)$ is irreducible, the only continuous linear operators in $\mathcal{H}$ commuting with all $U(x)$ are scalars, and conversely. We will return to this point in more detail in § 8.

(iv) *Spectral decomposition of a normal operator.* Let us study the classical case of a normal operator $N$ and of the algebra $\mathbf{A} = \mathbf{A}(N)$ generated by $N$ and $N^*$. The space $\widehat{\mathbf{A}}$ can be identified to the spectrum of $N$ by $\chi \mapsto \chi(N) = \zeta$. All measures, functions, integrals introduced in the general case

[84] The traditional version of Schur's lemma assumes that we are in dimension is finite and can be proved in an elementary manner. A less obvious version, which does not use Hilbert structures, applies to "semisimple" algebras, i.e. those in which every invariant subspace has an invariant complement.

are, therefore, defined on $Sp(N) = X \subset \mathbb{C}$, and $N$ is the operator $M(\varphi)$ for the function $\zeta$. Then relation (17) becomes

$$(22.18) \quad N^p N^{*q} = \int \zeta^p \overline{\zeta^q} dM(\zeta), \quad \text{or} \quad (N^p x | N^q y) = \int \zeta^p \overline{\zeta^q} d\mu_{x,y}(\zeta).$$

This result is due to von Neumann[85] except for notation: what I write $dM(\zeta)$ is usually written $dE(\zeta)$.

If $H$ is Hermitian, its spectrum is real and contained in $[m_H, M_H]$. In the former version of spectral decomposition, everything was reduced to Stieltjes integrals over $\mathbb{R}$ using projections

$$(22.19) \qquad E(\lambda) = M([m_H, \lambda]), \quad \lambda \in [m_H, M_H],$$

and right-continuous and increasing functions

$$(22.20) \qquad \mu_{x,x}(\lambda) = \mu_{x,x}([m_H, \lambda])$$

which define the right Radon measures $\mu_{x,x}$ of the general theory. Formula (17) is traditionally written

$$(22.21) \qquad f(H) = \int_{m_H}^{M_H} f(\lambda) dM(\lambda)$$

for every bounded Borel function $f$ on $[m_H, M_H]$. It was only really used for $f(\lambda) = \lambda^n$. One had to interpret as a Stieltjes integral in which the "measure" of an interval is an orthogonal projective operator on $\mathcal{H}$. For example, the measure of $]\alpha, \beta] \subset [m_H, M_H]$ is $E(\beta) - E(\alpha)$, i.e. $M(]\alpha, \beta])$ in the notation $M(\omega)$ of section (iii), but that of $]\alpha, \beta[$, i.e. $M(]\alpha, \beta])$, is defined by taking the limit of $E(\beta') - E(\alpha)$ as $\beta' < \beta$ tends to $\beta$, etc. These constructions contained many traps since no one, apart from F. Riesz, used the characterization of measures as linear functions. The GN theorem, that was beginning to be glimpsed in the case of the algebra $\mathbf{A}(H)$, was replaced by explicit arguments on operators.

If $U$ is unitary, then $U^* = U^{-1}$, whence $|\chi(U)| = 1$ for every character of $\mathbf{A}(U)$. So the spectrum of $U$ is contained in the unit circle $|\zeta| = 1$; thus one can interpret $M$ as the spectral measure on $\mathbb{T}$ and write the spectral decomposition of $U$ as

$$U^n = \int_{\mathbb{T}} u^n dM(u),$$

which is far better[86] than the formula

[85] Von Neumann considered operators that were not defined everywhere, a far more difficult case than the one considered here. His 1929 articles in Math. Annalen are perfect examples of axiomatic and abstract arguments and can still be read profitably. Those who criticize N. Bourbaki's taste for "modern" mathematics could also target the "father of computer science" ...

[86] Since it can be generalized to unitary representations of all locally compact commutative groups.

$$U^n = \int_0^{2\pi} e^{nit} dE(t)$$

that one still encounters.

## 23 – Self-Adjoint Operators

(i) *Inverse of an injective Hermitian operator.* Besides continuous Hermitian operators, there are non-continuous ones that are not defined everywhere. These are the *self-adjoint* operators; they are very important, in particular in the theory of partial differential equations. They are the subject (von Neumann) of a similar theory where the spectrum is an unbounded interval of $\mathbb{R}$, hence on which the function $\lambda$ is not necessarily integrable with respect to all measures $\mu_{x,x}$. It is easy to understand where they come from even without leaving an "abstract" framework – it suffices to study the inverse, not defined everywhere, of a Hermitian operator, which is what we are going to do –, but it is possible to come up with concrete examples: the operator in $L^2(\mathbb{R})$, which transforms every function $f(t)$ into $tf(t)$ is obviously not defined everywhere, though it is quite reasonable. Taking the Fourier transform one gets $f \mapsto 2\pi i f'$, an operator whose direct definition would be more tricky.

So let us consider a Hermitian, and hence continuous, operator $H$ on $\mathcal{H}$, and set $X = Sp(H)$ to be a compact subset of $\mathbb{R}$. $H$ is invertible if and only if the function $\lambda$ has a *continuous* inverse in $X$, which means that $0 \notin X$ since $X$ is closed. Introducing the measures $\mu_{x,y}$ of (22.21) then gives

$$(H^{-1}x|y) = \int_X \lambda^{-1} d\mu_{x,y}(\lambda)\,, \tag{23.1}$$

$$\|H^{-1}x\|^2 = \int_X \lambda^{-2} d\mu_{x,x}(\lambda) \tag{23.2}$$

for all $x, y \in \mathcal{H}$. But if $0 \in X$, there is no reason why the function $\lambda^{-1}$ should be integrable, much less why it should be square-integrable, with respect to the spectral measures $\mu_{x,y}$ of $H$, so that (1) and (2) need not be make sense.

The difficulty comes from the fact that $H$ may be neither injective, nor surjective.[87] The first one is not serious. The subspace $\text{Ker}(H)$ of $x$ for which $Hx = 0$, i.e. such that $(x|Hy) = 0$ for all $y$, is orthogonal to the subspace $\text{Im}(H)$. The orthogonal complement of $\text{Ker}(H)$, i.e. the closure of $\text{Im}(H)$, is a closed $H$-invariant subspace on which $H$ is injective. It is, therefore, reasonable, to work in this subspace, in other words, to suppose that $H$ is injective, which is what we will do in what follows. The subspace $\mathcal{D} = \text{Im}(H)$ is then everywhere dense in $\mathcal{H}$. By the way, note that, by (22.14), the injectivity of $H$ is equivalent to

$$\mu_{x,x}(\{0\}) = 0 \quad \text{for all } x \in \mathcal{H}\,.$$

[87] Counterexample on $\mathcal{H} = L^2(\mathbb{N})$: the operator defined by $Hf(n) = 0$ if $n = 0$, $Hf(n) = f(n)/n$ if $n \neq 0$.

In these conditions, $H$ has an inverse

$$H^{-1} : \mathcal{D} = \mathrm{Im}(H) \longrightarrow \mathcal{H}$$

given by

$$y = H^{-1}x \Longleftrightarrow x = Hy .$$

This operator, which is not defined everywhere and even less continuous, has very remarkable properties.

**Theorem 34.** *Let $H$ be an* injective *continuous Hermitian operator on a Hilbert space $\mathcal{H}$ and $H^{-1} : \mathcal{D} = \mathrm{Im}(H) \longrightarrow \mathcal{H}$ its inverse.*

(a) *The elements $x \in \mathcal{D}$ are characterized by the relation*

$$\int \lambda^{-2} d\mu_{x,x}(\lambda) < +\infty . \tag{23.3}$$

(b) *The following hold:*

$$\begin{aligned} H^{-1}Hx &= x \quad \text{for all } x \in \mathcal{H} , \\ HH^{-1}x &= x \quad \text{for all } x \in \mathcal{D} . \end{aligned}$$

(c) *Every continuous operator $A$ commuting with $H$ maps $\mathcal{D}$ to $\mathcal{D}$ and satisfies*

$$AH^{-1}x = H^{-1}Ax \quad \text{for all } x \in \mathcal{D} .$$

(d) *The function $\lambda^{-1}$ is integrable with respect to $\mu_{x,y}$ for all $x \in \mathcal{D}$ and $y \in \mathcal{H}$ and*

$$d\mu_{H^{-1}x,y}(\lambda) = \lambda^{-1} d\mu_{x,y}(\lambda) ,$$

$$(H^{-1}x|y) = \textstyle\int \lambda^{-1} d\mu_{x,y}(\lambda) \quad \text{for } x \in \mathcal{D} \quad \text{and } y \in \mathcal{H} . \tag{23.4}$$

(e) *The operator $H^{-1}$ is self-adjoint:*

$$(H^{-1}x|y) = (x|H^{-1}y) \quad \text{for } x, y \in \mathcal{D} ,$$

*and the only $y \in \mathcal{H}$ for which the map $x \longmapsto (H^{-1}x|y)$ from $\mathcal{D}$ to $\mathcal{H}$ is continuous are the elements $y \in \mathcal{D}$.*

(f) *For any Borel set $\omega \subset X$ whose complement is a* neighbourhood *of $0$ in $X$, $\mathcal{H}(\omega) \subset \mathcal{D}$, $H^{-1}$ maps $\mathcal{H}(\omega)$ to $\mathcal{H}(\omega)$ and the restriction of $H^{-1}$ to $\mathcal{H}(\omega)$ is continuous.*

*Proof of* (a). We begin by observe that, $\{0\}$ being a null set with respect to all spectral measures of $H$, the functions $\lambda^{-1}$ and $\lambda^{-2}$ are defined almost everywhere with respect to these measures. Let us then consider a vector $x = Hx' \in \mathrm{Im}(H)$, whence

$$d\mu_{x,x}(\lambda) = \lambda^2 d\mu_{x',x'}(\lambda)$$

by property (MS 4) of spectral measures [n° 22, (ii)]. As the function $\lambda^{-2}\lambda^2$ is almost everywhere equal to 1, and hence integrable with respect to $\mu_{x',x'}$, theorem 25 shows that $\lambda^{-2} \in L^1(X; \mu_{x,x})$, proving the necessity of (3).

Conversely, let us suppose that (3) holds and write $B^\infty(X)$ for the set of bounded Borel functions on $X$. It is everywhere dense in $L^2(X; \mu_{x,x})$. Let $\mathcal{H}(x)$ be the closure of the set of $M(f)x$ in $\mathcal{H}$, where $f \in B^\infty(X)$. As

$$\|M(f)x\|^2 = \int |f|^2 d\mu_{x,x}$$

by (22.9), the map $f \longmapsto M(f)x$ extends to an isomorphism $j$ from $L^2(X; \mu_{x,x})$ onto $\mathcal{H}(x)$. Relation $M(fg)x = M(f)M(g)x$, i.e. $j(fg) = M(f)j(g)$, which holds for all $f, g \in B^\infty(X)$ by n° 22, (ii), shows that $j$ transforms the multiplication operator $\varphi \longmapsto f\varphi$ in $L^2(X; \mu_{x,x})$ – a continuous operator since $f$ is bounded – into the operator $M(f)$ in $\mathcal{H}(x)$. Therefore,

$$j(f\varphi) = M(f)j(\varphi)$$

for all $f \in B^\infty(X)$ and all bounded and unbounded $\varphi \in L^2(X; \mu_{x,x})$.

This being so, let us suppose that $x$ satisfies (3). The function $\varphi(\lambda) = \lambda^{-1}$ is in $L^2(X; \mu_{x,x})$, the function $f(\lambda) = \lambda$ is bounded and Borel on $X$ and $f\varphi = 1$ almost everywhere with respect to $\mu_{x,x}$. So $M(f)j(\varphi) = M(f\varphi) = M(1) = x$. Thus there exists $x' = j(\varphi)$ such that $x = M(f)x' = Hx'$, which is what needed to be proved.

*Proof of* (b). Set theory, chap. I.

*Proof of* (c). If $A$ commutes with $H$ and if $x \in \mathcal{D}$, then there exists $x'$ such that $x = Hx'$, whence $Ax = Hy'$ where $y' = Ax'$. As a result, $Ax = y = Hy'$ is in $\mathcal{D}$ and $H^{-1}Ax = H^{-1}y = y' = Ax' = AH^{-1}x$.

*Proof of* (d). For $x \in \mathcal{D}$, we once again consider the isomorphism $j : L^2(X; \mu_{x,x}) \longrightarrow \mathcal{H}(x)$ used in point (a) of the proof. As was then seen, $x = j(\varphi)$, where $\varphi(\lambda) = \lambda^{-1}$ and $x = Hx'$, where $x' = j(f)$, $f(\lambda) = \lambda$. Then, for all $y \in \mathcal{H}$,

$$d\mu_{x,y}(\lambda) = d\mu_{Hx',y}(\lambda) = \lambda.d\mu_{x',y}(\lambda).$$

This relation obviously continues to hold if the measures and the function $\lambda$ are replaced by their absolute values [n° 17, (ii)]. As $\lambda^{-1}\lambda = 1$ is integrable with respect to $|\mu_{x',y}|$, theorem 25 shows that $\lambda^{-1}$ is integrable with respect to $|\mu_{x,y}|$ and that

$$\lambda^{-1}.d\mu_{x,y}(\lambda) = d\mu_{x',y}(\lambda) \quad \text{if } x' = H^{-1}x, \quad y \in \mathcal{H},$$

which is the first relation in (4). Integration of the function 1 with respect to these two measures leads to the second one.

Exchanging the roles of $x$ and $y$ and taking conjugate measures, the formula

$$\lambda^{-1}.d\mu_{x,y}(\lambda) = d\mu_{x,y'}(\lambda) \quad \text{if } y' = H^{-1}y\,, \quad x \in \mathcal{H}$$

is obtained in a similar way. Replacing $y$ by $y'$ in the last but one formula, once again thanks to theorem 25, which is in fact almost trivial in this case,

$$(*) \qquad \lambda^{-2}.d\mu_{x,y}(\lambda) = d\mu_{x',y'}(\lambda) \quad \text{if } x' = H^{-1}x \quad \text{and } y' = H^{-1}y\,.$$

*Proof of* (e). These relations show that

$$(H^{-1}x|y) = (x|H^{-1}y) \Longleftrightarrow \int \lambda^{-1}d\mu_{x,y}(\lambda) = \int \lambda^{-1}d\mu_{x,y}(\lambda)\,,$$

which is clear and shows that, for all $y \in \mathcal{D}$, the linear functional $x \longmapsto (H^{-1}x|y)$ is continuous on $\mathcal{D}$. Conversely suppose that, for some $y \in \mathcal{H}$, there is a vector $y'$ such that $(H^{-1}x|y) = (x|y')$ for all $x \in \mathcal{D}$, hence of the form $x = Hx'$. As $Hx \in \text{Im}(H) = \mathcal{D}$, $x$ may be replaced by $Hx$ in this relation, whence $(x|y) = (Hx|y') = (x|Hy')$. Since $\mathcal{D}$ is everywhere dense, one deduces that $y = Hy' \in \mathcal{D}$.

*Proof of* (f). If the open set $\omega$ avoids the neighbourhood of 0, then there is a number $m > 0$ such that $|\lambda| \geq m$ for all $\lambda \in \omega$. But for all $x \in \mathcal{H}(\omega)$, $\mu_{x,x}(X-\omega) = 0$ by (22.12). The bounded function $\lambda^{-2}$ on $X-\omega$, is, therefore, integrable with respect to $\mu_{x,x}$, whence $x \in \mathcal{D}$. Furthermore,

$$\begin{aligned}\|H^{-1}x\|^2 &= \int d\mu_{H^{-1}x,H^{-1}x}(\lambda) = \int \lambda^{-2}d\mu_{x,x}(\lambda) \quad \text{by } (*)\\ &\leq m^{-2}\int d\mu_{x,x}(\lambda) = m^{-2}\|x\|^2\,.\end{aligned}$$

Thus $H^{-1}$ is continuous on $\mathcal{H}(\omega)$. $H^{-1}$ maps $\mathcal{H}(\omega)$ to $\mathcal{H}(\omega)$ since $H^{-1}M(\omega) = M(\omega)H^{-1}$, which is a very particular case of (c) for $A = M(\omega)$, qed.

*Exercise 1.* Let **A** be a GN algebra in $\mathcal{H}$ and $\varphi$ a real Borel function on the spectrum of **A**. Find the assumptions needed to define $M(\varphi)$ as a self-adjoint operator.

(ii) *Canonical extension of a positive symmetric operator.* Point (e) of the statement implicitly defined the notion of a *self-adjoint* operator. To understand it, consider an everywhere dense subspace $\mathcal{D}$ of $\mathcal{H}$ and a *symmetric* map $S : \mathcal{D} \longrightarrow \mathcal{H}$ (I prefer keeping the term "Hermitian" for continuous operators), i.e. such that

$$(Sx|y) = (x|Sy) \quad \text{for } x, y \in \mathcal{D}\,.$$

For all $y \in \mathcal{D}$, the linear functional $x \mapsto (Sx|y)$ is therefore continuous on $\mathcal{D}$. But it may also be so for some $y \notin \mathcal{D}$. Let $\mathcal{D}^*$ be the set of such $y$. It is a subspace of $\mathcal{H}$ containing $\mathcal{D}$, and associating to each $y \in \mathcal{D}^*$, its unique vector $y^*$ such that $(Sx|y) = (x|y^*)$ for all $x \in \mathcal{D}$, we define a map

$$S^* : \mathcal{D}^* \longrightarrow \mathcal{H}$$

naturally called the *adjoint* map of $S$ extending $S$. $S$ is said to be *self-adjoint* if $\mathcal{D}^* = \mathcal{D}$, in which case $S$ does not admit any other symmetric extension apart from itself. This is exactly what proposition (e) of the previous theorem says.

When a symmetric operator $S$ is not self-adjoint, it is not always possible to find a self-adjoint operator extending it. I will not address this problem in the general case,[88] but in the far simpler and often used case, where $S$ is *positive*, i.e. satisfies $(Sx|x) \geq 0$ for all $x \in \mathcal{D}$. Then there is a *canonical* self-adjoint extension of $S$. It is a consequence of the next result:

**Theorem 35**[89]. *Let $\mathcal{H}$ be a Hilbert space, $\mathcal{D}$ an everywhere dense subspace of $\mathcal{H}$ and $S$ a positive symmetric operator defined on $\mathcal{D}$. Set*

$$m = \inf\,(Sx|x)\,/\,(x|x)\ . \tag{23.5}$$

*If $m > 0$, there is an injective, positive, continuous Hermitian operator $H$ in $\mathcal{H}$ having norm $\leq 1/m$ such that*

(a) *$HS = id$ on $\mathcal{D}$,*

(b) *$H$ commutes with every unitary operator $U$ such that*[90]

$$U(\mathcal{D}) \subset \mathcal{D}\,, \quad USx = SUx \quad \text{for all } x \in \mathcal{D}\,.$$

*If $m = 0$, then there is an injective, positive, continuous Hermitian operator $H$ in $\mathcal{H}$ having norm $\leq 1$ such that $HS = 1 - H$ on $\mathcal{D}$ and satisfying* (b).

The case $m = 0$ immediately reduces to the previous on by taking $S + 1$. Hence, one may suppose $m > 0$ and even $m = 1$.

The proof only uses the elementary results of n° 19. It consists in constructing a second Hilbert space $\mathcal{H}'$ and a bijection from $\mathcal{H}'$ onto a subspace of $\mathcal{H}$. There is a tendency to identify $\mathcal{H}'$ with this subspace, but it is then impossible to know whether its vectors should be considered to be elements of $\mathcal{H}$ or of $\mathcal{H}'$. This book not being aimed at experts, I will use somewhat clumsy but precise notation in order to avoid such possible confusion.

---

[88] See for example Dieudonné, XV.13 or any other treatise on Hilbert spaces.

[89] According to Béla v. Sz. Nagy, *Spektraldarstellung Linearer Transformationen des Hilbertschen Raumes* (Ergebnisse der Math., Bd. R, 1942, p. 35–36), a book that in earlier times was like the Bible for me, this theorem was first proved by von Neumann (1929), then by M. H. Stone in his book on Hilbert spaces (1932). The proof in the text is due to Hans Freudenthal (1936) and simplifies a proof by Kurt Friedrichs, *Spektraltheorie halbbeschrnkter Operatoren* (Math. Annalen, **109** and **110**, 1934). French PDE tradition attributes it to Gelfand, Lions, Kato, etc. Looking at it this way, it could also be attributed to me since I was using it in 1946–1947...

[90] If these properties hold, $U$ will be said to commute with $S$.

(i) Let us consider the positive Hermitian form

$$(x|y)' = (Sx|y) \tag{23.6}$$

on $\mathcal{D}$. It enables us [n° 19, (i)] to construct a Hilbert space $\mathcal{H}'$, whose norm will be written $\|x\|'$, and a canonical map

$$\pi : \mathcal{D} \longrightarrow \mathcal{H}'$$

from $\mathcal{D}$ onto an everywhere dense subspace $\mathcal{D}'$ of $\mathcal{H}'$. Hence

$$\big(\pi(x)\big|\pi(y)\big)' = (Sx|y) \quad \text{for } x, y \in \mathcal{D} \tag{23.6'}$$

and, by assumption (5) for $m = 1$,

$$\|x\| \leq \|\pi(x)\|' \quad \text{for } x \in \mathcal{D}\,.$$

Thus the map $\pi$ is injective. Furthermore, the set of $\pi(x)$ being everywhere dense in $\mathcal{H}'$, there is a unique continuous linear map

$$J : \mathcal{H}' \longrightarrow \mathcal{H}\,,$$

having norm $\leq 1$, such that

$$J\left[\pi(x)\right] = x \quad \text{for all } x \in \mathcal{D}\,.$$

(6') then shows that $(u|v)' = (SJu|Jv)$ for $u, v \in \mathcal{D}'$. But for given $u \in \mathcal{D}'$, both sides are continuous functions of $v \in \mathcal{H}'$ since $J$ is continuous. Hence, as $\mathcal{D}'$ is dense in $\mathcal{H}'$,

$$(u|v)' = \big(SJu\big|Jv\big) \quad \text{for } u \in \mathcal{D}'\,, \quad v \in \mathcal{H}'\,.$$

In particular, $Jv = 0$ implies $(u|v)' = 0$ for all $u \in \mathcal{D}'$ and so $v = 0$. As a result, $J$ is *injective.*

(ii) Since $J : \mathcal{H}' \longrightarrow \mathcal{H}$ is continuous, it has an adjoint [n° 19, (iii)]

$$J^* : \mathcal{H} \longrightarrow \mathcal{H}'$$

having norm $\leq 1$ like $J$. By definition,

$$(J^*x|v)' = (x|Jv) \quad \text{for } x \in \mathcal{H} \text{ and } v \in \mathcal{H}'\,,$$

which shows that $J^*x = 0$ implies $(x|Jv) = 0$ for all $v \in \mathcal{H}'$. Thus $x = 0$ since $J(\mathcal{H}')$ is everywhere dense in $\mathcal{H}$. The operator $J^*$ is, therefore, *injective* (as is the adjoint of any operator whose image is everywhere dense).

Replacing $x$ by $Sx$ and $v$ by $\pi(y)$ for $y \in \mathcal{D}$ in the previous relation, for all $y \in \mathcal{D}$,

$$\big(J^*Sx\big|\pi(y)\big)' = (Sx|J\pi(y)) = (Sx|y) = (\pi(x)|\pi(y))'\,.$$

So the set of $\pi(y)$ being everywhere dense in $\mathcal{H}'$, $J^*Sx = \pi(x)$ for all $x \in \mathcal{D}$. Hence

$$JJ^*Sx = x \quad \text{for all } x \in \mathcal{D}.$$

The operator

$$H = JJ^* : \mathcal{H} \longrightarrow \mathcal{H},$$

is continuous, has *injective* norm $\leq 1$ like $J$ and $J^*$ and satisfies condition (a) of the statement. $H$ is positive Hermitian like any operator of the form $AA^*$.

(iii) Let us consider a unitary operator $U$ on $\mathcal{H}$ which preserves $\mathcal{D}$ and commutes with $S$ on $\mathcal{D}$. Then, for $x, y \in \mathcal{D}$,

$$(\pi(Ux)|\pi(Uy))' = (SUx|Uy) = (USx|Uy) = (Sx|y) = (\pi(x)|\pi(y))' .$$

As a result, there is a unitary operator $U'$ on $\mathcal{H}'$ such that

$$\pi \circ U = U' \circ \pi .$$

By definition of $J$, it is written $J(U'y) = UJ(y)$ for all $y \in \mathcal{D}'$. The same is true (continuity) for $y \in \mathcal{H}'$, whence $JU' = UJ$. This implies

$$J^*U^* = U'^*J^* \quad \text{and so } U'J^* = J^*U$$

since $U^* = U^{-1}$ for all unitary operators. The fact that $U$ commutes with $H$ is then obvious. qed.

Let us now show how to deduce from the theorem a *canonical self-adjoint extension* of $S$, i.e. commuting with every unitary operator commuting with $S$.

If $m > 0$, it is the inverse $H^{-1}$ of $H$ defined in theorem 34. It is positive like $H$, defined on the image of $H$, hence on $\mathcal{D}$, and as $H^{-1}H = 1$, clearly $H^{-1}x = Sx$ for all $x \in \mathcal{D}$.

If $m = 0$, we start from the relation $HS = 1 - H$ and write the spectral decomposition

$$H = \int_X \lambda dM(\lambda) , \quad 1 - H = \int_X (1 - \lambda) dM(\lambda)$$

of $H$, where $X \subset [0, 1]$ is the spectrum of $H$. For $x \in \mathcal{D}$, $HSx$ is the image of $H$, hence also $x = HSx + Hx$. So $H^{-1}$ may be applied, the outcome being the relation

$$Sx = H^{-1}x - x . \tag{23.7}$$

As a result, the self-adjoint operator

$$H^{-1} - 1 = \int_X \left(\lambda^{-1} - 1\right) dM(\lambda)$$

extends $S$. It is defined on the same vectors as $H^{-1}$, i.e. on $x \in \mathcal{H}$ such that

$$\int_X \lambda^{-2} d\mu_{x,x}(\lambda) < +\infty .$$

*Exercise 2.* Suppose $\mathcal{H} = L^2(\mathbb{R})$, $\mathcal{D} = L(\mathbb{R})$ and $Sf = \varphi f$, where $\varphi \geq 1$ is such that $\varphi^2$ is locally integrable. Set out explicitly $\mathcal{H}'$, $J$, $J^*$ and $H$ in this case. Compute the self-adjoint extension of $S$ assuming only $\varphi \geq 0$.

**Corollary 1.** *Let $S$ be a non-trivial positive symmetric operator defined on an everywhere dense subspace $\mathcal{D}$ of $\mathcal{H}$. There is a non-zero positive, continuous Hermitian operator $S'$ on $\mathcal{D}$ such that:*
(a) $(S'x|x) \leq (Sx|x)$ *for all* $x \in \mathcal{D}$,
(b) $S'$ *commutes with all unitary operator commuting with* $S$.

Indeed, let us choose numbers $a$ and $b$ such that $0 < a < b < 1$ and consider the spectral projection

$$M(a,b) = M\left([a,b]\right) = \int_{[a,b]} dM(\lambda)$$

of the operator $H$ of the theorem. Since

$$\left(H^{-1} - 1\right) M(a,b) = \int_{[a,b]} \left(\lambda^{-1} - 1\right) dM(\lambda)$$

with $a > 0$, this operator can be extended by a *continuous* operator $S'$ on $\mathcal{H}$, which is obviously Hermitian positive and commutes with every unitary operator commuting with $S$.

$S'$ satisfies condition (b) – obvious – and condition (a) since

$$(S'x|x) = \int_{[a,b]} \left(\lambda^{-1} - 1\right) d\mu_{x,x}(\lambda) \quad \text{for } x \in \mathcal{H} ,$$
$$(Sx|x) = \int_{[0,1]} \left(\lambda^{-1} - 1\right) d\mu_{x,x}(\lambda) \quad \text{for } x \in \mathcal{D} .$$

On the other hand, the kernel of $S'$ is the set of $x$ such that

$$\|S'x\|^2 = \int_{[a,b]} \left(\lambda^{-1} - 1\right)^2 d\mu_{x,x}(\lambda) = 0 .$$

Since $0 < a < b < 1$, this means that $[a,b]$ is a null set with respect to $\mu_{x,x}$. If, for given $x \in \mathcal{D}$, this is the case for all $a$ and $b$, the measure $\mu_{x,x}$ is concentrated on the subset $\{0,1\}$ of the spectrum of $H$. As $H$ is injective, 0 is not an eigenvalue of $H$ and as a result, $\mu_{x,x}(\{0\}) = 0$. Thus $\mu_{x,x}([0,1[) = 0$, whence $Hx = x$. Therefore, $S' = 0$ may hold on $\mathcal{D}$ for all choices of $a$ and $b$ only if $H = 1$ on $\mathcal{D}$. But then relation $HS = 1 - H$ shows that $HS = 0$, and so $S = 0$ since $H$ is injective, qed.

**Corollary 2 (Schur's lemma II).** *Let $(\mathcal{H}, U)$ be an irreducible unitary representation of some* lcg $G$ *and $S$ a self-adjoint or positive symmetric operator defined on an everywhere dense subspace $\mathcal{D}$ of $\mathcal{H}$. Suppose $\mathcal{D}$ is $U(x)$-invariant and that $U(x)S = SU(x)$ in $\mathcal{D}$. Then $S$ is a scalar.*

If $S$ is positive symmetric, the canonical self-adjoint extension $S$ commutes with all $U(x)$, hence so do its spectral projections, which being continuous, are scalars, proving the result in this case.

If $S$ is not positive, but self-adjoint, consider $S + i$. As in n° 19, (iv),

$$\|(S+i)x\|^2 = \|Sx\|^2 + \|x\|^2 \quad \text{for all } x \in \mathcal{D};$$

Thus $S + i$ is injective. If $\lim(S + i)x_n = v$ exists for $x_n \in \mathcal{D}$, the sequence $(x_n)$ satisfies Cauchy's criterion, and so converges to some $x \in \mathcal{H}$. Thus $\lim Sx_n = y = v - ix$ exists. But then, for all $u \in \mathcal{D}$,

$$(Su|x) = \lim (Su|x_n) = \lim (u|Sx_n) = (u|y) \ .$$

Hence $x \in \mathcal{D}$ and $Sx = y$ since $S$ is self-adjoint. The limit $y + ix$ of the $(S+i)x_n$ is, therefore, $(S+i)x$, and so the subspace $\operatorname{Im}(S+i)$ is closed. Any vector $y$ orthogonal to all $(S + i)x$ satisfies $(Sx|y) = -i(x|y)$ for all $x \in \mathcal{D}$, and thus is in $\mathcal{D}$ since $S$ is self-adjoint, and since $Sy = iy$ then holds, it follows that $y = 0$. Finally, $\operatorname{Im}(S + i) = \mathcal{H}$. So $S + i$ is a bijection from $\mathcal{D}$ onto $\mathcal{H}$, its inverse being also continuous and having norm $\leq 1$. (See exercise below for the rest..).

That being so, if $S$ commutes with all $U(x)$, so does $S+i$, and hence also its inverse, which, being continuous, is a scalar by Schur's lemma I [n° 22, (iii), end]. Thus so is $S$, qed.

*Exercise 3* (von Neumann, 1930). Let $H$ be a self-adjoint operator defined on a subspace $\mathcal{D}$ of $\mathcal{H}$. We have just seen that $H + i : \mathcal{D} \longrightarrow \mathcal{H}$ is bijective and that the inverse map $(H + i)^{-1}$ from $\mathcal{H}$ to $\mathcal{D}$ is a continuous operator. (a) Show that the map $U : x \mapsto (H - i)(H + i)^{-1}x$ from $\mathcal{H}$ to $\mathcal{H}$ is unitary and that

$$1 - U = 2i(H + i)^{-1} .$$

(b) Let

$$U^n = \int_{\mathbb{T}} u^n.dM(u)$$

be the spectral decomposition of $U$. Show that the spectral manifold $\mathcal{H}(\{1\})$ is trivial. Denote by $dM(\lambda)$ the image of $dM(u)$ under the homeomorphism

$$u \longmapsto \lambda = i(1 + u)/(1 - u)$$

from $\mathbb{T} - \{1\}$ onto $\mathbb{R}$. Show that this gives a non-degenerate spectral measure on $\mathbb{R}$ such that

$$H = \int \lambda.dM(\lambda),$$

the elements $x \in \mathcal{D}$ being characterized by $\lambda^2 d\mu_{x,x}(\lambda) < +\infty$.

## 24 – Continuous Sum Decomposition

(i) *Virtual eigenvectors.* In finite dimension, any commutative GN algebra $\mathbf{A}$ is diagonalizable: if, for every character $\chi$ of $\mathbf{A}$, $\mathcal{H}(\chi)$ denotes the subspace of $x \in \mathcal{H}$ such that

$$(*) \qquad Tx = \chi(T)x \quad \text{for all } T \in \mathbf{A},$$

then the subspaces $\mathcal{H}(\chi)$ are pairwise orthogonal and $\mathcal{H}$ is the direct sum of these subspaces.

This is obviously no longer the case in infinite dimension – relation (*) usually has no solution $x \neq 0$ –, but we will see that, nonetheless, there is an analogous result provided "eigenvectors" not belonging to $\mathcal{H}$ are allowed, "direct integrals" or "continuous sums" are substituted to standard direct sums and $\mathcal{H}$ is assumed to be separable.

So let $\mathbf{A} \subset \mathcal{L}(\mathcal{H})$ be a commutative GN algebra $X$ its spectrum and let us suppose that $\mathcal{H}$ is separable. Once and for all, we choose linearly independent vectors $a_n$ whose linear combinations are everywhere dense in $\mathcal{H}$, and denote by $\mu_n$ the spectral measure $\mu_{x,x}$ for $x = a_n$. There are numbers $\alpha_n > 0$ such that

$$\sum \alpha_n \|a_n\|^2 = \sum \alpha_n \mu_n(X) < +\infty,$$

which enables us to define the measure $\mu = \sum \alpha_n \mu_n$ by $\mu(f) = \sum \alpha_n \mu_n(f)$. This is a trivial case of a continuous sum of measures. The measures $\mu_n$ are clearly bounded above by $\mu$, up to some factors, so that if $\omega \subset X$ is a Borel and $\mu$-null set, then

$$\|M(\omega)a_n\|^2 = \big(M(\omega)a_n \big| a_n\big) = \mu_n(\omega) = 0.$$

So $M(\omega) = 0$, and hence $(M(\omega)x|x) = 0$ as well for all $x$. As a result (LN), *all measures $\mu_{x,y}$ are absolutely continuous with respect to $\mu$ and relations $M(\omega) = 0$ and $\mu(\omega) = 0$ are equivalent.* Any measure $\mu \geq 0$ with this property is suitable for the following arguments.

Let us set

$$d\mu_{x,y}(\chi) = p_{x,y}(\chi)d\mu(x).$$

$p_{x,y} \in L^1(X, \mu) = L^1$ since $\mu_{x,y}$ is bounded. Functions $p_{x,y}$ clearly have the same formal properties as the measures $\mu_{x,y}$, for example

$$p_{y,x}(\chi) = \overline{p_{x,y}(\chi)}, \quad p_{x,x}(\chi) \geq 0, \tag{24.1}$$

but these relations hold *almost everywhere* only with respect to $\mu$. If they were true for *all* $\chi$ (which is the case if $\widehat{\mathbf{A}}$ is finite), the function $(x, y) \mapsto p_{x,y}(\chi)$ would be a positive Hermitian form on $\mathcal{H}$ with respect to all $\chi$. Taking the quotient by the subspace of $x$ such that $p_{x,x}(\chi) = 0$ and then its completion (n° 19, beginning) would lead to a Hilbert space $\mathcal{H}(\chi)$, with a canonical map from $\mathcal{H}$ to $\mathcal{H}(\chi)$ which it would be natural to write

$$x \longmapsto x(\chi);$$

$$p_{x,y}(\chi) = (x(\chi)|y(\chi))$$

would then hold and as a result,

$$(M(\varphi)x|y) = \int \varphi(\chi)\,(x(\chi)|y(\chi))\,d\mu(\chi)$$

for all $x, y$ and bounded measurable functions $\varphi$ on $X$. Thus the space $\mathcal{H}$ would seemingly be the "continuous sum" or "direct integral" of spaces $\mathcal{H}(\chi)$ with respect to $\mu$, operators $M(\varphi)$ likely mapping a function $x(\chi)$ to the function $\varphi(\chi)x(\chi)$ with, in particular, the relation

$$y = Tx \Longrightarrow y(\chi) = \chi(T)x(\chi) \quad \text{for all } T \in \mathbf{A}$$

analogous to (*)...

This far too naive argument needs to be corrected using the null-set elimination machinery. The method, due to von Neumann,[91] consists in coming back to the initial pseudo-basis $(a_n)$ and in assuming $x = a_i$, $y = a_j$ in (1). The corresponding functions $p_{x,y} = p_{i,j}$ being chosen once and for all, relations (1) hold outside null sets $N_{ij}$, and thus do so simultaneously outside a null set. The latter can even be chosen in such a way that for all $\chi \in X - N$, the matrix $(p_{i,j}(\chi))$ is positive Hermitian, in other words satisfies

$$\sum_{1 \leq i,j \leq n} p_{i,j}(\chi)\xi_i\overline{\xi_j} \geq 0 \tag{24.2}$$

for all integers $n$ and coefficients $\xi_1, \ldots, \xi_n \in \mathbb{C}$. Indeed, setting $x = \sum \xi_i a_i$, the left hand side of (2) is just $p_{x,x}(\chi)$, up to a null set, so that (2) holds outside a null set $N_{x,x}$. But to check (2) for given $n$ and $\chi$ and arbitrary coefficients $\xi_i$, by continuity, it suffices to do so when the set of $\xi_i$ is everywhere dense in $\mathbb{C}$, hence in a countably infinite number of cases. By making $n$ vary, one finally gets a countably infinite number of sets $N_{x,x}$ to be eliminated. Taking their union with the null set introduced above gives a null set $N$ outside which formulas (1) *and* (2) hold for *all* ordered pairs $(i, j)$.

[91] *On rings of operators. Reduction theory* (Annals of Math., **50**, 1949).

Let us denote by $\Lambda$ the (everywhere dense but not closed) vector subspace of $\mathcal{H}$ generated by all $a_n$. If $x, y \in \Lambda$, then

$$(24.3) \qquad x = \sum \xi_n a_n \,, \quad y = \sum \eta_n a_n$$

with *well-determined* coefficients since the $a_n$ have been assumed to be linearly independent. Hence

$$(24.4) \qquad p_{x,y}(\chi) = \sum \xi \bar{\eta}_j p_{i,j}(\chi) \text{ ae.}$$

The functions $p_{x,y}$ being determined up to null sets, for ordered pairs of form (3), (4) can be taken to *define* $p_{x,y}(\chi)$ for *all* $\chi \in X - N$. For these $\chi$, by (2), the map $(x, y) \mapsto p_{x,y}(\chi)$ is then a positive Hermitian form on the subspace $\Lambda$ of $\mathcal{H}$ consisting of vectors (3). Replacing all $p_{i,j}(\chi)$ by 0 on $N$, (4) may even be assumed to hold for all $\chi$.

Arguments similar to those used at the beginning of n° 19 lead to a genuine Hilbert space $\mathcal{H}(\chi)$ and to a linear map $x \mapsto x(\chi)$ from $\Lambda$ onto an everywhere dense subspace of $\mathcal{H}(\chi)$, a map for which, for all $\chi$,

$$(x(\chi)|y(\chi)) = p_{x,y}(\chi) \quad \text{for al } x, y \in \Lambda\,.$$

The left hand side is a $\mu$-integrable function. Since

$$\begin{aligned}(M(\varphi)x|y) &= \int \varphi(\chi) d\mu_{x,y}(\chi) = \int \varphi(\chi).p_{x,y}(\chi) d\mu(\chi) \\ &= \int \varphi(\chi) p_{x,y}(\chi).d\mu(\chi)\,,\end{aligned}$$

the formula

$$(24.5) \qquad (M(\varphi)x|y) = \int \varphi(\chi)\,(x(\chi)|y(\chi))\,d\mu(\chi)$$

becomes legitimate for $x, y \in \Lambda$ and for every bounded Borel function $\varphi$. In particular it shows that

$$(24.6) \qquad \|x\|^2 = \int \|x(\chi)\|^2\,d\mu(\chi)\,.$$

$\|x\|$ is thus the norm $N_2$ of the function $\|x(\chi)\|$, which is in $L^2(X, \mu)$ since $\|x(\chi)\|^2 = p_{x,x}(\chi)$ is in $L^1(X, \mu)$.

Let us now show that a function $x(\chi) \in \mathcal{H}(\chi)$, defined up to a null set, can be associated to each $x \in \mathcal{H}$ in such a way that (5) continues to hold for $x, y \in \mathcal{H}$. We start by remarking that, for all $x \in \mathcal{H}$, there exist $x_n \in \Lambda$ such that

$$x = \sum x_n \quad \& \quad \sum \|x_n\| < +\infty\,.$$

Since the second relation can be written

$$\sum \left( \int \|x_n(\chi)\|^2 \, d\mu(\chi) \right)^{1/2} < +\infty ,$$

theorem 6 of n° 4 applied for $p = 2$ to functions $\|x_n(\chi)\|$ shows that the series $\sum \|x_n(\chi)\|$ converges ae. and in $L^2(X;\mu)$ to a square integrable function $F(\chi)$ with respect to $\mu$. Thus the series $\sum x_n(\chi)$ converges ae. to some $x(\chi) \in \mathcal{H}(\chi)$ satisfying $\|x(\chi)\| \leq \sum \|x_n(\chi)\| = F(\chi)$ ae. Setting $s_p = x_1 + \ldots + x_p \in \Lambda$, $x(\chi) = \lim s_p(\chi)$ ae., the functions $\|s_p(\chi)\|$ being, moreover, in $L^2(X;\mu)$ and dominated by $F$. Therefore, the the function $\|x(\chi)\|$ is square integrable and its norm $N_2$ is bounded above by the sum of norms $N_2$ of functions $\|x_n(\chi)\|$. Replacing $x$ by $x - s_p$, this shows that

$$\lim_{p\infty} \int \|x(\chi) - s_p(\chi)\|^2 \, d\mu^*(\chi) = 0 .$$

It follows that the function $\chi \mapsto x(\chi)$ only depends on the vector $x \in \mathcal{H}$, up to a null set. Indeed, if $x = \sum x'_n$ is another way of approximating $x$ by $x'_n \in \Lambda$, the series $\sum x_n - x'_n$ converges to 0. The previous result then shows that

$$\lim_{p\infty} \int \|s_p(\chi) - s'_p(\chi)\|^2 \, d\mu(\xi) = 0 .$$

Hence a partial sequence tending ae. to 0 can be extracted from the sequence of functions $\|s_p(\chi) - s'_p(\chi)\|$, and as anyhow the complete sequences $s_p(\chi)$ and $s'_p(\chi)$ converge ae., this is sufficient to prove that

$$\sum x_p(\chi) = \sum x'_p(\chi) \text{ pp}$$

as expected.

This being settled, let us prove (5) for $x, y \in \mathcal{H}$. To this end, we choose sequences $s_n$ (resp. $t_n$) in $\Lambda$ satisfying the following conditions: (a) they converge to $x$ (resp. $y$) in $\mathcal{H}$, (b) $s_n(\chi)$ (resp. $t_n(\chi)$) converges to $x(\chi)$ (resp. $y(\chi)$) in $\mathcal{H}(\chi)$ for almost all $\chi$, (c) the functions $\|s_n(\chi)\|$ (resp. $\|t_n(\chi)\|$) are dominated by functions $F$ (resp. $G$) in $L^2(X;\mu)$. Then, by (5) applied to $\Lambda$ and the continuity of $M(\varphi)$,

$$(M(\varphi)x|y) = \lim \left(M(\varphi)s_p|t_p\right) = \lim \int \varphi(\chi) \left(s_p(\chi)|t_p(\chi)\right) d\mu(\chi) .$$

As the functions $(s_p(\chi)|t_p(\chi))$ are (Cauchy-Schwarz) dominated by $FG \in L^1(X;\mu)$ and converge ae. to $(x(\chi)|y(\chi))$, Lebesgue's theorem immediately least to (5) for $x, y$ and even (replace $\varphi$ by $\varphi\overline{\psi}$) to

$$(24.7) \qquad (M(\varphi)x|M(\psi)y) = \int \varphi(\chi)\overline{\psi(\chi)} \left(x(\chi)|y(\chi)\right) d\mu(\chi)$$

for all bounded and measurable $\varphi$ and $\psi$.

Let us now show that

$$(24.8) \qquad y = M(\varphi)x \Longrightarrow y(\chi) = \varphi(\chi)x(\chi) \text{ pp.}$$

and that in particular

$$y = Tx \Longrightarrow y(\chi) = \chi(T)x(\chi) \text{ ae.}$$

for all $T \in \mathbf{A}$. This result is essential if we want the $x(\chi)$ to resemble common eigenvectors of all $T \in \mathbf{A}$. Indeed, by (7), for $y = M(\varphi)x$,

$$\begin{aligned} 0 = \|M(\varphi)x - y\|^2 &= (M(\varphi)x|M(\varphi)x) - 2\,\mathrm{Re}\,(M(\varphi)x|y) + (y|y) \\ &= \int |\varphi(x)|^2 . \|x(\chi)\|^2 \, d\mu(\chi) - 2\,\mathrm{Re} \int \varphi(\chi)\,(x(\chi)|y(\chi))\, d\mu(\chi) \\ &\quad + \int \|y(\chi)\|^2 \, d\mu(\chi) \\ &= \int d\mu(\chi) \left[ \|\varphi(\chi)x(\chi)\|^2 - 2\,\mathrm{Re}\,((\varphi(\chi)x(\chi)|y(\chi)) + \|y(\chi)\|^2 \right] . \end{aligned}$$

The expression under the $\int$ sign is just $\|\varphi(\chi)x(\chi) - y(\chi)\|^2$, whence

$$\int \|\varphi(\chi)x(\chi) - y(\chi)\|^2 \, d\mu(\chi) = 0$$

and (8).

Finally the functions $x(\chi)$ thus obtained need to be *characterized*. The following two conditions answer the question:

(a) the function $(x(\chi)|y(\chi))$ is measurable for all $y \in \Lambda$,

(b) $\int \|x(\chi)\|^2 d\mu^*(\chi) < +\infty$.

Indeed, suppose that these two conditions hold. The function $\|x(\chi)\|$ is then measurable since

$$\|x(\chi)\| = \sup_y |(x(\chi)|y(\chi))| \,/\, \|y(\chi)\|$$

provided $y \in \mathcal{H}$ is made to vary in such a way that the set of $y(\chi)$ is everywhere dense in $\mathcal{H}(\chi)$. For this to be the case, it suffices to chose linear combinations of $a_n$ with "rational complex" coefficients. The set of these $y$ being countable, condition (a) immediately implies that the function $\|x(\chi)\|$ is measurable. Condition (b) then shows that it is in $L^2(X; \mu)$.

For two functions $x(\chi)$ and $y(\chi)$ satisfying (a), the inner product $(x(\chi)|y(\chi))$ is measurable because of the age-old identity

$$4B(x, y) = B(x + y, x + y) - \ldots$$

valid for all Hermitian forms. If $x$ and $y$ also satisfy (b), the function $(x(\chi)|y(\chi))$ is integrable (Cauchy-Schwarz). By (5) applied to $y \in \mathcal{H}$,

$$\left|\int (x(\chi)|y(\chi))\,d\mu(\chi)\right|^2 \leq \int \|x(\chi)\|^2\,d\mu(\chi).\int \|y(\chi)\|^2\,d\mu(\chi)$$
$$= \|y\|^2 \int \|x(\chi)\|^2\,d\mu(\chi)\,.$$

Hence there exists (n° 19, theorem 31) $x' \in \mathcal{H}$ such that

$$(x'|y) = \int (x(\chi)|y(\chi))\,d\mu(\chi) \quad \text{for all } y \in \mathcal{H}\,,$$

i.e.

$$\int (x(\chi) - x'(\chi)|y(\chi))\,d\mu(\chi) = 0 \quad \text{for all } y \in \mathcal{H}\,.$$

By (7), replacing $y$ by $M(\varphi)y$ multiplies $y(\chi)$ by $\varphi(\chi)$ , whence

$$\int \varphi(\chi)\,(x(\chi) - x'(\chi)|y(\chi))\,d\mu(\chi) = 0$$

for all $y \in \mathcal{H}$. As $\varphi$ is arbitrary, $(x(\chi) - x'(\chi)|y(\chi)) = 0$ ae.. Applying this result to the vectors $y = a_n$, by eliminating the null sets, we conclude that $x(\chi) = x'(\chi)$ ae., which completes the construction.

In conclusion, starting from the given algebra $\mathbf{A}$ in $\mathcal{H}$, it is possible to find a measure $\mu$ on $X$ and a decomposition of $\mathcal{H}$ into continuous sums (with respect to $\mu$) of Hilbert spaces $\mathcal{H}(\chi)$ so that all $T \subset \mathbf{A}$ multiply the "components" $x(\chi)$ for all $x \in \mathcal{H}$ in $\mathcal{H}(\chi)$ by $\widehat{T}(\chi)$, the $M(\varphi)$ multiplying them more generally by $\varphi(\chi)$. This is the perfect analogue of a simultaneous diagonalization, and like in finite dimension, this is the result that should be called the spectral decomposition of $\mathbf{A}$. The number $\dim \mathcal{H}(\chi)$, defined almost everywhere, plays the role of the "multiplicity" of the "common eigenvalue" $\chi$ of $T \in \mathbf{A}$.

The difference with the case of finite-dimensional spaces is the availability of a wide choice of measures $\mu$. The only canonical aspect of the construction is the class of $\mu$ in the sense of Lebesgue-Nikodym. Changing $\mu$ would modify $\mathcal{H}(\chi)$ only almost everywhere and would multiply $x(\chi)$ by scalars independent of $x$. This is the best that one can hope for in this abstract and general context.

In concrete examples, $\mathcal{H}$ is often a function space of $L^2$ type. Our hope is then to be able to consider the $\mathcal{H}(\chi)$ as spaces of eigenfunctions (or distributions) of operators $M(\varphi)$. This requires methods from analysis far beyond the framework of integration theory.

(ii) *Continuous sums of Hilbert spaces.* In all of the above, we have talked of continuous sums or of direct integrals of Hilbert spaces as physicists do, i.e. without mathematically defining this notion. We can now do so, and I will content myself with it in order not to lengthen the presentation excessively.

For this, let us take a locally compact space[92] $T$, a measure $\mu \geq 0$ on $T$ and, for all $t \in T$, a Hilbert space $\mathcal{H}(t)$. Direct sums of trivial spaces being of little interest, it is reasonable to suppose that $\mathcal{H}(t) \neq \{0\}$ almost everywhere. For the same reason, it is reasonable to suppose that the support[93] of $\mu$ is all of $T$.

As "measurable" or "square integrable" functions $x(t)$ with values in $\mathcal{H}(t)$ need to be defined, which on the fact of it has no meaning if the spaces $\mathcal{H}(t)$ are unrelated to each other, suppose given in advance a family $\Lambda$ of functions $x(t)$ with values in $\mathcal{H}(t)$, defined up to null sets, and satisfying the following conditions:

(VN 1) $\alpha x + \beta y \in \Lambda$ for all $x, y \in \Lambda$ and $\alpha, \beta \in \mathbb{C}$;
(VN 2) the function $(x(t)|y(t))$ is measurable for all $x, y \in \Lambda$;
(VN 3) there is a countable set $\mathcal{D} \subset \Lambda$ such that, for almost all $t$, the set of $x(t)$, $x \in \mathcal{D}$, is everywhere dense in $\mathcal{H}(t)$.

It is possible to dispense with countability in (VN 3) by having recourse to Lusin's theorem and to equimeasurable families, but this would needlessly complicate matters.

These assumptions enable us to define functions $x(t)$ that will be referred to as *measurable*: they are required to satisfy the condition that $(x(t)|y(t))$ is measurable for all $y \in \mathcal{D}$; by (VN 2), this is the case for all $x \in \Lambda$. If $x(t)$ is measurable, then clearly so is $\varphi(t)x(t)$ for any bounded measurable scalar function $\varphi$ and any limit-ae. of measurable functions is measurable. If $x(t)$ is measurable, then so is $\|x(t)\|$ by (VN 2) and (VN 3). If $x$ and $y$ are measurable, then so is the function

$$(x(t)|y(t)) = \|x(t) + y(t)\|^2 /4 - \ldots$$

A function $x(t)$ will be said to be *square integrable* if it is measurable and if

$$\int \|x(t)\|^2 \, d\mu(t) < +\infty . \tag{24.9}$$

The existence of such functions is immediate: choose a measurable function $x(t)$ which is not zero ae., replace $x(t)$ by $x(t)/\|x(t)\|$ if $x(t) \neq 0$ and by $0$ if $x(t) = 0$, and finally multiply the result by an arbitrary function $\varphi \in L^2(T; \mu)$.

This said, the *continuous sum* $\mathcal{H}$ of $\mathcal{H}(t)$ with respect to $\mu$ is, by definition, the set of classes of square integrable functions $x(t)$, equipped with the inner product

---

[92] In the chapter of his book on von Neumann algebras dedicated to this topic, Dixmier even considers the case of an "abstract" measure because the "spectrum" of a non-commutative GN algebra, equipped with its natural topology, is not always locally compact in a strict sense.

[93] Open sets such that $\mu(U) = 0$ that form an increasing filtering family, their union also being a null set. The support of $\mu$ is its complement. Saying that the support of $\mu$ is $T$ means that 0 is the only function $f \in L_+(T)$ such that $\mu(f) = 0$.

$$(x|y) = \int (x(t)|y(t))\, d\mu(t)\,.$$

By (8) and Cauchy-Schwarz, the integral converges for $x$ and $y$. One can then associate a continuous operator $M(\varphi)$ on $\mathcal{H}$ defined by

$$y = M(\varphi)x \quad \text{if} \quad y(t) = \varphi(t)x(t)$$

to any $\varphi \in L^\infty(T;\mu)$. Formulas (22.4) of spectral theory continue to hold here: $\|M(\varphi)\| \leq \|f\|_\infty$, where this is the norm of $f$ in $L^\infty(T;\mu)$, and in fact

$$\|M(\varphi)\| = \|f\|_\infty \tag{24.10}$$

as is easily seen.

By confining ourselves to continuous $\varphi$ on the Alexandrov compactification of $T$, we thus get a self-adjoint algebra $\mathbf{A}$ in $\mathcal{H}$ containing the unit operator. As the support of $\mu$ is assumed to be $T$, the norm $N_\infty$ of such a function is the same as its uniform norm $\|f\|$. Then because of $\|M(\varphi)\| = \|\varphi\|$, $\mathbf{A}$ is closed in the Banach space $\mathcal{L}(\mathcal{H})$, and so is a commutative GN algebra. As was seen at the end of n° 12, its spectrum is the compactification of $T$. If $T$ is compact, a case it can easily be reduced to, we thus clearly recover the spectral decomposition of $\mathbf{A}$ in the sense defined above. Hence decomposing a Hilbert space $\mathcal{H}$ into a continuous sum of spaces $\mathcal{H}(t)$ reduces to choosing a separable GN algebra $\mathbf{A}$ in $\mathcal{H}$ and to carrying out the spectral decomposition. For example, $\mathbf{A}$ may be assumed to be generated by a Hermitian operator $H$, in which case $T$ is the spectrum of $H$, which acts on each $\mathcal{H}(\lambda)$ as the scalar $\lambda$, like in finite dimension...

*Exercise 1.* Show that an equivalent definition[94] of direct integrals would consist in taking $T$, $\mu$, separable Hilbert spaces $\mathcal{H}$ and $\mathcal{H}(t)$, $t \in T$, and in assuming that a class of functions $x(t) \in \mathcal{H}(t)$ satisfying the following conditions is associated to each $x \in \mathcal{H}$: (i) linearity, (ii) $(x(t)|y(t))$ is integrable for all $x, y \in \mathcal{H}$ and $(x|y) = \int (x(t)|y(t))d\mu(t)$, (iii) every function $x(t)$ such that $(x(t)|y(t))$ is integrable for all $y \in \mathcal{H}$ corresponds to some $x \in \mathcal{H}$.

Finally, let us give some indications about "decomposable" operators in a direct integral. Let $A(t)$ be a function whose values are continuous linear operators in $\mathcal{H}(t)$. Assuming the functions $A(t)x(t)$ to be measurable for all $x \in \mathcal{D}$ and the function $\|A(t)\|$ to be bounded, there is a unique continuous operator $A$ on $\mathcal{H}$ such that

$$y = Ax \Longleftrightarrow y(t) = A(t)x(t) \text{ pp.}$$

"Decomposable" operators obtained thereby commute with all $M(\varphi)$. Conversely, any operator $A$ on $\mathcal{H}$ commuting with all $M(\varphi)$ is decomposable. On the other hand, suppose given a von Neumann algebra $\mathcal{A}$ in $\mathcal{H}$. For almost all $t$, there is a von Neumann algebra $\mathcal{A}(t)$ in $\mathcal{H}(t)$ such that the relation $A \in \mathcal{A}$

[94] See for example Richard Kadison and John R. Ringrose, *Fundamentals of the Theory of Operator Algebras* (AMS, 4 vol., 1999–2001), chap. 14.

is equivalent to $A(t) \in \mathcal{A}(t)$ ae. if and only if the algebra $\mathcal{Z}$ of operators $M(\varphi)$ is contained in the centre of $\mathcal{A}$. This is also true for the commutator algebra $\mathcal{A}'$, and $\mathcal{A}'(t) = \mathcal{A}(t)'$ ae. Etc...

(iii) *The $L^2$ space of the integral of a measure.* At the end of n° 17, we mentioned the following question: given an integral of measures

$$\nu = \int \mu_x d\lambda(x)\,,$$

under what condition is the $L^2$ space of $\nu$ the continuous sum (with respect to $\lambda$) of $L^2$ spaces for $\mu_x$? As we saw, the answer is simple if there are finitely many $\mu_x$: they must be pairwise disjoint. Let us now consider the general case keeping the notation of n° 14 and giving its precise meaning to the notion of "continuous sum".

(a) To begin with, it is prudent to suppose $X$ and $Z$ to be Polish. $L^2(Z;\nu)$ and $\mathcal{H}(x) = L^2(Z;\mu_x)$ are then separable. For almost all $x$, each $f \in L^2(Z;\nu)$ is square integrable with respect to $\mu_x$ (generalized LF), hence defines a vector $f_x \in \mathcal{H}(x)$. For $f, g \in L^2(Z;\nu)$, the function $(f_x|g_x)$ is integrable and

$$(f|g) = \int (f_x|g_x)\, d\lambda(x)\,. \tag{24.11}$$

The set $\Lambda$ of functions $x \mapsto f_x$, $f \in L(Z)$ obviously satisfies axioms (VN 1), (VN 2). To check (VN 3), we choose a countable subset $D$ of $L(Z)$ such that every $f \in L(Z)$ is the uniform limit of functions $f_n \in D$ vanishing outside a fixed compact set. The classes of $f \in D$ in the $L^2$ space of *all* measures on $Z$ are then everywhere dense, so that (VN 3) is satisfied by choosing functions $x \mapsto f_x$ for $f \in D$.

This enables us to define measurable functions with values in $\mathcal{H}(x)$ and hence the direct integral $\mathcal{H}$ of $\mathcal{H}(x)$. By (10), it is clear that, for any $f \in L^2(Z;\nu)$, the associated function $x \mapsto f_x$ is in $\mathcal{H}$ and that this gives an *isomorphism from $L^2(Z;\nu)$ onto a closed subspace $\mathcal{H}_0$ of $\mathcal{H}$*, but not necessarily onto $\mathcal{H}$ as already seen in n° 17.

(b) Let us now show that $\mathcal{H}_0 = \mathcal{H}$ if and only if, for all $f \in L^2(Z;\nu)$ and all bounded measurable scalar function $\varphi$, there exists $g \in L^2(Z;\nu)$ such that

$$g_x = \varphi(x) f_x \quad \text{ae. for } \lambda\,.$$

This means that *the subspace $L^2(Z;\nu) = \mathcal{H}_0$ of $\mathcal{H}$ must be stable under operators $M(\varphi)$*. This condition is clearly necessary.[95] To show that it

[95] It is not always easy to check in concrete examples, in particular in extensions of Fourier transforms to semisimple groups, when it is not immediately obvious what the operators $M(\varphi)$ are and, as far as I know, no author who has dealt with the topic has given them explicitly. If this is really the case, this means that the (generalized Plancherel) decomposition of the regular representation of the group into "continuous sums" of irreducible representations has not yet been *proved* in the strict sense of the word, even for groups such as $SL_2(\mathbb{R})$ (Harish-Chandra) or $SL_2(\mathbb{C})$ (Gelfand and Neumark)...

is sufficient, let $x \mapsto h_x$ be an element of $\mathcal{H}$ orthogonal to $\mathcal{H}_0$. By assumption, $\int (f_x|h_x)d\lambda(x) = 0$ for all $f \in \mathcal{H}_0$, hence also by assumption, $\int \varphi(x)(f_x|h_x)d\lambda(x) = 0$ for all bounded measurable $\varphi$, and so $h_x = 0$ ae. follows from (VN 3), qed.

(c) We suppose that this condition holds and let $X = A \cup B$ be a partition of $X$ into two measurable sets $A$ and $B = X - A$. Setting

$$\text{(24.12)} \qquad \nu_A = \int \mu_x.\chi_A(x)d\lambda(x)\,, \quad \nu_B = \int \mu_x.\chi_B(x)d\lambda(x)\,,$$

$\nu = \nu_A + \nu_B$ and the space $L^2$ of $\nu_A$ (resp. $\nu_B$) can be identified with the subspace of $f \in L^2(Z;\nu)$ such that $f_x = 0$ outside $A$ (resp. $B$). However, the operator $E_A = M(\chi_A)$ on a continuous sum of Hilbert spaces is, by definition, the orthogonal projection onto the subspaces of vectors whose "components" are trivial outside $A$. As a result, $E_A$ (resp. $E_B$) maps $L^2(Z;\nu)$ onto $L^2(Z;\nu_A)$ (resp. $L^2(Z;\nu_B)$). The subspace $\mathcal{H}_0$ of $\mathcal{H}$ is, therefore, stable under all operators $E_A$ and

$$\text{(24.13)} \qquad L^2(Z;\nu) = L^2(Z;\nu_A) \oplus L^2(Z;\nu_{X-A})\,, \quad \textit{direct}\ \text{sum}\,.$$

Conversely, this condition ensures that $L^2(Z;\nu) = \mathcal{H}$. Indeed, it shows that $L^2(Z;\nu)$ is stable under $M(\varphi)$ for any function $\varphi$ which is the linear combination of characteristic functions of measurable sets. Now, any bounded measurable function $\varphi$ is the uniform limit of a sequence of functions $\varphi_n$ of the previous type. As $\|M(\varphi)\| = \|\varphi\|_\infty$ for all bounded $\varphi$, the functions $M(\varphi_n)f \in \mathcal{H}_0$ converge in $\mathcal{H}$ to $g = M(\varphi)f$. Thus $g \in \mathcal{H}_0$ since $\mathcal{H}_0$ is closed. Condition (11), therefore, holds, whence $L^2(Z;\nu) = \mathcal{H}$ by section (b) above.

(d) Hence, $L^2(Z;\nu) = \mathcal{H}$ *if and only if* (13) holds for any measure subset $A$ of $X$. By n° 17, (iii), this means that

$$\text{(24.14)} \qquad \nu_A \text{ and } \nu_{X-A} \quad \text{are disjoint}$$

for all $A$. The generalization to finite or countable partitions of $X$ is obvious. It would be tempting to conjecture that this condition means that the measures $\mu_x$ and $\mu_y$ are disjoint whenever $x \neq y$. As will be shown in point (e) below, this is almost necessary, but the following counterexample shows that it is not sufficient.

*Exercise 2.* Take $X = Z = [0,1]$. Let $m$ denote the Lebesgue measure and $\epsilon_x$ the Dirac measure at $x$. Choose $\mu_x = m$ for $x = 0$, $\mu_x = \epsilon_x$ for $x > 0$ and $\lambda = m + \epsilon_0$. Show that, for $A = \{0\}$ and $B = ]0,1]$, $\nu_A = \nu_B = m$.

(e) In the general case, we let $D$ denote the diagonal of the product $X \times X$ and assume that (13) holds. Then *the measures* $\mu_x$ *and* $\mu_y$ *are disjoint for almost all* $(x,y) \in X \times X - D$. The condition is thus necessary, but not sufficient.

If $\nu_A$ and $\nu_B$ are disjoint, then indeed there is (LN) a Borel set $N \subset Z$ such that

$$\nu_A(Z - N) = 0\,, \quad \nu_B(N) = 0\,.$$

By theorem 22, or (13.7) applied to measures $\nu_A$ and $\nu_B$,

$$\text{(24.13')} \qquad \begin{aligned} \mu_x(Z - N) &= 0 \quad \text{for all } x \in A - N_A\,, \\ \mu_y(N) &= 0 \quad \text{for all } y \in B - N'_A \end{aligned}$$

where $N_A$ and $N'_A$ are $\lambda$-null sets. However, the set $N_{A,B}$ of $(x, y) \in A \times B$ such that the measures $\mu_x$ and $\mu_y$ are not disjoint is contained in $A \times N'_A \cup N_A \times B$, a null set by (LF). Hence

$$\lambda \times \lambda(N_{A,B}) = 0\,.$$

Since $X \times X - D$ is open in the product, it can be covered by open sets of the form $U \times V$, where $U$ and $V$ are open in $X$ and disjoint (Hausdorff). As $X \times X - D$ is locally compact and separable, and so the union of a sequence of compact sets, $X \times X - D$ is the union of sets $U_n \times V_n$ where, for each $n$, $U_n$ are $V_n$ are disjoint open sets. As

$$U_n \times V_n \subset U_n \times (X - U_n)\,,$$

relation (13) for $A = U_n$ and $B = X - U_n$ shows that the set of $(x, y) \in U_n \times V_n$ such that $\mu_x$ and $\mu_y$ are not disjoint is a null set with respect to $\lambda \times \lambda$, proving the result. This is the best that can be said since one can modify the measures $\mu_x$ on a null set without changing $\nu$.

*Exercise 3.* Let $G$ be a unimodular lcg acting properly on a space $Z$, $\nu$ a $G$-invariant measure on $Z$ and $\lambda$ the quotient measure [n° 15, (iv)]. Hence $\nu = \int \mu_x d\lambda(x)$ using notation (15.13). Show that the condition of point (d) holds in this case.

*Exercise 4.* (a) Let $\mu'$ and $\mu''$ be two positive measures on a locally compact space $Z$. For any $f \in L_+(Z)$, set

$$\mu(f) = \inf_{\substack{h \in L_+(Z) \\ h \leq f}} [\mu'(h) + \mu''(f - h)]\,.$$

Show that $\mu$ is the restriction to $L_+(Z)$ of a measure on $Z$ and that a positive measure is bounded above by $\mu'$ and $\mu''$ if and only if it is bounded above by $\mu$. Set $\mu = \inf(\mu', \mu'')$. (b) Suppose $Z$ to be separable. Show that in the previous formula, it is enough to make $h$ vary in a countable set independent of $\mu'$ and $\mu''$ . (c) Let $X$ and $Z$ be two separable locally compact sets, $\lambda$ a positive measure on $X$ and $(\mu'_x)$, $(\mu''_x)$ two families of positive measures on $Z$ measurable with respect to $\lambda$. Set $\mu_{x,y} = \inf(\mu'_x, \mu''_y)$ for $x, y \in X$. Show that, for all $f \in L(X \times X)$, the function $(x, y) \mapsto \mu_{x,y}(f)$ is measurable with respect to $\lambda \times \lambda$ and that the set of $(x, y)$ such that $\mu_{x,y} = 0$ is measurable.

# § 7. The Commutative Fourier Transform

## 25 – Convolution Product on a Lcg

To illustrate the general statements in integration theory and in particular the Lebesgue-Fubini theorem, let us begin by show how they can be used to extend to a lcg $G$ the convolution product which has already been encountered in Chap. VII (Fourier series) and even in Chap. II, n° 18, example 3 for discrete groups. In all that follows, unless otherwise mentioned, "measurable" or "integrable" over $G$ or over $G \times G$ will be with respect to a left invariant measure $dx$ or to the product measure $dxdy$. To simplify, $G$ may be assumed to be countable at infinity, though this assumption is not needed for any of the results.

(i) *Convolutions and representations.* Let us start by proving a reassuring result:

**Lemma 1.** *Let $f$ be a function on $G$ with values in a topological space. The function $(x,y) \mapsto f(xy)$ is measurable on $G\times G$ if and only if $f$ is measurable on $G$.*

Indeed, the most primitive version of LF shows that l'on a

$$\begin{aligned}\iint F(x,y)dxdy &= \iint F\left(x,y^{-1}\right)\Delta(y)dxdy = \int \Delta(y)dy \int F\left(x,y^{-1}\right)dx \\ &= \int \Delta(y)dy \int F\left(xy,y^{-1}\right)\Delta(y)^{-1}dx \\ &= \iint F\left(xy,y^{-1}\right)dxdy\end{aligned}$$

for all $F \in L(G \times G)$, where the function $\Delta$ is that of n° 15, (i). Therefore, the measure $dxdy$ is invariant under the homeomorphism $(x,y) \mapsto (xy, y^{-1})$, so that the latter transforms every measurable function into a measurable function. As it transforms $f(xy)$ into $f(x)$, $(x,y) \mapsto f(xy)$ is measurable if and only if so is $(x,y) \mapsto f(x)$, i.e. if and only if so is $f$, qed.

The most natural way of introducing the convolution product is to start with a representation $(\mathcal{H}, U)$ of $G$ [n° 15, (ii)], which need not be unitary nor bounded. The function $U(x)\mathbf{a}$ being continuous for all $\mathbf{a} \in \mathcal{H}$ and hence bounded on all compact sets $K \subset G$, (n° 15)

$$\sup_{x\in K} \|U(x)\| < +\infty\,. \tag{25.1}$$

The function $\|U(x)\| = \rho(x)$ is, moreover, lsc since it is the upper envelope of the continuous functions $x \mapsto \|U(x)\mathbf{a}\|/\|\mathbf{a}\|$. So, for any compact set $K \subset G$, there is point where $\rho$ is minimum [Chap. V, n° 10, property (vi)]. This minimum cannot be zero since the functions $U(x)$ are invertible, whence

(25.1')
$$\inf_{x \in K} \|U(x)\| > 0\,.$$

This being settled, the continuous function $x \mapsto U(x)\mathbf{a}$ can be integrated (n° 4) with respect to any complex measure $\mu$ such that

(25.2)
$$\int \rho(x) d|\mu|(x) < +\infty\,.$$

The vector

$$U(\mu)\mathbf{a} = \int U(x)\mathbf{a}.d\mu(x)$$

thus obtained depends linearly on $\mathbf{a}$, and

$$\| \int U(x)\mathbf{a}.d\mu(x) \| \leq \int \|U(x)\mathbf{a}\| .d|\mu|(x)$$

by (17.4), also holds for vector-valued functions. The operator $U(\mu)$ is thus continuous and

$$\|U(\mu)\| \leq \int \rho(x) d|\mu|(x)\,.$$

In particular, if the functions $U(x)$ are isometric, in which case $\mu$ must be a bounded measure, then $\|U(\mu)\| \leq \|\mu\|$, where $\|\mu\| = |\mu|(G)$ is the norm of $\mu$.

If $\lambda$ are $\mu$ are two measures satisfying (2), first of all (n° 4)

$$U(x)U(\mu)\mathbf{a} = U(x) \int U(y)\mathbf{a}.d\mu(y) = \int U(x)U(y)\mathbf{a}.d\mu(y)\,.$$

Thus

$$U(\lambda)U(\mu)\mathbf{a} = \int U(x)U(\mu)\mathbf{a}.d\lambda(x) = \int d\lambda(x) \int U(xy)\mathbf{a}.d\mu(y)\,,$$

and

(25.3)
$$U(\lambda)U(\mu)\mathbf{a} = \iint U(xy)\mathbf{a}.d\lambda(x)d\mu(y)\,.$$

Since $\rho(xy) \leq \rho(x)\rho(y)$, the double integral is bounded above by

$$\iint \rho(x)\rho(y).d|\lambda|(x)d|\mu|(y) = \int \rho(x)d|\lambda|(x). \int \rho(y)d|\mu|(y) < +\infty\,,$$

up the a factor $\|\mathbf{a}\|$, and LF justifies the formal calculation.

However, for any $f \in L(G)$, by (1'), the function $|f(x)|$ is dominated by $\|U(x)\|$, up to a constant factor. Hence, since $\|U(xy)\|$ is integrable with respect to $d|\lambda|(x)d|\mu|(y)$, so is $f(xy)$. Setting

$$\nu(f) = \iint f(xy)d\lambda(x)d\mu(y) \tag{25.4}$$

defines a linear functional on $L(G)$. It is positive if this is also the case of $\lambda$ and $\mu$. In the general case, $\lambda$ and $\mu$ are linear combinations of positive measures bounded above by $|\lambda|$ and $|\mu|$. This leads to the conclusion that (4) is a complex *measure.*

(ii) *Convolution of two measures.* Regardless of whether representations of $G$ are used or not, these arguments lead to the definition of the measure

$$\lambda * \mu : f \longmapsto \iint f(xy)d\lambda(x)d\mu(y) = \iint \epsilon_{xy}(f)d\lambda(x)d\mu(y)$$

provide only that

$$\iint |f(xy)|\, d|\lambda|(x)d|\mu|(y) < +\infty \tag{25.5}$$

for all $f \in L(G)$. For example no restrictions need be imposed if one of the measures has compact support because, if $\lambda$ is concentrated on a compact set, the function $y \mapsto \int |f(xy)|d|\lambda|(x)$ is in $L(G)$. The measure $\nu$ has compact support if this is also the case of $\lambda$ and $\mu$ since, denoting by $K$ and $H$ the compact subsets of $G$ on which $\lambda$ and $\mu$ are concentrated, $\nu(f) \neq 0$ only if the function $f(xy)$ is not identically zero on $K \times H$. This supposes that $f$ is not identically zero on the compact set $KH$ of $xy$ where $x \in K$ and $y \in H$. The measure $\nu$ is, therefore, concentrated on $KH$. In particular, if $\lambda$ and $\mu$ are the Dirac measures $\epsilon_a$ and $\epsilon_b$ at the points $a$ and $b$ of $G$, then

$$\epsilon_a * \epsilon_b = \epsilon_{ab}\,,$$

and, for any measure $\mu$, the measure

$$\epsilon_a * \mu : f \longmapsto \int f(ay)d\mu(y)\,,$$

is the image of $\mu$ under the translation $g \mapsto ag$. There is a similar result for $\mu * \epsilon_a$. Writing simply $\epsilon$ for the Dirac measure at $e$ reveals its role of unit element for the convolution product.

The definition also makes sense if $\lambda$ and $\mu$, hence $\lambda \times \mu$ as well, are bounded since, for $f \in L(G)$, the function $f(xy)$ is continuous and bounded on $G \times G$. The measure $\nu$ is then bounded and

$$\|\lambda * \mu\| \leq \|\lambda\|.\|\mu\| \tag{25.6}$$

since integral (5) is $\leq \|\lambda\|.\|\mu\|.\|f\|$.

As $\lambda * \mu$ is the image of $\lambda \times \mu$ under $(x, y) \mapsto xy$, theorem 23 of n° 14 can be applied. It follows that:

(a) a function $f$ defined on $G$ and with values in a topological space is measurable with respect to $\nu$ if and only the function $(x, y) \mapsto f(xy)$ is measurable with respect to the product measure $|\lambda| \times |\mu|$,

(b) a reasonable function $f$ with values in a Banach space is integrable with respect to $\lambda * \mu$ if and only if it satisfies (5),

(c) relation (4) continues to hold in this case. In particular, (3) shows that

$$U(\lambda * \mu) = U(\lambda)U(\mu)\,. \tag{25.7}$$

*Exercise 1.* Show that

$$\lambda * \mu = \int \epsilon_x * \mu.d\lambda(x) = \int \lambda * \epsilon_x.d\mu(x)$$

(apply the definition of n° 12) and compare the previous results to those obtained by directly applying the generalized LF theorem [n° 13, theorem 22, in cases where condition (a) holds].

Every measure is the continuous sum of punctual measures

$$\mu = \int \epsilon_x.d\mu(x)$$

and the convolution product, which from this point of view is just

$$\lambda * \mu = \iint \epsilon_x * \epsilon_y.d\lambda(x)d\mu(y)\,,$$

consists in extending by linearity group multiplication to these sums. This is the argument which, a century ago, led Schur to define "the algebra" of a finite group: it is a vector space over $\mathbb{C}$ (or any other commutative field) having as basis the elements of the group, i.e. the Dirac measures, and equipped with a an associative and linear multiplication whose restriction to the basis is the composition law of $G$. In the case of a general lcg, the analogue of Schur's construction is any one of the following: the algebra $\mathcal{M}^1(G)$ of bounded measures on $G$ or the algebra $L^1(G)$ equipped with the convolution defined below. The second choice is better than the first one which is far too extensive.

The convolution product is *bilinear* (obvious) and *associative*:

$$(\lambda * \mu) * \nu = \lambda * (\mu * \nu) \tag{25.8}$$

provided

$$\iiint |f(xyz)|\, d|\lambda|(x)d|\mu|(y)d|\nu|(z) < +\infty \tag{25.9}$$

for all $f \in L(G)$. Indeed then

$$\iint f(xyz)d\lambda(x)d\mu(y) = \int f(uz)d\lambda * \mu(u)$$

by LF, so that

$$\iiint f(xyz)d\lambda(x)d\mu(y)d\nu(z) = \iint f(uz)d\lambda * \mu(u)d\nu(z)$$

is the integral of $f$ with respect to the left hand side of (8). But it is also

$$\int d\lambda(x) \iint f(xyz)d\mu(y)d\nu(z) = \int d\lambda(x) \int f(xv)d\mu * \nu(v)\,,$$

whence (8). The order of integration being of no importance, under assumption (9), relation

$$\lambda * \mu * \nu = \int \lambda * \epsilon_x * \nu . d\mu(x)$$

and other similar ones follow.

The case of a *unitary* representation $(\mathcal{H}, U)$ leads to the definition of an operation $\mu \mapsto \tilde{\mu}$ on measures which plays an important role. It is obtained by calculating the adjoint of $U(\mu)$:

$$\begin{aligned}(U(\mu)^*\mathbf{a}|\mathbf{b}) = \overline{(U(\mu)\mathbf{b}|\mathbf{a})} &= \int \overline{(U(x)\mathbf{b}|\mathbf{a})}d\overline{\mu}(x) = \int \left(U\left(x^{-1}\right)\mathbf{a}|\mathbf{b}\right)d\overline{\mu}(x) \\ &= \int \left(U(x)\mathbf{a}|\mathbf{b}\right)d\tilde{\mu}(x)\,,\end{aligned}$$

where

$$d\tilde{\mu}(x) = d\overline{\mu}(x^{-1}) \tag{25.10}$$

is set to be the image under $x \mapsto x^{-1}$ of the conjugate $\overline{\mu}$. The notation $\mu^*$ would be better since

$$U(\mu)^* = U(\tilde{\mu})\,. \tag{25.11}$$

An immediate calculation shows that

$$(\lambda * \mu)^{\tilde{}} = \tilde{\mu} * \tilde{\lambda} \tag{25.12}$$

if the convolution of $\lambda$ and $\mu$ exists, and if

$$\tilde{\epsilon}_x = \epsilon_{x^{-1}} \quad \text{for all } x \in G\,.$$

(iii) *Convolution of a measure and a function.* We suppose $d\mu(x) = \varphi(x)dx$ where $\varphi$ is locally integrable. Then, formally,

$$\nu(f) = \iint f(xy).d\lambda(x)\varphi(y)dy = \iint f(xy).\varphi(y)d\lambda(x)dy$$
$$= \iint f(y)\varphi(x^{-1}y).d\lambda(x)dy\,.$$

Thus, permuting the names of the variables $x$ and $y$,

$$\nu(f) = \int f(x).\lambda * \varphi(x)dx\,,$$

where the function

$$\lambda * \varphi(x) = \int \varphi\left(y^{-1}x\right) d\lambda(y) \tag{25.13}$$

is, by definition, the convolution of the function $\varphi$ and the measure $\lambda$. For example,

$$\epsilon_y * \varphi(x) = \varphi\left(y^{-1}x\right)\,, \tag{25.14}$$

whence

$$\lambda * \varphi(x) = \int \epsilon_y * \varphi(x).d\lambda(y)\,.$$

No problems arise when $\varphi$ or $\lambda$ has compact support, or else when $\varphi$ is integrable and $\lambda$ bounded, in which case $\varphi * \lambda$ is integrable (since the measure $\lambda * \mu$ is bounded) and satisfies

$$\|\lambda * \varphi\|_1 \leq \|\lambda\|.\|\varphi\|_1\,.$$

If we now suppose that $d\lambda(x) = \varphi(x)dx$, then for any measure $\mu$, formally,

$$\nu(f) = \iint f(xy)\varphi(x)dxd\mu(y) = \iint f(x)\varphi\left(xy^{-1}\right)\Delta(y)^{-1}dxd\mu(y)\,.$$

So $d\nu(x) = \varphi * \mu(x)dx$, where

$$\varphi * \mu(x) = \int \varphi\left(xy^{-1}\right)\Delta\left(y^{-1}\right)d\mu(y)\,. \tag{25.13'}$$

In particular,

$$\varphi * \epsilon_y(x) = \varphi\left(xy^{-1}\right)\Delta(y)^{-1}\,. \tag{25.14'}$$

These factors $\Delta$ disappear if $G$ is unimodular.

**Theorem 36**. *If $\lambda$ is a* bounded *measure, then $\lambda * \varphi$ is defined for all $\varphi \in L^p(G)$ and all $p \in [1, +\infty]$, and is again in $L^p$ and*

$$\|\lambda * \varphi\|_p \leq \|\lambda\|.\|\varphi\|_p . \tag{25.15}$$

$\lambda$ and $\varphi$ may be supposed to be positive by replacing them with their absolute values. We first need to show that, for all $f \in L_+(G)$, the function $f(xy)$ is integrable with respect to $d\lambda(x)\varphi(y)dy$, i.e. (theorem 25) that $f(xy)\varphi(y)$ is integrable with respect to $d\lambda(x)dy$. First, it is measurable with respect to $d\lambda(x)dy$: the function $(x, y) \mapsto \varphi(x)$ is measurable since $\varphi$ is measurable with respect to $dx$ [n° 11, property (C7)], and so is $(x, y) \mapsto f(xy)$ since it is continuous. By LF [Theorem 19, (iv)], it therefore suffices to show that

$$\int d\lambda(x) \int f(xy)\varphi(y)dy < +\infty , \tag{25.16}$$

where *a priori* these are upper integrals. The integral with respect to $y$ poses no problem since $\varphi$ is locally integrable. As $L(G) \subset L^q(G)$, where $1/p+1/q = 1$, given left invariance of the measure needed to define $\|f\|_q$,

$$\int f(xy)\varphi(y)dy \leq \|f\|_q\|\varphi\|_p .$$

So the left hand side of (16) is $\leq \|\lambda\|.\|f\|_q\|\varphi\|_p < +\infty$, which proves the first result (corollary of theorem 30, n° 18). It remains to show that the measure $\lambda * \varphi$ is in fact defined by a function satisfying (15).

We start by observing that the measure $d\lambda(x)dy$ is invariant under the homeomorphism $(x, y) \mapsto (x, x^{-1}y)$ from $G \times G$ onto $G \times G$ since for all $F \in L(G \times G)$,

$$\iint F(x, y)d\lambda(x)dy = \int d\lambda(x) \int F(x, y)dy = \int d\lambda(x) \int F\left(x, x^{-1}y\right) dy .$$

As this homeomorphism transforms the function $f(xy)\varphi(y)$ appearing in (16) into $f(y)\varphi(x^{-1}y)$, if $\nu = \lambda * \varphi$, then

$$\begin{aligned} \nu(f) = \iint f(xy)\varphi(y)d\lambda(x)dy &= \iint f(y)\varphi\left(x^{-1}y\right) d\lambda(x)dy = \\ &= \int f(y)dy \int \varphi\left(x^{-1}y\right) d\lambda(x) . \end{aligned}$$

LF then shows that, with respect to the Haar measure, *the function $x \mapsto \varphi(x^{-1}y)$ is $\lambda$-integrable*[96] *for almost all $y$* where $f$ is not zero, hence almost

[96] This result always seems incredible when first encountered: what does measurability with respect to $dx$ have to do with measurability with respect to an almost arbitrary measure $\lambda$? But we do not claim that $\varphi(x^{-1}y)$ is $\lambda$-measurable with respect to *all* $y$, we only claim that this is the case *almost* always. Exercise: suppose $G = \mathbb{R}$ and $\lambda(f) = \sum f(\xi_n)/n^2$, where $n \mapsto \xi_n$ is a bijection from $\mathbb{N}$ onto $\mathbb{Q}$. Find an interpretation for this result in this case.

everywhere if $G$ is supposed to be countable at infinite (or almost everywhere on every compact set otherwise). In addition, this calculation shows that the function

$$\lambda * \varphi(y) = \int \varphi\left(x^{-1}y\right) d\lambda(x)\,,$$

defined almost everywhere, is locally integrable and that as was seen above,

$$|\nu(f)| \leq \int |\lambda * \varphi(y)| \,.\, |f(y)|\, dy \leq \|\lambda\|.\|f\|_q\|\varphi\|_p$$

for all $f \in L(G)$. Corollary of theorem 30, n° 18 finally shows that $\lambda * \varphi \in L^p(G)$ and satisfies (15), qed.

If $U_p(x)$ denotes the representation of $G$ by left translations on $L^p$ [n° 15, (ii)], the function $y \mapsto \varphi(x^{-1}y)$ is just $U_p(x)\varphi$. However, the general definition (4)

$$U_p(\lambda)\mathbf{a} = \int U_p(x)\mathbf{a}.d\lambda(x)$$

applies to the representation considered on $L^p(G)$. It is, therefore, "obvious" that, in this case,

$$U_p(\lambda)\varphi = \lambda * \varphi \tag{25.17}$$

for all $\varphi \in L^p$. This is correct, but the left hand side is the integral of the function $x \mapsto U_p(x)\varphi$ with values in $L^p$, defined by the general method of n° 4, (i), whereas the right hand side is defined in a completely different way. Hence a proof is needed.

For the convenience of calculations, generally speaking, we set

$$(\varphi|\psi) = \int \varphi(x)\overline{\psi(x)}dx \quad \text{if } \varphi \in L^p\,,\ \psi \in L^q\,.$$

To show that $\varphi = 0$, it suffices to show that $(\varphi|f) = 0$ for all $f \in L(G)$. So the proof reduces to checking that

$$(U_p(\lambda)\varphi - \lambda * \varphi|f) = 0 \quad \text{for } f \in L(G)\,.$$

As $\varphi \mapsto (\varphi|f)$ is a continuous linear functional on $L^p$, [n° 4, (i)]

$$\left(\int U_p(x)\varphi.d\lambda(x)\middle|f\right) = \int \left(U_p(x)\varphi|f\right) d\lambda(x)\,.$$

Hence by definition of $U_p(x)$, at least formally,

$$\begin{aligned}\left(\int U_p(x)\varphi|f\right)d\lambda(x) &= \int d\lambda(x) \int \varphi\left(x^{-1}y\right) \overline{f(y)}dy \\ &= \int \overline{f(y)}dy \int \varphi\left(x^{-1}y\right) d\lambda(x) \\ &= \int \lambda * \varphi(y).\overline{f(y)}dy = (\lambda * \varphi|f)\,.\end{aligned}$$

This calculation is justified by LF since we already know that under the assumptions made, the function $\overline{f}(y)\varphi(x^{-1}y)$ is integrable with respect to $d\lambda(x)dy$. Hence (17).

As $U(\lambda * \mu) = U(\lambda)U(\mu)$ in any representation, (17) shows that

$$(\lambda * \mu) * \varphi = \lambda * (\mu * \varphi)$$

for all bounded measures $\lambda$ and $\mu$ and the function $\varphi \in L^p(G)$. This also follows from (8).

Finally, the map $\mu \mapsto \tilde{\mu}$ defined in (10) can be restated when $d\mu(x) = \varphi(x)dx$. Clearly,

$$d\tilde{\mu}(x) = \overline{\varphi(x^{-1})}d\left(x^{-1}\right) = \tilde{\varphi}(x)dx\,,$$

where

$$\tilde{\varphi}(x) = \overline{\varphi(x^{-1})}/\Delta(x), \quad = \overline{\varphi(x^{-1})} \quad \text{if } G \text{ is unimodular}. \tag{25.18}$$

The map $\varphi \mapsto \tilde{\varphi}$ is isometric on $L^1(G)$, and even on all $L^p(G)$ if $G$ is unimodular. Relation (12) shows that

$$(\lambda * \varphi)\tilde{} = \tilde{\varphi} * \tilde{\lambda}\,. \tag{25.19}$$

(iv) *Convolution of two functions.* If $d\lambda(x) = \varphi(x)dx$ and $d\mu(x) = \psi(x)dx$ are absolutely continuous, whence $d|\lambda|(x)d|\mu|(y) = |\varphi(x)\psi(y)|dxdy$, then the existence of the convolution product means (theorem 25) that $f(xy)\varphi(x)\psi(y)$ is integrable over $G \times G$ for all $f \in L(G)$. Besides, it would be enough to check it when $f$, instead of being continuous, is the characteristic function of a compact set $K \subset G$, which would be reflected by the condition

$$\iint_{xy\in K} |\varphi(x)\psi(y)|\,dxdy < +\infty\,.$$

Then

$$\nu(f) = \iint f(xy)\varphi(x)\psi(y)dxdy = \iint f(x)\varphi(xy)\psi(y^{-1})dxdy\,.$$

Hence, if the measure $\nu$ is well-defined, the function $y \mapsto \varphi(xy)\psi(y^{-1})$ is (LF) integrable for almost all $x$ and its integral is a locally integrable function of $x$ since its product with any $f \in L(G)$ is integrable. This gives the definition of the convolution

$$\varphi * \psi(x) = \int \varphi(xy)\psi(y^{-1})dy = \int \varphi(y)\psi(y^{-1}x)dy \tag{25.20}$$

of two locally integrable functions and the formula

$$\nu(f) = \int f(x).\varphi * \psi(x)dx\,. \tag{25.21}$$

Instead of $y$, it is possible to take $y^{-1}$ in (20) provided $dy = d_l y$ is replaced by $dy^{-1} = \Delta(y)^{-1}dy$, a detail that should not be forgotten when $G$ is not unimodular.

There still remains to find assumptions under which integral (20) will be well-defined. This is obviously the case if one of the functions $\varphi$, $\psi$ is in $L(G)$ and the other one is integrable, the result being a continuous function with compact support if this is also the case of $\varphi$ and $\psi$. Theorem 36 leads to a less trivial result:

**Theorem 37.** *Let $G$ be a locally compact group. For all $\varphi \in L^1(G)$ and $\psi \in L^p(G)$, $1 \leq p \leq +\infty$, the function $y \mapsto \varphi(xy)\psi(y^{-1})$ is integrable for almost all $x \in G$, the function*

$$\int \varphi(xy)\psi(y^{-1})dy = \varphi * \psi(x)$$

*is in $L^p(G)$ and*

$$\|\varphi * \psi\|_p \leq \|\varphi\|_1.\|\psi\|_p\,. \tag{25.22}$$

If $G$ is *unimodular,* the product $\varphi * \psi$ exists when $\varphi \in L^p$, $\psi \in L^q$ with $1/p+1/q = 1$. To see this it suffices to verify that $f(xy)\varphi(x)\psi(y)$ is integrable over $G \times G$ for all $f \in L(G)$. The change of variable $(x, y) \mapsto (x, x^{-1}y)$ reduces the verification to the function $f(y)\varphi(x)\psi(x^{-1}y)$, and the change of variable $(x, y) \mapsto (yx, y)$ to $f(y)\varphi(yx)\psi(x^{-1})$. The function $x \mapsto \varphi(yx)$ is also in $L^p$ for all $y$, and $x \mapsto \psi(x^{-1})$ is again in $L^q(G)$ since $G$ is unimodular. Moreover, the norm of these functions is the same as that of $\varphi$ and $\psi$. Hence $\int |\varphi(yx)\psi(x^{-1})|dx \leq \|\varphi\|_p\|\psi\|_q$, and so the double integral is convergent, with here too relation (19).

For $p = 1$, (22) enables us to consider $L^1(G)$, equipped with the convolution product and the norm of $L^1$, as a complete normed algebra – except that it does not necessarily have a unit element. Such an element would indeed be a *function* $\epsilon \in L^1$ for which $\epsilon * f = f$ for all $f$, i.e. for which

$$\int f(xy)\epsilon\left(y^{-1}\right)dy = f(x) \quad \text{almost everywhere}\,.$$

But if $f \in L(G)$, both sides are continuous functions of $x$, so that the previous relation must hold for all $x$. So the measure $\epsilon(y^{-1})dy$ is the initial Dirac measure. It can be defined by an integrable function only if $G$ is *discrete.* Replacing $L^1$ by the product

$$\mathbf{A}^1(G) = \mathbb{C} \times L^1(G)$$

equipped with the obvious multiplication and the norm $\|(\alpha, f)\| = |\alpha| + \|f\|_1$ reduces the general case to a normed algebras with unit element $\epsilon$. This boils down to introducing measures of the form $\alpha d\epsilon(x) + f(x)dx$ with $\alpha \in \mathbb{C}$.

When $\varphi$ and $\psi$ are both integrable, relation (7) shows that

$$U(\varphi * \psi) = U(\varphi)U(\psi) \tag{25.23}$$

for any *bounded* representation $(\mathcal{H}, U)$ of $G$, i.e. such that $\sup \|U(x)\| < +\infty$. If $(\mathcal{H}, U)$ is *unitary,* (11) shows that

$$U(\varphi)^* = U(\tilde{\varphi})\,. \tag{25.23'}$$

It should be noted that, for $\varphi, \psi \in L(G)$,

$$\begin{aligned}\tilde{\psi} * \varphi(x) &= \int \tilde{\psi}(xy)\varphi\left(y^{-1}\right) dy = \int \overline{\psi\left(y^{-1}x^{-1}\right)}\Delta(xy)^{-1}\varphi\left(y^{-1}\right) dy \\ &= \int \overline{\psi\left(yx^{-1}\right)}\Delta(x)^{-1}\varphi(y)dy = \int \varphi(yx)\overline{\psi(y)}dy\end{aligned}$$

since $d(yx) = \Delta(x)dy$. For $x = e$, this gives the very useful relation

$$(\varphi|\psi) = \epsilon(\tilde{\psi} * \varphi) \tag{25.24}$$

which enables us to compute the inner product of $L^2(G)$, at least if $\varphi, \psi \in L(G)$. Similarly

$$\varphi * \tilde{\psi}(x) = \int \varphi(xy)\tilde{\psi}(y^{-1})dy = \int \varphi(xy)\overline{\psi}(y)\Delta(y)dy$$

and so

$$(\varphi|\psi) = \epsilon\left(\varphi * \tilde{\psi}\right) \quad \text{if } G \text{ is unimodular}\,. \tag{25.24'}$$

Therefore, the identity

$$\epsilon(f * g) = \epsilon(g * f) \quad \text{for } f, g \in L(G) \tag{25.25}$$

characterizes *unimodular* groups. Measures satisfying it are said to be *central* and will play an important role in n° 31.

(v) *Dirac sequences* . We saw how useful they were in Chap. V, n° 27 and again in Chap. VII in relation to Fourier sequences and integrals. In fact they can be defined in any locally compact space and, in the case of a group, they can be constructed using convolution products. Let us start with a general result:

**Lemma 2.** *Let $X$ be a locally compact space, $a$ a point of $X$ and $(\mu_n)$ a sequence*[97] *of bounded complex measures on $X$. Assume the following conditions hold:*

(D 1) $\sup \|\mu_n\| < +\infty$;
(D 2) $\lim \int d\mu_n(x) = 1$;
(D 3) for any compact neighbourhood $V$ of $a$,

$$\lim \int_{X-V} d|\mu_n|(x) = 0 \,. \tag{25.26}$$

*Then*

$$\lim \int f(x) d\mu_n(x) = f(a) \tag{25.27}$$

*for all Borel functions $f$, bounded and continuous at $a$. If the measures $\mu_n$ are carried by a fixed compact set $K$, (27) holds for any Borel function continuous at $a$ and bounded on $K$.*

Dividing $\mu_n$ by $\mu_n(1)$ which, by (D 2), tends to 1 reduces the proof to the case where $\mu_n(1) = 1$ for all $n$. The function $f$ being Borel and bounded, and hence integrable with respect to any bounded measure,

$$\mu_n(f) - f(a) = \mu_n\,[f - f(a)] = \mu_n(g)\,,$$

where $g$ satisfies the same assumptions as $f$, with moreover $g(0) = 0$. Then, for any compact neighbourhood $V$ of $a$,

$$\mu_n(g) = \int_V g(x) d\mu_n(x) + \int_{X-V} g(x) d\mu_n(x)\,.$$

The absolute value of the right hand side is bounded above by integral (26), up to the factor $\|g\|_\infty$, and so tends to 0 for all $V$. As $g$ is continuous and zero at $a$, $|g(x)| < r$ on $V$ if $V$ is sufficiently small. For such a neighbourhood $V$ and sufficiently large $n$, (D 1) shows that the left hand side is $\leq Mr$, where $M$ is a constant, whence (27). If there is a compact set $K \subset X$ such that $|\mu_n|(X - K) = 0$ for all $n$, it is then unnecessary to investigate what is happening in $X - K$, and replacing $\|g\|_\infty$ by the uniform norm $\|g\|_K$ on $K$, the argument remains the same, qed.

It should again be noted that the proof equally applies to functions with values in a Banach space. This can prove useful as we will see.

If $(\mathcal{H}, U)$ is a representation of some lcg $G$, this result can indeed be applied to the function $x \mapsto U(x)\mathbf{a}$ for all $\mathbf{a} \in \mathcal{H}$, since such a function

[97] This countability assumption could be avoided by considering a family of measures $\mu_V$ dependent on a sufficiently small neighbourhood $V$ of $e$ and by changing conditions (D 1) to (D 3) in an obvious manner.

is continuous and bounded on every compact set. As a result, *if* $(\mu_n)$ *is a Dirac sequence at* $a = e$, *then*

$$\lim U(\mu_n)\mathbf{a} = \mathbf{a} \tag{25.28}$$

*for all* $\mathbf{a} \in \mathcal{H}$.

Let us for example choose the representation given by left translations on $L^p(G)$ and apply (17). We find that

$$\text{l.i.m.}^p \mu_n * \varphi = \varphi \quad \text{for all } \varphi \in L^p(G)\,. \tag{25.29}$$

If $d\mu_n(x) = f_n(x)dx$, where the functions $f_n$ are in $L_+(X)$, with integrals equal to 1, and for large $n$, vanish outside an arbitrary compact neighbourhood of the unit element, then the conditions of lemma 2 hold. Thus

$$\text{l.i.m.}^p f_n * \varphi = \varphi\,. \tag{25.29'}$$

But in this case, the functions $f_n * \varphi$ are *continuous*. Indeed, for all $f \in L(X)$,

$$f * \varphi(x) = \int f(xy)\varphi(y^{-1})dy\,.$$

As $x$ tends to a limit $x_0$, the function $y \mapsto f(xy)$ converges uniformly to $y \mapsto f(xy_0)$ while remaining zero outside a fixed compact set over which $\varphi$ is integrable. This enables to pass to the limit in the integral and so gives a systematic way of approximating any $f \in L^p$ by continuous functions.

These results admit several variants that one learns as and when required.

Till n° 28, we will content ourselves with commutative lcgs.

## 26 – Fourier Transform[98] on $L^1(G)$

(i) *Characters of a commutative* lcg. Though generally speaking the algebra $L^1(G)$ does not have any unit element, we will say that a *character* of $L^1(G)$ is a homomorphism $\chi$ from $L^1(G)$ to $\mathbb{C}$ satisfying $\chi(\epsilon) = 1$ if $G$ is discrete; $\chi$ may, on the contrary, be considered to be the trivial homomorphism when $G$ is not discrete. I will denote by $X(G)$ the set of these characters and by $\infty$ the trivial character. Every character $\chi$ of $L^1$ then arises from the character $(\lambda, f) \mapsto \lambda + \chi(f)$ of the algebra $\mathbf{A}^1(G) = \mathbb{C} \times L^1(G)$, and so is continuous. At the same time, $X(G)$ is just the spectrum of the normed algebra $\mathbf{A}^1(G)$.

[98] All following results are already in André Weil's book, but he used what was then known (L. Pontrjagin) on the construction of general commutative lcgs based on "classical" groups. Henri Cartan and the author's article, *Analyse harmonique et théorie de la dualité dans les groupes abéliens localement compacts* (Ann. scient. de l'École normale supérieure, 1947) gives a direct presentation of the theory starting from the Gelfand-Raïkov theorem on the existence of irreducible unitary representations for every lcg and using functions of positive type which we will discuss in the next §.

*Example* 1. If $G = \mathbb{R}$, the Fourier transform immediately provides us with such homomorphisms, namely

$$f \mapsto \widehat{f}(y) = \int f(x)\overline{\mathbf{e}(xy)}dx\,,$$

where recall that $\mathbf{e}(x) = \exp(2\pi ix)$. If $\mathbb{R}$ is replaced by a Cartesian space $E$ with dual $E^*$, then likewise, setting

$$\mathbf{e}(x,y) = \exp\left(2\pi i\,\langle x,y\rangle\right)\,, \tag{26.1}$$

where $\langle x, y\rangle = y(x)$ is the value of the linear functional $y \in E^*$ at $x \in E$, every $y \in E^*$ defines a character

$$f \mapsto \widehat{f}(y) = \int f(x)\overline{\mathbf{e}(x,y)}dx$$

of $L^1(E)$, where $dx$ is "the" Lebesgue measure on $E$ used to define the convolution product

$$f * g(x) = \int f(x-y)g(y)dy\,.$$

**Lemma 1.** *Let $\chi$ be a homomorphism from $L^1(G)$ to $\mathbb{C}$ which is not identically zero. Then there is a unique continuous function $\chi(x)$ on $G$ such that*

$$\chi(xy) = \chi(x)\chi(y) \tag{26.2}$$

*and for which*

$$\chi(f) = \int f(x)\overline{\chi(x)}dx \tag{26.3}$$

*for all $f \in L^1(G)$. Then $|\chi(x)| = 1$ and conversely, every bounded and measurable solution of (2) defines a character of $L^1(G)$.*

The existence of a bounded and measurable function $\chi(x)$ satisfying (3) follows from the duality of $L^p$ (n° 18, theorem 30). By definition of $f * g$, relation $\chi(f * g) = \chi(f)\chi(g)$ can then be written

$$\iint \overline{\chi(xy)}f(x)g(y)dxdy = \iint \overline{\chi(x)\chi(y)}f(x)g(y)dxdy\,.$$

This is in particular the case for $f, g \in L(G)$. If $\chi$ is shown to be almost everywhere equal to a *continuous* function, it will obviously satisfy the first relation (1), $|\chi(x)| = 1$ following from the fact that, for any $x$, the set of numbers $\chi(x)^n = \chi(x^n)$ must be bounded.

But as the previous equality holds for all $f, g \in L(G)$,

$$\int g(y)\overline{\chi(xy)}dy = \int g(y)\overline{\chi(x)\chi(y)}dy = \overline{\chi(x)}\chi(g)$$

necessarily holds for almost all $x$. The left hand side is the convolution of a locally integrable function and of a continuous function with compact support $y \mapsto g(y^{-1})$. So it is a continuous function of $x$. Thus the result follows by choosing $g$ in such a way that $\chi(g) \neq 0$. The converse is obvious, qed.

A bounded, continuous solution of (2) which is identically zero is, by definition, a *character* of the group $G$; $\widehat{G}$ will denote the set of characters of $G$. Hence

$$X(G) = \begin{array}{ll} \widehat{G} & \text{if } G \text{ is discrete} \\ \widehat{G} \cup \{\infty\} & \text{otherwise}. \end{array} \tag{26.4}$$

Following André Weil, the elements of $\widehat{G}$ are often written by letters such as $\widehat{x}$, $\widehat{y}$, $\widehat{g}$, etc, the value of the character $\widehat{x}$ at $x \in G$ being written $\langle x, \widehat{x}\rangle = \widehat{x}(x)$. To avoid confusion with the notation of linear algebra – also found in Fourier transforms: example 1 above –, I will instead use the notation

$$\mathbf{e}(x, \chi) = \chi(x) \tag{26.5}$$

which recalls that of Chap. VII.

*Example* 2. For $G = \mathbb{R}$, we recover the exponentials $\mathbf{e}(xy)$ of Fourier transforms. There are no other solutions given the characterization of exponential functions by their functional equation (Chap. IV). In the more general case of a Cartesian space $E$, we clearly only get the functions $\mathbf{e}(x, y)$ of example 1. If $G = \mathbb{T}$, the multiplicative group of complex numbers with absolute value 1, we obtain the exponentials $\mathbf{e}_n(u) = u^n$ of the theory of Fourier series (Chap. VII, § 1), etc.

As the product of two characters is again a character, and so is the inverse of a character, multiplication defines a commutative group structure on $\widehat{G}$. Hence

$$\mathbf{e}(xy, \chi) = \mathbf{e}(x, \chi)\mathbf{e}(y, \chi), \quad \mathbf{e}(x, \chi\chi') = \mathbf{e}(x, \chi)\mathbf{e}(x, \chi'), \tag{26.6}$$

$$\mathbf{e}(x, \chi)^{-1} = \mathbf{e}\left(x, \chi^{-1}\right) = \overline{\mathbf{e}(x, \chi)}, \quad |\mathbf{e}(x, \chi)| = 1. \tag{26.6'}$$

*Example* 3. For $G = \mathbb{R}$, the characters are parameterized by real numbers and multiplication in $\widehat{G}$ becomes addition in $\mathbb{R}$. For $G = \mathbb{T}$, they are parameterized by $n \in \mathbb{Z}$ and $\widehat{G}$ is the additive group $\mathbb{Z}$. For $G = \mathbb{Z}$, the characters are functions $n \mapsto u^n$, where $u \in \mathbb{T}$, and, as a group, $\widehat{G}$ is isomorphic to $\mathbb{T}$. By the way, note that in these three classical cases, $\widehat{\widehat{G}} = G$ provided $\widehat{G}$ is equipped with the usual topology.

*Exercise 1.* Determine $\widehat{G}$ when $G$ is the multiplicative group $\mathbb{C}^*$.

In the light of these examples, in the general case, we associate a *Fourier transform*

$$(26.7) \qquad \widehat{f}(\chi) = \int f(x)\overline{\mathbf{e}(x,\chi)}dx$$

to all functions $f \in L^1(G)$. By definition of characters,

$$(26.7') \qquad \widehat{f}(\chi) = \chi(f)$$

and so, trivially, $\widehat{f * g} = \hat{f}\hat{g}$ and $\widehat{\bar{f}} = \overline{\hat{f}}$. Also note the very useful formula

$$(26.8) \qquad f * \chi(x) = \int f(y)\chi\left(xy^{-1}\right)dy = \widehat{f}(\chi)\chi(x)$$

valid for all $f \in L^1(G)$.

(ii) *The topology on the dual group.* In the standard theory, the Fourier transform of an integrable function is continuous and tends to 0 at infinity (Chap. VII, n° 27, theorem 23). To generalize this theorem of Riemann-Lebesgue proved in Chap. VII by methods special to $\mathbb{R}$, we are going to transform it into the *definition* of the topology of $\widehat{G}$. We will thus avoid proving it, even if it means checking whether this topology is indeed the usual one in standard cases.

Thus first of all it means equipping $\widehat{G}$ with a topology that will make the Fourier transforms $\widehat{f}(\chi) = \chi(f)$ continuous. The simplest way of proceeding is to do the bare minimum to ensure this result. However, $\widehat{G}$ only differs from the spectrum $X(G)$ of $\mathbf{A}^1(G)$ by the possible addition of the character $(\lambda, f) \mapsto \lambda$. So the solution is to equip $\widehat{G}$ with the weak topology of the dual of $\mathbf{A}^1(G)$, i.e. of the dual $L^\infty$ of $L^1$: a neighbourhood of a character $\chi_0 \in \widehat{G}$ must contain the intersection of a *finite* number of subsets of $\widehat{G}$ each defined by a relation of the form

$$|\chi(f) - \chi_0(f)| < r, \quad \text{i.e.} \quad \left|\widehat{f}(\chi) - \widehat{f}(\chi_0)\right| < r.$$

Now, the set $X(G)$ of characters of $\mathbf{A}^1(G)$ is compact with respect to the weak topology (n° 20, end). Since by lemma 1 or relation (4), $\widehat{G}$ can be identified with $X(G)$ if $G$ is discrete and with $X(G) - \{\infty\}$ otherwise, $\widehat{G}$ is a *locally compact* space in all cases (and compact if $G$ is discrete, in which case it is not necessary to check that the Fourier transforms tend to 0 at infinity!). If $G$ is not discrete, the compact subsets of $\widehat{G}$ are the closed subsets of $X(G)$ which do not contain 0. Now, any relation of the form $|\widehat{f}(\chi)| \geq r$ with $r > 0$ defines a closed subset of $X(G)$ which does not contain 0. Thus relation $|\widehat{f}(\chi)| \geq r$ defines a *compact* subset of $\widehat{G}$, and so the function $\widehat{f}$ tends to

0 at infinity[99] on $\widehat{G}$ as expected. This result, trivial except for vocabulary, explains the notation $\infty$.

**Lemma 2**. *The weak topology is identical to the topology of compact convergence on $\widehat{G}$, t*

(This shows that, for $G = \mathbb{R}$, this is the usual topology of $\mathbb{R}$). First, compact convergence implies weak convergence in the unit ball of $L^\infty$. Indeed, for $f \in L^1$ and $\varphi, \varphi_0$ in the unity ball of $L^\infty$,

$$\left| \int_G f(x)\varphi(x)dx - \int_G f(x)\varphi_0(x)dx \right| \leq \int_G |f(x)| . |\varphi(x) - \varphi_0(x)|\, dx$$
$$= \int_K |f(x)| . |\varphi(x) - \varphi_0(x)|\, dx + \int_{G-K}$$

for every compact set $K \subset G$. For all $r > 0$, $K$ can be chosen so that the integral of $|f|$ over $G - K$ is $< r$. Then the latter integral is $\leq 2r$ for all $\varphi$ and $\varphi_0$. If moreover $|\varphi(x) - \varphi_0(x)| \leq r$ on $K$, in other words if $\varphi$ is sufficiently near $\varphi_0$ with respect to the topology of compact convergence, the former one is $\leq r\|f\|_1$. Hence the result.

Conversely, to prove that, weak convergence in $\widehat{G}$ implies compact convergence, we use a general result:

**Lemma 3**. *Let $E$ be a Banach space, $E'$ its topological dual, $B'$ the unit ball*[100] *of $E'$ and $K$ a compact subset of $E$. Weak convergence in $B'$ is equivalent to uniform convergence in $K$.*

We need to show that, for all $f_0 \in B'$ and all $r > 0$, there exists a neighbourhood $V$ of $f_0$ in $B'$ equipped with the weak topology such that

$$f \subset V \Longrightarrow |f(\mathbf{x}) - f_0(\mathbf{x})| < r \quad \text{for all } \mathbf{x} \in K\,.$$

$K$ being compact, for all $r > 0$ there are finitely many $\mathbf{x}_i \in K$ such that the balls $B(\mathbf{x}_i, r) \subset E$ cover $K$. By definition, the set of $f \in B'$ such that $|f(\mathbf{x}_i) - f_0(\mathbf{x}_i)| < r$ for all $i$ is a neighbourhood $V$ of $f_0$ with respect to the weak topology. For all $\mathbf{x} \in K$, there exists $i$ such that $\|\mathbf{x} - \mathbf{x}_i\| < r$. For $f \in V$, all of

$$|f(\mathbf{x}_i) - f_0(\mathbf{x}_i)| < r\,, \quad |f(\mathbf{x}_i) - f(\mathbf{x})| \leq r\,, \quad |f_0(\mathbf{x}) - f_0(\mathbf{x}_i)| \leq r$$

[99] Recall that a function $f$ defined on a locally compact space $X$ tends to 0 at infinity if the set $|f(x)| \geq r$ is contained in a compact set for all $r > 0$. Taking the Alexandrov compactification $X \cup \{\infty\}$ of $X$ and setting $f(\infty) = 0$, this means that $f$ is continuous at $\infty$ since its open neighbourhoods are by definition the complements of the compact subsets of $X$.

[100] If $E$ is a Banach space, any continuous linear functional $f$ on $E$ has a norm

$$\|f\| = \sup |f(x)| / \|x\|\,,$$

so that the set $E'$ of these linear functionals is also a normed space, which is obviously complete.

hold. The last two inequalities follow from the essential assumption that $\|f\| \leq 1$. Thus

$$f \in V \Longrightarrow |f(\mathbf{x}) - f_0(\mathbf{x})| \leq 3r$$

for all $\mathbf{x} \in K$, qed.

Coming back to the proof of lemma 2, as a general rule let us set that $f_x(y) = f(xy)$. Then relation (8) becomes

$$\widehat{f}(\chi)\chi(x) = f * \chi(x) = \int f(xy)\overline{\chi(y)}dy = \chi(f_x). \tag{26.9}$$

As $f_x$ is the image of $f$ under the translation operator $U(x^{-1})$ on $L^1$, the map $x \mapsto f_x$ from $G$ to the Banach space $L^1$ is continuous [n° 15, (ii)]. For any compact set $K \subset G$, the set $K_f$ of $f_x$, where $x \in K$, is, therefore, a compact subset of $L^1$. Hence, if a varying $\chi \in \widehat{G}$ converges weakly to a limit $\chi_0 \in \widehat{G}$, lemma 2 then implies the compact convergence of functions $\widehat{f}(\chi)\chi(x)$ to $\widehat{f}(\chi_0)\chi_0(x)$. As $\widehat{f}(\chi)$ tends to $\widehat{f}(\chi_0)$ and as $f$ can be chosen so that $\widehat{f}(\chi_0) \neq 0$, we conclude that $\chi(x)$ converges uniformly to $\chi_0(x)$ on $K$, proving lemma 2.

This result shows that the topology of $\widehat{G}$ is compatible with the group structure of $\widehat{G}$.

The set of functions $\widehat{f}$ is an *algebra* of continuous functions on $X(G)$ since $\widehat{f * g} = \widehat{f}\widehat{g}$. Replacing $f$ by $\tilde{f}$ replaces $\widehat{f}$ by the conjugate function. The functions $\widehat{f}$ separate the points of $\widehat{G}$. It would even be enough to only consider $f \in L(G)$. Functions of the form $\chi \mapsto \widehat{f}(\chi) + \lambda$, where $\lambda$ is an arbitrary constant – i.e. the Gelfand transforms of the elements of $\mathbf{A}^1(G)$ –, separate the points of $X(G)$. Hence they enable us to verify that the assumptions of the Stone-Weierstrass theorem hold. As a result, Functions of the form $\widehat{f}(\chi) + \lambda$ are everywhere dense in the space of continuous functions on $X(G)$. But to approximate a function vanishing at $0 = \infty$, i.e. a function of $L_\infty(\widehat{G})$, there is no need to add constants. Thus:

**Lemma 4.** *Fourier transforms of $f \in L^1(G)$ are everywhere dense in the space $L_\infty(\widehat{G})$ of continuous functions tending to 0 at infinity on $\widehat{G}$.*

We have already seen that if $G$ is discrete, then $\widehat{G}$ is compact. Let us show that *if $G$ is compact, $\widehat{G}$ is discrete.* To this end, we use the *orthogonality relations*

$$\int \chi(x)\overline{\chi'(x)}dx = 1 \quad \text{if } \chi = \chi', \quad = 0 \text{ if } \chi \neq \chi' \tag{26.10}$$

for the characters of a compact group, already encountered in Chap. VII when $G = \mathbb{T}$. This assumes that the invariant measure of $G$ is normalized, thus making (10) obvious for $\chi = \chi'$. For $\chi \neq \chi'$, we compute

$$\chi' * \chi(x) = \int \chi'\left(xy^{-1}\right)\chi(y)dy = \chi'(x)\left(\chi|\chi'\right).$$

As $\chi' * \chi = \chi * \chi'$, $\chi'(x)(\chi|\chi') = \chi(x)(\chi|\chi')$, and as two characters are proportional only if they are the same, the expected result follows. This being settled, it remains to check that the left hand side of (10) is a *continuous* function of $(\chi, \chi')$ on $\widehat{G} \times \widehat{G}$. This is clearly since the topology of $\widehat{G}$ is that of uniform convergence in $G$.

To summarize:

**Theorem 38.** *Let $G$ be a locally compact commutative group. The weak topology of the dual of $L^1$ coincides with the topology of compact convergence on the multiplicative group $\widehat{G}$ of characters of $G$. The Fourier transform of every $f \in L^1(G)$ is continuous and tends to 0 at infinity on $\widehat{G}$. Any continuous function tending to 0 at infinity on $\widehat{G}$ is the uniform limit of functions $\widehat{f}$, where $f \in L^1(G)$.*

(iii) *The canonical homomorphism $G \longrightarrow \widehat{\widehat{G}}$.* Since, for given $x$, the function $\chi \mapsto \chi(x) = \mathbf{e}(x, \chi)$ is continuous, by (6), it is a character of $\widehat{G}$, and hence an element of $\widehat{\widehat{G}}$. This gives a homomorphism of groups

$$j : G \longrightarrow \widehat{\widehat{G}} \tag{26.11}$$

for which, by definition,

$$\mathbf{e}(x, \chi) = \mathbf{e}\left[\chi, j(x)\right] . \tag{26.12}$$

In this relation, the left hand side refers to the duality between $G$ and $\widehat{G}$, the right hand one to the duality between $\widehat{G}$ and $\widehat{\widehat{G}}$. Let us show that $j$ is continuous, in other words that, if some varying $x \in G$ converges to some $a \in G$, then $\mathbf{e}(x, \chi) = \chi(x)$ converges uniformly to $\chi(a)$ on every compact set $K \subset \widehat{G}$.

Indeed, by (9),

$$\left|\widehat{f}(\chi)\chi(x) - \widehat{f}(\chi)\chi(a)\right| = |\chi(f_x - f_a)| \leq \|f_x - f_a\|_1$$

for all $\chi \in \widehat{G}$ and $f \in L^1$. As $x \mapsto f_x$ is continuous, the right hand side is arbitrarily small for $x$ sufficiently near $a$. So, for all $r > 0$, there is a neighbourhood $V$ of $a$ such that $x \in V$ implies

$$\left|\widehat{f}(\chi)\chi(x) - \widehat{f}(\chi)\chi(a)\right| < r \quad \text{for all } \chi \in \widehat{G} .$$

To deduce that, for all $r > 0$,

$$|\chi(x) - \chi(a)| < r \quad \text{for all } \chi \in K$$

if $x$ is sufficiently near $a$, it suffices to show that there exists $f \in L^1$ such that

(26.13) $$\inf_{\chi\in K}\left|\widehat{f}(\chi)\right| > 0\,.$$

This is computation rule (R 3) of Chap. III, n° 7 concerning uniform convergence. But using the function $\tilde{f}(x)$ defined at the end of n° 24, (iii), a trivial calculation shows that the Fourier transforms of $f$ and $\tilde{f}$ are mutually conjugate. As a result,

(26.14) $$f = \sum \tilde{f}_i * f_i \Longrightarrow \widehat{f}(\chi) = \sum \left|\widehat{f}_i(\chi)\right|^2 \geq 0\,.$$

This said, for all $\chi \neq 0$, there exists $f$ such that $\widehat{f}(\chi) \neq 0$. One may assume that $|\widehat{f}(\chi)| > 1$ in the neighbourhood of $\chi$, and so cover the compact set $K$ with finitely many open sets $\{|f_i(\chi)| > 1\}$. The function $f = \sum \tilde{f}_i * f_i$ is then $> 1$ on $K$, qed.

## 27 – Fourier Transform on $L^2(G)$

(i) *The algebra* $\mathbf{A}(G)$ *and its characters.* We consider the regular representation $(\mathcal{H}, U)$ of $G$,where $\mathcal{H} = L^2(G)$ and where $U(x)$ transforms $f \in \mathcal{H}$ into

(27.1) $$U(x)f : y \longmapsto f\left(x^{-1}y\right)\,.$$

As was seen in (25.17), for $f \in L^1$, the integrals

(27.2) $$U(f) = \int U(x)f(x)dx$$

are the convolution operators

(27.3) $$U(f)g = f * g$$

on $\mathcal{H}$. They satisfy

(27.4) $$U(f * g) = U(f)U(g)\,, \quad U(f+g) = U(f) + U(g)$$

and also

(27.5) $$U(f)^* = U(\tilde{f})\,.$$

These formal properties show that the set $\mathbf{A}(G)$ of limits *in norm* of operators of the form $U(f) + \lambda 1$ is a commutative GN algebra, not to be confused with the algebra $\mathbf{A}^1(G)$ used above. The Fourier transform on $L^2$ will be obtained by setting out clearly the results of n° 22 in this case; this is in particular the case of Plancherel's formula

$$\int |f(x)|^2\, dx = \int |\widehat{f}(\chi)|^2 d\chi\,,$$

which obviously assumes that the invariant measure $d\chi$ of $\widehat{G}$ has been conveniently chosen.

Let us begin with the characters $\chi$ of $\mathbf{A}(G)$. Knowing them on an everywhere dense subspace is sufficient, for example on the subspace of operators $U(f) + \lambda 1$, and as $\chi(\lambda 1) = \lambda$, it is sufficient to know $\chi$ on all $U(f)$. Since $\|U(f)\| \leq \|f\|_1$, every character $\chi$ of $\mathbf{A}(G)$ defines a character of $L^1(G)$, which will also be denoted by $\chi$, such that

$$\chi(f) = \widehat{f}(\chi) = \chi\left[U(f)\right] . \tag{27.6}$$

This leads to an *injective* map from $\mathbf{A}(G)$ to $X(G)$ which, as will be seen later, is a homeomorphism.

For the moment, let us draw an important consequence of (6):

**Theorem 39**. *The Fourier transform $f \mapsto \widehat{f}$, $f \in L^1(G)$, is injective. For any pair of distinct elements $x, y$ of $G$, there is a character $\chi$ of $G$ such that $\chi(x) \neq \chi(y)$.*

If $\widehat{f}(\chi) = 0$ for all $\chi \in \widehat{G}$, then, by (6), $\chi[U(f)] = 0$ must perforce hold for any character $\chi$ of $\mathbf{A}(G)$. But in a GN algebra, the Gelfand transform is injective (theorem 33). Hence $U(f) = 0$, i.e. $f * g = 0$ for all $g \in L^2$. Thus replacing $g$ by a Dirac sequence and using (25.29) gives $f = 0$.

The second proposition, obvious for the groups mentioned in the examples of the previous n°, reduces to showing that for $x \neq e$, there exists $\chi$ such that $\chi(x) \neq 1$. But (26.9) can also be written

$$\chi(f)\chi(x) = \chi\left(f_x\right) , \tag{27.7}$$

where $f_x(y) = f(x^{-1}y)$. If $\chi(x) = 1$ for all $\chi$, then $\chi(f - f_x) = 0$ for all $\chi \in \widehat{G}$ and $f \in L(G)$, whence it would follow from the first proposition of the theorem that $f - f_x = 0$ for all $f \in L(G)$, an absurdity if $x \neq e$.

The map $\widehat{\mathbf{A}}(G) \longrightarrow X(G)$ defined by (6) is clearly continuous with respect to the weak topologies of the duals of $\mathbf{A}(G)$ and $L^1$. As it is injective, it is, therefore, a homeomorphism from $\mathbf{A}(G)$ onto a compact subset of $X(G)$. In fact:

**Lemma 1**. *The map $\widehat{\mathbf{A}}(\widehat{G})$ to $X(G)$ is surjective.*

The proof consists of two parts, the first one comprising computations without any immediately clear connections with the statement.

(a) We choose a character $\chi \in \widehat{G}$. For all $f \in L^1$, the function

$$\chi' \longmapsto \widehat{f}\left(\chi^{-1}\chi'\right) = \int f(x)\chi(x)\overline{\chi'(x)}dx$$

is the Fourier transform of the function

$$V(\chi)f : x \longmapsto f(x)\chi(x) = f(x)\mathbf{e}(x,\chi) . \tag{27.8}$$

Thus, considering $\chi'$ a linear functional on $L^1$,

(27.9) $$\chi'[V(\chi)f] = \hat{f}(\chi^{-1}\chi') .$$

Operator (8) of multiplication by a function with absolute value 1 can be defined on every $L^p$ and in particular on $L^2$, where it is unitary and satisfies

$$V(\chi)^* = V(\overline{\chi}) = V(\chi)^{-1} .$$

In fact the map e $\chi \mapsto V(\chi)$ is a unitary representation of $\widehat{G}$ on $L^2(G)$.

The product of $V(\chi)$ and some $U(f)$ can be easily found:

$$V(\chi)U(f)g(x) = \chi(x).f * g(x) = \chi(x) \int f(xy^{-1})\, g(y)dy$$
$$= \int \chi(xy^{-1})\, f(xy^{-1})\chi(y)g(y)dy .$$

This function is obtained by applying the operator $V(\chi)$ to $g$, then the convolution operator by $\chi(x)f(x)$, i.e. by function (8), on the function obtained. As a result, $V(\chi)U(f) = U[V(\chi)f]V(\chi)$, i.e.

(27.10) $$U[V(\chi)f] = V(\chi)U(f)V(\chi)^{-1} \quad \text{for all } \chi \in \widehat{G} .$$

It follows that

(27.11) $$T \in \mathbf{A}(G) \Longrightarrow V(\chi)TV(\chi)^{-1} \in \mathbf{A}(G) .$$

This is shown by (9) for $T = U(f)$, hence for $T = \lambda 1 + U(f)$, and in the general case can be deduced by passing to the limit in norm.

(b) We are now ready to prove the lemma. Let $\chi'$ be a non-trivial character of $\mathbf{A}(G)$ on $L^1$. Map (11) being a ring homomorphism, by (10), the map

$$\chi'' : T \longmapsto \chi'\left[V(\chi)TV(\chi)^{-1}\right]$$

is also a non-trivial character of $\mathbf{A}(G)$ on $L^1$. Identifying $\chi'$ and $\chi''$ with the corresponding characters of $L^1$, if $T = U(f)$, then (10) shows that, for all $\chi \in G$, $V(\chi)TV(\chi)^{-1}$ corresponds to the function $V(\chi)f$. So by (9),

$$\chi''(f) = \chi'[V(\chi)f] = \hat{f}(\chi^{-1}\chi') .$$

Hence if a character $\chi'$ of $G$ stems from a character of $\mathbf{A}(G)$, so does $\chi^{-1}\chi'$ for all $\chi \in \widehat{G}$. As a result, any character of $G$ stems from a character of $\mathbf{A}(G)$, qed.

*Exercise*[101] 1. Show that the set of operators $uV(\chi)U(x)$, where $u$ is a scalar with absolute value 1 and where $x \in G$, $\chi \in \widehat{G}$, is a group with the usual multiplication. What the group law is thereby obtained on $\mathbb{T} \times G \times \widehat{G}$?

---

[101] Trivial starting point from which in 1964 André Weil drew far less trivial consequences (*Collected Works*, volume 3, pp. 143–211). For $G = \mathbb{R}$, the question is related to the *Heisenberg group*, i.e the set of real $3 \times 3$ upper triangular matrices with 1's on the diagonal, whose Plancherel formula I had computed in 1949 (Journal de Liouville, 1951). Dieudonné transformed part of Weil's article into exercises 1 to 8 of XXII.15.

(ii) *Spectral decomposition of the regular representation.* We are now able to apply the general constructions of n° 22 to the algebra$\mathbf{A}(G)$. Before doing so, we set out clearly the GN theorem in this case.

For all $T \in \mathbf{A}(G)$, by lemma 1, the Gelfand transform of the general case can be identified with a continuous function $\widehat{T}(\chi)$ on $X(G)$ such that

$$\widehat{T}(\chi) = \widehat{f}(\chi) \quad \text{if } T = U(f)\,, \tag{27.12}$$

on the understanding that $\widehat{f}(0)$ is set to be equal to 0 if $\chi$ is trivial on $U(f)$ for all $f \in L^1(G)$. The GN theorem tells us that the map $T \mapsto \widehat{T}$ is an isomorphism for all structures. The functions $\widehat{T}$ are, therefore, *the* continuous functions on $X(G)$, i.e. functions of the form $\varphi + Cte$, where $\varphi \in L_\infty(\widehat{G})$, the Fourier transforms of $f \in L^1(G)$ being functions belonging to $L_\infty(\widehat{G})$.

On the other hand, n° 22 enables us to associate to any pair of functions $p, q \in \mathcal{H} = L^2(G)$ a measure $\mu_{p,q}$ on $X(G)$ such that

$$(Sp|Tq) = \int_{X(G)} \widehat{S}(\chi)\overline{\widehat{T}(\chi)}d\mu_{p,q}(\chi) \tag{27.13}$$

for all $S, T \in \mathbf{A}(G)$. For $S = U(f)$ and $T = U(g)$ or 1, $\widehat{S} = \widehat{f}$ and $\widehat{T} = \widehat{g}$ or 1. Thus relations

$$(f * p|g * q) = \int \widehat{f}(\chi)\overline{\widehat{g}(\chi)}d\mu_{p,q}(\chi) \tag{27.14}$$

$$(f * p|q) = \int \widehat{f}(\chi)d\mu_{p,q}(\chi) \tag{27.14'}$$

which hold for all $f, g \in L^1$. All these integrals are *a priori* extended to $X(G)$ and not only to $\widehat{G}$.

These measures enable us to associate an operator $M(\varphi)$ on $L^2(G)$, given by

$$(M(\varphi)p|q) = \int_{X(G)} \varphi(\chi)d\mu_{p,q}(\chi)\,, \tag{27.15}$$

to "every" bounded (for example Borel) function $\varphi$ on $X(G)$. These operators commute with all $T \in \mathbf{A}(G)$, in particular with all $U(f)$, and commute pairwise as in all spectral decompositions. As was seen in n° 22,

$$\begin{gathered} M(\alpha\varphi + \beta\psi) = \alpha M(\varphi) + \beta M(\psi)\,, \\ M(\varphi\psi) = M(\varphi)M(\psi)\,, \quad M(\overline{\varphi}) = M(\varphi)^*\,, \end{gathered} \tag{27.16}$$

and also, by (14'),

$$M(\widehat{f}) = U(f) \quad \text{for all } f \in L^1\,. \tag{27.17}$$

In particular, any reasonable subset $\omega \subset X(G)$ defines a projection operator $M(\omega)$ on $L^2$. Computing the operator $M$ associated to the product of the characteristic function of $\omega$ and a bounded function $\varphi$ shows that

$$(27.18) \qquad (M(\omega)M(\varphi)p|q) = \int_\omega \varphi(\chi)d\mu_{p,q}(\chi)$$

for all bounded $\varphi$ and $p, q \in L^2(G)$. Let us for example take $\omega = \{\infty\}$ by assuming that $G$ is not discrete. For $\varphi = \widehat{f}$, the right hand side of (12) is $\widehat{f}(\infty)\mu_{p,q}(\omega) = 0$. Hence, as in this case $M(\varphi)p = U(f)p = f * p$, $(M(\omega)U(f)p|q) = 0$. So $q \in L^2$ being arbitrary, this proves that, for all functions $f \in L^1$ and $p \in L^2$, $f * p$ is a zero of $M(\omega)$. But in $L^2$, every function $p$ can be approximated by such convolutions using Dirac sequences. So $M(\omega) = 0$, which shows that $\{\infty\}$ is null with respect to all $\mu_{p,q}$. *Hence in (13) it is enough to integrate over* $\widehat{G}$ or, equivalently, to regard all $\mu_{p,q}$ as bounded measures on $\widehat{G}$.

One can also choose the function $\chi \mapsto \chi(x)$ for some $x \in G$. It is continuous and bounded. Denoting by $M(x)$ the corresponding operator,

$$(M(x)M(\varphi)p|q) = \int \chi(x)\varphi(\chi)d\mu_{p,q}(\chi)$$

for all bounded and Borel $\varphi$ on $\widehat{G}$, in particular for $\varphi = 1$. Let us show that $M(x) = U(x)^{-1}$. For $\varphi = \widehat{f}$, a trivial calculation gives

$$\chi(x)\widehat{f}(\chi) = \int f(xy)\overline{\chi(y)}dy\,.$$

This is the Fourier transform of the function $U(x)^{-1}f = f_x : y \mapsto f(xy)$. So, as $M(\varphi) = U(f)$ in this case,

$$M(x)U(f) = U(f_x) = \int f(xy)U(y)dy = U(x)^{-1}U(f)\,,$$

whence $M(x) = U(x)^{-1}$. Therefore,

$$(27.19) \qquad (U(x)M(\varphi)p|q) = \int \overline{\mathbf{e}(x,\chi)}\varphi(\chi)d\mu_{p,q}(\chi)$$

for any bounded and Borel function $\varphi$ on $\widehat{G}$, especially for $\varphi = 1$.

(iii) *The invariant measure on the dual.* To simplify notation, set

$$L^{12} = L^{12}(G) = L^1(G) \cap L^2(G)\,,$$

A confusion with the space $L^p$ for $p = 10 + 2 = 7 + 5$ being unlikely. The Fourier transforms of $f \in L^{12}$ are everywhere dense in $L_\infty(\widehat{G})$ since $L^{12}$

contains $L(G)$. The set they form is also a subalgebra of $L_\infty(\widehat{G})$ with respect to the usual multiplication, since $L^{12}$ is a subalgebra of $L^1$ with respect to the convolution product thanks to relations

$$L^1 * L^1 \subset L^1 , \quad L^1 * L^2 \subset L^2 .$$

The measures $\mu_{p,q}$ associated to the functions of $L^{12}$ satisfy an identity that will be the starting point of the proof of Plancherel's theorem:

**Lemma 2**. *For all* $f, g, p, q \in L^{12}(G)$,

$$\widehat{f}(\chi)\overline{\widehat{g}(\chi)}d\mu_{p,q}(\chi) = \widehat{p}(\chi)\overline{\widehat{q}(\chi)}d\mu_{f,g}(\chi) . \tag{27.20}$$

As (20) is an identity between two *measures*, we need to show that

$$\int \varphi(\chi)\widehat{f}(\chi)\overline{\widehat{g}(\chi)}d\mu_{p,q}(\chi) = \int \varphi(\chi)\widehat{p}(\chi)\overline{\widehat{q}(\chi)}d\mu_{f,g}(\chi) \tag{27.20'}$$

for all $\varphi \in L(\widehat{G})$ or, more generally, Borel and bounded. This means that

$$(M(\varphi\widehat{f}\,\overline{\widehat{g}})p|q) = (M\,(\varphi\widehat{p}\overline{\widehat{q}})\,f|g)$$

or, by (16) and (17), that

$$(M(\varphi)U(f)p|U(g)q) = (M(\varphi)U(p)f|U(q)g) .$$

This is obvious since, under the assumptions of the lemma,

$$U(f)p = U(p)f = f * p$$

because $G$ is commutative.

**Lemma 3**. *For* $p, q \in L^{12}(G)$, *the measure* $d\mu_{p,q}$ *does not depend on the function* $p * \tilde{q}$.

Indeed by (17), under the assumptions of the lemma, for $f \in L(G)$

$$\mu_{p,q}(\widehat{f}) = (M(\widehat{f})p|q) = (U(f)p|q) = (f * p|q) .$$

So by (25.24'),

$$\mu_{p,q}(\widehat{f}) = (f * p) * \tilde{q}(e) = f * (p * \tilde{q})\,(e) ,$$

associativity being justified since $f$, $p$ and $q$ are integrable. As $\widehat{f}$ determines $f$ (theorem 39) and as all $\widehat{f}$ are everywhere dense in $L_\infty(\widehat{G})$, the bounded measure $\mu_{p,q}$ only depends on $p * \tilde{q}$, which proves the lemma.

**Theorem 40.** *The invariant measure on $\widehat{G}$ can be chosen in such a way that*

$$d\mu_{p,q}(\chi) = \widehat{p}(\chi)\overline{\widehat{q}(\chi)}d\chi \tag{27.21}$$

*for all $p, q \in L^{12}(G)$.*

To simplify calculations, I will denote by $\Lambda$ the set of functions of the form $P(\chi) = \widehat{p}(\chi)\overline{\widehat{q}(\chi)}$ on $G$, where $p, q \in L^{12}(G)$. Lemma 3 enables us to associate the measure

$$d\mu_P(\chi) = d\mu_{p,q}(\chi)$$

to such a function $P$. Then the following properties hold:

($\Lambda$ 1): $PQ \in \Lambda$ for all $P, Q \in \Lambda$;
($\Lambda$ 2): for all $P, Q \in \Lambda$,

$$\int P(\chi)d\mu_Q(\chi) = \int Q(\chi)d\mu_P(\chi)\,; \tag{27.22}$$

($\Lambda$ 3): for every compact set $K \subset \widehat{G}$, there exists $P \in \Lambda$ such that $P(\chi) \neq 0$ for all $\chi \in K$.

The first one follows from the fact that $L^{12} * L^{12} \subset L^{12}$. The second one is just lemma 2. To obtain the third one, observe that, the Fourier transforms of $f \in L(G)$ being everywhere dense on $L_\infty(\widehat{G})$, there exists $p \in L(G)$ such that $|\widehat{p}(\chi)| \geq 1$ on $K$. The function $P(\chi) = |\widehat{p}(\chi)|^2$ answers the question.

This being settled, let us show that there is a measure $\mu$ on $\widehat{G}$ such that $d\mu_P(\chi) = P(\chi)d\mu(\chi)$ for all $P \in \Lambda$, i.e.

$$\mu_P(\varphi) = \mu(\varphi P) \quad \text{for } \varphi \in L(\widehat{G}) \text{ and } P \in \Lambda\,. \tag{27.23}$$

As (22) can be formally written as

$$P(\chi)^{-1}d\mu_P(\chi) = Q(\chi)^{-1}d\mu_Q(\chi)\,, \tag{*}$$

"obviously" the measure $P(\chi)^{-1}d\mu_P(\chi)$ does not depend on $P$, which gives the expected measure $\mu$. But both sides of (*) are well-defined only if the functions $P$ and $Q$ do not vanish.[102] Arguments, therefore, need to be less brutal.

Therefore, let us associate the open set $\Omega_P = \{P(\chi) \neq 0\}$ to each $P \in \Lambda$ and identify $L(\Omega_P)$ with the subspace of $f \in L(\widehat{G})$ whose support is contained in $\Omega_P$. We define a measure $\nu_P$ on $\Omega_P$ – and not on $\widehat{G}$ – setting

[102] The reader desiring an easy proof for $G = \mathbb{R}^n$ will observe that, in this case, the function $p(x) = \exp(-\pi\|x\|^2)$, obviously in $L^{12}$, does not vanish and is equal to its Fourier transform (here $\widehat{G} = G$). For $P(\chi) = \widehat{p}(\chi)\overline{\widehat{p}(\chi)} = p(\chi)^2$, the left hand side of (*) is well-defined. Denoting it by $d\mu(\chi)$, both sides of (14) can then be divided by $P(\chi)$, and so $d\mu_{p,q}(\chi) = \widehat{p}(\chi)\overline{q(\chi)}d\mu(\chi)$ for all $p, q \in L^{12}$, qed.

$$\nu_P(\varphi) = \int \varphi(\chi)P(\chi)^{-1}d\mu_P(\chi)$$

for all $\varphi \in L(\Omega_P)$. This is well-defined since the function being integrated vanishes outside a compact subset of $\Omega_P$. To show that the result does not depend on $P$, we consider another function $Q \in \Lambda$ and suppose that the support $K$ of $\varphi$ is contained in $\Omega_P \cap \Omega_Q$. As $PQ$ does not vanish on $\Omega_P \cap \Omega_Q$, $\varphi = \psi PQ$ for some $\psi \in L(X, K)$. Thus by (22),

$$\nu_P(\varphi) = \mu_P(\psi Q) = \mu_Q(\psi P) = \nu_Q(\varphi) .$$

As ($\Lambda$ 3) implies that the open sets $\Omega_P$ cover $\widehat{G}$, the existence of $\mu$ will be a consequence of a general result which enables us to construct a measure by "gluing" together measures defined on open sets:

**Lemma 4.** *Let $X$ be a locally compact space and $(\Omega_i)_i \in I$ an open cover of $X$. Suppose that a measure $\mu_i$ is given on each $\Omega_i$. There is a measure $\mu$ on $X$ such that*

$$\mu(f) = \mu_i(f) \tag{27.24}$$

*for all $f \in L(X)$ with support contained in $\Omega_i$ if and only if, for any ordered pair $(i, j)$,*

$$\mu_i(f) = \mu_j(f) \tag{27.25}$$

*for all $f \in L(X)$ with support $K \subset \Omega_i \cap \Omega_j$.*

(25) is clearly necessary. If this condition holds and if some $f \in L(X)$ vanishes outside a compact set $K$, there exist $p_i \in L(X)$ whose supports are contained in the sets $\Omega_i$, which are trivial for almost all $i$, and such that $\sum p_i = 1$ on $K$. If some $q_i \in L(X)$ satisfy the same conditions, then by the Lemma's assumptions $\mu_i(fp_iq_j) = \mu_j(fp_iq_j)$. Hence

$$\sum_i \mu_i(fp_i) = \sum_{i,j} \mu_i(fp_iq_j) = \sum_{i,j} \mu_j(fp_iq_j) = \sum_j \mu_j(fq_j) .$$

Therefore, the expression $\mu(f) = \sum \mu_i(fp_i)$ only depends on $f$. It is the expected measure $\mu$.

We have thus shown that there exists a (unique) measure $\mu$ on $\widehat{G}$ such that

$$(M(\varphi)p|q) = \mu_{p,q}(\varphi) = \int \varphi(\chi)\widehat{p}(\chi)\overline{\widehat{q}(\chi)}d\mu(\chi) \tag{27.26}$$

for all functions $\varphi \in L(\widehat{G})$ and $p, q \in L^{12}(G)$. It is positive since the left hand side is $\geq 0$ for $\varphi \geq 0$ and $p = q$. The *invariance* of $\mu$ remains to be shown. .

First of all, (26) means that the measures $d\mu_{p,q}(\chi)$ and $\widehat{p}(\chi)\overline{\widehat{q}(\chi)}d\mu(\chi)$ are identical for all $p, q \in L^{12}(G)$. Since the measure $d\mu_{p,p}(\chi) = |\widehat{p}(\chi)|^2 d\mu(\chi)$ is bounded, this shows (n° 18, corollary of theorem 30) that $\widehat{p} \in L^2(\widehat{G};\mu)$ for all $p \in L^{12}(G)$. Formula (26) thus holds for any bounded continuous (and even Borel: theorem 25) function $\varphi$ on $\widehat{G}$, and in particular if $\varphi(\chi) = \widehat{f}(\chi)$ for some $f \in L^{12}(G)$. In this case, (26) becomes

(27.27) $$(f * p|q) = \int \widehat{f}(\chi)\widehat{p}(\chi)\overline{\widehat{q}(\chi)}d\mu(\chi)$$

for $f, p, q \in L^{12}$. As in general

$$\widehat{p}\left(\chi\chi_0^{-1}\right) = \int p(x)\overline{\mathbf{e}\left(x, \chi\chi_0^{-1}\right)}dx = \int p(x)\chi_0(x)\overline{\mathbf{e}(x,\chi)}dx\,,$$

making a translation act on the function being integrated in (27) amounts to multiplying $f$, $p$ and $q$ by the same character $\chi_0$ de $G$. Denoting this operation by $f \mapsto f'$,

$$f' * p'(x) = \int \chi_0\left(xy^{-1}\right) f\left(xy^{-1}\right)\chi_0(y)p(y)dy = \chi_0(x).f * p(x)\,,$$

i.e. $(f * p)' = f' * p'$. Therefore, since $f \mapsto f'$ is obviously unitary,

(27.28) $$(f' * p'|q') = (f * p|q)\ .$$

This proves that integral (27) does not change if the integrated function is made to undergo a translation $\chi \longmapsto \chi\chi_0^{-1}$ or else that any image $\mu'$ of $\mu$ under a translation also satisfies (27). So

$$\int \widehat{f}(\chi)\widehat{p}(\chi)\overline{\widehat{q}(\chi)}d\mu(\chi) = \int \widehat{f}(\chi)\widehat{p}(\chi)\overline{\widehat{q}(\chi)}d\mu'(\chi)$$

for all $p, q \in L^{12}(G)$. As the functions $\widehat{f}$ are everywhere dense on $L_\infty(\widehat{G})$, the equality

$$\widehat{p}(\chi)\overline{\widehat{q}(\chi)}d\mu(\chi) = \widehat{p}(\chi)\overline{\widehat{q}(\chi)}d\mu'(\chi)$$

between bounded measures follows, whence $\mu = \mu'$ because of ($\Lambda$ 3). This completes the proof of theorem 40.

**Corollary 1 (Plancherel).** *There is a unique isomorphism $f \mapsto \widehat{f}$ from $L^2(G)$ onto $L^2(\widehat{G})$ which reduced to the Fourier transform on $L^{12}(G)$.*

The proof reduces to showing that the set of $\widehat{f}$, $f \in L^{12}$, is everywhere dense on $L^2(\widehat{G})$. Otherwise, there is a non-zero function $F \in L^2(\widehat{G})$ orthogonal to all $\widehat{f}$, in particular to all $P \in \Lambda$ used in the proof of theorem 40, whence

$$\int F(\chi)\widehat{p}(\chi)\overline{\widehat{q}(\chi)}.d\chi = 0$$

for all $p, q \in L^{12}(G)$. As $F$ and $\widehat{p}$ are in $L^2(\widehat{G})$, the measure $F(\chi)\widehat{p}(\chi)d\chi$ is *bounded*, and the previous relation becomes (theorem 25)

$$\int \overline{\widehat{q}(\chi)}.F(\chi)\widehat{p}(\chi)d\chi = 0\,.$$

The functions $\widehat{q}$ being everywhere dense on $L_\infty(\widehat{G})$, the *measure* $F(\chi)\widehat{p}(\chi)d\chi$ is zero, and as the open sets $\{\widehat{p}(\chi) \neq 0\}$ cover $\widehat{G}$, the same holds for $F(\chi)d\chi$, whence $F(\chi) = 0$ ae., qed.

**Corollary 2**. *The Fourier transform of the product $fg$ of two square integrable functions is the convolution $\widehat{f} * \widehat{g}$ of their Fourier transforms.*

Let us prove this by replacing $g$ by its conjugate. The Fourier transform of the convolution is the function

$$\chi \longmapsto \int f(x)\overline{g(x)\mathbf{e}(x;\chi)}dx\,.$$

That of the function $g(x)\mathbf{e}(x;\chi)$ is clearly the function

$$\chi' \longmapsto \widehat{g}\left(\chi\chi'^{-1}\right)\,.$$

Plancherel's formula thus shows that the Fourier transform of $fg$ is

$$\chi \longmapsto \int \widehat{f}\,(\chi')\,\overline{\widehat{g}\,(\chi\chi'^{-1})}d\chi'\,,$$

qed.

(iv) *Fourier inversion formula and biduality.*

**Theorem 41**. *Let $f$ be a function belonging either to $L^1(G)$ or to $L^2(G)$, and whose Fourier transform is integrable; then*

$$f(x) = \int \mathbf{e}(x;\chi)\widehat{f}(\chi)d\chi \quad \text{almost everywhere} \tag{27.29}$$

*and the right hand side is a continuous function on $G$.*

(a) If $f \in L^1(G)$, then the operator $U(f) : g \mapsto f * g$ is in the algebra $\mathbf{A}(G)$ and

$$(U(f)p|q) = \int \widehat{f}(\chi)\widehat{p}(\chi)\overline{\widehat{q}(\chi)}d\chi \tag{27.30}$$

for all $p, q \in L^2(G)$. Hence, if $p, q \in L(G)$, then

$$(U(f)p|q) = \iint \widehat{f}(\chi)\widehat{p}(\chi)\overline{q(x)}\mathbf{e}(x;\chi)dxd\chi\,,$$

the function being integrated being in $L^1(G\times\widehat{G})$ since $\widehat{f}$ and $q$ are integrable and $\widehat{p}$ bounded. If this result is applied to the functions $p\in L_+(G)$ with integrals equal to 1 and vanishing outside smaller and smaller neighbourhoods of $e$, then the function $\widehat{p}(\chi)$ converges uniformly to 1 on every compact set, with $|\widehat{p}(\chi)|\leq 1$. Hence passing to the limit in the integral is possible. On the left hand side, $U(f)p=f*p$ converges to $f$ in $L^1(G)$, and as $q$ is assumed to be in $L(G)$, passing to the limit on the left hand side is possible. So

$$\int f(x)\overline{q(x)}dx=\iint \widehat{f}(\chi)\overline{q(x)}\mathbf{e}(x;\chi)dxd\chi$$

for all $q\in L(G)$. The double integral does not pose any problems because of the assumptions on $\widehat{f}$. Thus the inversion formula readily follows.

(b) Case where $f\in L^2(G)$. For all $g\in L(G)$, (Plancherel)

$$f*\tilde{g}(x)=(f|U(x)g)=\int\widehat{f}(\chi)\overline{\widehat{g}(\chi)}\mathbf{e}(x,\chi)d\chi\,.$$

Let us apply this identity to some $g_n\geq 0$ with integral 1 and converging to the Dirac measure at $e$. The function appearing in the integral tends to $\widehat{f}(\chi)\mathbf{e}(x,\chi)$ while remaining $\leq|\widehat{f}(\chi)|$. Therefore, the second integral tends (dominated convergence) to $\int\widehat{f}(\chi)\mathbf{e}(x,\chi)$. On the left hand side,

$$f*\tilde{g}_n(x)=U(\tilde{g}_n)f$$

and so, by (25.29),

$$f=\text{l.i.m.}^2 f*\tilde{g}_n(x)\,.$$

The result then follows from Riesz-Fischer.

**Theorem 42**. *The canonical homomorphism $j:G\longrightarrow\widehat{\widehat{G}}$ is bijective.*

The Fourier transform on $G$ is an isomorphism $f\mapsto\widehat{f}$ from $L^2(G)$ onto $L^2(\widehat{G})$. Likewise, the Fourier transform on $\widehat{G}$ is an isomorphism onto $L^2(\widehat{\widehat{G}})$ mapping $\widehat{f}$ onto $\widehat{\widehat{f}}$. This gives an isomorphism $f\mapsto\widehat{\widehat{f}}$ from $L^2(G)$ onto $L^2(\widehat{\widehat{G}})$. If $\widehat{f}\in L^{12}(\widehat{G})$, the functions $f$ and $\widehat{\widehat{f}}$ are continuous and, by the previous theorem, coincide on $G$ if the Fourier transform on $\widehat{G}$ is defined accordingly. The theorem will follow once we have shown that, under the assumption $G\neq\widehat{\widehat{G}}$, there exists a function $f\neq 0$ such that $\widehat{f}\in L^{12}(\widehat{G})$ and $\widehat{\widehat{f}}=0$ on $G$.

For this purpose, we consider two functions $p, q \in L(\widehat{\widehat{G}})$ and the corresponding functions $\varphi, \psi \in L^2(\widehat{G})$. Plancherel's formula for $\widehat{G}$ (corollary 2 above) shows that $p * q$, calculated in $\widehat{\widehat{G}}$, is the Fourier transform of $\varphi\psi$. This function is integrable and, since the Fourier transforms of $\varphi$ and $\psi$ are integrable, $\varphi$ and $\psi$ are given by the inversion formula applied to $\widehat{G}$. As a result, $\varphi$ and $\psi$ are continuous and bounded, and so $\varphi\psi \in L^{12}(\widehat{G})$. By the previous theorem applied to $G$, the function $f \in L^2(G)$ such that $\widehat{f} = \varphi\psi$ is, therefore, continuous on $G$ and $\widehat{\widehat{f}} = p * q$. If $G \neq \widehat{\widehat{G}}$, $p$ and $q$ can easily be chosen in such a way that $p * q$ is zero on $G$, but not on all of $\widehat{\widehat{G}}$: if $A$ and $B$ are the supports of $p$ and $q$, it suffices for the intersection of $AB$ and the closed set $G \subset \widehat{\widehat{G}}$ to be empty. This condition is satisfied by choosing $A$ to be a disjoint compact subset of $G$ with non-empty interior, and $B$ to be a sufficiently small compact neighbourhood of the unit, qed.

*Exercise 1.* Generalize the Poisson summation formula.

*Exercise 2.* Show that

$$d\mu_{p,q}(x) = \widehat{p}(x)\overline{\widehat{q}(x)}dx$$

for all $p, q \in L^2(G)$.

## § 8. Unitary Representations of Locally Compact Groups

### 28 – Further Representation Theory

As was mentioned in n° 15 and 23, a representation of a lcg $G$ is a homomorphism $x \mapsto U(x)$ from $G$ to the group of invertible continuous operators of a Banach space $\mathcal{H}$ such that the map

$$x \longmapsto U(x)\mathbf{a}$$

is continuous for all $\mathbf{a} \in \mathcal{H}$. n° 25, (i) enabled us to associate to every complex measure $\mu$ on $G$ a continuous operator $U(\mu)$ given by

$$U(\mu)\mathbf{a} = \int U(x)\mathbf{a}.d\mu(x)\,, \tag{28.1}$$

at least if $\int ||U(x)||d|\mu|(x) < +\infty$, an empty condition when $\mu$ has compact support, and in particular when $d\mu(x) = f(x)dx$ for some $f \in L(G)$. The corresponding operator is then denoted by $U(f)$. n° 25 shows that

$$U(\mu * \nu) = U(\mu)U(\nu) \tag{28.2}$$

if these operators are well-defined, and especially if $\mu$ and $\nu$ have compact support. In particular,

$$U(f * g) = U(f)U(g) \tag{28.2'}$$

for all $f, g \in L(G)$. The next result, where we only consider functions $f \in L(G)$, though easy is very useful:

**Lemma 1.** *Let $(\mathcal{H}, U)$ be a representation of $G$. Every $\mathbf{a} \in \mathcal{H}$ is the limit of vectors of the form $U(f)\mathbf{a}$. A closed vector subspace of $\mathcal{H}$ is invariant under all $U(x)$, if and only if it is so under all $U(f)$. A continuous operator commutes with all $U(x)$ if and only if it commutes with all $U(f)$. If $(\mathcal{H}, U)$ is unitary, the set of $U(f)$ is everywhere dense with respect to the ultrastrong topology on the von Neumann algebra they generate.*

The first proposition was proved at the end of n° 25: given measures of the form $f_n(x)dx$, with $f_n \geq 0$ having integral equal to 1 and vanishing outside compact on decreasingly small neighbourhoods of $e$, $\lim U(f_n)\mathbf{a} = \mathbf{a}$ for all $\mathbf{a}$. This also implies the last proposition: indeed if $\mathcal{A}$ denotes the von Neumann algebra generated by all $U(f)$, i.e. the set of operators which commute with all operators commuting with all $U(f)$, then [n° 19, (vii)] the set of $U(f) + \lambda 1$ is known to be everywhere dense in $\mathcal{A}$ with respect to the ultrastrong topology; to show that so is the set of $U(f)$, it suffices [n° 19, (vii), lemma 9] to check that 0 is the the only element of $\mathcal{H}$ mapped to 0 by all $U(f)$, which is obvious.

On the other hand, the formula $U(f)\mathbf{a} = \int U(x)\mathbf{a}.f(x)dx$ shows that $U(f)\mathbf{a}$ is a limit of linear combinations of vectors $U(x)\mathbf{a}$. Any closed subspace invariant under all $U(x)$ is, therefore, invariant under all $U(f)$. The relation

$$U(x)U(f)\mathbf{a} = U(\epsilon_x)\,U(f)\mathbf{a} = U(\epsilon_x * f)\,\mathbf{a}$$

shows that any subspace, not necessarily closed, invariant under all $U(f)$, is invariant under all $U(x)$.

If a continuous linear operator $A$ on $\mathcal{H}$ commutes with all $U(x)$, the linearity of the integral of vector functions [n° 4, formula (*****)] shows that

$$AU(f) = \int AU(x)f(x)dx = \int U(x)Af(x)dx = U(f)A\,.$$

Conversely, if $A$ commutes with all $U(f)$, for any $x \in G$, we choose $f_n$ such that the measure $f_n(y)dy$ converges to the Dirac measure at $x$. Then

$$U(x)A\mathbf{a} = \lim U(f_n)\,A\mathbf{a} = A.\lim U(f_n)\,\mathbf{a} = AU(x)\mathbf{a}\,,$$

qed.

In the case of a unitary representation, formulas (25.10) and (25.11) hold for any bounded measure $\mu$:

$$U(\mu)^* = U(\tilde{\mu}) \quad \text{où } \tilde{\mu}(x) = \overline{d\mu(x^{-1})}\,, \tag{28.3}$$

$$U(f)^* = U(\tilde{f}) \quad \text{où } \tilde{f}(x) = \overline{f(x^{-1})}/\Delta(x) \tag{28.3'}$$

for $f \in L^1(G)$. The set of operators $U(f)$ is, therefore, a self-adjoint algebra in $\mathcal{H}$. Conversely, a unitary representation of $G$ can be obtained from operators $U(f)$:

**Lemma 2.** *Let $\mathcal{H}$ be a Hilbert space and $f \mapsto U(f)$ a homomorphism from the algebra $L(G)$ to the algebra $\mathcal{L}(\mathcal{H})$. Suppose that*

$$U(f)^* = U(\tilde{f})\,, \quad \|U(f)\| \le \|f\|_1 \quad \text{for all } f \in L(G)$$

*and that the relation $U(f)\mathbf{a} = 0$ for all $f \in L(G)$ implies $\mathbf{a} = 0$. Then there is a unique unitary representation $x \mapsto U(x)$ of $G$ in $\mathcal{H}$ such that*

$$U(f) = \int U(x)f(x)dx \quad \text{for all } f \in L(G)\,.$$

First, the subspace $\mathcal{H}_0$ generated by all vectors $U(f)\mathbf{a}$ is everywhere dense in $\mathcal{H}$ because a vector orthogonal to $\mathcal{H}_0$ is a zero of all $U(f)^*$, hence of all $U(f)$, and so is zero.

To define $U(x)$ for $x \in G$, we first do so on $\mathcal{H}_0$. The formula to be used is obvious:

$$U(x)\sum U(f_i)\,\mathbf{a}_i = \sum U(\epsilon_x * f_i)\,\mathbf{a}_i\,.$$

A small calculation immediately shows that the norms of the two sums are equal, which gives bijective and isometric $U(x) : \mathcal{H}_0 \longrightarrow \mathcal{H}_0$ such that $U(xy) = U(x)U(y)$. They can be extended to unitary operators on $\mathcal{H}$. We check that the maps $x \mapsto U(x)\mathbf{a}$ are continuous and that the representation $(\mathcal{H}, U)$ satisfies the required condition by using the simplest properties of the convolution product on $L(G)$.

$L(G)$ could be replaced by $L^1(G)$ in this statement.

Both versions of Schur's lemma proved at the end of n° 22, (iii) and in n° 23, (iii), corollary 2 of theorem 35 can be applied to irreducible unitary representations. The next version is also very useful:

**Schur's Lemma III.** *Let $(\mathcal{H}, U)$ and $(\mathcal{H}', U')$ be two irreducible unitary representations of $G$ and $T : \mathcal{H} \longrightarrow \mathcal{H}'$ a continuous linear map such that $TU(x) = U'(x)T$ for all $x \in G$. If $T \neq 0$, the two given representations are equivalent.*[103]

To prove proposition (b), observe that $T^*T : \mathcal{H} \longrightarrow \mathcal{H}$ commutes with all $U(x)$, and so is a scalar which can be assumed to be equal to 1. The map $T$ is then isometric, and so its image is a closed subspace of $\mathcal{H}'$. As it is invariant and non-trivial, $T$ is bijective, qed.

A trivial corollary: Every irreducible unitary representation of a commutative lcg $G$ has dimension one, i.e. is a group character. Hence Plancherel's formula can be regarded as a decomposition of the regular representation into a continuous sum of irreducible representations of dimension one:

$$\int_G (U(x)f|g) = \int_{\widehat{G}} \left(U(x;\chi)\widehat{f}(\chi)|\widehat{g}(\chi)\right) d\chi\,.$$

On the left hand side, $U(x)$ is the translation operator which, in $L^2(G)$, transforms the function $f(y)$ into the function $f(x^{-1}y)$, On the right hand side, we equip $\mathbb{C}$ with the inner product $(u|v) = u\overline{v}$, Finally, for $\chi \in \widehat{G}$, $U(x;\chi)$ is the multiplication operator by $\chi(x) = \mathbf{e}(x;\chi)$ on $\mathbb{C}$. By the way, note that in the formula

$$\widehat{f}(\chi) = \int f(x)\overline{\chi(x)}dx = \int U\left(x^{-1};\chi\right) f(x)dx$$

$\widehat{f}(\chi)$ appears as the operator $U(f)$ associated to the conjugate representation $\overline{\chi}$ of $\chi$. In the case of a non-commutative lcg, to obtain formulas compatible with the previous one, one would need to define operators $U(f)$ by $U(f) = \int U(x^{-1})f(x)dx$. The "Fourier transform" of $f$ would then be the function which associates to each class of irreducible unitary representations of $G$ the corresponding operator $U(f)$. But this definition of $U(f)$ would lead to the relation $U(f * g) = U(g)U(f)$ which, on a commutative group, assuredly reduces to the formula $(f * g)\widehat{\ } = \hat{f}\hat{g}$, but which nobody wants in the general case.

[103] i.e. there is an isomorphism from $\mathcal{H}$ onto $\mathcal{H}'$ transforming $U(x)$ into $U'(x)$.

*Exercise 1* (Gelfand and Neumark). (a) Let $f$ be a measurable function on $\mathbb{R}$. Show that the function $(t, x) \mapsto f(tx)$ is measurable on $\mathbb{R}^* \times \mathbb{R}$. (b) Suppose that, for all $a \neq 0$, there is a null set $N(a)$ such that

$$f(ax) = f(x) \quad \text{for all } x \notin N(a)\,.$$

Show that $f$ is almost everywhere equal to a constant [using property (C1) of n° 10, show that, for almost all $x$, $f(ax) = f(x)$ for almost all $a$]. (c) In $L^2(\mathbb{R})$, consider the unitary operators

$$U(a, b) : f(x) \longmapsto a^{1/2+it} f(ax + b)$$

with given $t \in \mathbb{R}$, and varying $a \in \mathbb{R}^*$, $b \in \mathbb{R}$. Using a Fourier transform show that any continuous operator commuting with these operators is a scalar. (d) Use this result to construct irreducible representations of the group $x \mapsto ax + b$ ($a \in \mathbb{R}^*$, $b \in \mathbb{R}$) on $L^2(\mathbb{R})$. What about the case where $a > 0$?

## 29 – Fourier Transform on a Compact Group

(i) *Irreducible representations of central groups.* Let $G$ be a lcg. Schur's lemma shows that the centre $Z$ of $G$ acts by scalars on every irreducible unitary representation $(\mathcal{H}, U)$ of $G$:

$$U(z) = \alpha(z)1$$

where $\alpha$ is a character of $Z$. So the coefficients

$$\varphi_{a,b}(x) = (U(x)\mathbf{a}|\mathbf{b})$$

of the representation are solutions of the functional equation

$$f(xz) = f(x)\alpha(z)\,.$$

It follows that the functions $|f(x)|$ are in fact defined on the quotient group $G/Z$. Hence, if the latter is compact, in which case $G$ is said to be a *central group*, they can be integrated over $G/Z$.

Here too, Schur's lemma is going to prove a basic property of these groups using arguments which, in a slightly less simple version, will lead us to the Bargmann orthogonality relations in n° 32.[104]

[104] As regards my own case, a few months *after* having proved the Bargmann relations in 1947, I realized how the same type of arguments led to a direct proof of this theorem which all specialists, few in number at the time, obviously knew (at least in the case of a compact group). The "result" was new inasmuch as in 1927 the inventors of compact group theory, Fritz Peter and Hermann Weyl, were only considering finite dimensional irreducible representations. But everyone could prove it from their results (see Dieudonné, XXI.4, whose notation complicates a quite straightforward question). This proof is now attributed to Leopoldo Nachbin, an excellent Brazilian mathematician who took the trouble to publish it in 1961.

**Theorem 43.** *Every irreducible unitary representation of a central* lcg $G$ *is finite-dimensional. If* $(\mathcal{H}, U)$ *and* $(\mathcal{H}', U')$ *are two non-equivalent irreducible unitary representations of* $G$ *defining the same character of the centre* $Z$ *of* $G$*, then*

$$(29.1) \qquad \int_{G/Z} (U(g)\mathbf{a}|\mathbf{b})\,\overline{(U'(g)\mathbf{a}'|\mathbf{b}')}dg = 0$$

*for all* $\mathbf{a}, \mathbf{b} \in \mathcal{H}$ *and* $\mathbf{a}', \mathbf{b}' \in \mathcal{H}'$*. If* $(\mathcal{H}, U) = (\mathcal{H}', U')$*, then*

$$(29.2) \qquad \int_{G/Z} (U(g)\mathbf{a}|\mathbf{b})\,\overline{(U'(g)\mathbf{a}'|\mathbf{b}')}dg = \dim(\mathcal{H})^{-1}\,(\mathbf{a}|\mathbf{a}')\,\overline{(\mathbf{b}|\mathbf{b}')}$$

*where* $dg$ *is the normalized Haar measure of* $G/Z$*.*

For given $\mathbf{b} \in \mathcal{H}$ and $\mathbf{b}' \in \mathcal{H}'$, we set

$$T(\mathbf{x}, \mathbf{x}') = \int_{G/Z} (U(g)\mathbf{x}|\mathbf{b})\,\overline{(U'(g)\mathbf{x}'|\mathbf{b}')}dg\,.$$

This gives

$$(29.3) \qquad |T(\mathbf{x}, \mathbf{x}')| \leq \|\mathbf{b}\|.\|\mathbf{b}'\|.\|\mathbf{x}\|.\|\mathbf{x}'\|\,,$$

and so [n° 19, corollary of theorem 31] a continuous linear map $T$ from $\mathcal{H}$ to $\mathcal{H}'$, satisfying $\|T\| \leq \|\mathbf{b}\|.\|\mathbf{b}'\|$, and such that

$$(T\mathbf{x}|\mathbf{x}') = \int_{G/Z} (U(g)\mathbf{x}|\mathbf{b})\,\overline{(U'(g)\mathbf{x}'|\mathbf{b}')}dg\,.$$

The invariance of the measure shows that $(TU(h)\mathbf{x}|U'(h)\mathbf{x}') = (T\mathbf{x}|\mathbf{x}')$ for all $h \in G$. So, if the given representations are not equivalent, then $T = 0$ (Schur's lemma III), which proves the first proposition of the theorem.

In the second case, $T$ is a scalar $T(\mathbf{b}', \mathbf{b})$, whence

$$T(\mathbf{b}', \mathbf{b})\,(\mathbf{x}|\mathbf{x}') = \int_{G/Z} (U(g)\mathbf{x}|\mathbf{b})\,\overline{(U'(g)\mathbf{x}'|\mathbf{b}')}dg\,.$$

(3) again shows that $T(\mathbf{b}', \mathbf{b}) = (T\mathbf{b}'|\mathbf{b})$, where $T \in \mathcal{L}(\mathcal{H})$, and the invariance of the measure again shows that $T$ commutes with all $U(g)$, and so is a scalar $\lambda$. Thus

$$\lambda\,(\mathbf{b}'|\mathbf{b})\,(\mathbf{x}|\mathbf{x}') = \int_{G/Z} (U(g)\mathbf{x}|\mathbf{b})\,\overline{(U(g)\mathbf{x}'|\mathbf{b}')}dg\,.$$

Then let $\mathbf{e}_1, \mathbf{e}_2, \ldots$ be a finite or infinite orthonormal basis of $\mathcal{H}$. The functions

$$u_{ij}(g) = (U(g)\mathbf{e}_i|\mathbf{e}_j)$$

are pairwise orthogonal in $L^2(G)$. Using Kronecker's index, we even have

$$(29.4) \qquad (u_{ij}|u_{pq}) = \delta_{ip}\delta_{jq}\lambda\,,$$

where the scalar product is obviously calculated on $G/Z$. So $\|u_{ij}\|_2^2 = \lambda$. Since $\|\mathbf{x}\|^2 = \sum |(\mathbf{x}|\mathbf{e}_i)|^2$ for all $\mathbf{x}$,

$$\sum_{i,j\leq n} |u_{ij}(g)|^2 = \sum_{i,j\leq n} |(U(g)\mathbf{e}_i|\mathbf{e}_j)|^2 \leq \sum_{i\leq n} \|U(g)\mathbf{e}_i\|^2\,.$$

This result equals $n$ since all $U(g)$ are unitary. Integrating the left hand side shows that $n^2\lambda \leq n$, and so $n \leq \lambda^{-1}$, which proves that $\dim(\mathcal{H}) \leq \lambda^{-1}$. For $n$ maximum, the inequality is replaced by an equality and thus $\lambda = \dim(\mathcal{H})^{-1}$, qed.

The orthogonality relations can also be written differently by considering the integral

$$\int Tr[AU(g)]\overline{Tr[BU(g)]}dg\,.$$

As a function of the ordered pair $(A, B)$, it is a positive Hermitian form on $\mathcal{L}(\mathcal{H})$, and equals $Tr(ASB^*)$ for some unique $S \in \mathcal{L}(\mathcal{H})$. As above, the invariance of the measure shows that $S$ commutes with all $U(x)$, and so is a scalar $\lambda$. For $A$ and $B$ choosing operators whose matrices, with respect to an orthonormal basis of $\mathcal{H}$, only contain one non-zero entry, we recover the orthogonality relations of the theorem and $\lambda = \dim(\mathcal{H})^{-1}$; and

$$(29.5) \qquad \int Tr[AU(g)]\overline{Tr[BU(g)]}dg = \dim(\mathcal{H})^{-1}\, Tr(AB^*)\,.$$

(ii) *Central functions on a compact group.* The functions $f(g) = Tr[U(g)]$ satisfy $f(xy) = f(yx)$, i.e. are *central functions* on $G$. Over a compact group, the formula

$$(29.6) \qquad f^\natural(g) = \int f(s^{-1}gs)ds$$

transforms every $f \in L(G)$ into a continuous central function equal to $f$ if $f$ is central, and non-zero if $f$ is $> 0$. Proving the formulas

$$(29.7) \qquad (f^\natural)^\natural = f^\natural\,, \quad (f|g) = (f^\natural|g) \quad \text{if } g \text{ is central}\,,$$

$$(29.7') \qquad (f^\natural * g)^\natural = f^\natural * g^\natural$$

using the Haar measure will be a simple exercise for the reader. Clearly,

$$(29.8) \qquad U(f^\natural) = \int U(s)U(f)U(s)^{-1}.ds$$

on *any* finite or infinite-dimensional continuous linear representation $(\mathcal{H}, U)$ of $G$, for all functions such that $\int \|U(x)\|.|f(x)|dx < +\infty$.

The group $G$ acts on the space $G$ by the map $(s, g) \longmapsto sgs^{-1}$. Being compact, $G$ acts properly [n° 15, (iii), lemma 3]. The quotient space $X$ of conjugacy classes of $G$ is, therefore, separated and even compact. Lemma 4 of n° 15, (iv) on quotient measures shows that if the central functions are considered to be defined on the quotient space, the map $f \longmapsto f^\natural$ transforms $L(G)$ into $L(X)$. Computing the quotient measure when $G$ is a Lie group is a famous exercise using the structure of semisimple Lie algebras and enables us to compute explicitly the characters of $G$ (H. Weyl[105]).

Using central functions, we get the Dirac sequences at the origin. Indeed, transforming by (6) all $f \geq 0$ with integral 1 of a Dirac sequence at the origin gives positive functions $f^\natural$ with integral 1. If $f$ vanishes outside a neighbourhood $W$ of $e$, $f^\natural$ vanishes outside the union of $sWs^{-1}$. So the proof reduces to showing that *for any open neighbourhood $V$ of $e$, there is a neighbourhood $W$ such that $sWs^{-1} \subset V$ for all $s$.* To this end, we associate to each open neighbourhood $W$ of $e$ the set $W'$ of $s$ such that $W \subset sVs^{-1}$. As $V$ and $W$ are open, so is $W'$. As $sVs^{-1}$ is a neighbourhood of $e$ for all $s$, every $s \in G$ belongs to some $W'$. Since the open sets $W'$ cover the compact space $G$, there are neighbourhoods $W_1, \ldots, W_n$ such that $W'_p$ cover $G$. This means that, for all $s \in G$, there exists $p$ such that $sVs^{-1} \supset W_p$. Hence

$$sVs^{-1} \supset W_1 \cap \ldots \cap W_n = W\,,$$

qed. This result shows that, for $\varphi \in L^2(G)$,

$$f * \varphi = 0 \quad \text{for all central } f \implies \varphi = 0\,. \tag{29.9}$$

The map $\natural$ can be extended to all $L^p(G)$ and is still defined *almost everywhere* by relation (6). To begin with,

$$\iint f(sgs^{-1})dsdg = \int f(g)dg$$

for all $f \in L(G)$. This means that the image of the measure $dsdg$ under the map $(s, g) \longmapsto sgs^{-1}$ is the invariant measure of $G$. So theorem 23 of n° 14 can be used: for all $f \in L^p(G)$, the function $(s, g) \longmapsto f(sgs^{-1})$ is in $L^p(G \times G)$, and so is integrable. As a result, $\int |f(sgs^{-1})|ds < +\infty$ for almost all $g$, which makes integral (6) well-defined almost everywhere. Finally

$$\iint \left|f\left(sgs^{-1}\right)\right|^p ds = \int |f(g)|^p dg$$

shows that $\|f^\natural\|_p \leq \|f\|_p$. This is obtained by extending the map $f \longmapsto f^\natural$ from $L(G)$ to $L(G)$ by continuity to any $L^p(G)$. $Z(G)$ will denote the set of central functions in $L(G)$ and $Z^p(G)$ the set of central functions in $L^p(G)$.

[105] The general formula can be found in Dieudonné, *Eléments d'analyse*, chap. XXI, p. 106, the main problem being that notations are hard to understand.

On the other hand, $\tilde{f}$ is clearly central if so is $f$. This is also the case of $f * g$ if so are $f$ and $g$. An easy calculation shows that central functions are characterized by the fact that

$$f * g = g * f \quad \text{for all } g \in L^1(G) .$$

In $L^2(G)$, the set of operators

$$L(f) : g \longmapsto f * g , \quad f \in Z^1(G)$$

is, therefore, a self-adjoint commutative algebra whose elements commute with all left and right translations

$$L(x) : g \longmapsto \epsilon_x * g , \quad R(x) : g \longmapsto g * \epsilon_x$$

as well as with all left and right convolution operators. So do the operators of the Gelfand-Neumark algebra $\mathbf{Z}(G)$ generated by these $L(f)$, i.e. the limits *in norm* of operators $L(f) + \alpha 1$, where $\alpha \in C$ and $f \in Z^1(G)$, and of the corresponding von Neumann algebra[106] $\mathcal{Z}(G)$, i.e. the (weak, strong, ultrastrong) limits of the same operators [n° 19, (vii)]. It is not necessary to add scalar operators to the set of operators $L(f)$ since using Dirac sequences shows that the unit operator is the ultrastrong limit[107] of operators $L(f)$.

As an interlude, let us show that $\mathcal{Z}(G)$ is in fact the set of *all* operators commuting with all right and left translations. Indeed let $\mathcal{L}(G) = \mathcal{L}$ be the von Neumann algebra generated by all $L(x)$ and $\mathcal{R}(G) = \mathcal{R}$ the algebra generated by all $R(x)$. Clearly, $\mathcal{L} \subset \mathcal{R}'$ and $\mathcal{R} \subset \mathcal{L}'$. In fact, $\mathcal{L} = \mathcal{R}'$ and $\mathcal{R} = \mathcal{L}'$. We will generalize this result [n° 31, (ii)] to all unimodular groups later. To see this, let us consider operators $A \in \mathcal{L}'$ and $B \in \mathcal{R}'$. $A$ commutes with all $L(x)$, so with $L(f)$ for all $f \in L^1(G)$, hence for all $f \in L^2(G)$ since $G$ is compact. So, for $f, g \in L^2(G) \subset L^1(G)$,

$$A(f * g) = AL(f)g = L(f)Ag = f * Ag$$

and similarly $B(f * g) = Bf * g$; whence

$$AB(f * g) = A(Bf * g) = Bf * Ag = B(f * Ag) = BA(f * g) .$$

These calculations are well-defined in a *compact* group since the convolution product of two functions in $L^2(G)$ is again in $L^2(G)$. It is in fact a continuous function. As the set of convolutions $f * g$ is everywhere dense in $L^2$, it follows that $AB = BA$ for all $A \in \mathcal{L}'$ and $B \in \mathcal{R}'$. Hence, using the notation of von N [n° 19, (vii)], $\mathcal{L}' \subset \mathcal{R}''$ and $\mathcal{R}' \subset \mathcal{L}''$. But by his density theorem, $\mathcal{R}'' = \mathcal{R}$ and $\mathcal{L}'' = \mathcal{L}$. So

[106] Henceforth, italicized capital letters will be used for von Neumann algebras and bold capitals for GN algebras.

[107] By lemma 6 of n° 19, (vii), it suffices to shows that if a function $g \in L^2(G)$ is a zero of all operators $L(f)$, then $g = 0$. This is obvious thanks to Dirac sequences.

$$\mathcal{R}(G)' = \mathcal{L}(G), \quad \mathcal{L}(G)' = \mathcal{R}(G).$$

We can now come back to the algebra $\mathcal{Z}(G)$ generated by the operators $L(f) = R(f)$, for central $f$. We need to show that *every* operator $A$ commuting with all left and right translations, i.e. such that $A \in \mathcal{L} \cap \mathcal{R}$, is in $\mathcal{Z}(G)$. The functions $f \in Z^1(G)$ being characterized by the relation $L(x)R(x)f = f$, $Af$ is central if so is $f$. Let us then choose a Dirac sequence consisting of measures $f_n(x)dx$, where the functions $f_n$ are central. The sequence of $L(f_n)$ converges weakly to the unit operator, so that the sequence of $AL(f_n)$ converges weakly to $A$. But since $A \in \mathcal{R}$, $AL(f_n)g = A(f_n * g) = Af_n * g = L(Af_n)g$ for all $g \in L^2$. As all $Af_n$ are central, $L(Af_n) \in \mathcal{Z}(G)$, so that $A$ is the limit of operators belonging to $\mathcal{Z}(G)$, and so is in $\mathcal{Z}(G)$. Thus, indeed

$$\mathcal{Z}(G) = \mathcal{L}(G) \cap \mathcal{R}(G).$$

In the case of a general lcg, an algebra $\mathbf{Z}(G)$ generated by convolutions can always be defined using integrable central functions. But the only *integrable* central function on $G$ may be the trivial one, for example if $G$ is a semisimple non-compact Lie group such as $SL_2(R)$. Then $Z^1(G) = \{0\}$. In this case, the algebra $\mathcal{Z}(G)$ is defined by the previous formula,[108] but it is then impossible to replace $\mathcal{Z}(G)$ by a GN algebra whose spectrum would be much more reasonable than the totally discontinuous one of $\mathcal{Z}(G)$. This is where the difficulty of the general case lies.

Using $\mathbf{Z}(G)$ in the same way as the algebra $\mathbf{A}(G)$ of the commutative case [n° 25, (i)], we will obtain the entire theory of compact groups.[109] To understand the method – there are other ones –, it is useful to start with the proof of some formulas related to irreducible representations.

By Schur's lemma, the operators $U(f)$, $f \in Z^1(G)$, on any irreducible representation $(\mathcal{H}, U)$, are scalars. Setting $U(f) = \chi(f)1$,

$$\chi(f) = \dim(\mathcal{H})^{-1}\, Tr[U(f)] = Sp[U(f)].$$

This convenient notation (*Spur* = trace in German) restricts the legions of factors $\dim(\mathcal{H})$ that clutter most presentations. $Sp(1) = 1$ and as $Sp$ is a linear functional on $\mathcal{L}(\mathcal{H})$,

$$Sp[U(f)] = \int Sp[U(x)]f(x)dx = \int f(x)\chi(x)dx,$$

where

[108] $\mathcal{Z}(G)$ is the von N algebra generated by $\mathbf{Z}(G)$ if and only if every neighbourhood of $e$ in $G$ contains an invariant neighbourhood under inner automorphisms. Apart from central groups, about the only ones to have this property are the discrete groups.

[109] The reader is advised to interpret all the following calculations and results in the case where $G$ is commutative.

(29.10) $$\chi(x) = Sp[U(x)] = \dim(\mathcal{H})^{-1}\, Tr[U(x)]$$

is a central function proportional to what is usually called the *character* of $(H, U)$, namely the function $Tr[U(x)]$. As $\chi(e) = 1$, I will call (10) the *normalized character* of the representation considered. Obviously $\tilde{\chi} = \chi$. As there is a relation analogous to (5) for inequivalent representations, except that the right hand side is then zero, putting $A = B = 1$ shows that the (standard or normalized) *characters of two inequivalent representations are orthogonal.*

The map

$$f \longmapsto \chi(f) = \int f(x)\chi(x)dx$$

is a character of the normed algebra $Z^1(G)$ which, in fact, can be extended to $\mathbf{Z}(G)$ since, all irreducible representations $(\mathcal{H}, U)$ being realizable on[110] $L^2(G)$, $U(f)$ is the restriction to $\mathcal{H}$ of the convolution operator $L(f)$ by $f$ whose norm on $\mathcal{H}$ is bounded above by its norm on $L^2(G)$. Thus $\|U(f)\| \leq \|L(f)\|$. Relation (10) can be generalized to all $f \in L^1(G)$ in the form

$$Sp[U(f)] = \chi(f^\natural),$$

since, by invariance of the trace,

$$Sp[U(f)] = Sp\left[U(s)U(f)U\left(s^{-1}\right)\right]$$

for all $s \in G$. Thus integrating with respect to $s$ and using (8) lead to the formula. We also write down the relation

$$\chi * f(x) = \int Sp\left[U\left(xy^{-1}\right)\right] f(y)dy = Sp\left[U(x)U(f')\right],$$

where $f'(y) = f(y^{-1})$. If $f$ is central, $U(f')$ reduces to the scalar

$$Sp\left[U(f')\right] = \int Sp\left[U\left(y^{-1}\right)\right] f(y)dy = (f|\chi).$$

Thus

(29.11) $$\chi * f = (f|\chi)\chi \quad \text{if } f \in Z^1(G)$$

and in particular

(29.11') $$\chi * \chi = (\chi|\chi)\chi, \quad \chi * \chi' = 0 \quad \text{if} \chi \neq \chi'.$$

[110] Choose a non-zero vector $\mathbf{a}$ in $\mathcal{H}$ and associate the function

$$g \longmapsto \left(U(g^{-1})\mathbf{x}|\mathbf{a}\right)$$

to each $\mathbf{x} \in \mathcal{H}$. This map transforms the operator $\mathbf{x} \longmapsto U(s)\mathbf{x}$ into the left translation $L(s)$.

We prove the second relation by first observing that, by (11), $\chi * \chi' = (\chi|\chi')\chi = 0$, then that $\chi$ and $\chi'$ are orthogonal as was seen above.

Normalized characters have a simple functional equation. For all $x$, the operator $\int U(uxu^{-1})du$ commutes with the representation, and so reduces to the scalar

$$Sp\left[\int U\left(uxu^{-1}\right)du\right] = \int Sp\left[U\left(uxu^{-1}\right)\right]du = \int Sp\left[U(x)\right]du = \chi(x)\,.$$

Hence

$$\int U\left(uxu^{-1}\right)du = Sp\left[U(x)\right]1 \quad \text{for all } x \in G\,. \tag{29.12}$$

We then deduce that

$$\int \chi\left(uxu^{-1}y\right)du = Sp\left[\int U\left(uxu^{-1}\right)U(y)du\right] = Sp[U(x)]\,Sp[U(y)]\,,$$

i.e. that

$$\int \chi\left(uxu^{-1}y\right)du = \chi(x)\chi(y)\,. \tag{29.13}$$

Setting $U(x) = A$, relation (12) becomes

$$\int U(u)AU(u)^{-1}du = Sp(A) \tag{29.14}$$

and in fact holds for all $A \in \mathcal{L}(\mathcal{H})$, since any operator $A$ is a linear combination of operators $U(x)$ by von Neumann's density theorem, which is almost trivial and purely algebraic in finite dimension ("Burnside's lemma"). We could have also observed that the left hand side commutes with all $U(x)$, and so reduces to a scalar necessarily equal to $Sp[\int U(u)AU(u)^{-1}du] = Sp(A)$.

We will again encounter these formula for the simple reason that, like in the case of a commutative group, (classes of) of irreducible representations of $G$ correspond bijectively to the characters of $\mathbf{Z}(G)$.

(iii) *Spectral decomposition of* $\mathbf{Z}(G)$. Let us denote the compact spectrum of $\mathbf{Z}(G)$ by $X(G)$ and let $\widehat{G}$ be the set of $\chi \in X(G)$ that are not identically zero on $Z^1(G)$. Since $\widehat{G}$ is the complement of at least one element of $X(G)$ – obviously $X(G) = \widehat{G}$ if $G$ is finite –, it is a locally compact space (in fact, discrete as will be seen) for which the forgotten character of $\mathbf{Z}(G)$ plays the role of the "point at infinity". For $\chi \in X(G)$ and $f \in L^1(G)$,

$$\left|\chi\left(L(f^\natural)\right)\right| \leq \left\|L\left(f^\natural\right)\right\| \leq \left\|f^\natural\right\|_1 \leq \|f\|_1\,.$$

So there is (duality between $L^1$ and $L^\infty$) a bounded measurable function $\chi(x)$ such that

$$\chi\left[L(f^\natural)\right] = \int f(x)\chi(x)dx$$

for all $f \in L^1(G)$. The function $\chi$ may be assumed to be central – otherwise, replace it with $\chi^\natural$ –, which determines it completely since then

$$\int f(x)\chi(x)dx = \int f^\natural(x)\chi(x)dx = \chi\left[L(f^\natural)\right]. \tag{29.15}$$

It is zero if $\chi$ is the point at infinity of $\widehat{G}$.

For all $f \in L^1(G)$, we will set

$$\chi(f) = \int f(x)\chi(x)dx = \chi(f^\natural) = (f|\overline{\chi}). \tag{29.16}$$

For given $f$, we get a continuous function $\chi \longmapsto \chi(f)$ on $\widehat{G}$ tending to 0 at infinity by definition of the topology of $X(G)$. The equality

$$\chi(f * g) = \chi(f)\chi(g)$$

holds if $f$ *or* $g$ is central because of the second relation (7).

If $\chi$ is a character of $\mathbf{Z}(G)$, so is the conjugate function

$$\overline{\chi}(f) = \overline{\chi(\overline{f})}.$$

Let us now show that, for all $\chi \in \widehat{G}$,

$$\chi * \chi(x) = (\chi|\chi)\chi(x) \tag{29.17}$$

(*a priori* almost everywhere) and more generally that

$$f * \chi(x) = (f|\chi)\chi(x) \tag{29.18}$$

for all $f \in Z^1(G)$. To this end, let us set $g'(x) = g(x^{-1})$ for all functions $g$. Clearly, $(g * h)' = h' * g'$ since $G$ is unimodular. Moreover, $\int g(x)h(x)dx = g' * h(e)$ for all $g, h$ and in particular $\chi(f) = f' * \chi(e)$. Hence, setting $\chi * f = \varphi$, i.e. a central function like $f$ and $\chi$, for any function $h \in L^1(G)$,

$$\begin{aligned}\int h(x).\varphi(x)dx &= h' * \varphi(e) = h' * f * \chi(e) = (h' * f')' * \chi(e) = \chi(h * f') \\ &= \chi(h)\chi(f') = \chi(f')\int h(x).\chi(x)dx\end{aligned}$$

since $f'$ is central, whence $\varphi = \chi(f')\chi = (f|\chi)\chi$ almost everywhere, qed.

As the function $\chi * \chi$ is continuous, $\chi(x)$ may be supposed to be continuous, which determines it everywhere. It then satisfies (17) and (18) including with respect to a null measure, and as $\chi * \chi(e) = (\chi|\chi)$, we deduce that $\chi(e) = 1$.

*Exercise 1.* Using the fact that

$$\chi\left(f^\natural * g^\natural\right) = \chi\left(f^\natural\right)\chi\left(g^\natural\right) = \chi(f)\chi(g)$$

for all $f$ and $g \in L(G)$, show that

$$\int \chi\left(uxu^{-1}y\right) du = \chi(x)\chi(y)\,. \tag{29.19}$$

If $\chi$ and $\chi'$ are distinct characters, (18) shows that $\chi * \chi'$ is proportional to $\chi$ *and* $\chi'$. As two proportional characters of a normed algebra must be identical,

$$\chi * \chi' = 0\,, \quad (\chi|\chi') = 0\,. \tag{29.20}$$

However, for all $f \in Z^1(G)$, the function $(f|\chi)$ is continuous on $\widehat{G}$. This is the case of $\chi \longmapsto (\chi'|\chi)$ for all $\chi' \in \widehat{G}$. Since this function vanishes everywhere except at $\chi = \chi'$, the subspace $\widehat{G} = X(G) - \{\infty\}$ is *discrete*.

The spectral decomposition

$$T = \int_{X(G)} \widehat{T}(\chi) dM(\chi)\,, \quad T \in \mathbf{Z}(G)\,,$$

of n° 22 is then particularly simple. First, for each $\chi \in \widehat{G}$, by (22.14), the spectral manifold $L^2(G;\chi)$ associated to the open set $\{\chi\} \subset X(G)$ is the set of $\varphi \in L^2(G)$ such that

$$T\varphi = \chi(T)\varphi \quad \text{for all } T \in \mathbf{Z}(G),$$

or even only for $T = L(f)$ where $f \in Z^1(G)$. In this case, $\chi(T) = \chi(f) = (f|\overline{\chi})$ by (16), whence

$$f * \varphi = \chi(f)\varphi = (f|\overline{\chi})\varphi\,. \tag{29.21}$$

If $f = \chi' \in \widehat{G}$, then $\chi(f) = \int \chi'(x)\chi(x)dx = 0$ except if $\chi' = \overline{\chi}$, in which case $\chi(f) = (\chi|\chi)$. So

$$\overline{\chi} * \varphi = (\chi|\chi)\varphi \tag{29.21'}$$

in $L^2(G;\chi)$. Conversely, (21') implies (21) since

$$(\chi|\chi) f * \varphi = f * (\overline{\chi} * \varphi) = (f * \overline{\chi}) * \varphi = (f|\overline{\chi})\overline{\chi} * \varphi = \chi(f)(\chi|\chi)\varphi$$

by (18) for $\overline{\chi}$. Hence

$$\varphi \in L^2(G;\chi) \Longleftrightarrow \overline{\chi} * \varphi = (\chi|\chi)\varphi\,. \tag{29.22}$$

Given (17) for $\overline{\chi}$, $\overline{\chi} \in L^2(G;\chi)$.

According to the general results of n° 22, (iii), the spectral manifold associated to the open and discrete subset $\widehat{G} = X(G) - \{\infty\}$ of $X(G)$, is the Hilbert direct sum of the spaces $L^2(G;\chi)$ for all $\chi \in \widehat{G}$. Therefore, like in any GN algebra, the spectral manifold associated to $X(G) = \widehat{G} \cup \{\infty\}$, namely $L^2(G)$ is the direct sum of the spaces $L^2(G;\chi)$ and of the spectral manifold associated to $\{\infty\}$. It is trivial since its elements satisfy $L(f)\varphi = 0$ for all $f \in Z^1(G)$, whence $\varphi = 0$ by (9). So the spectral manifold associated to $\widehat{G}$ is $L^2(G)$, and so

$$L^2(G) = \bigoplus_{\chi \in \widehat{G}} L^2(G;\chi)\,, \tag{29.23}$$

which is a Hilbert direct sum.

Let us compute $E(\chi)$, the orthogonal projection operator onto $L^2(G;\chi)$. The operator $f \longmapsto \overline{\chi} * f$ is Hermitian and proportional to its square by (17). By (22), it reduces to the scalar $(\chi|\chi)$ on $L^2(G;\chi)$ and it obviously vanishes on the other spectral manifolds $L^2(G;\chi')$. So it is proportional to $E(\chi)$, and as $E(\chi)$ needs to equal 1 on $L^2(G;\chi)$,

$$E(\chi)f = (\chi|\chi)^{-1}.\overline{\chi} * f \tag{29.24}$$

necessarily holds for all $f \in L^2(G)$.

Permuting $\chi$ and $\overline{\chi}$ in (23), for all $f \in L^2(G)$,

$$f = \sum (\chi|\chi)^{-1}.\chi * f\,, \tag{29.25}$$

i.e. a series of pairwise orthogonal functions, and

$$(f|g) = \sum (\chi|\chi)^{-2}(\chi * f|\chi * g) \tag{29.25'}$$

for all $f$ and $g$. Summation is over $\chi \in \widehat{G}$, series (25') converging unconditionally (Chap. II, n° 15).

(iv) *Characters of* $\mathbf{Z}(G)$ *and irreducible representations.* As operators $T \in \mathbf{Z}(G)$ commute with all the right and left translations, the spaces $L^2(G;\chi)$ are bi-invariant. Let us show that, the right and left translation operators on $L^2(G;\chi)$ define an *irreducible* representation of $G \times G$.

Indeed, let $A$ be an operator on $L^2(G;\chi)$ commuting with these translations. Extend it to $L^2(G)$ by requiring it to be zero on the other spaces $L^2(G;\chi')$. This gives an operator $T$ belonging to the von N algebra $\mathcal{L}(G) \cap \mathcal{R}(G) = \mathcal{Z}(G)$ generated by $\mathbf{Z}(G)$ [end of section (ii)]. As any $T$ in $\mathbf{Z}(G)$, or in $\mathcal{Z}(G)$, reduces to a scalar on $L^2(G;\chi)$, so does $A$, which implies irreducibility.

$G \times G$ being compact, theorem 42 shows that *the subspaces* $L^2(G;\chi)$ *are finite-dimensional.* Then let $\mathcal{H}$ be a minimal non-trivial left invariant subspace of $L^2(G;\chi)$. Denoting by $U(x) : \varphi \longmapsto \epsilon_x * \varphi$ the restriction of $L(x)$ to

$\mathcal{H}$, the representation $(\mathcal{H}, U)$ of $G$ is irreducible. For all $f \in Z^1(G)$, the operator $U(f) = \int U(x)f(x)dx$, which is the restriction of $L(f) = L(x)f(x)dx$ to $\mathcal{H}$, reduces to a scalar, namely

$$\lambda(f) = Sp[U(f)] = \int f(x)\lambda(x)dx\,,$$

where

$$\lambda(x) = Sp[U(x)]$$

is the normalized character of the representation $(\mathcal{H}, U)$ defined in section (ii). As every $T \in \mathbf{Z}(G)$ reduces to the scalar $\chi(T)$ on $L^2(G;\chi)$ and so perforce in $\mathcal{H}$, $\lambda(f) = \chi[L(f)] = \int f(x)\chi(x)dx$. Thus

$$\chi(x) = Sp[U(x)] \tag{29.26}$$

for any irreducible component $(\mathcal{H}, U)$ of the left regular representation of $G$ on $L^2(G;\chi)$. *Hence the algebra* $\mathbf{Z}(G)$ *only has characters defining the irreducible representations of* $G$ (and the character "at infinity"). Exercise 1 was therefore superfluous since we already knew (13).

We can now reconstruct $L^2(G;\chi)$ using $(\mathcal{H}, U)$. To begin with, by (24), $L^2(G;\chi)$ is the set of functions of the form $f * \overline{\chi}$. However, for all $f \in L^2(G)$,

$$f * \overline{\chi}(x) = \int f(xy)\chi(y)dy = \int f(xy)\, Sp[U(y)]dy$$

and so

$$f * \overline{\chi}(x) = Sp[U(x)^{-1}U(f)]\,. \tag{29.27}$$

As every $A \in \mathcal{L}(\mathcal{H})$ is in some $U(f)$ because of irreducibility, it follows that $L^2(G;\chi)$ *is the set of functions of the form*

$$x \longmapsto Sp\left[AU(x)^{-1}\right]\,.$$

The preceding calculations assume that a minimal left invariant subspace $\mathcal{H}$ has been chosen, but changing $\mathcal{H}$ replaces the representation $(\mathcal{H}, U)$ by an equivalent representation since, as was seen above, the characters of two inequivalent irreducible representations are orthogonal. In conclusion, *the subspaces* $L^2(G;\chi)$ *correspond bijectively to irreducible representations of* $G$.

The equality

$$(f|g) = \sum(\chi|\chi)^{-2}(\chi * f|\chi * g)\,.$$

can now be written differently. Calculations leading to (27) also shows that

$$f * \chi(x) = Sp\left[U(x)U(f')\right]\,,$$

where $f'(x) = f(x^{-1})$. So, using the orthogonality relations in the form (5),

$$\begin{aligned}(\chi * f|\chi * g) &= \int Sp\,[U(x)U(f')]\,\overline{Sp\,[U(x)U(g')]}dx \\ &= \dim(\mathcal{H})^{-2}\,Sp\,[U(f')U(g')^*] = (\chi|\chi)\,Sp\,[U(f')U(g')^*]\end{aligned}$$

because (5) for $A = B = 1$ also shows that

$$(\chi|\chi) = \dim(\mathcal{H})^{-2}\,.$$

Hence

$$\begin{aligned}(\chi|\chi)^{-2}(\chi * f|\chi * g) &= (\chi|\chi)^{-1}\,Sp\,[U(f')U(g')^*] \\ &= \dim(\mathcal{H})^2\,Sp\,[U(f')U(g')^*] \\ &= \dim(\mathcal{H})\,Tr\,[U(f')U(g')^*]\,.\end{aligned}$$

This gives Plancherel's formula (it is better to say Peter-Weyl's, even if they argued in terms of matrices instead of operators)

$$(f|g) = \sum \dim(\mathcal{H})\,Tr\,[U(f')U(g')^*]\,, \tag{29.28}$$

where the summation is over the irreducible classes of representations. Besides, as $(f'|g') = (f|g)$, $f'$ and $g'$ can also be replaced by $f$ and $g$ on the right hand side.

To obtain a formula which, for a commutative $G$, reduces to those of the previous §, for each $\chi \in \widehat{G}$, we could choose an irreducible representation

$$x \longmapsto U(x;\chi)$$

with character $\chi$ in the space $\mathcal{H}(\chi)$ and set

$$\widehat{f}(\chi) = \int U(x;\chi)^* f(x)dx = U(f';\chi)\,. \tag{29.29}$$

For every $f \in L^1(G)$; this Fourier transform of $f$, a function whose values are *operators*, reduces for central $f$ to the scalar$(f|\chi)$. Denoting by $\dim(\chi)$ the dimension of $\mathcal{H}(\chi)$, (28) becomes

$$(f|g) = \sum \dim(\chi)\,Tr\left[\widehat{f}(\chi)\widehat{g}(\chi)^*\right] \tag{29.28'}$$

for $f, g \in L^2(G)$. This immediately gives formulas

$$g(x) = f(axb) \Longrightarrow \widehat{g}(\chi) = U(b,\chi)\widehat{f}(\chi)U(a,\chi)\,, \tag{29.30}$$

$$(\widehat{f * g})(\chi) = \widehat{g}(\chi)\widehat{f}(\chi)\,. \tag{29.31}$$

*Exercise 2.* Prove an analogue of the Fourier inversion formula.

*Exercise 3.* Consider the von Neumann algebra $\mathcal{Z}(G)$. Every $T \in \mathcal{Z}(G)$ acts on each subspace $L^2(G;\chi)$ by a scalar $\widehat{T}(\chi)$. Show that this gives all the functions such that

$$\sup |\widehat{T}(\chi)| < +\infty .$$

To conclude, let us show how a unitary representation $(\mathcal{H}, U)$ of $G$ can be decomposed.[111] Since, by (17), the function $(\chi|\chi)^{-1}\chi$ is Hermitian and identical to its convolution square, the operator

$$E(\chi) = (\chi|\chi)^{-1} \int U(x)\overline{\chi(x)}dx = (\chi|\chi)^{-1}U(\overline{\chi}) \tag{29.32}$$

is a projection onto a subspace $\mathcal{H}(\chi)$. Since $\chi * \chi' = 0$ if $\chi \neq \chi'$, these subspaces are pairwise orthogonal, and invariant since the characters are central functions. For $\mathbf{a}, \mathbf{b} \in \mathcal{H}$, age-old calculations give

$$(\chi|\chi)^{-1} \int (U(xy^{-1})\mathbf{a}|\mathbf{b})\, \chi(y)dy = (U(x)E(\chi)\mathbf{a}|\mathbf{b}) = (U(x)E(\chi)\mathbf{a}|E(\chi)\mathbf{b}) .$$

If $\mathbf{a} \in \mathcal{H}(\chi)$, the coefficient

$$(U(x)\mathbf{a}|\mathbf{b}) = \varphi_{\mathbf{a},\mathbf{b}}(x) \tag{29.33}$$

of $(\mathcal{H}, U)$ thus satisfies $(\chi|\chi)^{-1}.\chi * \varphi_{\mathbf{a},\mathbf{b}} = \varphi_{\mathbf{a},\mathbf{b}}$, and so is in $L^2(G;\chi)$. Therefore, as $\mathbf{b}$ varies in $\mathcal{H}$, functions (33) remain in a finite-dimensional subspace of $L^2(G)$, in fact of $L(G)$. It follows that the subspace $\mathcal{H}(\mathbf{a})$ of $\mathcal{H}$ generated by all $U(x)\mathbf{a}$ is also finite-dimensional. As it is invariant, it is the direct sum of minimal non-trivial invariant subspaces on each of which the operators $U(x)$ act irreducibly. The representations thus obtained on $\mathcal{H}(\mathbf{a})$ all belong to the class $\chi$ considered since there operators (32) are equal to 1. As for the subspace $\mathcal{H}(\chi)$, it can be infinite-dimensional, but as the orthogonal complement of some $\mathcal{H}(\mathbf{a})$ in $\mathcal{H}(\chi)$, it is also invariant. $\mathcal{H}(\chi)$ is the Hilbert direct sum (in several ways!) of subspaces $\mathcal{H}(\mathbf{a})$, hence of finite-dimensional irreducible subspaces, each of which correspond to a representation with character $\chi$.

It remains to be shown that $\mathcal{H}$ is a Hilbert direct sum of the subspaces $\mathcal{H}(\chi)$. This is an exercise whose details can surely be omitted at this point of the presentation.

(v) *Easy generalizations.* The reader, having probably observed similarities between the methods used for commutative groups and compact groups, will suspect that it should be possible to address both cases simultaneously. There are three ways at our disposal to achieve this. I will only briefly outline them.

[111] Apart from orthogonality questions, the following arguments apply to all continuous representations in a Banach space.

(a) The obvious method[112] consists in generalizing calculations and results to central groups ($G/Z$ compact) and, for every character $\alpha$ of the centre $Z$ of $G$, in replacing $L^2(G)$ by the space $L^2(G/Z;\alpha)$ of mod $Z$ square integrable solutions of $f(xz) = f(x)\alpha(z)$. There is still a map $f \longmapsto f^\natural$ from $L^1(G)$ onto its centre $Z^1(G)$, given by

$$f^\natural(x) = \int_{G/Z} f\left(uxu^{-1}\right) du\,.$$

The convolution products by $f \in Z^1(G)$ obviously preserve the space $L^2(G/Z;\alpha)$ on which an *ersatz* of the convolution can be defined by setting

$$f *_Z g(x) = \int_{G/Z} f\left(xy^{-1}\right) g(y)dy\,.$$

In this space, (ordinary) convolutions by $f \in Z^1(G)$ generate an algebra $\mathbf{Z}(G/Z;\alpha)$ with the same properties as the algebra $\mathbf{Z}(G)$ of the compact case. As above, one can show that the characters of this algebra are the functions $\chi(x) = Sp[U(x)]$ of the irreducible representations $(\mathcal{H}, U)$ on which the centre $Z$ de $G$ acts by $\alpha$. Denoting by $L^2(G;\chi)$ the subspace of functions $Tr[AU(x)^{-1}]$, where $A \in \mathcal{L}(\mathcal{H})$, $L^2(G/Z;\alpha)$ is shown to be a Hilbert direct sum of these subspaces. If, for any character $\chi$, $(U(x;\chi), \mathcal{H}(\chi))$ denotes a representation with normalized character $\chi(x) = Sp[U(x;\chi)$, the projection of $f \in L^2(G/Z;\alpha)$ onto $L^2(G;\chi)$ is the function

$$x \longmapsto \dim(\chi) \int_{G/Z} Tr\left[U(xy^{-1};\chi)\right] f(y)dy\,.$$

On the other hand, associating the function

$$f_\alpha(x) = \int f(xz)\overline{\alpha(z)}dz$$

to each $f \in L(G)$ gives functions $f_\alpha$ everywhere dense in $L^2(G;\alpha)$. The projection of $f_\alpha$ onto $L^2(G;\chi)$ is the function

$$\begin{aligned}\dim(\chi) \textstyle\int_{G/Z} dy \int_Z Tr\left[U(xy^{-1};\chi)\right] f(yz^{-1})\alpha(z)dz &= \\ = \dim(\chi) \textstyle\int dy \int Tr\left[U(x(yz)^{-1};\chi)\right] f(yz)dz &= \\ = \dim(\chi) \textstyle\int_G Tr\left[U(xy^{-1};\chi)\right] f(y)dy\,,\end{aligned}$$

i.e. setting

$$\widehat{f}(\chi) = \int U(x;\chi)^* f(x)dx$$

[112] S. Grosser and M. Moskowitz, *Harmonic analysis on central topological groups* (Trans. AMS, **156**, 1971, pp. 419–454). Also Calvin Moore, *Groups with finite-dimensional irreducible representations* (idem, **166**, 1972, pp. 401–410).

as in the compact case, this is the function $\dim(\chi)\ Tr[U(x;\chi)\widehat{f}(\chi)]$. There is again an expansion

$$(29.34)\qquad (f_\alpha|g_\alpha) = \sum_{\chi=\alpha \text{ sur } Z} \dim(\chi)\ Tr\left[\widehat{f}(\chi)\widehat{g}(\chi)^*\right]$$

for $f, g \in L(G)$. But Plancherel's formula for $Z$ shows that the inner product $(f|g)$ on $L^2(G)$ is given by

$$(f|g) = \int_{G/Z} dx \int_Z f(xz)\overline{g(xz)}dz = \int_{G/Z} dx \int_{\widehat{Z}} f_\alpha(x)\overline{g_\alpha}(x)d\alpha\,,$$

i.e. by

$$(29.35)\qquad (f|g) = \int (f_\alpha|g_\alpha)d\alpha\,.$$

Hence, given (34),

$$(f|g) = \int_{\widehat{Z}} d\alpha \sum_{\chi=\alpha \text{ sur } Z} \dim(\chi)\ Tr[f(\chi)\widehat{g}(\chi)^*]\,.$$

This can be simplified further. In the locally compact space $\widehat{G}$ of the characters of $Z^1(G)$, the relation $\{\chi = \chi' \text{ on } Z\}$ is an equivalence relation whose classes are *discrete* sets. The quotient space can be canonically identified to the dual $\widehat{Z}$ of $Z$. A measure $d\chi$ can then be *defined* on $\widehat{G}$ by setting

$$\int_{\widehat{G}} \varphi(\chi)d\chi = \int_{\widehat{Z}} d\alpha \sum_{\chi=\alpha \text{ on } Z} f(\chi)$$

for all $f \in L(\widehat{G})$. This is a very simple example of a continuous sum of discrete measures (n° 13). This definition leads to the final formula

$$(29.36)\qquad \int_G f(x)\overline{g(x)}dx = \int_{\widehat{G}} Tr\left[\widehat{f}(\chi)\widehat{g}(\chi)^*\right] \dim(\chi)d\chi\,.$$

It unifies the results obtained for commutative groups and compact groups. Though the subject is not of much use, proving all these formulas *without* having recourse to we already know about commutative or compact groups would be one of the best exercises for the reader to familiarize himself with these methods.

(b) A significantly more difficult generalization[113] as it uses G. W. Mackey's theory of induced representations consists in supposing that $G$ contains a normal commutative closed subgroup (but not necessarily central) $Z$ such that $G/Z$ is compact.

[113] Kleppner and R. L. Lipsman, *Plancherel formula for group extensions* (Ann. sc. École normale sup., **5** (1972) and **6** (1973)).

(c) The third and an extremely ingenious and much more useful method is Gelfand's *theory of spherical functions.* Suppose we are given a compact subgroup $K$ in $G$ such that the convolution product is commutative on the set of functions such that $f(uxv) = f(x)$ for all $u, v \in K$. A trivial example: $G = K \times K$, the compact subgroup being the diagonal of the product; the solutions of $f(uxv, uyv) = f(x, y)$ can be identified to the central functions on $K$. A far less trivial example: suppose there is an "involution" $x \mapsto x'$ in $G$ satisfying

$$(x')' = x\,, \quad (xy)' = y'x'\,, \quad x' = x^{-1} \quad \text{for all } x \in K\,,$$

and that all $x \in G$ can be written as $x = kp$ with $k \in K$ and $p' = p$. For $G = GL_n(\mathbb{R})$, $K$ is the orthogonal group and $x'$ the transpose of the matrix $x$. The situation is the same in every real semisimple Lie groups.[114] It then only requires one line of calculations to check that the convolution product, restricted to $K$-bi-invariant functions, is commutative for the simple reason that we then have $f(x') = f(x)$. In all cases, the set of integrable $K$-bi-invariant functions is again a commutative normed algebra whose characters, in other words the functions said to be *spherical*, are the (necessarily bi-invariant) solutions of the functional equation

$$\int f(xky)dk = f(x)f(y)\,.$$

There is still a Plancherel's formula not for all of $L^2(G)$ – it would be miraculous – but, for want of anything better, for the subspace of $L^2(G)$ generated by the images of bi-invariant functions under right and left translations. Only irreducible unitary representations of $G$ admitting a $K$-invariant vector are involved in this formula. This vector is necessarily unique up to a scalar. This method was adopted by Dieudonné in his volume 6, and I could have done the same.[115] Regardless of its merit, it seemed more reasonable to me not to throw into disarray readers who are not expected to already know everything. In addition, the real problems – explicit calculations of irreducible spherical functions using integral formulas, explicit calculations of Plancherel's formula – cannot be solved by this type of *abstract nonsense*, however ingenious it

[114] As a first approximation, these are the closed subgroups of $GL_n(\mathbb{R})$ satisfying the following conditions: (i) $G$ is algebraic, i.e. defined by polynomial equations, (ii) Every commutative closed normal subgroup of $G$ is finite. Examples: $SL_n(\mathbb{R})$, the group of automorphisms of symmetric or alternate bilinear forms (orthogonal or symplectic groups), the subgroups of $SL_n(\mathbb{C})$ leaving a symmetric bilinear form or a Hermitian form invariant, etc. There are also "exceptional groups" that cannot be easily defined... General semisimple groups are coverings of semisimple linear groups. The most simple general definition uses the Lie algebra of $G$.

[115] I presented it at the Séminaire Bourbaki (1958) adding to Gelfand's short note in *Doklady* a Plancherel formula which did not appear there. My talk, or that of Serge Lang, $SL_2(\mathbb{R})$ (Springer, 1985), Chap. IV, is probably easier to read than that of Dieudonné's...

may be. They have been completely solved by Harish-Chandra[116] in articles that display his genius, a term that should not be used lightly, and his prodigious obstinacy.

We are now going to drop the subject of compact groups and briefly outline possible (or impossible) extensions to general lcgs.[117]

## 30 – Measures and Functions of Positive Type

(i) *Measures of positive type.* Let $G$ be a locally compact group. As was seen in n° 25, the inner product on $L^2(G)$ is given by

$$(f|g) = \epsilon(\tilde{g} * f)\,, \tag{30.1}$$

where $\epsilon$ is the Diract measure at $e$. It follows that the map

$$(f, g) \longmapsto \epsilon(\tilde{g} * f) \tag{30.2}$$

is a positive Hermitian form on $L(G)$, and one obviously obtains $L^2(G)$ by applying the general method of n° 19 to it .

It is thus necessary to more generally consider measures $\mu$ on $G$ having this property, namely measures of positive type (notation: $\mu \gg 0$). Such a measure enables us to define a unitary representation of $G$, which, for $\mu = \epsilon$, is just the "left" regular representation of $G$ on $L^2(G)$.

Indeed, if $\mu \gg 0$, then taking the quotient and the completion, the general method of n° 19, (i) leads to a Hilbert space $\mathcal{H}(\mu)$ and to a canonical map $f \mapsto f_\mu$ from $L(G)$ onto an everywhere dense subspace of $\mathcal{H}(\mu)$. By definition,

$$(f_\mu|g_\mu) = \mu(\tilde{g} * f) = \int f(yx)\overline{g(y)}d\mu(x)dy \tag{30.3}$$

follows from (28.3'). If $f$ and $g$ are made to undergo the same *left* translation, the function $\tilde{g} * f$ does not change. So, if $f_\mu = 0$, neither do any of images of $f$ under left translations. Hence, for each $x \in G$ there is a linear operator $L_\mu(x)$ defined on the subspace of $f_\mu$ by

$$g = \epsilon_x * f \Longrightarrow g_\mu = L_\mu(x)f_\mu\,. \tag{30.4}$$

It is bijective, preserves the inner product and can be extended to all of $\mathcal{H}(\mu)$. This gives a unitary representation of $G$ on $\mathcal{H}(\mu)$. To check that the maps

[116] About Harish-Chandra, the "second Ramanujan" as he is now called in India, see the articles by Borel, Helgason, Langlands and Varadarajan in Robert S. Doran & V.S. Varadarajan, eds., *The Mathematical Legacy of Harish-Chandra* (Amer. Math. Soc. 2000). India has produced other first-class mathematicians, but only rarely does one meet there or elsewhere Harish's level.

[117] For the rest of this chapter, it will be useful to refer to the second part of Dixmier's more complete presentation, *$C^*$-algebras* (North-Holland, 1972), which has the inconvenience or advantage, depending on one's point of view, of being based on the general theory of GN algebras. See also Nolan R. Wallach, *Real Reductive Groups* II (Academic Press, 1992), Chap. 14.

$x \mapsto L_\mu(x)\mathbf{a}$ are continuous for all $\mathbf{a} \in \mathcal{H}(\mu)$, it suffices to do so for all $\mathbf{a}$ everywhere dense, for example, for $f_\mu$, where $f \in L(G)$, and only at $x = e$. As

$$\|L_\mu(x)f_\mu - f_\mu\|^2 = \mu\left[(\epsilon_x * f - f)^{\sim} * (\epsilon_x * f - f)\right],$$

it suffices to check that $x \mapsto \mu(\tilde{f} * \epsilon_x * f)$ is continuous.

Every $f \in L(G)$ defines an operator

$$L_\mu(f) = \int L_\mu(x)f(x)dx \tag{30.5}$$

on $\mathcal{H}(\mu)$. By definition,

$$\begin{aligned}(L_\mu(f)g_\mu|h_\mu) &= \int (L_\mu(x)g_\mu|h_\mu)\, f(x)dx = \mu\left(\tilde{h} * \epsilon_x * g\right) f(x)dx \\ &= \mu\left(\tilde{h} * f * g\right) = ((f * g)_\mu|h_\mu)\end{aligned}$$

[n° 25, (ii), exercise 1]. We then deduce that

$$h = f * g \Longrightarrow h_\mu = L_\mu(f)g_\mu \tag{30.6}$$

for $f, g, h \in L(G)$.

This result could be interpreted by first observing that, by Cauchy-Schwarz, in the algebra $L(G)$, the relation $\mu(\tilde{f} * f) = 0$ is equivalent to $\mu(\tilde{g} * f) = 0$ for all $g \in L(G)$, whence $\mu(\tilde{f} * \tilde{g} * g * f) = 0$. The set of $f$ such that $f_\mu = 0$ is, therefore, a *left ideal* $I(\mu)$ of $L(G)$. But if $I$ is a left ideal of a ring $A$, $A$ can be made to act on the quotient $A/I$ since, for $f, g \in A$, the coset of $f * g$ mod $I$ only depends on the coset of $g$. Formula (6) falls within this algebraic framework.

The Hermitian symmetry of $\mu(\tilde{g} * f)$ shows that

$$\mu\left(\tilde{g} * f\right) = \overline{\mu(\tilde{f} * g)}\,.$$

Making the measure $g(x)dx$ converge to the Dirac measure at $e$ gives $\mu(f) = \overline{\mu(\tilde{f})}$. Thus

$$\tilde{\mu} = \mu \quad \text{if} \quad \mu \gg 0\,. \tag{30.7}$$

When $G$ is a Lie group,[118] one can consider $C^\infty$ functions on $G$, and thus define the Schwartz space $\mathcal{D}(G)$ of $C^\infty$ functions with compact support, hence also distributions on $G$ by generalizing in an obvious manner what has been said in Chap. VII for $\mathbb{R}$. For $f, g \in \mathcal{D}(G)$, $f * g \in \mathcal{D}(G)$, which enables us to define *distributions of positive type* by requiring that $\mu(\tilde{g} * f)$ be a positive

[118] i.e. equipped with a $C^\infty$ manifold structure such that the map $(x, y) \mapsto xy^{-1}$ is $C^\infty$.

Hermitian form on $\mathcal{D}(G)$. The construction of the space $\mathcal{H}(\mu)$ and of the unitary representation $x \mapsto L_\mu(x)$ can be generalized to this case.

(ii) *Case of a commutative group.* Since the representation defined by $\epsilon$ leads to Plancherel's formula, one expects a similar result for all measures $\mu \gg 0$, namely:

**Theorem 44**. *Let $G$ be a commutative* lcg *and $\mu$ a measure of positive type on $G$. Then there is a unique positive measure $\widehat{\mu}$ on $\widehat{G}$ such that*

$$\mu(f * \tilde{g}) = \int \widehat{f}(\chi)\overline{\widehat{g}(\chi)}d\widehat{\mu}(\chi) \tag{30.8}$$

*for all $f, g \in L(G)$.*

The proof is almost the same as that of theorem 40. Consider the set $\mathbf{A}(\mu)$ in $\mathcal{H}(\mu)$ of limits in norm of operators of the form $L_\mu(f) + \lambda 1$; it is a commutative GN algebra and the proof reduces to setting out explicitly its spectral decomposition. To start with, as

$$\|L_\mu(f)\| \leq \|f\|_1\,,$$

any character of $\mathbf{A}(\mu)$ defines a (possibly trivial) character of $L^1(G)$, which gives a homeomorphism from the spectrum $\mathbf{A}(\mu)$ onto a compact subspace $X_\mu(G)$ of $X(G) = G$ or $G \cup \{\infty\}$. The Gelfand transform $T \mapsto \widehat{T}$ for $\mathbf{A}(\mu)$ associates the restrictions to $X_\mu(G)$ of the Fourier transforms $\widehat{f}(\chi)$ to operators $L_\mu(f)$. Hence by the general theory of n° 22, (ii) et (iii), for $f, p, q \in L(G)$,

$$(L_\mu(f)p_\mu|q_\mu) = \int \widehat{f}(\chi)d\mu_{p,q}(\chi)\,,$$

where $\mu_{p,q}$ is, in theory, a measure on the compact set $X_\mu(G)$, and so can be identified to a measure on $X(G)$ concentrated on $X_\mu(G)$. Any bounded Borel function $\varphi$ on $X(G)$ defines an operator $M(\varphi)$ on $\mathcal{H}(\mu)$ such that

$$(M(\varphi)p_\mu|q_\mu) = \int \varphi(\chi)d\mu_{p,q}(\chi)\,.$$

As was seen using formula (27.18) in the case of the regular representation, $\{\infty\}$ is a null set with respect to the measures $\mu_{p,q}$, which, therefore, become bounded on $\widehat{G}$.

Then, replacing $\varphi$ by $\widehat{f}\overline{\widehat{g}}$, where $f, g \in L(G)$, in the previous formula,

$$\begin{aligned}\int \widehat{f}(\chi)\overline{\widehat{g}(\chi)}d\mu_{p,q}(\chi) &= (L_\mu(g)^* L_\mu(f)p_\mu|q_\mu) \\ &= \mu(\tilde{q} * \tilde{g} * f * p) = \mu(\tilde{g} * \tilde{q} * p * f) \\ &= \int \widehat{p}(\chi)\overline{\widehat{q}(\chi)}d\mu_{f,g}(\chi)\,.\end{aligned}$$

We then deduce that

$$\widehat{f}(\chi)\overline{\widehat{g}(\chi)}d\mu_{p,q}(\chi) = \widehat{p}(\chi)\overline{\widehat{q}(\chi)}d\mu_{f,g}(\chi)$$

for all $f, g, p, q \in L(G)$. The end of the proof is now obvious.

This theorem shows that $\widehat{f} \in L^2(\widehat{G}; \widehat{\mu})$ for all $f \in L(G)$ and that the map $f \mapsto \widehat{f}$ can be extended (modulo taking the quotient) to an isometric map of $\mathcal{H}(\mu)$ in $L^2(\widehat{G}; \widehat{\mu})$. As in the classical case ($\mu = \epsilon$), it is bijective (exercise!).

By definition, the measure $\widehat{\mu}$ is the *Fourier transform* of $\mu$. Plancherel's traditional formula thus says that the Fourier transform of the measure $\epsilon$ is an invariant measure of $\widehat{G}$.

If $G$ is of the form $\mathbb{R}^p \times \mathbb{T}^q$, in which case $\widehat{G} = \mathbb{R}^p \times \mathbb{Z}^q$, the arguments of theorem 44 associate a *measure* $\widehat{\mu} \geq 0$ on $\widehat{G}$ to each *distribution* $\mu \gg 0$ on $G$, with formula (8) holding for $f, g \in \mathcal{D}(G)$. We leave it to the reader to fill in the gaps in this proof if he feels the need to do so. The space $\mathcal{S}(G)$ of rapidly decreasing $C^\infty$ functions on $G$ can also be defined over such a group, i.e. functions for which the product of any derivative and a polynomial function tends to 0 at infinity. Hence the notion of tempered distributions on $G$ as in Chap. VII. If such a distribution $\mu$ is of positive type, $\mathcal{D}(G)$ can be replaced by $\mathcal{S}(G)$ in the previous arguments and formula (8) thereby obtained for all $f, g \in \mathcal{S}(G)$. It follows that the Fourier transform of a tempered distribution of positive type is a positive tempered measure, which means that there is a polynomial function $p$ on $\widehat{G}$ such that the measure $p(\chi)d\widehat{\mu}(\chi)$ is bounded.

Measures $\widehat{\mu}$ obtained from distributions of positive type are not the most general positive measures on $\widehat{G}$. To obtain these, $\mathcal{D}(G)$ or $\mathcal{S}(G)$ would need to be replaced by the algebra (with respect to the convolution product) $\mathcal{W}(G)$ of functions whose Fourier transform has compact support (for $G = \mathbb{R}$, see the Paley-Wiener theorem in Chap. VIII, n° 12). It is then possible to associate to any measure $\nu \geq 0$ on $\widehat{G}$ a linear functional $\mu : f \mapsto \nu(\widehat{f})$ on $\mathcal{W}(G)$ for which

$$\mu(\tilde{f} * f) = \nu(|\widehat{f}|^2) \geq 0\,.$$

In fact all this falls within the scope of a general theorem which Dieudonné (XV.9) strangely called "Plancherel-Godement" as he had heard me present it around 1950. As it was an almost trivial generalization of the method used by Cartan and myself to present the Fourier transform on commutative groups, I did not publish the result.

*Exercise 1.* Suppose that $d\mu(x) = f(x)dx$ for some $f \in L^2(G)$. Show that $d\widehat{\mu}(\chi) = \widehat{f}(\chi)d\chi$ and that $\widehat{f}(\chi) \geq 0$ almost everywhere.

(iii) *Functions of positive type.* Among measures of positive type on a lcg, there are measures $d\mu(x) = \varphi(x)dx$, where $\varphi$ is locally integrable. $\varphi$ is then said to be of positive type. However, by (6),

$$\mu\left(\tilde{g} * f\right) = \int f(yx)\overline{g(y)}\varphi(x)dxdy = \int \varphi\left(y^{-1}x\right) f(x)\overline{g(y)}dxdy\,.$$

Hence the condition

$$\int \varphi\left(y^{-1}x\right) f(x)\overline{f(y)}dxdy \geq 0 \tag{30.9}$$

valid for all $f \in L(G)$. As will be seen, a general procedure for obtaining *continuous* functions of positive type consists in starting from a unitary representation $(\mathcal{H}, U)$ of $G$ and in considering a diagonal coefficient

$$\varphi(x) = (U(x)\mathbf{a}|\mathbf{a})\ , \quad \mathbf{a} \in \mathcal{H}\,, \tag{30.10}$$

of the latter. Then indeed, for $d\mu(x) = \varphi(x)dx$,

$$\begin{aligned}\mu\left(\tilde{g} * f\right) &= \iint \left(U\left(y^{-1}x\right)\mathbf{a}|\mathbf{a}\right) f(x)\overline{g(y)}dxdy \\ &= \iint (U(x)\mathbf{a}|U(y)\mathbf{a})\, f(x)\overline{g(y)}dxdy\,,\end{aligned}$$

whence

$$\mu(\tilde{g} * f) = (U(f)\mathbf{a}|U(g)\mathbf{a}). \tag{30.11}$$

As the left hand side is, by definition, $(f_\mu|g_\mu)$, the representation of $G$ defined by $\varphi(x)dx$ can be obtained up to equivalence in the following manner: $\mathcal{H}(\mu)$ is the closure of the subspace of $U(f)\mathbf{a} \in \mathcal{H}$, operators $L_\mu(x)$ being the restrictions of $U(x)$ to this subspace. In particular, if the representation $(\mathcal{H}, U)$ is irreducible, it is equivalent to the representation defined by $\mu$.

**Lemma 1 (Gelfand-Raikov).** *Every locally integrable function $\varphi$ of positive type* bounded in the neighbourhood of unity *is almost everywhere equal to a bounded continuous function. There is a unitary representation $(\mathcal{H}, U)$ of $G$ such that $\varphi(x) = (U(x)\mathbf{a}|\mathbf{a})$ for some $\mathbf{a} \in \mathcal{H}$. It is unique up to equivalence if the linear combinations of the $U(x)\mathbf{a}$ are required to be everywhere dense in $\mathcal{H}$.*

Here too, the result is obtained by using the unitary representation defined by $d\mu(x) = \varphi(x)dx$. Indeed, for $f, g \in L(G)$,

$$(f_\mu|g_\mu) = \mu\left(\tilde{g} * f\right) = \iint \varphi\left(y^{-1}x\right) f(x)\overline{g(y)}dxdy\,. \tag{30.12}$$

By assumption, $|\varphi(x)| \leq M$ ae. in a neighbourhood of 0 which can be supposed to be of the form $K^{-1}K$, where $K$ is a compact neighbourhood of $e$. If $f$ vanishes outside $K$ and if we put $g = f$ in (12), we get the inequality

$$\|f_\mu\| \leq M \int |f(x)|\, dx\,.$$

Then, choosing functions $f \geq 0$ with integral equal to 1 and vanishing outside $K$, we suppose that the measure $f(x)dx$ converges to a Dirac measure at $e$.

As the function

$$\int \varphi\left(y^{-1}x\right)\overline{g(y)}dy = \int \varphi\left(y^{-1}\right)\overline{g(xy)}dy = \overline{g} * \varphi(x)$$

is continuous for $g \in L(G)$, $(f_\mu|g_\mu)$ converges to the value of this function at $x = e$. Since $\|f_\mu\| \leq M$ for the functions $f$ considered, at the limit,

$$\left|\int \varphi\left(y^{-1}\right)\overline{g(y)}dy\right| \leq M\,\|g_\mu\| \quad \text{for all } g \in L(G)\,.$$

So there is a vector $\mathbf{a} \in \mathcal{H}(\mu)$ such that

$$(\mathbf{a}|g_\mu) = \int \varphi\left(y^{-1}\right)\overline{g(y)}dy$$

for all $g \in L(G)$ and, replacing $g(y)$ by $g(xy)$,

$$\int \varphi\left(y^{-1}\right)\overline{g(xy)}dy = \left(\mathbf{a}|L_\mu\left(x^{-1}\right)g_\mu\right) = (L_\mu(x)\mathbf{a}|g_\mu)\ ;$$

but then (12) becomes

$$(f_\mu|g_\mu) = \int (L_\mu(x)\mathbf{a}|g_\mu)\,f(x)dx = (L_\mu(f)\mathbf{a}|g_\mu)\ ,$$

whence $f_\mu = L_\mu(f)\mathbf{a}$ for all $f \in L(G)$. So

$$\begin{aligned}\iint \varphi\left(y^{-1}x\right)f(x)\overline{g(y)}dxdy &= (f_\mu|g_\mu) = (L_\mu(f)\mathbf{a}|L_\mu(g)\mathbf{a})\\ &= \iint (L_\mu(x)\mathbf{a}|L_\mu(y)\mathbf{a})\,f(x)\overline{g(y)}dxdy\end{aligned}$$

for all $f, g \in L(G)$. We then deduce that

$$\varphi\left(y^{-1}x\right) = (L_\mu(x)\mathbf{a}|L_\mu(y)\mathbf{a}) = (L_\mu(y^{-1}x)\mathbf{a}|\mathbf{a})$$

almost everywhere on $G \times G$. Thus

$$\varphi(x) = (L_\mu(x)\mathbf{a}|\mathbf{a})$$

for almost all $x$.

Conversely, if $\varphi(x) = (U(x)\mathbf{a}|\mathbf{a})$ for a unitary representation $(\mathcal{H}, U)$ and some $\mathbf{a} \in \mathcal{H}$, $\mathcal{H}$ may be assumed to be generated by all $U(x)\mathbf{a}$. Then, as was seen above, for the measure $d\mu(x) = \varphi(x)dx$ and $f, g \in L(G)$,

$$(U(f)\mathbf{a}|U(g)\mathbf{a}) = \mu\left(\tilde{g} * f\right) = (f_\mu|g_\mu)\ . \tag{30.13}$$

Therefore, there is an isomorphism from $\mathcal{H}(\mu)$ onto $\mathcal{H}$ which, for all $f \in L(G)$, maps $f_\mu$ onto $U(f)\mathbf{a}$. So the representation defined by $\mu$ is equivalent to $(\mathcal{H}, U)$, which proves the lemma.

In particular if $(\mathcal{H}, U)$ and $(\mathcal{H}', U')$ are two unitary representations admitting "generators" $\mathbf{a}$ and $\mathbf{a}'$, and if $(U(x)\mathbf{a}|\mathbf{a}) = (U'(x)\mathbf{a}'|\mathbf{a}')$ for all $x$, then all given representations are equivalent. Corollary: two *irreducible* representations having a common diagonal coefficient are equivalent.

Supposing that $\varphi$ is continuous, it is clear that $\varphi(e) \geq 0$, $|\varphi(x)| \leq \varphi(e)$ and

$$\varphi\left(x^{-1}\right) = \overline{\varphi(x)}$$

for all $x$. Hence $\tilde{\varphi}(x) = \Delta(x)\varphi(x)$, i.e. $\tilde{\varphi} = \varphi$ when $G$ *unimodular.*

On the other hand, under the same assumptions, relation (13) shows that if a measure $\nu$ of positive type is $\ll \mu$, then $\nu(\tilde{f} * f) \leq \|U(f)\mathbf{a}\|^2$ and so

$$|\nu\left(\tilde{g} * f\right)| \leq \|U(f)\mathbf{a}\| \, . \, \|U(g)\mathbf{a}\| \, .$$

The $U(f)\mathbf{a}$ being everywhere dense in $\mathcal{H}$, there is (n° 19, corollary of theorem 31) a Hermitian operator $H$ satisfying $0 \leq H \leq 1$ and

$$\nu\left(\tilde{g} * f\right) = (HU(f)\mathbf{a}|U(g)\mathbf{a}) \, .$$

The left hand side being invariant if $f$ and $g$ are made to undergo the same left translation, $H$ commutes with all $U(f)$ and all $U(x)$. Hence if $(\mathcal{H}, U)$ is irreducible, then $H$ is a scalar and $\nu$ is proportional to $\mu$.

Conversely suppose that this condition holds for a measure $\mu$ of the form $\varphi(x)dx$, where $\varphi$ is continuous of positive type. If a Hermitian operator $H$ on $\mathcal{H}$, satisfying $0 \leq H \leq 1$, commutes with all $U(x)$, so does $H^{1/2}$ and the function

$$\varphi_H(x) = (U(x)H\mathbf{a}|\mathbf{a}) = \left(U(x)H^{1/2}H^{1/2}\mathbf{a}|\mathbf{a}\right) = \left(U(x)H^{1/2}\mathbf{a}|H^{1/2}\mathbf{a}\right)$$

is of positive type, as well as $\varphi_{1-H} = \varphi - \varphi_H$. Therefore, $\varphi_H \ll \varphi$, and so $\varphi_H = c\varphi$ for some scalar $c \geq 0$. It readily follows that $H$ is a scalar.

Let us say that a continuous function of positive type is *elementary* if it is a "diagonal" coefficient of an *irreducible* unitary representation. The previous arguments prove the next statement:

**Lemma 2.** *Let $\varphi$ be a continuous function of positive type. $\varphi$ is elementary if and only if every continuous function of positive type $\psi \ll \varphi$ is proportional to $\varphi$. Every measure $\mu$ of positive type such that $\mu \ll \varphi$ is then proportional to the measure $\varphi(x)dx$.*

An easy calculation shows that, for any finite number of $x_i \in G$ and scalars $\xi_i \in \mathbb{C}$, in notation (10),

$$\sum \varphi\left(x_j^{-1}x_i\right)\xi_i\bar{\xi}_j = \left\|\sum \xi_i U\left(x_i\right)\mathbf{a}\right\|^2 \geq 0 \, . \tag{30.14}$$

This is the original definition of (continuous) functions of positive type used in the 1930s by Salomon Bochner for $G = \mathbb{R}$, and adopted by everyone, in particular by probabilists studying stationary stochastic processes. The equivalence of (14) and (9) can be easily shown. Indeed, if $K$ is a compact set outside which the continuous functions $f$ and $g$ appearing in (9) vanish, then $K$ can be partitioned into finitely many measurable sets $A_i$ on which $f$ and $g$ are constant up to $r > 0$. Choosing $x_i$ in $A_i$ and denoting the extension of the integral of $f$ (resp. $g$) to $A_i$ by $\xi_i$ (resp. $\eta_i$), $\varphi$ being continuous, the difference between integral (9) and the sum $\sum \varphi(x_j^{-1}x_i)\xi_i\eta_j$ can clearly be made to be arbitrarily small. Hence the result.

For $G = \mathbb{R}$, F. Riesz[119] was the first to see the relations between continuous functions of positive type and unitary representations of $\mathbb{R}$ studied by Marshall Stone. This is where Gelfand and Raikov (1943) probably got the idea of generalizing the result to arbitrary lcg; this was my case a year later. This point of view cannot be found in Weil's book, but it contains the generalization of Bochner's essential result to all commutative lcgs:

**Theorem 45 (S. Bochner).** *A continuous function $\varphi$ on a commutative lcg $G$ is of positive type if and only if there is a bounded positive measure*[120] *$d\widehat{\varphi}(\chi)$ on $\widehat{G}$ such that*

$$\varphi(x) = \int \mathbf{e}(x,\chi)d\widehat{\varphi}(\chi) \quad \text{for all } x \in G. \tag{30.15}$$

*The measure $\widehat{\varphi}$ is unique.*

Indeed, applying theorem 44 to the measure $d\mu(x) = \varphi(x)dx$, we get a measure $\widehat{\mu} \geq 0$ on $\widehat{G}$ such that

$$\int \widehat{f}(\chi)\overline{\widehat{g}(\chi)}d\widehat{\mu}(\chi) = \mu(f * \tilde{g}) = \iint \varphi\left(y^{-1}x\right) f(x)\overline{g(y)}dxdy$$

for all $f, g \in L(G)$. To show that $\widehat{\mu}$ is bounded, we put $f = g$ and make the measure $f(x)dx$ tend to the Dirac measure at $e$. The second integral tends to $\varphi(e)$ since $\varphi$ is continuous and bounded. On the left hand side, $|\widehat{f}(\chi)|^2$ converges uniformly to 1 on every compact set while staying dominated by a constant. So the left hand side tends to $\widehat{\mu}(1)$. Thus $\widehat{\mu}(1) = \varphi(e) < +\infty$. To get (15), we make $g(y)dy$ tend to a Dirac measure at $e$ and $f(x)dx$ to a Dirac measure at a point $a \in G$. $\widehat{f}(\chi)$ converges to $\mathbf{e}(a,\chi)$ and $\widehat{g}(\chi)$ to 1, while staying dominated by a constant function. This enables us to pass to the limit in the first integral. In the middle expression, the limit is $\varphi(a)$, giving (15). The converse consists in calculating $\mu(f * \tilde{g})$ from (15).

*Exercise 2.* State the theorem explicitly for $G = \mathbb{Z}$ (The G. Herglotz "moment problem") and $G = \mathbb{T}$.

[119] *Über Sätze von Stone und Bochner* (Acta Szeged, 1933, pp. 184–198)

[120] So the notation $\widehat{\varphi}$ does not denote a function, though, in some cases, the measure $d\widehat{\varphi}(\chi)$ is absolutely continuous with respect to $d\chi$.

*Exercise 3.* (a) Show that the Fourier transform (in the sense of Plancherel's formula) of a square integrable*continuous* function $\varphi$ of positive type is positive, in $L^{12}(\widehat{G})$ and that

$$\varphi(x) = \int \mathbf{e}(x,\chi)\widehat{\varphi}(\chi)d\chi\,.$$

(b) Show that there is a unique function $\psi \in L^2(G)$ of positive type such that $\varphi = \psi * \psi$. (c) Let $(\mathcal{H}, U)$ be the unitary representation defined by $\varphi$ (The Gelfand-Raikov lemma). Show that it is equivalent to the representation on $L^2(\widehat{G}; \widehat{\varphi})$ associating the operator of multiplication by $\mathbf{e}(x;\chi)$ to each $x \in G$. (d) Denote by $\mathcal{H}(\psi)$ the smallest closed invariant subspace of $L^2(G)$ containing $\psi$. Show that $(\mathcal{H}, U)$ is equivalent to the representation by translations on $\mathcal{H}(\psi)$.

*Exercise 4.* Let $(\mathcal{H}, U)$ be a unitary representation of $G$. Using Bochner's theorem, show there is spectral measure $M$ on $\widehat{G}$, in the sense of n° 22, such that[121]

$$U(x) = \int \mathbf{e}(x,\chi)dM(\chi)$$

for all $x \in G$ and $U(f) = M(\widehat{f})$ for all $f \in L^1(G)$. What about the case $G = \mathbb{Z}$?

*Exercise 5.* (a) Let $G$ be a lcg and $K \subset L^\infty(G)$ the convex set of continuous functions of positive type such that $\varphi(e) \leq 1$. Show that $K$ is compact with respect to the weak topology of $L^\infty(G)$ (use lemma 1 and the fact that, in the topological dual of a Banach space, the unit ball is weakly compact). (b) A point of a convex set $K$ is said to be *extremal* if it is not in the interior of any line segment contained in $K$. Using lemma 2, show that the extremal points of $K$ are precisely the elementary $\varphi$ such that $\varphi(e) = 1$ and the function 0.

In a locally convex topological vector space, for example $L^\infty(G)$ equipped with the weak topology, every *compact* convex set is the smallest closed convex set containing all its extremal points (Krein-Milman theorem). The previous exercise then enables us to show the existence of "many" elementary functions of positive type and even that, for all $x \in G$ such that $x \neq e$, there

[121] For $G = \mathbb{R}$, it is Stone's theorem referred to in the title of F. Riesz's article mentioned above. It extension to the general case was independently published by M. A. Neumark (1943), W. Ambrose (1944) and myself (CR de l'Académie des Sciences, June 1944) at a time when, as I have already mentioned somewhere, Soviet (or American) journals did not reach France. Mine is even now not cited, perhaps because it was published in CRAS, a journal little read by mathematicians. Some have criticized CRAS for not having *referees.* As most mathematicians are familiar with the distinction between a necessary and a sufficient condition, this argument does not prove anything: to decide whether a two page article is interesting or not, it should be read. Let me add that my note already contained much of the formalism of spectral measures given in n° 22.

is an irreducible unitary representation $(\mathcal{H}, U)$ of $G$ such that $U(x) \neq 1$: This is Gelfand-Raikov's result and their proof. To go further, two methods are available. The general theorem for $G$ can be used. Choquet's theorem for compact convex: for all $\varphi \in K$ such that $\varphi(e) = 1$, there is a positive measure $m$ on $K$, of total weight 1, *concentrated on the set of elementary functions* and such that

$$\varphi(x) = \int \psi(x) d\mu(\psi) \quad \text{for all } x\,.$$

This result generalizes Bochner's theorem, except for the uniqueness of the measure $\mu$. Otherwise, using von Neumann's methods described in n° 24, one can show that any unitary representation of $G$ is a continuous sum of irreducible representations. But this result from pure measure theory is of little interest. In concrete cases, one always expects to find a much more natural or canonical decomposition, not to mention the fact that, for some bizarre groups, a given representation can be decomposed into irreducible representations in two completely different ways.

## 31 – Quasi-Regular Representations of a Unimodular Group

(i) *Central measures of positive type.*[122] As was seen in section (i) above, the regular representation of $G$ by left translations

$$L(x) : f \longmapsto \epsilon_x * f \tag{31.1}$$

on $L^2(G)$ is obtained by applying the method valid for any measure $\mu \gg 0$ to the Diract measure $\epsilon$. But right translations defined by

$$R(x) : f \longmapsto f * \epsilon_x \tag{31.1'}$$

define a second representation on $L^2(G)$[123] commuting with the previous one. It is unitary because, $G$ being unimodular, the measure $\epsilon$ satisfies the identity

$$\epsilon(f * g) = \epsilon(g * f)\,, \tag{31.2}$$

equivalent to

$$\int f(x) g\left(x^{-1}\right) dx = \int g(x) f\left(x^{-1}\right) dx\,.$$

[122] For this section, see my *Mémoire sur la théorie des caractères dans les groupes localement compacts unimodulaires* (Journal de math. pures et appliquées, XXX, 1951, pp. 1–110).

[123] As $f * \epsilon_x$ is the function $y \mapsto f(yx^{-1})$, $R(xy) = R(y)R(x)$, so that we get a representation of the "opposite" group, obtained by replacing the given composition law $(x, y) \mapsto xy$ by $(x, y) \mapsto yx$. The map $x \mapsto R(x^{-1})$ is the one that would define a representation of $G$. Denoting by $G'$ the "opposite" group to $G$, the map $(x, y) \mapsto L(x)R(y)$ is, therefore, a – very particular – unitary representation of the group $G \times G'$.

Then

$$\begin{aligned}(R(x)f|g) &= \epsilon\,(\tilde{g} * f * \epsilon_x) = \epsilon\,(\epsilon_x * \tilde{g} * f) \\ &= \epsilon[(g * \epsilon_{x^{-1}})^{\sim} * f] = \left(f|R\left(x^{-1}\right)g\right) ,\end{aligned}$$

whence the result. Moreover, $L^2(G)$ contains an "involution"

$$S : f \longmapsto \tilde{f}$$

for which

$$SL(x)S^{-1} = R\left(x^{-1}\right) . \tag{31.3}$$

(2) says that the measure $\epsilon$ is *central.* An immediate calculation shows that central measure are invariant under inner automorphisms $x \mapsto sxs^{-1}$ de $G$, and conversely.

Similarly, for any central measure $\mu$ of positive type, there are two unitary representations $L_\mu$ and $R_\mu$ and an involution on the space $\mathcal{H}(\mu)$. Indeed the inner product $(f_\mu|g_\mu) = \mu(\tilde{g} * f)$ is invariant under left and right translations $f \mapsto \epsilon_x * f$ and $f \mapsto f * \epsilon_x$; hence two representations $L_\mu(x)$ and $R_\mu(x)$. They commute like they do on $L^2(G)$. For the same reason,

$$\mu\,(\tilde{g} * f) = \overline{\mu(\tilde{f} * g)} \quad (\text{car } \mu \gg 0) = \overline{\mu(g * \tilde{f})} = \mu\,(f * \tilde{g}) ,$$

which proves that $f \mapsto \tilde{f}$ induces a quotient map and defines an involution $S$ on $\mathcal{H}(\mu)$ such that

$$g = \tilde{f} \Longrightarrow g_\mu = Sf_\mu . \tag{31.4}$$

Relation (3) obviously holds in this setup.

Let us denote by $\mathbf{A}(\mu)$ the image of $L(G)$ under $f \mapsto f_\mu$. It is the quotient of $L(G)$ by the left ideal $I(\mu)$ consisting of $f$ such that $\mu(\tilde{f} * f) = 0$. As $\mu$ is central, this ideal is stable under $f \mapsto \tilde{f}$, and so is a two-sided ideal[124] of $L(G)$. This provides an *algebra* structure on $\mathbf{A}(\mu) = L(G)/I(\mu)$. Hence, denoting multiplication in $\mathbf{A}(\mu)$ by a point to avoid any confusion,

$$f_\mu.g_\mu = (f * g)_\mu , \tag{31.5}$$

$$S\left(f_\mu.g_\mu\right) = Sg_\mu.Sf_\mu . \tag{31.6}$$

Except in some cases, in general the $f_\mu$ cannot be interpreted as functions on $G$. The symbol used to multiply them is, therefore, not a convolution product in the usual sense.

Over a Lie group, $L(G)$ could be replaced by the Schwartz space $\mathcal{D}(G)$ and the same constructions carried out for all central *distributions* of positive type. We could even go further by considering positive Hermitian

[124] In a ring $A$, a two-sided ideal is a subgroup $I$ of the additive group such that $axb \in I$ for all $a, b \in A$ and $x \in I$. Relations $x = x' \bmod I$ and $y = y' \bmod I$ then imply $xy = x'y' \bmod I$, defining a multiplication – a ring structure – on $A/I$.

forms $\mu(f,g)$ defined on a vector subspace $\mathfrak{a}$ of $L^1(G)$ whose elements are "sufficiently regular" functions – whatever that means... – satisfying some obvious conditions:[125]

(BT 1) $\mathfrak{a}$ is invariant under the maps $f \mapsto \tilde{f}$, $f \mapsto \epsilon_x * f$ for all $x$ and $f \mapsto g * f$ for all $g \in \mathfrak{a}$ [so that $\mathfrak{a}$ is a subalgebra of $L^1(G)$],
(BT 2) $\mathfrak{a}$ is everywhere dense in $L^1(G)$,
(BT 3) the function $\mu(\epsilon_x * f, g)$ is continuous for all $f, g \in \mathfrak{a}$ and

$$(31.7)\quad \mu(h * f, g) = \int \mu(\epsilon_x * f, g)\, h(x) dx$$

for all $h \in L(G)$,
(BT 4) $\mu(\tilde{f}, \tilde{g}) = \mu(g, f)$ for all $f, g \in \mathfrak{a}$.

There is no need to go further to find examples of such objects: take $G = \mathbb{R}$, for $\mathfrak{a}$ take the set $W(\mathbb{R})$ [not contained in $L(G)$] of integrable functions whose Fourier transforms have *compact supports* as in n° 30, (ii) and set $\mu(f,g) = \int \widehat{f}(t)\overline{\widehat{g}(t)} d\lambda(t)$, where $\lambda$ is a positive arbitrary measure on $\mathbb{R}$. If $\mu$ is a central distribution of positive type on a Lie group $G$, choose $\mathfrak{a}$ to be Schwartz space $\mathcal{D}(G)$ and set $\mu(f,g) = \mu(\tilde{g} * f)$.

In the general case, specifying a function $\mu$ satisfying the stated conditions enables us to define a Hilbert space $\mathcal{H}(\mu)$, a canonical map $f \mapsto f_\mu$ from $\mathfrak{a}$ to $\mathcal{H}(\mu)$, an involution $S$, two representations $L_\mu$ and $R_\mu$ of $G$ and of the opposite group $G'$, and finally a multiplication in the image $\mathbf{A}(\mu)$ of $\mathfrak{a}$ in $\mathcal{H}(\mu)$. Like Dixmier, we will call such a Hermitian form $\mu$ a *bitrace* on $G$.

(ii) *The commutation theorem.* "Two-sided representations" of a unimodular group defined by bitraces have a property well-known (and trivial) in the case of the regular representation of a finite group which was proved in n° 29, (ii) for the regular representation of a compact group:

**Theorem[126] 46.** *Let $\mu$ be a bitrace on a locally compact unimodular group $G$ and $\mathcal{L}(\mu)$ [resp. $\mathcal{R}(\mu)$] the algebra of continuous operators on the space $\mathcal{H}(\mu)$ commuting with right (resp. left) translations. Then*

$$(31.8)\qquad \mathcal{L}(\mu) = \mathcal{R}(\mu)', \quad \mathcal{R}(\mu) = \mathcal{L}(\mu)'$$

*and*

$$(31.9)\qquad SAS^{-1} = A^* \quad \text{for all } A \in \mathcal{L}(\mu) \cap \mathcal{R}(\mu)\,.$$

[125] See R. Godement, *Séminaire Bourbaki*, Mars 1951, and *Théorie des caractères II* (Annals of Math., **59**, 1954, pp. 63–85), where spaces $\mathfrak{a}$ of bounded measures are considered instead. The definition adopted here consists in replacing $\mathfrak{a}$ by its intersection with $L^1(G)$ and is suffices for our purpose.

[126] For the regular representation, see I. E. Segal (Annals of Math., **51**, 1950, pp. 293–298) and for the general case, R. Godement (J. de Math. pures et appliquées, XXX, 1951, pp. 1–110) and notes in CRAS (November 1949).

Since $\mathcal{A}'' = \mathcal{A}$ for any von Neumann algebra [n° 19, (vii)], relations (8) are equivalent. As $\mathcal{R}(\mu)$ contains all $R(x)$, it is clear that $\mathcal{R}(\mu)' \supset \mathcal{L}(\mu)$ and $\mathcal{L}(\mu)' \supset \mathcal{R}(\mu)$. So it suffices to show that all $A \in \mathcal{L}(\mu)$ commute with all $B \in \mathcal{R}(\mu)$.

For $f \in \mathfrak{a}$, the definition algebra of $\mu$, consider the operators

$$L_\mu(f) = \int L_\mu(x) f(x) dx \,, \quad R_\mu(f) = \int R_\mu(x) f(x) dx \,.$$

Formula (31.7), or (30.4) in the case of a measure, shows that, in notation (5),

$$L_\mu(f) g_\mu = (f * g)_\mu = f_\mu . g_\mu$$

and similarly,

$$R_\mu(f) g_\mu = (g * f)_\mu = g_\mu . f_\mu \,.$$

The assumption implies that $A$ commutes with all $R_\mu(f)$ and $B$ with all $L_\mu(f)$, and so that

$$A\,(g_\mu . f_\mu) = A g_\mu . f_\mu \,, \quad B\,(f_\mu . g_\mu) = f_\mu . B g_\mu \tag{31.10}$$

for $f, g \in \mathfrak{a}$. The "obvious" proof of the theorem, used for compact groups, would consist in writing that

$$\begin{aligned} BA\,(f_\mu . g_\mu) &= B\,(A f_\mu . g_\mu) = A f_\mu . B g_\mu \,, \\ AB\,(f_\mu . g_\mu) &= A\,(f_\mu . B g_\mu) = A f_\mu . B g_\mu \,, \end{aligned}$$

but neither $Af_\mu$ nor $Bg_\mu$ are in general in $\mathbf{A}(\mu)$, and *a priori*, their product is not well-defined, even if $G = \mathbb{R}$ and $\mu = \epsilon$.

It is, therefore, necessary to use somewhat less simple arguments. The theorem being only a particular case of a much more general theorem with exactly the same proof, but more applications, we will work in the latter framework.

We start from the following three data:[127]

(i) an algebra $\mathbf{A}$ with or without unit element whose multiplication will be denoted by $f.g$,
(ii) a semi-linear map ("involution") $S$ on $\mathbf{A}$ such that $S^2 = 1$,
(iii) an inner product $(f|g)$ on $\mathbf{A}$ such that $(f|f) = 0$ implies $f = 0$.

[127] The notation used ($f, g, \varphi$, etc.) aims at highlighting the similarity with the regular representation case, where Roman letters denote functions of $L(G)$, Greek letters a priori arbitrary functions on $L^2(G)$. For what follows, see R. Godement, *Théorie des caractères. I. Algèbres unitaires* (Annals of Math., **59**, 1954, pp. 47–62) and a talk of March 1951 in séminaire Bourbaki. At the time, exchanging ideas with Jacques Dixmier was very helpful for me. See his article in *Compositio Mathematica* of 1952 and the two books already referred to.

Set $\|f\| = (f|f)^{1/2}$ and denote by $\mathcal{H}$ the Hilbert space completion of $\mathbf{A}$, which, therefore, is $\mathcal{H}(\mu)$ in the case of the theorem and $L^2(G)$ if $\mu = \epsilon$. The map $S$ can be extended to $\mathcal{H}$. I will in general write $Sf = \tilde{f}$.

$\mathbf{A}$ will be said to be a *Hilbert algebra* if it satisfies the following conditions:

(AH 1) $Sf.Sg = S(g.f)$ for $f, g \in \mathbf{A}$

(AH 2) $(f.g|h) = (g|Sf.h)$ for $f, g, h \in \mathbf{A}$;

(AH 3) $(f|g) = (Sg|Sf)$ for $f, g \in \mathbf{A}$;

(AH 4) for all $f \in \mathbf{A}$, there is an upper bound $\|f.g\| \leq M(f)\|g\|$;

(AH 5) the products $f.g$ are everywhere dense in $\mathbf{A}$ (hence in $\mathcal{H}$).

These axioms are satisfied by the algebra $\mathbf{A}(\mu)$ associated to a bitrace $\mu$. The first one is obvious. The second one can be written $\tilde{h}*(f*g) = (\tilde{f}*h)*g$. (AH 3) follows from (BT 4). (AH 4) is clear since $f_\mu . g_\mu = L_\mu(f)g_\mu$ with a continuous operator $L_\mu(f)$ on $\mathcal{H}(\mu)$, as on any unitary representation. Finally, (AH 5) follows from the fact that, for any unitary representation $(\mathcal{H}, U)$, every $x \in \mathcal{H}$ is in the closure of the set of vectors $U(f)x$ where $f \in L^1(G)$. However, $\mathfrak{a}$ is everywhere dense in $L^1(G)$.

We next consider the general "abstract" situation. For all $f \in \mathbf{A}$, multiplication in $\mathbf{A}$ leads to maps

$$L(f)g = f.g, R(f)g = g.f \tag{31.11}$$

from $\mathbf{A}$ to $\mathcal{H}$. By (AH 4), the maps $L(f)$ extend to *continuous* operators on $\mathcal{H}$. (AH 2) says that

$$L(f)^* = L(\tilde{f}) \quad \text{for } f \in \mathbf{A}. \tag{31.12}$$

On the other hand, (AH 1), (AH 2) and (AH 3) show that

$$SL(\tilde{f})S^{-1}g = S(\tilde{f}.\tilde{g}) = g.f = R(f)g,$$

i.e. that

$$SL(\tilde{f})S^{-1} = R(f), \tag{31.13}$$

and so all $R(f)$ also extend to continuous operators on $\mathcal{H}$. Trivially (associativity),

$$L(f)R(g) = R(g)L(f), \quad L(f)g = R(g)f$$

for $f, g \in \mathbf{A}$. These operators enable us to consider $\mathcal{H}$ as a "bimodule" over the ring $\mathbf{A}$: for this it suffices to set

$$f.\varphi = L(f)\varphi, \quad \varphi.f = R(f)\varphi \quad \text{for } f \in \mathbf{A}, \quad \varphi \in \mathcal{H}. \tag{31.14}$$

Associativity follows from $L(fg) = L(f)L(g), R(fg) = R(g)R(f)$. For the regular representation on $L^2(G)$ we thereby obtain convolution products, and this example shows (if $G$ is not compact) that, without assumptions on $\varphi$ and $\psi$, the product $\varphi.\psi$ of two elements of $\mathcal{H}$ is not well-defined (or, if it is from the point of view of group theory, does not belong to $\mathcal{H}$).

To get theorem 46, it, therefore, suffices to prove the next result:

**Theorem 47**. *Let* $\mathbf{A}$ *be a Hilbert algebra and* $\mathcal{H}$ *its completion. If* $A$ *and* $B$ *are continuous operators on* $\mathcal{H}$ *such that*

$$AR(f) = R(f)A\,, \quad BL(f) = L(f)B$$

*for all* $f \in \mathbf{A}$, *then* $AB = BA$.

The first step of the proof consists in associating to each $\varphi \in \mathcal{H}$ a linear map $R(\varphi) : \mathbf{A} \longrightarrow \mathcal{H}$ by setting

$$R(\varphi)f = L(f)\varphi = f.\varphi \quad \text{for } f \in \mathbf{A}\,. \tag{31.15}$$

$\varphi$ will be said to be right *moderate* if $R(\varphi)$ is continuous, which the case of $f \in \mathbf{A}$ by (AH 3), and the extension of $R(\varphi)$ to $\mathcal{H}$ will also be denoted by $R(\varphi)$. The left moderate $\varphi$ are defined in the same way. Hence in this case, there is a continuous operator $L(\varphi)$ satisfying

$$L(\varphi)f = R(f)\varphi \quad \text{for all } f \in \mathbf{A}\,. \tag{31.15'}$$

**Lemma 1**. *If* $\varphi \in \mathcal{H}$ *is right moderate, so is* $B\varphi$ *and*

$$R(B\varphi) = BR(\varphi)\,.$$

Indeed, for all $f \in \mathbf{A}$,

$$BR(\varphi)f = BL(f)\varphi = L(f)B\varphi\,.$$

As $B$ and $R(\varphi)$ are continuous, the map $f \mapsto L(f)B\varphi$ is continuous, so that $B\varphi$ is right moderate, qed.

Similarly, if $\varphi$ is left moderate, so is $A\varphi$ and

$$L(A\varphi) = AL(\varphi)\,.$$

**Lemma 2**. *If* $\varphi$ *is right moderate, then so is* $\tilde{\varphi}$ *and*

$$R(\tilde{\varphi}) = R(\varphi)^*\,.$$

Indeed by (15), for $f, g \in \mathbf{A}$,

$$\begin{aligned}(R(\varphi)f|g) &= (L(f)\varphi|g) = (\varphi|L(\tilde{f})g) = (\varphi|\tilde{f}.g) = (\tilde{g}.f|\tilde{\varphi}) \\ &= (L(\tilde{g})f|\tilde{\varphi}) = (f|L(g)\tilde{\varphi}) = (f|R(\tilde{\varphi})g)\,.\end{aligned}$$

This proves the formula and shows that $\tilde{\varphi}$ is right moderate.
Similarly,

$$L(\tilde{\varphi}) = L(\varphi)^*$$

if $\varphi$ is left moderate.

**Lemma 3.** *Any right moderate $\varphi \in \mathcal{H}$ is left moderate and*

$$SR(\varphi)S^{-1} = L(\tilde{\varphi}) = L(\varphi)^*$$

For $f \in \mathbf{A}$,

$$\begin{aligned} SR(\varphi)S^{-1}f &= SR(\varphi)\tilde{f} = SL(\tilde{f})\varphi = R(f)S^{-1}\varphi \quad \text{by (10)} \\ &= R(f)\tilde{\varphi} = L(\tilde{\varphi})f \end{aligned}$$

by (15'), so that $\tilde{\varphi}$ is left moderate. Hence so is $\varphi = (\tilde{\varphi})^{\sim}$ by lemma 2, qed.
Lemmas 2 and 3 show that the notions of left and right moderation are the same. Hence we will talk of moderate elements without any further detail.

**Lemma 4.** *If $\varphi$ and $\psi$ are moderate, then*

$$L(\varphi)\psi = R(\psi)\varphi\,. \tag{31.16}$$

For $f \in \mathbf{A}$,

$$\begin{aligned} (L(\varphi)\psi|f) &= (\psi|L(\tilde{\varphi})f) \quad \text{(lemma 2)} \\ &= (\psi|R(f)\tilde{\varphi}) = (R(\tilde{f})\psi|\tilde{\varphi}) = (SL(f)S^{-1}\psi|S\varphi) \quad \text{by (13)} \\ &= (\varphi|L(f)\tilde{\psi}) = (\varphi|R(\tilde{\psi})f) = (R(\psi)\varphi|f) \end{aligned}$$

by lemma 2, qed.
We are now ready to show that $AB = BA$. For $f, g \in \mathbf{A}$, $Af$ and $Bg$ are moderate by lemma 1. So

$$\begin{aligned} AB(f.g) &= ABL(f)g = AL(f)Bg = L(Af)Bg \quad \text{(lemma 1)} \\ &= R(Bg)Af \quad \text{(lemma 4)} \quad = BR(g)Af \quad \text{(lemma 1)} \\ &= BAR(g)f = BA(f.g)\,; \end{aligned}$$

(AH 4) then shows that $AB = BA$, qed.
The equality

$$SAS^{-1} = A^* \tag{31.17}$$

remains to be proved for any continuous operator $A$ commuting with all $L(f)$ and all $R(f)$, i.e. belonging to the common centre of the von Neumann algebras $\mathcal{L}$ and $\mathcal{R}$ associated to $\mathbf{A}$. For $f, g, h \in \mathbf{A}$,

$$\begin{aligned}(SAS^{-1}(f.g)|h) &= (\tilde{h}|A(\tilde{g}.\tilde{f})) = (\tilde{h}|AL(\tilde{g})\tilde{f}) = \\ &= (\tilde{h}|L(\tilde{g})A\tilde{f}) = (L(g)\tilde{h}|A\tilde{f}) = (R(\tilde{h})g|A\tilde{f}) = \\ &= (g|R(h)A\tilde{f}) = (g|AR(h)\tilde{f}) = (g|R(Ah)\tilde{f}) \quad \text{(lemma 1)} \\ &= (g|L(\tilde{f})Ah) \quad \text{(lemma 4)} = (L(f)g|Ah) = (f.g|Ah)\,.\end{aligned}$$

Thus (17) follows since the $f.g$ are everywhere dense. qed.

Finally, note that the set of operators $L(f)$ where $f \in \mathcal{H}$ is moderate is a *two-sided ideal* $\mathcal{L}_2$ of the von Neumann algebra $\mathcal{L}$ (the notation will be explained later). Indeed, lemma 1 shows that it is a left ideal. On the other hand, keeping the notation of lemma 1,

$$R(\varphi)B = [B^*R(S\varphi)]^* = R\,(B^*S\varphi)^* = R\,(SB^*S\varphi)\;.$$

The result follows by permuting left and right.

**Corollary**. *The set of* $L(f)$ [resp. $R(f)$], $f \in \mathbf{A}$, *is ultrastrongly everywhere dense in the von Neumann algebra* $\mathcal{L}$ (resp. $\mathcal{R}$) *of continuous operators commuting with* $R(f)$ [resp. $L(f)$] *for any* $f \in \mathbf{A}$.

Indeed the theorem shows that $\mathcal{L} = \mathcal{R}'$, so that [n° 19, (vii), density theorem] the algebra of $L(f) + \lambda 1$ is ultrastrongly dense in $\mathcal{L}$. To show that that so are the $L(f)$, it suffices [n° 19, (vii), lemma 6] to show that, for $\varphi \in \mathcal{H}$, for all $f \in \mathbf{A}$, relation $L(f)\varphi = 0$ implies $\varphi = 0$. As

$$(L(f)\varphi|g) = \left(\varphi|\tilde{f}.g\right)\,,$$

this is axiom (AH 5) of Hilbert algebras, qed.

Lemma 4 enables us to define in the set $\widehat{\mathbf{A}}$ of moderate $\varphi \in \mathcal{H}$ a multiplication extending that considered in $\mathbf{A}$, namely

$$\varphi.\psi = L(\varphi)\psi = R(\psi)\varphi\,. \tag{31.18}$$

$\varphi.\psi$ can even be defined if only one of the factors is moderate: set

$$\varphi.\psi = L(\varphi)\psi \quad \text{or} \quad R(\psi)\varphi \tag{31.18'}$$

according to whether $\varphi$ or $\psi$ is moderate. This multiplication is associative in the set of moderate elements of $\mathcal{H}$, which thereby becomes an algebra, and transforms $\mathcal{H}$ into a bimodule over it. Indeed if $\varphi$ and $\psi$ are moderate, then for all $\theta \in \mathcal{H}$

$$\begin{aligned}(\varphi.\psi)\,.\theta &= L\,(\varphi.\psi)\,\theta = L\,(L(\varphi)\psi)\,\theta \\ &= L(\varphi)L(\psi)\theta \quad \text{(lemma 1)} \quad = \varphi.\,(\psi.\theta)\;.\end{aligned}$$

In particular, $L(\varphi.\psi) = L(\varphi)L(\psi)$ and $R(\varphi.\psi) = R(\psi)R(\varphi)$. On the other hand, the involution $\varphi \mapsto \tilde{\varphi}$ preserves moderate elements and by lemma 3,

$$\widetilde{(\varphi.\psi)} = SL(\varphi)\psi = SL(\varphi)S^{-1}S\psi = R(\tilde{\varphi})\tilde{\psi} = \tilde{\psi}.\tilde{\varphi}\,.$$

The algebra $\widehat{\mathbf{A}}$ – nothing to do with a dual group or a spectrum – of moderate elements clearly satisfies conditions (AH 1),..., (AH 5), and so is a Hilbert algebra. Applying the previous construction to it does not give anything more. If the only moderate elements of $H$ are the maps $f \in \mathbf{A}$, $\mathbf{A}$ will be said to be a *maximal* (or *full* in Dixmier) Hilbert algebra.

For example, let us take $\mathbf{A} = L(G)$, where $G$ is commutative, whence $\mathcal{H} = L^2(G)$. By Fourier transform, the moderate $\varphi \in \mathcal{H}$ are the functions for which the map $\widehat{f} \mapsto \widehat{\varphi}\widehat{f}$, defined on the Fourier transforms of $f \in L(G)$, is continuous on $L^2(\widehat{G})$. This obviously means that $\widehat{\varphi}$ is *bounded*. So in this case, the maximal algebra $\widehat{\mathbf{A}}$ is the set of $f \in L^2(G)$ with bounded Fourier transform. If unimodular $G$ is not commutative, then $\widehat{\mathbf{A}}$ is the set of $\varphi \in L^2(G)$ such that $f \mapsto \varphi * f$ (or $f \mapsto f * \varphi$: lemma 3) maps $L^2(G)$ to $L^2(G)$.

If a Hilbert algebra $\mathbf{A}$ is *irreducible*, i.e. if $\mathcal{H}$ does not contain any non-trivial right and left invariant closed subspace, the algebra $\mathcal{Z} = \mathcal{L} \cap \mathcal{R}$ of operators commuting with all $L(f)$ and all $R(f)$ is reduced to scalars. As $\mathcal{Z}$ is the centre of $\mathcal{L}$, algebras $\mathcal{L}$ and $\mathcal{R}$ are *factors* in the sense of von Neumann theory [n° 19, (vii)].

*Exercise 1.* Let $\mathcal{H}'$ be a closed subspace of $\mathcal{H}$ invariant under all $L(f)$ and all $R(f)$ and let $E$ be the projection onto $\mathcal{H}'$. Show that $\mathbf{A}' = \mathbf{A} \cap \mathcal{H}' = E(\mathbf{A}')$ is everywhere dense in $\mathcal{H}'$ and is a maximal Hilbert algebra in $\mathcal{H}'$ if $\mathbf{A}$ is maximal in $\mathcal{H}$.

(iii) *Traces on a Hilbert algebra.* Let us first consider a central measure $d\mu(x) = \varphi(x)dx$ defined by a *continuous* function $\varphi$ such that $\varphi(xy) = \varphi(yx)$. This simple particular case rarely occurs except for commutative, compact or discrete groups. We denote the algebra of moderate elements of $\mathcal{H}(\mu) = \mathcal{H}$ by $\mathbf{A}$. There is [n° 30, (iii), proof of the Gelfand-Raikov lemma] a vector $u \in \mathcal{H}$ such that

$$\varphi(x) = (L_\mu(x)u|u)\ , \quad f_\mu = L_\mu(f)u$$

for all $x \in G$ and all $f \in L(G)$, whence $f_\mu = f_\mu.u$ by definition (18') of the multiplication. It follows that $R(u) = 1$, and so that $u$ is a moderate element of $\mathcal{H}(\mu)$. Since then $v.u = R(u)v = v$, the vector $u$ is a right unit element of the algebra $\mathbf{A}$, and since $v.u = v$ implies $\tilde{u}.\tilde{v} = \tilde{v}$, $u$ is in fact a unit element of $\mathbf{A}$ and $\tilde{u} = u$. The most obvious example is that of the measure $\epsilon$ on a discrete group $G$. As $L(u) = R(u) = 1$, for any operator $A \in \mathcal{L}$ and any $f \in \mathbf{A}$,

$$Af = A(u.f) = AR(f)u = R(f)Au = Au.f = a.f\,. \tag{31.19}$$

The element $Au = a \in \mathcal{H}$ defines $A = L(a)$ and is moderate since $A$ is continuous. The von Neumann algebra $\mathcal{L}$ is, therefore, the set of $L(a)$, $a \in \mathbf{A}$.

For all $A \in \mathcal{L}$, the formula

$$Tr(A) = (Au|u) = (a|u) \tag{31.20}$$

defines a linear functional on $\mathcal{L}$ with all the properties of a trace. Indeed, if $A = L(a)$ and $B = L(b)$ where $a, b \in \mathbf{A}$, then

$$Tr(B^*A) = (Au|Bu) = (a|b) = (\tilde{b}|\tilde{a}) = Tr(AB^*)\,,$$

whence

$$Tr(AB) = Tr(BA)\,. \tag{31.21}$$

Obviously

$$Tr(A^*) = \overline{Tr(A)}\,, \quad Tr(H) \geq 0 \quad \text{if } H^* = H \geq 0\,. \tag{31.22}$$

If moreover $0 = Tr(H) = (Hu|u) = \|H^{1/2}u\|^2$, then $H^{1/2} = L(H^{1/2}u) = 0$ and so $H = 0$; trace (20) is then said to be *faithful.*

If $(H_n)$ is an increasing sequence of positive Hermitian operators on $\mathcal{L}$ and is bounded above [n° 19, (v), lemma 8], then the definition of $\sup H_n$ shows that

$$Tr(\sup H_n) = \sup Tr(H_n)\,. \tag{31.23}$$

Finally, it is clear that the initial function $\varphi(x) = (L_\mu(x)u|u)$ becomes

$$\varphi(x) = Tr[L_\mu(x)] \tag{31.24}$$

and that

$$(f_\mu|g_\mu) = Tr[L_\mu(g)^*L_\mu(f)] \tag{31.25}$$

for all $f, g \in L(G)$. Besides this is only a different way of stating relation

$$Tr[L(b)^*L(a)] = (a|b) \tag{31.25'}$$

obtained above as an aside. All this generalizes in an obvious way to unitary algebras *with unit element* and is directly connected to the theory of von Neumann algebras of finite type due to Dixmier and to their wonderful properties.[128]

In the general case of a maximal algebra $\mathbf{A}$ without unit element, the two-sided ideal $\mathcal{L}_2$ of $\mathcal{L}$ defined above is the set of $L(f), f \in \mathbf{A}$. For $A, B \in \mathcal{L}_2$, in accordance with (25'), we set

$$Tr(B^*A) = (f|g) \quad \text{if } A = L(f)\,, \quad B = L(g)\,,$$

[128] For example, Dixmier has shown that, if $\mathcal{Z}$ is the centre of such an algebra $\mathcal{A}$, then there is a map $T \mapsto T^\natural$ from $\mathcal{A}$ onto $\mathcal{Z}$ with the formal properties of the similar map with respect to a compact group, that $T^\natural$ is the only element of $\mathcal{Z}$ belonging to the closed convex envelope of the set of $UTU^{-1}$, where $U$ is unitary in $\mathcal{A}$, and that the traces on $\mathcal{A}$ are the functions $T \mapsto \chi(T^\natural)$ where $\chi$ is a character of $\mathcal{Z}$.

where $f, g \in \mathbf{A}$. The result only depends on the operator $B^*A = L(\tilde{g}.f)$ since, for $u, v \in \mathbf{A}$,

$$(B^*Au|v) = (L(\tilde{g}.f)\,u|v) = (L(f)u|L(g)v) = (R(u)f|R(v)g)\ .$$

Choosing $u_n$ and $v_n$ such that $R(u_n)f$ and $R(v_n)g$ converge to $f$ and $g$ in $\mathcal{H}$ (density theorem), $(f|g) = \lim(B^*Au_n|v_n)$, whence the result. $Tr(B^*A)$ is clearly a positive Hermitian form on $\mathcal{L}_2$. Equipping $\mathcal{L}_2$ with this inner product and with the obvious involution gives a Hilbert algebra isomorphic to $\mathbf{A}$, for good reason... As above, the relation $Tr(AB) = Tr(BA)$ is equivalent to the formula $(f|g) = (\tilde{g}|\tilde{f})$.

One can also define the trace of an operator $H$ in the set $\mathcal{L}_+$ of positive Hermitian operators on $\mathcal{L}$. Set

$$Tr(H) = Tr\left(H^{1/2}H^{1/2}\right) \quad \text{if } H^{1/2} \in \mathcal{L}_2\,, \tag{31.26}$$

$$Tr(H) = +\infty \quad \text{if } H^{1/2} \notin \mathcal{L}_2\,. \tag{31.26'}$$

$Tr(H) \geq 0$ and $Tr(H) = 0$ if and only if $H = 0$.

**Theorem 48.** *The function $Tr(H)$ has the following properties:*

(a) $Tr(UHU^*) = Tr(H)$ *for any unitary operator* $U \in \mathcal{L}$;
(b) $0 \leq H' \leq H$ implies $Tr(H') \leq Tr(H)$.
(c) $Tr(H' + H'') = Tr(H') + Tr(H'')$.
(d) *If $H_1 \leq H_2 \leq \ldots$ is an increasing sequence*[129] *bounded above in $\mathcal{L}_+$, then*

$$Tr(\sup H_n) = \sup Tr(H_n)\ . \tag{31.27}$$

(a) The operator $UH^{1/2}U^*$ is positive Hermitian and $U^*U = 1$ since $U$ is isometric. So $(UH^{1/2}U^*)^2 = UH^{1/2}U^*UH^{1/2}U^* = UHU^*$ and

$$UH^{1/2}U^* = (UHU^*)^{1/2}\ .$$

This being so, we first suppose that $Tr(H) < +\infty$, whence $H^{1/2} = L(a)$ for some moderate $a \in \mathcal{H}$. Using relations $UL(f) = L(Uf)$ and $L(f)^* = L(\tilde{f}) = L(Sf)$ of section (ii), it follows that

$$\begin{aligned} UH^{1/2}U^* &= UL(a)U^* = L(Ua)U^* = [UL(Ua)^*]^* \\ &= [UL(SUa)]^* = L(USUa)]^* = L(SUSUa)\,. \end{aligned}$$

As $SUSUa$ is moderate, $UHU^*$ has finite trace and

$$Tr(UHU^{-1}) = (SUSUa|SUSUa) = (a|a) = Tr(H)$$

[129] The argument equally applies to an increasing filtering family. See Dixmier's book on von Neumann algebras.

since $U$ and $S$ are isometric.

On the other hand, $H = U^*(UHU^*)U$. Conversely, if $U$ and $U^*$ are isometric, in other words if $U$ is *unitary*, then $Tr(H) < +\infty$ since $Tr(UHU^*) < +\infty$, qed.

(b) One may assume that $Tr(H) < +\infty$, and so $H^{1/2} = L(u)$ for some $u \in \mathbf{A}$. As $\|H'^{1/2}f\|^2 = (H'f|f) \leq (Hf|f) = \|u.f\|^2$, there is a continuous operator $A'$ on $\mathcal{H}$ such that $H'^{1/2}f = A'H^{1/2}f$ for all $f \in \mathbf{A}$. Since this relation only determines $A'$ on the image of $H^{1/2}$, image whose orthogonal is $\mathcal{N} = \mathrm{Ker}(H^{1/2})$, $A'$ can be required to be trivial on $\mathcal{N}$. The operator $A'$ in then fully determined and satisfies $H'^{1/2} = A'H^{1/2} = H^{1/2}A'^*$ since $H' = H'^*$. This relation shows that the image of $H'^{1/2}$ is contained in the image of $H^{1/2}$. Denoting its closure by $\mathcal{M}$, i.e. the subspace orthogonal to $\mathcal{N}$, one sees that $A'$ maps $\mathcal{M}$ to $\mathcal{M}$ and $\mathcal{N}$ to $\mathcal{N}$.

Let us show that $A' \in \mathcal{L}$, i.e. that it commutes with all $R(g)$, $g \in \mathbf{A}$.

$$R(g)A'H^{1/2}f = R(g)H'^{1/2}f = H'^{1/2}R(g)f = A'H^{1/2}R(g)f = A'R(g)H^{1/2}f$$

holds for all $f \in \mathbf{A}$, so that $R(g)A'$ and $A'R(g)$ coincide on $\mathcal{M}$. As $R(g)$ commutes with $H$, it preserves $\mathcal{N}$; $R(g)A'$ and $A'R(g)$ are, therefore, trivial on $\mathcal{N}$. Thus $R(g)A' = A'R(g)$ as desired.

This said, by lemma 1 about Hilbert algebras, $H'^{1/2} = A'H^{1/2} = A'L(u) = L(A'u)$. As a result,

$$Tr(H') = \|A'u\|^2 \leq \|u\|^2 = Tr(H)$$

since the norm of $A$ is $\leq 1$, qed.

(c) As $H' + H'' = H$ is an upper bound of $H'$ and $H''$, the trace of $H'$ and $H''$ may be assumed to be finite. So $H'^{1/2} = L(u')$ and $H''^{1/2} = L(u'')$ for some $u', u'' \in \mathbf{A}$. As was mentioned above, there are operators $A', A'' \in \mathcal{L}$ that vanish on the subspace $\mathcal{N}$ of vectors orthogonal to $\mathrm{Im}(H^{1/2})$, preserving the orthogonal $\mathcal{M}$ of $\mathcal{N}$, and such that

$$H'^{1/2} = A'H^{1/2} = H^{1/2}A'^*, \quad H''^{1/2} = A''H^{1/2} = H^{1/2}A''^*. \tag{31.28}$$

By lemma 1,

$$H' = H^{1/2}A'^*H'^{1/2} = H^{1/2}A'^*L(u') = H^{1/2}L(v') \quad \text{where } v' = A'^*u'.$$

Likewise, $H'' = H^{1/2}L(v'')$, where $v'' = A''^*u''$. So

$$H = H^{1/2}L(v) \quad \text{where } v = A'^*u' + A''^*u''.$$

All vectors $u', \ldots, v''$ are moderate, and so are in $\mathbf{A}$.

The previous relation shows that the operator

$$T = H^{1/2} - L(v) = H^{1/2} - A'^*H'^{1/2} - A''^*H''^{1/2} \tag{31.29}$$

is a zero of $H^{1/2}$. Since $\mathcal{M}$ and $\mathcal{N}$ are invariant under $H^{1/2}$, $A'$ and $A''$, hence also under their adjoint maps, so are they under $T$ as well. Now, $H^{1/2}$

is injective on the orthogonal subspace $\mathcal{M}$ to $\mathrm{Ker}(H^{1/2})$. So $H^{1/2} = L(v)$ on $\mathcal{M}$. The images of $H'^{1/2}$ and $H''^{1/2}$ being contained in $\mathcal{M}$ as was seen in the previous lemma, these operators act trivially on $\mathcal{N}$, whence $T = 0$.

Thus $H^{1/2} = L(v)$, which shows that $H$ has finite trace, and by definition,

$$Tr(H) = \|v\|^2, \quad Tr(H') = \|u'\|^2, \quad Tr(H'') = \|u''\|^2.$$

However, the relation $H = H^{1/2}(A'^*A' + A''^*A'')H^{1/2}$ shows that

$$\|f\|^2 = \|A'f\|^2 + \|A''f\|^2 \tag{31.30}$$

for all $f$ of the form $H^{1/2}g$, and so for all $f \in \mathcal{M}$. On the other hand,

$$L(u') = A'H^{1/2} = A'L(v) = L(A'v)$$

(lemma 1), whence $u' = A'v$ by axiom (AH 5) of Hilbert algebras. Likewise, $u'' = A''v$. Relation (30), therefore, proves that $Tr(H) = Tr(H') + Tr(H'')$ if $\mathcal{M}$ is shown to contain $v$. But $\mathcal{M}$ is the closure of the set of vectors $H^{1/2}f = L(v)w = R(w)v$ with $w \in \mathbf{A}$ since $L(v)$ is continuous, and we know that the unit operator is a strong limit of operators $R(w)$, qed.

(d) The first expression being greater than the second one, it suffices to prove the inverse inequality when the second expression is finite. Thus $H_n^{1/2} = L(a_n)$ with $a_n \in \mathbf{A}$ such that $\|a_n\|^2 = Tr(H_n)$. Set $H = \sup H_n$. We know that $(H_n)$ [n° 19, (v), lemma 8] converges strongly to $H$ and [n° 22, (i), exercise 2] that $H_n^{1/2}$ converges strongly to $H^{1/2}$. So, for $x, y \in \mathbf{A}$,

$$(H^{1/2}x|y) = \lim (H_n^{1/2}x|y) = \lim (a_n.x|y) = \lim (a_n|y.\tilde{x}).$$

As $\sup \|a_n\| < +\infty$ and as the $y.\tilde{x}$ are everywhere dense in $\mathcal{H}$, $(a_n)$ converges weakly to some $a \in \mathcal{H}$ such that

$$(H^{1/2}x|y) = (a|y.\tilde{x}) = (a.x|y)$$

for all $x, y \in \mathbf{A}$. As $H^{1/2}$ is continuous, $a$ is moderate and $H^{1/2} = L(a)$, whence $Tr(H) = \|a\|^2 \le \sup \|a_n\|^2 = \sup Tr(H_n)$, qed.

*Exercise 2.* For all $H \in \mathcal{L}_+$, there exists $H' \in \mathcal{L}_+$ such that

$$H' \le H, \quad 0 < Tr(H') < +\infty.$$

[Set $H' = H^{1/2}L(\tilde{f}.f)H^{1/2}$ for some conveniently chosen $f \in \mathbf{A}$].

*Exercise 3.* $Tr(A^*A) = Tr(AA^*)$ for all $A \in \mathcal{L}$.

In general there are many other functions on $\mathcal{L}_+$ having properties (a),..., (d) of the statement: setting $\mathcal{Z} = \mathcal{L} \cap \mathcal{R}$ and $\mathcal{Z}_+ = \mathcal{Z} \cap \mathcal{L}_+$, any $A \in \mathcal{Z}_+$ defines such a function, namely

$$Tr_A(H) = Tr(AH). \tag{31.31}$$

This is well-defined since, $A$ and $H$ being positive Hermitian and commuting, $AH \in \mathcal{L}_+$ [n° 22, (i), exercise 1]. If $H^{1/2} = L(a)$ for some moderate $a \in \mathcal{H}$, then $(AH)^{1/2} = A^{1/2}H^{1/2} = L(A^{1/2}a)$, and since $A^{1/2}a$ is moderate, it follows that

$$Tr_A(H) = (A^{1/2}a|A^{1/2}a) \leq \|A\| \, Tr(H) .$$

As shown by Dixmier, this gives *all* traces having properties (a),..., (d) of the theorem and that of exercise 3. Dixmier has also shown that any "reasonable" trace on a von Neumann algebra can be obtained in this manner from a unitary algebra; see this books.

(iv) *Case of a commutative group.* In the case of a commutative lcg, one can add to theorem 44 thanks to the Fourier transform. We will confine ourselves to the case of the measure $\mu = \epsilon$. The reader can easily address the general case by the same method.

**Theorem 49.** *Let $G$ be a locally compact commutative group and $T$ a continuous operator on $L^2(G)$ commuting with the translations. Then there is a function $\widehat{T} \in L^\infty(\widehat{G})$ such that $T$ is the operator $\widehat{f} \longmapsto \widehat{T}\widehat{f}$.*

The set $\mathcal{R}(G)$ of these operators is a von Neumann algebra to which theorem 46 can be applied for $\mu = \epsilon$: since $G$ is commutative, $\mathcal{R}(G) = \mathcal{R}(G)'$. Let $\mathcal{M}(\widehat{G})$ be the image of $\mathcal{R}(G)$ under the Fourier transform. It it the commutator of the algebra $\mathbf{A}$ of operators $M(p) : f \longmapsto pf$, where $p$ is a finite linear combination of characters of $\widehat{G}$. Hence $\mathcal{M}(\widehat{G}) = \mathbf{A}'$, and so $\mathcal{M}(\widehat{G}) = \mathcal{M}(\widehat{G})' = \mathbf{A}''$ using von N's notation. Since $\mathbf{A}$ is a self-adjoint algebra containing 1, $\mathbf{A}$ is everywhere dense in $\mathcal{M}(\widehat{G})$ with respect to the ultrastrong topology [density theorem, Chap. XI, n° 19, (vii)].

This means that, if $(e_i)$ is an orthonormal basis[130] of $L^2(\widehat{G})$, then, for all $r > 0$, there is a linear combination $p$ of characters of $\widehat{G}$ such that

$$\sum \|Te_i - pe_i\|_2^2 < r^2 .$$

So there is as sequence of functions $p_n$ for which

$$Te_i(\chi) = \text{l.i.m.}^2 \, p_n(\chi)e_i(\chi) \quad \text{for all } i .$$

By the Riesz-Fischer theorem, for each $i$, there is a subsequence for which the previous relation holds in the sense of convergence almost everywhere. As the set of indices $i$ is countable, the same subsequence can be assumed to be suitable for all $i$, in other words that

$$Te_i(\chi) = \lim p_n(\chi)e_i(\chi) \text{ ae.}$$

for all $i$.

[130] To simplify the proof, it will be assumed to be countable, which amounts to supposing that $G$ is separable.

However, $\widehat{G}$ is the union of sets $\{e_i(\chi) \neq 0\}$, up to a null set, since otherwise there would be functions $\neq 0$ on $L^2(\widehat{G})$ orthogonal to all $e_i$. As a result, the limit

$$\widehat{T}(\chi) = \lim p_n(\chi)$$

exists almost everywhere. As $Te_i(\chi) = \widehat{T}(\chi)e_i(\chi)$ ae. for all $i$, $Tf(\chi) = \widehat{T}(\chi)f(\chi)$ ae. for any finite linear combination of $e_i$. But any $f \in L^2(\widehat{G})$ is a limit in mean, hence also almost everywhere, of such combinations. So $Tf(\chi) = \widehat{T}(\chi)f(\chi)$ ae. for all $f \in L^2(\widehat{G})$. It remains to observe that, the operator $T$ being continuous, $\|\widehat{T}\|_\infty = \|T\|$ necessarily holds.

The reader will easily show that if $T$ is positive Hermitian, then the function $\widehat{T}$ is positive and that $Tr(T) = \int \widehat{T}(\chi)d\chi$.

(v) *Characters of a locally compact group.* Let $\mu$ be a bitrace on a unimodular lcg $G$, $(\mathcal{H}(\mu), L_\mu, R_\mu)$ the two-sided representation of $G$ and $\mathbf{A}(\mu)$ the maximal Hilbert algebra defined by $\mu$. For all $f \in L(G)$, the image $f_\mu$ of $f$ in $\mathcal{H}(\mu)$ is in $\mathbf{A}(\mu)$. For $f, g \in L(G)$, by definition,

$$\mu(f,g) = (f_\mu|g_\mu) = Tr\,[L_\mu(g)^* L_\mu(f)]$$

where $Tr$ is the canonical trace (theorem 48) on the von Neumann algebra $\mathcal{L}(\mu)$ generated by the $L_\mu(f)$ or $L_\mu(x)$. $\mu$ will be said to be a *character* of $G$ if the algebra $\mathcal{Z}(\mu)$ of operators commuting with all $L_\mu(f)$ and $R_\mu(f)$ is reduced to scalars, in other words if the algebras $\mathcal{L}(\mu)$ and $\mathcal{R}(\mu)$ are factors in the sense of von Neumann. The representation $(x,y) \longmapsto L(x)R(y)$ of $G \times G$ on $\mathcal{H}(\mu)$ is, therefore, irreducible and conversely.

As shown by von N., there are four large classes of factors. He distinguished between them by investigating the closed subspaces $\mathcal{E}$ of $\mathcal{H}$ such that the corresponding projection $P_E$ is in the algebra $\mathcal{A}$ considered, and by considering equivalent two subspaces $\mathcal{E}'$ and $\mathcal{E}''$ when there is an operator on $\mathcal{A}$ mapping $\mathcal{E}'$ isometrically onto $\mathcal{E}''$. The equivalence classes of these subspaces are characterized by a number $\dim(\mathcal{E}) \in [0, +\infty]$ having the formal properties of a *dimension*. Then, up to a constant factor, the possibilities are as follows:

Case ($\mathrm{I}_n$), $n \in \mathbb{N} \cup \{+\infty\}$: the dimension takes values $0, 1, \ldots, n$;
Case ($\mathrm{II}_1$): it takes all values in the interval $[0, 1]$;
Case ($\mathrm{II}_\infty$): it takes all values in the interval $[0, +\infty]$;
Case (III): it only takes values 0 and $+\infty$.

This classification applies to characters of a unimodular lcg. Case (III) does not occur, but it is encountered in the regular representation of some non-unimodular groups.

In the case of the algebra $\mathcal{A} = \mathcal{L}(\mathcal{H})$, the von N dimension is the usual one and is just $Tr(P_E)$, where $P_E$ is the orthogonal projection onto $\mathcal{E}$ and where $Tr(H) = \sum(He_i|e_i)$ is the usual trace calculated in an orthonormal basis. Nonetheless, the notion of the trace of an operator hardly appears in von N.'s articles. Partly influenced by representations defined by central measures of positive type (the case of compact groups was already indicative), Dixmier reinterpreted von N.'s theory in terms of traces and freed it from its restriction to factors. In the case of a factor, Dixmier showed that, up to a constant factor, there is a unique function $Tr$ defined on Hermitian elements $\geq 0$ of $\mathcal{A}$, satisfying theorem 48 and such that $\dim(\mathcal{E}) = Tr(P_E)$.

Like elementary functions of positive type [n° 30, (iii), lemma 2], characters can be defined by an "extremal" property. To see this, it suffices to observe that, for every bitrace $\mu$, bitraces $\mu' \ll \mu$ correspond bijectively to operators $H \in \mathcal{Z}(\mu)$ such that $0 \leq H \leq 1$, the bitrace defined by such an $H$ being

$$\mu_H(f,g) = Tr\left[L_\mu(g)^* H L_\mu(f)\right].$$

Conversely, every bitrace $\mu' \ll \mu$ defines an operator $H$ by

$$(Hf_\mu | g_\mu) = \mu'(f,g).$$

The proof is the same as that of lemma 2 of n° 30. The conclusion that follows is that a bitrace $\mu$ is a character if and only if every bitrace $\mu' \ll \mu$ is proportional to $\mu$, which is equivalent to saying that characters of $G$ correspond to extremal generators of the convex cone of bitraces. In this general situation, there is unfortunately no analogue of the Krein-Milman existence theorem, let alone of Choquet's theorem, on *compact* convex sets.

Nonetheless, the theory of direct integrals of Hilbert spaces of n° 24 applies to the algebra $\mathcal{Z}(\mu)$ or to any Gelfand-Naimark subalgebra $\mathbf{Z} \subset \mathcal{Z}(\mu)$ such that $\mathbf{Z}'' = \mathcal{Z}(\mu)$. Indeed, using separability assumptions, there is then a decomposition

$$\mathcal{H}(\mu) = \int \mathcal{H}(\zeta) d\lambda(\zeta),$$

where integration is over the spectrum of $\mathbf{Z}$. There then exist operators $T(\zeta)$ in $\mathcal{H}(\zeta)$, defined up to a null set, corresponding to every bounded operator $T$ commuting with $\mathbf{Z}$, i.e. with $\mathcal{Z}(\mu)$. In particular, for almost every $\zeta$, the decomposition of operators $L_\mu(x)$, $L_\mu(f)$, $R_\mu(x)$, $R_\mu(f)$ and $S$, leads to similar operators $L(x;\zeta)$, etc. in $\mathcal{H}(\zeta)$. Likewise, in almost all $\mathcal{H}(\zeta)$, there are algebras $\mathcal{L}(\zeta)$ and $\mathcal{R}(\zeta)$ generated respectively by the $L(f;\zeta)$ and $R(f;\zeta)$ corresponding to the algebras $\mathcal{L}(\mu)$ and $\mathcal{R}(\mu)$ [n° 24, (ii), end]. The algebras $\mathcal{L}(\zeta)$ and $\mathcal{R}(\zeta)$ are *factors* for almost all $\zeta$ and the canonical trace $Tr$ on $\mathcal{L}(\mu)$ can be computed by the relation

$$Tr(T^*T) = \int Tr\left[T(\zeta)^* T(\zeta)\right] d\lambda(\zeta), \tag{31.32}$$

where, for each $\zeta$, a convenient function $Tr$ is chosen on the factor $\mathcal{L}(\zeta)$. For all $T = L_\mu(f)$ where $f$ is in the domain of definition of the initial bitrace $\mu$, this gives or almost all $\zeta$ a bitrace $f \longmapsto Tr[L(f;\zeta)L(g;\zeta)^*]$ f that is the character of $G$. The subject is dealt with in detail by Dixmier in chap. 8, 17, 18 of his *$C^*$-algebras*. See also Lajos Pukánszky, *Characters of connected Lie groups* (AMS, 1999) and Chap. 14, Abstract Representation Theory, of Nolan R. Wallach, *Real Reductive Groups* II (Academic Press, 1992).

Case ($II_1$) corresponds to irreducible Hilbert algebras with unit element or, in group theory, to central continuous functions of positive type studied at the beginning of section (iii) above. Consider for example the regular representation of a discrete group all of whose conjugacy classes are infinite.

In case ($II_\infty$), thanks to his theory of induced representations, G. W. Mackey obtained examples for groups seemingly quite simple, the semidirect products[131] of commutative groups.

(vi) *Characters of type (I).* Groups that have been by far the most studied are those whose characters are of type (I): all *linear algebraic* groups fall in this case; so do real semisimple Lie groups with infinite centre, a fundamental result proved by Harish-Chandra, and all solvable real algebraic groups, i.e. whose matrices are, if necessary by passing to $\mathbb{C}$, simultaneously triangularizable (Dixmier, Kirillov, etc.); the others are, as a first approximation, semidirect products of a solvable group and a semisimple group, which enabled Dixmier to obtain the general result. Actually, for these groups, characters can be shown to be *distributions* and, in the semisimple case, *locally integrable functions*, a very hard result of Harish-Chandra's.

Let $\mathbf{A}$ be a unitary algebra and $\mathcal{H}$ its completion. Assuming $\mathbf{A}$ to be irreducible, the von N. algebras $\mathcal{L}$ and $\mathcal{R}$ are factors. Let us suppose they are of type (I) – as they are isomorphic by $T \longmapsto STS^{-1}$, it suffices that one of them be so – and let $Tr$ be the trace on $\mathcal{R}$, normalized so as to take values $0, 1, 2, \ldots$ We choose a projection $P \in \mathcal{R}$ such that $Tr(P) = 1$ and let $\mathcal{E} = \mathrm{Im}(P)$. Since $P \in \mathcal{R}$, the subspace $\mathcal{E}$ is invariant under $T \in \mathcal{L}$. These act irreducibly on $\mathcal{E}$. Indeed, if a closed subspace $\mathcal{E}' \subset \mathcal{E}$ is $\mathcal{L}$-invariant, the corresponding projection $P'$ is in $\mathcal{R}$ and satisfies $0 \leq P' \leq P$. So $Tr(P') \leq Tr(P)$, whence $Tr(P') = 0$ or 1. In the first case, $\mathcal{E}'$ is trivial; in the second, $Tr(P - P') = 0$. Thus $P' = P$ and $\mathcal{E}' = \mathcal{E}$, which gives the result. It follows that the restrictions of operators $T \in \mathcal{L}$ to $\mathcal{E}$ are everywhere dense in the algebra $\mathcal{L}(\mathcal{E})$ of all continuous operators on $\mathcal{E}$. But as $\mathcal{L}$ is closed with respect

[131] $G$ is said to be a semidirect product of two groups $H$ and $N$ if $H$ and $N$ are closed subgroups of $G$, if $N$ is invariant in $G$, and if the map $(h,n) \longmapsto hn$ is bijective. An easy example: the Euclidean group in $\mathbb{R}^n$ is the semidirect product of the group of rotations about the origin and the group of translations. An example difficult to analyze: the group of transformations

$$(z, z') \longmapsto (e(\alpha t)z + u, e(\alpha' t)z' + u') ,$$

of $\mathbb{C}^2$, where $t \in \mathbb{R}$, $u, u' \in \mathbb{C}$ and where $\alpha, \alpha'$ are non-trivial real constants with irrational ratio; $G$ is the semidirect product of $\mathbb{R}$ and the additive group $\mathbb{C}^2$. In fact, all possible "pathological" circumstances seem to be encountered in groups of this type.

to all von N. topologies and preserves $\mathcal{E}$, the set of restrictions of operators $T \in \mathcal{L}$ to $\mathcal{E}$ is identical to $\mathcal{L}(\mathcal{E})$. On the other hand, if $T \in \mathcal{L}$ is trivial on $\mathcal{E}$, it is trivial on the closed subspace generated by $A(\mathcal{E})$, where $A \in \mathcal{R}$. Now, this closed subspace is bi-invariant, and so is identical to $\mathcal{H}$ since $\mathbf{A}$ is irreducible. As a consequence, the map associating to each $T \in \mathcal{L}$ its restriction to $\mathcal{E}$ is an isomorphism from $\mathcal{L}$ onto $\mathcal{L}(\mathcal{E})$. Since, there is a unique trace function on $\mathcal{L}$, up to a constant factor, it is necessarily the usual trace function of $\mathcal{L}(\mathcal{E})$. In conclusion, for some $T \in \mathcal{L}$, the condition $Tr(T^*T) < +\infty$ means that the restriction of $T$ to $\mathcal{E}$ is a *de Hilbert-Schmidt* operator.

In the case of Hilbert algebras associated to groups, one readily deduces the next result:

**Theorem 50**. *Let $\mu$ be a character of type* (I) *of a unimodular group $G$, defined on a subalgebra $\mathfrak{a}$ of $L^1(G)$. There is an irreducible unitary representation $(\mathcal{E}, U)$ of $G$ having the following properties:*
(a) *$U(f) = \int U(x)f(x)dx$ is a Hilbert-Schmidt operator for all $f \in \mathfrak{a}$,*
(b) *for all $f, g \in \mathfrak{a}$*

$$\mu(f,g) = Tr\left[U(f)U(g)^*\right]. \tag{31.33}$$

This property shows that, from the point of view of harmonic analysis, in some sense groups of type (I) generalize compact groups. It would be wrong to infer conjectures optimistic about difficulties involved in their study (classification of irreducible representations, explicit computation of characters and of Plancherel's formula, etc.): fifty-five years of efforts (1947–2002) have not been enough to solve all these problems, in particular the first one, in the case of semisimple groups.

The fact that real semisimple groups (Harish-Chandra, 1950–52) or over a local field (J. Bernstein, 1974) have property (a) of the theorem follows from a far more important theorem. For the sake of simplicity, let us suppose that the centre of $G$ is finite. Then there are pairwise conjugate maximal compact subgroups in $G$,[132] and if $K$ is one of them, one can show that, for any irreducible unitary representation $(\mathcal{H}, U)$ of $G$ and any type $\mathfrak{d}$ of irreducible representations of $K$, the multiplicity of $\mathfrak{d}$ in the given representation of $G$ is $\leq \dim(\mathfrak{d})$. If $\chi$ is a character of $K$, normalized by the condition that $\chi * \chi = \chi$, and if

$$U(\overline{\chi}) = \int U(k)\overline{\chi(k)}dk$$

[n° 29, (iii), end, where this operator is written $E(\chi)$], then the subspace $\mathcal{H}(\chi) = \mathrm{Im}(U(\overline{\chi}))$ has dimension $\dim(\chi)^2$, where $\dim(\chi)$ denotes the dimension of the irreducible representation having character $\chi$. However, for any $f \in L^1(G)$ such that

[132] Typical examples: the group $SO(n)$ in $SL_n(\mathbb{R})$, the group $SU(n)$ in $SL_n(\mathbb{C})$, etc.

$$(31.34) \qquad f(x) = \chi * f * \chi'(x) = \iint f(kxk')\overline{\chi(k)\chi'(k')}dkdk' ,$$

$U(f) = U(\overline{\chi})U(f)U(\overline{\chi'})$. The rank of the operator $U(f)$ is, therefore, finite. But using the theory of compact groups of n° 29 it is easy to see that finite sums of functions of the previous type are everywhere dense in $L^1(G)$. Hence it follows that the set $\mathfrak{a}$ of $f \in L^1(G)$ for which $U(f)$ is a Hilbert-Schmidt operator satisfies conditions of section (i) of this n°. This enables us to associate a character $\mu(f,g) = Tr[U(f)U(g)^*]$, obviously of type (I), to the given representation.

Conversely, any character $\mu$ of $G$ is obtained in this manner. Indeed, let $(\mathcal{H}, L, R, S)$ be the two-sided representation of $G$ associated to $\mu$. The map $(x, x') \longmapsto L(x)R(x')$ is an irreducible representation of the semisimple group $G \times G$ in which $K \times K$ is a maximal compact subgroup. The characters of $K \times K$ are obviously the functions $(x, x') \longmapsto \chi(x)\chi'(x')$ where $\chi$ and $\chi'$ are characters of $K$, and for the corresponding projections,

$$E(\chi, \chi') = \iint L(k)R(k')\chi(k)\chi'(k')dkdk' = L(\chi)R(\chi') .$$

We then choose some $\chi'$ such that $R(\chi') \neq 0$ and set $\mathcal{E} = \mathrm{Im}(R(\chi'))$. This closed subspace is invariant under the von Neumann algebra $\mathcal{L}$. Denoting the restriction of $L(x)$ to $\mathcal{E}$ by $U(x)$, we get a unitary representation $(\mathcal{E}, U)$ of $G$. Obviously, $\mathcal{E}(\chi) = \mathrm{Im}[L(\chi)R(\chi')]$ holds for the latter. But as $G \times G$ is semisimple, the general result stated above implies that this subspace is finite-dimensional. We then deduce that that there are several operators $T \in \mathcal{L}$ whose restrictions to $\mathcal{E}$ are Hilbert-Schmidt operators, and as $\mathcal{L}$ is a factor, this is enough to show that $\mathcal{L}$, and so $\mu$, is of type (I).

Once again, I refer to Dixmier's books for additional information or more general proofs.

## 32 – Discrete Components of the Regular Representation

Given a unimodular group $G$, the basic problem is to decompose the regular representation of $G$ into a direct integral of irreducible representations. Such a decomposition can include a "discrete spectrum" and a "continuous spectrum". As in the theory of compact groups where the second one does not exist, the first one is obtained by using irreducible representations that can be embedded in $L^2(G)$, in other words the *minimal* left – or right (at choice) – invariant closed subspaces of $L^2(G)$. Such a representation is, by definition a *discrete component* of the regular representation. They are sufficient for the decomposition of the regular representation if $G$ is compact, but do not exist if $G = \mathbb{R}$. Hence it may be thought that they also do not exist in the general case. Using explicit calculations, one of Einstein's assistants in Princeton, Valentine Bargmann, found[133] that the group $SL_2(\mathbb{R})$ does have some, as

[133] *Irreducible unitary representations of the Lorentz group* (Ann. of Math., **48**, 1947, pp. 568–640). The Lorentz group itself is isomorphic to the quotient of $SL^2(\mathbb{C})$ by

will be seen in Chap. XII, n° 16 and 30. In this case, the decomposition of the regular representation, i.e. Plancherel's formula for $G$, contains both "Fourier" series and integrals. More precisely, Bargmann found that there are irreducible representations $(\mathcal{H}, U)$ some of whose coefficients are square integrable, and checked that they satisfy orthogonality relations similar to those of compact group theory. As shown by Harish-Chandra, these representations are also encountered in the case of semisimple groups having a maximal compact torus, for example the symplectic group $Sp_n(\mathbb{R})$. They are realized on square integrable holomorphic functions spaces and are essential in his general Plancherel formula. They are also encountered in the case of semisimple $p$-adic groups and in that of associated adelic groups, which are lcgs but not Lie groups, and they exist for some nilpotent Lie groups.[134]

In fact, Bargmann already knew, and Harish-Chandra insisted on this point in the general case, that it is essential to consider not only $SL_2(\mathbb{R})$, but its universal cover $G$, a group that cannot be embedded in a linear group. Like any Lie group, $SL_2(\mathbb{R})$ is isomorphic to the quotient of its universal cover by a discrete subgroup $Z$ of the latter's centre. In this case, $Z$ is isomorphic to $\mathbb{Z}$, and so is infinite.

More generally, let $G$ be a lcg, $Z$ its centre and $(\mathcal{H}, U)$ a unitary representation of $G$. If $(\mathcal{H}, U)$ is irreducible, there is a character $\alpha$ of $Z$ such that

$$U(z) = \alpha(z)1 \quad \text{for all } z \in G. \tag{32.1}$$

The coefficients

$$\varphi_{a,b}(x) = (U(x)a|b) \tag{32.2}$$

of $(\mathcal{H}, U)$, therefore, satisfy the functional equation

its centre $\{1, -1\}$. Around the same period, using integral methods, Gelfand and Naimark obtained Plancherel's formula for $SL_2(\mathbb{C})$ then for $SL_n(\mathbb{C})$; see my two talks in the Séminaire Bourbaki of 1948–49. Harish-Chandra quickly generalized their Plancherel formula to all complex semisimple groups. Bargmann was only using infinitesimal methods, which, at the time, assumed that some coefficients of the representations were analytic functions on the group; but like a good physicist, Bargmann did not let this "detail" stop him. The existence of several "analytic" vectors, i.e. for which the function $x \longmapsto U(x)a$ is analytic, was shown shortly after (see Chap. XII, n° 26). Bargmann's and GN's articles are still worth reading. The explicit Plancherel formula for $SL_2(\mathbb{R})$, more difficult to obtain than for $SL_2(\mathbb{C})$, is also due to Harish-Chandra (1950) ; see Serge Lang, $SL_2(\mathbb{R})$ (Springer, 1985).

[134] C. C. Moore and J. A. Wolf, *Square integrable representations of nilpotent Lie groups* (Trans. AMS, **185** (1973), pp. 445–462). A Lie group $G$ is nilpotent if there are closed normal subgroups $Z_1 \subset Z_2 \subset \ldots \subset Z_n = \{e\}$ in $G$ such that, for all $p$, the image of $Z_p$ in $G/Z_{p-1}$ is in the centre of $G/Z_{p-1}$, i.e. if

$$x \in G \quad \& \quad y \in Z_p \Longrightarrow xyx^{-1}y^{-1} \in Z_{p-1}.$$

Example: any closed subgroup of the of triangular matrices with diagonal 1 (unipotent matrices).

$$\varphi(xz) = \varphi(x)\alpha(z)\,. \tag{32.3}$$

If $Z$ is infinite, they cannot be square integrable over $G$; but as $|\varphi(xz)| = |\varphi(x)|$, they can be square integrable over the quotient group $G/Z$. The representation will be said to be *square integrable* if all its coefficients are square integrable mod $Z$. The set of classes of measurable solutions $\varphi$ of (3) such that

$$\int_{G/Z} |\varphi(x)|^2\, dx < +\infty \tag{32.4}$$

is a Hilbert space $L^2(G/Z;\alpha)$ with inner product

$$(f|g) = \int_{G/Z} f(x)\overline{g(x)}dx\,. \tag{32.5}$$

Since $Z$ is in the centre of $G$, left and right translations$f \longmapsto \epsilon_x * f$ and $f \longmapsto f * \epsilon_x$ clearly act on this space, and so does the involution $f \longmapsto \widetilde{f}$. This gives a "two-sided representation" which will be seen to be identical to the two-sided representation defined by a central measure $\mu$ of positive type. One could call it the regular representation defined by $\alpha$. All this has already been seen in the context of "central groups" [n° 29, (iii)].

For any $f \in L(G)$, the function

$$f_\alpha(x) = \int f(xz)\overline{\alpha(z)}dz \tag{32.6}$$

obviously satisfies (3) and thereby gives all the continuous solutions of (3) having compact support mod $Z$. The proof, left to the reader, consists in imitating that of n° 15, (iv) related to the unit character. As the reader will also easily check by imitating standard integration theory, the space $L(G/Z;\alpha)$ of these continuous solutions with compact support mod $Z$ is everywhere dense in $L^2(G/Z;\alpha)$. All limit theorems valid for the usual spaces $L^2(G/Z)$, and especially Riesz-Fischer, can be applied to $L^2(G/Z;\alpha)$. Finally, as in the case of central groups, the regular representation of $G$ on $L^2(G)$ is clearly the continuous sum of its representations on $L^2(G/Z;\alpha)$: this is relation (29.35).

For $f, g \in L(G)$,

$$\begin{aligned}(f_\alpha|g_\alpha) &= \int_{G/Z} dx \iint f(xz_1)\overline{g(xz_2)\alpha(z_1)}\alpha(z_2)dz_1dz_2 = \\ &= \int \overline{\alpha(z)}dz \int_{G/Z} dx \int f(xz_2z)\overline{g(xz_2)}dz_2\,.\end{aligned}$$

But integrating over $Z$ then over $G/Z$ amounts to integrating over $G$. As a result,

$$(f_\alpha|g_\alpha) = \int \overline{\alpha(z)}dz \int_G f(xz)\overline{g(x)}dx = \int \widetilde{g} * f(z)\overline{\alpha(z)}dz\,.$$

Hence associating the measure $\mu$ on $G$ defined by

$$\mu(f) = \int_Z f(z)\overline{\alpha(z)}dz \tag{32.7}$$

to the character $\alpha$ of $Z$, we at last get

$$(f_\alpha|g_\alpha) = \mu(\widetilde{g} * f)\,. \tag{32.8}$$

This shows that the obviously central measure $\mu$ is of positive type and that $L^2(G/K;\alpha)$ is isomorphic to the space $\mathcal{H}(\mu)$ of section (i), the isomorphism transforming $f_\alpha$ into $f_\mu$ for all $f \in L(G)$ being compatible with representations of $G$ on these two spaces and transforming the algebra $L(G/Z;\alpha)$ into the Hilbert algebra $\mathbf{A}(\mu)$. Multiplication in the latter is defined by (31.7): $f_\mu . g_\mu = (f * g)_\mu$ for $f, g \in L(G)$. A small calculation readily shows that, for $f, g \in L(G/Z;\alpha)$, this product is just the "convolution mod $Z$"

$$f_\mu . g_\mu(x) = f *_Z g(x) = \int_{G/Z} f(xy)g(y^{-1})dy\,. \tag{32.9}$$

**Theorem 51 (Bargmann orthogonality relations).** *Let $(\mathcal{H}, U)$ be an irreducible unitary representation of a locally compact unimodular group $G$ and $Z$ the centre of $G$. Set $U(z) = \alpha(z)1$ for all $z \in Z$.*

(i) *If the representation $(\mathcal{H}, U)$ has a non-trivial square integrable coefficient over $G/Z$, it is square integrable.*

(ii) *For such a representation, there exists a scalar $\lambda > 0$ such that*

$$\int_{G/Z} (U(g)a|b)\,\overline{(U(g)a'|b')}dg = \lambda\,(a|a')\,\overline{(b|b')} \tag{32.10}$$

*for all $a, a', b, b' \in \mathcal{H}$.*

(iii) *For all $f \in L(G)$, $U(f)$ is a Hilbert-Schmidt operator everywhere dense in the Hilbert space $\mathcal{L}_2(\mathcal{H})$.*

(iv) *The coefficients of two inequivalent square integrable irreducible unitary representations corresponding to the same character $\alpha$ of $Z$ are orthogonal.*

(v) *Every square integrable representation such that $U(z) = \alpha(z)1$ is a discrete component of the regular representation of $G$ on $L^2(G/Z;\alpha)$ and conversely.*

We start by observing that for any representation $(\mathcal{H}, U)$ of $G$, coefficients (2) satisfy

$$\varphi_{a,b}\left(y^{-1}gx\right) = \varphi_{U(x)a,U(y)b}(g) \tag{32.11}$$

and

$$\overline{\varphi_{a,b}(g)} = \varphi_{b,a}\left(g^{-1}\right)\,. \tag{32.12}$$

So, if $G$ is unimodular, then

$$(32.13) \qquad (\varphi_{a,b})^{\tilde{}} = \varphi_{b,a}\,.$$

*Proof of* (i) *and* (ii). Let us suppose that $\varphi_{a',a} \in L^2(G/Z;\alpha)$ for a pair of vectors $a, a' \in \mathcal{H}$. The set $\mathcal{D}$ of $u \in \mathcal{H}$ such that $\varphi_{u,a}$ has the same property is invariant by (11), and so is everywhere dense in $\mathcal{H}$. We equip $\mathcal{D}$ with the inner product

$$(32.14) \qquad (u|v)' = (u|v) + (\varphi_{u,a}|\varphi_{v,a})\,,$$

the right hand side being calculated in $L^2(G/Z;\alpha)$, and let $\mathcal{H}'$ be the completion of $\mathcal{D}$. For all $x \in G$, $U(x)$, is clearly unitary with respect to this inner product. If $(u_n)$ is a Cauchy sequence in $\mathcal{D}$ with respect to inner product (14), clearly, $\lim u_n = u$ exists in $\mathcal{H}$ and the coefficients $\varphi$ corresponding to ordered pairs $(u_n, a)$ converge in mean in $L^2(G/Z;\alpha)$. Now, they converge everywhere to $\varphi_{u,a}$ in $G$ since $(u_n)$ converges to $u$ in $\mathcal{H}$. As a result (Riesz-Fischer),

$$\text{l.i.m.}^2\, \varphi_{u_n,a} = \varphi_{u,a} \quad \text{in } L^2(G/Z;\alpha)\,.$$

So $u \in \mathcal{D}$ and as $(u_n)$ converges to $u$ in $\mathcal{H}'$, $\mathcal{D} = \mathcal{H}'$.

Since $(u|u) \le (u|u)'$, the canonical injection $J$ from $\mathcal{H}'$ to $\mathcal{H}$ is continuous, hence has an adjoint $J^* : \mathcal{H} \longrightarrow \mathcal{H}' = \mathcal{D}$. $JJ^* : \mathcal{H} \Longrightarrow \mathcal{H}$ clearly commutes with $U(x)$, hence is a scalar, so that $J^*$ is proportional to an isometric operator. Thus its image is a closed invariant subspace contained in $\mathcal{D}$, whence $\mathcal{D} = \mathcal{H}$. It first follows that $\varphi_{u,a} \in L^2(G/Z;\alpha)$ for all $u \in \mathcal{H}$, hence, applying arguments to $u$, also that $\varphi_{u,v} \in L^2(G/Z;\alpha)$ for all $u$ and $v$, and then that, for given $a$, $J^*$ is unitary up to a constant factor. Hence so is $J$, which gives relation of the form

$$(32.15) \qquad (\varphi_{u,a}|\varphi_{u',a}) = \lambda(a)\,(u|u')\,.$$

For $u = u'$, the left hand side does not change if $a$ and $u$ are permuted. Thus $\lambda(a)(u|u) = \lambda(u)(a|a)$, and so $\lambda(a) = \lambda(a|a)$ where $\lambda$ is a constant. Then (15) becomes

$$(32.15') \qquad (\varphi_{u,a}|\varphi_{u',a}) = \lambda\,(a|a)\,(u|u')\,.$$

Therefore, (10) follows by $4(a|b) = (a+b|a+b) - \ldots$

*Proof of* (iii). Let $(e_i)$ be an orthonormal basis for $\mathcal{H}$. Setting

$$(32.16) \qquad \varphi_{ij}(x) = (U(x)e_i|e_j)\,,$$

for all $f \in L(G)$,

$$(U(\overline{f})e_i|e_j) = \int_G \overline{f(x)} f_{ij}(x) = \int_{G/Z} dx \int_Z \overline{f(xz)} f_{ij}(x)\alpha(z)dz\,.$$

Thus

$$(U(\overline{f})e_i|e_j) = (\varphi_{ij}|f_\alpha) . \tag{32.17}$$

By function (16), this is an inner product on $L^2(G/Z;\alpha)$. However, (10) shows that the functions $\varphi_{ij}$ are pairwise orthogonal and have the same norm on $L^2(G/Z;\alpha)$. Therefore, they are part of an orthonormal basis, up to a factor, and so $\sum |(\varphi_{ij}|\psi)|^2 \leq \|\psi\|^2$ for all $\psi \in L^2(G;\alpha)$; in particular

$$\sum |(U(f)e_i|e_j)|^2 \leq \|f_\alpha\|^2 < +\infty \tag{32.18}$$

by (17). This shows that for all $f$, $U(f)$ is an operators of $HS$. To prove that they are everywhere dense in the space of operators $HS$, it suffices to show that, for all $S \in \mathcal{L}_2(\mathcal{H})$ and all $r > 0$, there exists $f \in L(G)$ such that $\|U(f) - S\|_2 < r$, where

$$\|A\|_2 = \left(\sum \|Ae_i\|^2\right)^{\frac{1}{2}} = Tr(A^*A)^{\frac{1}{2}}$$

is the norm on $\mathcal{L}_2(\mathcal{H})$. This inequality written

$$\sum \|U(f)e_i - Se_i\|^2 < r^2$$

will follow from von Neumann's ultrastrong density theorem [n° 19, (vii)].

To prove it, we first recall that as the commutator algebra of $U(f)$ reduces to scalars, $\mathcal{L}(\mathcal{H})$ is the bi-commutator of the self-adjoint algebra of $U(f)$. So the operators $U(f)$ are dense in $\mathcal{L}(\mathcal{H})$ with respect to the ultrastrong topology [n° 19, (vii), lemma 9]. For all $g \in L(G)$, let us set $a_i = U(g)e_i$. Since $\sum \|a_i\|^2 < +\infty$, the map $T \longmapsto (\sum \|Ta_i\|^2)^{\frac{1}{2}}$ is one of the seminorms defining the ultrastrong topology of $\mathcal{L}(\mathcal{H})$. So there exists $f \in L(G)$ such that

$$\|U(f)U(g)e_i - SU(g)e_i\|_2 = \sum \|U(f)a_i - Sa_i\|^2 < r^2 .$$

As $U(f)U(g) = U(f*g)$, the operators $U(f)$ enable us to approximate $SU(g)$ in $\mathcal{L}_2(\mathcal{H})$. However, for all $S$ of type $HS$, the map $T \longmapsto ST$ from the space $\mathcal{L}(\mathcal{H})$, equipped with the ultrastrong topology, to the Hilbert space $\mathcal{L}_2(\mathcal{H})$ is continuous. On the other hand, we know that the operator 1 is the ultrastrong limit of operators $U(g)$. So $S$ can be approximated with operators $SU(g)$, hence by operators $U(f)$. Thus the $U(f)$ are dense in $\mathcal{L}_2(\mathcal{H})$.

*Proof of* (iv). See theorem 41 on compact groups.

*Proof of* (v). Let us suppose that $(\mathcal{H}, U)$ is irreducible and square integrable, choose some non-trivial $a \in \mathcal{H}$ and associate the function $\varphi_{u,a}(x) = (U(x)u|a)$ to each $u \in \mathcal{H}$. By orthogonality relations,

$$(\varphi_{u,a}|\varphi_{v,a}) = \lambda\,(a|a)\,(u|v) . \tag{32.19}$$

For given $a$, the set of functions $\varphi_{u,a}$ is, therefore, a closed subspace $\mathcal{H}(a)$ of $L^2(G/Z;\alpha)$. Replace $u$ by $U(x)u$ by replacing $\varphi_{u,a}(g)$ by $\varphi_{u,a}(gx)$. Then the map $u \longmapsto \varphi_{u,a}$ transforms representation $U$ of $G$ on $\mathcal{H}$ into the right regular representation on $L^2(G/Z;\alpha)$, so that $(\mathcal{H}, U)$ is a discrete component of the latter.

Note that the orthogonality relations lead to a formula for convolution products mod $Z$ of two coefficients:

$$\begin{aligned} \varphi_{u,a} *_Z \varphi_{v,a}(x) &= \int (U(xg)u|a) \left(U\left(g^{-1}\right) v|a\right) dg = \\ &= \int \left(U(g)u|U\left(x^{-1}\right) a\right) \overline{(U(g)a|v)} dg = \\ &= \lambda(u|a) \left(v|U(x)^{-1}a\right) , \end{aligned}$$

whence

$$\varphi_{u,a} *_Z \varphi_{v,a} = \lambda(u|a)\varphi_{v,a} \,. \tag{32.20}$$

The convolution of two functions of $\mathcal{H}(a)$ is again in $\mathcal{H}(a)$.

Conversely, let $\mathcal{E}$ be a closed subspace of $L^2(G/Z;\alpha)$ invariant under right translations $R(x)$ and minimal. The proof reduces to showing that the functions

$$(u|R(x)v) = \int_{G/Z} u(y)\overline{v\left(yx^{-1}\right)} dy = \widetilde{v} *_Z u(x) \tag{32.21}$$

are square integrable mod $Z$ for all $u, v \in \mathcal{E}$, and that for this to be the case so is one of the coefficients.

Let $\mathbf{A} = L(G/Z;\alpha)$ denote the Hilbert algebra associated to the measure $\alpha(z)dz$ on $G$ and, as in n° 31, $\widehat{\mathbf{A}}$ be the corresponding maximal Hilbert algebra. It is the set of $u \in L^2(G/Z;\alpha) = \mathcal{H}$ such that the maps

$$R(u) : f \longmapsto f.u \,, \qquad L(u) : f \longmapsto u.f$$

from $\mathbf{A}$ to $\mathcal{H}$ are continuous with respect to the $L^2$ norm. Besides, as was seen in the proof of theorem 47 of n° 31, it suffices that this holds for one of them. If $v_n \in \mathbf{A}$ converge in $\mathcal{H}$, by (9) the functions $v_n.u = v_n *_Z u$ clearly converge everywhere and even uniformly on $G$. At the limit, we thus again get

$$v.u = R(u)v = v *_Z u \tag{32.22}$$

for all $v \in \mathcal{H}$. Replacing $v$ by $\widetilde{v}$, (21) shows that

$$\widetilde{v}.u(g) = (u|R(x)v)$$

for all $u \in \widehat{\mathbf{A}}$ and $v \in \mathcal{H}$. The left hand side being in $L^2$, the proof, therefore, reduces to showing that $\mathcal{E}$ contains a non-trivial $u \in \widehat{\mathbf{A}}$. However, $\mathcal{E}$ being right invariant, the projection $P$ of $L^2(G/Z;\alpha)$ onto $\mathcal{E}$ commutes with $R(u)$ for all $u \in \mathbf{A}$, hence also for all $u \in \widehat{\mathbf{A}}$. Then, for $u, v \in \widehat{\mathbf{A}}$,

$$PL(v)u = PR(u)v = R(u)Pv = L(Pv)u\,.$$

As $PL(v)$ is continuous, so is $L(Pv)$, which shows that $Pv \in \widehat{\mathbf{A}}$ for all $v \in \widehat{\mathbf{A}}$. Hence the moderate elements of $\mathcal{E}$ are everywhere dense in $\mathcal{E}$, qed.

*Exercise 1.* Show that $\mathcal{E} \subset \widehat{\mathbf{A}}$.

*Exercise 2.* Let $(\mathcal{H}, U)$ be a square integrable irreducible representation such that $U(z) = \alpha(z)1$. Show that, if the rank of $A, B \in \mathcal{L}(\mathcal{H})$ is finite, then

$$\int_{G/Z} Tr\,[U(x)A]\,\overline{Tr\,[U(x)B]}dx = \lambda\; Tr\,(AB^*)\;.$$

Associate the function $\varphi_A(x) = \lambda^{-\frac{1}{2}}\; Tr[U(x)A]$ to each $A$ of finite rank. Show that $A \longmapsto \varphi_A$ extends to an isomorphism from the Hilbert space $\mathcal{L}_2(\mathcal{H})$ onto a *minimal* closed bi-invariant subspace $\mathcal{M}$ of $L^2(G/Z;\alpha)$ and that every isomorphism from $(\mathcal{H}, U)$ onto a minimal right (or left) invariant closed subspace of $L^2(G/Z;\alpha)$ maps $\mathcal{H}$ to $\mathcal{M}$. Show that if $G$ is compact one recovers the subspaces $L^2(G;\chi)$ of n° 29, (ii).

*Exercise 3.* (a) Show that a function $\varphi \in \mathcal{H} = L^2(G/Z;\alpha)$ is of positive type if and only if the operator $L(\varphi) : \mathbf{A} \longrightarrow \mathcal{H}$ is positive symmetric. Assume this condition holds in what follows. (b) Let

$$H = \int_0^{+\infty} \lambda dM(\lambda)$$

be the self-adjoint canonical extension of $S = L(\varphi)$ [theorem 35 of n° 23, (ii)], whence $S = H$ in $\mathbf{A}$. Let $\omega$ be a compact set contained in $\mathbb{R}_+^*$ and $M(\omega)$ the corresponding spectral projection [notation of n° 22, (iii)]. Show that $M(\omega)$ belongs to the von Neumann algebra $\mathcal{L}$ and that $M(\omega)L(\varphi)$ is continuous. Deduce that $\varphi' = M(\omega)\varphi$ is moderate and that $0 \ll \varphi' \ll \varphi$. (c) Assume that $\varphi$ is continuous and elementary [n° 30, (iii)]. Show that $\varphi$ is moderate, that $\varphi *_Z \varphi$ is proportional to $\varphi$ and that $L(\varphi)$ is proportional to a projection. Let $\mathcal{E} = \mathrm{Im}(L(\varphi))$. (d) Show that the representation of $G$ on $\mathcal{E}$ by right translations is equivalent to the irreducible representation associated to $\varphi$ [n° 30, (iii)]. (e) Show that all elements of $\mathcal{E}$ are moderate and that the function $(R(x)u|v)$ is in $L^2(G/Z;\alpha)$ for all $u, v \in \mathcal{E}$. Deduce the orthogonality relations for any irreducible representation having a square integrable *diagonal* coefficient.[135]

[135] Theorem 51 has a curious history. Immediately after having read Bargmann's article, I published a proof of it (CRAS, September and October 1947) which assumes $Z$ is finite using moderate functions in $L^2(G)$ and theorem 35 of n° 23; this is the one proposed in exercise 3. In 1956, Harish-Chandra (Œuvres complètes, II, pp. 90–107) generalized the result to semisimple Lie groups without any assumptions on $Z$ and supposing that a not necessarily diagonal coefficient is square integrable. He refers to my notes and says that my proof *could perhaps be modified* so as to apply to the case of an infinite $Z$. His proof uses very special

*Exercise 4* [Dixmier, *C*-algebras*, n° 13.8]. More generally show that if a function $\varphi \in L^2(G/Z;\alpha)$ is *continuous* and of positive type, there is a function $\psi \in L^2(G/Z;\alpha)$ such that $\varphi = \widetilde{\psi} *_Z \psi$. What about the case of a commutative group?

---

properties of semisimple groups, but his far harder aim was the explicit computation of the characters of the representations considered. Proposition (i) of the theorem can also be proved using the closed graph theorem; see for example A. Borel, *Représentations de groupes localement compacts* (Springer, Lecture Notes, **276**, 1972, pp. 56–57), Dieudonné XXI.4, exercises 5 to 9 and N. Wallach, *Real Reductive Groups* (Academic Press, 1988), chap. I, which do not mention the inventor of the method.

# XII – The Garden of Modular Delights or The Opium of Mathematicians

*§ 1. Infinite Series and Products in Number Theory – § 2. The series $\sum 1/\cos \pi nz$ and $\sum \exp(\pi i n^2 z)$ – § 3. The Dirichlet series $L(s;\chi)$ – § 4. Elliptic Functions – § 5. $SL_2(\mathbb{R})$ as a Locally Compact Group – § 6. Modular Functions: the Classical Theory – § 7. Fuchsian Groups – § 8. Hecke Theory – § 9. $SL_2(\mathbb{R})$ as a Lie Group*

## § 1. Infinite Series and Products in Number Theory

### 1 – The Mellin Transform of a Fourier Transform

There are more interesting relations between Mellin and Fourier transforms (Chap. VIII, n° 13) than those arising from a trivial variable change in the integration. At the very least, they apply to all functions $f$ *defined, continuous and integrable over* $\mathbb{R}$ (with respect to the usual measure $dx$) *and whose Fourier transforms are themselves integrable*, in other words to functions $f \in F^1(\mathbb{R})$. The Fourier inversion formula [Chap. VII, § 6, theorem 26 or Chap. XI, n° 27, (iv), theorem 41] applies to these functions. Setting $d^*x = dx/|x|$ as before, one can associate functions

$$(1.1) \quad \Gamma_f^0(s) = \int_{\mathbb{R}^*} f(x)|x|^s d^*x\,, \quad \Gamma_f^1(s) = \int_{\mathbb{R}^*} f(x)|x|^s \operatorname{sgn}(x) d^*x\,,$$

to such a function as well as to its Fourier transform. These are in fact the Mellin transforms of the functions $f_+(x) = f(x) + f(-x)$ and $f_-(x) = f(x) - f(-x)$ and they are equal to $2\Gamma_f(s)$ (resp. 0 ) if $f$ is even (resp. odd) and conversely. As $f$ is continuous, its convergence in the neighbourhood of 0 is guaranteed for $\mathrm{Re}(s) > 0$; and as $f$ is integrable with respect to $dx$, convergence at infinity is guaranteed if the function $|x|^{s-1}$ is bounded there, hence for $\mathrm{Re}(s) < 1$. The functions (1) are, therefore, defined and holomorphic at least on the strip

$$(1.2) \qquad 0 < \mathrm{Re}(s) < 1\,.$$

If $f$ is in the Schwartz space $\mathcal{S}(\mathbb{R})$, $f_+$ and $f_-$ are also in $\mathcal{S}$. So their Mellin transforms are meromorphic on $\mathbb{C}$, possibly with simple poles at the points

$n \leq 0$, and are rapidly decreasing at infinity on any vertical strip of finite width (Chap. VIII, § 3, n° 13, theorem 14). In fact, the functions $\Gamma_f^0(s)$ and $\Gamma_f^1(s)$ have simple poles at even negative integers in the former case and at odd ones in the latter one: consider the asymptotic expansions of $f_+$ and $f_-$ at the origin.

The basic result is that, for $f \in F^1(\mathbb{R})$, *the ratios*

$$\Gamma_{\hat{f}}^0(1-s)/\Gamma_f^0(s) \qquad \text{and} \qquad \Gamma_{\hat{f}}^1(1-s)/\Gamma_f^1(s)$$

*are independent of* $f$ and can be explicitly calculated. This can be generalized to much more complicated situations than that of the very trivial field of real numbers and arises out[1] of John Tate's Ph.D. thesis; around 1950, in Harvard, he used calculations of this type to study zeta functions of algebraic number fields in the "adelic" framework. There are also non-commutative generalizations, for example by replacing the field $\mathbb{R}$ by the ring $M_n(\mathbb{R})$ of real $n \times n$ matrices, the multiplicative group $\mathbb{R}^*$ by the group $GL_n(\mathbb{R})$ of invertible matrices and the function $x \longmapsto |x|^s$ by $g \longmapsto |\det(g)|^s$ in the simplest case.

To unify calculations, it is useful to temporarily set

$$\chi(x) = |x|^s \quad \text{or} \quad |x|^s \operatorname{sgn}(x) \tag{1.3}$$

according to cases. Hence $\chi(xy) = \chi(x)\chi(y)$, and

$$\Gamma_f(\chi) = \int f(x)\chi(x)d^*x\,, \tag{1.4}$$

where integration is over $\mathbb{R}^*$. We will also set

$$\chi^*(x) = \chi(x)^{-1}|x|\,,$$

which corresponds to the shift from $s$ to $1-s$.

Therefore the result we have in mind means that, for all $f, g \in F^1(\mathbb{R})$,

$$\Gamma_f\left(\chi^*\right)\Gamma_{\hat{g}}(\chi) = \Gamma_g\left(\chi^*\right)\Gamma_{\hat{f}}(\chi)\,. \tag{1.5}$$

$0 < \operatorname{Re}(s) < 1$ needs to be assumed, but the result holds in all of $\mathbb{C}$ by analytic extension if $f$ and $g$ are in $\mathcal{S}(\mathbb{R})$ and even under broader assumptions.

A formal calculation reduces the proof to showing that

$$\iint f(x)\hat{g}(y)\chi^*(x)\chi(y)d^*xd^*y = \iint g(x)\hat{f}(y)\chi^*(x)\chi(y)d^*xd^*y\,. \tag{*}$$

The change of variable $y \longmapsto xy$ in the integration with respect to $y$ transforms relation (*) into

$$\iint f(x)\hat{g}(xy)\chi(y)dxd^*y = \iint g(x)\hat{f}(xy)\chi(y)dxd^*y\,.$$

[1] Like the method used in § 3 for the series $L(s;\chi)$.

So showing that

$$(**) \qquad \int f(x)\hat{g}(xy)dx = \int g(x)\hat{f}(xy)dx$$

is sufficient, but we have known this relation for a long time.[2]

These operations remain to be justified. Since, for $0 < \mathrm{Re}(s) < 1$, $f(x)\chi^*(x)$ and $\hat{g}(x)\chi(x)$ are integrable with respect to $d^*x$, the product $f(x)\chi^*(x)\hat{g}(y)\,\chi(y)$ is integrable with respect to the product measure $d^*xd^*y$, and so is the function obtained by permuting $f$ and $g$. Hence it is possible to compute both sides of (*) by first integrating with respect to $y$ for given $x$, which entitles us to make the variable change $y \longmapsto xy$, then with respect to $x$: Lebesgue-Fubini theorem.

Relation (5) being proved, the ratios $\Gamma_f^0(1-s)/\Gamma_{\hat{f}}^0(s)$ and $\Gamma_f^1(1-s)/\Gamma_{\hat{f}}^1(s)$ remain to be explicitly calculated assuming $f$ even in the former case, odd in the latter one. The same then holds for $\hat{f}$ and the functions $\Gamma_f^0$, etc. reduce to $\Gamma_f$, up to a factor of 2. The easiest is to choose a particular function $g$ whose Fourier transform is known.

In the first case, we can for example choose

$$(1.6) \qquad g(x) = \left(1+x^2\right)^{-1}, \quad \text{and so } \hat{g}(y) = \pi \exp\left(-2\pi|y|\right)$$

by (8.12) of Chap. VIII for $w = 1$. Here it is necessary to assume $0 < \mathrm{Re}(s) < 1$ in order to get convergent integrals in the formulas

$$\Gamma_g^0(1-s) = \int_{\mathbb{R}^*} \frac{|x|^{1-s}}{1+x^2} d^*x = 2\int_0^{+\infty} \frac{x^{-s}}{1+x^2} dx\,,$$

$$\Gamma_{\hat{g}}^0(s) = 2\pi \int_0^{+\infty} \exp(-2\pi x)x^s d^*x\,.$$

The first integral is just function (15.2) of Chap. VIII for $1-s$. So by VIII, (15.10),

$$(1.7) \qquad \Gamma_g^0(1-s) = \pi/\cos(\pi s/2)\,.$$

The change of variable $x \longmapsto x/2\pi$ shows that on the other hand

$$\Gamma_{\hat{g}}^0(s) = (2\pi)^{1-s}\Gamma(s)$$

by definition of the $\Gamma$ function. For this choice of $g$ and for $\chi(x) = |x|^s$, $f \in F^1$, relation (5) becomes

$$(1.8) \qquad (2\pi)^{1-s}\Gamma(s)\Gamma_f^0(1-s) = \frac{\pi}{\cos(\pi s/2)}\Gamma_{\hat{f}}^0(s)\,.$$

[2] Relation (**) not being exact if integration is restricted to be over $x > 0$, this calculation shows that in the definition of Mellin transforms used here it is necessary to integrate over $\mathbb{R}^*$ and not simpy over $\mathbb{R}_+^*$ as in Chap. VIII.

If $f$ is even, $\Gamma_f^0$ and $\Gamma_{\hat{f}}^0$ can be replaced in the formula by the Mellin transforms $\Gamma_f$ and $\Gamma_{\hat{f}}$. This result assumes $0 < \mathrm{Re}(s) < 1$ but by analytic extension remains valid if the Mellin transforms of $f$ and $\hat{f}$ are meromorphic on $\mathbb{C}$, for example if $f \in \mathcal{S}(\mathbb{R})$. When one obtains such a formula, it is prudent to at least check that both sides have the same poles.

We could have also chosen the function

$$g(x) = \exp\left(-\pi x^2\right) = \hat{g}(x) \tag{1.9}$$

(Chap. V, n° 25, example 2 and Chap. VII, n° 28). In this case, the change of variable $\pi x^2 = y$ for which $d^*x = \frac{1}{2}d^*y$ and $x = (y/\pi)^{\frac{1}{2}}$ gives

$$\Gamma_g^0(x) = 2\int_0^{+\infty} e^{-\pi x^2} x^s d^*x = \int_0^{+\infty} e^{-y}\,(y/\pi)^{s/2}\,d^*y = \pi^{-s/2}\Gamma(s/2)\,.$$

Formula (5) then becomes

$$\pi^{-s/2}\Gamma(s/2)\Gamma_f^0(1-s) = \pi^{-(1-s)/2}\Gamma\left[(1-s)/2\right]\Gamma_{\hat{f}}^0(s)\,. \tag{1.10}$$

Applying it to function (6), one deduces that

$$\pi^{-s/2}\Gamma(s/2)\pi/\cos(\pi s/2) = \pi^{-(1-s)/2}\Gamma\left[(1-s)/2\right](2\pi)^{1-s}\Gamma(s)\,.$$

As relation $\Gamma(s)\Gamma(1-s) = \pi/\sin \pi s$ shows that

$$\pi/\cos(\pi s/2) = \Gamma\left[(1-s)/2\right]\Gamma\left[(1+s)/2\right]\,,$$

(10) is equivalent to

$$\Gamma(s) = \pi^{-\frac{1}{2}}2^{s-1}\Gamma(s/2)\Gamma\left[(1+s)/2\right]\,. \tag{1.11}$$

This result is itself equivalent to the *duplication formula*

$$\Gamma(2s) = \pi^{-\frac{1}{2}}2^{2s-1}\Gamma(s)\Gamma\left(s+\frac{1}{2}\right)$$

of Chap. VIII, (10.5.9) and we thereby get a more natural proof of it.

Relation (10) has the benefit of being perfectly symmetric in $s$ and $1-s$. Multiplying both sides by $\Gamma\left(\frac{1}{2}+s/2\right)$ and taking account of the duplication formula and Euler's relation, it can (exercise!) be expressed in the simpler form

$$\Gamma_f^0(1-s) = 2(2\pi)^{s-1}\sin(\pi s/2)\Gamma(1-s)\Gamma_{\hat{f}}^0(s)\,. \tag{1.10'}$$

In the case of $\Gamma_f^1(s)$, choose for example

$$g(x) = x\exp\left(-\pi x^2\right) = -g(-x)\,.$$

This is more or less the derivative of the Gaussian function, so that

$$\hat{g}(y) = -iy\exp\left(-\pi y^2\right) = -ig(y).$$

Thus

$$\Gamma_g^1(s) = \int_{-\infty}^{+\infty} \exp\left(-\pi x^2\right)|x|^{s+1}d^*x = \pi^{-(s+1)/2}\Gamma\left[(1+s)/2\right].$$

As $\Gamma_{\hat{g}}^1 = -i\Gamma_g^1$, this time

$$(1.12) \qquad -i\pi^{-(s+1)/2}\Gamma\left[(1+s)/2\right]\Gamma_f^1(1-s) = \pi^{-(2-s)/2}\Gamma(1-s/2)\Gamma_{\hat{f}}^1(s)$$

or

$$\Gamma\left[(1+s)/2\right]\Gamma_f^1(1-s) = i\pi^{s-\frac{1}{2}}\Gamma(1-s/2)\Gamma_{\hat{f}}^1(s).$$

Multiplying by $\Gamma[(1-s)/2]$, by the relation on complements and (11),

$$\Gamma_f^1(1-s) = i\pi^{s-3/2}\cos(\pi s/2)\Gamma(1-s)\pi^{\frac{1}{2}}2^s\Gamma_{\hat{f}}^1(s).$$

Thus finally the formula

$$(1.12') \qquad \Gamma_f^1(1-s) = 2i(2\pi)^{s-1}\cos(\pi s/2)\Gamma(1-s)\Gamma_{\hat{f}}^1(s)$$

similar to (10'). The poles $s = 1, 3, \ldots$ of $\Gamma(1-s)$ are canceled out by the zeros of $\cos(\pi s/2)$, so that the left hand side has poles at $s = 2, 4, \ldots$ as it should since $f$ is odd.

*Exercise.* Let $\mathcal{S}'$ be the space of tempered distributions on $\mathbb{R}$ (Chap. VII, § 6, n° 32). (a) Find a reasonable definition of meromorphic functions with values in $\mathcal{S}'$. (b) For non-negative integers $s$, define the distributions $|x|^s d^*x$ and $|x|^s \operatorname{sgn}(x)d^*x$ and compute their Fourier transforms. (c) Regarding $s \longmapsto |x|^s d^*x$ as a meromorphic function with values in $\mathcal{S}'$, calculate its residues at $0, -1, \ldots$

Finally, we could choose the function

$$f(x) = 1/\operatorname{ch}\pi x = \hat{f}(x)$$

of Chap. VIII, n° 15, but the outcome would be quite different. It is obviously in the Schwartz space. As it is even, it suffices to calculate

$$\begin{aligned}\Gamma_f(s) &= 2\int_0^{+\infty}\frac{x^s d^*x}{e^{\pi x}+e^{-\pi x}} = 2\int\frac{e^{-\pi x}x^s d^*x}{1+e^{-2\pi x}} = \\ &= 2\int e^{-\pi x}x^s d^*x\sum(-1)^n e^{-2n\pi x} = \\ &= 2\sum(-1)^n\int e^{-(2n+1)\pi x}x^s d^*x = \\ &= 2\pi^{-s}\Gamma(s)\sum_{\mathbb{N}}(-1)^n/(2n+1)^s.\end{aligned}$$

The first series expansion is justified since $e^{-\pi x} < 1$, and so is the permutation with the integral for $\text{Re}(s) > 1$ because

$$\sum \int \left| (-1)^n e^{-(2n+1)\pi x} x^s \right| d^*x < +\infty$$

(Chap. XI, n° 4, theorem 5, a result that will be constantly used in this type of calculations).

The series obtained can also be written

$$L(s, \chi) = \sum (-1)^n/(2n+1)^s = \sum \chi(n)/n^s\,, \tag{1.13}$$

where the "character mod 4" $\chi$, unrelated to functions (3), is given by

$$\chi(n) = \begin{array}{rl} 1 & \text{if } n = 1 \pmod 4 \\ 0 & \text{if } n = 0 \text{ or } 2 \pmod 4 \\ -1 & \text{if } n = 3 \pmod 4\,. \end{array} \tag{1.14}$$

It resembles the Riemann series $\zeta(s) = \sum 1/n^s$ . Relation

$$\Gamma_f(s) = 2\Gamma(s)\pi^{-s} L(s; \chi) \tag{1.15}$$

shows that the function $L(s;\chi) = \Gamma_f(s)/2\Gamma(s)\pi^s$ extends analytically to all of $\mathbb{C}$ like $\Gamma_f(s)$, which has simple poles at the points $s = -1, -2, -5, \ldots$ As $1/\Gamma(s)$ is an entire function with zeros at these values of $s$, the series $L(s)$ is the restriction of an *entire* function to the half-plane $\text{Re}(s) > 1$. And as $2\Gamma_f = \Gamma_f^0 = \Gamma_{\hat{f}}^0$ since $f$ is even and identical to its Fourier transform, (10) leads to a functional equation for $L(s)$, which can, as usual, be simplified by some gamma function calculations: the function

$$(\pi/4)^{-(1+s)/2}\Gamma\left[(s+1)/2\right] L(s;\chi) \tag{1.16}$$

is invariant under $s \longmapsto 1 - s$. This result will be generalized in § 3, n° 10.

## 2 – The Functional Equation of the $\zeta$ Function

This topic has already been addressed in Chap. VIII, n° 13, but let us present it again for the convenience of the reader. We start with the Poisson summation formula $\sum f(n) = \sum \hat{f}(n)$, which is at least valid for all $f \in \mathcal{S}(\mathbb{R})$, and apply it to the function $y \longmapsto f(x^{-1}y)$ for given $x \neq 0$. A trivial calculation then shows that

$$\sum f\left(x^{-1}n\right) = |x| \sum \hat{f}(xn)\,.$$

Hence, setting

$$\theta_f^*(x) = \sum_{n \neq 0} f(nx) = \theta_f^*(-x)\,, \tag{2.1}$$

(2.2) $$f(0) + \theta_f^*\left(x^{-1}\right) = \hat{f}(0)|x| + |x|\theta_{\hat{f}}^*(x)\,.$$

We are going to deduce a relation between Mellin transforms of $\theta_f^*$ and of $\theta_{\hat{f}}^*$, but first their behaviour at the origin and at infinity need to be studied.

**Lemma 1.** *For all $f \in \mathcal{S}(\mathbb{R})$, the function $\theta_f^*$ is $C^\infty$ on $\mathbb{R}^*$ and rapidly decreasing at infinity and so are all its derivatives.*

As, for all $N \in \mathbb{N}$, there is an upper bound $|x^N f(x)| \leq c_N$ valid in all of $\mathbb{R}$ [the left hand side is continuous on $\mathbb{R}$ and bounded for large $|x|$, hence on $\mathbb{R}$], it follows that, for $N \geq 2$,

$$\left|\theta_f^*(x)\right| \leq c_N \sum |nx|^{-N} = c'_N |x|^{-N}\,.$$

Hence it is rapidly decreasing at infinity. To show that $\theta_f^*$ is $C^\infty$, it suffices to prove that the derived series $\sum n^p f^{(p)}(nx) = x^{-p}\sum g_p(nx)$, where $g_p(x) = x^p f^{(p)}(x)$, converge uniformly in every compact subset of $\mathbb{R}^*$, which is obvious since $g_p \in \mathcal{S}(\mathbb{R})$. The previous calculation then shows that the first derived series of $\theta_f^*(x)$ and hence, by induction, the following ones, are rapidly decreasing at infinity, qed.

Formula (2) now gives the behaviour of $\theta_f^*$ in the neighbourhood of $x = 0$, namely

$$\theta_f^*(x) = \hat{f}(0)|x|^{-1} - f(0) + O\left(|x|^N\right)$$

for all $N$ by lemma 1. This can be written as an unlimited asymptotic expansion whose only non-trivial trivial terms are the first two. Hence the methods and results of Chap. VIII, n° 13 can be applied to the Mellin transform

(2.3) $$\xi_f(s) = \int_0^{+\infty} \theta_f^*(x) x^s d^*x\,, \quad \mathrm{Re}(s) > 1\,.$$

It is holomorphic on the indicated half-plane; it extends to a meromorphic function on $\mathbb{C}$ whose only singularities are simple poles at $s = 1$ and $s = 0$, the corresponding residues being

(2.4) $$\mathrm{Res}\,(\xi_f, 0) = -f(0)\,, \quad \mathrm{Res}\,(\xi_f, 1) = \hat{f}(0)\,;$$

finally $\xi_f(s)$ and all its derivatives are rapidly decreasing at infinity on any vertical strip of finite width.

Moreover, we have

(2.5) $$\xi_f(s) = \xi_{\hat{f}}(1-s)\,.$$

To see this, let us consider the entire function

(2.6) $$\xi_f^+(s) = \int_1^{+\infty} \theta_f^*(x) x^s d^*x\,.$$

For $\text{Re}(s) > 1$, by (2),

$$\begin{aligned}\xi_f(s) - \xi_f^+(s) &= \int_0^1 \theta_f^*(x) x^s d^*x = \int_1^{+\infty} \theta_f^*\left(x^{-1}\right) x^{-s} d^*x = \\ &= \int_1^{+\infty} \left[\hat{f}(0)x - f(0) + x\theta_{\hat{f}}^*(x)\right] x^{-s} d^*x = \\ &= \hat{f}(0) \int_1^{+\infty} x^{-s} dx - f(0) \int_1^{+\infty} x^{-s-1} dx + \\ &+ \int_1^{+\infty} \theta_{\hat{f}}^*(x)] x^{1-s} d^*\,, x\,,\end{aligned}$$

all the integrals being convergent. Hence finally

$$\xi_f(s) = \xi_f^+(s) + \xi_{\hat{f}}^+(1-s) - \left[\frac{f(0)}{s} + \frac{\hat{f}(0)}{1-s}\right]. \tag{2.7}$$

As $\hat{\hat{f}}(x) = f(-x)$, replacing $f$ by $\hat{f}$ and $s$ by $1-s$ clearly permutes the first two and the last two terms on the right hand side, whence (5). Relation (7) confirms that $\xi_f(s)$ extends analytically to a meromorphic function with at most simple poles at 0 and 1.

However, for $\text{Re}(s) > 1$,

$$\begin{aligned}\xi_f(s) &= \sum_{n\neq 0} \int_0^{+\infty} f(nx)|x|^s d^*x = \sum_{n\geq 1} \int_{-\infty}^{+\infty} = \\ &= \sum_{n\geq 1} n^{-s} \int f(x)|x|^s d^*x = \Gamma_f^0(s)\zeta(s)\,,\end{aligned}$$

where $\Gamma_f^0$ is defined by (1.1) and where $\zeta(s) = \sum 1/n^s$ is the Riemann function. To justify these formal calculations, showing

$$\sum \int |f(nx)|\,.\,|x^s|\, d^*x < +\infty$$

is sufficient, but this is clear for $\text{Re}(s) > 1$.

This being proved, relation (7) becomes

$$\Gamma_f^0(s)\zeta(s) = \Gamma_{\hat{f}}^0(1-s)\zeta(1-s) \tag{2.8}$$

provided both sides are defined by analytic extension since without relation (7), the left hand side would suppose that $\text{Re}(s) > 1$ and the right hand one that $\text{Re}(s) < 0$.

**Theorem 1.** *The function*

$$\xi(s) = \pi^{-s/2}\Gamma(s/2)\zeta(s)\,, \quad \mathrm{Re}(s) > 1 \tag{2.9}$$

*extends analytically to a meromorphic function on* $\mathbb{C}$ *whose only singularities are simple poles at* $s = 1$ *and* $s = 0$*, where its residues are equal to 1 and* $-1$*. It satisfies the functional equation*

$$\xi(1-s) = \xi(s)\,. \tag{2.10}$$

*The* $\zeta$ *function extends analytically to all of* $\mathbb{C}$*, except for a simple pole at* $s = 1$ *where*

$$\mathrm{Res}(\zeta, 1) = 1\,. \tag{2.11}$$

$\zeta(s) = 0$ *for* $s = -2, -4, \ldots$ .

The assertions about $\xi(s)$ follow from properties of $\xi_f$ by choosing $f(x) = \exp(-\pi x^2) = \hat{f}(x)$, whence $\Gamma_f(s) = \frac{1}{2}\pi^{-s/2}\Gamma(s/2)$. On the other hand, thanks for example to the infinite product expansion of Euler's function, we know that $1/\Gamma(s)$ is an entire function of $s$. We also know that the only singularities of $\Gamma(s)$ simple poles at points $s = 0, -1, -2, \ldots$, which are, therefore, simple zeros of $1/\Gamma(s)$. In

$$\zeta(s) = \frac{\pi^{s/2}}{\Gamma(s/2)}\xi(s)\,,$$

the simple pole of $\xi(s)$ at $s = 0$ is canceled out by the simple zero of $\Gamma(s/2)$ at this point, so that $\zeta(0) \neq 0$. The fact that $\zeta(s) = 0$ for $s = -2, -4, \ldots$ follows from the analogous property of $\Gamma(s/2)$. At $s = 1$, $\Gamma(s/2) = \pi^{\frac{1}{2}}$, and as $\xi$ has a simple pole with residue 1 at this point, in the neighbourhood of $s = 1$,

$$\zeta(s) = [1+?(s-1)+\ldots]\,[1/(s-1)+\ldots]\,.$$

Hence the existence of a simple pole and relation (11), qed.

*Exercise 1.* Show that

$$\zeta(1-2n) = -b_{2n}/2n$$

for $n = 1, 2, \ldots$, where the $b_{2n}$ are the Bernoulli numbers of Chap. VI, n° 12. [Use functional equation (10), relation (13.7)

$$2\zeta(2n) = (-1)^{n+1}(2\pi)^{2n}b_{2n}/(2n)!$$

of Chap. VI, formula

$$\Gamma\left(\frac{1}{2}-n\right)\Gamma\left(\frac{1}{2}+n\right) = (-1)^n\pi$$

and the duplication formula of the $\Gamma$ function].

*Exercise 2* (Riemann). Show that $\Gamma(s)\zeta(s)$ is the Mellin transform of the function $1/(e^t - 1)$ and that

$$2\pi i\zeta(s)/\Gamma(1-s) = \int_{-\infty}^{(0+)} \left(e^{-z} - 1\right)^{-1} z^{s-1} dz$$

for all $s \in \mathbb{C}$, where integration is over the same path as for the Hankel integral for the $\Gamma$ function (Chap. VIII, n° 10, (iii)). What is the connection with previous exercise?

As all authors of textbooks on analytic function theory remind their readers, Riemann, the first to have proved theorem 1 using the Jacobi function[3]

$$\theta(ix) = \sum \exp\left(-\pi n^2 x\right) ,$$

conjectured in 1859 that the only non-trivial zeros (i.e. other than $-2, -4, \ldots$) of his function are located on the line $\text{Re}(s) = \frac{1}{2}$. For more than a century, many eminent mathematicians have made significant efforts to prove this. Amateurs, who have written thousands of wrong proofs of Fermat's theorem, do not tackle Riemann's hypothesis – too technical, whereas Fermat's theorem can be regarded as elementary arithmetic ...–, neither do professional lacking unlimited confidence in their own ability.[4] For a long time, it was hoped standard methods of analytic function theory would work. They have led to significant results, but not to *the* result. For example, G. H. Hardy showed in 1914 that the $\zeta$ function has infinitely many zeros on the "critical line" $\text{Re}(s) = \frac{1}{2}$, which could explain his comments on the intellectual level of mathematics applied to artillery and his conviction that his mathematics will not be useful as it neither *tends to accentuate the existing inequalities in the distribution of wealth, [n]or promotes directly the destruction of human life.*

See my Postface to vol. II. In an article on mathematics in *Le Monde* on May 25th, 2000, exalting applications without any discriminations, by a curious coincidence, I found the following passage:

> Thus, one should reflect on the unfortunate comments of the English mathematician Godfrey Hardy, who considered that real mathematics would never have any military applications. If only because, through cryptography, the theory of prime numbers plays an essential role in the world of intelligence and in that of the internet.

These comments seem to me to be far more "unfortunate" than Hardy's declaration.

---

[3] Generally speaking, we set $\theta(z) = \sum \exp(\pi i n^2 z)$ for $\text{Im}(z) > 0$.

[4] In his youth, André Weil hoped to prove it before its centenary date and to publish it in 1959. He burst out laughing the day when, after the publication of his complete works, I remarked to him that if he finally found a fifteen page proof, Springer-Verlag would be obliged to add a very thin volume to its edition.

(1) Though *Le Monde* twice dedicated two pages to "the year of mathematics", the cited passage is the one and unique reference to their military use. Is this topic so irrelevant – or so dangerous? – that it is only worth five ironic lines?

(2) M. Augereau forgets to mention the word "directly" that Hardy had prudently employed, which enables him, as well as most commentators, to get away with ridiculing him. When Hardy wrote this in 1915, the chemistry of toxic gases (chlorine, phosgene, mustard gas, etc.) was starting to play a *direct* role in warfare thanks to future German Nobel laureates (Fritz Haber, James Franck, Otto Hahn, Gustav Hertz) and to dozens of less well-known chemists, who were readily imitated by their French and British colleagues.[5] Haber's *initiative* caused an enormous scandal; its author, fearing he would be judged by the Allies, temporarily fled to Switzerland in 1918. In 1919, he received the chemistry Nobel prize for his pre-1914 research on the direct synthesis of ammonia, which was doubly scandalizing since the first use of the new technique, developed by BASF in 1912, saved Germany from an armament crisis[6] that would have been fatal for the country long before 1918. To this day, public opinion continues to disapprove the use of these armaments, gas warfare having given rise after 1918 to the first "anti-Science" comments in some circles. Grass-root veterans of my youth, knowing very well that they and their German counterparts had been in the same drunken boat together, did not feel any hatred for them; but according to them, Haber should have been shot in 1918.

People who ridicule Hardy should learn about the context. Haber's chemistry is altogether a different matter from using the Arabs' trigonometry to locate enemy batteries or inventing a cryptography system allowing the military of all nations to protect their little secrets...

(3) That the new cryptography is of interest to the "world of intelligence", which I mentioned (Chap. V, § 5, note 36) without waiting for M. Augereau and the mathematicians he interviewed, is in no way obvious. Upon its invention by Diffie and Hellman in 1976, the principle of public-key cryptography was made... public,[7] after which Rivest, Samir and Adelman invented and they too made public[8] their system based on the factorization of very large integers; it was commercialized by a company, RSA Systems, which today is worth billions of dollars on Wall Street. It stands to reason that a cryptography system can really be of use to a government agency – in this case, the American National Security Agency – only if it has exclusive rights to it, particularly concerning an almost indecipherable system. Making it available for everyone was a first in the history of the subject, and from the point of view of the NSA, was extremely absurd since the Soviets, who were not "bad at Maths", could use it. All this was done despite opposition from the NSA and a 1951 law that would

[5] The main source for the topic is L. F. Haber, *The Poisonous Cloud* (Oxford UP, 1986), whose author, the son of the chemist, is a specialist of the history of the chemical industry.

[6] After the war, the new technique made it possible to introduce large quantities of chemical nitrogen fertilizers – this was in theory BASF's original purpose – and to almost multiply the world population by four in one century. Chemists have "revolutionized society" long before von Neumann and computer scientits. Vaclav Smil, *Enriching the Earth. Fritz Haber, Carl Bosch, and the Transformation of World Food Production* (MIT Press, 2001).

[7] Including in France, thanks to the translation (*Pour la Science*, October 1979) of one of Hellman's article in *Scientific American*.

[8] *Communications of the ACM*, **21**, 1978, pp. 120–126.

have allowed it to "classify" these new methods,[9] obstacles which resistance from scientific circles and the enormous power of *business* successfully overcame. As recently written by an American journalist, the outcome[10] is that the efficiency of the NSA, submerged under a flow of messages especially due to the development of the internet, has significantly diminished, the process being heightened by the development of fibre optics.

(4) If Hardy must be contradicted, it would have been better to remark that making the new methods of cryptography available to international trade and money launderers is not such as to reduce *existing inequalities in the distribution of wealth.*

Computers have made it possible to calculate more than a billion zeros of the $\zeta$ function, all on the critical line, i.e. whose real part is 0,5000... with a very large but finite number of zeros. This keeps calculators busy and were it needed, would reassure mathematicians; but it does not prove anything. Everyone has always believed in Fermat's theorem, but the problem was proving it, not checking it numerically. This proved to be tremendously more interesting than the statement itself. This will obviously also be the case of the Riemann hypothesis and it is mainly for this reason that it fascinates those who hope to be able to justify it.

Similarities between the Riemann function and series associated to algebraic curves have now convinced specialists that methods of algebraic geometry will lead to the proof, as was the case with Ramanujan's conjecture (Chap. IV, § 3, n° 20) and with Fermat's theorem. It may well be so. Others consider combining them to those of non-commutative harmonic analysis, i.e. to generalizations of Fourier transforms to non-commutative groups like $SL_2(\mathbb{R})$. In such cases, the best is to do like the English: *wait and see.*

Riemann showed that his hypothesis, whose statement is at a first glance anecdotal, was connected to the statistical distribution of prime numbers. Using prime number and logarithm tables, Gauss and Legendre had conjectured that, if $\pi(x)$ denotes the number of prime numbers $p < x$, then

$$\pi(x) \sim x/\log x \tag{2.12}$$

and what is even somewhat better,

$$\pi(x) \sim li(x) = \int_2^x dt/\log t \quad (\text{"logarithmic integral"}). \tag{2.13}$$

*Exercise.* Show that

$$li(x) \approx x/\log x - x/\log^2 x + x/\log^3 x - \ldots \qquad \text{when } x \longrightarrow +\infty .$$

[9] Among the articles I collected at the time, I find Duncan Campbell, *Whose eyes on secret data?* (New Scientist, 2 March 1978), *Cryptology: A Secret Meeting at IDA?* (Science, **200**, 14 April 1978), *Intelligence Agency Chief Seeks "Dialogue" with Academics* (Science, **202**, 27 October 1978), *Prior Restraints on Cryptography Considered* (Science, **208**, 27 June 1980), *Cryptography: A New Clash Between Academic Freedom and National Security?* (Science, **209**, 29 août 1980), *High Technology: Back in the Bottle?* (Technology Review, août/septembre 1981), Christopher Paine, *Admiral Inman's tidal wave* (The Bulletin of the Atomic Scientists, March 1982).

[10] Seymour M. Hersh, *The Intelligence Gap*, The New Yorker, 6 December 1999, by the author of a famous book on the "My Laï massacre" during the Vietnam war.

In 1896 Jacques Hadamard and Charles de la Vallée-Poussin independently obtained proofs of these results,[11] related to the non-existence of zeros of $\zeta(s)$ on the line $\mathrm{Re}(s) = 1$, and they have been much improved since. But the Riemann hypothesis is equivalent to

$$\pi(x) = li(x) + O\left(x^{\frac{1}{2}+\epsilon}\right) \qquad \text{for all } \epsilon > 0\,,$$

a far stronger result than the previous one. With their probabilistic conception of real mathematics, obviously, cryptography experts do not hesitate to make use of it.

## 3 – Weil's method for the Function $\eta(z)$

In Chap. IV, § 3 of n° 20 entitled "strange identities", we introduced the function

$$\eta(z) = e^{\pi iz/12} \prod_{n\geq 1} (1 - \exp(2\pi inz)) = q^{1/12} \prod (1 - q^{2n} \,, \tag{3.1}$$

where $q$ is set to be

$$q = \exp(\pi iz)\,, \qquad q^s = \exp(\pi is)$$

for $\mathrm{Im}(z) > 0$ and $s \in \mathbb{C}$. These are simple abbreviations, the variable still being $z$ and not $q$. As was seen then, the infinite product is absolutely convergent for $\mathrm{Im}(z) > 0$ since $|q| < 1$. Moreover, since the series $\sum \exp(2\pi inz) = \sum e(nz)$ is dominated in the half-plane $y \geq r$ by the convergent series $\sum \exp(-2\pi r)^n$ for all $r > 0$, the infinite product is normally convergent in such a half-plane. So function (1) is holomorphic on the half-plane $\mathrm{Im}(z) > 0$ (Chap. VII, § 4, n° 20, theorem 18). In Chap. IV, we stated the two functional equations satisfied by the $\eta$ function. We are now going to show how the Mellin transform simplifies their proof using a method [12] which amounts to reversing the arguments leading to the functional equation of the zeta function; this is a great exercise. A proof using the series expansion

$$\eta(z) = \exp(\pi iz/12) \sum (-1)^n \exp\left[n(n+3)\pi iz\right]$$

and the Poisson summation formula will be given later (§ 3, n° 7).

**Theorem 2 (Dedekind).** *The following relations hold:*

$$\eta(z+1) = \exp(\pi i/12)\eta(z)\,, \tag{3.2'}$$

$$\eta(-1/z) = (z/i)^{\frac{1}{2}}\eta(z)\,. \tag{3.2''}$$

The first relation is obvious. By analytic extensions, it is sufficient to prove the second one for purely imaginary $z = iy$ with $y > 0$. The $\eta$ function, being

---

[11] See for example Freitag & Busam, Chap. VII, § 6, or the last chapter in Serge Lang's *Complex Analysis* (Springer-Verlag, 4th. ed., 1999). All that was known before 1978 can be found in K. Prachar, *Primzahlverteilung* (2. Aufl., Springer-Verlag).

[12] André Weil, *Sur une formule classique* (J. Math. Soc. Japan, vol. 20, 1967, pp. 400–403). This type of method was in fact invented in 1936 by the great expert of the theory of modular functions, Erich Hecke. We could even go back to Riemann, with his proof of the functional equation of the $\zeta(s)$ function.

represented by an absolutely convergent product all of whose terms are non-zero, has no zeros (Chap. IV, § 3, n° 17, theorem 13) and for $z = iy$, all its terms are $> 0$. Hence its logarithm may be considered. It is somewhat simpler to study the function

$$f(y) = -\pi y/12 - \log \eta(iy) = -\sum \log\left(1 - \exp(-2\pi y)^n\right) =$$

$$(3.3) \qquad = \sum_{m,n\geq 1} \exp(-2\pi y)^{mn}/m\,,$$

the logarithmic series being justified since $\exp(-2\pi y) < 1$. By the general theorems (Chap. II, § 2, n° 15 and 18), the double series (3) with positive terms obviously converges unconditionally. So the terms can be arbitrarily arranged (associativity).

The Mellin transform of $f$ is

$$\begin{aligned} \Gamma_f(s) = \varphi(s) &= \int y^s d^*y \sum m^{-1} \exp(-2\pi mny) = \\ &= \sum \int m^{-1} \exp(-2\pi mny) y^s d^*y = \\ &= (2\pi)^{-s} \Gamma(s) \sum 1/m^{s+1} n^s\,, \end{aligned}$$

whence

$$(3.4) \qquad \varphi(s) = (2\pi)^{-s} \Gamma(s) \zeta(s) \zeta(s+1)\,.$$

This formal calculation is justified by the convergence of the series whose general term is

$$\int \left| m^{-1} \exp(-2\pi mny) y^s \right| d^*y = (2\pi)^{-s} \Gamma(\sigma)/m^{\sigma+1} n^\sigma\,,$$

where $\sigma = \mathrm{Re}(s)$, i.e. the product of two absolutely convergent series for $\mathrm{Re}(s) > 1$.

As the $\zeta$ function has a rather remarkable functional equation, namely

$$(3.5) \qquad \xi(1-s) = \xi(s) \quad \text{where } \xi(s) = \pi^{-s/2} \Gamma(s/2) \zeta(s)\,,$$

$\varphi(s)$ needs to be compared to the function

$$\xi(s)\xi(s+1) = \pi^{-s-\frac{1}{2}} \Gamma(s/2) \Gamma\left[(s+1)/2\right] \zeta(s) \zeta(s+1)\,,$$

However, the duplication formula (1.11) of the gamma function

$$(3.6) \qquad \Gamma(s/2) \Gamma\left((s+1)/2\right) = \pi^{\frac{1}{2}} 2^{1-s} \Gamma(s)$$

shows that

$$(3.7) \qquad \xi(s)\xi(s+1) = 2(2\pi)^{-s} \Gamma(s) \zeta(s) \zeta(s+1)\,.$$

Hence

$$(3.8) \qquad \varphi(s) = \frac{1}{2} \xi(s) \xi(s+1)\,, \quad \text{whence } \varphi(s) = \varphi(-s)\,.$$

Since $\xi(s)$ is meromorphic and its only singularities are simple poles at $s = 0$ and $s = 1$, $\varphi(s)$ extends analytically to all of $\mathbb{C}$, except for simple poles at 1 and $-1$ and a double pole at $s = 0$ with residues

$$\text{(3.9)} \qquad \mathrm{Res}(\varphi, 1) = \frac{1}{2}\xi(2)\,\mathrm{Res}(\xi, 1) = \frac{1}{2}\xi(2) = \frac{1}{2}\pi^{-1}\Gamma(1)\zeta(2) = \pi/12$$

since $\zeta(2) = \pi^2/6$, and so, by symmetry,

$$\text{(3.9')} \qquad \mathrm{Res}(\varphi, -1) = -\pi/12\,.$$

In the neighbourhood of $s = 0$, there are series expansions

$$\xi(s) = -1/s + c_0 + c_1 s + \ldots\,,$$
$$\xi(s+1) = \xi(-s) = 1/s + c_0 - c_1 s + \ldots\,,$$

whence

$$\text{(3.9'')} \qquad \varphi(s) = -1/2s^2 + a_0 + a_2 s^2 + \ldots$$

whose coefficients are of no relevance for the rest of this book.

Let us now show that the Mellin inversion formula

$$\text{(3.10)} \qquad 2\pi i f(y) = \int_{\mathrm{Re}(s)=\sigma>1} \varphi(s) y^{-s} dy$$

can be applied to $f$. Indeed, the function $t \longmapsto \varphi(\sigma + it)$ is the Fourier transform of the integrable function $u \longmapsto f[\exp(2\pi u)]\exp(2\pi\sigma u)$. But as $\xi(s)$ is rapidly decreasing at infinity on every vertical, so is $\varphi(s) = \frac{1}{2}\xi(s)\xi(s+1)$. The Fourier inversion formula then proves (10) as in the proof of theorem 14 of Chap. VIII, n° 13.

To exploit relation $\varphi(-s) = \varphi(s)$, (10) needs to be compared with the similar integral along the vertical $\mathrm{Re}(s) = -\sigma$, i.e. the residue formula needs to be applied to the contour bounded by verticals $\sigma$ and $-\sigma$ and horizontals $T$ and $-T$. At the limit, these do not contribute to the integral since $|y^{-s}| = y^{-\sigma}$ is independent of $T$, whereas $\varphi(\sigma \pm iT)$ is rapidly decreasing. On the other hand, the integral over $\mathrm{Re}(s) = -\sigma$ reduces to an integral over $\mathrm{Re}(s) = \sigma$ by $s \longmapsto -s$. Setting

$$\psi(s) = \varphi(s)y^{-s} = \varphi(-s)y^{-s}\,,$$

therefore, gives

$$\begin{aligned} 2\pi i \sum \mathrm{Res}(\psi) &= \int_{\mathrm{Re}(s)=\sigma} [\psi(s) - \psi(-s)]\, ds = \\ &= \int_{\mathrm{Re}(s)=\sigma} \left[y^{-s}\varphi(s) - y^{s}\varphi(s)\right] ds = \\ \text{(3.11)} \qquad &= 2\pi i\left[f(y) - f(1/y)\right]\,, \end{aligned}$$

whence

$$f(1/y) = f(y) - \sum \mathrm{Res}(\psi)\,.$$

The residues of $\psi(s) = \varphi(s)y^{-s}$ at poles 1, 0 and $-1$ remain to be calculated. As 1 and $-1$ are simple poles of $\varphi$, by (9) and (9')

$$\mathrm{Res}(\psi, 1) = \pi/12y\,, \qquad \mathrm{Res}(\psi, -1) = -\pi y/12\,.$$

In the neighbourhood of 0, by (9")

$$\varphi(s) = -1/2s^2 + a_0 + \ldots ,$$
$$y^{-s} = \exp(-s \log y) = 1 - s \log y + \ldots ,$$

whence

$$\mathrm{Res}(\psi, 0) = \frac{1}{2} \log y .$$

So

$$f(1/y) = f(y) - \pi/12y + \pi y/12 - \frac{1}{2} \log y ,$$

i.e.

$$f(1/y) + \pi y^{-1}/12 = f(y) + \pi y/12 - \frac{1}{2} \log y . \tag{3.12}$$

Since, by (1) and (3), $\eta(iy) = \exp[-f(y) - \pi y/12]$, it follows that

$$\eta(i/y) = y^{\frac{1}{2}} \eta(iy) .$$

Thus, by analytic extension, we get the strange identity

$$\eta(-1/z) = (z/i)^{\frac{1}{2}} \eta(z)$$

that had been prematurely announced at the end of Chap. IV, § 3 in the hope of inspiring new Ramanujans.

*Exercise.* Let $w$ be a complex number such that $|w| < 1$, $w \neq 1$. Set

$$f(y) = -\log\left(1 - we^{-y}\right) = \sum w^n e^{-ny}/n .$$

Show that $f$ satisfies the assumptions of Chap. VIII, n° 13, theorem 14 and deduce that the function

$$\Gamma(s) \sum w^n/n^{s+1}$$

extends analytically to $\mathbb{C}$.

Several infinite products similar to the Dedekind function are encountered in the theory of elliptic or modular functions. Like the latter, they all have simple functional equations that can be proved in the same way.

*Exercise 1: functional equation of* $F(z) = q^{-1/24} \prod (1 + q^{2n-1})$. The Mellin transform of

$$f(y) = \pi y/24 - \log F(iy) = \sum (-1)^m e^{-(2n-1)m\pi y}/m \tag{3.13}$$

is given by

$$\begin{aligned} \varphi(s)/\pi^{-s}\Gamma(s) &= \sum (-1)^m/(2n-1)^s m^{s+1} = \\ &= \sum (-1)^m/m^{s+1} \sum 1/(2n-1)^s . \end{aligned} \tag{3.14}$$

Using (7) and

$$\sum 1/(2n-1)^s = \zeta(s) - \sum 1/(2n)^s = \left(1 - 2^{-s}\right) \zeta(s) ,$$

one deduces that

$$\begin{aligned}\varphi(s) &= \pi^{-s}\Gamma(s)\zeta(s)\zeta(s+1)\left(1-2^{-s}\right)\left(2^{-s}-1\right) = \\ &= \frac{1}{2}\xi(s)\xi(s+1)\left(1-2^{-s}\right)\left(1-2^{s}\right) = \varphi(-s)\,.\end{aligned} \tag{3.15}$$

Like $\xi$, the function $\varphi$ is rapidly decreasing on every vertical strip since the functions $2^s$ and $2^{-s}$ are bounded there. The function

$$\psi(s) = \varphi(s)y^{-s} = \frac{1}{2}\xi(s)\xi(s+1)\left(1-2^{s}\right)\left(1-2^{-s}\right)y^{-s} \tag{3.16}$$

being integrated is, like $\varphi$, rapidly decreasing on all vertical strips since the factor $y^{-s}$ is at most of polynomial growth. Hence the Mellin inversion formula applies, and calculating as in (11) gives

$$f(1/y) = f(y) - \sum \mathrm{Res}(\psi)\,,$$

the residues being with respect to poles contained between verticals $\sigma > 1$ and $-\sigma$. But (24) shows that, like the function $\xi(s)\xi(s+1)$, $\psi$ has at most poles at points 1, 0 and $-1$. The corresponding residues remain to be calculated.

As the poles of $\frac{1}{2}\xi(s)\xi(s+1)$ at 1 and $-1$ are simple, it suffices to multiply the residues by the values of the function $(1-2^s)(1-2^{-s})y^{-s}$ at 1 and $-1$, i.e. by $-1/2y$ at $s=1$ and $-y/2$ at $s=-1$, whence

$$\mathrm{Res}(\psi,1) = -\pi/24y\,, \qquad \mathrm{Res}(\psi,-1) = \pi y/24\,. \tag{3.17}$$

The function $\frac{1}{2}\xi(s)\xi(s+1)$ has a double pole at $s=0$. It is canceled out by the double zero of the factor $(1-2^s)(1-2^{-s})$, thus the corresponding residue is trivial.

So finally,

$$f(1/y) = f(y) - \pi y/24 + \pi/24y\,. \tag{3.18}$$

One, therefore, concludes that *the function*

$$F(z) = q^{-1/24}\prod\left(1+q^{2n-1}\right) \quad \text{where } q = \exp(\pi i z)$$

*satisfies the functional equation*

$$F(-1/z) = F(z) \tag{3.19}$$

and trivially,

$$F(z+2) = q^{-1/12}F(z)\,.$$

*Exercise 2*: functional equation of $G(z) = q^{-1/24}\prod(1-q^{2n-1})$. The Mellin transform of

$$f(y) = \pi y/24 - \log G(iy) = \sum e^{-(2n-1)m\pi y}/m$$

is

$$\begin{aligned}\varphi(s) &= \pi^{-s}\Gamma(s)\sum 1/m^{s+1}(2n-1)^s = \left(2^s-1\right)(2\pi)^{-s}\,\Gamma(s)\zeta(s)\zeta(s+1) = \\ &= \frac{1}{2}\left(2^s-1\right)\xi(s)\xi(s+1)\,,\end{aligned}$$

from which one concludes that

$$\varphi(-s) = -2^{-s}\varphi(s)\,.$$

The residues of the function $\psi(s) = y^{-s}\varphi(s)$ are $\pi/12y$ at $s=1$, $-\frac{1}{2}\log 2$ at $s=0$ [since $2^s = \exp(s\log 2) = 1 + s\log 2 + \ldots$] and $\pi y/24$ at $s=-1$. Here method (11) gives

$$f(y) + f(2/y) = \pi/12y + \pi y/24 - \frac{1}{2}\log 2\,,$$

and so, replacing $y$ with $2y$,

$$G(2z)G(-1/z) = \sqrt{2} \tag{3.20}$$

by analytic extension. We could have deduced this result from the functional equation of the Dedekind function by observing that

$$G(2z) = \eta(z)/\eta(2z)\,.$$

*Exercise 3*: functional equation of $H(z) = q^{1/24}\prod(1-(-q)^n)$. We start from

$$f(y) = -\pi y/24 - \log H(iy) = \sum(-1)^{mn}\exp(-\pi mny)/m \quad (y>0)\,.$$

A few calculations then show that the Mellin transform of this function is

$$\varphi(s) = \pi^{-s}\Gamma(s)\sum(-1)^{mn}/m^{s+1}n^s = \frac{1}{2}\xi(s)\xi(s+1)a(s)a(-s)\,,$$

where $a(s) = 2^{s/2} + 1 - 2^{-s/2}$. So it is invariant under $s \longmapsto -s$. Calculating the residues and using the inverse Mellin transform like in previous exercises lead to the functional equation

$$H(-1/z) = (z/i)^{\frac{1}{2}}H(z)\,.$$

Observing that

$$\eta\,(1/2 + z/2) = \exp(\pi i/24)\,H(z)\,,$$

the relation obtained means that *the* $\eta(z)$ *function satisfies*

$$\eta\,(1/2 - 1/2z) = (z/i)^{\frac{1}{2}}\eta\,(1/2 + z/2)\,. \tag{3.21}$$

With a few lines of calculations, one can deduce this relation from the functional equations of $\eta(z)$:

$$\begin{aligned}\eta\,((z-1)/2z) &= \exp(\pi i/12)\eta\,(-(z+1)/2z) = \\ &= \exp(\pi i/12)\,[2z/i(z+1)]^{\frac{1}{2}}\,\eta\,[2z/(z+1)] = \\ &= \exp(\pi i/12)\,[2z/i(z+1)]^{\frac{1}{2}}\,\eta\,[2-2/(z+1)] = \\ &= \exp(\pi i/4)\,[2z/i(z+1)]^{\frac{1}{2}}\,\eta\,[-2/(z+1)] = \\ &= \exp(\pi i/4)\,[2z/i(z+1)]^{\frac{1}{2}}\,[(z+1)/2i]^{\frac{1}{2}}\,\eta\,[(z+1)/2]\,.\end{aligned}$$

This proves (21) provided

$$\exp(\pi i/4)\,[2z/i(z+1)]^{\frac{1}{2}}\,[(z+1)/2i]^{\frac{1}{2}} = (z/i)^{\frac{1}{2}}$$

holds. Despite appearances, this is not obvious[13] since, by (14.12) of Chap. VIII,

$$(*) \qquad (w_1 w_2)^{\frac{1}{2}} = w_1^{\frac{1}{2}} w_2^{\frac{1}{2}} \iff |\operatorname{Arg}(w_1) + \operatorname{Arg}(w_2)| < \pi\,,$$

which excludes formal calculations available for integer exponents. However, like $z$, the points $(z+1)/2$ and $2z/(z+1)$ are in the upper half-plane, so that their arguments are between 0 and $\pi$. The points

$$w_1 = 2z/i(z+1) = -2iz/(z+1) \quad \text{and} \quad w_2 = (z+1)/2i = -i(z+1)/2$$

occurring in (15) are, therefore, in the half-plane $\operatorname{Re}(w) > 0$. So $|\operatorname{Arg}(w_1)| < \pi/2$ and $|\operatorname{Arg}(w_2)| < \pi/2$. Thus, condition (*) holds. Hence

$$\left[2z/i(z+1)\right]^{\frac{1}{2}} \left[(z+1)/2i\right]^{\frac{1}{2}} = (-z)^{\frac{1}{2}}$$

as expected. But as $\operatorname{Arg}(-z) \in \,]-\pi, 0[\,$, $\operatorname{Arg}(z/i) = \pi/2 + \operatorname{Arg}(-z)$ has to be chosen in such a way that it stays in the interval $]-\pi, +\pi[$, whence $(z/i)^{\frac{1}{2}} = (-z)^{\frac{1}{2}} \exp(\pi i/4)$, qed.

*Exercise 4*: functional equation of $P(z) = q^{1/24} \prod(1+q^{2n})$. This time we set

$$f(y) = \log \prod \left(1 + e^{-2\pi n y}\right) = \sum (-1)^{m-1} \exp(-2\pi m n y)/m\,,$$

and so, for the Mellin transform

$$\varphi(s) = (2\pi)^{-s} \Gamma(s) \sum (-1)^{m-1}/m^{s+1} n^s = \frac{1}{2}\left(1 - 2^{-s}\right) \xi(s)\xi(s+1)$$

and $\varphi(-s) = -2^s \varphi(s)$. The residues of $\varphi(s) y^{-s}$ are $\pi/48y$ at $s = 1$, 0 at $s = 0$ and $\pi y/24$ at $s = -1$. The Mellin inversion then shows that

$$f(y) + f(1/2y) = \pi/48y + \pi y/24\,,$$

whence

$$(3.22) \qquad P(z)P(-1/2z) = 1\,.$$

---

[13] More generally, if $a, b, c, d$ are integers such that $ad - bc = 1$, setting $\gamma(z) = (az+b)/(cz+d)$, we get

$$\eta\left[\gamma(z)\right] = \epsilon(\gamma)(cz+d)^{\frac{1}{2}} \eta(z)\,,$$

but the explicit calculation of the constant $\epsilon(\gamma)$ encounters arithmetical problems which, however "elementary", were not solved before 1931 (Hans Rademacher) or, with a simpler formula, 1954 (Hans Petersson), despite the fact that the $\eta$ function was known for a long time. See for example M. I. Knopp, *Modular Functions in Analytic Number Theory* (Markham, 1970). A similar problem arises for the $\theta(z)$ function. It was solved by Erich Hecke in 1944; see for example Neal Koblitz, *Introduction to Elliptic Curves and Modular Functions* (Springer-Verlag, 2d ed., 1993), pp. 148–152.

To conclude this n°, let us note the similarity between

$$\eta(z) = q^{1/12} \prod (1 - q^{2n})$$

and the function

$$\prod (1 - q^{2n})^{-1} = \sum p(n) q^{2n} = q^{1/12} \eta(z)^{-1}$$

of partition theory (Chap. IV, § 3, n° 20; the variable $q$ of Chap. IV has been replaced by $q^2$). A detailed study of the behaviour of $\eta(z)$ in the neighbourhood of the real axis leads to precise information about the asymptotic behaviour of $p(n)$ for large $n$, for example the asymptotic formula

$$p(n) \sim \frac{1}{4(3n)^{\frac{1}{2}}} \exp\left[(2n/3)^{\frac{1}{2}} \pi\right],$$

due to Hardy and Ramanujan, then an expansion of $p(n)$ into a convergent series due to Hans Rademacher, the author of what still is one of the most beautiful books on analytic number theory.[14]

[14] Dieudonné gives, obviously without proofs, a very very good idea of the subject in the *Dictionnaire des Mathématiques, algèbre, analyse, géométrie* (Paris, Encyclopaedia Universalis and Albin Michel, 1997) and, probably wising to give educated populations an idea of the power of mathematics, writes the 27 digits of $p(721)$. Let us also draw attention to the excellent article by Jean-Luc Verley on analytic functions in the same volume. Except for some details, the articles of the *Dictionnaire* can be found in the many volumes of the *Encyclopaedia Universalis.*

## § 2. The series $\sum 1/\cos \pi nz$ and $\sum \exp\left(\pi i n^2 z\right)$

### 4 – The series $\sum 1/\cos \pi nz$

We showed in Chap. VIII, n° 15 that the function $1/\operatorname{ch} \pi x$ is equal to its Fourier transform. A change of variable then shows that, for all $t > 0$, the Fourier transform of $x \longmapsto 1/\operatorname{ch}(\pi x/t)$ is $x \longmapsto t/\operatorname{ch}(\pi xt)$. This function being in the Schwartz space $\mathcal{S}(\mathbb{R})$, the Poisson summation formula can be applied to it; thus

$$\varphi(1/t) = t\varphi(t) \quad \text{where } \varphi(t) = \sum_{\mathbb{Z}} 1/\operatorname{ch}(\pi nt) \tag{4.1'}$$

and more generally,

$$\sum 1/\operatorname{ch}\left[\pi(u+n)t^{-1}\right] = t\sum \mathbf{e}(nu)/\operatorname{ch}(\pi nt) \tag{4.1''}$$

for all $u \in \mathbb{R}$.

Let us now consider the analogous series

$$f(z) = \sum 1/\cos(\pi nz)\,. \tag{4.2}$$

We begin by showing that it converges normally in all half-planes of the form $\operatorname{Im}(z) \geq r > 0$. Indeed, in this half-plane

$$\begin{aligned} 2\left|\cos(\pi nz)\right| &= \left|e^{\pi(ny-inx)} + e^{\pi(-ny+inx)}\right| \geq \\ &\geq \left|e^{\pi|n|r} - e^{-\pi|n|r}\right| \geq e^{\pi|n|r} - 1\,. \end{aligned}$$

The convergent series $\sum 1/(e^{\pi|n|r} - 1)$ thus dominates series (2) in the half-plane considered.

We now now that $f$ satisfies two simple functional equations. First,

$$f(z+2) = f(z)\,. \tag{4.3}$$

On the other hand, function (2) is holomorphic on the half-plane $\operatorname{Im}(z) > 0$ and reduces to $\varphi(t)$ for $z = it$. Relation (1') meaning that

$$f(-1/z) = (z/i).f(z) \tag{4.4'}$$

for imaginary $z$ and both sides being analytic on the half-plane $\operatorname{Im}(z) > 0$, (4') holds on the latter.

Similarly, relation (1") shows that, for $u \in \mathbb{R}$,

$$\sum 1/\cos\left[-\pi(u+n)/z\right] = (z/i)\sum \mathbf{e}(nu)/\cos(\pi nz)\,. \tag{4.4''}$$

Relations (3) and (4') resemble those proved in Chap. VII, n° 28 for the Jacobi function

$$\theta(z) = \sum \exp\left(\pi i n^2 z\right) = 1 + 2\left(q + q^4 + q^9 + q^{16} + \ldots\right),$$

where

$$q = \exp(\pi iz)\,.$$

Here too, the series converges for $\mathrm{Im}(z) > 0$. Applying the Poisson summation formula and the fact that the function $x \longmapsto \exp(-\pi x^2)$ equals its Fourier transform, we have already shown that

$$\theta(-1/z) = (z/i)^{\frac{1}{2}}\theta(z)\,,$$

which is the reason why Riemann used it to prove the functional equation of his $\zeta(s)$ series. Obviously,

$$\theta(z+2) = \theta(z)$$

holds as well. Thus the function $\theta(z)^2$ satisfies (3) and (4'). In fact,

$$f(z) = \theta(z)^2\,. \tag{4.5}$$

The proof of this equality is going to take us directly to the theory of modular functions in the next n°.

Like the theta series, $f$ has an easily computable Fourier series expansion (Chap. VII, § 4, n° 17). Indeed

$$f(z) = 1 + 4\sum_{m\geq 1}\frac{q^m}{1+q^{2m}} = 1 + 4\sum_{m\geq 1} q^m \sum_{k\geq 0}(-1)^k q^{2km} =$$

$$= 1 + 4\sum_{m\geq 1, k\geq 0}(-1)^k q^{(2k+1)m} = 1 + 4\sum_{n\geq 1} a_n q^n \tag{4.6}$$

with a series which theorem 7 of of Chap. II, n° 15 tells us converges unconditionally. The coefficient $a_n$ is the sum of the numbers $(-1)^k$ for all ordered pairs $(k, m)$ such that $k \geq 0$, $(2k+1)m = n$, But the existence of an integer $m$ such that $(2k+1)m = n$ means that $2k+1$ is an odd divisor of $n$. We infer from this that

$$\begin{aligned} a_n = {} & \text{number of divisors of } n \text{ of the form } 4k+1 - \\ & - \text{number of divisors of } n \text{ of the form } 4k+3\,, \end{aligned} \tag{4.7}$$

the number $1 (k = 0)$, and possible also $n$, always occurring among the enumerated divisors. Easy calculations then show that

$$f(z) = 1 + 4\left(q + q^2 + q^4 + 2q^5 + q^8 + q^9 + 2q^{10} + 2q^{13} + \ldots\right)\,.$$

For $n = 12$ for example, odd divisors are $1 \equiv 1(\bmod 4)$ and $3 \equiv 3(\bmod 4)$, and so $a_n = 0$. For $n = 25$, these are 1, 5 and 25, all $\equiv 1(\bmod 4)$, and so $a_{25} = 3$. For $n = 50$, the odd divisors are the same, thus again $a_{50} = 3$.

To make conjecture (5) plausible, let us calculate the expansion of $\theta(z)^2$.

$$\theta(z) = \sum_{\mathbb{Z}} q^{x^2} = 1 + 2\sum_{x\geq 1} q^{x^2}$$

and so

$$\theta(z)^2 = 1 + 4\sum_{x\geq 1} q^{x^2} + 4\sum_{x,y\geq 1} q^{x^2+y^2} = 1 + 4\sum_{n\geq 1} b_n q^n\,.$$

If $n$ is not a square, $b_n$ is clearly the number of ordered pairs $(x, y)$ with integers $x, y \geq 1$ such that $x^2 + y^2 = n$; if $n$ is a square, 1 needs to be added to this number to obtain $b_n$, which amounts to admitting the ordered pair $(x, 0)$ if $n = x^2$. Hence, in all cases, $b_n$ is the number of ordered pairs $(x, y)$ such that

(4.8) $$x^2 + y^2 = n\,, \quad x \geq 1\,, \quad y \geq 0\,.$$

Again easy calculations then show that the beginning of the expansion is the same as that of $f$. For $n = 12$, (8) has no solution, whence $b_{12} = 0$. For $n = 25 = 5^2 = 3^2 + 4^2 = 4^2 + 3^2$, we find $b_{25} = 3$. For $n = 50$,

$$50 = 1^2 + 7^2 = 7^2 + 1^2 = 5^2 + 5^2\,,$$

whence $b_{50} = 3$. The reader having access to a computer will be able to check further if, like the new cryptographers, he considers it important to replace *obsolete* deductive mathematics of the past by new inductive or experimental mathematics.[15] Hence equality $f(z) = \theta(z)^2$.

Admitting it without proof, we get a classical result of arithmetic: *the number of solutions of* (8) *is given by formula* (7).

There are elementary proofs of this result, for example in Hardy & Wright. Conversely, (5) can be deduced from the classical result.[16]

## 5 – The Identity $\sum 1/\cos \pi nz = \theta(z)^2$

As for proving directly that $f(z) = \theta(z)^2$, as stated above, it is a problem in the theory of modular functions. In fact, apart from the infinite product expansion of the $\theta$ that has not yet been proved, everything can be proved "without any knowledge" of this theory, or almost without. Some of its features will be presented in § 5.

(i) *The fundamental domain of* $\Gamma(\theta)$. First, the function

$$g(z) = f(z)/\theta(z)^2$$

is clearly invariant under the maps

(5.1) $$T : z \longmapsto z + 2\,, \qquad S : z \longmapsto -1/z$$

from the half plane

$$P = \{\mathrm{Im}(z) > 0\}$$

to itself, hence also under every map $T^p S^q T^r \ldots$, where $p, q, r, \ldots$ are arbitrary rational integers (in fact it suffices to give the exponents of $S$ the values 0 and 1 since $S^2 = id$). These form a group $\Gamma(\theta)$ of conformal representations

$$\gamma(z) = (az + b)/(cz + d)$$

of $P$ on itself, with coefficients $a, b, c, d \in \mathbb{Z}$ such that $ad - bc = 1$, like $S$ and $T$. Multiplication of transformations of this type is similar to that of the corresponding

[15] See Jacques Stern, *La science du secret* (Paris, Odile Jacob, 1998), Chapter VI.

[16] This is what Freitag and Busam do, in exercises of their chapter VII, § 3, *Funktionentheorie*, albeit they write $f(z)$ in the less spectacular form that can be obtained by permuting the summations with respect to $k$ and $m$ in the first double series of equation (6), which conceals the $\cos \pi nz$. Their method consists in comparing the Mellin transforms of $f(z)$ and $\theta(z)^2$, which are quite well-known Dirichlet series. We will return to this later. The method used here, directly taken from Freitag and Busam (chap. VII, 1.8), is standard in the theory of modular functions. I did not find anywhere calculations leading directly to the functional equation of $f(z)$ and to the behaviour of $f(1 - 1/z)$ at infinity, but then the literature is considerable.

matrices[17] $(a\ b|c\ d)$. The latter form the *modular group*. Transformations $\gamma \in \Gamma(\theta)$ may be showed to be characterized by the additional condition

$$a+b+c+d \equiv 0 \pmod 2,$$

but it is less obvious and useless. When studying elliptic functions in $\mathbb{C}$ with fundamental periods $\omega_1$ and $\omega_2$ (Chap. II, § 3, n° 23), it is not necessary to consider them on all of $\mathbb{C}$. It suffices to consider them on the parallelogram $P$ generated by these two vectors since any $z \in \mathbb{C}$ can be transformed into $P$ by adding periods. We thus need to imitate this procedure by replacing $\mathbb{C}$ and the group of periods by the half-plane $P$ and the group $\Gamma(\theta)$, i.e. to construct a subset $\mathcal{F}$ of $P$, of the simplest form possible, such that, for any $z \in P$, there exists a transformation $\gamma \in \Gamma(\theta)$ that maps $z$ to a point of $\mathcal{F}$. This amounts to requiring that

$$P = \bigcup_{\gamma \in \Gamma(\theta)} \gamma(\mathcal{F}). \tag{5.2}$$

The ideal would be to make sure that the sets $\gamma(\mathcal{F})$ are pairwise disjoint, but this is impossible since there may exist $\gamma \in \Gamma(\theta)$ such that $\gamma(z) = z$, for example $S$ which leaves the point $i$ fixed. It is, however, possible to choose $\mathcal{F}$ in such a way that (i) $\mathcal{F}$ is the closure of some open set, (ii) the intersection of two images $\gamma'(\mathcal{F})$ and $\gamma''(\mathcal{F})$ only contains subsets of their boundaries. This is what happens for elliptic functions. The next lemma describes such a *fundamental domain* of the group $\Gamma(\theta)$ [Chap. XI, n° 15, (vii)].

**Lemma**. *The set $\mathcal{F}$ defined by inequalities*

$$0 \leq \mathrm{Re}(z) \leq 2, \qquad |z| \geq 1, \qquad |z-2| \geq 1. \tag{5.3}$$

*satisfies (2). So does the set $F$ defined by*

$$|\mathrm{Re}(z)| \leq 1, \qquad |z| \geq 1. \tag{5.3'}$$

To see this, we first note that if $\gamma = (a\ b|c\ d)$ is a real matrix with determinant $ad - bc = 1$, then

$$\mathrm{Im}\left[\gamma(z)\right] = \frac{1}{2i}\left(\frac{az+b}{cz+d} - \frac{a\bar{z}+b}{c\bar{z}+d}\right) = \frac{\mathrm{Im}(z)}{|cz+d|^2}. \tag{5.4}$$

This formula shows that $\gamma$ maps $P$ to $P$, and since so does the inverse $\gamma^{-1} = (d\ -b|-c\ a)$, any matrix of this form defines a conformal representation of the half-plane $P$ onto itself.

This being so, for given $z \in P$, let us consider the imaginary parts of the points $\gamma(z)$ for $\gamma \in \Gamma(\theta)$. As both $c$ and $d$ are non-trivial integers, the numbers $cz + d$ appearing in (4) belong to the lattice generated by 1 and $z$ in $\mathbb{C}$. Hence $c$ and $d$ can be so chosen that $|cz + d|$ is minimal, hence that $\mathrm{Im}[\gamma(z)]$ is maximal. As the translation $T : z \longmapsto z + 2$ leaves $\mathrm{Im}(z)$ invariant, multiplying $\gamma$ with a power of $T$, one may even suppose that $|\,\mathrm{Re}\,\gamma(z)| \leq 1$. Since $S \in \Gamma(\theta)$, the image $z' = \gamma(z)$ of $z$ obtained thereby then satisfies

$$|\mathrm{Re}\,(z')| \leq 1, \qquad \mathrm{Im}\,(-1/z') \leq \mathrm{Im}\,(z').$$

[17] In what follows, the notation $(a\ b|c\ d)$ will sometimes denote the $2 \times 2$ matrix whose first row is $a\ b$ and the second one $c\ d$.

But $\mathrm{Im}(-1/z') = \mathrm{Im}(z')/|z'|^2$. Therefore, both

$$|\mathrm{Re}\,(z')| \le 1\,, \qquad |z'| \ge 1$$

hold, whence (3').

We proceed in the same way to reduce to inequalities (3). For given $z \in P$, choose $\gamma$ such that $\mathrm{Im}(\gamma z)$ is maximal. So as above, $|cz' + d| \ge 1$ for any matrix in the group. Using $z \longmapsto z + 2$, transform $z'$ into the strip bounded by the verticals of 0 and 2. Relation $\mathrm{Im}(-1/z') \le \mathrm{Im}(z')$ shows that $|z'| \ge 1$, inequality $|z' - 2| \ge 1$ being obtained in a similar way using $TS : z \longmapsto (z-1)/(z-2)$, qed.

Pursuing these calculations further, the pairwise intersection of the images $\gamma(\mathcal{F})$ could be shown to only contain boundary points, but this is not needed for what follows.

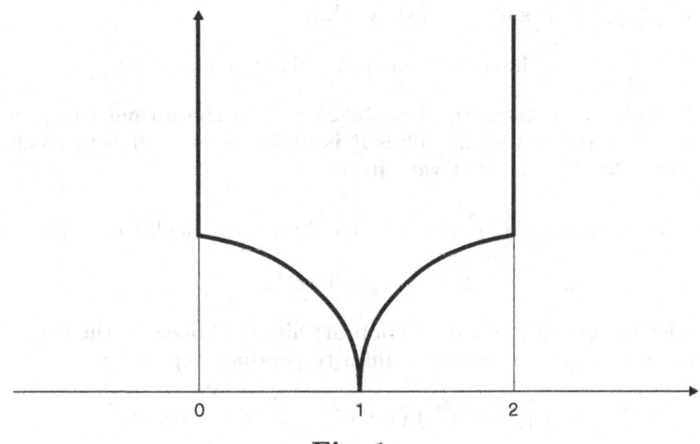


**Fig. 1.**

The above figure represents set (3). Observing that any homography $z \longmapsto (az+b)/(cz+d)$ with real coefficients and determinant 1 transforms a circle centered on the real axis or possibly at the limit of a vertical into a figure of the same type, its images under $\gamma \in \Gamma(\theta)$ are easily obtained. The images of (3') form a tessellation of the half-plane $P$ with closed triangles bounded by circular arcs centered on the real axis, or with verticals, and accumulating on the real axis. The reader can practise drawing the tessellation obtained from the set $\mathcal{F}$ defined by (3) or (3').

(ii) *A general method.* Let us consider a function $g(z)$ in $P$ which is holomorphic everywhere and invariant under the group $\Gamma(\theta)$. We intend to search for conditions enabling us to affirm that $g(z) = 1$. The analogous problem in the theory of elliptic functions is easy to solve (Chap. VII, § 4, n° 18): if an elliptic function $g$ is holomorphic everywhere, the continuous function $|g(z)|$ has a maximum in a period parallelogram since it is compact, hence by periodicity in $\mathbb{C}$. The result follows.

Likewise, in the case at hand, it would suffice to show that $|g(z)|$ has a maximum in the closed set $\mathcal{F}$. The latter is not compact but can be made to be so by removing all $z$ such that either $\mathrm{Im}(z) > c$ for given $c > 0$, or $|z - 1| < \epsilon$ for given $\epsilon > 0$. So, to show that $g = 1$, it suffices to show that $g(z)$ tends to 1 as $z \in \mathcal{F}$ tends either to infinity or 1. Indeed, if this is the case, then, for all $r > 0$, inequality $|g(z) - 1| \ge r$ gets rid of all points $z \in \mathcal{F}$ too near 1 or infinity, hence defines a compact set

$K(r) \subset \mathcal{F}$. If $g-1$ were not identically zero, then $K(r)$ would be non-empty for sufficiently small $r$ and there would be some $a \in K(r)$ such that

$$|g(z)-1| \leq |g(a)-1|$$

for all $z \in K(r)$, hence for all $z \in \mathcal{F}$, thus by "periodicity" for all $z \in P$, a contradiction.

So it suffices to analyze the behaviour of $g(z)$ either for large $\mathrm{Im}(z)$ or for $z \in \mathcal{F}$ near 1. In the second case, we set

$$z = 1 - 1/\zeta, \quad \text{whence } \zeta = 1/(1-z).$$

The half-circles $|z-1| = 1$ and $|z-2| = 1$ bounding $\mathcal{F}$ are transformed into the verticals $\mathrm{Re}(z) = \frac{1}{2}$ and $\mathrm{Re}(z) = -\frac{1}{2}$, so that making $z$ tend to 1 on $\mathcal{F}$ reduces to making $\zeta$ tend to infinity on the strip bounded by these verticals. Thus it follows that to prove $g(z) = 1$, it suffices to show that

$$\lim g(z) = \lim g(1-1/z) = 1 \tag{5.5}$$

as $z$ tends to infinity on the strip $0 \leq \mathrm{Re}(z) \leq 2$ in the former case, on the strip $-\frac{1}{2} \leq \mathrm{Re}(z) \leq \frac{1}{2}$ in the latter one. Thus it is all the more sufficient to check (5) as $\mathrm{Im}(z)$ tends to $+\infty$. This is what we will do.

(iii) *The identity* $f(z)/\theta(z)^2 = 1$. Let us then come back to the function

$$g(z) = f(z)/\theta(z)^2,$$

invariant under the group $\Gamma(\theta)$. It is holomorphic on $P$ because the theta series has no zeros there thanks to a miraculous infinite product expansion

$$\theta(z) = \prod_{n\geq 1} \left(1-q^{2n}\right)\left(1+q^{2n-1}\right)^2, q = \exp(\pi i z),$$

a more general form of which will be proved in the next n°.

To study the behaviour of $g$ as $\mathrm{Im}\, z$ tends to infinity, i.e. as $q$ tends to 0, observe that

$$\theta(z) = 1 + 2\left(q+q^4+q^9+\ldots\right), \quad f(z) = 1 + \sum_{n>0} a_n q^n.$$

The first condition (5) thus trivially holds. The second one requires some calculations. Summation is over all $n \in \mathbb{Z}$ in what follows.

$$\theta(z+1) = \sum \exp\left(\pi i n^2 z + \pi i n\right)$$

since $n$ and $n^2$ are either both odd or both even. As our study concerns $\theta(1-1/z)$, the Poisson summation formula which we have successfully used for the $\theta(z)$ series is again required. The Fourier transform of $t \longmapsto \exp(\pi i t^2 z)$ being $u \longmapsto (z/i)^{-\frac{1}{2}} \exp(-\pi i u^2/z)$, the general formula

$$\sum \varphi(t+n) = \sum \hat{\varphi}(n) \exp(2\pi i n t)$$

shows that

$$\sum \exp\left(-\pi i n^2/z + 2\pi i n t\right) = (z/i)^{\frac{1}{2}} \sum \exp\left[\pi i z(t+n)^2\right]. \tag{5.6}$$

For $t = \frac{1}{2}$, this gives

$$\theta(1-1/z) = (z/i)^{\frac{1}{2}} \sum \exp\left[\pi i z \left(n + \frac{1}{2}\right)^2\right] = (z/i)^{\frac{1}{2}} q^{\frac{1}{4}} \sum q^{n(n+1)} =$$

$$= 2(z/i)^{\frac{1}{2}} q^{\frac{1}{4}} \left(1 + q^2 + q^6 + \ldots\right).$$

Hence

$$\theta(1-1/z)^2 \sim 4(z/i)q^{\frac{1}{2}} \quad \text{as } \mathrm{Im}(z) \longrightarrow +\infty. \tag{5.7}$$

If the same result is obtained for $f(z) = \sum 1/\cos \pi nz$, this will prove the second condition (5), and hence that $f(z) = \theta(z)^2$.

To this end, we need to study the behaviour of

$$f(1-1/z) = \sum \frac{(-1)^n}{\cos(-\pi n/z)} = \sum \frac{\mathrm{e}(n/2)}{\cos(-\pi n/z)}$$

as $\mathrm{Im}(z)$ tends to $+\infty$. The formula

$$\sum 1/\cos\left[\pi(u+n)z\right] = (z/i)^{-1} \sum \mathrm{e}(nu)/\cos(-\pi n/z) \tag{4.4"}$$

obtained above is going to lead to the result: putting $u = -\frac{1}{2}$ gives

$$f(1-1/z) = (z/i) \sum 1/\cos\left[\pi\left(n - \frac{1}{2}\right) z\right] =$$

$$= 2(z/i) \sum_{\mathbb{Z}} 1 \Big/ \left(q^{n-\frac{1}{2}} + q^{\frac{1}{2}-n}\right) = 4(z/i) q^{\frac{1}{2}} \sum_{n \geq 0} \frac{q^n}{1 + q^{2n+1}}.$$

The latter series converges normally for $|q| \leq r < 1$. So it is possible to pass to the limit term by term as $q$ tends to 0. The term $n = 0$ tends to 1 and the others to 0, so that the sum of the series tends to 1. Hence

$$f(1-1/z) \sim 4(z/i)q^{\frac{1}{2}} \tag{5.8}$$

in accordance to (7). The result follows.

**Theorem 3.** *For* $\mathrm{Im}(z) > 0$,

$$\sum_{\mathbb{Z}} 1/\cos \pi nz = \theta(z)^2. \tag{5.9}$$

Other authors prefer writing

$$\left(\sum_{\mathbb{Z}} q^{n^2}\right)^2 = 1 + 4 \sum_{1}^{\infty} \frac{q^n}{1 + q^{2n}}, \quad |q| < 1. \tag{5.9'}$$

## 6 – The Infinite Product of the Function $\theta(u, z)$

The previous proof is based on the equality

(6.1) $$\theta(z) = \prod_{n\geq 1} \left(1 - q^{2n}\right)\left(1 + q^{2n-1}\right)^2 , \quad q = \exp(\pi i z),$$

which is necessary (or sufficient – but no other proof is known) to show that $\theta(z)$ does not have any zeros in the half-plane. The calculations of n° 3 and the arguments used above to show that $\sum 1/\cos \pi n z = \theta(z)^2$ enable us to prove (1).

Once again, let us consider the functions

$$\eta(z) = q^{1/12} \prod \left(1 - q^{2n}\right),$$
$$F(z) = q^{-1/24} \prod \left(1 + q^{2n-1}\right)$$

of n° 3. Equality (1) which we need to prove becomes

(6.1') $$\theta(z) = \eta(z) F(z)^2 .$$

As

$$\begin{aligned} \eta(z+1) &= \exp(\pi i/12)\eta(z), & \eta(-1/z) &= (z/i)^{\frac{1}{2}}\eta(z) \\ F(z+2) &= \exp(-\pi i/12)F(z), & F(-1/z) &= F(z), \\ \theta(z+2) &= \theta(z), & \theta(-1/z) &= (z/i)^{\frac{1}{2}}\theta(z), \end{aligned}$$

The ratio $\theta(z)/\eta(z)F(z)^2$ is clearly invariant under $z \longmapsto z+2$ and $z \longmapsto -1/z$, hence under the group $\Gamma(\theta)$ of the function $\theta$. The denominator, a convergent infinite product, has no zeros in $P$. Like in the proof of the previous theorem, to prove (10'), it will, therefore, be sufficient to show that, as $\operatorname{Im} z$ tends to infinity,

(6.2') $$\theta(z) \sim \eta(z)F(z)^2 ,$$
(6.2") $$\theta(1 - 1/z) \sim \eta(1 - 1/z)F(1 - 1/z)^2 .$$

The first point is clear: as $z \in \mathcal{F}$ tends to infinity, i.e. as $q$ tends to 0, $\theta(z) \sim 1$ and

$$\eta(z) \sim q^{1/12}, \qquad F(z) \sim q^{-1/24}$$

since the infinite product tends to 1.

To obtain (2"), we first observe that, by (5.7),

$$\theta(1 - 1/z) \sim 2(z/i)q^{1/4} .$$

On the other hand,

$$\eta(1 - 1/z) = \exp(\pi i/12)\eta(-1/z) = \exp(\pi i/12)(z/i)^{\frac{1}{2}}\eta(z)$$

and so

$$\eta(1 - 1/z) \sim \exp(\pi i/12)(z/i)^{\frac{1}{2}} q^{1/12} .$$

Finally, to obtain the behaviour of $F(1 - 1/z)$, we use the relation

$$F(z+1) = \exp(-\pi i/24)G(z),$$

where $G(z) = q^{-1/24} \prod(1 - q^{2n-1})$ is the function of exercise 2 of n° 3. Formula (3.20) then shows that

$$F(2z+1)F(1-1/z) = \sqrt{2}\exp(-\pi i/12).$$

As $F(2z+1) = e^{-(2z+1)\pi i/24} \prod(1 - q^{4n-2})$ and hence as

$$F(2z+1) \sim \exp(-\pi i/24)q^{-1/12},$$

it follows that

$$F(1-1/z) \sim \sqrt{2}\exp(-\pi i/24)q^{1/12}$$

and as a result,

$$\eta(1-1/z)F(1-1/z)^2 \sim \theta(1-1/z),$$

proving (1).

*Exercise.* (1) is equivalent to

$$\theta(z)\eta(z) = \exp(\pi i/12)\eta\left[(z+1)/2\right]^2. \tag{6.3}$$

As (1) is only a particular case of a much more general formula whose classic proof uses far more "elementary" methods, the previous proof is likely to make experts scream; but I am not writing for them. Moreover, (1) is going to provide us with a very quick proof of the general formula:

**Theorem 4 (Jacobi, 1828).** *For $|q| < 1$ and $w \in \mathbb{C}^*$,*

$$\sum_{\mathbb{Z}} q^{m^2} w^m = \prod_{n \geq 1} \left(1 - q^{2n}\right)\left(1 + q^{2n-1}w\right)\left(1 + q^{2n-1}w^{-1}\right). \tag{6.4}$$

In what follows, $J(q,w)$ will denote the Jacobi series and $A(q,w)$ the Abel infinite product like in Remmert, *Funktionentheorie 2*, pp. 22–27, which presents the classic proof that I will not follow, except for its quasi-trivial part. Mine (?) will show the reader getting to grips with these questions that he has not worked in vain by following the calculations of n° 3.

By Weierstrass' general theorems on normally convergent infinite products (Chap. VII, n° 20, theorem 18), for given $q$, $A(q,w)$ is clearly a holomorphic function of $w$ on $\mathbb{C}^*$, so has a Laurent series expansion

$$A(q,w) = \sum_{\mathbb{Z}} a_n(q)w^n.$$

Replacing $w$ by $q^2 w$ in the product $A(q,w)$, removes the factor $1+qw$ and adds a factor $1+q^{-1}w^{-1} = q^{-1}w^{-1}(1+qw)$. So

$$A\left(q, q^2 w\right) = q^{-1}w^{-1}A(q,w), \tag{6.5}$$

i.e.

$$\sum a_n(q)q^{2n}w^n = \sum a_n(q)q^{-1}w^{n-1} = \sum a_{n+1}(q)q^{-1}w^n.$$

We deduce that $a_{n+1}(q) = q^{2n+1}a_n(q)$, then that

$$a_n(q) = q^{n^2} a_0(q), \tag{6.6}$$

first for $n \geq 0$, in fact for all $n$ since $a_{-n}(q) = a_n(q)$ because of the invariance of the infinite product under $w \longmapsto w^{-1}$. Hence

$$A(q,w) = a_0(q)J(q,w) \tag{6.7}$$

for all $w \in \mathbb{C}^*$ and $|q| < 1$.

For $w = 1, A(q,w)$ and $J(q,w)$ reduce to the two sides of identity (1), and so are equal. Thus

$$A(q,1) = a_0(q)A(q,1)\,.$$

As $|q| < 1$, none of the factors of the infinite product $A(q,1)$ is zero, so that $A(q,1) \neq 0$ (Chap. IV, n° 17, theorem 13). Thus $a_0(q) = 1$, qed.

The $\theta(z)$ series is that value for $u = 0$ of the function

$$\theta(u,z) = \sum_{\mathbb{Z}} e^{\pi i n^2 z + 2\pi n u} = \sum_{\mathbb{Z}} q^{n^2} w^n\,, \tag{6.8}$$

where $q = \exp(\pi i z)$, $w = \mathbf{e}(u)$ for $\mathrm{Im}(u) > 0$, $u \in \mathbb{C}$. We will again come across it in the theory of elliptic functions.

As was mentioned in Chap. IV, this history of this formula is somewhat strange. Jacobi had discovered before 1828 the significance of the left hand side series, Abel having, for his part, used the right hand side infinite product. When Jacobi published his formula, Gauss wrote to him that he was familiar with his result ever since 1808, greatly outraging Legendre who accused Gauss of wanting to appropriate Jacobi's result. Nonetheless, Gauss (1777–1855) was right as was realized when his secret paper were published in 1868. He had became interested in these question even before 1800, used to write down his results every day in his diary and, Houzel relates that,[18] he had obtained most of Abel and Jacobi's infinite product expansions of elliptic functions as well as many more results of the same type. One wonders why he never published them.

To prove (4), there are methods that do not require algebraic calculations on series.[19] G. H. Hardy and E. M. Wright classic *An Introduction to the Theory of Numbers*, as well as Remmert, give the simplest proof. Its starting point, replacing $w$ by $q^2 w$, is already in Gauss, and as was seen, immediately leads to relation (6). We then check through ingenious calculations that

$$A(q,i) = A\left(q^4,-1\right)\,, \qquad J(q,i) = J\left(q^4,-1\right)\,,$$

from which we deduce that $a_0(q^4) = a_0(q)$. As $a_0(q)$ is obviously a power series in $q$, we conclude that $a_0(q) = a_0(0) = A(0,w)/J(0,w) = 1$.

Let us give two striking consequences of (4).

**Corollary 1 (Euler).**

$$\prod_{n\geq 1}(1-q^n) = \sum_{\mathbb{Z}}(-1)^m q^{(3m+1)m/2}\,. \tag{6.9}$$

Replace $q$ by $q^{3/2}$ and $w$ by $-q^{\frac{1}{2}}$ in (2).

---

[18] See his article on the history of elliptic functions in Jean Dieudonné, *Abrégé d'histoire des mathématiques* (Hermann, 1978), vol. 2, pp. 34–40)

[19] See for example G. E. Andrews, Proc. AMS, **16**, 1965, pp. 333–334.

The previous formula can also be written as

$$\eta(z) = \sum_{\mathbb{Z}} (-1)^m \exp\left[(6m+1)^2 \pi i z/12\right]$$

and can be used to prove the functional equation of the $\eta$ function in just a few lines; see n° 8.

**Corollary 2 (Jacobi).**

$$\prod_{n\geq 1} (1-q^n)^3 = \sum_{\mathbb{Z}} (-1)^m m q^{m(m+1)/2} =$$

$$= \sum_{m\geq 0} (-1)^m (2m+1) q^{m(m+1)/2}. \tag{6.10}$$

Replacing $q$ by $q^{\frac{1}{2}}$ and $w$ by $-q^{\frac{1}{2}}w$ in (4) gives

$$\sum (-w)^m q^{m(m+1)/2} = \prod (1-q^n)(1-q^n w)\left(1-q^{n-1}w^{-1}\right) =$$
$$= \left(1-w^{-1}\right)\prod (1-q^n)(1-q^n w)\left(1-q^n w^{-1}\right),$$

and so

$$\frac{1}{w-1}\sum (-w)^m q^{m(m+1)/2} = w^{-1}\prod_{n\geq 1} (1-q^n)(1-q^n w)\left(1-q^n w^{-1}\right).$$

As $w$ tends to 1, the right hand side, which converges normally in every annulus $0 < r \leq |w| \leq R < +\infty$, tends to product (10). The left hand side tends to the derivative of the Laurent series at $w = 1$, qed.

## 7 – The Reciprocity Law for Gauss Sums

(i) *Cauchy's method.* As was seen in the proof of (6.1), the behaviour of the $\theta(z)$ function in the neighbourhood of $z = 1$ on the real axis in given by

$$\theta(1-1/z) \sim 2(z/i)^{1/2} e^{\pi i z/4}, \tag{7.1}$$

as $z$ tends to infinity in the fundamental domain $\mathcal{F}$. More generally, one can work in the neighbourhood of a rational point on the real axis and thus prove analytically one of Gauss' most famous discoveries. Cauchy was the first to notice this.

For this, we set $z = 2p/q + it$ where $p$ and $q$ are non-trivial rational integers and where $t$ tends to 0 through positive values or, more generally, so that $-1/z$ tends to infinity on $\mathcal{F}$. As $-1/z \sim i/t$, this means that $\mathrm{Re}(1/t)$ tends to $+\infty$.

This said,

$$\theta(2p/q + it) = \sum \exp\left(2\pi i n^2 p/q - \pi n^2 t\right).$$

As $\exp(2\pi i z) = \mathbf{e}(z) = \mathbf{e}(z+1)$, the value of $\exp(2\pi i n^2 p/q) = \mathbf{e}(n^2 p/q)$ only depends on the class of $n \bmod q$. So we also have

$$\theta(2p/q + it) = \sum_{n \bmod q} \mathbf{e}\left(n^2 p/q\right) \sum_m \exp\left[-\pi (n+mq)^2 t\right]. \tag{7.2}$$

The Poisson summation formula shows that, for all $n$ and $\mathrm{Re}(t) > 0$,

$$\sum \exp\left[-\pi(n+mq)^2 t\right] = \left(q^2 t\right)^{-\frac{1}{2}} \sum \exp\left(-\pi m^2/q^2 t + 2\pi i m n/q\right) ,$$

where the argument of $(q^2 t)^{-\frac{1}{2}}$ must be chosen between $-\pi/4$ and $+\pi/4$. As

$$\left|\exp\left(-\pi m^2/q^2 t + 2\pi i m n/q\right)\right| = \exp\left[-\pi m^2 \,\mathrm{Re}(1/t)/q^2\right] ,$$

the series obtained converges normally in $\mathrm{Re}(1/t) \geq r$ for all $r > 0$. It is, therefore, possible to pass to the limit term by therm as $\mathrm{Re}(1/t)$ increases indefinitely. However, except for the term $m = 0$, all the terms of the series then tend to 0. Thus, for given $n$,

$$\text{(7.3)} \qquad \sum \exp\left[-\pi(n+mq)^2 t\right] \sim \left(q^2 t\right)^{-\frac{1}{2}} \quad \text{as } \mathrm{Re}(1/t) \longrightarrow +\infty .$$

Substituting this result in (2), we get

$$\text{(7.4)} \qquad \theta\left(2p/q + it\right) \sim G(p,q)\left(q^2 t\right)^{-\frac{1}{2}} ,$$

where

$$\text{(7.5)} \qquad G(p,q) = \sum_{n \bmod q} \mathbf{e}\left(n^2 p/q\right)$$

is the *Gauss sum.* To exploit the relation $\theta(-1/z) = (z/i)^{\frac{1}{2}}\theta(z)$, let us set $z = 2p/q + it$ and $-1/z = -q/2p + iu$. A few calculations show that, as $\mathrm{Re}(1/t)$ tends to $+\infty$, so does

$$\mathrm{Re}(1/u) = 4\,\mathrm{Re}(1/t)p^2/q^2 ,$$

with moreover $4p^2 u \sim q^2 t$. Instead of (2), we write

$$\begin{aligned}\theta\left(-q/2p + iu\right) &= \sum_{\mathbb{Z}} \exp\left(-\pi i n^2 q/2p - \pi n^2 u\right) = \\ &= \sum_{n \bmod 2p} \exp\left(-\pi i n^2 q/2p\right) \sum_m \exp\left[-\pi(n+2pm)^2 u\right] .\end{aligned}$$

By (3) for $q = 2p$, the sum over $m$ is equivalent to $(4p^2 u)^{-\frac{1}{2}}$ as $\mathrm{Re}(1/u)$ tends to $+\infty$. As $4p^2 u \sim q^2 t$, setting

$$\text{(7.6)} \qquad G'(q,p) = \sum_{n \bmod 2p} \exp\left(\pi i n^2 q/2p\right) = \sum \mathbf{e}\left(n^2 q/4p\right) ,$$

it, therefore, follows that

$$\theta\left(-q/2p + iu\right) \sim G'(-q,p)\left(q^2 t\right)^{-\frac{1}{2}} .$$

But the left hand side of the relation found is just $\theta(-1/z) = (z/i)^{\frac{1}{2}}\theta(z)$, where $z = 2p/q + it$. Thus, by (4),

$$\theta\left(-q/2p + iu\right) \sim \left(t + 2p/iq\right)^{\frac{1}{2}} G(p,q)\left(q^2 t\right)^{-\frac{1}{2}}$$

as well. As $t$ tends to 0 from the right of the imaginary axis, $z/i$ tends to $2p/iq$. As $(z/i)^{\frac{1}{2}}$ is defined by choosing $\mathrm{Arg}(z/i)$ between $-\pi/2$ and $+\pi/2$ for $\mathrm{Im}(z) > 0$, we have to choose

$$\mathrm{Arg}(2p/iq) = -\pi/2 \quad \text{if } p/q > 0\,, \quad = +\pi/2 \quad \text{if } p/q < 0\,.$$

Hence

$$(2p/iq)^{\frac{1}{2}} = |2p/q|^{\frac{1}{2}} \exp\left[-\pi i.\,\mathrm{sgn}(pq)/4\right] =$$

$$= |p/q|^{\frac{1}{2}} \left[1 - \mathrm{sgn}(pq)i\right]\,. \tag{7.7}$$

Comparing the two asymptotic evaluations found for $\theta(-q/2p + iu)$, we finally get the equality

$$|p|^{-\frac{1}{2}} G'(-q, p) = |q|^{-\frac{1}{2}} G(p, q) \left[1 - \mathrm{sgn}(pq)i\right]\,. \tag{7.8}$$

In particular, this result applies for $p = 1$. Then $G'(-q, p) = 1 + \exp(-\pi iq/2) = 1 + i^{-q}$, and so

$$|q|^{-\frac{1}{2}} G(1, q) = \left(1 + i^{-q}\right) / \left[1 - \mathrm{sgn}(q)i\right] \tag{7.9}$$

and (Gauss)

$$q^{-\frac{1}{2}} G(1, q) = \frac{1 + i^{-q}}{1 + i^{-1}} = \begin{array}{ll} 1 + i & \text{if } q \equiv 0 \bmod 4 \\ 1 & \text{if } q \equiv 1 \bmod 4 \\ 0 & \text{if } q \equiv 2 \bmod 4 \\ i & \text{if } q \equiv 3 \bmod 4\,. \end{array} \quad (q > 0) \tag{7.9'}$$

In the general case to which we now return, sum (6) can be calculated differently. For this observe that, *if $p$ is odd* and $\neq 1$, then each class mod $2p$ is obtained once and only once by setting $n = 2u + pv$ where $u$ varies mod $p$ and $v \bmod 2$, which transforms the general term of (6) into $\mathbf{e}(u^2 q/p)\mathbf{e}(v^2 pq/4)$. The sum over $u \bmod p$ equals $G(q, p)$ and the sum over $v \bmod 2$ equals $1 + i^{pq}$. Hence

$$G'(q, p) = G(q, p)\left(1 + i^{pq}\right)\,, \tag{7.10}$$

a result which, *for $p$ and $q$ odd*, can also be written

$$G'(q, p) = G(q, p)\left[1 + \epsilon(pq)i\right] \tag{7.10'}$$

by setting

$$\epsilon(n) = 1 \quad \text{if } n \equiv 1 \bmod 4\,, \quad = -1 \quad \text{if } n \equiv 3 \bmod 4\,.$$

By the way, notice the useful relations

$$n \equiv \epsilon(n) \bmod 4\,, \qquad \epsilon(-n) = -\epsilon(n)\,,$$

$$\epsilon(mn) = \epsilon(m)\epsilon(n)\,, \qquad 1 + i^n = 2^{\frac{1}{2}}\left[1 + \epsilon(n)i\right]\,.$$

Comparing (8) and (10) and observing that

$$G(-q, p) = \overline{G(q, p)}\,,$$

it follows that, for $p$ and $q$ odd,

(7.11) $$|p|^{-\frac{1}{2}}\overline{G(q,p)} = |q|^{-\frac{1}{2}}G(p,q)\,[1-\mathrm{sgn}(pq)i]\,/\,[1-\epsilon(pq)i]\ .$$

Before deducing the arithmetic consequences of point (iii), we give a very different method to obtain these formulas.

(ii) *The Dirichlet method.* These formulas have many proofs, including of course arithmetic ones. Dirichlet had the idea of applying his theorem on Fourier series (Chap. VII, § 3, n° 11, Theorem 7 bis) to the function with period 1 on $\mathbb{R}$ given by

(7.12) $$f(t) = \sum_{0\le k\le q-1} \mathbf{e}\left[(t+k)^2p/q\right] \quad \text{for } 0<t<1\,.$$

As $t\in\,]0,1[$ tends to 0 or 1, its limit values equal $G(p,q)$, and since it has right and left derivatives everywhere,

(7.13) $$G(p,q) = \sum \hat{f}(n)\,,$$

the sum of the series being the limit of its partial symmetric sums. We are going to deduce relation (8), which is not essential in the proof of reciprocity formula (11).

To make the clever calculations of the general case more understandable, let us first suppose $p=q=1$, so that $G(p,q)=1$. Here,

$$\hat{f}(-n) = \int_0^1 \mathbf{e}\left(t^2+nt\right)dt = \mathbf{e}\left(-n^2/4\right)\int_0^1 \mathbf{e}\left[(t+n/2)^2\right]dt =$$
$$= \mathbf{e}\left(-n^2/4\right)\int_{n/2}^{(n+2)/2} \mathbf{e}\left(t^2\right)dt\,.$$

So, by Dirichlet's theorem on Fourier series,

$$1 = \sum_n \mathbf{e}\left(-n^2/4\right)\int_{n/2}^{(n+2)/2} \mathbf{e}\left(t^2\right)dt\,.$$

Setting $n=2m$ or $2m+1$ according to the case, this becomes

$$1 = \sum_{\mathbb{Z}}\int_m^{m+1} \mathbf{e}\left(t^2\right)dt - i\sum_{\mathbb{Z}}\int_{\frac{1}{2}+m}^{\frac{1}{2}+m+1} \mathbf{e}\left(t^2\right)dt\,.$$

Setting $t^2 = u$ reduces to proof to integrating the functions $|u|^{-\frac{1}{2}}\cos(2\pi u)$ and $|u|^{-\frac{1}{2}}\sin(2\pi u)$ over $\mathbb{R}$, which leads to (not absolutely) convergent integrals (Chap. V, § 7, n° 24, Theorem 23), thereby giving the formula

(7.14) $$\int_{\mathbb{R}} \exp\left(2\pi i t^2\right)dt = 1/(1-i) = (1+i)/2$$

which is going to prove useful later. The reader will deduce the Fresnel integrals $\int\cos(2\pi t^2)dt$ and $\int\sin(2\pi t^2)dt$ extended to $\mathbb{R}_+$.

*Exercise 1.* Recover (14) by integrating the function $\exp(2\pi i z^2)$ along the closed contour consisting of the interval $[0,R]$ of $\mathbb{R}$, of the arc $0<\mathrm{Arg}(z)<\pi/4$ of the circle centered at 0 of radius $R$, and of the line segment connecting the endpoint of this arc to the origin. Jordan's lemma (Chap. VIII, n° 8, (iv)) will be useful.

*Exercise 2.* Recover (14) starting from the integral of $\exp(\pi i t^2 z)$, $\mathrm{Im}(z) > 0$, and passing to the limit. Show that

$$\int \hat{f}(t) \exp\left(\pi i t^2 z\right) dt = (z/i)^{\frac{1}{2}} \int f(t) \exp\left(-\pi i t^2/z\right) dt$$

for all $f \in \mathcal{S}(\mathbb{R})$ and $\mathrm{Im}(z) > 0$. Deduce that the Fourier transform of the distribution[20] $\exp(\pi i t^2)dt$ is $e^{-\pi i/4} \exp(-\pi i t^2)dt$.

Let us now consider the general case.

$$\hat{f}(-n) = \sum_k \int_0^1 \mathbf{e}\left[nx + (x+k)^2 p/q\right] dx = \int_0^q \mathbf{e}\left(nx + x^2 p/q\right) dx$$

since $k$ varies from 0 to $q-1$. Second degree trinomial theory then shows that

$$\hat{f}(-n) = \mathbf{e}\left(-n^2 q/4p\right) \int_{nq/2p}^{q+nq/2p} \mathbf{e}\left(x^2 p/q\right) dx\,.$$

As the exponential only depends on the class of $n \bmod 2p$,

$$\begin{aligned} G(p,q) &= \sum_{n \bmod 2p} \mathbf{e}\left(-n^2 q/4p\right) \sum_m \int_{mq+nq/2p}^{(m+1)q+nq/2p} \mathbf{e}\left(x^2 p/q\right) dx = \\ &= \sum_{n \bmod 2p} \mathbf{e}\left(-n^2 q/4p\right) \int_{\mathbb{R}} \mathbf{e}\left(x^2 p/q\right) dx\,. \end{aligned}$$

So, setting $t = p/q$,

$$G(p,q) = G'(-q,p) \int_{\mathbb{R}} \mathbf{e}(tx^2) dx\,, \tag{7.15}$$

where $G'(q,p)$ is sum (6). By (14), the integral equals $|p/q|^{-\frac{1}{2}}/[1 - \mathrm{sgn}(pq)i]$ and we recover relation (8) obtained by Cauchy's method.

(iii) *The quadratic reciprocity law.*[21] If $p$ is a *prime* number, then the ring $\mathbb{F}_p = \mathbb{Z}/p\mathbb{Z}$ of integers $\bmod p$ is a field, the multiplicative group $G(p)$ then being the set of its non-trivial elements. As it is commutative, its image under the map $x \longrightarrow x^2$ is a subgroup $G(p)^+$ whose number of elements is equal to the quotient of $\mathrm{Card}[G(p)] = |p| - 1$ by the number of solutions of $x^2 = 1$, namely 2 if $p$ is *odd* as will be assumed. Integers that are not divisible by $p$ fall into two disjoint sets: the set $G(p)^+$ of *quadratic remainders* $\bmod p$, and the set $G(p)^-$ of non-squares. Then, for $n \in \mathbb{Z}$ non-multiple of $p$, we set

$$\left(\frac{n}{p}\right) = \left(\frac{n}{|p|}\right) = \begin{array}{rl} +1 & \text{if } n \in G(p)^+ \\ 0 & \text{if } n = 0 \bmod p \\ -1 & \text{if } n \in G(p)^-\,. \end{array} \tag{7.16}$$

[20] See Chap. VII, n° 32. Generally speaking, $\varphi(t)dt$ denotes the distribution $f \longrightarrow \int f(t)\varphi(t)dt$.

[21] This section assumes that the reader is familiar with the definitions of n° 9.

Since there are only two cosets modulo $G(p)^+$ in the group $G(p)$, the *Legendre symbol* that has just been defined clearly satisfies

$$\left(\frac{mn}{p}\right) = \left(\frac{m}{p}\right)\left(\frac{n}{p}\right) \tag{7.17}$$

for all $m, n \in \mathbb{Z}$. Hence it is a character mod $p$ in the sense of n° 9, (iii).

$G(p)^+$ and $G(p)^-$ having the same number of elements, the cardinality of $G(p)^+$ is $(|p| - 1)/2$. So[22]

$$n^{(|p|-1)/2} = 1 \quad \text{for all } n \in G(p)^+ , \tag{7.18}$$

as well as

$$n^{|p|} \equiv n \bmod p \quad \text{for all } n \in \mathbb{Z} .$$

This is "Fermat's little theorem" used by cryptographers.

In fact, relation (18) characterizes squares since, being an algebraic equation of degree $(|p| - 1)/2$, it cannot have more that $(|p| - 1)/2$ roots in the *field* $\mathbb{F}_p$.

(18) enables us to determine whether $-1$ is or is not a quadratic remainder mod $p$. The first case occurs if and only if $(-1)^{(|p|-1)/2} = 1$ in $\mathbb{Z}/p\mathbb{Z}$, hence in fact in $\mathbb{Z}$ since $p$ is odd. In other words,

$$\left(\frac{-1}{p}\right) = (-1)^{(|p|-1)/2} = \epsilon\left(|p|\right) = \epsilon(p)\,\mathrm{sgn}(p) . \tag{7.19}$$

This result will prove useful. Recall that $\epsilon(n) = 1$ or $-1$ according to whether odd $n$ equals $= 1$ or $3 \bmod 4$, so that $n = \epsilon(n) \bmod 4$.

Let us now return to the Gauss sum $G(p, q)$ by assuming that $p$ and $q$ are *odd primes*. Setting $\mathbf{e}(z) = \exp(2\pi i z)$,

$$G(p, q) = \sum_{n \bmod q} \mathbf{e}\left(n^2 p/q\right) = 1 + 2 \sum_{G(q)^+} \mathbf{e}(mp/q) \tag{7.20}$$

since any quadratic remainder can be written in two ways as $n^2$. We then distinguish between the two cases.

If $p \in G(q)^+$, the map $m \longmapsto mp$ permutes the elements of the group $G(q)^+$. Hence

$$G(p, q) = 1 + 2 \sum_{G(q)^+} \mathbf{e}(m/q) . \tag{7.21}$$

This result is independent of $p$, and so is equal to its value for $p = 1$. Thus

$$G(p, q) = G(1, q) \quad \text{if } p \in G(q)^+ . \tag{7.22}$$

---

[22] If $G$ is finite group and if $x \in G$, the powers $x^n$ are not pairwise distinct. Thus there is some $n > 0$ such that $x^n = e$, the unit element, and the smallest $n$ with this property is obvious the cardinality of the subgroup $G(x)$ generated by $x$. As $G$ is the union of cosets $G(x)y$ containing the same number of elements as $G(x)$, the cardinality of $G$ is a multiple of the cardinality of $G(x)$. Thus $x^{\mathrm{Card}(G)} = e$ for all $x \in G$.

If, however, $p \in G(q)^-$, then $m \longmapsto mp$ is a bijection from $G(q)^+$ onto $G(p)^-$ and

$$G(p,q) = 1 + 2\sum_{G(q)^-} \mathbf{e}(m/q) =$$

$$= 1 + 2\sum_{G(q)} \mathbf{e}(m/q) - 2\sum_{G(q)^+} \mathbf{e}(m/q)\,. \tag{7.23}$$

The sum over $G(q)$ would be zero if $m$ varied over all of $\mathbb{Z}/p\mathbb{Z}$ (sum of the roots of unity $q^e$); as summation is over all $m \neq 0$, it equals $-1$. Taking (21) and (22) into account leads to

$$G(p,q) = -G(1,q) \quad \text{if } p \in G(q)^-\,. \tag{7.24}$$

Both cases are unified by

$$G(p,q) = \left(\frac{p}{q}\right) G(1,q)\,. \tag{7.25}$$

*Exercise.* Set

$$\chi(n) = \left(\frac{n}{q}\right)\,.$$

Show that

$$G(p,q) = \sum_{n \bmod q} \chi(n)\mathbf{e}(np/q) = G(\chi,q)$$

in the notation of n° 9, (iv) and deduce (24).

Let us now come back to relation (11). For $n$ odd, $i^n = \epsilon(n)i$. Thus (11) now becomes

$$|p|^{-\frac{1}{2}}\left(\frac{q}{p}\right)\overline{G(1,p)} = |q|^{-\frac{1}{2}}\left(\frac{p}{q}\right)G(1,q)\frac{1-\operatorname{sgn}(pq)i}{1-\epsilon(pq)i}\,.$$

$$|p|^{-\frac{1}{2}}\overline{G(q,p)} = |q|^{-\frac{1}{2}}G(p,q)\left[1-\operatorname{sgn}(pq)i\right]/\left[1-\epsilon(pq)i\right]\,. \tag{7.11'}$$

But, we saw above that for an odd prime $q$,

$$|q|^{-\frac{1}{2}}G(1,q) = \left(1+i^{-q}\right)/\left[1-\operatorname{sgn}(q)i\right] = \left[1-\epsilon(q)i\right]/\left[1-\operatorname{sgn}(q)i\right]$$

and that there is also a similar relation for $p$. It then follows that

$$\left(\frac{p}{q}\right)\left(\frac{q}{p}\right) = \frac{1-\epsilon(q)i}{1-\operatorname{sgn}(q)i}\,\frac{1+\operatorname{sgn}(p)i}{1+\epsilon(p)i}\,\frac{1-\operatorname{sgn}(pq)i}{1-\epsilon(pq)i}\,. \tag{7.25}$$

The right hand side being equal to $+1$ or $-1$, it suffices to calculate its argument mod $2\pi$. For $\alpha = +1$ or $-1$, the argument of $1+\alpha i$ equals $\alpha\pi/4$. Mod $2\pi$, the argument of the right hand side, therefore, equals $k\pi/4$, where

$$k = -\epsilon(q) + \operatorname{sgn}(p) - \operatorname{sgn}(pq) + \operatorname{sgn}(q) - \epsilon(p) + \epsilon(pq) =$$
$$= \left[\epsilon(p)-1\right]\left[\epsilon(q)-1\right] - \left[\operatorname{sgn}(p)-1\right]\left[\operatorname{sgn}(q)-1\right]\,.$$

Since $\epsilon(p) = p \bmod 4$, we finally get the *quadratic reciprocity law*

$$\left(\frac{p}{q}\right)\left(\frac{q}{p}\right) = (-1)^{(p-1)(q-1)/4}$$

which holds for positive odd primes $p$ and $q$ or only when they are not both negative. The result changes sign otherwise. There are easy generalizations where $p$ and $q$ are no longer assumed to be prime. They can be found in all number theory textbooks.

Legendre guessed the formula and Gauss attained instant fame by proving it. Finding generalizations, for example for rings of algebraic integers, or other proofs was a national sport for the German dynasty created by Gauss until the rest of the world, from Japanese Takagi in 1920 to Chevalley some ten years later, discovered the topic, and after 1935, expanded it. Governed by a High Commissioner rigorously monitoring the alignment of the Great Pyramids, it is today one of the most respected domains in mathematics.

## § 3. The Dirichlet Series $L(s;\chi)$

### 8 – The Functional Equation of $\eta(z)$: bis

Consider once again the formula

$$\eta(z) = \exp(\pi iz/12)\sum_{n\geq 1}\left(1-q^{2n}\right) = \sum_{\mathbb{Z}}(-1)^m \exp\left[(6m+1)^2\pi iz/12\right] =$$

$$(8.1)\qquad = \sum_{\mathbb{Z}}\exp\left[(12m+1)^2\pi iz/12\right] - \sum_{\mathbb{Z}}\exp\left[(12m+7)^2\pi iz/12\right]$$

which follows from the Jacobi identity (Corollary 1). Let $\chi(n)$ denote the function whose values, depending only on the class of $n \bmod 12$, are given by the following table:

| $n$ | 0 | 1 | 2 | 3 | 4 | 5 | 6 | 7 | 8 | 9 | 10 | 11 |
|---|---|---|---|---|---|---|---|---|---|---|---|---|
| $\chi(n)$ | 0 | 1 | 0 | 0 | 0 | −1 | 0 | −1 | 0 | 0 | 0 | 1 |

;

Clearly, $\chi(n) \neq 0$ if and only if $n$ is coprime to 12 and

$$(8.2)\qquad \chi(mn) = \chi(m)\chi(n)\,, \quad \chi(-n) = \chi(n)$$

for all $m, n \in \mathbb{Z}$. With this notation, (1) can also be written

$$(8.3)\qquad \eta(z) = \frac{1}{2}\sum_{\mathbb{Z}}\chi(n)\exp\left[\pi in^2 z/12\right].$$

If indeed $\chi(n) = 1$, then either $n = 12m+1$, or $n = 12m+11 = -(12m'+1)$ with $m' = -m-1$. Because of the $\frac{1}{2}$ factor, the contribution of these values of $n$ to series (3) is equal to the first series (1). If, however, $\chi(n) = -1$, then either $n = 12m+7$, or $n = 12m+5 = -(12m'+7)$ with $m' = -m-1$. Hence the second series (1).

That being so, let us write

$$\sum_{\mathbb{Z}}\chi(n)\exp\left[\pi in^2 z/12\right] = \sum_{a \bmod 12}\chi(a)\sum_{\mathbb{Z}}\exp\left[\pi i(a+12n)^2 z/12\right]$$

and apply the Poisson summation formula to each subseries. After a few lines of calculations, we find

$$\sum \exp\left[\pi i(a+12n)^2 z/12\right] = (12z/i)^{-\frac{1}{2}} \sum \exp\left[-\pi i n^2/12z + \pi i a n/6\right].$$

In consequence,

$$(12z/i)^{\frac{1}{2}}\eta(z) = \sum_{a \bmod 12} \chi(a) \sum_{\mathbb{Z}} \exp\left[-\pi i n^2/12z + \pi i a n/6\right].$$

As the factor $\exp(\pi i a n/6) = \mathbf{e}(an/12)$ only depends on the class of $a \bmod 12$, the result can also be written as

$$(12z/i)^{-\frac{1}{2}}\eta(z) = \sum G(\chi,n)\exp\left(-\pi i n^2/12z\right) \tag{8.4}$$

setting

$$G(\chi,n) = \sum \chi(a)\mathbf{e}(an/12) = \sum \chi(a)\exp(\pi i a n/6), \tag{8.5}$$

where summation is extended to the four classes $a = 1, 5, 7$ and $11 \bmod 12$. The result only depends on the class of $n \bmod 12$ since $\mathbf{e}(z+1) = \mathbf{e}(z)$.

Setting $\omega = \exp(\pi i/6)$, which implies $\omega^6 = -1$ and $\omega^{12} = 1$, and taking into account the values of $\chi(n)$, we get

$$\begin{aligned} G(\chi,n) &= \omega^n - \omega^{5n} - \omega^{7n} + \omega^{11n} = \omega^n + \omega^{-n} + (-1)^{n+1}\left(\omega^n + \omega^{-n}\right) = \\ &= 2\left[1 + (-1)^{n+1}\right]\cos \pi n/6. \end{aligned}$$

Like $\chi(n)$, the result is trivial if $n$ is even and, if $n = 2m+1$, it equals $4\cos n\pi/6$. As $\cos\pi/6 = \cos 11\pi/6 = \sqrt{3}/2$, $\cos 3\pi/6 = \cos 9\pi/6 = 0$ and $\cos 5\pi/6 = \cos 7\pi/6 = -\sqrt{3}/2$, it finally follows that

$$G(\chi,n) = \sqrt{12}\chi(n) \tag{8.6}$$

for all $n$.

Hence relation (4) can now be written as

$$\eta(z) = (12z/i)^{-\frac{1}{2}} 12^{\frac{1}{2}} \sum \chi(n)\exp\left[-\pi i n^2/12z\right] = (z/i)^{-\frac{1}{2}}\eta(-1/z),$$

where the summation is over $\mathbb{Z}$, thereby leading to a functional equation of the Dedekind function by a much simpler method than Weil's; but it assumes the Jacobi identity.

By the way, note that, by (3), the Mellin transform of $t \longmapsto \eta(it^2) - 1$ is the series

$$\frac{1}{2}\sum_{\mathbb{Z}} \chi(n) \int \exp\left(-\pi n^2 t^2/12\right) t^s d^*t = (\pi/12)^{-s/2}\Gamma(s/2)\sum_{n\geq 1}\chi(n)/n^s.$$

The relation $\eta(it^{-2}) = t\eta(it^2)$ immediately gives a functional equation for it in $s \longmapsto 1-s$.

## 9 – Arithmetic Interlude

(i) *Quotient rings.* A *ring* (commutative in what follows) with unit element is a set $A$ with two operations $(x,y) \longmapsto x+y$ and $(x,y) \longmapsto xy$ satisfying all algebraic rules, including the existence of elements 0 and 1 that had better be assumed to be distinct short of a fascination for properties of rings with no elements or only one. So called elementary arithmetics is the study of the ring $\mathbb{Z}$. A *field* is a ring in which every $x \neq 0$ has an inverse and where $1 \neq 0$, so that a field always has at least two elements, the set $\{0,1\}$ equipped with the obvious rules is effectively a field with two elements.

If $A'$ and $A''$ are two rings, a ring structure can be defined on the Cartesian product $A' \times A''$ by setting

$$(x', x'') + (y', y'') = (x' + x'', y' + y'') \,, \quad (x', x'') \,.\, (y', y'') = (x'x'', y'y'') \,.$$

Identifying every $x' \in A'$ with the ordered pair $(x', 0)$ and every $x'' \in A''$ with the the ordered pair $(0, x'')$, then gives $(x', x'') = x' + x''$, but the product

$$x'x'' = (x', 0)\,(0, x'') = (0, 0) = 0$$

in $A' \times A''$ is not equal to

$$(x', x'') = (x', 1)\,(1, x'') \,.$$

An *ideal* of a ring $A$ is an additive subgroup $I$ of $A$ such that

$$x \in A \quad \& \quad y \in I \Longrightarrow xy \in I \,.$$

For example, the set $xA$ of multiples $xy$ of $x \in A$. These are the only ones in $\mathbb{Z}$, for if, in a non-trivial ideal $I$ of $\mathbb{Z}$, $d$ denotes its smallest element $> 0$ (such an element exists since $x \in I \Longrightarrow -x \in I$) and if every $x \in \mathbb{Z}$ is written as

$$x = dq + r \quad \text{with } q \in \mathbb{Z} \,, \quad 0 \le r < d \,,$$

the remainder $r = x - dq$ of the Euclidean division of $x$ by $d$, which is in $I$ if $x \in I$, is trivial as it is $< d$; whence $I = d\mathbb{Z}$. Clearly,

$$d\mathbb{Z} \supset a\mathbb{Z} \Longleftrightarrow d|a \,,$$

where we use conventional notation to say that $d$ divides $a$.

This result shows that, for all $a, b \in \mathbb{Z}$,

$$a\mathbb{Z} \cap b\mathbb{Z} = m\mathbb{Z} \,,$$

where $m$ is the *lcm* of $a$ and $b$: the common multiples of $a$ and $b$ are precisely the multiples of $m$.

As the sum $a\mathbb{Z} + b\mathbb{Z}$ is also an ideal of $\mathbb{Z}$, likewise

$$a\mathbb{Z} + b\mathbb{Z} = d\mathbb{Z}$$

for some number $d = au + bv$. As $d\mathbb{Z}$ contains $a\mathbb{Z}$ and $b\mathbb{Z}$, $d$ is a common divisor of $a$ and $b$. Any other common divisor divides $au + bv$, hence divides $d$. In other words $d$ is the *gcd* of $a$ and $b$, and is denoted $d = (a, b)$. It is easily seen that $ab = md$. If $(a, b) = 1$, $a$ and $b$ are said to be *coprime*. Equivalently: all $x \in \mathbb{Z}$ is of the form

$$x = au + bv \qquad \text{(Bezout's identity)} \,, \tag{9.1}$$

whereas, in the general case, this property assumes that $x$ is a multiple of $(a,b)$. In all cases,

$$a = da', \quad b = db' \quad \text{with } (a',b') = 1.$$

If $I$ is an ideal of a ring $A$, the relation $x - y \in I$ is written $x \equiv y \bmod I$, The *coset* $\bmod I$ of $x \in A$ is the set $x + I$ of $x + y$ where $y \in I$. For $x, y \in A$, it is clear that the cosets $\bmod I$ of $x + y$ and $xy$ only depend on the cosets of $x$ and $y \bmod I$. This defines an addition and a multiplication in the set $A/I$ of cosets $\bmod I$, which enables us to talk of the *quotient ring* $A/I$.

The coset of $a \in A$ is invertible in $A/I$ if and only if there exists $u \in A$ such that $au \equiv 1 \bmod I$. Then $x \equiv aux \bmod I$ for all $x \in A$. Hence $A = aA + I$, and conversely. In the case $A = \mathbb{Z}$, $I = b\mathbb{Z}$, this means that there are integers $u$ and $v$ such that $au + bv = 1$, in other words that $(a,b) = 1$. Like in any ring, the set of invertible elements of $A/I$ is a multiplicative group $(A/I)^*$. The groups

$$G(m) = (\mathbb{Z}/m\mathbb{Z})^*$$

play a particularly important role in arithmetics.

$A/I$ is a field if and only if $A = aA + I$ for all $a$ not belonging to $I$. This means that the only ideals $J$ containing $I$ are $I$ and $A$ since $J \supset aA + I$ for all $a \in J$. $I$ is then said to be a *maximal ideal* of A. In the case $A = \mathbb{Z}$, $I = pZ$, the ideals $J = d\mathbb{Z}$ containing $I$ correspond to the divisors of $p$. So the ideal $p\mathbb{Z}$ is maximal if and only if $p$ is a *prime*. Equivalently: if $p$ divides $xy$, it divides $x$ or $y$. Indeed this relation means that, in the quotient ring $\mathbb{Z}/p\mathbb{Z}$, the relation $xy = 0$ implies $x = 0$ or $y = 0$, in other words that $\mathbb{Z}/p\mathbb{Z}$ is an *integral ring* or that, for all $x \neq 0$ in $\mathbb{Z}/p\mathbb{Z}$, the map $y \longmapsto xy$ is injective. As $\mathbb{Z}/p\mathbb{Z}$ is a finite set, this map is bijective and $x$ is invertible in $\mathbb{Z}/p\mathbb{Z}$. The quotients $\mathbb{F}_p = \mathbb{Z}/p\mathbb{Z}$, $p = 2, 3, 5, 7, 11, \ldots$ are the simplest examples of finite fields.

All integers are products of prime factors. This is written as

$$n = \epsilon \prod p^{v_p(n)}$$

with a factor $\epsilon \in \{-1, 1\}$ and exponents $v_p(n) \in \mathbb{N}$ all of which except a finite number are trivial. For the *gcd* (resp. *lcm*) of two integers $m$ and $n$, the exponent $p$ is the smallest (resp. the largest) of all integers $v_p(m)$ and $v_p(n)$. So any prime factor of the *lcm* is a prime factor of $m$ or of $n$ (or both).

If $I$ and $J$ are ideals of a commutative ring $A$ and if $I \supset J$, relation $x \equiv y \bmod I$ implies $x \equiv y \bmod J$. Hence the coset $\bmod J$ of $x \in A$ only depends on its coset $\bmod I$. This gives a "canonical" map

$$A/I \longrightarrow A/J$$

which is a ring homomorphism. If $I, J, K$ are ideals such that $I \subset J \subset K$, the map $A/I \longrightarrow A/K$ is clearly composed of maps $A/I \longrightarrow A/J$ and $A/J \longrightarrow A/K$.

If $m = m'm''$ in $\mathbb{Z}$, we get a canonical ring homomorphism

$$\mathbb{Z}/m'm''\mathbb{Z} \longrightarrow \mathbb{Z}/m'\mathbb{Z} \times \mathbb{Z}/m''\mathbb{Z} \tag{9.2}$$

by associating to the class $\bmod m$ of every $x \in \mathbb{Z}$ its classes $\bmod m'$ and $m''$.

**Lemma 1**. *Homomorphism (2) is injective if and only if $(m', m'') = 1$; it is then bijective.*

The injectivity of (2) means that any common multiple of $m'$ and $m''$ is a multiple of $m'm''$, hence that $m'm''$ is the *lcm* of $m'$ and $m''$. As this *lcm* is always equal to $m'm''/(m',m'')$, condition $(m',m'')=1$ follows. Conversely if the latter holds, the surjectivity of (2) remains to be proved, in other words that, for all $a',a''\in\mathbb{Z}$, there exists $x\in\mathbb{Z}$ such that both $x\equiv a' \bmod m'$ and $x\equiv a'' \bmod m''$ hold. However, by Bezout's theorem, there are integers $u',u''$ such that

$$a'-a''=m''u''-m'u'.$$

The number $x=a'+m'u'=a''+m''u''$ then answers the question.

(ii) *The groups* $G(m)$ ; *characters* $\bmod m$. For any integer $m\geq 2$, as was mentioned above, we set

$$G(m)=(\mathbb{Z}/m\mathbb{Z})^*.$$

This is the multiplicative group of classes of coprimes to $m$, for example the classes of $1,5,7$ and $11$ if $m=12$. The canonical map $A/I\longrightarrow A/J$ defined in the general case for the ideals $I$ and $J\supset I$ induces a homomorphism from the multiplicative group $(A/I)^*$ to the multiplicative group $(A/J)^*$. For $A=\mathbb{Z}$, there is a more precise result:

**Lemma 2**. *Let $m$ and $m'$ be two integers such that $m'|m$. The canonical homomorphism $G(m)\longrightarrow G(m')$ is surjective.*

The proof reduces to showing that any $a'$ such that $(m',a')=1$ is the class $\bmod m'$ of some $a$ such that $(m,a)=1$, i.e. that there exists $x$ such that $(m,a'+xm')=1$. This means that no prime divisor $p$ of $m$ should divide $a'+xm'$. If $p$ divides $m'$, it divides $xm'$ but not $a'$ since $(a',m')=1$. For these primes, the condition thus holds for all $x$. If, on the other hand $p$ does not divide $m'$, there are two possible cases. If $p$ divides $a'$, any $x$ not divisible by $p$ is suitable. If $p$ does not divide $a'$, any $x$ divisible by $p$ is suitable. So the proof reduces to including in the list of prime factors of $x$ all the prime factors of $m$ dividing $a'$ (they do not divide $m'$) and to excluding from it those that divide $a'$, which is possible since these sets of primes are disjoint, qed.

Let us suppose that $m=m'm''$. Homomorphism (2), namely $x\longmapsto(x',x'')$, defines a homomorphism from $G(m)$ to the multiplicative group $A^*$ of invertible elements of the ring $\mathbb{Z}/m'\mathbb{Z}\times\mathbb{Z}/m''\mathbb{Z}=A$. The unit element of $A$ being $(1,1)$, its invertible elements are the ordered pairs $(x',x'')$ with $x'\in G(m')$, $x''\in G(m'')$. If, for all $x\in\mathbb{Z}/m\mathbb{Z}$, $x'$ and $x''$ denote the images of $x$ in $\mathbb{Z}/m'\mathbb{Z}$ and $\mathbb{Z}/m''\mathbb{Z}$, homomorphism (2) transforms $x$ into $(x',x'')\in A$ and the map

$$G(m)\longrightarrow G\left(m'\right)\times G\left(m''\right) \tag{9.3}$$

thus obtained is obviously a homomorphism from the group $G(m)$ to the Cartesian product of the groups $G(m')$ and $G(m'')$. As the multiplicative groups of two isomorphic rings are isomorphic, lemma 2 shows that (3) *is an isomorphism if and only if* $(m',m'')=1$. As $(x',x'')=(x',1)(1,x'')$ in the Cartesian product $A$, the result can also be formulated as follows:

**Lemma 3**. *Let $m'$ and $m''$ be two coprimes and $m=m'm''$. Let $G'(m)$ (*resp. *$G''(m)$) be the subgroup of $x\equiv 1 \pmod{m''}$ (*resp. *$\pmod{m'}$) in $G(m)$. Then every $x\in G(m)$ is uniquely expressible as the product of an element of $G'(m)$ and of an element of $G''(m)$ and these subgroups are isomorphic to $G(m')$ and $G(m'')$.*

This enables us to calculate the number $\varphi(m)$ of elements of $G(m)$, known as the *Euler indicator* of $m$. To start with, it is clear that $\varphi(m'm'') = \varphi(m')\varphi(m'')$ if $(m', m'') = 1$. Obviously, for a prime $m = p$,

$$\varphi(p) = p - 1 = m(1 - 1/p)\,.$$

If $m = p^r$ with $p$ prime, $G(m)$ decomposes into classes of numbers not divisible by $p$. There are $p^r - p^{r-1}$ of them, and so once again $\varphi(m) = m(1 - 1/p)$. In the general case, $m$ has a prime factor decomposition and we get a product

$$\varphi(m) = m \prod (1 - 1/p) \tag{9.4}$$

extended over all prime divisors of $m$.

If, instead of only considering $G(m)$, one also consider $\mathbb{Z}/m\mathbb{Z}$, the elements $x$ can be classified according to the value of $d = (x, m)$, For any divisor $d$ of $m$, these $x$ are the classes $\bmod\, m$ of the numbers $dy$ with $(y, m/d) = 1$. As the relation $dy \equiv 0 (\bmod\, m)$ is equivalent to $y \equiv 0 (\bmod\, m/d)$, it follows that the number of such classes is equal to $\varphi(m/d)$. Hence the formula

$$\sum_{d|m} \varphi(m/d) = m = \sum_{d|m} \varphi(d)\,. \tag{9.5}$$

For an integer $m \neq 0, 1, -1$, the *character* $\bmod\, m$ is said to be any function

$$\chi : G(m) \longrightarrow \mathbb{C}$$

satisfying

$$\chi(xy) = \chi(x)\chi(y)\,, \quad \chi(1) = 1\,. \tag{9.6'}$$

Let us extend the definition by setting

$$\chi(x) = 0 \quad \text{if } (x, m) \neq 1 \tag{9.6''}$$

so that relation (6') continues to hold for all $x$ and $y$.

$$|\chi(n)| = 1 \quad \text{if } (m, n) = 1 \tag{9.7}$$

necessarily holds since the image of $G(m)$ under $\chi$ is a finite subgroup of $\mathbb{C}^*$. Its elements are roots of unity since the powers of an element of a finite group are not pairwise distinct. The simplest character is $x \longmapsto 1$; it is called the *unit character*. If $\chi'$ and $\chi''$ are two characters, so is their product $x \longmapsto \chi'(x)\chi''(x)$.

If $m'$ is a divisor of $m$ and if $\pi : G(m) \longrightarrow G(m')$ is the canonical map, every character $\chi' \bmod\, m'$ defines a character

$$\chi = \chi' \circ \pi \quad \text{de } G(m)\,.$$

Do not forgot to set $\chi(x) = 0$ for $(x, m) \neq 1$. It is then clear that[23]

$$(x, m) = 1 \quad \& \quad x \equiv 1 \bmod m' \Longrightarrow \chi(x) = 1\,. \tag{9.8}$$

Conversely, let us suppose that this condition holds. As the homomorphism $\pi : G(m) \longrightarrow G(m')$ is surjective and as (8) means that $\chi(x) = 1$ for $x$ such that $\pi(x) = 1$, relation (6') shows that $\pi(x) = \pi(y)$ implies $\chi(x) = \chi(y)$, so that $\chi$ is

[23] Be careful with the condition $(x, m) = 1$. A coprime $x$ to $m'$ is not necessarily coprime to $m$. Equality $\chi = \chi' \circ \pi$ no longer holds in $G(m)$.

given by the composition of $\pi$ and a well-determined character $\chi'$ of $G(m')$. $\chi$ will then be said to *arise* from a character $\bmod m'$ or, simply, by abuse of language, that $\chi$ is a character $\bmod m'$. If $m'|m''|m$, then clearly $\chi$ also arises from a character $\chi'' \bmod m''$, namely that given by the composition of $\chi'$ and of the canonical map from $G(m'')$ onto $G(m')$.

A character $\bmod m$ is said to be *primitive* if it can be obtained in this way from a non-trivial divisor of $m$. This means that, for divisors $m'$ of $m$, it never reduces to 1 on the kernel of the homomorphism $G(m) \longrightarrow G(m')$, in other words that, for every divisor $m' \neq m$ de $m$, there exists $x$ satisfying

$$(x, m) = 1\,, \quad x \equiv 1 \bmod m'\,, \quad \chi(x) \neq 1\,. \tag{9.9}$$

In the notation of lemma 3, this means that, for non-trivial decompositions $m = m'm''$, $\chi$ never reduces to the unit on $G'(m)$.

**Lemma 4.** *Let $\chi$ be a character* $\bmod m$ *and $m', m''$ divisors of $m$ such that $\chi$ is both a character* $\bmod m'$ *and* $\bmod m''$. *Let $d$ be the gcd of $m'$ and $m''$. Then $\chi$ is a character* $\bmod d$.

Let $\mu$ be the *lcm* of $m'$ and $m''$; it divides $m$, so that $\chi$ follows from the composition either of maps

$$G(m) \longrightarrow G(\mu) \longrightarrow G(m') \longrightarrow \chi'\,,$$

or of analogous maps for $m''$ and $\chi''$, where $\chi'$ and $\chi''$ are the characters of $G(m')$ and $G(m'')$. As $\chi$ cannot arise from two different characters of $G(\mu)$ $G(m) \longrightarrow G(\mu)$ being surjective, it is enough to consider $G(\mu)$, in other words to suppose that $m$ is the *lcm* of $m'$ and $m''$, whence $m = m'm''/d$.

By (9), the proof then reduces to showing that

$$(x, m) = 1 \quad \& \quad x \equiv 1 \bmod d \Longrightarrow \chi(x) = 1\,.$$

But, by Bezout, $x \equiv 1 \bmod d$ means there are $u'$ and $u''$ such that $x = 1 + m'u' + m''u''$. $(m, 1+m'u') = 1$ for if a prime $p$ divides $m$ and $1+m'u'$, it could not divide $m'$ or it would divide 1, thus also $m''$ since $m$ is the *lcm* of $m'$ and $m''$, hence would divide $1+m'u'+m''u'' = x$, which is impossible as $(x, m) = 1$. This being so, since $x$ and $1 + m'u'$ are coprime to $m$ and equal $\bmod m''$, $\chi(x) = \chi(1 + m'u')$ by (8) applied to $m''$. But as $1 + m'u'$ is coprime to $m$ and $\equiv 1 \bmod m'$, $\chi(1 + m'u') = 1$ by (8) applied to $m'$, qed.

This implies that, if $d$ is the *gcd* of all divisors $m'$ of $m$ such that $\chi$ is a character $\bmod m'$, then $\chi$ arises from an obviously primitive character $\bmod d$. The numbers $m'$ considered are then clearly all the possible multiples of $d$ dividing $m$. The number $d$ is called the *conductor* of $\chi$; it is the unique divisor $d$ of $m$ for which $\chi$ arises from a primitive character $\bmod d$.

(iii) *Orthogonality relations.* The definition of characters $\bmod m$ can be generalized to all groups $G$: a character of $G$ is a homomorphism $\chi$ from $G$ to the multiplicative group $\mathbb{C}^*$. The simplest case is that of a finite commutative group $G$. To address it, one could invoke the general theory of Chap. XI, n° 26 and 27, but it is much simpler to argue directly.

Let us consider the (finite-dimensional) Hilbert space $L^2(G)$, i.e. the vector space of functions $f : G \longrightarrow \mathbb{C}$ equipped with the inner product

$$(f|g) = \mathrm{Card}(G)^{-1} \sum f(x)\overline{g(x)}\,. \tag{9.10}$$

In $L^2(G)$, the translation operator $U_a$, transforming the function $f(x)$ into the function $U_a f(x) = f(ax)$ for all $f \in L^2(G)$, can be associated to all $a \in G$. Trivial calculations show that

$$(U_a f, U_a g) = (f, g)$$

for all $a$, $f$ and $g$, as well as

$$U_a U_b = U_{ab}$$

for all $a, b \in G$. The former relation says that all $U_a$ are unitary and hence diagonalizable. Since $G$ is commutative, the latter one implies that they they can be simultaneously diagonalized, more precisely that there is an orthogonal basis of $L^2(G)$ whose elements are the eigenvectors of all $U_a$. If $\chi \in L^2(G)$ is such an element, relation $U_a\chi = \lambda(a)\chi$ with $\lambda(a) \in \mathbb{C}$ means that $\chi(ax) = \lambda(a)\chi(x)$ for all $a$ and $x$. Thus $\chi(a) = \chi(e)\lambda(a)$ and so $\chi(a)\chi(x) = \chi(e)\chi(ax)$. As $\chi$ is not the trivial function, one can always assume that $\chi(e) = 1$, in which case it follows that

$$\chi(xy) = \chi(x)\chi(y)\,.$$

The image of $G$ under $\chi$ being a finite group in $\mathbb{C}^*$, $|\chi(x)| = 1$ for all $x$ and so

$$(\chi|\chi) = 1$$

since out of precaution, we introduced a factor $\mathrm{Card}(G)^{-1}$ in the definition of the inner product. Finally, there *is an orthonormal basis of* $L^2(G)$ *composed of* $n = \mathrm{Card}(G)$ *characters of* $G$.

Besides, the vectors of the basis $\chi_i, 1 \leq i \leq \mathrm{Card}(G)$ that have been found are the only characters of $G$. Indeed, almost by definition, a character $\chi'$ is a common eigenvector of all $U_a$. If $\chi' \neq \chi_i$ for some $i$, there exists $x$ such that $\chi'(x) \neq \chi_i(x)$, so that $\chi'$ and $\chi_i$ are eigenvectors for $U_x$ with distinct eigenvalues . Hence they are orthogonal. As a non-trivial element of $L^2(G)$ cannot be orthogonal to all basis vectors, the conclusion follows.

The equalities

$$(\chi|\chi') = \begin{matrix} 1 & \text{if } \chi = \chi', \\ 0 & \text{if } \chi \neq \chi' \end{matrix} \tag{9.11}$$

are the *orthogonality relations* for characters.

*Exercise 1.* For every function $f \in L^2(G)$, set

$$\hat{f}(\chi) = (f|\chi)\,.$$

Without using § 7 du Chap. XI nor the Lebesgue integral prove that

$$f(x) = \mathrm{Card}(G)^{-1} \sum \hat{f}(\chi)\chi(x)$$

for all $x \in G$ and that

$$(f|g) = \mathrm{Card}(G)^{-1} \sum \hat{f}(\chi)\overline{\hat{g}(\chi)} = \left(\hat{f}|\hat{g}\right)$$

for all $f, g$. Analogy with Fourier series?

*Exercise 2.* The product of two characters of $G$ being also a character of $G$, a multiplication can be defined on the set $\hat{G}$ of characters of $G$. It transforms $\hat{G}$ into a commutative group, the *dual* of $G$. Show that every character of $\hat{G}$ is of the form

$\hat{x} : \chi \longmapsto \chi(x)$ for some unique $x \in G$. Deduce that $x \longmapsto \hat{x}$ is an isomorphism from $G$ onto $\hat{\hat{G}}$. Write down the orthogonality relations for the characters of $\hat{G}$.

*Exercise 3.* Let $H$ be a subgroup of $G$ and $\chi$ a character of $H$. Let $L^2(G;\chi)$ be the space of functions $f$ on $G$ such that $f(xh) = f(x)\chi(h)$ for all $h \in P$, equipped with the inner product

$$(f|g) = \mathrm{Card}(G/H)^{-1} \sum_{xH} f(x)g(x)^*\,,$$

where summation is over all cosets mod $H$ since the function being "integrated" is constant on these classes. Show that the characters of $G$ equal to $\chi$ on $H$ form an orthogonal basis for $L^2(G;\chi)$. Deduce that the restriction homomorphism $\hat{G} \longrightarrow \hat{H}$ is surjective. Generalize to all locally compact commutative groups.

(iv) *Gauss sums.* One can construct a theta series

$$\theta_f(x;\chi) = \sum \chi(n) f\left(nx/m^{\frac{1}{2}}\right) = \chi(-1)\theta_f(-x;\chi) \tag{9.12}$$

from a character $\chi \bmod m$ and a function $f \in \mathcal{S}(\mathbb{R})$ as was done in n° 2 to prove the functional equation of the $\zeta$ function or, less trivially, in n° 8 to prove that of the $\eta$ function. As $\chi(-n) = \chi(-1)\chi(n)$, the formula only involves the function $\frac{1}{2}[f(x) + \chi(-1)f(-x)]$; so it is reasonable to confine ourselves functions $f$ for which $f(-x) = \chi(-1)f(x)$. The reason for the factor $m^{\frac{1}{2}}$ will become clear later. In the meantime to make calculations easier, we set

$$g(x) = f\left(x/m^{\frac{1}{2}}\right), \quad \text{whence } \hat{g}(x) = m^{\frac{1}{2}} \hat{f}\left(xm^{\frac{1}{2}}\right).$$

Grouping together the terms of series (19) belonging to a same class mod $m$ shows that

$$\theta_f(x;\chi) = \sum_{\mathbb{Z}/m\mathbb{Z}} \chi(a) \sum_{p\in\mathbb{Z}} g\left[(a+pm)x\right].$$

The Poisson summation formula can then be applied to the function $t \longmapsto g[(a + tm)x]$. Three lines of calculations lead to

$$\theta_f(x;\chi) = (mx)^{-1} \sum_{a,n \bmod m} \hat{g}(n/mx)\chi(a)\mathbf{e}(an/m) =$$

$$= (mx)^{-1} \sum \Gamma_m(n,\chi)\hat{g}(n/mx) \tag{9.13}$$

with *Gauss sums*

$$\Gamma_m(n,\chi) = \sum_{a \bmod m} \chi(a)\mathbf{e}(an/m) \tag{9.14}$$

that most authors simply denote by $G(n,\chi)$. The function $\mathbf{e}(z) = \exp(2\pi i z)$ has period 1, which justifies the summation mod $m$, which in fact is extended to all $a \in G(m)$ for otherwise $\chi(a) = 0$. We also set

$$\Gamma_m(\chi) = \Gamma_m(1,\chi) = \sum \chi(a)\mathbf{e}(a/m)\,. \tag{9.15}$$

The functional equation of the series $\theta(x;\chi)$ being primarily based on properties of Gauss sums, in this section we will prove those that will be essential for us, and even a bit more. The main result is point (a 1) of the next theorem:

**Theorem 5.** *Let $\chi$ be a character* $\bmod m$ *and $n$ an integer. Set*

$$d = (m,n)\,, \quad m = dm'\,, \quad n = dn'\,. \tag{9.16}$$

(a) *In the three following cases*

$$\Gamma_m(n,\chi) = \Gamma_m(\chi)\overline{\chi(n)}\ : \tag{9.17}$$

- (a 1) *for all $n$ if $\chi$ is primitive;*
- (a 2) *for $(m,n) = 1$ for all $\chi$;*
- (a 3) *$\chi$ is not a character* $\bmod m'$.

(b) *If $\chi$ arises from a character $\chi'$* $\bmod m'$, *then*

$$\Gamma_m(n,\chi) = \frac{\varphi(m)}{\varphi(m')}\Gamma_{m'}(\chi')\chi'(n')\,. \tag{9.17'}$$

(14) shows that first of all

$$\Gamma_m(n,\chi)\chi(y) = \sum \chi(xy)\mathbf{e}(xn/m) \quad \text{for all } y \in \mathbb{Z}\,.$$

$y$ may be supposed[24] to be in $G(m)$ for otherwise the previous relation would reduce to $0 = 0$. Let $z$ be the inverse of $y$ in $G(m)$. The map $x \longmapsto xy = x'$ permutes the elements of $G(m)$, and as $x = x'z$, this change of variable in the Gauss sum shows that

$$\Gamma_m(n,\chi)\chi(y) = \sum \chi(x')\mathbf{e}\left(x'zn/m\right) = \Gamma_m(zn,\chi)\,.$$

Several cases that need to be considered

*If $n$ is coprime to $m$*, one may choose $y = n$, in which case $zn = 1 \bmod m$. Then $\chi(y) = \chi(n)$, and so since $|\chi(n)| = 1$,

$$\Gamma_m(n,\chi) = \Gamma_m(\chi)\overline{\chi(n)} \quad \text{if } (n,m) = 1\,, \tag{9.18}$$

which proves proposition (a 2) in the statement. For example, in the extreme case where $\chi$ is a unit character,

$$\Gamma_m(n,1) = \sum_{a \in G(m)} \mathbf{e}(an/m)$$

has the same value for all exponents $n$ coprime to $m$.

*If $n$ is not coprime to $m$*, use (16), which implies $(m',n') = 1$. As $\mathbf{e}(an/m) = \mathbf{e}(an'/m')$ only depends on the class of $a \bmod m'$, the terms of $G(m)$ belonging to the same class $\bmod m'$ can be grouped together in (14), before summing over all classes $\bmod m'$. Hence

$$\Gamma_m(n,\chi) = \sum_{x \bmod m'} \mathbf{e}\left(xn'/m'\right) \sum_{a \equiv x \bmod m'} \chi(a)\,. \tag{9.19}$$

[24] In this type of calculations, it is convenient not to make any difference between an integer and its class $\bmod m$; the context always indicates the correct interpretation.

Since $\chi(a) \neq 0$ assumes $(m, a) = 1$, the partial sum over all $a$ is extended to all $a \in G(m)$ with the same image as $x$ in $G(m')$. These are $xh$ for $h$ in the kernel $H$ of the homomorphism $G(m) \longrightarrow G(m')$, whence

$$\sum_{a \equiv x \bmod m'} \chi(a) = \chi(x) \sum_{h \in H} \chi(h) .$$

The restriction of $\chi$ to the subgroup $H$ being a character of $H$, two cases are possible:

*If $\chi$ is not a character* $\bmod m'$, the restriction of $\chi$ to $H$ is a non-trivial character of $H$. So all partial sums (18) of the Gauss sum $\Gamma_m(\chi, n)$ are trivial by orthogonality relations (11). Hence $\chi(n) = 0$, and thus we also have

$$\Gamma_m(n, \chi) = \Gamma_m(\chi)\overline{\chi(n)} , \tag{9.20}$$

which proves propositions (a 1) and (a 3) of the theorem in this case.

*If $\chi$ is a character* $\bmod m'$, then $\chi = 1$ on $H$ and the sum of all $\chi(h)$ equals $\mathrm{Card}(H) = \mathrm{Card}[G(m)]/\,\mathrm{Card}[G(m')] = \varphi(m)/\varphi(m')$. Then the value of the partial sum over all $a \equiv x \bmod m'$ is $\mathrm{Card}(H)\chi(x)\mathbf{e}(xn'/m')$ and

$$\Gamma_m(n, \chi) = \varphi(m)\,/\varphi(m') \sum \chi(x)\mathbf{e}\,(xn'/m') , \tag{9.21}$$

where summation is $\bmod m'$ over all $x \in G(m)$. But if $\chi'$ is the character of $G(m')$ from which $\chi$ arises and each $x \in G(m)$ is replaced by its image $x'$ in $G(m')$, then $\chi(x) = \chi'(x')$ and $\mathbf{e}(xn'/m') = \mathbf{e}(x'n'/m')$ since $x' = x \bmod m'$. Thus relation (21) becomes

$$\begin{aligned}\Gamma_m(n, \chi) &= \varphi(m)\,/\varphi(m') \sum_{x' \in G(m')} \chi'(x')\mathbf{e}\,(x'n'/m') = \\ &= \varphi(m)\,/\varphi(m')\; \Gamma_{m'}\,(n', \chi') .\end{aligned}$$

However, by definition of $m'$, $(m', n') = 1$. So (18) can be applied to the new situation. Hence

$$\Gamma_m(n, \chi) = \Gamma_{m'}(\chi')\overline{\chi'(n')}\varphi(m)/\varphi(m') , \tag{9.22}$$

with

$$\Gamma_{m'}(\chi') = \sum_{G(m')} \chi'(x')\mathbf{e}\,(x'/m') , \tag{9.23}$$

which proves proposition (b) of the theorem.

(v) *Case of the unit character.* As a useful illustration, for the unit character $\chi$ take: $\chi(x) = 1$ if $x \in G(m)$, $\chi(x) = 0$ otherwise. Then $\chi' = 1$ as well, hence $\chi'(n') = 1$ since $(m', n') = 1$. Setting

$$\Gamma_m(n) = \Gamma_m(n, 1) = \sum_{\substack{(a,m)=1 \\ a \bmod m}} \mathbf{e}(na/m) = \Gamma_m(-n) \tag{9.24}$$

and

$$\mu(m) = \Gamma_m(1)$$

in order to simplify notation, we therefore find

$$\Gamma_m(n) = \frac{\varphi(m)}{\varphi(m/d)}\mu(m/d) \qquad \text{if } d = m/(m,n)\,. \tag{9.25}$$

Sums (24) satisfy

$$\sum_{d|m} \Gamma_{m/d}(n) = \begin{cases} m & \text{if } m|n\,, \\ 0 & \text{otherwise}\,. \end{cases} \tag{9.26}$$

Indeed if the restriction $(a,m) = 1$ is omitted in sum (24), we would get the sum $\sum \mathbf{e}(na/m)$ extended to all $a \in \mathbb{Z}/m\mathbb{Z}$. As $a \longmapsto \mathbf{e}(na/m)$ is an additive character which reduces to the unit only if $m|n$, the result obtained would be equal to the right hand side of (26). But, in the sum extended to all roots of unity, let us group together the terms according to the value $d$ of $(a,m)$. Setting $m = dm'$ and $a = da'$, we get $(a',m') = 1$ and $\mathbf{e}(na/m) = \mathbf{e}(na'/m')$. So the contribution from these classes equals $\Gamma_{m/d}(n)$. As $d$ can take any value, including $d = m$ which corresponds to the term $a = 0$ of (24), (26) follows.

For $n = 1$, (26) becomes

$$\sum_{d|m} \mu(m/d) = \sum_{d|m} \mu(d) = \begin{cases} 1 & \text{if } m = 1\,, \\ 0 & \text{if } m > 1\,. \end{cases} \tag{9.26'}$$

To compute this *Möbius function*, named after the inventor of the strip, first observe that $\mu(1) = 1$ by (24) and

$$\mu(p) = -1 \quad \text{if } p \text{ is prime} \tag{9.27.1}$$

since 1 and $p$ are the only divisors of $p$. If $m = p^r$, the divisors of $m$ are $1, p, \ldots, p^r$, and so

$$\mu\left(p^r\right) = 0 \quad \text{if } r \geq 2\,. \tag{9.27.2}$$

In the general case, first observe that

$$(m', m'') = 1 \Longrightarrow \mu\left(m'm''\right) = \mu\left(m'\right)\mu\left(m''\right)\,. \tag{9.27.3}$$

Since $d = d'd''$, we can identify divisors of $m = m'm''$ with ordered pairs of divisors of $m'$ and $m''$. If the relation we need to show is proved for all $< m = m'm''$, then $\mu(d) = \mu(d')\mu(d'')$ except possibly if $d = m$, in which case $d' = m'$, $d'' = m''$. But then (26) becomes

$$\mu(m) + \sum_{\substack{d'|m' \\ d''|m''}} \mu(d')\mu(d'') = 0$$

provided the pair $m', m''$ of the sum is omitted. On the other hand, multiplying relations (26) for $m'$ and $m''$, the same relation is recovered except that $\mu(m)$ is replaced by $\mu(m')\mu(m'')$. Hence (27.3) holds by induction on $m = m'm''$.

To summarize:

$$\mu(m) = \begin{cases} 1 & \text{if } m = 1\,, \\ (-1)^k & \text{if } m = p_1 \ldots p_k \text{ with distinct } p_i \\ 0 & \text{in other cases}\,. \end{cases} \tag{9.28}$$

The Möbius function – as well as many other analogues – is related to the Riemann $\zeta$ function. Indeed, let us compute

$$\sum 1/n^s \sum_{m\geq 1} \mu(m)/m^s = \sum_{m,n\geq 1} \mu(m)/(mn)^s$$

by grouping together terms according to the value $k$ of $mn$. They are obtained by choosing an arbitrary divisor $d$ of $k$ and by setting $m = d$, $n = k/d$. The total coefficient of $1/k^s$ in the product of the two series is, therefore, the sum $\sum \mu(d)$ extended to divisors of $k$, which is trivial if $k \geq 2$ and equal to 1 for $k = 1$. As a result,

$$1/\zeta(s) = \sum \mu(n)/n^s\,. \tag{9.29}$$

(Alternatively, we could have used (28) to get the infinite product of the $\zeta$ function directly). This supposes $\mathrm{Re}(s) > 1$ if we are keen on convergent series. But the computation is purely formal and some authors even turn this into a theory analogous to that of formal series $\sum a_n X^n$ and, for example, define some sort of associative convolution product

$$f * g(m) = \sum_{d|m} f(d)g(m/d) \tag{9.30}$$

applicable to functions on integers $n \geq 1$. At least it has the not so worthwhile merit of satisfying the identity

$$\sum_{m\geq 1} f(m)/m^s \sum_{m\geq 1} g(m)/m^s = \sum_{m\geq 1} f * g(m)/m^s\,, \tag{9.31}$$

which can be written

$$\zeta_f \zeta_g = \zeta_{f*g}$$

using obvious notation. In particular let us take $g(m) = 1$ for all $m$, whence

$$f * g(m) = \sum f(m/d) = f'(m)\,.$$

As $\zeta_g(s) = \zeta(s)$, it follows that $\zeta_f \zeta = \zeta_{f'}$, and so

$$\zeta_f(s) = \zeta_{f'}(s)/\zeta(s) = \zeta_{f'}(s)\zeta_\mu(s) = \zeta_{f'*\mu}(s)\,.$$

This implies the *Möbius inversion formula:*

$$f'(m) = \sum f(m/d) \Longleftrightarrow f(m) = \sum \mu(d) f'(m/d)\,. \tag{9.32}$$

We next consider the equality $\sum \varphi(m/d) = m$ satisfied by Euler's indicator. As the function $g$ remains the same, $\varphi * g(m) = m$ for all $m$. However, the function $\zeta_h$ associated to $h(m) = m$ is $\zeta(s-1)$. Hence $\zeta_\varphi(s)\zeta(s) = \zeta(s-1)$, and so

$$\sum_{m\geq 1} \varphi(m)/m^s = \zeta(s-1)/\zeta(s)\,. \tag{9.33}$$

We can at last apply the method to the function $f(m) = \Gamma_m(n) = \Gamma_m(|n|)$ for given $n$. Here $f'(m) = m$ if $m|n$, $= 0$ otherwise. Thus, setting

$$\sigma_s(n) = \sum_{d|n} d^s = \sum_{d|n} (n/d)^s = n^s\sigma_{-s}(n)$$

for all $n > 0$ gives

$$\zeta(s)\zeta_f(s) = \zeta_{f'}(s) = \sum_{m|n} (|n|/m)^{1-s} = \sigma_{1-s}(|n|) \ .$$

Therefore, we finally get the formula

$$\sum_{m\geq 1} \Gamma_m(n)m^{-s} = \sigma_{1-s}(|n|)/\zeta(s) = n^{s-1}\sigma_{s-1}(|n|)/\zeta(s) \tag{9.34}$$

which will be used to get the Fourier series expansion of the "reduced" Eisenstein and Maaß series.

## 10 – The Series $\theta_f(x;\chi)$ and $L(s;\chi)$

(i) *Functional equation of* $\theta_f(x;\chi)$. We can now come back to the series

$$\theta_f(x;\chi) = \sum \chi(n)f\left(nx/m^{\frac{1}{2}}\right)$$

defined in (9.12) and to relation (9.13)

$$mx\theta_f(x;\chi) = \sum \Gamma_m(n,\chi)\hat{g}\left(n/mx\right)$$

with $g(x) = f\left(x/m^{\frac{1}{2}}\right)$ and $\hat{g}(x) = m^{\frac{1}{2}}\hat{f}\left(xm^{\frac{1}{2}}\right)$. It can be written

$$mx\theta_f(x;\chi) = m^{\frac{1}{2}}\sum \Gamma_m(n,\chi)\hat{f}\left(n/m^{\frac{1}{2}}x\right) .$$

If $\chi$ is a *primitive* character mod $m$, theorem 5 enables us to replace $\Gamma_m(n,\chi)$ by $\Gamma_m(\chi)\overline{\chi(n)}$. So, replacing $x$ by $1/x$ and setting

$$\epsilon(\chi) = \Gamma_m(\chi)/m^{\frac{1}{2}} , \tag{10.1}$$

$$\theta_f\left(1/x;\chi\right) = \epsilon(\chi)x\theta_{\hat{f}}\left(x;\overline{\chi}\right) . \tag{10.2}$$

Replacing $f$ by $\hat{f}$ in (1), replaces the function $\hat{f}$ by $x \longmapsto f(-x)$. So, if $\chi$ is also replaced by the conjugate character $\overline{\chi}$, then

$$\theta_{\hat{f}}\left(x;\overline{\chi}\right) = \epsilon\left(\overline{\chi}\right)x^{-1}\theta_f\left(-1/x;\chi\right) .$$

Thus, substituting in (1),

$$\theta_f\left(1/x;\chi\right) = \epsilon(\chi)\epsilon\left(\overline{\chi}\right)\theta_f\left(-1/x;\chi\right) .$$

As $\chi(-n) = \chi(-1)\chi(n)$,

$$\theta_f(-x;\chi) = \chi(-1)\theta_f(x;\chi) \tag{10.3}$$

for all $x$ and $f$. Since the functions $\theta_f(x;\chi)$ are not all trivial for given $\chi$, it follows that

(10.4) $$\epsilon(\chi)\epsilon(\overline{\chi}) = \chi(-1)$$

for every primitive character. But

$$\Gamma_m(\overline{\chi}) = \sum \overline{\chi(a)}\mathbf{e}(a/m) = \sum \overline{\chi(a)\mathbf{e}(-a/m)} = \overline{\Gamma_m(-1,\chi)} = \chi(-1)\overline{\Gamma_m(\chi)}\,.$$

Hence

(10.5) $$\epsilon(\overline{\chi}) = \chi(-1)\overline{\epsilon(\chi)}$$

as well, so that (4) becomes

(10.6) $$|\epsilon(\chi)| = 1$$

or, taking (2) into account,

(10.6') $$|\Gamma_m(\chi)| = m^{\frac{1}{2}}\,.$$

As an example let us choose the even function $1/\operatorname{ch}\pi x$, which is identical to its Fourier transform and suppose that $\chi$ is primitive and $\chi(-1) = +1$. Then, by (1), the function

(10.7) $$f(z;\chi) = \sum \chi(n)/\cos\left(\pi nz/m^{\frac{1}{2}}\right), \quad \operatorname{Im}(z) > 0$$

satisfies

(10.8) $$f(-1/z;\chi) = \epsilon(\chi)(z/i)f(z;\overline{\chi})$$

for pure imaginary $z = ix$, hence on all of the half-plane, which generalizes the formulas found in n° 4 in the case $\chi(n) = 1$ for all $n \in \mathbb{Z}$.

We could also have chosen the function $x \longmapsto \exp(\pi i x^2 z)$ where $\operatorname{Im}(z) > 0$. Its Fourier transform is $x \longmapsto (z/i)^{-\frac{1}{2}}\exp(-\pi i x^2/z)$ and writing (1) for $x = 1$, we conclude that the function

(10.9) $$\theta(z;\chi) = \sum_{n\in\mathbb{Z}} \chi(n)\exp\left(\pi i n^2 z/m\right) = \theta(z+2m;\chi)$$

satisfies

(10.10) $$\theta(-1/z;\chi) = \epsilon(\chi)(z/i)^{\frac{1}{2}}\theta(z;\overline{\chi})$$

by assuming $\chi$ to be *primitive and even.*

(ii) *The series $L(s,\chi)$.* To generalize the calculations of § 1, n° 2 concerning the Riemann series to the series $\theta_f(x;\chi)$, it is necessary to study their asymptotic behaviour as $|t|$ tends to $+\infty$ or .

Assuming $f(-x) = \chi(-1)f(x)$ and noting that $\chi(0) = 0$ for any character modulo an integer, $m > 1$,

(10.11) $$\frac{1}{2}\theta_f(x;k) = \sum_{n\geq 1} \chi(n)g(nx)\,,$$

where $g(x) = f\left(x/m^{\frac{1}{2}}\right)$. As $g$ is rapidly decreasing at infinity and as $|\chi(n)| = 1$, by the lemma of § 1, n° 2, the result is rapidly decreasing at infinity.

In the neighbourhood of 0, the relation

$$\theta_f(x;\chi) = (mx)^{-1}\sum \Gamma_m(n,\chi)\hat{g}\,(n/mx)\,,$$

where summation is over $n \in \mathbb{Z}$, shows that the left hand side is the sum of a function $O(|x|^N)$ for all $N$ and of the additional term

$$(mx)^{-1}\hat{g}(0)\Gamma_m(0,\chi)\,.$$

But $\Gamma_m(0,\chi) = \sum \chi(a)$, where summation is over all $a \in G(m)$. This is the inner product (9.8) on $L^2[G(m)]$ of $\chi$ and the unit character. If $\chi$ is the unit character, then $\Gamma_m(0,\chi) = \varphi(m)$, i.e. the number of elements of $G(m)$. Otherwise, the result is trivial because of the orthogonality relations. Getting rid of the case of the unit character, which anyhow is not primitive, shows that the function $\theta_f(x;\chi)$ is $O(|x|^N)$ for all $N$ in the neighbourhood of 0. As a result, the integral defining its Mellin transform converges for all $s \in \mathbb{C}$ and so is an *entire* function of $s$ (Chap. VIII, n° 13).

A formula calculation shows it is equal to

$$\int_0^{+\infty} \theta_f(x;\chi)x^s d^*x = 2\sum_{n\geq 1}\chi(n)\int_0^{+\infty} f\left(nx/m^{\frac{1}{2}}\right)x^s d^*x =$$

$$= m^{s/2}\Gamma_f(s)L(s;\chi)\,, \tag{10.12}$$

where $\Gamma_f(s)$ is the Mellin transform of $f$ and where the Dirichlet series

$$L(s;\chi) = \sum_{n\geq 1}\chi(n)/n^s \tag{10.13}$$

converges absolutely for $\mathrm{Re}(s) > 1$. In this domain, term by term integration is justified because of

$$\sum \int \left| f\left(nx/m^{\frac{1}{2}}\right)x^s\right| d^*x < +\infty\,,$$

which amounts to the convergence of the Riemann series since the integral defining the Mellin transform of $f \in \mathcal{S}(\mathbb{R})$ converges absolutely on $\mathrm{Re}(s) > 0$. Hence this calculation shows that, if $\chi$ is not the unit character, the product $m^{s/2}\Gamma_f(s)L(s;\chi)$ is the restriction of an *entire* function of $s$ to the half-plane $\mathrm{Re}(s) > 1$.

As this assumes $f(-x) = \chi(-1)f(x)$, one can for example choose

$$f(x) = \exp\left(-\pi x^2\right) = \hat{f}(x) \qquad \text{if } \chi(-1) = +1\,, \tag{10.14'}$$

$$f(x) = \pi x\exp\left(-\pi x^2\right) = i\hat{f}(x) \qquad \text{if } \chi(-1) = -1\,. \tag{10.14''}$$

As shown by the calculations of § 1, n° 1, the first case leads to the function

$$\Lambda(s;\chi) = (\pi/m)^{-s/2}\Gamma(s/2)L(s;\chi)\,, \quad \chi(-1) = +1\,; \tag{10.15'}$$

and in the second one, multiplying the result by the cosmetic factor $m^{\frac{1}{2}}$ gives

$$\Lambda(s;\chi) = (\pi/m)^{-(1+s)/2}\Gamma\left[(1+s)/2\right]L(s;\chi)\,, \quad \chi(-1) = -1\,. \tag{10.15''}$$

As $1/\Gamma(s)$ is an entire function, so is the series $L(s;\chi)$. If $\chi$ is even, it has zeros at $s = 0, -2, \ldots$ like $1/\Gamma(s/2)$. If $\chi$ is odd, it has zeros at $s = -1, -3, \ldots$ like $1/\Gamma[(1+s)/2]$.

Taking into account the functional equation

$$\theta_f(1/x;\chi) = \epsilon(\chi)x\theta_f\left(x;\overline{\chi^*}\right),$$

which holds if $\chi$ is primitive, finally leads to the next result:

**Theorem 6.** *Let $m$ be an integer $> 1$ and $\chi$ a* primitive *character* mod $m$. *Then the function*

$$L(s;\chi) = \sum \chi(n)/n^s$$

*is the restriction of an entire function to the half-plane* $\mathrm{Re}(s) > 1$, *and the function $\Lambda(s;\chi)$ defined by (15') or (15") satisfies the functional equation*

$$\Lambda(1-s;\chi) = \epsilon(\chi)\Lambda\left(s;\overline{k}\right) \quad \text{if } \chi(-1) = +1\,, \tag{10.16'}$$

$$\Lambda(1-s;\chi) = -i\epsilon(\chi)\Lambda\left(s;\overline{k}\right) \quad \text{if } \chi(-1) = -1\,. \tag{10.16"}$$

As a first example, let us take $m = 3$ and the character $\chi(n) = 1$ if $n = 1 \bmod 3$, $\chi(n) = -1$ if $n = 2 \bmod 3$. This is the only possibility in this case. Then

$$L(s;\chi) = \sum 1/(3n+1)^s - \sum 1/(3n+2)^s \tag{10.17}$$

where summation is over all $n \in \mathbb{N}$, including 0. As

$$\Gamma_3(\chi) = \sum \chi(a)\mathbf{e}(a/m) = \exp(2\pi i/3) - \exp(4\pi i/3) = 3^{\frac{1}{2}}i\,,$$

$\epsilon(\chi) = i$. Hence, since $\overline{\chi} = \chi$, (16") shows that

$$\Lambda(1-s;\chi) = \Lambda(s;\chi)\,,$$

where

$$\Lambda(s;\chi) = (\pi/3)^{-(1+s)/2}\Gamma\left[(1+s)/2\right]L(s;\chi)\,.$$

The reader who is not very familiar with these calculations should completely redo them in this particular case.

For $m = 4$ and the character used at the end of n° 1 of § 1,

$$\Gamma_4(\chi) = \mathbf{e}(1/4) - \mathbf{e}(3/4) = i - i^3 = 2i = 4^{\frac{1}{2}}i\,,$$

whence $\epsilon(\chi) = i$. As $\chi(-1) = -1$,

$$\Lambda(s;\chi) = (\pi/4)^{-(1+s)/2}\Gamma[(1+s)/2]L(s;\chi)\,.$$

Relation (16") shows that $\Lambda(s;\chi) = \Lambda(1-s;\chi)$, as obtained in n° 1.

For $m = 12$ and the character $\chi$ used in n° 8, formula (8.6) shows that $\Gamma_{12}(\chi) = 12^{\frac{1}{2}}$, whence $\epsilon(\chi) = +1$. As $\chi$ is even, formula (16') applies. Here

$$L(s;\chi) = \sum 1/(12n+1)^s - 1/(12n+5)^s - 1/(12n+7)^s + \\ +1/(12n+11)^s\,, \tag{10.18}$$

where summation is over $n \geq 0$.

Like the Riemann $\zeta$ function, the functions $L(s;\chi)$ have an infinite product expansion

$$L(s;\chi) = \prod [1 - \chi(p)/p^s]^{-1}\,, \quad \mathrm{Re}(s) > 1\,, \tag{10.19}$$

extended to all prime numbers not dividing $m$ since $\chi(p) = 0$ for the others. Indeed, for all $p$,

$$[1 - \chi(p)/p^s]\, L(s;\chi) = \sum \chi(n)/n^s - \sum \chi(pn)/(pn)^s\,,$$

which removes from the series all terms for which $p$ divides $n$. The end of the argument ends is like in the case of the zeta function. (19) shows that the series $L(s;\chi)$ are $\neq 0$ for $\mathrm{Re}(s) > 1$.

(19) does not suppose $\chi$ to be primitive. If $f$ is the conductor of $\chi$, i.e. the least divisor $f$ of $m$ such that $\chi$ is a character mod $f$, and if $\chi'$ denotes the primitive character mod $f$ from which $\chi$ arises, expansions (15) of $L(s;\chi)$ and $L(s;\chi')$ are the same except that factors corresponding to $p$ coprime to $f$ but not to $m$ appear in $L(s;\chi')$, but equal 1 in $L(s;\chi)$. Hence

$$L\,(s;\chi') = L(s;\chi) \prod \left[1 - \chi'(p)/p^s\right]^{-1}\,, \tag{10.20}$$

where the product is extended to all $p$ dividing $m$ but not $f$. As theorem 6 applies to the left hand side, it can be deduced that, even if $\chi$ is a primitive character, the $L(s;\chi)$ series extends analytically.

Finally note that there is a property of the series $L$ distinguishing them from the Riemann $\zeta$ function: *if $\chi$ is not the unit character, the series $L(s;\chi)$ converges uniformly*[25] *in the half-plane* $\mathrm{Re}(s) \geq \sigma$ *for all* $\sigma > 0$ whereas, for the $\zeta$ function, $\sigma > 1$ would need to be assumed.

These series having been invented by Dirichlet, it is not surprising that the result can be obtained by the arguments used in Chap. III, § 3, n° 11 to prove the convergence criterion by the same author. Recall that it is about a numerical series $\sum u_n v_n$ whose partial sums are assumed to be bounded and the sequence $v_n > 0$ to tend decreasingly to 0. These assumptions are too restrictive. In fact, $u_n$ and $v_n$ may be taken to be complex, in which case the assumptions to be checked are

(1) the sums $s_n$ are bounded,
(2) $\lim v_n = 0\,,$
(3) $\sum |v_{n+1} - v_n| < +\infty\,.$

Indeed let us set $t_n = u_1 v_1 + \ldots + u_n v_n$. The proof reduces to showing that, for all $\epsilon > 0$, $|t_q - t_{p-1}| < \epsilon$ for sufficiently large $p$ and $q$. To this end, as in Chap. III, let us write

$$t_q - t_{p-1} = -s_{p-1}v_p + s_p\,(v_p - v_{p+1}) + \ldots + s_{q-1}\,(v_{q-1} - v_q) + s_q v_q\,.$$

Assumption (1) provides an upper bound

$$\begin{aligned} |t_q - t_{p-1}| \leq M\,|v_p| + M\,|v_q| + M\,(|v_p - v_{p+1}| + \ldots +\\ + |v_{q-1} - v_q|)\,, \end{aligned} \tag{10.21}$$

[25] Recall that, for a series of functions, uniform convergence is by definition equivalent to that of its partial sums. It is obviously guaranteed by normal convergence, but does not hold for the $L$-series. In the case of a series of holomorphic functions, uniform convergence ensures that the sum also converges uniformly and its derivatives can be computed by differentiating the given series term by term.

assumption (2) shows that the first two terms tend to 0 and assumption (3) that so does the third one, whence the general theorem.

In the particular case at hand, set $u_n = \chi(n)$ and $v_n = 1/n^s$. The first assumption holds . As $\chi(n+km) = \chi(n)$, the sum of $\chi(n)$ as $n$ varies from $km$ to $(k+1)m$ is indeed equal to the sun of $\chi(x)$ extended to $x \in G(m)$, and so is trivial since $\chi$ is not a unit character. Hence setting $n = mq + r$ with $0 \leq r < q$, $s_n = s_r$. As a consequence, there are finitely many possible values for $s_n$ and a constant $M$ independent of the parameter $s$.

To show that convergence is uniform in the half-plane $P$ defined by $\mathrm{Re}(s) \geq \sigma$ with $\sigma > 0$, it is necessary to show that, for all $\epsilon > 0$, there is an integer $N$ independent of $s$ such that

$$s \in P \quad \& \quad p, q > N \Longrightarrow |t_q - t_{p-1}| < \epsilon .$$

As $|v_n| = 1/n^{\mathrm{Re}(s)} \leq 1/n^\sigma$, the first two terms of the right hand side of (21) do not create any problems. Finding an upper bound for the differences

$$v_n - v_{n+1} = n^{-s} - (n+1)^{-s} = n^{-s}\left[1 - (1+1/n)^{-s}\right]$$

amounts to finding an upper bound for $|(1+z)^{-s} - 1|$ for $z = 1/n$. Newton's binomial formula shows that $(1+z)^{-s} - 1 = -sz + O(1/z^2)$ as $z$ tends to 0, whence $v_n - v_n + 1 = O(1/n^{s+1})$. This result is sufficient to prove convergence, but not that the latter is *uniform*. The derivative of the function $t \longmapsto (1+tz)^{-s}$ being $-sz(1+tz)^{-s-1}$, TF shows that

$$\left|(1+z)^{-s} - 1\right| = \left|-sz \int_0^1 (1+tz)^{-s-1} dt\right| \leq |sz| . \sup_{0 \leq t \leq 1} |1+tz|^{-\mathrm{Re}(s)-1} .$$

For $z = 1/n$, $|1+tz|$ varies between 1 and $1 + 1/n$. As the exponent is $< 0$, sup is attained at $t = 0$. Thus

$$|v_n - v_{n+1}| \leq |s| n^{-\mathrm{Re}(s)-1} \leq |s| n^{-\sigma-1}$$

in the half-plane $P$ and so

$$|v_p - v_{p+1}| + \ldots + |v_{q-1} - v_q| \leq |s| \left(p^{-\sigma-1} + \ldots + q^{-\sigma-1}\right) .$$

As the series $\sum 1/n^{\sigma+1}$ converges, Cauchy's criterion for it gives the expected uniform upper bound.

## § 4. Elliptic Functions

### 11 – Liouville's Theorems

Lett $\omega_1$ and $\omega_2$ be two complex numbers whose ratio is not real and $L$ the lattice of points $\omega = n_1\omega_1 + n_2\omega_2$ with $n_1, n_2 \in \mathbb{Z}$. An elliptic function attached to the lattice $L$ is a *meromorphic* function on $\mathbb{C}$ satisfying

$$f(u + \omega) = f(u)$$

for all $\omega \in L$, for example the series $\sum 1/(u-\omega)^k$ with $k > 2$ and

$$\wp_L(u) = 1/u^2 + \sum \left[1/(u-\omega)^2 - 1/\omega^2\right]$$

of Chap. II, n° 23, which we will encounter again later. The general properties of these functions are set out in theorems due to Liouville (1844), Hermite and Weierstrass, now easy to prove. If $f$ is elliptic, then obviously so is $f(u+a)$ for all $a$, as well as the derivatives of $f$. Any function $P(f_1, \ldots, f_n)/Q(f_1, \ldots, f_n)$, where $P$ and $Q$ are polynomials and the $f_i$ are elliptic functions is again an elliptic function. In fact, it will be shown further down that any elliptic functions is a rational function of $\wp$ and $\wp'$.

This theory is the simplest non-trivial illustration of the expansions on Riemann surfaces given in Chap. X. First of all, the quotient $S(L) = \mathbb{C}/L$ of the additive group $\mathbb{C}$ by the discrete subgroup $L$ can be considered as a *compact group* [Chap. XI, n° 15, (vi)]. It is compact since there are compact sets $K \subset \mathbb{C}$ having non-trivial intersections with all classes mod $L$, for example the parallelogram generated by $\omega_1$ and $\omega_2$. To define a Riemann surface structure on $S(L)$ , it must be equipped with an atlas $(U_i, \varphi_i)$ whose chart changing maps are holomorphic (Chap. X, n° 1). To this end, we consider open sets $W$ in $\mathbb{C}$ whose images $W + \omega$ under translations by elements of $L$ are pairwise disjoint. For all $z \in \mathbb{C}$, any sufficiently small disc centered at $z$ clearly answers the question. Since the canonical map $\pi$ from $\mathbb{C}$ onto $\mathbb{C}/L$ transforms every open set into an open set [Chap. XI, n° 15, (iii)], it maps each $W$ homeomorphically onto an open subset of $\mathbb{C}/L$. Hence there is an inverse homeomorphism $\varphi$ from $\pi(W)$ onto $W$, and thus a chart $\pi(W)$. If $W$ and $W'$ are two such open subsets and $\varphi, \varphi'$ the corresponding maps, for all $\zeta \in \pi(W) \cap \pi(W')$, there exists $\omega \in L$ such that $\varphi'(\zeta) = \varphi(\zeta) + \omega$, and as there clearly exists only one $\omega$ mapping a sufficiently small disc centered at $\varphi(\zeta)$ onto a disc centered at $\varphi'(\zeta)$, the change of chart map reduces to a translation $z \longmapsto z + \omega$ on every sufficiently small open subset of $\pi(W) \cap \pi(W')$, qed.

With this definition, holomorphic functions on an open subset $U$ of $\mathbb{C}/L$ are the same as holomorphic functions defined on the $L$-invariant subset $\pi^{-1}(U)$. The notion of poles, of the order of a pole or of a zero, etc. is the same as in $\mathbb{C}$ and in $\mathbb{C}/L$. In particular, *elliptic functions are just the meromorphic functions on* $\mathbb{C}/L$.

Henceforth, when used without further explanations, the letter $\omega$ will denote an arbitrary element of $L$.

**Theorem 7 (Liouville, Hermite).** *All entire elliptic functions are constants. For any non-constant elliptic function* $f$,

$$\sum v_a(f) = 0 , \qquad \sum \operatorname{Res}(f, a) = 0 .$$

Needless to say that summations are extended to all classes mod $L$. The former result is obvious even without knowing Cauchy theory: it suffices to expand the given function into a Fourier series with respect to one of the periods (Chap. VII, n° 17) and to write the relations that coefficients have to satisfy so that it admits the other period. The latter, as indeed the former, is just a restatement of theorem 1 of Chap. X, n° 1.

**Corollary 1.** *Let* $f$ *and* $g$ *be two elliptic functions such that* $v_a(f) = v_a(g)$ *for all* $a \in \mathbb{C}$. *Then* $f$ *and* $g$ *are proportional.*

For $f/g$ is an elliptic function without any zeros or poles.

**Corollary 2.** *There are no elliptic functions of order* 1.

For such a function would have a unique simple pole with non-trivial residue.

**Corollary 3.** *An elliptic function has as many zeros as poles.*

The zeros and poles are obviously counted mod $L$ and their multiplicities taken account of. The total number of zeros or poles, or more generally of the equation $f(u) = c$, is called the *order* of $f$.

Applying this corollary to $f - c$, where $c$ is a constant, shows that $f$ is of order $p$, and the equation $f(u) = c$ has $p$ roots mod $L$, as usual taking their multiplicities into account. For example, an elliptic function of minimal possible order, namely 2, has only the following possibilities:

(i) two simple poles and two simple zeros,
(ii) a double pole and two simple zeros,
(iii) two simple poles and a double zero,
(iv) a double pole and a double zero.

On a Riemann surface, it is necessary to consider meromorphic differential forms as well as functions. If $\varpi$ is such a form on $S(L)$, its inverse image $\varpi \circ \pi$ is clearly (Chap. IX, n° 16) a meromorphic differential form $f(z)dz$ invariant under $z \longmapsto z + \omega$, which is equivalent to saying that $f$ is an elliptic function. As, in the neighbourhood of every point of $\mathbb{C}$, the form $f(z)dz$ expresses $\varpi$ in a local chart of $\mathbb{C}/L$ at $\pi(z) = a$, the order $v_a(\varpi)$ at $a$ defined in Chap. X, n° 1, is just $v_a(f)$. Hence $v(\varpi) = v(f) = 0$, and as the genus $g$ of a compact Riemann surface is defined by the relation $v(\varpi) = 2g - 2$, we conclude that *the Riemann surface* $\mathbb{C}/L$ *is of genus* 1. One can show that all compact Riemann surfaces of genus 1 are obtained in this way.

Finally note that $\mathbb{C}$ can be considered to be the *universal covering* space of the Riemann surface $\mathbb{C}/L$ (Chap. X, n° 3).

## 12 – Elliptic Functions and Theta Series

There has always been two possible points of view concerning the construction of elliptic functions: that of Abel and Jacobi, historically the first one, on the use of theta series and infinite products representing them and that of Eisenstein and Weierstrass, based on partial fractions expansions. Most textbook authors follow the second one, which is easier as it spares readers Abel and Jacobi's calculations. This advantage is in fact an inconvenience, for as was seen in the preceding §§ these calculations have other uses. When they have been understood, they lead to existence theorems more quickly than Weierstrass' methods. We will therefore present both points of view starting with that of Abel and Jacobi in this n°.

(i) *Abel's theorem.* Abel and Jacobi infinite products studied in § 2, n° 6 enable us to find all elliptic functions through an extraordinarily simple procedure.

**Theorem 8.** *Let* $a_1, \ldots, a_n$ *be the zeros and* $b_1, \ldots, b_n$ *the poles* mod $L$ *of an elliptic function* $f$*, counted with their multiplicities. Then*

$$\sum a_k \equiv \sum b_k \bmod L\,. \tag{12.1}$$

*Conversely, relation (1) ensures the existence of an elliptic function* $f$ *whose only zeros and poles* mod $L$ *are the points* $a_k$ *and*$b_k$*. This function is unique up to a constant.*

*Necessity of the condition.* Let us set $g = f'/f$ and $h(u) = ug(u)$. The function $g$ being elliptic, $h(u+\omega) = h(u) + \omega g(u)$. Consider the period "parallelogram" given above. It is somewhat deformed so that $f$ does not have any poles or zeros on its boundary.

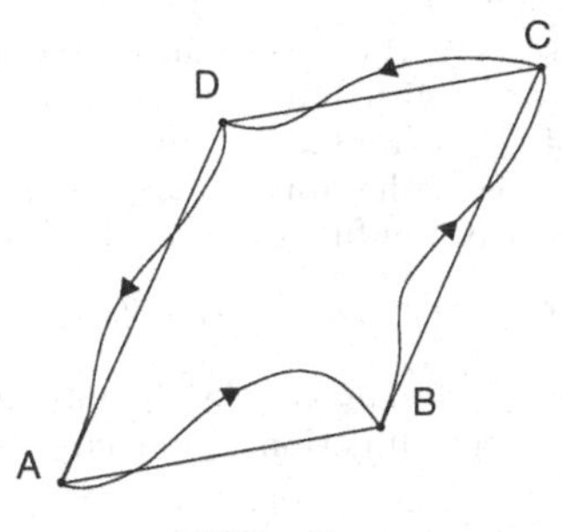


**Fig. 2.**

As

$$\int_{AB} h(u)du + \int_{CD} h(u)du = \int_{AB} h(u)du + \int_{BA} h\left(u + \omega_2\right) du =$$

$$= -\omega_2 \int_{AB} g(u)du$$

and as there is a similar relation between integrals along $BC$ and $DA$,

$$\int_{ABCDA} h(u)du = -\omega_1 \int_{BC} df/f - \omega_2 \int_{AB} df/f\,. \tag{12.2}$$

Since $f(A) = f(B) = f(C) = f(D)$, the next lemma applies:

**Lemma.** *Let $\mu : [0,1] \longrightarrow \mathbb{C}$ be a $C^1$ path and $f$ a holomorphic function on an open set containing $\mu(I)$ and without zeros in $\mu(I)$. Suppose that the values of $f$ at the endpoints are equal. Then*

$$\int_{\mu} df/f \in 2\pi i\mathbb{Z}\,.$$

By definition, the integral to be calculated is obtained by integrating along the inverse image $I = [0,1]$ under $\mu$ of the differential form $df/f$. Setting $\varphi(t) = f[\mu(t)]$, it is, therefore, a question of integrating $d\varphi/\varphi = \varphi'(t)dt/\varphi(t)$ from 0 to 1. However, the map $t \longmapsto \varphi(t)$ is a path $\gamma$ in $\mathbb{C}$, and the integral to be calculated is, by definition, just that of $dz/z$ along $\gamma$, i.e. the variation of a uniform branch of $\mathcal{L}og\, z$ along $\gamma$ [Chap. VIII, § 2, n° 4, (i)]. If $\varphi(0) = f[\mu(0)] = f[\mu(1)] = \varphi(1)$, the path $\gamma$ is closed. So this variation is an integer multiple of $2\pi i$, qed.

This being settled, (2) shows that the integral of $h$ along $ABCDA$ is of the form $n_1\omega_1 + n_2\omega_2$, up to a factor $2\pi i$, i.e. is in $L$. But the poles of $h$ in the interior of this contour are the zeros and poles of $f$. At a point $a$ where the series expansion of $f$ starts with a term in $(u-a)^k$, that of $f'/f$ starts with a term $k/(u-a)$ and that of $uf'(u)/f(u)$ with the term $ka/(u-a)$, whence $\mathrm{Res}(h,a) = ka$. So the sum of the residues of $h$ in $P(a)$ equals the sum of the zeros $a_k$ minus that of the poles $b_k$, repeated as many times as their multiplicities since each point $a$ where the order of $f$ is $k$, which can be positive of negative, contributes an amount equal to $ka$ to the result, proving the necessity of (1).

*Sufficiency of the condition.* The solution is clearly unique by corollary a of theorem 7.

To simplify calculations a bit, the proof can be reduced to the case where the lattice $L$ is generated by 1 an $z$ with $\mathrm{Im}(z) > 0$. To see this, it suffices to observe that the imaginary parts of $\omega_1/\omega_2$ and $\omega_2/\omega_1$ are of opposite signs, which enables us to assume that $z = \omega_2/\omega_1$ is in the upper half-plane. Then the elliptic functions of the given lattice $L$ are clearly the functions $f(\omega_1 u)$, where

$$f(u+1) = f(u+z) = f(u) .$$

On the other hand, the given $a_k$ and $b_k$ are clearly only involved through their class mod $L$. To construct the expected function $f$, one can, therefore, suppose that

$$\sum a_k = \sum b_k . \tag{12.3}$$

This being so, let us set

$$q = \exp(\pi i z) , \quad w = \mathbf{e}(u) = \exp(2\pi i u) \tag{12.4}$$

and start from the Jacobi identity (n° 6, theorem 4)

$$\begin{aligned} \theta(u;z) &= \sum e^{\pi i n^2 z + 2\pi i n u} = \sum q^{n^2} w^n = \\ &= \prod_{n>0} \left(1-q^{2n}\right)\left(1+q^{2n-1}w\right)\left(1+q^{2n-1}w^{-1}\right) . \end{aligned} \tag{12.5}$$

A trivial calculation using the series shows that

$$\theta(u+1,z) = \theta(u;z) , \quad \theta(u+z) = q^{-1}\mathbf{e}(-u)\theta(u;z) . \tag{12.6}$$

Since $\theta(u,z)$ is, moreover, an entire function of $u$ (Chap. VII, § 4, n° 20, theorem 18), clearly, for all $c \in \mathbb{C}$, the function

$$f(u) = \prod \theta\left(u - a_k + c; z\right) \Big/ \prod \theta\left(u - b_k + c; z\right) \tag{12.7}$$

is meromorphic and

$$f(u+1) = f(u) , \quad f(u+z) = \mathbf{e}\left(\sum a_k - \sum b_k\right) f(u) = f(u)$$

by (3). The points $a_k$ and $b_k$ remain to be shown to be the only zeros and poles of $f$, counting multiplicities.

However, the zeros of the infinite product $\theta(u,z)$ are the values of $u$ where $w = q^{2n-1}$ for some $n \in \mathbb{Z}$, i.e. such that

$$u \equiv \omega_0 \bmod L \quad \text{where} \quad \omega_0 = \frac{1}{2}(1+z) .$$

For two different factors of the infinite product to be simultaneously zero would require $q^{2n-1}w = 1$ for two (not necessarily positive) values of $n$. But then $w$ would be a root of unity, and so $|q| = 1$, which is impossible. As moreover the roots of the equation $\mathbf{e}(u) = c$ are simple, *the zeros of $\theta(u,z)$ are $u \equiv \omega_0 \bmod L$ and are all simple.*

Choosing $c = \omega_0$ shows that the zeros of the numerator (resp. denominator) of the right hand side of (7) are, modulo $L$, the points $a_k$ (resp. $b_k$) with the correct multiplicities. The zeros of the numerator cannot cancel out those of the denominator since all $a_k$ are distinct from all $b_h$ mod $L$. So the function

$$f(u) = \prod \theta\left(u - a_k + \omega_0; z\right) \Big/ \prod \theta\left(u - b_k + \omega_0; z\right) \tag{12.8}$$

answers the question upon the choice of (3), which is always possible, for $a_k$ and $b_h$, qed.

Using the function $\theta_1(u;z)$ defined in (iii) further down, the result can be put in the much simpler form

$$\prod \theta_1(u-a_k;z)\Big/\prod \theta_1(u-b_k;z)\,, \tag{12.9}$$

which Hermite availed of in the 1860s. He was one of the first to systematically apply Cauchy's methods.

Without more ado, some authors attribute the previous theorem to Abel, who used an infinite product very similar to that of Jacobi. Abel and Jacobi, and Gauss before them without publishing, could obviously show that formula (8) provides *some* elliptic functions and, through explicit calculations, those that were already known. But how could they have proved that it gives *all* of them in the 1820s, when Cauchy's theory, barely invented, were practically unknown, starting with the precise notion of an analytic or a holomorphic or a meromorphic function? For example, Christian Houzel says[26] that « *without in any way using Cauchy's methods to study one variable complex functions* », in 1844 Liouville proved that a doubly periodic functions has as many zeros as poles, that the sum of its zeros is equal to the sum of its poles and that if $f$ is an elliptic function of order 2, any « *meromorphic* » function admitting the same periods is of the form $P(f)+Q(f)f'$, where $p$ and $Q$ are rational fractions. He also says that Hermite was the first who introduced in 1848 Cauchy's methods by integrating along the boundary of a period parallelogram, Cauchy showing in 1851 that all this follows easily from his residue theory. In fact, referring to Cauchy was not necessary to implicitly ensure that, in the neighbourhood of each point $a \in \mathbb{C}$, elliptic functions are power series in $z-a$ possibly divided by a power of $z-a$: this is how a rational fraction behaves, a fact then long well known, and this is what Liouville implicitly implied when talking of « functions of $x+\sqrt{-1}y$ ». For example to show that (8) represents all elliptic functions, it then suffices to show that an elliptic function without poles is a constant. As I mentioned in Chap. VII, n° 18 this Liouville knew how to do using Fourier series. As for Abel and Jacobi, they seemingly found it sufficient to have shown that their miraculous infinite products made it possible to compute all elliptic integrals and functions that they give rise to.

Finally, note that the Jacobi series has the advantage over infinite products of converging far more rapidly because the exponents of $q$ are the squares 1, 4, 9, 16, etc.. This makes numerical computations easier for practical purposes. Jacobi was the first to have observed this in the case of the "Poinsot type" movement of a heavy solid body with a fixed point, a problem of mechanics leading directly to elliptic integrals.

(ii) *General theta functions.* The previous proof shows the importance of having available *entire* functions, the "general theta functions", which, instead of being doubly periodic, satisfy relations of the form

$$f(u+1)=\mathbf{e}(au+b)f(u)\,,\quad f(u+z)=\mathbf{e}(cu+d)f(u) \tag{12.10}$$

with given constants $a,b,c,d$. There is then obviously a more general relation

$$f(u+\omega)=\mathbf{e}\left(a_\omega u+b_\omega\right)f(u)$$

[26] *in* Jean Dieudonné, *Abrégé d'histoire des mathématiques 1700–1900* (Hermann, 1978), tome II, pp. 21–22.

for all $\omega \in L$, so that the zeros of such a function can be divided into finitely many classes mod $L$. The same is also true for all products and quotients of meromorphic functions satisfying identities such as (10).

If $P(u) = au^2 + bu + c$ is a second degree polynomial, then

$$P(u+1) = P(u) + 2au + a + b\,, \quad P(u+z) = P(u) + 2auz + az^2 + bz\,,$$

so that $f(u) = \exp[P(u)]$ is a theta function, so-called trivial for obvious reasons. It has no zeros, and this property characterizes the trivial theta functions since (10) shows that $(f'/f)'$ is an elliptic function without any poles and so is constant if $f$ has no zeros. It is then immediate that multiplying a solution of (10) by a trivial theta function, one may reduce to the case where

$$f(u+1) = f(u)\,, \quad f(u+z) = \mathbf{e}(cu+d)f(u) \tag{12.10'}$$

with coefficients $c$ and $d$ different from above, which makes calculations somewhat easier and as will be seen, enables us to determine all the solutions of the problem.

Replacing $u$ with $u+1$ in the second relation (10), we first observe that the function $\mathbf{e}(cu+d)$ must be of period 1. Hence it is necessary to assume $c = -p \in \mathbb{Z}$.

On the other hand, the first relation (10') means (Chap. VII, § 4, n° 17) that $f$ is a Fourier series

$$f(u) = \sum_{\mathbb{Z}} a_n \mathbf{e}(nu) \tag{12.11}$$

converging normally in every compact set. The second relation (10') is equivalent to

$$\sum a_n q^{2n} \mathbf{e}(nu) = \mathbf{e}(d) \sum a_n \mathbf{e}\left[(n-p)u\right] = \mathbf{e}(d) \sum a_{n+p} \mathbf{e}(nu)\,.$$

So

$$a_{n+p} = \mathbf{e}(-d) q^{2n} a_n$$

and more generally, after a short calculation,[27]

$$a_{n+kp} = a_n \mathbf{e}(-kd) q^{k(k-1)p+2kn}$$

for all not positive and negative $k \in \mathbb{Z}$. Decomposing series (11) into partial sums

$$f_n(u) = \sum_{r \equiv n \bmod p} a_r \mathbf{e}(ru) = \mathbf{e}(nu) \sum_{k \in \mathbb{Z}} a_{n+kp} \mathbf{e}(pku)\,, \tag{12.12}$$

gives

$$f_n(u) = \mathbf{e}(nu) \sum_{k \in \mathbb{Z}} \mathbf{e}(-kd) q^{k(k-1)p+2kn} \mathbf{e}(kpu) = \mathbf{e}(nu) \sum q^{pk^2} \mathbf{e}(kv)\,,$$

up to a factor $a_n$, and so

$$f_n(u) = \mathbf{e}(nu) \sum_{k \in \mathbb{Z}} q^{pk^2} \mathbf{e}(kv) = \mathbf{e}(nu)\theta(v; pz)\,, \tag{12.13}$$

[27] There are many trivial but tedious calculations in this theory that I will not give details of, leaving it to the reader to do so.

where

$$(12.14) \qquad v = pu - u_0 + nz\,, \qquad u_0 = d + pz/2\,.$$

This calculation however assumes that

$$\sum_{k \in \mathbb{Z}} |q|^{pk^2+(2n-p)k} < +\infty\,.$$

Since, for large $|k|$, the exponent $|q|$ is equivalent to $pk^2$, the series converges in the same conditions as

$$\sum \exp\left(\pi i p z k^2\right)\,,$$

i.e. for $\mathrm{Im}(pz) > 0$. As $\mathrm{Im}(z) > 0$, *it is, therefore, necessary to assume* $p \geq 1$ *to get non-trivial solutions.*

As the Jacobi series satisfies (9), it is conversely to see, after some short calculations, that function (13) satisfies

$$(12.15) \qquad f(u+1) = f(u)\,, \qquad f(u+z) = \mathbf{e}(-pu+d)f(u)$$

for all $n$. The coefficients $a_n$ $(0 \leq n \leq p-1)$ can be chosen arbitrarily, from which one deduces that, for $c = -p$, the solutions of (10) form a $p$-dimensional complex vector space. In fact, this assumes that the $p$ functions $f_n(u)$ are known to be linearly independent, but (12) shows that the same function $\mathbf{e}(ku)$ never occurs in two different series (12). Linear independence then follows from the fact that a (real or complex) Fourier series can be identically zero only if all its coefficients are zero.

For $p = 1$, the previous calculations shows that all solutions of

$$f(u+1) = f(u)\,, \qquad f(u+z) = \mathbf{e}(-u+d)f(z)$$

are proportional to $\theta(u - u_0; z)$, where $u_0 = d + z/2$.

(iii) *Metamorphoses of the Jacobi series.* The Jacobi series has fascinated dozens of mathematicians in the 19th and also in the 20th century. Starting with Jacobi himself,[28] they have proved numerous identities where it occurs. Jacobi, and much before him Gauss in his secret papers, singled out four particularly important functions (in all cases summation is over $n \in \mathbb{Z}$):

---

[28] The reader can refer to his memoir *Ueber unendliche Reihen deren Exponenten zugleich in zwei verschiedenen quadratischen Formen enthalten sind* (Œuvres complètes, tome 2, pp. 217–288). Dozens of identities between infinite products and series can be found in it. They are collected at the end of the article in three typographically very dense pages of results.

$$\begin{aligned}
\theta_1(u;z) &= -i\exp(\pi iu+\pi iz/4)\theta\left[u+(z+1)/2;z\right] \\
&= -i\sum(-1)^n\exp\left[\pi i\left(n+\tfrac{1}{2}\right)^2z+(2n+1)\pi iu\right] \\
&= 2q^{1/4}\left(\sin\pi u-q^{1.2}\sin 3\pi u+q^{2.3}\sin 5\pi u-\ldots\right) \\
\theta_2(u;z) &= \exp(\pi iu+\pi iz/4)\theta(u+z/2;z) \\
&= \sum\exp\left[\pi i\left(n+\tfrac{1}{2}\right)^2z+(2n+1)\pi iu\right] \\
&= 2q^{1/4}\left(\cos\pi u+q^{1.2}\cos 3\pi u+q^{2.3}\cos 5\pi u+\ldots\right) \\
\theta_3(u;z) &= \theta(u;z) \\
&= \sum\exp\left(\pi in^2z+2\pi inu\right) \\
&= 1+2\left(q\cos 2\pi u+q^4\cos 4\pi u+q^9\cos 6\pi u+\ldots\right) \\
\theta_4(u;z) &= \theta\left(u+\tfrac{1}{2};z\right) \\
&= \sum(-1)^n\exp\left(\pi in^2z+2\pi inu\right) \\
&= 1-2\left(q\cos 2\pi u-q^4\cos 4\pi u+q^9\cos 6\pi u-\ldots\right).
\end{aligned}$$

In my preface to vol. I I wrote that I refused to economize paper. There are nevertheless limits to be respected, and as a simple but efficient way of limiting the length of text is to replace detailed proofs by exercises, here are some. They will be referred to later.

*Exercise 1.* Prove the following relations, where $w=\mathbf{e}(u)$:

$$\begin{aligned}
\theta_1(u;z) &= q^{1/4}\prod_{n\geq 1}\left(1-q^{2n}\right)\left(1-q^{2n}w\right)\left(1-q^{2n-2}w^{-1}\right); \\
\theta_2(u;z) &= q^{1/4}\prod_{n\geq 1}\left(1-q^{2n}\right)\left(1+q^{2n}w\right)\left(1+q^{2n-2}w^{-1}\right); \\
\theta_3(u;z) &= \prod_{n\geq 1}\left(1-q^{2n}\right)\left(1+q^{2n-1}w\right)\left(1+q^{2n-1}w^{-1}\right); \\
\theta_4(u;z) &= \prod_{n\geq 1}\left(1-q^{2n}\right)\left(1-q^{2n-1}w\right)\left(1-q^{2n-1}w^{-1}\right).
\end{aligned} \tag{12.16}$$

*Exercise 2.* The table below gives the zeros of functions modulo periods, the factors by which $u\longmapsto u+1$ or $u\longmapsto u+z$ multiplies them by and what they become under $z\longmapsto -1/z$.

| | zéros | $u\longmapsto u+1$ | $u\longmapsto u+z$ | $z\longmapsto -1/z$ |
|---|---|---|---|---|
| $\theta_1(u;z)$ | $0$ | $-1$ | $-\mathbf{e}(-u-z/2)$ | $-i(z/i)^{\frac{1}{2}}\exp\left(\pi iu^2z\right)\theta_1(uz;z)$ |
| $\theta_2(u;z)$ | $1/2$ | $-1$ | $+\mathbf{e}(-u-z/2)$ | $(z/i)^{\frac{1}{2}}\exp\left(\pi iu^2z\right)\theta_4(uz;z)$ |
| $\theta_3(u;z)$ | $(1+z)/2$ | $+1$ | $+\mathbf{e}(-u-z/2)$ | $(z/i)^{\frac{1}{2}}\exp\left(\pi iu^2z\right)\theta_3(uz;z)$ |
| $\theta_4(u;z)$ | $z/2$ | $+1$ | $-\mathbf{e}(-u-z/2)$ | $(z/i)^{\frac{1}{2}}\exp\left(\pi iu^2z\right)\theta_2(uz;z)$ |

Concerning the effect of $z \longmapsto -1/z$, use the Poisson formula

$$\sum f(t+n) = \sum \hat{f}(n)\mathbf{e}(nt) \tag{12.17}$$

and the fact that the Fourier transform of $\exp(\pi i t^2 z)$ is $(z/i)^{-\frac{1}{2}} \exp(-\pi i t^2/z)$ for $\mathrm{Im}(z) > 0$ (Chap. VII, § 6, n° 28).

Show that the functions $u \longmapsto \theta_k(u; z)$ are even for $k \geq 2$ and that $\theta_1(u; z)$ is odd.

The importance of these series is that their zeros are obvious and they all transform under $u \longmapsto u+1$ or $u \longmapsto u+z$ in the same way, up to sign.

*Exercise 3.* Define the Germans' *Thetanullwerte* by

$$\theta_k(z) = \theta_k(0; z) \quad \text{for } 2 \leq k \leq 4 \,.$$

Show that

$$\begin{aligned} \theta_2(z) &= 2q^{1/4} \prod \left(1-q^{2n}\right)\left(1+q^{2n}\right)^2 &&= \sum \exp\left[\pi i \left(n+\tfrac{1}{2}\right)^2 z\right], \\ \theta_3(z) &= \prod \left(1-q^{2n}\right)\left(1+q^{2n-1}\right)^2 &&= \sum \exp\left(\pi i n^2 z\right) = \theta(z), \\ \theta_4(z) &= \prod \left(1-q^{2n}\right)\left(1-q^{2n-1}\right)^2 &&= \sum (-1)^n \exp\left(\pi i n^2 z\right) \end{aligned} \tag{12.18}$$

and that

$$\theta_4(z)\theta_2(z/2) = 2\eta(z)\eta(z/2)\,, \tag{12.19}$$

where $\eta(z)$ is the Dedekind function (n° 3).

*Exercise 4.* Let $\theta_k^{(n)}(z)$ denote the $n$th derivative of l $u \longmapsto \theta_k(u; z)$ for $u = 0$; an abusive but traditional notation. Show that

$$\theta_1'(z) = \pi q^{1/4} \sum_{\mathbb{Z}} (-1)^n (2n+1) q^{n(n+1)} \tag{12.20}$$

and that

$$\theta_1'(z) = 2\pi\eta(z)^3 = 2\pi q^{1/4} \prod \left(1-q^{2n}\right)^3 , \tag{12.21'}$$

$$= \pi\theta_2(z)\theta_3(z)\theta_4(z)\,. \tag{12.21''}$$

Reduce (21') to identity (6.10) of § 2 and, for (21"), use the identity

$$\prod \left(1+q^n\right)\left(1-q^{2n-1}\right) = 1\,, \tag{12.22}$$

which is proved by multiplying the left hand side by $\prod(1-q^n)$.

*Exercise 5.* Justify the table below, which gives the effect of $z \longmapsto z+1$ and $z \longmapsto -1/z$ on the functions concerned:

| | $z \longmapsto z+1$ | $z \longmapsto -1/z$ |
|---|---|---|
| $\theta_1'(z)$ | $i^{\frac{1}{2}}\theta_1'(z)$ | $(z/i)^{3/2}\theta_1'(z)$ |
| $\theta_2(z)$ | $i^{\frac{1}{2}}\theta_2(z)$ | $(z/i)^{\frac{1}{2}}\theta_4(z)$ |
| $\theta_3(z)$ | $\theta_4(z)$ | $(z/i)^{\frac{1}{2}}\theta_3(z)$ |
| $\theta_4(z)$ | $\theta_3(z)$ | $(z/i)^{\frac{1}{2}}\theta_2(z)$ |

where $i^{\frac{1}{2}} = \exp(\pi i/4)$.

*Exercise 6.* Recover formulas of exercises 1 to 4 of §1, n° 3.

*Exercise 7.* Prove (21') and (21") by investigating the behaviour of both sides under $z \longmapsto z+2$ and $z \longmapsto -1/z$ and by showing that the ratio of one side of (21') or of (21") to the other satisfies condition (5.5) of § 2, n° 5.

*Exercise 8.* Using the same method show that

$$\theta_3(z)^4 = \theta_2(z)^4 + \theta_4(z)^4 \tag{12.23}$$

or that

$$q\left(1 + 2q^{1.2} + 2q^{2.3} + \ldots\right)^4 + \left(1 - 2q + 2q^4 - 2q^9 + \ldots\right)^4 =$$
$$= \left(1 + 2q + 2q^4 + 2q^9 + \ldots\right)^4 .$$

Check the plausibility of this relation by a direct calculation of the first few terms of both sides.

*Exercise 9.* Using the fact that, for $c = -p$, the set of solutions of (10) has dimension $p$, show that the four functions $\theta_k(u;z)^2$ generate a 2-dimensioinal vector space. Using the fact that two general theta functions satisfying the same functional equations and having the same zeros are proportional, show that

$$\theta_4(z)^2\theta_4(u;z)^2 = \theta_3(z)^2\theta_3(u;z)^2 - \theta_2(z)^2\theta_2(u;z)^2\,, \tag{12.24'}$$
$$\theta_4(z)^2\theta_1(u;z)^2 = \theta_2(z)^2\theta_3(u;z)^2 - \theta_3(z)^2\theta_2(u;z)^2\,, \tag{12.24"}$$

by first showing that the right hand side of (24') is of the form $c(z)\theta_4(u;z)^2$ and by calculating $c(z)$ using (23).

*Exercise 10.* Show that

$$\left(\frac{\theta_1'(z)\theta_4(u;z)}{\theta_4(z)\theta_1(u;z)}\right)^2 - \left(\frac{\theta_1'(z)\theta_2(u;z)}{\theta_2(z)\theta_1(u;z)}\right)^2 = \pi^2\theta_3(z)^4\,, \tag{12.25'}$$

$$\pi^2\theta_3(z)^4 = \theta_4''(z)/\theta_4(z) - \theta_2''(z)/\theta_2(z)\,. \tag{12.25"}$$

First check that the left hand side of (25') is independent of $u$. To calculate its value, put $u = \frac{1}{2}$ and use (21"). To get (25"), use the Taylor series of the functions $u \longmapsto \theta_k(u;z)$ at $u = 0$.

*Exercise 11.* Show that (notation of exercises 3 and 4)

$$\theta_k''(z) = 4\pi i.d\theta_k(0;z)/dz\,.$$

Using (25"), deduce that

$$\left(\sum_{\mathbb{Z}} q^{n^2}\right)^4 = 1 + 8\sum_{m\geq 1} mq^m/\left(1-q^m\right) - 8\sum_{m\geq 1} 4mq^4m/\left(1-q^4m\right) =$$
$$= 1 + 8\sum c_n q^n\,,$$

where $c_n$ is the sum of the positive divisors of $n$ that are not multiples of 4. Show (Lagrange, and proved again by Jacobi) that *every rational integer is the sum of four squares.*[29]

[29] There is a very different proof using Eisenstein series, as well as a result related to sums of 8 squares; see for example Freitag and Busam, *Funktionentheorie*, Chap. VII, §1.

Many other properties of theta series can be found in Rademacher, *Topics in Analytic Number Theory* (Springer, 1973), Chap. 10 and 11, in Whittaker & Watson, *Modern Analysis*, where I found these exercises, in David Mumford, *Tata Lectures on Theta* I (Birkhäuser, 1983), in Henry McKean & Victor Moll, *Elliptic Curves* (Cambridge UP, 1997) and in a number of earlier authors. Nowadays, mostly analogous series in several variables are studied. They have been used since Riemann in algebraic function theory and in other questions, for example in analytic theory of quadratic forms with integer coefficients, not to mention the extraordinary identities given recently by Macdonald, some of which we now describe.

They provide series expansions for all functions $\eta(z)^N$, where $\eta(z)$ is the Dedekind function (n° 3) and $N$ the dimension of a complex semisimple Lie algebra or, equivalently, of a compact subgroup of some $GL_n(\mathbb{R})$ with finite centre. How to classify them and study them using the theory of "root systems" has long been known. This can for example be learnt in N. Bourbaki's books on Lie algebras, in Dieudonné, vol. 5 and many other books. The simple ones consist of four series of classical groups (unitary, orthogonal with even or odd numbers of variables, and symplectic groups) and five exceptional groups with a complicated description. In the next formulas, summation is over integer vectors $v$ satisfying certain conditions and $\|v\|^2$ denotes the sum of the squares of the coordinates of $v$. A constant $c_0$ appears in each formula; it could be explicitly given by taking the constant term into account. We set $q = \exp(\pi i z)$.

For the unitary group with $n+1$ variables and $n$ even,

$$\eta(z)^{n(n+2)} = c_0 \sum_v \sum_{i<j} (v_i - v_j)\, q^{\|v\|^2/(2n+2)},$$

where summation is extended to all $v = (v_0, \ldots, v_n) \in \mathbb{Z}^{n+1}$ for which $\sum v_i = 0$ and $v_i \equiv i \bmod n+1$.

The other classical groups give similar formulas with $N = n(2n+1)$ or $n(2n-1)$.

Using the five exceptional groups, it is possible to expand $\eta(z)$ to the power of 14, 52, 78, 133 and 248. For example,

$$\eta(z)^{248} = c_0 \sum_{u,v} \sum_{i<j} (v_i - v_j) \sum_{i<j<k} (u + v_i + v_j + v_k)\, q^{(\|v\|^2 - u^2)/60},$$

where summation is over $u \in \mathbb{Z}$ and vectors $v = (v_1, \ldots, v_9) \in \mathbb{Z}^9$ satisfying

$$u \equiv 8 \bmod 30, \quad v_i \equiv i \bmod 30\, (1 \leq i \leq 8), \quad v_9 = 0, \quad 3u + \sum v_i = 0.$$

Macdonald gives other curious formulas, for example

$$\eta(z)^7 \eta(3z)^7 = c_0 \sum_v \prod_i v_i \sum_{i<j} (v_i - v_j)\, q^{\|v\|^2/12},$$

where summation is over $v \in \mathbb{Z}^3$ such that $v_i \equiv i \bmod 6$ and $\sum v_i = 0$.

Those who think that the age of beautiful formulas is over are mistaken.

## 13 – Eisenstein and Weierstrass' Point of View

(i) *Convergence of Eisenstein series.* Obtaining functions defined on a group $G$, not necessarily commutative, and *right invariant* under a countable subgroup[30] $\Gamma$ of $G$, i.e. satisfying

$$f(x\gamma) = f(x) \quad \text{for all } x \in G \quad \text{and all } \gamma \in \Gamma\,,$$

consists in choosing a function $\varphi$ on $G$ and in setting

$$f(x) = \sum \varphi(x\gamma)\,. \tag{13.1}$$

Indeed

$$f\left(x\gamma'\right) = \sum \varphi\left(x\gamma'\gamma\right) \tag{13.2}$$

then, and since, by definition of a group, the map $\gamma \longmapsto \gamma'\gamma$ is a permutation of the elements of $\Gamma$, relation (1) trivially follows. For example, this is what the transformation of a function $f$ on $\mathbb{R}$ into a function $\sum f(x+n)$ in the Poisson summation formula is all about.

A slightly more general situation consists in supposing that $\Gamma$ acts on a set $X$ and to then search for $\Gamma$-*invariant*-functions on $X$, i.e. satisfying

$$f(\gamma x) = f(x)$$

for all $\gamma \in \Gamma$ and $x \in X$. (1) is then replaced by

$$f(x) = \sum \varphi(\gamma x)\,. \tag{13.1'}$$

This argument nevertheless supposes that series (1) or (1') converges unconditionally, as this is the only assumption under which it can undergo the same transformations as a finite sum.

Hence, if $G$ is chosen to be the additive group $\mathbb{C}$ and $\Gamma$ the period lattice $L$, then this leads to the construction of elliptic functions of the form

$$\sum f(u-\omega)\,, \tag{13.3}$$

where summation is extended over all periods $\omega \in L$. The series first needs to be made unconditionally convergent, which supposes that $f$ tends to 0 sufficiently rapidly at infinity. As the aim is to get holomorphic or meromorphic functions, $f$ must satisfy the same assumption. By Liouville's theorem $f$ cannot be supposed to be entire. Hence $f$ must be admitted to have at least one pole at a point $a$, in which case if series (3) converges in a reasonable way, it will have poles at all points $a+\omega$.

Functions $f(u) = 1/u^k$ where $k$ is an integer $> 0$ are the simplest ones satisfying these conditions. They lead to the series

$$\wp_k(u) = \sum 1/(u-\omega)^k\,. \tag{13.4}$$

These converge unconditionally for $k \geq 3$ as shown by the following argument (Chap. II, § 3, n° 23).

[30] In practice, the group $G$ is always equipped with a locally compact topology (or even with a differentiable structure) and $\Gamma$ is always a discrete subgroup of $G$. See Chap. XI, n° 15, (vi).

Let us start by considering a compact set $K \subset \mathbb{C}$ and the general term of (4) when $u$ remains in $K$. To analyze the behaviour of series (3), terms for which $u-\omega$ is 0 in $K$ must first be omitted, i.e. the points $\omega \in K$. As there are finitely many of them, the sum of these exceptional terms is a rational function of $u$. For the other terms, the relation $|\omega| - |u| \leq |u - \omega| \leq |u| + |\omega|$ shows that, for sufficiently large $|\omega|$ $\frac{1}{2}|\omega| \leq |u - \omega| \leq 2|\omega|$ for all $u \in K$. Normal convergence in all compact subsets of (4) is equivalent to the unconditional convergence of $\sum 1/|\omega|^k$.

To determine the suitable values of $k$ – here $k$ need not be an integer – we use the partition of the period lattice $L$ into sets $L_n$ of $8n$ elements, namely containing the points $\omega \in L$ located on the contour of the parallelogram $P_n$, image of the fundamental parallelogram under the homothety with centre $O$ and ratio $n$. For these $\omega \in L_n$, $rn \leq |\omega| \leq Rn$ with constants $r, R > 0$ independent of $n$. However, by the general theorems (Chap. II, § 2, n° 18, associativity theorem 13), the unconditional convergence of the series $\sum 1/|\omega|^k$ is equivalent to

$$\sum_{n>0} \sum_{\omega \in L_n} 1/|\omega|^k < +\infty .$$

The partial sum extended to $L_n$ having the same order of magnitude as $n/n^k = 1/n^{k-1}$, the question reduces to the convergence of the Riemann series, whence the condition $k > 2$.

These arguments still apply if in (3), instead of $1/u^k$, one chooses a rational function $f$ satisfying

$$f(u) \asymp 1/u^k \quad \text{at infinity with}, k \geq 3$$

hence of the form $p(u)/q(u)$ with $d^\circ(q) \geq d^\circ(p) + 3$. The general term $f(u - \omega)$ is again of the same order of magnitude as $1/|\omega|^k$ and, for large $|\omega|$, there is again an upper bound $|f(u-\omega)| < M_K/|u-\omega|^k < M'_K|\omega|^k$ valid for all $u$ in a given compact set $K$. If $K$ is chosen to be an arbitrary closed disc, apart from finitely many terms with a pole in $K$, we get a normally convergent holomorphic functions series in $K$ and so a holomorphic sum in the interior of $D$. In conclusion:

**Theorem 9.** *Let $f$ be a rational function such that $f(u) \asymp 1/u^k$ at infinity, with $k > 3$. Then the series*

$$f^L(u) = \sum f(u - \omega) \tag{13.5}$$

*converges normally in every compact set and its sum is an elliptic function whose poles are,* mod $L$, *those*[31] *of $f$, with the same polar parts.*

This is in particular the case of the functions $\wp_k$ defined by (4): they have a pole of order $k$ at lattice points and are holomorphic elsewhere. But the theorem applies to several other rational functions other than $1/u^k$, for example to $1/(u - a)(u - b)(u - c)$ where $a$, $b$, $c$ are distinct mod $L$.

(ii) *The Weierstrass $\wp$-function.* The series $\sum 1/(u-\omega)^2$ diverges and so cannot be used to construct an elliptic function with double poles at lattice points. As was shown in Chap. II, n° 23, the modified series[32]

[31] To get the polar part of $f^L$ at a point $a$, the polar parts of $f$ at all points $a + \omega$ must be added, so that $f^L$ may well not have a pole at $a$. Hence, strictly speaking, the poles of $f$ would need to be assumed to be distinct mod $L$.

[32] In formula (6) and many similar ones, the series are obviously extended to non-trivial periods. Some authors use the symbol $\sum'$ to indicate this explicitly. I prefer to trust the reader's common sense.

$$\wp(u) = 1/u^2 + \sum \left[1/(u-\omega)^2 - 1/\omega^2\right] \tag{13.6}$$

converges because, for large $|\omega|$,

$$\begin{aligned} 1/(u-\omega)^2 &= \omega^{-2}\left(1 + 2u/\omega + 3u^2/\omega^2 + \ldots\right) = \\ &= 1/\omega^2 + \omega^{-3}u\left(2 + 3u/\omega + \ldots\right). \end{aligned}$$

If $u$ remains in a compact set $K \subset \mathbb{C}$, then $|u/\omega| \leq \frac{1}{2}$ except for a finite and fixed set of periods. Hence, for all $\omega$ not in $K$, there is an upper bound

$$\left|1/(u-\omega)^2 - 1/\omega^2\right| \leq M/|\omega|^3$$

with a constant $M$ independent of $u \in K$. Series (6) from which the finitely many terms for which $\omega \in K$ have been removed is thus normally convergent in $K$, which shows that its sum is meromorphic on $\mathbb{C}$, with double poles at all lattice points.

Though correct, the relation $\wp(u+\omega) = \wp(u)$ is no longer obvious. A direct proof was given in Chap. II, but as a similar problem arises in other cases, the situation may as well be generalized:

**Lemma.** *Let $\Gamma$ be a group acting on the set $X$, $f$ a function defined on $X$ and $c$ a function defined on $\Gamma$. Suppose the following conditions hold:*

(i) *the series*

$$g(x) = \sum \left[f(\gamma x) - c(\gamma)\right] \tag{13.7}$$

*converges unconditionally for all $x \in X$;*

(ii) *the series $\sum[c(\gamma\gamma_0) - c(\gamma)]$ converges unconditionally for all $\gamma_0 \in \Gamma$;*

(iii) *for all $\gamma_0 \in \Gamma$ of finite order and all $\gamma \in \Gamma$,*[33]

$$\lim_{|n|=+\infty} c\left(\gamma\gamma_0^n\right) = 0. \tag{13.8}$$

*Then $g(\gamma x) = g(x)$ for all $\gamma \in \Gamma$ and all $x \in X$.*

As $\gamma_0 \in \Gamma$ is given, first of all,

$$\begin{aligned} g\left(\gamma_0 x\right) &= \sum \left[f\left(\gamma\gamma_0 x\right) - c(\gamma)\right] = \\ &= \sum \left[f\left(\gamma\gamma_0 x\right) - c\left(\gamma\gamma_0\right)\right] + \sum \left[c\left(\gamma\gamma_0\right) - c(\gamma)\right]. \end{aligned}$$

This formal calculation is justified by assumptions (i) and (ii). The map $\gamma \longmapsto \gamma\gamma_0$ being a permutation of the elements of $\Gamma$, the first series has sum $g(x)$. So the proof reduces to showing that

[33] If $\gamma$ is an element of a group $\Gamma$, the set of powers $\gamma^n$ ($n \in \mathbb{Z}$) of $\gamma$ can, depending on the case, be finite or infinite. The order of $\gamma$ is by definition the cardinal of this set. Example: in the multiplicative group $\mathbb{T}$ of complex numbers with absolute value 1, the roots of unity have finite order and the other elements infinite order. In the modular group $SL_2(\mathbb{Z})$ of $2\times 2$ matrices with integer entries and determinant 1, which acts on the upper half-plane by $z \longrightarrow (az+b)/(cz+d)$, the matrix $S = \begin{pmatrix} 0 & -1 \\ 1 & 0 \end{pmatrix}$ has order 2 and the matrix $T = \begin{pmatrix} 1 & 1 \\ 0 & 1 \end{pmatrix}$ infinite order. They define the transformations $z \longrightarrow -1/z$ and $z \longrightarrow z+1$ that we have already extensively used.

$$(13.9) \qquad \sum [c(\gamma\gamma_0) - c(\gamma)] = 0$$

for all $\gamma_0$.

This result generalizes the "obvious" formula

$$(13.10) \qquad \sum_{n\in\mathbb{Z}} [f(n+1) - f(n)] = 0$$

whose proof is immediate in the case that is going to be useful to obtain the Lemma. If the series converges reasonably, its sum is the limit of its partial symmetric sums $f(n+1) - f(-n)$. This gives (10) *if* $f(n)$ *tends to* 0 as $|n|$ increases indefinitely.[34]

Another useful case is that of a periodic function $f$, i.e. satisfying $f(n+p) = f(n)$ for a given integer $p$. As the general term remains unchanged under $n \longmapsto n+p$, formula (10) is not well-defined in this case and it needs to be replaced by

$$(13.11) \qquad \sum_{n \bmod p} [f(n+1) - f(n)] = 0\,.$$

The proof reduces to checking that

$$f(1) - f(0) + f(2) - f(1) + \ldots + f(p) - f(p-1) = 0\,,$$

which is obvious since $f(0) = f(p)$.

Coming back to (10), let $\Gamma_0$ denote the set of $\gamma_0^n$, $n \in \mathbb{Z}$. Writing $\gamma\Gamma_0$ for the set of elements $\gamma\gamma'$ with $\gamma' \in \Gamma_0$ (coset mod $\Gamma_0$), we get a partition of $\Gamma$. The associativity theorem for unconditional convergence then shows that series (10) becomes

$$\sum_{\gamma\Gamma_0} \sum_{\gamma'\in\gamma\Gamma_0} [c(\gamma'\gamma_0) - c(\gamma')]\,,$$

where summation is first over elements $\gamma'$ of some coset $\gamma\Gamma_0$, then over the set of these cosets. As $\gamma'' \longmapsto \gamma\gamma''$ is a bijection from $\Gamma_0$ onto $\gamma\Gamma_0$, the previous series can also be written

$$\sum_{\gamma\Gamma_0} \sum_{\gamma''\in\Gamma_0} [c(\gamma\gamma''\gamma_0) - c(\gamma\gamma'')]\,.$$

Hence it suffices to show that

$$\sum [c(\gamma\gamma''\gamma_0) - c(\gamma\gamma'')] = 0\,,$$

where summation is over $\gamma'' \in \Gamma_0$, i.e. over the set of powers of $\gamma_0$. Setting $\gamma'' = \gamma_0^n$ and $f(n) = c(\gamma\gamma_0^n)$, the general term of the series becomes $f(n+1) - f(n)$. This reduces the proof to (10) if $\gamma_0$ has infinite order since the elements $\gamma\gamma_0^n$ of $\Gamma$ are then pairwise distinct, and assumption (iii) of the lemma ensures the conclusion. If $\gamma_0$ has finite order $p$, then[35] $\gamma_0^p = e$, the identity element. Hence $f(n+p) = f(n)$, which reduces the proof to (11), proving the lemma.

---

[34] In Bernoulli's days, it would probably have been noticed that each term $f(n)$ occurs twice with opposite signs (for example if $f(n) = n\ldots$). The argument of the text shows that sum (10) equals $f(+\infty) - f(-\infty)$ if $f(n)$ tends to finite limits as $n$ tends to $+\infty$ or $-\infty$.

[35] If an element $g$ of a group $G$ has finite order, its powers are pairwise distinct, and so $g^a = g^b$ with distinct integers $a$ and $b$. Hence there are integers $p > 0$ such that $g^p = e$. If $p$ is the smallest one among them, the elements $g^0 = e, g, \ldots, g^{p-1}$ are pairwise distinct and the other powers $g^n$ reduce to them by replacing $n$ with the remainder of its division by $p$. The number $p$ is thus the order of $g$.

Example: The function

$$f(z) = 1/z + \sum [1/(z+n) - 1/n]$$

on $\mathbb{C} - \mathbb{Z}$ has period 1: make the additive group $\mathbb{Z}$ act on $\mathbb{C}$ by translations and set $c(n) = 1/n$ if $n \neq 0$, $c(0) = 0$. The function $f$ is meromorphic on $\mathbb{C}$, its poles are obvious and $f(z) - \pi \operatorname{cotg} \pi z$ is entire with period 1. To show that $f(z) = \pi \operatorname{cotg} \pi z$, the difference would need to be shown to tend to 0 at infinity, a non-obvious exercise.

Coming back to the Weierstrass function (6), the lemma can be applied by choosing $X = \mathbb{C}$ with its periods removed since the series is not strictly speaking well-defined if $u \in L$, and $\Gamma$ to be the additive group $L$ acting on $X$ by translations $u \longmapsto u + \omega$. The simplest function $c$[36] is then given by $c(\omega) = 1/\omega^2$ if $\omega \neq 0$, $c(0) = 0$. As was seen earlier, assumption (i) holds. Assumption (ii) means that

$$\sum \left|1/\left(\omega + \omega_0\right)^2 - 1/\omega^2\right| < +\infty\,,$$

where the two terms that are not well-defined are omitted. This is already known since the general term is $O(1/\omega^3)$. Finally, assumption (iii) means that $\lim 1/(\omega + n\omega_0)^2 = 0$ for $\omega_0 \neq 0$, which is clear. Formula (6) thus defines an elliptic function. We already know that its only singularities are double poles at lattice points.

This property more or less characterizes the function $\wp$. Indeed if $\varphi(u)$ is an elliptic function whose only singularities are double poles at points of $L$, there is an expansion $\varphi(u) = a/u^2 + b + \ldots$ in the neighbourhood of the origin because the residues of $\varphi$ are zero by theorem 7. The elliptic function $\varphi(u) - a\wp(u)$ is then entire, hence constant. In conclusion:

**Theorem 10.** *The series*[37]

$$\wp_L(u) = 1/u^2 + \sum \left[1/(u-\omega)^2 - 1/\omega^2\right] \tag{13.12}$$

*converges normally in every compact set and its sum is an elliptic function whose only singularities are double poles at lattice points. Any other elliptic function with this property is of the form $a\wp_L(u) + b$ where $a$ and $b$ are constants.*

As was seen in Chap. II, n° 23, and as is now obvious because of Weierstrass' theorem on holomorphic function series (term by term differentiation), for $k \geq 3$, the functions

$$\wp_k(u) = \sum 1/(u-\omega)^k$$

are proportional to the derivatives of $\wp$:

$$\wp'(u) = -2\wp_3(u)\,, \quad \wp''(u) = 2.3\wp_4(u), \ldots\,. \tag{13.13}$$

Note that as $u$ tends to 0, the terms of series (12) with the term $1/u^2$ removed tend to 0. As this series converges normally in the neighbourhood of the origin, its sum also tends to 0. Therefore, the function $\wp$ being even, there is a series expansion of the form

[36] The choice $c(0) = 0$ as well as that of $c(e)$ in the general lemma are is irrelevant. Modifying $c(\gamma)$ for a *finite* set of values of $\gamma$ adds a constant to function (7), so does replacing $c(\gamma)$ by $c(\gamma) + c'(\gamma)$ where $\sum |c'(\gamma)| < +\infty$.

[37] The notation $\wp_L$ is used to point out the dependence with respect to the lattice $L$. It will always be omitted when there is no ambiguity.

$$\wp(u) = 1/u^2 + cu^2 + \dots .$$

Calculating its coefficients is easy: the series appearing on the right had side of (12) being normally convergent in the neighbourhood of 0, the power series of the functions it is composed of can be added term by term.[38] However, for $\omega \neq 0$,

$$1/(u-\omega)^2 = \omega^{-2}(1-u/\omega)^{-2} = \omega^{-2}\sum (n+1)u^n/\omega^n =$$
$$= \sum (n+1)u^n/\omega^{n+2} .$$

Thus

$$\wp_L(u) = 1/u^2 + \sum a_{2n}u^{2n} \tag{13.14'}$$

with

$$a_{2n} = (2n+1)\sum 1/\omega^{2n+2} = (2n+1)G_{2n+2}(L) , \tag{13.14''}$$

where summation is over non-trivial periods. As shown much earlier (Chap. II, n° 23, Chap. VII, n° 18), as a result, the function $\wp$ satisfies the differential equation

$$\wp'(u)^2 = 4\wp(u)^3 - 20a_2\wp(u) - 28a_4 = 4\wp(u)^3 - g_2\wp(u) - g_3$$

in conventional notation. The proof consists in showing that the polar parts of both sides coincide at the origin, and that, as a consequence, the difference between the two sides is an elliptic function without any poles, hence constant, and in checking that it is trivial by calculating the terms independent of $u$ in the power series of both sides of (14").

In the case where the lattice $L$ is generated by 1 and $z$, with $\mathrm{Im}(z) > 0$, let $\wp(u;z)$ denote the corresponding function $\wp_L$. Then

$$a_{2n} = (2n+1)\sum 1/(cz+d)^{2n+2} = (2n+1)G_{2n+2}(z) ,$$

where summation is over all ordered pairs of integers $(c,d) \neq (0,0)$. As will be seen, the *Eisentein series* $G_{2n}(z)$ play a basic role in the theory of modular functions because of the functional equation

$$G_n[(az+b)/(cz+d)] = (cz+d)^n G_n(z)$$

satisfied by all transformations of $SL_2(\mathbb{Z})$. The function $\wp(u;z)$ remains unchanged if $z$ is replaced by $z+1$, but an immediate calculation shows that

$$z^{-2}\wp(-u/z; -1/z) = \wp(u;z) . \tag{13.15}$$

(iii) *The series* $\sum \pi^2/\sin^2 \pi(u+nz)$ *and* $G_2(z)$. For a less trivial example of the construction of elliptic functions that makes elliptic functions out of functions

[38] The coefficients of the power series of a holomorphic function in the neighbourhood of 0 are its successive derivatives at the origin. However, one can pass to the limit in the derivatives of a *uniformly* convergent series of holomorphic functions (Weierstrass). The power series of a uniformly convergent limit (resp. series) is thus given by the formal calculation that would have seemed obvious to Euler...

that are not, consider the function $\pi^2/\sin^2 \pi u$. It has period 1 and double poles at points $n \in \mathbb{Z}$. In a given compact set $K$, the terms of the series

$$f(u;z) = \sum \pi^2/\sin^2 \pi(u+nz) \tag{13.16}$$

are all holomorphic except the finitely many that have a pole in $K$. Hence, if (16) converges normally in all of $K \subset \mathbb{C} - L$, then the result will be of the form

$$f(u;z) = a(z)\wp(u;z) + c(z)$$

with the constants appearing in the Laurent series

$$f(u;z) = a(z)/u^2 + c(z) + \dots$$

of $f$ at the origin.

To analyze convergence, observe that setting as usual $q = \exp(\pi i z)$ and $w = \exp(2\pi i u)$,

$$f(u;z) = -4\pi^2 \sum_{\mathbb{Z}} 1 \Big/ \left(q^n w^{\frac{1}{2}} - q^{-n} w^{-\frac{1}{2}}\right)^2 .$$

If $u$ stays in a compact subset of $\mathbb{C} - L$, then $r \leq |w| \leq R$ with $0 < r < R < +\infty$, and so $|q^n w^{\frac{1}{2}} - q^{-n} w^{-\frac{1}{2}}| \geq |r^{\frac{1}{2}} |q|^n - R^{\frac{1}{2}} |q|^{-n}| \neq 0$ since $|q| < 1$. As $n$ tends to $+\infty$, $q^n$ tends to 0 and $|q|^{-n}$ to $+\infty$. Thus

$$\left|q^n w^{\frac{1}{2}} - q^{-n} w^{-\frac{1}{2}}\right| \geq \frac{1}{2} R^{\frac{1}{2}} |q|^{-n} \quad \text{for large } n,$$

and so $|1/(q^n w^{\frac{1}{2}} - q^{-n} w^{-\frac{1}{2}})^2| \leq 2R^{-\frac{1}{2}} |q|^n$, which shows that the series with terms $n \geq 0$ converges normally. The proof remains the same for the series extended to $n \leq 0$.

To evaluate the constants $a(z)$ and $c(z)$, observe that, for any normally convergent series of meromorphic functions, the Laurent series of the sum can be formally calculated from those of the terms of the series. For $n \neq 0$, the term $\pi^2/\sin^2 \pi(u+nz)$ contributes its value $\pi^2/\sin^2 \pi nz$ at $u = 0$ to the calculation of $c(z)$ and nothing to the calculation of $a(z)$. For $n = 0$, we use the formula

$$\pi^2/\sin^2 \pi u = \pi^2 \left(\pi u - \pi^3 u^3/6 + \dots\right)^{-2} = u^{-2} + \pi u^2/3 + \dots,$$

which gives $a(z) = 1$ and a contribution $\pi^2/3$ to the calculation of $c(z)$, so that

$$c(z) = \pi^2/3 + \sum_{n \neq 0} \pi^2/\sin^2 \pi nz .$$

We thus get the formula

$$\wp(u;z) = \sum_{\mathbb{Z}} \pi^2/\sin^2 \pi(u+nz) - \pi^2/3 - \sum_{n\neq 0} \pi^2/\sin^2 \pi nz . \tag{13.17}$$

It is not unrelated to that of Weierstrass. Indeed[39]

[39] Differentiate the relation

$$\pi \operatorname{cotg} \pi u = \frac{1}{u} + \sum_{n \neq 0} \left(\frac{1}{u-n} + \frac{1}{n}\right).$$

$$\pi^2/\sin^2 \pi u = \sum_{\mathbb{Z}} 1/(u-m)^2$$

for non-integer $u \in \mathbb{C}$, and so

$$f(u;z) = \sum_{n\in\mathbb{Z}} \sum_{m\in\mathbb{Z}} 1/(u-m-nz)^2 \ .$$

We recover the series $\sum 1/(u-\omega)^2$ except for a "detail": its unconditional sum is not well-defined, whereas a convergent result is obtained by *first* summing over $m$ *then* over $n$. We can then write

$$\wp(u;z) = \sum_n \sum_m 1/(u-m-nz)^2 - \pi^2/3 -$$

$$-\sum_{n\neq 0} \sum_m 1/(m+nz)^2 \ . \tag{13.18}$$

Grouping together similar terms of both series, we find

$$\wp(u;z) = 1/u^2 + \sum_{\omega\neq 0} \left[1/(u-\omega)^2 - 1/\omega^2\right] - \pi^2/3 + \sum_{m\neq 0} 1/m^2 \ .$$

This is Weierstrass' formula since $\sum 1/m^2 = \pi^2/3$. (18) is generally written as

$$\wp(u;z) = \sum_n \sum_m 1/(u-m-nz)^2 - G_2(z) \tag{13.19}$$

with a function

$$G_2(z) = \sum_{\substack{n \quad m \\ (m,n)\neq(0,0)}}\sum 1/(m+nz)^2 = \pi^2/3 + \sum_{n\neq 0} \pi^2/\sin^2 \pi nz \tag{13.20}$$

similar to the series $G_{2n}(z)$ ($n \geq 2$), but which should be used cautiously since the double series does not converge unconditionally. The next section will show that permuting the summation order actually changes the result.

(iv) *Relation between* $\wp$ *and* $\theta_1$ *functions* . Suppose the lattice $L$ is generated by 1 and $z$ with $\mathrm{Im}(z) > 0$. To construct the Weierstrass function let us return to the power series $\theta_k(u;z)$ introduced in n° 2, (iii). We know their zeros (all simple because of infinite product expansions) and their transformation formulas under $u \longmapsto u+1$ and $u \longmapsto u+z$ (n° 12, *Exercise* 2). These immediately show that the functions

$$f_k(u) = \theta_k(u;z)^2/\theta_1(u;z)^2 \quad (k=2,3,4)$$

are elliptic and their only singularities are double poles at the period lattice points. Hence

$$f_k(u) = a_k(z)\wp(u;z) + b_k(z) = a_k(z)/u^2 + b_k(z) + \dots \ ,$$

with coefficients depending only on $z$. These are calculated by considering the Laurent series of the functions $f_k$ at the origin. Setting

$$\theta_k^*(u;z) = \theta_k(u;z)/\theta_k(z) \qquad \text{for } 2 \leq k \leq 4 \, ,$$
$$\theta_1^*(u;z) = \theta_1(u;z)/\theta_1'(z)$$

in order to get functions in the neighbourhood of 0 equivalent to 1 for $k \geq 2$ and to $1/u$ for $k = 1$, we find

$$\wp(u;z) = \theta_k^*(u;z)^2 / \theta_1^*(u;z)^2 - \theta_k''(z)/\theta_k(z)\,,$$

where recall that $\theta_k''(z)$ denotes the second derivative at $u = 0$ of the function $u \longmapsto \theta_k(u;z)$. This result is not unrelated to *Exercise* 9 of n° 12.

Another method amounts to observing that, given a general theta function $f(u)$ satisfying relations of the form

$$f(u+1) = \exp(az+b)f(u)\,, \quad f(u+z) = \exp(cz+d)f(u)\,,$$

its logarithmic derivative $g = f'/f$ satisfies

$$g(u+1) = g(u) + a\,, \quad g(u+z) = g(u) + c\,,$$

so that $h = (f'/f)'$ is an elliptic function.

For example let us choose the odd function $\theta_1(u;z)$ which is non-trivial at lattice points. Writing

$$\theta_1(u;z) = a(z)u + b(z)u^3 + \ldots$$

in the neighbourhood of 0 gives $g(u) = 1/u + \ldots$ and hence $h(u) = -1/u^2 + \ldots$ So there is a relation of the form

$$\left[\theta_1'(u;z)/\theta_1(u;z)\right]' = -\wp(u;z) + c(z)$$

with a function $c(z)$ which, as will be seen, is the same as the one used to obtain (17). Since

$$f(u) = \theta_1(u;z) = \prod \left(1-q^{2n}\right)\left(1-q^{2n}w\right)\left(1-q^{2n-2}w^{-1}\right)$$

with $w = \exp(2\pi i u)$ and differentiating with respect to $u, w' = 2\pi i w$, $(w^{-1})' = -2\pi i w^{-1}$,

$$\begin{aligned} f'(u)/f(u) = -2\pi i \sum_{n\geq 1} q^{2n}w/\left(1-q^{2n}w\right) + \\ +2\pi i \sum_{n\geq 0} q^{2n}w^{-1}/\left(1-q^{2n}w^{-1}\right)\,. \end{aligned} \tag{13.21}$$

Differentiating again with respect to $u$, it follows that

$$\begin{aligned} \left(f'(u)/f(u)\right)' = 4\pi^2 \sum q^{2n}w/\left(1-q^{2n}w\right)^2 + \\ + 4\pi^2 \sum q^{2n}w^{-1}/\left(1-q^{2n}w^{-1}\right)^2 \end{aligned} \tag{13.21'}$$

and

$$\begin{aligned} \left(f'(u)/f(u)\right)' &= 4\pi^2 \sum_{n\geq 1} 1 \Big/ \left(q^n w^{\frac{1}{2}} - q^{-n}w^{-\frac{1}{2}}\right)^2 + \\ &+4\pi^2 \sum_{n\geq 0} 1 \Big/ \left(q^n w^{-\frac{1}{2}} - q^{-n}w^{\frac{1}{2}}\right)^2 = \\ &= 4\pi^2 \sum_{n\geq 1} 1/[2i\sin\pi(u+nz)]^2 + \\ &+4\pi^2 \sum_{n\geq 0} 1/[2i\sin\pi(-u+nz)]^2\,. \end{aligned}$$

So finally,

$$\left[\theta_1'(u;z)/\,\theta_1(u;z)\right]' = -\sum_{\mathbb{Z}} \pi^2/\sin^2\pi(u+nz)\,.$$

This is function (16). Using (17), $c(z) = -G_2(z)$ follows and so

$$\left[\theta_1'(u;z)/\,\theta_1(u;z)\right]' = -\wp(u;z) - G_2(z)\,. \tag{13.22}$$

This result enables us to calculate the effect of $z \longmapsto -1/z$ on the function $G_2(z)$. Let us denote by $h(u;z)$ the left hand side and, in (22), replace $z$ by $-1/z$. Taking into account the relation

$$\theta_1(u;-1/z) = -i(z/i)^{\frac{1}{2}}\exp(\pi i u^2 z)\theta_1(uz;z)$$

and the factor $z^2$ which occurs when the last term of the right hand side is differentiated twice with respect to $u$, we get

$$h\,(u;-1/z) = z^2 h(uz;z) + 2\pi i z = z^2\left[\wp(uz;z) + G_2(z)\right] + 2\pi i z\,.$$

However, by (15), $\wp(u;-1/z) = z^2\wp(uz;z)$. So relation (22) for $-1/z$ becomes

$$z^2\left[\wp(uz;z) + G_2(z)\right] - 2\pi i z = z^2\wp(uz;z) + G_2\left(-1/z\right)\,,$$

given the expected formula

$$G_2(-1/z) = z^2 G_2(z) - 2\pi i z\,. \tag{13.23}$$

It is due to Eisenstein who gave a direct elementary proof,[40] but full of traps. Definition (20) shows that

$$G_2\left(-1/z\right) = z^2\sum_n\sum_m 1\,/(n-mz)^2 = z^2\sum_m\sum_n 1\,/(m+nz)^2\,,$$

(the term $m = n = 0$ is obviously omitted), so that we again get the series $G_2(z)$, up to the factor $z^2$, but *the order of summation has be changed.* This operation, which would not change anything if the double series converged unconditionally, is not innocuous: indeed (23) can be put in the form

$$\sum_m\sum_n 1\,/(m+nz)^2 = \sum_n\sum_m 1\,/(m+nz)^2 - 2\pi i/z\,. \tag{13.23'}$$

*Exercise 1.* Show that, applying the relation $1/(1-z) = \sum z^m$, which supposes $|z| < 1$, to the terms of the two series appearing in (21) gives a double series converging unconditionally if and only if $u$ is in the horizontal strip

$$B: -\operatorname{Im}(z) < \operatorname{Im}(u) < 0\,.$$

Permuting the summation order, show that

$$\wp(u;z) = -G_2(z) + 4\pi^2\sum_{m\neq 0}\frac{mq^m}{q^m - q^{-m}}\mathbf{e}(mu)\,. \tag{13.24}$$

[40] See for example Freitag and Busam, pp. 390–393.

This is the Fourier series expansion of the function $u \longmapsto \wp(u;z)$.[41] What is the Fourier series of $\wp(u)$ in the strip $p\,\mathrm{Im}(z) < \mathrm{Im}(u) < (p+1)\,\mathrm{Im}(z)$, where $p \in \mathbb{Z}$?

*Exercise 2.* Using the identity

$$\sum_{n\geq 1} nx^n/(1-x^n) = \sum_{m,n\geq 1} nx^{mn} = \sum_{m\geq 1} x^m(1-x^m)^2$$

valid for $|x| < 1$, as well as the infinite product of the function $\eta(z)$, show that

$$G_2(z) = -4\pi i\eta'(z)/\eta(z)\,, \tag{13.25}$$

$\eta$ being the Dedekind function. Deduce[42] that

$$G_2(-1/z) = z^2G_2(z) - 2\pi iz \Longleftrightarrow \eta(-1/z) = (z/i)^{\frac{1}{2}}\eta(z)\,.$$

(v) *Elliptic functions with given simple poles.* They can be obtained either using Abel and Jacobi's infinite products, or the Eisenstein-Weierstrass method. We still assume that $L$ is generated by 1 and $z$.

The first one, presented above, is the simplest: we fix the zeros $a_k$ and the poles $b_k$ $(1 \leq k \leq n) \bmod L$ beforehand and consider the function

$$\varphi(u) = \prod \theta_1(u-a_k)/\theta_1(u-b_k)\,. \tag{13.26}$$

It clearly answers the question if it is elliptic. Invariance under $u \longmapsto u+1$ is obvious. Due to formulas of n° 12, exercise 2, $u \longmapsto u+z$ multiplies the function by a factor

$$\prod \mathbf{e}[-(u-a_k)-z/2]\Big/\prod \mathbf{e}[-(u-b_k)-z/2] = \mathbf{e}\left(\sum a_k - \sum b_k\right)\,.$$

Therefore, supposing $\sum a_k \equiv \sum b_k \bmod \mathbb{Z}$ is sufficient to obtain the result. As we anyhow need to suppose $\sum a_k \equiv \sum b_k \bmod L$, the condition can always be assumed to hold.

This method allows us to impose poles and zeros, but not residues. It is possible to construct an elliptic function with simple poles $\bmod L$ at given pairwise distinct points $a_1, \ldots, a_n$, with given residues $\varrho_k$ at these points, but we then have no information about its zeros: everything cannot be imposed at the same time. Hermite only remarks that, if

$$\sum \varrho_k = 0\,,$$

a compulsory condition by theorem 8, then the function

$$C + \sum \varrho_k\theta_1'(u-a_k;z)/\theta_1(u-a_k)\,,$$

where $C$ is a constant, clearly answers the question (i.e. is elliptic) thanks to the transformation formulas given in the exercises of n° 12.

---

[41] It holds in strip (24) where it is holomorphic of period 1 and diverges elsewhere.

[42] This is for instance what Neal Koblitz does in *Introduction to Elliptic Curves and Modular Forms* (Springer, 1993, 2d. ed.), p. 121 to get (23).

The lattice $L$ now being arbitrary, Weierstrass starts from the rational function

$$f(u) = \sum \varrho_k / (u - a_k) \tag{13.27}$$

satisfying the imposed conditions but which is not elliptic and makes it elliptic using the series

$$f^L(u) = \sum f(u - \omega) = \sum\sum \varrho_k / (u - \omega - a_k) \ . \tag{13.28}$$

For it to converge, one must make sure that $f(u)$ decreases sufficiently rapidly at infinity. At a first glance, $f(u) \sim c/u$ where $c = \sum \varrho_k$. Hence one must assume that $\sum \varrho_k = 0$, a not very surprising condition. Then for large $\omega$,

$$\begin{aligned}\varrho_k (u - \omega - a_k)^{-1} &= -\varrho_k \omega^{-1} [1 + (u - a_k)/\omega]^{-1} = \\ &= \varrho_k \left[1/\omega - (u - a_k)/\omega^2 + (u - a_k)^2/\omega^3 - \ldots\right] .\end{aligned}$$

Adding the results for $1 \leq k \leq n$, the terms in $1/\omega$ disappear since $\sum \varrho_k = 0$ and the coefficient of $1/\omega^2$ reduces to $\sum \varrho_k a_k$. To make sure that (28) converges, it is also necessary to suppose

$$c = \sum \varrho_k a_k = 0 ; \tag{13.29}$$

which does not augur well. Let us in particular suppose that $n = 2$. The two residues are opposites and (29) means that $a_1 = a_2$, in which case the function $f$, hence also $f^L$, is identically zero... Nonetheless, there are elliptic functions with only two given simple poles $a_1$ and $a_2$ as can be seen by using theta series.

The general lemma used above to construct the Weierstrass function explains the mystery. For $n = 2$, $f(u) = 1/(u - a_1) - 1/(u - a_2)$, up to a constant factor, and so for given $u$ and large $\omega$,

$$f(u - \omega) = c/\omega^2 + O\left(1/\omega^3\right)$$

with $c = a_1 - a_2$. Thus the solution, for $c \neq 0$, is to define $f^L$ not by (28), which diverges, but by

$$\begin{aligned}f_L(u) &= f(u) + \sum \left[f(u - \omega) - c/\omega^2\right] = \\ &= \sum \left[f(u - \omega) - c(\omega)\right] ,\end{aligned} \tag{13.30}$$

where $c(\omega) = c/\omega^2$ if $\omega \neq 0$, $c(0) = 0$. Conditions (i), (ii) and (iii) of the general lemma can be readily checked and this gives the expected elliptic function.

More generally, when starting from the function

$$f(u) = \sum \varrho_k / (u - a_k) \quad \text{with } \sum \varrho_k = 0 ,$$

the necessity of condition (27) to ensure convergence of the series $\sum f(u - \omega)$ is dealt with in a similar way. As we saw, for large $\omega$,

$$f(u - \omega) = c/\omega^2 + O\left(1/\omega^3\right) \quad \text{where } c = \sum a_k \varrho_k .$$

Hence we apply the general lemma by choosing the function

$$c(\omega) = c/\omega^2 \quad \text{if } \omega \neq 0, \quad c(0) = 0 ,$$

which leads to a function

$$f^L(u) = f(u) + \sum_{\omega \neq 0} \left[f(u-\omega) - c/\omega^2\right] \tag{13.31}$$

with simple poles at $a_k + \omega$, and given residues $\varrho_k$, and holomorphic elsewhere. It is unique, up to an additive constant.

(vi) *The functions* $\zeta_L$ *and* $\sigma_L$. Let us continue our variations on the same the with Weierstrass and start from the function $f(u) = 1/u$. The series $1/(u-\omega)$ does not converge. For large $\omega$,

$$\begin{aligned} 1/(u-\omega) &= -1/\omega(1-u/\omega) = -\left(1/\omega + u/\omega^2 + u^2/\omega^3 + \ldots\right) = \\ &= -1/\omega - u/\omega^2 + O\left(1/\omega^3\right), \end{aligned}$$

so that the series

$$\begin{aligned} \zeta_L(u) &= 1/u + \sum \left[1/(u-\omega) + 1/\omega + u/\omega^2\right] = \\ &= 1/u + \sum \left(u^2/\omega^3 + u^3/\omega^4 + \ldots\right) = \\ &= 1/u + \sum_{n \geq 2} G_{2n}(L) u^{2n-1} \end{aligned} \tag{13.32}$$

converges and represents an *odd* meromorphic function whose singularities are obvious. This why this type of formula is important. Because of the presence of the term $u/\omega^2$, depending on $u$, the general lemma cannot be applied to prove invariance under $u \longmapsto u + \omega$. Term by term differentiation gives

$$-\zeta_L'(u) = \wp_L(u). \tag{13.33}$$

Thus, for any period,

$$\zeta_L(u+\omega) = \zeta_L(u) + \eta_L(\omega) \tag{13.34}$$

with a constant

$$\eta_L(\omega) = 2\zeta_L(\omega/2) \quad \text{if } \omega/2 \notin L \tag{13.34'}$$

since $\zeta_L$ is odd. Obviously,

$$\eta_L(\omega' + \omega'') = \eta_L(\omega') + \eta_L(\omega'')$$

for all periods $\omega'$ and $\omega''$.

*Exercise 3.* If $\mathrm{Im}(\omega_2/\omega_1) > 0$, then $\omega_2\eta_L(\omega_1) - \omega_1\eta_L(\omega_2) = 2\pi i$.

The function $\zeta_L$ enables us to put formula (25) in the form

$$f_L(u) = \sum \varrho_k \zeta_L(u - a_k) \tag{13.35}$$

because the sum of the terms $\varrho_k/\omega$ which should appear in the last sum of (25) is 0 since $\sum \varrho_k = 0$. By relation (34), the periodicity of the right hand side of (35) is obvious.

Another function related to $\wp_L$ and $\zeta_L$ is the infinite product

$$\sigma_L(u) = u \prod_{\omega \neq 0} (1 - u/\omega) \exp\left(u/\omega + u^2/2\omega^2\right) = -\sigma_L(-u) \tag{13.36}$$

invented by Weierstrass to obtain an entire function with simple zeros at all points of the lattice $L$ and $\neq 0$ elsewhere. Convergence is immediate since the general term is

$$\begin{aligned}(1-u/\omega)\left(1+u/\omega+u^2/2\omega^2+\ldots\right)\left(1+u^2/2\omega^2+\ldots\right) &= \\ = (1-u/\omega)\left(1+u/\omega+u^2/\omega^2+\ldots\right) &= 1+O\left(u^3/\omega^3\right).\end{aligned}$$

Differentiating (3) logarithmically gives

$$\sigma_L'(u)/\,\sigma_L(u)=\zeta_L(u) \tag{13.37}$$

and as a result,

$$\wp_L=-\left(\sigma_L'/\sigma_L\right)'. \tag{13.38}$$

Comparison with (22) suggests a simple relation between functions $\sigma$ and $\theta_1$ and, at the very least, an infinite product expansion of $\sigma$[43] very different from (36)...

## 14 – Elliptic Integrals

(i) *The field of elliptic functions.* Weierstrass was apparently the first to obtain the next result:

**Theorem 11.** *Every elliptic function $f$ is a rational function of $\wp$ and $\wp'$.*

This means that there is a function $R(X,Y)=P(X,Y)/Q(X,Y)$ of two variables, where $P$ and $Q$ are polynomial, such that

$$f(u)=R\left(\wp(u),\wp'(u)\right) \tag{14.1}$$

for all $u\in\mathbb{C}$ for which the relation is well-defined, i.e. except for a discrete set of values of $u$.

If the function $f$ has poles not belonging to the period lattice $L$, then let $a_k$ $(1\leq k\leq m)$ denote their representatives mod $L$ and let $r_k$ be their multiplicities. As the function $\wp=\wp_L$ is holomorphic at these points, the elliptic function

$$g(u)=f(u)\prod\left[\wp(u)-\wp\left(a_k\right)\right]^{r_k}$$

is holomorphic outside $L$. If it is holomorphic at $u=0$, hence on all of $\mathbb{C}$, it is constant and the theorem is proved.

So let us suppose this is not the case. There is a Laurent series

$$g(u)=c_r/\,u^r+c_{r-1}/u^{r-1}+\ldots$$

in the neighbourhood of 0 with $c_r\neq 0$ and $r\geq 2$ since an elliptic function cannot have a single simple pole mod $L$. If $r=2$, the function $g(u)-c_2\wp(u)$ has, at most, a simple pole, and so is again constant. If $r\geq 3$, there are two possibilities. If $r=2m$, the order of the pole of $g$ can be reduced by subtracting the function $c_r\wp(u)^m$. If $r$ is odd, then $r=3m+2n$ with $n\in\{0,1\}$. As $\wp'(u)=-2/u^3+\ldots$, it is obviously possible to choose a constant $c$ such that $g(u)-c\wp'(u)^m\wp(u)^n$ has a pole of order $<r$ at the origin. Induction on $r$ implies that a polynomial in $\wp$ and $\wp'$ canceling out the pole of $g$ can be subtracted from $g$, thus making $g$ constant, qed.

[43] See for example Joseph H. Silverman, *Advanced Topics in the Arithmetic of Elliptic Curves* (Springer, 1994), p. 53.

Since $\wp'^2$ is a polynomial in $\wp$, this argument shows that every elliptic function is actually of the form

$$f = R_1(\wp) + \wp' R_2(\wp) \tag{14.2}$$

where the $R_i$ are rational functions of one variable.

Even if theorem 11 does not hold for all elliptic functions, it is far from being restricted to the function $\wp$. One could for example use a function with only two simple poles. This is form (2) of $R(\wp, \wp')$ which is so because of its simplicity.

This can be interpreted in the algebraic language of commutative field theory. For a given lattice $L$, let $K$ denote the set of corresponding elliptic functions. They can be added, multiplied and divided, and these operations, applicable to all meromorphic functions on a given domain, obviously satisfy the axioms of a commutative field. Among the functions $f \in K$ there are rational functions of $\wp$, which form a subfield $\mathbb{C}(\wp)$ of $K$ containing $\mathbb{C}$. The element $\wp$ of $\mathbb{C}(\wp)$ is obviously transcendent over $\mathbb{C}$, i.e. does not satisfy any algebraic equation $\sum a_n \wp^n = 0$ with coefficients in $\mathbb{C}$: such a relation would force the Weierstrass function to take only finitely many values.

The differential equation of the function $\wp$ shows that, on the other hand, $\wp'$ is algebraic over $\mathbb{C}(\wp)$: $\wp'^2 \in \mathbb{C}(\wp)$. In fact, every $f \in K$ is algebraic over $\mathbb{C}(\wp)$. To see this, it suffices to write $f$ and $f^2$ as in (2) and to remove $\wp'$ from the two relations that arise. This clearly gives an equation $af^2 + bf + c = 0$ with $a, b, c \in \mathbb{C}(\wp)$. $a$, $b$, $c$ can be assumed to be polynomials in $\wp$ (get rid of denominators), which means that there is always a non-trivial algebraic relation $P(f, \wp) = 0$ between $f$ and $\wp$. More generally:

**Corollary 1.** *Let $f$ and $g$ be two elliptic functions having the same periods. There is a non-trivial polynomial $P \in \mathbb{C}[X, Y]$ such that $P(f, g) = 0$.*

It could be tempting to prove this by constructing a polynomial $P$ for which the elliptic function $P[f(u), g(u)]$ does not have poles. This would require intricate calculations since $f$ and $g$ cannot have an arbitrary number of poles randomly distributed in $\mathbb{C}$ , modulo the period lattice. The *abc* of commutative field theory gets totally rid of these difficulties and makes the corollary a very particular of a much more general and purely algebraic result.[44]

**Corollary 2.** *Every elliptic function satisfies a differential polynomial equation $P(f, f') = 0$.*

(ii) *The Riemann surface of the field of elliptic functions.* We saw in n° 11 that elliptic functions of the lattice $L$ can be identified with meromorphic functions on the compact Riemann surface $\mathbb{C}/L$. On the other hand, the functions $\wp$ and $\wp'$ are connected by algebraic equation (13.14"), which will be written

$$\zeta^2 - 4z^3 + g_2 z + g_3 = 0 \tag{14.3}$$

[44] For this consider a field $K$ and a subfield $k$ of $K$ (here take $k = \mathbb{C}$) and assume there are $d$ elements $x_i$ of $K$ (here take $d = 1$ and $x_1 = \wp$) such that every element of $K$ is algebraic over the subfield $k(x_1, \ldots, x_d)$ generated by $k$ and the elements $x_i$. This general theorem asserts that there always is a non-trivial algebraic relation $P(y_1, \ldots, y_{d+1}) = 0$ with coefficients in $k$ between $d + 1$ elements $y_j$ of $K$. The reader will find the proof – consisting essentially of linear algebra – in Serge Lang, *Algebra* and several other textbooks. In fact, §§ 9 and 26 to 29 of my *Cours d'algèbre* and the corresponding exercises ought to be enough.

here so as to follow the notation of Chap. X, n° 4. There is a compact Riemann surface $\hat{X}$ connected to this equation, which except for critical points of (3), is the set of solutions $(z, \zeta)$ of this equation. As $(\wp(u), \wp'(u))$ clearly satisfies (3) for $u \notin L$, the next result arises naturally:

**Theorem 12.** *The map*

$$j : u \bmod L \longmapsto (\wp(u), \wp'(u)) \tag{14.4}$$

*can be extended to an isomorphism from the Riemann surface $\mathbb{C}/L$ onto the algebraic Riemann surface $\hat{X}$ associated to equation (3).*

There would be nothing to prove if we already knew that, when the fields of meromorphic functions of two compact Riemann surfaces are isomorphic, so are the two surfaces considered. But we will have to forego this general theorem, construct the compact Riemann surface of (3) following step by step n° 4 of Chap. X, then define $j$ for all $u \in \mathbb{C}$ and check that, this gives a holomorphic bijection from $\mathbb{C}/L$ onto $\hat{X}$, whose inverse is also holomorphic. In fact, we will construct a holomorphic surjective map $j$ from $\mathbb{C}$ onto $\hat{X}$ such that $j(u) = j(v)$ if and only if $u \equiv v \bmod L$. Passing to the quotient will give the expected isomorphism.

In accordance with Chap. X, n° 4, we start by removing from $\mathbb{C}$ of the values of $z$ where the equation in $\zeta$ has a multiple root. These are the roots usually denoted by $e_1$, $e_2$, $e_3$, of the equation

$$4z^3 - g_2 z - g_3 = 0\,. \tag{14.5}$$

They correspond to numbers $u \in \mathbb{C}$ where $\wp'(u) = 0$, As $\wp'$ is an odd holomorphic elliptic function outside $L$, $\wp'(u) = 0$ if $u \equiv -u \not\equiv 0 \bmod L$, hence if $u = \omega/2$ for a period $\omega$ such that $\omega/2 \notin L$. So, denoting by $\omega_1$ and $\omega_2$ a basis of $L$,

$$u \in \left\{\omega_1/2, \omega_2/2, (\omega_1 + \omega_2)/2\right\} \bmod L\,. \tag{14.6}$$

These three numbers being distinct $\bmod L$,

$$e_1 = \wp(\omega_1/2)\,,\quad e_2 = \wp(\omega_2/2)\,,\quad e_3 = \wp\left[(\omega_1 + \omega_2)/2\right] = -e_1 - e_2 \tag{14.7}$$

since the sum of the roots of (4) is zero. These roots of (4) are simple. Indeed if for example $e_1 = e_2$, then the elliptic function $f(u) = \wp'(u)/2[\wp(u) - e_1]$ would clearly satisfy the equality $f(u)^2 = \wp(u) - e_3$, which is impossible since then $f$ would have a single simple pole.

To construct $\hat{X}$ and the map $\pi : \hat{X} \longrightarrow \hat{\mathbb{C}}$, we start by constructing the open subset $X$ of $\hat{X}$ located on top of $B = \mathbb{C} - \{e_1, e_2, e_3\}$. Topologically it is the set of solutions $P = (z, \zeta) \in \mathbb{C}^2$ of (3) for which $z \in B$, the map $\pi$ being given on $X$ by $\pi(z, \zeta) = z$.

Let us show that, for such an ordered pair, there exists a unique $u \in \mathbb{C} \bmod L$ for which $z = \wp(u)$, $\zeta = \wp'(u)$. The function $\wp$ being elliptic of order 2, the equation $\wp(u) = z$ has two simple roots or one double root, modulo $L$ (corollary of theorem 7). In the second case, $\wp'(u) = 0$, and so $z \in \{e_1, e_2, e_3\}$, a case temporarily excluded. As moreover $\wp$ is even and $\wp'$ odd and as $z$ determines $\zeta$ up to sign, the result follows.

Let $\pi_1$, $\pi_2$ and $\pi_3$ denote points of $\mathbb{C}$ where $\wp$ takes values $e_1$, $e_2$ and $e_3$, $B'$ the set of $u \in \mathbb{C}$ distinct from $\pi_1$, $\pi_2$, $\pi_3$ and 0, $\bmod L$, and let $\Omega$ be the image of $B'$ in $\mathbb{C}/L$, an open subset whose complement consists of the classes of $\pi_1$, $\pi_2$, $\pi_3$ and 0. We associate the element $j(u)$ of $X$ given by (4) to all $u \in B'$ . Taking the quotient $\bmod L$, $j$ defines a bijection from $\Omega$ onto $X$.

Let us show that $j$ is an isomorphism between the non-compact Riemann surfaces $X$ and $\Omega$. Let $P = (a, \alpha)$ be a point of $X$. In the neighbourhood of $P$, the function $q(z, \zeta) = z - a$ is a conformal representation of a neighbourhood of $P$ on a neighbourhood of $a$ since we are on top of a critical point of (3). On the other hand, if $b \in \mathbb{C}$ is chosen so that $a = \wp(b)$ and $\alpha = \wp'(b)$, the function $\wp$ is a conformal representation of any sufficiently small neighbourhood of $b$ on a neighbourhood of $a$ since $\wp'(b) \neq 0$. The composition of $u \longmapsto \wp(u)$ and the inverse of $(z, \zeta) \longmapsto z - a$ shows that $u \longmapsto (\wp(u), \wp'(u)) = j(u)$ is clearly also a conformal representation of a neighbourhood of $b$ on a neighbourhood of $(a, \alpha)$. Thus there is an isomorphism between $\Omega$ and $X$.

$j$ now needs to be defined at the four points of $\mathbb{C}/L$ that are not in $\Omega$, i.e. at $u \equiv \pi_1, \pi_2, \pi_3$ or $0 \bmod L$.

Let us first analyze the surface $\hat{X}$ on top of a neighbourhood of $a = e_1 = \wp(\pi_1)$. Let (notation of Chap. X, n° 4) $D(a)^*$ be a pointed disc centered at $a$ not containing any of the other points $e_j$. The inverse image $\pi^{-1}[D(a)^*] \subset X$ is a covering of order 2 of $D(a)^*$. Let us show it is connected. Otherwise, it would be possible to define two uniform branches $f(z)$ of equation (2) (Chap. X, theorem 5 for $k = 1$) in the neighbourhood of $a$, i.e two solutions of

$$f(z)^2 = 4(z - a)(z - e_2)(z - e_3)$$

holomorphic for sufficiently small $|z - a| > 0$. Such a function $f$ would be bounded in the neighbourhood of $a$, hence holomorphic on a *closed* disc centered at $a$ and zero at $a$: this is absurd since the right hand side has a *simple* zero at $a$.

Hence this shows that there is a unique point $P_i$ of $\hat{X}$ on top of each critical point $a = e_i$. When $P \in X$ converges to $P_i$, the ordered pair $(z, \zeta) = P$ converges in $\mathbb{C}^2$ to $(e_i, 0)$, which enables us to identify $P_i$ with $(e_i, 0) = (\wp(\pi_i), \wp'(\pi_i))$. Thus in accordance to (4), we set $j(\pi_i) = P_i$. This defines $j$ on $\mathbb{C} - L$, i.e. on $\mathbb{C}/L$ with the class of 0 removed. $j$ is clearly continuous.

To show that $j$ is an isomorphism in $\pi_i$, one needs to use a local uniformizer $q_i$ at $P_i$. Now, we know [Chap. X, eq. (4.2)] that it can be chosen in such a way that

$$q_i(z, \zeta)^2 = z - e_i = \wp(u) - \wp(\pi_i) \quad \text{if } (z, \zeta) = j(u)\,. \tag{14.8}$$

As $\wp'(\pi_i) = 0$, on the other hand, in the neighbourhood of $\pi_i$,

$$\wp(u) - \wp(\pi_i) = (u - \pi_i)^2 g(u)\,,$$

where $g$ is defined and holomorphic on a neighbourhood of $\pi_i$ and non-trivial at $\pi_i$, and so $q_i[j(u)]^2 = (u - \pi_i)^2 g(u)$ in the neighbourhood of $\pi_i$. As $g(\pi_i) \neq 0$, there is a holomorphic function such that $g(u) = f(u)^2$ on a neighbourhood of $\pi_i$. It is unique up to sign. Both sides of

$$q_i\,[j(u)]^2 = (u - \pi_i)^2 f(u)^2 \tag{14.9}$$

are holomorphic on a pointed disc $D(\pi_i)^*$ since $j$ is holomorphic except at the points $\pi_i$. Hence, as $D(\pi_i)^*$ is connected, if need be replacing $f$ by $-f$,

$$q_i\,[j(u)] = (u - \pi_i)\, f(u)\,. \tag{14.10}$$

It follows that the left hand side is holomorphic on a neighbourhood of $\pi_i$, which shows that $j$ is holomorphic as a map from $\mathbb{C} - L$ to the Riemann surface $\hat{X}$. Therefore, as $j$ is clearly injective, it is an isomorphism from $\mathbb{C} - L$ onto $\hat{X}$ with the points projecting onto $\infty \in \hat{\mathbb{C}}$ removed.

To finish the construction of $j$, let us analyze the surface $\hat{X}$ on top of the point $\infty$ of the Riemann sphere. To reduce to finite values of $z$ and $\zeta$, we set $z = 1/z'$, $\zeta = 1/\zeta'$, which transforms equation (3) into

$$z'^3 - \zeta'^2 \left(4 - g_2 z'^2 - g_3 z'^3\right) = 0 \tag{14.11}$$

which has to studied in a neighbourhood of $z' = 0$, a value where it has the double root $\zeta' = 0$. The point at infinity is, therefore, a critical point of the initial equation (3). If there were two points of $\hat{X}$ on top of $\infty$, the algebraic function $\zeta' = \mathcal{F}(z')$ defined by (11) would decompose into two uniform branches $f(z')$ and $-f(z')$ in a neighbourhood of 0, and by (11), one would have $f(z')^2 = z'^3 g(z')$ with $g(0) = 1/4$. This is obviously impossible. So there is a unique point $P_\infty$ of $\hat{X}$ projecting onto the point $z = \infty$ of $\hat{\mathbb{C}}$, with a local uniformizer $q_\infty$ satisfying $q_\infty(z', \zeta')^2 = z'$ as above or, coming back to the initial variables,

$$q_\infty(z, \zeta)^2 = 1/z\,,$$

which holds for large $|z|$.

To define $j$ at 0, it is natural to set $j(0) = P_\infty$. We still need to show that $j$ is an isomorphism from a neighbourhood $D(0)$ of 0 onto a neighbourhood of $P_\infty$. However, $j(u) = (\wp(u), \wp'(u))$ on $D(0)^*$, and so

$$q_\infty \left[j(u)\right]^2 = 1/\wp(u)\,.$$

As 0 is a double pole of $\wp$, $1/\wp(u) = u^2 f(u)^2$ where $f$ is holomorphic on $D(0)$ and not zero at 0. The arguments used for neighbourhoods of the other critical points imply that $q_\infty[j(u)] = uf(u)$, which ends the proof of theorem 12.

(iii) *Addition formula.* Corollary 1 of theorem 11 enables us to show for example that, if $f(u)$ is an elliptic function, then for all $n \in \mathbb{Z}$, there is a non-trivial algebraic relation between functions $f(u)$ and $f(nu)$: it suffices to observe that $u \longmapsto f(nu)$ admits the same periods (as well as others...) as $f(u)$. Ever since the second half of the 19th century, the study of these relations has given rise to research that has powerfully contributed to the development of the theory of algebraic equations, in particular Galois theory. Similarly, for all $a \in \mathbb{C}$, there is a relation between $f(u)$ and $f(u+a)$, in other words an "addition theorem". Abel and Jacobi were familiar with these results. Particular cases were known even before them. These had been obtained by explicit calculations sometimes close to wizardry.[45]

If $f$ is the $\wp$-function of the lattice $L$ considered, the addition theorem is obtained very simply. To calculate it, suppose for the moment that $2a$ is not in $L$. This condition ensures two propositions used in the proof: (a) the only zeros of the function $\wp(u) - \wp(a)$ are the obvious ones, namely the points $a$ and $-a \bmod L$; (b) $\wp'(a) \neq 0$ since $a$ has just been shown to be a simple zero of $\wp(u) - \wp(a)$.

This being so, the only singularities of the function $\wp(u-a)$ are double poles at points $\omega + a$, where $\omega \in L$. In the neighbourhood of such a point, it is the sum of $1/(u-a)^2$ and of a power series in $(u-a)^2$ without any constant term. As on the other hand $\wp$ is holomorphic at $a$, $\wp(u) - \wp(a) = \wp'(a)(u-a)[1 + \ldots]$, with $\wp'(a) \neq 0$ as has just been shown. So the elliptic function

$$g(u) = \wp(u-a) - \wp'(a)^2 / \left[\wp(u) - \wp(a)\right]^2 \tag{14.12}$$

[45] See Houzel's article in Dieudonné's *Abrégé d'histoire des mathématiques* tome 2, in particular pp. 8–11 that are difficult to understand without referring to sources since they present little else but results formulated in the notation of the time.

has at most a simple pole at $a$. Its other possible poles are the zeros of $\wp(u) - \wp(a)$, i.e. the points $u = a$ or $u = -a \bmod L$ by (a), and these are double poles since $\wp'(a) \neq 0$. But as $\wp'(u)$ is odd, $\wp'(u) + \wp'(a) = 0$ for $u = -a$. To remove the double pole at $a$, it is, therefore, sufficient to replace function (12) by

$$(14.13) \qquad h(u) = \wp(u-a) - \left[\wp'(u) + \wp'(a)\right]^2 \Big/ 4\left[\wp(u) - \wp(a)\right]^2 .$$

Since, in the neighbourhood of $u = a$,

$$\left[\wp'(u) + \wp'(a)\right]^2 \Big/ 4\left[\wp(u) - \wp(a)\right]^2 = \left[2\wp'(a) + \ldots\right]^2 \Big/ 4\left[\wp'(a)(u-a) + \ldots\right]^2 =$$
$$= (u-a)^{-2} + \ldots ,$$

and $\wp(u-a) = (u-a)^{-2} + \ldots$, like $g$, the new function $h$, has at most a simple pole at $a$.

Thus everything has been done to remove the pole of $g$ at $-a$, but replacing $g$ by $h$ introduces a singularity at $u = 0$, which now needs to be investigated by writing down the first few terms of the Laurent series of the functions concerned at the origin.

Since $\wp(u) = u^{-2} + a_2 u^2 + \ldots$ in the neighbourhood of 0,

$$\wp(u) - \wp(a) = u^{-2}\left[1 - \wp(a)u^2 + g_2 u^4 + \ldots\right] ,$$
$$\left[\wp(u) - \wp(a)\right]^2 = u^{-4}\left[1 - 2\wp(a)u^2 + \ldots\right] ,$$
$$(14.14) \qquad \left[\wp(u) - \wp(a)\right]^{-2} = u^4\left[1 + 2\wp(a)u^2 + \ldots\right] .$$

Similarly, $\wp'(u) = -2u^{-3} + 2a_2 u + \ldots$, and so

$$\wp'(u) + \wp'(a) = -2u^{-3}\left[1 - \frac{1}{2}\wp'(a)u^3 - g_2 u^4 + \ldots\right]$$

et

$$(14.15) \qquad \left[\wp'(u) + \wp'(a)\right]^2 \Big/ 4 = u^{-6}\left[1 - \wp'(a)u^3 + \ldots\right] .$$

Multiplying (14) and (15) sidewise, it follows that

$$\left[\wp'(u) + \wp'(a)\right]^2 \Big/ 4\left[\wp(u) - \wp(a)\right]^2 = u^{-2}\left[1 + 2\wp(a)u^2 + \ldots\right] =$$
$$= u^{-2} + 2\wp(a) + \ldots .$$

As $\wp(u-a) = \wp(-a) + u\wp'(-a) + \ldots = \wp(a) - \wp'(a)u + \ldots$,

$$(14.16) \qquad h(u) = -u^{-2} - \wp(a) + \ldots .$$

Thus the function $h(u) + \wp(u)$ is holomorphic at the origin, and so at points of $L$. Outside $L$, it has the same singularities as $h$, hence at most simple poles at points $u = a \bmod L$. In other words, $h + \wp$ is an elliptic function of order $\leq 1$, and so is constant. However, (8) shows that $h(u) + \wp(u) = -\wp(a) + \ldots$ in the neighbourhood of 0. Therefore, $h(u) + \wp(u) = -\wp(a)$, and taking (13) into account, we get

$$(14.17) \qquad \wp(u-a) = \left[\wp'(u) + \wp'(a)\right]^2 \Big/ 4\left[\wp(u) - \wp(a)\right]^2 - \wp(u) - \wp(a) .$$

This result is generally written in its most symmetric form

$$\wp(u+v)+\wp(u)+\wp(v)=\frac{1}{4}\left(\frac{\wp'(u)-\wp'(v)}{\wp(u)\wp(v)}\right)^2. \tag{14.18}$$

The proof of (18) was nonetheless based on the assumption that $a$, or $-v$ in the notation of (18), is not a half-period. This restriction is irrelevant since, for given $u$, both sides of (10) are meromorphic functions of $v$. If relation (18) holds outside a discrete subset of $\mathbb{C}$, it does so for every $v$ where both sides are well-defined, in other words provided none of the three numbers $u, v, u+v$ is a period.

For $u = v$, the fraction appearing in (18) obviously needs to be calculated by passing to the limit, which leads to the duplication formula

$$\wp(2u)+2\wp(u)=\wp''(u)^2/4\wp'(u)^2. \tag{14.19}$$

Calculating $\wp''$ by differentiating the differential equation of the $\wp$-function gives

$$\wp(2u)=\frac{\left[\wp(u)^2+g_2/4\right]^2+2g_3\wp(u)}{4\wp(u)^3-g_2\wp(u)-g_3}.$$

The geometric interpretation of (18) consists in considering the curve in $\mathbb{C}^2$ having equation

$$y^2=4x^3-g_2x-g_3. \tag{14.20}$$

The differential equation of the $\wp$-function gives a parametric representation

$$u\longmapsto(\wp(u),\wp'(u)) \tag{14.21}$$

of it, as was was seen in the proof of theorem 12. If $y = mx + p$ is the equation of a line in $\mathbb{C}^2$, the $x$-coordinates of its intersection points with the curve are the roots of the equation

$$4x^3-g_2x-g_3-(mx+p)^2=0.$$

Hence, if $A = (a, a')$, $B = (b, b')$ and $C = (c, c')$ are these three points, then $a + b + c = m^2/4$. Next let $u$, $v$ and $w$ be parameters, determined $\bmod L$, of the points $A$, $B$ and $C$. The slope of the line $AB$ being

$$m=(b'-a')/(b-a)=[\wp'(v)-\wp'(u)]/[\wp(v)-\wp(u)],$$

we get

$$\wp(u)+\wp(v)+\wp(w)=\frac{1}{4}[\wp'(v)-\wp'(u)]^2/[\wp(v)-\wp(u)]^2.$$

Comparing with the addition formula, we conclude that $\wp(w) = \wp(u + v)$, whence $w = \pm(u+v)$. As $u$, $v$ and $w$ can be arbitrarily permuted, $u+v+w = 0$. Therefore, the addition formula says that *the points with parameter $u$, $v$ and $-(u+v)$ of curve (20) are aligned.*

These calculations assume that none of the points $u$, $v$, $u+v$ is in $L$. In particular there is no point $O$ on the curve such that $A + O = O$ for all $A$ since it would correspond to the value $u = 0$ of the parameter. To get rid of this restriction, we need to work in the complex projective plane $P_2(\mathbb{C})$ and replace (19) by the equation

$$4x^3-y^2t-g_2xt^2-g_3t^3=0 \tag{14.22}$$

where $x, y, t$ denote the standard homogeneous coordinates [Chap. IX, § 4, n° 11, (iii)] in $P_2(\mathbb{C})$, which adds a point "at infinity" to the curve having homogeneous coordinates $(0, 1, 0)$. Map (21) can then be defined for all $u \in \mathbb{C}$: if $u$ is not in $L$, it associates the point with homogeneous coordinates $(\wp(u), \wp'(u), 1)$ to $u$. In the neighbourhood of 0, we for example replace these homogeneous coordinates by $(u^3\wp(u), u^3\wp'(u), u^3)$, and find the point $(0, -2, 0)$ at the limit, equivalently, the point $(0, 1, 0)$ at infinity of the projective curve. Map (21) becomes injective when it is interpreted as a map from $\mathbb{C}/L$ to $P_2(\mathbb{C})$. Its image is precisely curve (22). All this is simply a restatement of theorem 12. As the composition law $(u, v) \longmapsto u + v$ in $\mathbb{C}$ passes to the quotient by the subgroup $L$, it induces a commutative group structure on $\mathbb{C}/L$, hence also on curve (14). Its identity element is then the already $O$ with homogeneous coordinates $(0, 1, 0)$ we have already found, i.e. the only intersection point of (21) with the line at infinity of $P_2(\mathbb{C})$ and hence an inflection point of the curve. If $A$, $B$ are two points of the curve, the point $C = A + B$ given by the composition law is just the symmetric point with respect to $Ox$ of the third intersection point of the line $AB$ (or, if $A = B$, of the tangent at $A$) with the curve.

*Exercise 4.* Give a geometric interpretation for the equality

$$A + (B + C) = (A + B) + C\,.$$

To go further than this brief overview, it is necessary to go deeper into the much more general theory of Abelian varieties; a far too extensive program.

# § 5. $SL_2(\mathbb{R})$ as a Locally Compact Group

## 15 – Subgroups, Invariant Measure

(i) *Actions of $SL_2(\mathbb{R})$ on the half-plane.* In the theory of modular or more generally automorphic functions, one uses the group[46] $G = SL_2(\mathbb{R})$ of matrices

$$g = \begin{pmatrix} a & b \\ c & d \end{pmatrix} \quad \text{with } a, b, c, d \in \mathbb{R}, \quad ad - bc = 1,$$

and so

$$g^{-1} = \begin{pmatrix} d & -b \\ -c & a \end{pmatrix}.$$

Setting

$$gz = (az + b)/(cz + d)$$

in order to simplify notation and observing that

$$\text{(15.1)} \qquad \operatorname{Im}(gz) = |cz + d|^{-2} \operatorname{Im}(z)$$

has the same sign as $\operatorname{Im}(z)$, one sees that $G$ acts on the upper half-plane

$$P : \operatorname{Im}(z) > 0$$

by conformal representations $z \longmapsto gz$. These are the only ones: this is an easy classic result. For all $z, z' \in P$, there clearly exists $g \in G$ such that $z' = gz$. In fact, only matrices with $c = 0$ need be considered.

If $\Gamma$ is an arbitrary subgroup of $G$, the *stabilizer* of any $z \in P$ in $\Gamma$ is the subgroup $\Gamma_z$ of $\gamma \in \Gamma$ such that $\gamma z = z$. For $\Gamma = G$, I will rather denote it by $K_z$. The reason for this will readily become apparent. Writing $z = xi$ for $x \in G$, one need only search for elements $g$ such that $gxi = xi$, i.e. such that $x^{-1}gx \in K_i$. This subgroup is the set of matrices such that $a = d$, $c = -b$, hence of the form

$$\text{(15.2)} \qquad k(t) = \begin{pmatrix} \cos 2\pi t & \sin 2\pi t \\ -\sin 2\pi t & \cos 2\pi t \end{pmatrix}$$

with real $t$. $K_i$, denoted simply $K$, is compact and isomorphic to $\mathbb{R}/\mathbb{Z}$ or $\mathbb{T}$. So in the general case,

$$K_z = xKx^{-1} \quad \text{if } z = xi.$$

This is a compact subgroup like $K$.

As every $z \in P$ is the image of some $i$ under a unique $g \in G \bmod K$, $P$ may be identified with the homogeneous space $G/K$, even from a topological point of view. Thus $G$ acts properly on $P$, and more generally so does any closed subgroup $\Gamma$, for example discrete, of $G$ (Chap. XI, n° 15, theorem 24). In particular, the quotient space $\Gamma\backslash P$ is locally compact and for all compact sets $A, B \subset P$, the set of $\gamma \in \Gamma$ such that $\gamma A \cap B \neq \emptyset$ is compact, and so finite if $\Gamma$ is discrete. The theory of automorphic functions would be impossible without these two results.

(ii) *Automorphic forms as functions on $G$.* It is convenient to set

[46] Some authors prefer to use the group $GL_2^+(\mathbb{R})$ of matrices with determinant $> 0$.

(15.3) $$J(g;z) = cz + d\,.$$

Thus $d(gz)/dz = J(g;z)^{-2}$, $\mathrm{Im}(gz) = |J(g,z)|^{-2}\,\mathrm{Im}(z)$, and

(15.4) $$J(g'g;z) = J(g';gz)J(g;z).$$

This last formula is basic. As $J(e;z) = 1$, it follows that

(15.4') $$J(g^{-1};gz) = J(g;z)^{-1}\,.$$

These functions are needed for functional equations of automorphic functions. Given a discrete subgroup $\Gamma$ de $G$, for example the modular group $SL_2(\mathbb{Z}) = G(\mathbb{Z})$ of integer matrices,[47] I will say that a function $f(z)$ – not necessarily holomorphic or meromorphic, except in the classical theory – satisfying

(15.5) $$f(\gamma z) = J(\gamma,z)^r f(z) \quad \text{for all } \gamma \in \Gamma$$

is a *generalized automorphic form* of integer weight $r$ for $\Gamma$. For $G(\mathbb{Z})$, the most obvious examples are the functions

$$G_{2n}(z) = \sum (cz+d)^{-2n}$$

of the theory of elliptic functions (§ 3, n° 13).

Associating the function

(15.6) $$L_r(x)f : z \longmapsto J(x^{-1};z)^{-r} f(x^{-1}z)$$

to every $f(z)$ in $P$ and to every $x \in G$, for each $r \in \mathbb{Z}$, one gets linear operators $L_r(x)$ in the vector space of numerical functions defined on $P$. By definition, for $x, y \in G$,[48]

$$\begin{aligned} L_r(x)L_r(y)f(z) &= J\left(x^{-1};z\right)^{-r} L_r(y)f\left(x^{-1}z\right) = \\ &= J\left(x^{-1};z\right)^{-r} J\left(y^{-1};x^{-1}z\right)^{-r} f\left(y^{-1}x^{-1}z\right) = \\ &= J\left(y^{-1}x^{-1};z\right)^{-r} f\left(y^{-1}x^{-1}z\right) = L_r(xy)f(z)\,, \end{aligned}$$

whence

(15.7) $$L_r(x)L_r(y) = L_r(xy)\,.$$

If $\Gamma$ is a discrete subgroup of $G$, the functional equation

$$f(\gamma z) = (cz+d)^r f(z)$$

---

[47] The reasons for the notation $G(\mathbb{Z})$ are as follows. Let $G \subset SL_n(\mathbb{C})$ be a group of matrices $g = (g_{ij})$ defined by the condition $\det(g) - 1 = 0$ and polynomial equations $p_i(g) = 0$ with coefficients in $\mathbb{Z}$. For every commutative ring $A$ containing $\mathbb{Z}$, one can then consider the set $G(A) \subset SL_n(A)$ of solutions with coefficients $g_{ij} \in A$. It is a subgroup of $SL_n(A)$, a result which is less obvious in the general case than in those we will encounter here. The case of an algebraic subgroup of $GL_n$ reduces to the previous one by associating the "diagonal" matrix $(x, \det(x)^{-n})$ of order $n+1$ to every matrix $x \in GL_n$. For a ring $A$ not containing $\mathbb{Z}$, for example a field whose characteristic $p \neq 0$ divides the coefficients of the $p_i$, it may be that $G(A) = SL_n(A)$ or $GL_n(A)$...

[48] Recall that, as in Chap. IX, a notation like $L_r(g^{-1})f(z)$ denotes the value of the function $L_r(g^{-1})f$ at the $z$.

then becomes

$$L_r(\gamma)f = f \quad \text{for all } \gamma \in \Gamma.$$

All this looks like translation operators of group theory, and in fact, easily reduces to them. For example, for $r = 0$, one can associate the function

$$f_0(g) = f(z) \quad \text{if } z = gi$$

on $G$ to every function $f$ on $P$. This gives all the solutions of

$$\varphi(gk) = \varphi(g)$$

and clearly,

$$f(\gamma z) = f(z) \Longleftrightarrow f_0(\gamma g) = f_0(g).$$

For $r \neq 0$, we associate the function

$$f_r(g) = L_r\left(g^{-1}\right) f(i) = J(g;i)^{-r} f(gi) = (ci+d)^{-r} f\left(\frac{ai+b}{ci+d}\right) \tag{15.8}$$

to $f$, where the $L_r$ are defined as above. For $x \in G$, let us set $f' = L_r(x)f$ and let $f'_r$ be the function corresponding to $f'$ on $G$. Applying (8) to $f'$ and using (7) gives

$$f'_r(g) = L_r\left(g^{-1}\right) f'(i) = L_r\left(g^{-1}\right) L_r(x) f(i) = L_r\left(g^{-1}x\right) f(i) = f_r\left(x^{-1}g\right).$$

So the map $f \longmapsto f_r$ transforms the operator $L_r(x)$ into the left translation operator

$$L(x)\varphi = \epsilon_x * \varphi \quad g \longmapsto \varphi\left(x^{-1}g\right), \tag{15.9}$$

which brings us back to a situation familiar in group theory and to the "commutative diagram"

$$\begin{array}{ccc} f_r & \xrightarrow{\quad L(x) \quad} & f'_r \\ \uparrow & & \uparrow \\ f & \xrightarrow{\quad L_r(x) \quad} & f' \end{array}$$

Since the relation $L_r(x)f = f$ is equivalent to $L(x)f_r = f_r$, for all $\gamma \in G$,

$$f'(z) = J(\gamma;z)^{-r} f(\gamma z) \Longleftrightarrow f'_r(g) = f_r(\gamma g). \tag{15.10}$$

For every subgroup $\Gamma$ of $G$, going from $f$ to $f_r$ thus transforms the solutions of (5) into *left $\Gamma$-invariant* functions.

Independently of any subgroup $\Gamma$, function (8) has another quasi-invariance property. Let us consider the subgroup $K$ in $G$. For matrix (2) with parameter $t$,

$$J(k;i) = \exp(-2\pi i t) = \mathbf{e}(-t); \tag{15.11}$$

the definition of $f_r$ then shows that

$$f_r(gk) = J(gk;i)^{-r} f(gki) = J(g;i)^{-r} J(k;i)^{-r} f(gi) = J(k;i)^{-r} f_r(g).$$

So

$$f_r(gk) = f_r(g)\chi_r(k) \quad \text{for } k \in K, \quad g \in G, \tag{15.12}$$

where the functions

$$\chi_r(k) = \mathbf{e}(rt) = J(k;i)^{-r} \quad \text{for } k = k(t) \tag{15.13}$$

are characters of the compact group $K$. Reversing the order of calculations shows that conversely, for every function $\varphi$ of weight $r$ on $G$, where a *function of weight* $r$ on $G$ designates a solution of (12), conversely there is a function

$$f(z) = J(g;i)^r \varphi(g) \qquad (z = gi)$$

on the half-plane $P$ such that $\varphi = f_r$. It will sometimes be written $\varphi_P(z)$. Clearly, the set $\mathcal{F}_r(G)$ of functions of weight $r$ on $G$ is stable under left translations and every function $\varphi$ on $G$ at least slightly reasonable can be represented by a series of functions of weight $r$: it suffices to expand $k \longmapsto \varphi(gk)$ as a Fourier series. Let $\mathcal{H}_r(G)$ denote the set of functions of weight $r$ induced by holomorphic functions on $P$. They will sometimes be called *holomorphic functions of weight* $r$ *on* $G$.

We start from a generalized automorphic form $f$ of weight $r$ for a subgroup $\Gamma$. The corresponding function $f_r$ then satisfies

$$f_r(\gamma g k) = f_r(g)\chi_r(k)$$

and this relation is precisely functional equation (5).

We have thereby obtained a result which, judging from some recent books, does not seems to be yet adopted by all experts of the classical theory though it has been known at least since 1950. Despite being easy – it would take more to trivialize the theory of automorphic functions – it ought to be stated:

**Theorem 13.** *Let $r$ be a rational integer and to every function $f$ defined on the upper half-plane $P$ let us associate the function on $G = SL_2(\mathbb{R})$ defined by*

$$f_r(g) = J(g;i)^{-r} f(gi)\,.$$

*The functions obtained are characterized by the relation*

$$\varphi(gk) = \varphi(g)\chi_r(k) \quad \text{for all } k \in K\,.$$

*For all $\gamma \in G = SL_2(\mathbb{R})$, the functional equation*

$$f(\gamma z) = J(\gamma; z)^r f(z) \tag{15.14}$$

*is equivalent to*

$$f_r(\gamma g) = f_r(g)\,. \tag{15.15}$$

In the most classical theory, $\Gamma$ is the subgroup $SL_2(\mathbb{Z})$ of integer matrices which, as already mentioned, I will often denote by $G(\mathbb{Z})$. But there are many other interesting groups $\Gamma$, in particular the *congruence subgroups* of the modular group $G(\mathbb{Z})$. To define them, we choose an integer $N > 1$ and, for any matrix $\gamma \in G(\mathbb{Z})$, let $\gamma_N$ denote the matrix obtained by reducing the entries of $\gamma \bmod N$. This gives rise to a homomorphism

$$SL_2(\mathbb{Z}) \longrightarrow SL_2(\mathbb{Z}/N\mathbb{Z})$$

which is surjective (exercise). Its kernel, written $\Gamma(N)$, is thus the set of $\gamma$ such that $\gamma \equiv 1 (\bmod N)$. This being so, the congruence subgroups of $G(\mathbb{Z})$ are those containing some $\Gamma(N)$. The group $\Gamma(\theta)$ of the Jacobi function, generated by $z \longmapsto -1/z$ and $z \longmapsto z + 2$, contains $\Gamma(2)$.

As $G(\mathbb{Z})$ contains the matrix $-1$, condition (14) clearly has no solution $f \neq 0$ if $r$ is odd. But the group $\Gamma(N)$ does not contain the matrix $-1$ for $N \geq 3$, so that $r$ can take any values.

There is also the much more general class of Poincaré's Fuchsian groups, which will be defined later. They are not characterized by arithmetic conditions, so that in their case, results easily proved for $G(\mathbb{Z})$ – convergence of Eisenstein series, construction of fundamental domains, etc. – require more complicated arguments. This inconvenience is in fact an advantage for those who want to understand the subject as it gets rid of ad hoc arguments that cannot be generalized.

(iii) *Subgroups of $SL_2$.* Clearly,

$$J(g;z) = 1 \text{ for all } z \Longleftrightarrow g = \begin{pmatrix} 1 & u \\ 0 & 1 \end{pmatrix} = x(u) \tag{15.16}$$

for some $u \in \mathbb{R}$. The set of these matrices is a subgroup $U$ of $G$ which acts by translations $z \longmapsto z + u$ on the half-plane. Instead of $x(u)$ we will often write $u$, in other words, make no distinction between the matrix $x(u)$ and its parameter $u$.

Clearly, by (3),

$$J\left(g'g;z\right) = J(g;z) \text{ for all } z \Longleftrightarrow g' \in U\,, \tag{15.17}$$

or

$$J\left(g';z\right) = J(g;z) \text{ for all } z \Longleftrightarrow g' \in Ug\,. \tag{15.17'}$$

Hence the function $z \longmapsto J(g;z)$ only depends on the coset $Ug$ of $g$ modulo $U$ and, in fact, characterizes it.

We will also constantly need the subgroup[49] $B$ of triangular matrices ($c = 0$). It acts on $P$ by similitudes $z \longmapsto az + b$ with $a$ real $> 0$ and $b \in \mathbb{R}$. If $A$ denotes the subgroup of matrices

$$h = \begin{pmatrix} t & 0 \\ 0 & t^{-1} \end{pmatrix} = h(t)\,, \quad t \neq 0\,, \tag{15.18}$$

and $A_+$ the subgroup $t > 0$ of $A$, then clearly, any $b \in B$ can be written in a unique way as $b = hu$ with $h \in A$ and $u \in U$ and $hUh^{-1} = U$ for all $h \in A$:

$$\begin{pmatrix} t & 0 \\ 0 & t^{-1} \end{pmatrix} \begin{pmatrix} 1 & u \\ 0 & 1 \end{pmatrix} \begin{pmatrix} t^{-1} & 0 \\ 0 & t \end{pmatrix} = \begin{pmatrix} 1 & \alpha(h)u \\ 0 & 1 \end{pmatrix} \quad \text{where } \alpha(h) = t^2\,. \tag{15.19}$$

The subgroups $gBg^{-1}$ of $G$ are called *Borel subgroups* of $G$. As any matrix in $GL_n(\mathbb{R})$ *with real eigenvalues* is well known to be the conjugate of a triangular matrix, it follows that any $g \in G$ satisfying this condition belongs to a Borel subgroup.

Introducing the matrix[50]

$$w = \begin{pmatrix} 0 & -1 \\ 1 & 0 \end{pmatrix} = S \tag{15.20}$$

[49] Borel (Armand)'s initial, who made much use of it in far less trivial situations. Incidentally, let me mention his *Automorphic Forms on $SL_2(\mathbb{R})$* (Cambridge Tracts, 1997), a presentation of the new theory (Maaß-Selberg) of non-holomorphic automorphic functions for a Fuchsian group from a purely *group-theoretic* viewpoint.

[50] The character $S$ is used in the theory of modular functions $(z \longmapsto -1/z)$; the character $w$, Hermann Weyl's initial, refers to the theory of semisimple groups.

which acts by $z \longmapsto -1/z$ on $P$, we get

$$G = B \cup BwB \tag{15.21}$$

(*Bruhat decomposition* in a trivial case).[51] Since $B = AU$ and $Aw = wA$, every $g \notin B$ can even be written as

$$g = uwb \quad \text{with } u \in U\,, \quad b \in B\,.$$

An elementary calculation shows that this decomposition is unique. Analogous subgroups are encountered in all semisimple Lie groups and all linear algebraic groups over an arbitrary field.

*Exercise 1.* Every matrix $g$ such that $c \neq 0$ can be written in a unique way as

$$g = x\left(u'\right)\begin{pmatrix}1 & 0\\ c & 1\end{pmatrix}x\left(u''\right)\,. \tag{15.22}$$

Let us give some other ways of writing $g \in G$ using subgroups already defined. The first one is the *Iwasawa decomposition* in a trivial case. It consists in writing that

$$g = bk = uhk$$

with $b \in B$, $h \in A$, $u \in U$ and $k \in K$. Indeed

$$x + iy = \begin{pmatrix}1 & x\\ 0 & 1\end{pmatrix}\begin{pmatrix}y^{\frac{1}{2}} & 0\\ 0 & y^{-\frac{1}{2}}\end{pmatrix}i = uhi \tag{15.23}$$

for all $z = x + iy \in P$, so that setting $gi = z$, $g$ is the product of the previous matrix, which is in $B$, and a matrix leaving the point $i$ fixed, so in $K$. (23) can also be written

$$x(u)hki = u + \alpha(h)i \tag{15.23'}$$

for $u \in U$, $h \in A$, $k \in K$, where $\alpha(h) = t^2$ already occurred in (19). The definition of $\alpha$ can be conveniently generalized to all of $G$ by setting

$$\alpha(g) = \mathrm{Im}(gi) = J(g;i)^{-2}\,. \tag{15.24}$$

Obviously

[51] For a considerably less trivial example, see my *Cours d'Algèbre*, § 15, exercise 23. In my young days, when traumatizing secondary school children still went on, we used to learn that

$$\frac{ax+b}{cx+d} = a/c - \frac{(ad-bc)/c^2}{x+d/c}\,,$$

from which we deduced that every homographic transformation decomposes into two translations, a homothety and the inversion $x \longmapsto -1/x$. Exercises 20, 21 and 22 will be useful for other similar algebraic properties of $SL_2$ (commutator group, normal subgroups, definition by generators and relations). These exercises are inspired from C. Chevalley who proved all this for "split" semisimple algebraic groups over a commutative field in his famous article in *Tohoku Math. Journal* and thus started a theory that has in particular led to enormous progress in our knowledge of finite groups.

$$\alpha(uhgk) = \alpha(h)\alpha(g) \tag{15.25}$$

for all $u, h, g, k$ and, for $z = g'i$,

$$J(g; z) = J(g; g'i) = J(gg'; i)/J(g'; i),$$

whence

$$J(g; z)^{-2} = \alpha(gg')/\alpha(g') \quad \text{for } z = g'i. \tag{15.26}$$

Two other occasionally useful decompositions consist in putting every $g \in G$ in the form

$$g = kp = k_1 h k_2 \tag{15.27}$$

with $k, k_1, k_2 \in K$, $p$ symmetric positive definite and $h \in A_+$. For this observe that if $g'$ denotes the transpose of $g$, then the matrix $g'g$ is symmetric positive definite, and so $g'g = p^2$ where $p$ is symmetric positive definite.[52] Then the matrix $k = p^{-1}g$ satisfies $k'k = 1$, proving the first result. It shows that $P$ could be identified with the space of symmetric positive definite matrices by associating to a matrix $p$ of this type the point $z = pi$ of $P$.

On the other hand, any symmetric real matrix can be diagonalized over $\mathbb{R}$ using an orthonormal basis change in the Euclidean space it acts on. This means that $p = k_2^{-1} h k_2$ with $k_2$ orthogonal, $h \in A$ and even $h \in A_+$ if $p$ is positive definite. The first relation (27) thus implies the second one.

(iv) *Fixed points and eigenvalues.* The eigenvalues of $g \in G$ are the roots of the equation

$$t^2 - Tr(g)t + 1 = 0.$$

They are real and distinct if $|\,Tr(g)| > 2$; $g$ is then said to be *hyperbolic*. In this case, $\mathbb{R}^2$ has a basis of eigenvectors of $g$, so that hyperbolic matrices are the conjugates in $G$ of the matrices $h \in A$ other than $\epsilon = \pm 1$.

If $|\,Tr(g)| = 2$, $g$ is said to be *parabolic*. So $g$ has a double eigenvalue equal to 1 or $-1$. Then there is a vector $a \in \mathbb{R}^2$ such that $g(a) = \epsilon a$ with $\epsilon = \pm 1$. Choosing $b \in \mathbb{R}^2$ non-proportional to $a$, $g(b) = ua + \epsilon b$ for some $u \in \mathbb{R}$. Thus $g$ is the conjugate of a matrix $u \in U$, up to the factor $\epsilon$. Therefore, in both cases, $g$ belongs to a Borel subgroup of $G$.

If $|\,Tr(g)| < 2$, $g$ is said to be *elliptic*. Its eigenvalues are imaginary and $g$ can no longer be triangularized or diagonalized over $\mathbb{R}$ in this case. To bring back $g$ in a standard subgroup of $G$, we look for the *fixed points* of $g$ in $P$, defined for all $g \in G$ by $gz = z$. Assume that $g \neq \pm 1$. The solutions are the roots of the equation

$$cz^2 + (d - a)z - b = 0.$$

If $g$ est elliptic, then $c \neq 0$. This is then a quadratic equation with discriminant $Tr(g)^2 - 4 < 0$ and hence two distinct conjugate imaginary roots. One of them being in $P$, an elliptic matrix is thus the conjugate of an element of $K$ and conversely.

[52] For any real symmetric positive definite matrix $p$, there exists a unique matrix $p^{\frac{1}{2}}$ of the same type such that $p = \left(p^{\frac{1}{2}}\right)^2$. Existence: diagonalize $p$. Uniqueness: diagonalize $p^{\frac{1}{2}}$.

(v) *Invariant measure.* There is a left invariant measure $dg$ on $G$ like in the case of a locally compact group. It is actually also right invariant for the simple reason that there is no non-trivial homomorphism from $G$ to a commutative group.[53]

Since $P = G/K$ where $K$ is compact, there is a $G$-invariant measure $m$ on $P$; it is unique up to a constant factor and

$$\int \varphi(g)dg = \int dm(z) \int \varphi(gk)dk \quad (z = gi) \tag{15.28}$$

for any function $\varphi \in L(G)$, or more generally integrable. It can also be obtained by transforming every function $f(z)$ on $P$ into the function $f_0(g) = f(gi)$ and setting

$$\int f(z)dm(z) = \int f_0(g)dg \tag{15.28'}$$

where $dg$ is a left invariant measure on $G$. The decomposition $G = A_+UK$ or, equivalently, the map $b \longmapsto bi$, enables us to identify $B_+ = A_+U$ with $P = G/K$, which transforms the action of $B_+$ on $P$ into left translations on $B_+$. Considered as a measure on $B_+$, $m$ must therefore be the left invariant measure $db$ of the subgroup $B_+$. But for $B = AU$, one readily checks that it is possible to assume

$$\int \varphi(b)db = \iint \varphi(hu)dhdu$$

where $dh$ and $du$ are the obvious invariant measures on $A$ and $U$. Using calculations similar to (23), it can be deduced that

$$dm(z) = y^{-2}dxdy\,. \tag{15.29}$$

Its $G$-invariance can be directly checked using the change of variable formula for multiple integrals. In fact $dm(z)$ is associated to the invariant differential form $\omega = y^{-2}dz \wedge d\bar{z}$.

Formula (28') defines a positive linear functional on the functions $f(z)$. It is invariant under the operator $L_0(x)$ for all $x \in G$. To obtain non-trivial integrals invariant under all $L_r(x)$, it suffices to integrate $|f_r(g)|^p$ $(1 \leq p < +\infty)$ so as to obtain strictly $K$-invariant functions on $G$.

As it is then sufficient to integrate over $B$, the result is obtained by observing that, by definition (9) of $f_r$,

$$f_r(b) = y^{r/2}f(z) \quad \text{if } z = bi\,,$$
$$|f_r(g)| = y^{r/2}\,|f(z)| \quad \text{if } z = gi\,.$$

Hence

$$\int_P y^{r/2}\,|f(z)|\,dm(z) = \int_G |f_r(g)|\,dg = \|f_r\|_1\,. \tag{15.30}$$

This expression is invariant under operators $L_r(x)$. The same is true for the integral

$$(f'|f'')_r = \int_P y^r f'(z)\overline{f''(z)}dm(z) = (f'_r|\,f''_r)\,, \tag{15.30'}$$

[53] An algebraic proof would require showing that, as an abstract group, $G$ is generated by its commutators $xyx^{-1}y^{-1}$.

the inner product on $L^2(G)$. Hence generally speaking, one can define spaces $L_r^p(P)$ by the condition that

$$\int \left|y^{r/2} f(z)\right|^p dm(z) < +\infty .$$

Hans Petersson[54] introduced or used these notions without referring to the notion of a quotient measure. He defined the inner product of two automorphic forms $f'$ and $f''$ of weight $r$ by integrating over a fundamental domain of $\Gamma$ in $P$. Weil's book, published in 1940 and which German specialists of the time did not read, gave the correct solution: every discrete subgroup $\Gamma$ of $G$ acts properly on $G$ or $P$, $\Gamma\backslash G$ is locally compact, and so is $\Gamma\backslash P = \Gamma\backslash G/K$ (Chap. XI, n° 15, theorem 24). Associating the function

$$\varphi^\Gamma(g) = \sum \varphi(\gamma g) \tag{15.31}$$

to every $\varphi \in L(G)$, which gives a map from $L(G)$ onto $L(\Gamma\backslash G)$, defines a "natural" measure $\lambda$ on $X = \Gamma\backslash G$ by setting

$$\int_{\Gamma\backslash G} \varphi^\Gamma(x) d\lambda(x) = \int_G \varphi(g) dg . \tag{15.32}$$

We will always write this formula in the abusive but convenient form

$$\int_{\Gamma\backslash G} \varphi^\Gamma(g) dg = \int_G \varphi(g) dg . \tag{15.32'}$$

If one insists on working in the upper half-plane and on defining the inner product of two automorphic forms $f'$ and $f''$ of weight $r$ for $\Gamma$ in a natural way, then let

$$(f'|\,f'') = \int_{\Gamma\backslash G} f'_r(g)\overline{f''_r(g)} dg = \int_{\Gamma\backslash G} y^r f'(z)\overline{f''(z)} dm(z) . \tag{15.33}$$

Formula (32) holds for every function $f \in L^1(G)$: then the series $\sum f(\gamma g)$ converges absolutely almost everywhere and its sum is in $L^1(\Gamma\backslash G)$. This is a particularly simple application of the generalized Lebesgue-Fubini theorem [Chap. XI, n° 13, (ii)]: denoting by $\pi$ the canonical map from $G$ onto $\Gamma\backslash G$ and associating the measure

$$\mu_x : f \longmapsto \sum f(\gamma g) \quad \text{where } \pi(g) = x$$

to each $x \in \Gamma\backslash G$, relation (32) means that the measure $dg$ of $G$ is the integral of the measures $\mu_x$ with respect to the measure $\lambda$. Theorem 22 of Chap. XI, n° 13, (ii) then shows that $f \in L^1(G)$ implies that

$$\sum |f(\gamma g)| < +\infty \quad \text{almost everywhere}$$

(and, in good cases, everywhere), that the function $f^\Gamma(g) = \sum f(\gamma g)$ is integrable over $\Gamma\backslash G$ and that (32) again holds.

(vi) *The point of view of the unit disc.* It is sometimes useful to rewrite the definitions and results of this n° differently using the conformal representation

[54] *ber eine Metrisierung der ganzen Modulformen* (Jahresber. Deutschen Math. Verein., **49**, 1939, pp. 49–75).

$z \longmapsto (z-i)/(z+i) = \zeta$ of $P$ on the unit disc $D$. Easy calculations show that $SL_2(R)$ is replaced by the group $G'$ of matrices

$$(15.34) \qquad \begin{pmatrix} a & b \\ \bar{b} & \bar{a} \end{pmatrix} \qquad \text{such that } a\bar{a} - b\bar{b} = 1\,,$$

$K$ becoming the group $K'$ of rotations about 0. More precisely, if one chooses a matrix $s$ *with determinant* 1 such that $z = s\zeta$ (which determines it up to the factor $\pm 1$), the map

$$(15.35) \qquad g \longmapsto g' = s^{-1}gs$$

is an isomorphism from $G$ onto $G'$ transforming every subgroup $\Gamma$ of $G$ into a subgroup $\Gamma'$ of $G'$. In particular, the matrix $k(t) \in K$ defined in (2) is transformed into

$$(15.36) \qquad k(t)' = k'(t) = \begin{pmatrix} \mathbf{e}(t) & 0 \\ 0 & \mathbf{e}(-t) \end{pmatrix}$$

as can be easily seen.

On the other hand,

$$(15.37) \qquad \operatorname{Im}(z) = \left(1 - \zeta\bar{\zeta}\right) / \left|1 - \zeta\right|^2 .$$

Since $z = i(\zeta + 1)/(1 - \zeta)$,

$$dz = 2id\zeta/(1-\zeta)^2\,, \quad dz \wedge d\bar{z} = 4d\zeta \wedge d\bar{\zeta} / \left|1-\zeta\right|^4 .$$

Multiplication by $y^{-2} = |1-\zeta|^4(1-\zeta\bar{\zeta})^{-2}$ gives a measure on $D$ invariant under group (34), namely

$$(15.38) \qquad dm(\zeta) = 4\left(1 - |\zeta|^2\right)^{-2} d\xi d\eta\,,$$

the image of $dm(z)$.

The functional equation

$$(15.39) \qquad f(\gamma z) = J(\gamma; z)^r f(z)$$

of automorphic forms for the discrete group $\Gamma$ can be easily reworded. Setting $J(g;z) = cz + d$ for any complex matrix $g$ with determinant 1 and all $z \in \mathbb{C}$, the identities

$$J\left(gg';z\right) = J\left(g;g'z\right) J\left(g';z\right)\,, \quad d(gz) = J(g;z)^{-2}dz$$

continue to hold. Using (35), relation (39) becomes

$$\begin{aligned} f\left(s\gamma'\zeta\right) = f(\gamma s\zeta) = J(\gamma;s\zeta)^r f(s\zeta) = J(\gamma s;\zeta)^r J(s;\zeta)^{-r} f(s\zeta) = \\ = J\left(s\gamma';\zeta\right)^r J(s;\zeta)^{-r} f(s\zeta) = \\ = J\left(s;\gamma'\zeta\right)^r J\left(\gamma';\zeta\right)^r J(s;\zeta)^{-r} f(s\zeta)\,. \end{aligned}$$

Associating the function

$$(15.40) \qquad f_D(\zeta) = J(s;\zeta)^{-r} f(s\zeta) = J(s;\zeta)^{-r} f(z) = J(s^{-1};z)^r f(z)$$

to $f(z)$, which is holomorphic if and only if so is $f$, we thereby get the functional equation

(15.41) $$f_D(\gamma'\zeta) = J(\gamma';\zeta)^r f_D(\zeta)\,,$$

which replaces (39). Note that, for all $f$,

(15.41') $$f_D(\zeta) = (z+i)^r f(z)\,,$$

up to a factor $\pm\sqrt{2}$.

Everything that has been said above for *integer* $r$ of operators $L_r(g)$ in the space of functions on $P$ can be trivially transposed. The formulas are the same. On the other hand, (37) and (40) show that

$$y^{r/2}f(z) = (1-\zeta\overline{\zeta})^{r/2}(1-\zeta)^{-r}J(s,\zeta)^r f_D(\zeta)\,.$$

As $s$ has determinant 1,

$$J(s;\zeta)^{-2} = dz/d\zeta = 2i(1-\zeta)^{-2}\,.$$

So $J(s;\zeta) = (1-\zeta)/(2i)^{\frac{1}{2}}$. This is the value of $i^{\frac{1}{2}}$ that should be used depending on the matrix $s$ chosen. Hence

(15.42) $$y^{r/2}f(z) = c_r(1-\zeta\overline{\zeta})^{r/2}f_D(\zeta)\,,$$

where $c_r = 2^{r/2}(i^{\frac{1}{2}})^r$.

In theory, all this assumes $r$ is an integer. For non-integer real $r$, (40) can still be used to associate a function $f_D$ defined on $D$ to every function defined on $P$: it suffices to choose once and for all a uniform branch of $J(s;\zeta)^{-r}$ on $D$. Relation (42) continues to hold if $c_r$ is suitably defined. In particular, for all $p \geq 1$,

(15.43) $$\int |y^{r/2}f(z)|^p dm(z) = \int |(1-\zeta\overline{\zeta})^{r/2}f_D(\zeta)|^p dm(\zeta)$$

(up to a constant factor). For $p = +\infty$, we replace the integrals by the norms of uniform convergence.

## 16 – The Discrete Series of Representations of $SL_2(\mathbb{R})$

In Chap. VIII, n° 12, we defined a space[55] $\mathcal{H}^1_r(P)$, for all real $r = 2k$. It is the set of holomorphic functions on $P$ such that

$$\|f_r\|_1 = \int y^{r/2}|f(z)|dm(z) < +\infty\,.$$

One can more generally define spaces $\mathcal{H}^p_r(P)$ by setting

[55] The notation adopted here is different from those of chap. VIII: the measure written $dm(z)$ in Chap. VIII is $dxdy$. For what follows, the results of Chap. VIII need to be applied to the measure $d\mu(z) = y^{k-2}dxdy = y^{r/2}dm(z)$ in the notation adopted here, so that here the function $\varrho(y)$ of Chap. VIII is $y^{k-2}$. Finally I write $\mathcal{H}^1_r(P)$ for the space denoted by $\mathcal{H}^1_k(P)$ in chap. VIII, and $\mathcal{H}^1_r(G)$ for the space of $f_r$ where $f \in \mathcal{H}^1_r(P)$. This is a closed subspace of $L^1(G)$. In the maps that interest us here, $r$ is an integer, but everything remains valid for all real numbers $r$, provided (34) is taken as the definition of the symbol $\|f_r\|$. The case of a non-integer real exponent $r$ occurs when $SL_2(\mathbb{R})$ is replaced by its universal covering.

$$\|f_r\|_p = \left(\int \left|y^{r/2}f(z)\right|^p dm(z)\right)^{1/p} \qquad (p < +\infty)\,,$$

$$\|f_r\|_\infty = \sup \left|y^{r/2}f(z)\right| < +\infty\,.$$

For *integer* $r$, the spaces which correspond to them on $G$ by theorem 13 of n° 15, (ii) are

$$\mathcal{H}_r^p(G) = \mathcal{H}_r(G) \cap L^p(G)\,.$$

For all $r \in \mathbb{R}$, $\mathcal{H}_r^p(P)$ is *complete* and in fact convergence in mean implies compact convergence in $P$ [Chap. VIII, n° 4, (iv)], or in $G$ if $r \in \mathbb{Z}$. For integer $r$, these spaces play an important role in the theory of automorphic functions as well as for harmonic analysis on the group $G$. The same is true for non-integer $r$ – for some time now, experts have been devoting much energy to the study of modular forms of weight 1/2, 1 and 3/2 –, but then we must accept to replace $G$ by its universal covering, a topic that will not be broached in this chapter.[56] Nonetheless, for a real $r$, it should be observed that the formula

$$L_r(g^{-1})f(z) = J(g;z)^{-r}f(gz)\,,$$

though it is not well-defined for all $g \in G$, does not raise any difficulties for $g \in B_+$: it suffices to set

$$L_r(b^{-1})f(z) = t^r f(t^2z + u) \quad \text{for} \quad b = \begin{pmatrix} t & t^{-1}u \\ 0 & t^{-1} \end{pmatrix}\,, \qquad t > 0 \tag{16.1}$$

defining as usual $t^r = \exp(r \log t)$. We thereby get a representation of $B_+$ on each $\mathcal{H}_r^p(P)$. This remark will prove useful later, so will the expression

$$\|f\|_2^2 = \int |f_r(b)|^2 db$$

of the norm on $\mathcal{H}_r^2(P)$. This equality is obvious since $dg = dbdk$ for $g = bk$.

We start by recalling and completing the properties of functions of $\mathcal{H}_r^1(P)$.

(i) *Integrable holomorphic functions on the half-plane.* Let us make the weaker assumption

$$\int_{\mathrm{Im}(z)<T} y^{r/2}\,|f(z)|\, dm(z) < +\infty \quad \text{for all } T > 0\,. \tag{16.2}$$

[56] See a detailed construction in R. Godement, *Introduction à la théorie des groupes de Lie* (Springer, 2003), § 2, n° 7. An element of the covering is an ordered pair $g = (g, \omega)$ where the continuous and real-valued function $\omega(z)$ satisfies

$$J(g;z) = |J(g;z)|\mathbf{e}[\omega(z)]\,,$$

and so is one of the uniform branches in $P$ of the "function" $\mathrm{Arg}(cz+d)$, up to the factor $2\pi$. Multiplication is defined by

$$(g',\omega')\,(g'',\omega'') = (g,\omega) \quad \text{where } g = g'g''\,, \quad \omega(z) = \omega'\,(g''z) + \omega''(z)\,.$$

$G$ cannot be realized as a matrix group. On forms of weight 1/2, see the somewhat opaque Chap. IV of Neal Koblitz, *Introduction to Elliptic Curves and Modular Forms* (2d ed., Springer, 1993).

As was seen in Chap. VIII by comparing it to the integral of $f$ over a horizontal strip of finite width, the series $\sum f(z+n)$ and hence all its derived series converge normally in all compact subsets of $P$. The function $x \longmapsto f(x+iy)$ and all its derivatives are therefore integrable and tend to 0 at infinity. The Fourier transform of $f(x+iy)$ is continuous and $O(t^{-N})$ for all $N > 0$, and thus is integrable. The Fourier inversion formula can, therefore, be applied. But integrating along the contour of a horizontal rectangle whose vertical sides tend to infinity, we had shown that the integral

$$\hat{f}(t) = \int_{\mathrm{Im}(z)=y} f(z)\mathbf{e}(-tz)dz = \mathbf{e}(-ity)\int f(x+iy)\mathbf{e}(-tx)dx \tag{16.3}$$

does not depend on $y$. Then the inversion formula becomes

$$f(z) = \int \hat{f}(t)\mathbf{e}(tz)dt \tag{16.4}$$

and shows that $f$ is a complex Fourier transform. Finally, the Poisson summation formula applied to $x \longmapsto f(x+iy)$ shows that

$$\sum f(z+n) = \sum \hat{f}(n)\mathbf{e}(nz)\,, \tag{16.5}$$

both series converging normally in every horizontal strip of finite width. All this remains necessarily true if $f \in \mathcal{H}_r^1(P)$.

Moreover, by (3),

$$\left|\hat{f}(t)\right| \exp(-2\pi t y) \leq \int |f(x+iy)|\, dx\,,$$

whence

$$\int_0^T \left|\hat{f}(t)\right| \exp(-2\pi t y) y^{k-2} dy \leq \int_{y\leq T} \left|y^{r/2} f(z)\right| dm(z) < +\infty\,.$$

If $T < +\infty$, the left hand side is finite if and only if $k > 1$, i.e. if $r > 2$, a condition which is thus necessary for the existence of functions $f \neq 0$ satisfying (2). If $T = +\infty$, the integral in $y$ diverges for $t \leq 0$, so that convergence also supposes that $\hat{f}(t) = 0$ for all $t \leq 0$ since $\hat{f}$ is continuous. Hence finally,

$$f(z) = \int_0^{+\infty} \hat{f}(t)\mathbf{e}(tz)dt \quad \text{for all } f \in \mathcal{H}_r^1(P)\,. \tag{16.6}$$

For $r > 2$, as we have already checked that the functions $(z-\overline{w})^{-p}$ are in $\mathcal{H}_r^1(P)$ for all $w \in P$ provided $p > r$, it finally follows that $\mathcal{H}_r^1(P) \neq \{0\}$ if and only if $r > 2$.

To go further than these results of Chap. VIII, let us start by finding upper bounds for $\hat{f}(t)$ and $f(z)$ for $f \in \mathcal{H}_r^1(P)$. Since $y^{r/2}f(z)$ is integrable [with respect to $dm(z)$, as usual in this context] and $\mathbf{e}(tz)$ is bounded by 1 in $P$ for $t \geq 0$, one can compute

$$\begin{aligned}\int y^{r/2} f(z)\overline{\mathbf{e}(tz)}dm(z) &= \int y^{r/2-1}d^*y \int f(x+iy)\exp(-2\pi t y)\mathbf{e}(-tx)dx = \\ &= \hat{f}(t)\int y^{r/2-1}\exp(-4\pi t y)d^*y = \\ &= (4\pi t)^{1-r/2}\Gamma\left(r/2-1\right)\hat{f}(t)\,.\end{aligned}$$

The left hand side being bounded above by[57] $\|f_r\|_1 = \int y^{r/2}|f(z)|dm(z)$, we successively deduce that

$$t^{1-r/2}\left|\hat{f}(t)\right| \leq c_1(r)\,\|f_r\|_1\,,$$

$$|f(z)| \leq \int \left|\hat{f}(t)\mathbf{e}(tz)\right| dt \leq c_1(r)\,\|f_r\|_1 \int t^{r/2}\exp(-2\pi ty)d^*t\,, \tag{16.7}$$

$$y^{r/2}\,|f(z)| \leq c_2(r)\,\|f_r\|_1 \tag{16.8}$$

with constants

$$c_1(r) = \frac{(4\pi)^{r/2-1}}{\Gamma(r/2-1)}\,,\quad c_2(r) = (2\pi)^{-r/2}\Gamma(r/2)c_1(r)\,.$$

Since $y^{r/2}f(z)$ is bounded on $P$, one can also deduce that

$$\mathcal{H}_r^1(P) \subset \mathcal{H}_r^p(P) \quad \text{for all } p \geq 1\,. \tag{16.9}$$

For *integer* $r$, $|f_r(g)| = |J(g;i)^{-r}f(z)| = y^{r/2}|f(z)|$, so that $f_r$ is bounded on $G$. Similarly,

$$\|f_r\|_\infty \leq c_2(r)\,\|f_r\|_1 \tag{16.8'}$$

and, instead of (9),

$$\mathcal{H}_r^1(G) \subset \mathcal{H}_r^p(G) \quad \text{for } 1 \leq p \leq +\infty\,. \tag{16.9'}$$

(8') also shows that, in $\mathcal{H}_r^1(G)$, convergence in mean implies uniform convergence on $G$ and not only on every compact subset.

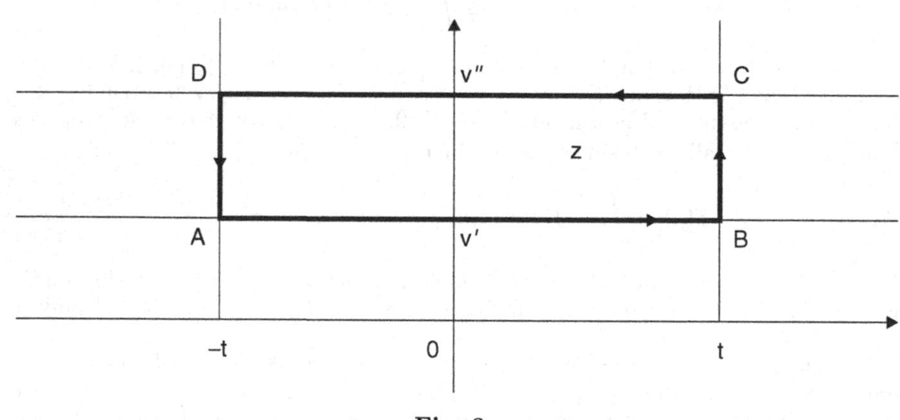


**Fig. 3.**

*Exercise 1* [direct proof of (8)]. (a) Applying Cauchy's formula to the contour $ABCD$ above and making $t$ tend to $+\infty$, show that

[57] I use this convenient notation despite the fact that the function $f_r(g)$ is not well-defined for non-integer $r$.

$$2\pi i f(z) = \int_{\mathrm{Im}(w)=v'} \frac{f(w)}{w-z} dw - \int_{\mathrm{Im}(w)=v''} \frac{f(w)}{w-z} dw$$

for $0 < v' < \mathrm{Im}(z) < v''$, the integrals being convergent. Let $c'$ and $c''$ be the values, independent of $v'$ and $v''$, of these two integrals. (b) Let $B$ be the horizontal strip $v_1'' \leq \mathrm{Im}(w) \leq v_2''$. Show that, for $\mathrm{Im}(z) < v_1'' < v_2''$,

$$\left| \int_B \frac{v^{r/2} f(w)}{w-z} dm(w) \right| \leq \frac{\|f_r\|_1}{v_1'' - \mathrm{Im}(z)} .$$

Computing the left hand side in terms of $c''$, $v_1''$ and $v_2''$, show that $c'' = 0$ and that

(16.10) $$2\pi i f(z) = \int_{\mathrm{Im}(w)=v} \frac{f(w)}{w-z} dw \quad \text{for } 0 < v < \mathrm{Im}(z) .$$

(c) For $0 < T < \mathrm{Im}(z)$, compute the integral

$$\int_{v \leq T} \frac{v^{r/2} f(w)}{w-z} dm(w)$$

and deduce that $y^{r/2} f(z)$ is bounded on $P$.

For $p = 2$, (9) proves the convergence of the integral

$$\|f_r\|_2^2 = \int y^r |f(z)|^2 dm(z) = \int y^{r-1} d^*y \int |f(x+iy)|^2 dx$$

for all $f \in \mathcal{H}_r^1(P)$. However, for given $y$, $f(x+iy)$ is bounded by (8), and integrable, and so is in $L^2(\mathbb{R}) \cap L^1(\mathbb{R})$. Its Fourier transform, namely $\hat{f}(t)\mathbf{e}(ity)$, is, therefore, in $L^2(\mathbb{R})$. Plancherel's formula then shows that

$$\begin{aligned} \int y^r |f(z)|^2 dm(z) &= \iint y^{r-1} \exp(-4\pi ty) \left|\hat{f}(t)\right|^2 dt d^*y = \\ &= \iint y^{r-1} \exp(-4\pi y) \left|\hat{f}(t)\right|^2 t^{1-r} dt d^*y = \\ &= (4\pi)^{1-r} \Gamma(r-1) \int \left|\hat{f}(t)\right|^2 t^{1-r} dt . \end{aligned}$$

This calculation is justified by LF since it only concerns positive functions. It follows that, for all $f \in \mathcal{H}_r^1(P)$,

(16.11) $$\int t^{1-r} \left|\hat{f}(t)\right|^2 dt < +\infty ,$$

which improves (7) but without characterizing the functions $\hat{f}$ (see theorem 15 below). Finally, the identity $4(f|g) = \ldots$ shows that, more generally, for $f, g \in \mathcal{H}_r^1(P)$,

$$\begin{aligned} (f_r | g_r) &= \int y^r f(z)\overline{g(z)} dm(z) = \\ &= c_3(r) \int \hat{f}(t)\overline{\hat{g}(t)} t^{1-r} dt \end{aligned}$$

(16.12) $$\text{where } c_3(r) = (4\pi)^{1-r} \Gamma(r-1) .$$

(ii) *The spaces* $\mathcal{H}_r^p$ *of the unit disc.* We saw at the end of the previous n°, whose notation we keep, that associating the function

$$f_D(\zeta) = J(s;\zeta)^{-r} f(s\zeta) = J(s;\zeta)^{-r} f(z) = J(s^{-1};z)^r f(z)$$

on the unit disc $D$ to every function $f$ on $P$ gives

$$y^{r/2} f(z) = \left(1 - |\zeta|^2\right)^{r/2} f_D(\zeta),$$

up to a constant factor. So, for $p < +\infty$,the space $\mathcal{H}_r^p(P)$ is transformed into the space $\mathcal{H}_r^p(D)$ of holomorphic functions on $|\zeta| < 1$ satisfying

$$\int \left|\left(1 - |\zeta|^2\right)^{r/2} \varphi(\zeta)\right|^p dm(\zeta) < +\infty \tag{16.13}$$

where $dm(\zeta)$, proportional to $(1 - |\zeta|^2)^{-2} d\xi d\eta$, is an invariant measure on $D$. In polar coordinates, this becomes

$$\iint |\varphi(\zeta)|^p \left(1 - \varrho^2\right)^{\frac{1}{2}pr-2} \varrho d\varrho d\theta < +\infty, \tag{16.13'}$$

where integration with respect to $p$ is extended to $(0,1)$. As

$$\int_0^1 \left(1 - \varrho^2\right)^{\frac{1}{2}pr-2} \varrho d\varrho < +\infty \Longleftrightarrow pr > 2,$$

for all $r > 2/p$, integer or not, $\mathcal{H}_r^p(D)$ contains every function $\varphi(\zeta)$ bounded on the unit disc. Coming back to the half-plane, this means that, *for* $r > 2/p$, *the space* $\mathcal{H}_r^p(P)$ *contains every function* $f(z)$ *such that* $(z+i)^r f(z)$ *is bounded*, for example the function $(z+i)^{-r}$, a result that can be easily verified directly. Hence $\mathcal{H}_r^p(P) \neq \{0\}$ for all real exponent $r > 2/p$.

This argument supposes $p < +\infty$. For $p = +\infty$, condition (13) is replaced by

$$\sup \left(|1 - \zeta|^2\right)^{r/2} |\varphi(\zeta)| < +\infty, \tag{16.13''}$$

whence $\mathcal{H}_r^\infty(D) \neq \{0\}$ for all $r \geq 0$, including $r = 0$.

Conversely, let us suppose that $\mathcal{H}_r^p(D) \neq \{0\}$. Though the operator $L_r(g)$ is not well-defined for all $g \in G'$ if $r$ is not an integer, for all $\varphi \in \mathcal{H}_r^p(D)$ and all $t \in \mathbb{R}$, the function $\varphi[\mathbf{e}(t)\zeta]$ is clearly still in $\mathcal{H}_r^p(D)$. This gives a representation of the group of rotations on $\mathcal{H}_r^p(D)$, which is obviously continuous and isometric. It can be decomposed into one-dimensional representations,[58] so that there are functions $\varphi \neq 0$ in $\mathcal{H}_r^p(D)$ such that

$$\varphi\left[\mathbf{e}(t)\zeta\right] = \mathbf{e}(t)^m \varphi(\zeta)$$

for some $m \in \mathbb{Z}$. Since $\varphi$ is holomorphic, this requires $\varphi(\zeta) = c\zeta^m$ and $m \geq 0$. Substituting in (13') gives

[58] This was shown (for every compact group) at the end of n° 29, (ii) of Chap. XI for unitary representations, but the proof and the result are easily generalizable to all representations $(\mathcal{H}, U)$ on an arbitrary Banach space: use operators $\int U(k)\chi(k)dk$, where $\chi$ varies in the set of characters of $K$.

$$\int \left(1-\varrho^2\right)^{\frac{1}{2}pr-2} \varrho^{pm+1} d\varrho < +\infty .$$

As $m \geq 0$, by (13"), the converse reduces to the condition $r > 2/p$ if $p < +\infty$, and to $r \geq 0$ if $p = +\infty$. As a result:

**Theorem 14**. *The space $\mathcal{H}_r^p(P)$ is non-trivial if and only if $r > 2/p$ when $p < +\infty$, or $r \geq 0$ when $p = +\infty$.*

We had already obtained this result for $p = 1$ using Fourier transforms, but less easily.

When $r$ is an *integer*, one can use the representation $L_r$ of $G'$ on $\mathcal{H}_r^p(D)$. For the matrix $k' = \mathrm{diag}(\mathbf{e}(t), \mathbf{e}(-t))$ which corresponds to $k(t) \in K$ and for $\varphi(\zeta) = a_m\zeta^m$,

$$L_r\left(k'^{-1}\right)\varphi(\zeta) = J\left(k';\zeta\right)^{-r}\varphi\left(k'\zeta\right) = \mathbf{e}(t)^r f\left[\mathbf{e}(2t)\zeta\right] = \mathbf{e}(t)^{2m+r}\varphi(\zeta)$$

or, coming back to $\mathcal{H}_r^p(P)$, $L_r[k(t)^{-1}]f(z) = \mathbf{e}(t)^{2m+r}f(z)$. So setting $n = 2m + r$ and $\chi_n[k(t)] = \mathbf{e}(nt)$ in accordance to (15.13),

$$L_r\left(k^{-1}\right)f = \chi_n(k)f \quad \text{with } n \geq r\,,\ n \equiv r \pmod 2\,. \tag{16.14}$$

For all $n$ of the form indicated, (14) has a unique solution, up to a constant factor. By (15.41') for $\varphi(\zeta) = \zeta^m$, it is the function

$$f(z) = (z+i)^{-r}\left(\frac{z-i}{z+i}\right)^{\frac{1}{2}(n-r)} = (z-i)^{\frac{1}{2}(n-r)}(z+i)^{-\frac{1}{2}(n+r)}\,. \tag{16.15}$$

As was seen above, the functions $f \in \mathcal{H}_r(G)$ represented on the unit disc by a monomial in $\zeta$ are characterized by a character $\chi$ of $K$ such that

$$f(kx) = \chi(k)f(x)\,, \qquad \text{i.e. } \chi * f = f\,.$$

For all $f \in \mathcal{H}_r(G)$, the function $\chi * f$ trivially satisfies this condition and on $D$ obviously corresponds to one of the terms of the power series of $f_D(\zeta)$, or to 0. Hence

$$f(x) = \sum \chi * f(x) \quad \text{for all } f \in \mathcal{H}_r(G)\,,$$

the series converging normally on every compact subset of $G$ as does the power series of $f_D$ on every compact subset of $D$.

This argument also shows that the subspace of solutions of $\chi * f = f$ has dimension 0 or 1 for all $\chi$. For $\chi = \chi_r$, it is obviously generated by the function $\varpi_r$ which on $D$ correspond to the function 1. It is, therefore, characterized by the relations

$$\varpi_r \in \mathcal{H}_r(G)\,, \quad \chi_r * \varpi_r = \varpi_r\,, \quad \varpi_r(e) = 1\,.$$

For $f \in \mathcal{H}_r(G)$, let us then consider the function

$$\chi_r * f(x) = \int f(kx)\overline{\chi_r(k)}dk\,.$$

It trivially satisfies the first two conditions imposed on $\varpi_r$, and so is proportional to it. For $x = e$, the previous integral equals $f(e)$ since $f$ has weight $r$. Thus it equals $f(e)\varpi_r(x)$ for all $x$. Replacing $f$ by $L(y^{-1})f \in \mathcal{H}_r(G)$ gives

$$\int f(ykx)\overline{\chi(k)}dk = f(y)\varpi_r(x) \quad \text{for all } f \in \mathcal{H}_r(G)\,.$$

As $k$ varies, the images $kx(0)$ of the point $0 \in D$ under $kx$ describe a circle centered at 0 and passing through $x(0)$, and the $ykx(0)$ describe the image of this circle under $y$, thus a circle whose interior contains the point $y(0)$. So the previous formula essentially means that integrating $f_D(\zeta)$ along this circle gives the value of $f_D$ at $y(0)$. As for Baron Cauchy, he did not only integrate over circles...

Thus, all this is simply a clever re-writing of standard and easy properties of power series. The group $G$ is too simple for the power of these methods to be gauged. If $G$ is only semisimple, with a maximal compact (non-commutative) subgroup $K$ so that $G/K$ has a complex analytic structure, then one should refer to Harish-Chandra's marvelous and far less than semi-simple articles on *Discrete series for semi-simple Lie groups* which can be found in his *Collected Works* (Springer).

(iii) *A theorem of Paley-Wiener type for* $\mathcal{H}_r^2(P)$. Fourier transforms have enabled us to associate a function $\hat{f}(t)$ with complex Fourier transform $f$ to each $f \in \mathcal{H}_r^1(P)$ In section (i) we obtained properties, but not any characterization, of these functions. One can also associate Fourier transforms $\hat{f}(t)$ to functions $f \in \mathcal{H}_r^2(P)$, and this time characterize them fully, but proofs are more difficult.

**Theorem 15**. *For all $f \in \mathcal{H}_r^2(P)$, $r > 1$, there is a measurable function $\hat{f}(t)$ zero for $t \leq 0$, satisfying*

$$\int \left|\hat{f}(t)\right|^2 t^{1-r} dt < +\infty \tag{16.16}$$

*and for which*

$$f(z) = \int_0^{+\infty} \hat{f}(t)\mathbf{e}(tz)dt\,. \tag{16.17}$$

$$\int y^r f(z)\overline{g(z)}dm(z) = (4\pi)^{1-r}\Gamma(r-1)\int_0^{+\infty} \hat{f}(t)\overline{\hat{g}(t)}t^{1-r}dt \tag{16.18}$$

*for all $f, g \in \mathcal{H}_r^2(P)$.*

*Conversely, integral (17) converges for all solutions of (16) and belongs to $\mathcal{H}_r^2(P)$.*

Since $\iint y^{r-2}|f(z)|^2 dxdy < +\infty$, there is a null set $N(f)$ such that

$$\int |f(x+iy)|^2\, dx < +\infty \quad \text{for } y \notin N(f)\,.$$

The function $f_y(x) = f(x+iy)$ then has a Fourier transform

$$\hat{f}_y(t) = \text{l.i.m.}^2 \int_{-n}^{n} f(x+iy)\mathbf{e}(-tx)dx \tag{16.19}$$

in $L^2(\mathbb{R})$, and Plancherel's formula shows that

$$\int |f(x+iy)|^2\, dx = \int \left|\widehat{f_y}(t)\right|^2 dt\,.$$

Let us show that there is function $\hat{f}(t)$ such that, for all $y \notin N(f)$,

$$\hat{f}_y(t) = \exp(-2\pi ty)\hat{f}(t) \quad \text{for almost all } t \in \mathbb{R}\,. \tag{16.20}$$

To this end, we set

$$\hat{f}_{y,n}(t) = \int_{-n}^{n} f(z)\mathbf{e}(-tz)dx$$

and show that

$$\lim_{n\infty} \left[\hat{f}_{a,n}(t) - \hat{f}_{b,n}(t)\right] = 0 \tag{16.21}$$

for all $a, b > 0$ and $t \in \mathbb{R}$.

**Lemma 1.** *For all functions $f \in \mathcal{H}_r^p(P)$, $p < +\infty$,*

$$\lim_{|x|\infty} f(x+iy) = 0$$

*uniformly on every compact subset of $\mathbb{R}_+^*$.*

Indeed let us suppose that $z$ stays in a strip $B : 0 < a \leq y \leq b < +\infty$ and consider a strip $B' : 0 < a' \leq y \leq b' < +\infty$, with $a' < a$ and $b < b'$. Let $K' \subset B'$ be the set of $\zeta = \xi + i\eta$ such that

$$-1 \leq \xi \leq 1\,, \qquad a' \leq \eta \leq b'$$

and set $K = K' \cap B$. There is [Chap. VIII, n° 4, (iv)] a constant $c_p$ such that

$$|f(z)|^p \leq c_p \int_{K'} |f(\zeta)|^p \, dm(\zeta) \quad \text{for all } z \in K$$

and all holomorphic functions on $B'$. Letting the same horizontal translation $z \longmapsto z+u$ act on $f$, $K$ and $K'$ replaces the left hand side with $f(z+u)$ and on the right hand side, $K'$ with $K'+u$. Thus

$$|f(x+iy)|^p \leq c_p \int_{K'+x} |f(\zeta)|^p \, dm(\zeta) \quad \text{for } a \leq y \leq b\,.$$

As the integral $\int |f(\zeta)|^p dm(\zeta)$ extended to all of the strip $B'$ converges, the right hand side clearly tends to 0 as $|x|$increases indefinitely, qed.

Relation (21) readily follows from the lemma. Integration of $f(z)\mathbf{e}(-tz)$ along the rectangle bounded by horizontals $a$ and $b$ and verticals $n$ and $-n$ shows that $\hat{f}_{a,n}(t) - \hat{f}_{b,n}(t)$ is indeed the difference between the integrals of $f(z)\mathbf{e}(-tz)$ along the verticals $n$ and $-n$ of strip $B$. As $|\mathbf{e}(-tz)| = \exp(2\pi ty)$ is bounded on $B$, the product $f(z)\mathbf{e}(-tz)$ converges uniformly to 0 as $z$ tends to infinity in strip $B$, whence (21).

We can now return to the proof of (20). For all $y \notin N(f)$, relation (19) can also be written

$$\hat{f}_y(t) = \text{l.i.m.}^2 \; \hat{f}_{y,n}(t) \exp(-2\pi ty)\,;$$

for $a, b \notin N(f)$, the difference $\hat{f}_a(t) - \hat{f}_b(t)$ is thus the limit in mean of a sequence converging *everywhere* to 0, and so (Riesz-Fischer) $\hat{f}_a(t) = \hat{f}_b(t)$ for almost all $t$, whence (20).

This being so, (Plancherel and LF)

$$\|f\|_2^2 = \iint y^{r-2}|f(z)|^2 dxdy = \int y^{r-2}dy \int \left|\hat{f}(t)\right|^2 \exp(-4\pi ty)dt =$$

$$= \int \left|\hat{f}(t)\right|^2 dt \int y^{r-1} \exp(-4\pi ty) d^*y\,. \tag{16.22}$$

In theory, these are upper integrals. The convergence of the integral in $y$ requiring $t > 0$, it follows that

$$\hat{f}(t) = 0 \quad \text{for almost all } t \leq 0$$

[and no longer for *all* $t < 0$ as was the case in $\mathcal{H}_r^1(P)$], then that

$$(4\pi)^{1-r}\Gamma(r-1)\int \left|\hat{f}(t)\right|^2 t^{1-r}dt = \|f\|_2^2 < +\infty\,,$$

which proves (16), as well as (18) for $f = g$, and so for arbitrary $f$ and $g$ by $4(f,g) = \ldots$.

As $\hat{f}(t)t^{(1-r)/2}$ and $t \longmapsto \mathbf{e}(tz)t^{-(1-r)/2}$ are square integrable over $t \geq 0$ for $\mathrm{Im}(z) > 0$ (and $r > 1$), integral (17) obviously converges. By (20), the Fourier transform $\hat{f}_y(t) = \hat{f}(t)|\mathbf{e}(tz)|$ is integrable for $y \notin N(f)$; since $x \longmapsto f(x+iy)$ is continuous, the Fourier inversion formula

$$f(x+iy) = \int \hat{f}_y(t)\mathbf{e}(tx)dt = \int \hat{f}(t)\mathbf{e}(tz)dt$$

is justified for $y \notin N(f)$. But the second integral is a continuous function of $z$ (dominated convergence). So (17) holds for *all* $z$.

Conversely, starting from a solution of (16), the same arguments show that integral (17) converges, represents a holomorphic function and (22) that $f \in \mathcal{H}_r^2(P)$, qed.

Multiplying the functions $\hat{f}(t)$ by $t^{(1-r)/2}$, the following statement follows: for all $r > 1$, the map associating the function

$$z \longmapsto \int t^{(r-1)/2}\varphi(t)\mathbf{e}(tz)dt$$

to all $\varphi \in L^2(\mathbb{R}_+)$ is an isomorphism from $L^2(\mathbb{R}_+)$ onto $\mathcal{H}_r^2(P)$. The typical Paley-Wiener theorem, which they in effect proved and which is given by all authors, corresponds to the case $r = 1$, excluded here because $\mathcal{H}_r^2(P) = \{0\}$ for $r \leq 1$. It characterizes complex Fourier transforms of $\varphi \in L^2(\mathbb{R}_+)$. The reader will easily find the statement and proof based on previous arguments.

(iv) *The kernel function of* $\mathcal{H}_r^2(P)$. The (complete) Hilbert space $\mathcal{H}_r^2(P)$ is non-trivial if and only if $r > 1$. On it, convergence in norm implies simple (and even compact) convergence. So, for all $z \in P$, there is (Chap. XI, n° 19, theorem 31) a function $K_z \in \mathcal{H}_r^2(P)$ such that

$$f(z) = (f\,|K_z) \quad \text{for all } f \in \mathcal{H}_r^2(P)\,. \tag{16.23}$$

It can be determined in several ways.

Setting

$$\overline{K_r(z,w)} = K_z(w) \quad \text{for } \mathrm{Im}(z) > 0\,, \quad \mathrm{Im}(w) > 0\,,$$

gives

$$f(z) = \int v^r K_r(z,w)f(w)dm(w) \quad (w = u+iv)\,. \tag{16.24}$$

where $K_r$ is the *kernel function* of $\mathcal{H}_r^2(P)$, a notion invented by Stefan Bergman in the 1920s- 30s and applicable to any Hilbert space whose elements are holomorphic functions, as well to many other situations (elliptic partial differential equations).

For given $z$, $K_r(z,w)$ is clearly a holomorphic function of $\overline{w}$. Relation (23) shows that $(K_z|K_w) = K_z(w)$. Thus

$$K_r(z,w) = (K_w, K_z) = K_r\overline{(w,z)}\,, \tag{16.25}$$

so that $K_r(z,w)$ is holomorphic at $z$ for given $w$.

First method. We now consider the representation $L_r$ of the subgroup $B_+$ on $\mathcal{H}^2_r(P)$ defined at the start of this n°. It is obviously unitary. By (15.6), replacing $f$ by $L_r(b^{-1})$ for $b \in B_+$ gives

$$J(b,z)^{-r} f(bz) = \left(L_r(b)^{-1} f|K_z\right) = \left(f|L_r(b)K_z\right)\,.$$

This can also be written

$$(f|K_{bz}) = f(bz) = \overline{J(b;z)}^{r}\,(f|L_r(b)K_z)\,.$$

Comparing with (23) for $bz$ and taking into account the fact that $K_z$ is determined by (23), it follows that

$$K_{bz} = \overline{J(b,z)}^{r} . L_r(b) K_{bz}\,, \tag{16.26}$$

which is equivalent to

$$K_r(bz,bw) = J(b,z)^r\,\overline{J(b,w)}^{r}\,K_r(z,w) \tag{16.26'}$$

or to $K_r(tz+\xi, tw+\xi) = t^{-r}K_r(z,w)$ for all $t > 0$ and all $\xi \in \mathbb{R}$. Setting $H(z,w) = (z-\overline{w})^r K_r(z,w)$ gives a holomorphic function in $z$ and antiholomorphic in $w$ such that $H(tz+\xi, tw+\xi) = H(z,w)$. So

$$H(z,w) = H\left[i, (w-x)/y\right]\,,$$

an expression which, for given $w$, must be holomorphic in $z$. However, $H(i,w)$ is antiholomorphic in $w$. Hence its differential with respect to $w$ is proportional to $d\overline{w}$. For given $w$, the differential of the function $z \longmapsto H[i,(w-x)/y]$ with respect to $z$ is thus proportional to $d[(\overline{w}^* - x)/y]$ and not to $dz$ if $H(i,w)$ is not a constant. Thus the only possibility is that $H(i,w)$, and so $H(z,w)$, is constant. Therefore,

$$K_r(z,w) = c_r\,(z-\overline{w})^{-r} \tag{16.27}$$

and

$$f(z) = c_r \int (z-\overline{w})^{-r}\, f(w) v^r\, dm(w)\,. \tag{16.28}$$

By a strange coincidence, function (27) appears among those which, in Chap. VIII, n° 12, (iv), eq. (12.23), have proved useful to show that $\mathcal{H}^1_r(P) \neq \{0\}$ if and only if $r > 2$: indeed we had shown that the function $(z-\overline{w})^{-p}$ is in $\mathcal{H}^1_r(P)$ if and only if both $\mathrm{Re}(p) > 1$ and $\mathrm{Re}(p) > r/2$, hence for $p = r > 2$. For $r > 2$, function (27) is, therefore, in $\mathcal{H}^1_r(P)$ and not only in $\mathcal{H}^2_r(P)$, and as $\mathcal{H}^1_r(P) \subset \mathcal{H}^2_r(P)$, it follows that, for $r > 2$, relation (23) applies to all $f \in \mathcal{H}^1_r(P)$ – and to all $f \in \mathcal{H}^2_r(P)$ if $r > 1$.

Second method. Show that, for all $z \in P$, there is a function $g_z \in \mathcal{H}^2_r(P)$ such that

$$\overline{\widehat{g_z}(t)} = t_+^{r-1}\mathbf{e}(tz) \tag{16.29}$$

and that $f(z) = \int f(w)\overline{g_z(w)}dm(w)$, up to a constant factor. Calculating $g_z(w)$, show that

$$K_r(z,w) = \frac{r-1}{4\pi}\,(z-\overline{w})^{-r}\,. \tag{16.30}$$

Third method. When $r$ is an *integer*, it is also possible to work in $G$. To do this, we replace $\mathcal{H}_r^2(P)$ with $\mathcal{H}_r^2(G) = \mathcal{H}_r(G)\cap L^2(G)$. As the space of holomorphic functions of weight $r$, it is closed in $L^2(G)$ and $\varphi \longmapsto \varphi(x)$ is a continuous linear functional for all $x \in G$. Hence there is a unique function $\omega_r \in \mathcal{H}_r^2(G)$ such that

$$\varphi(e) = (\varphi|\omega_r)$$

and so, replacing $\varphi$ by $L(x^{-1})\varphi$,

$$\varphi(x) = \left(L(x)^{-1}\varphi|\omega_r\right) = \left(\varphi|L(x)\omega_r\right)\,. \tag{16.31}$$

For $\varphi = \omega_r$, (31) becomes

$$\omega_r(x) = \left(\omega_r|L(x)\omega_r\right)\,,$$

which shows that $\omega_r$ is a function *of positive type* on $G$ [Chap. XI, n° 30, (iii)], hence satisfies $\omega_r(x^{-1}) = \overline{\omega_r(x)}$. Then, by (31),

$$\varphi(x) = \int \varphi(y)\overline{\omega_r\left(x^{-1}y\right)}dy = \int \varphi(y)\omega_r\left(y^{-1}x\right)dy\,,$$

in other words,

$$\varphi = \varphi * \omega_r\,. \tag{16.32}$$

This is relation (24) for $G$. Once again everything is easier on $G$ than on $P$ or $D$.

Furthermore, (27) follows readily from this. Indeed, the function $\omega_r = \widetilde{\omega}_r$ satisfies $\chi_r * \omega_r = \omega_r$. Relations (14) and (15) of section (ii) then readily show that, the corresponding function on the unit disc is the monomial of degree 0 in $\zeta$, hence that, for a constant $c_r$, $\omega_r$ corresponds to $c_r(z+i)^{-r}$.

(v) *The holomorphic discrete series of irreducible representations of $SL_2(\mathbb{R})$.*

**Theorem 16 (V. Bargmann).** *For all reals $r > 1$, the representation $L_r$ of $B_+$ on $\mathcal{H}_r^2(P)$ is irreducible. For integer $r$, the unitary representation $L_r$ of $SL_2(\mathbb{R})$ on $\mathcal{H}_r^2(P)$ is irreducible and square integrable.*

Let us first suppose that $r$ is an integer. To prove irreducibility, it suffices to show that any operator $A$ in $\mathcal{H}_r^2(G)$ commuting with left translations $L(x)$ is a scalar. It then amounts to showing that the function $\omega' = A\omega_r$ is proportional to $\omega_r$, since relation $A\omega_r = \lambda\omega_r$ implies $AL(x)\omega_r = \lambda L(x)\omega_r$. Thus the result follows since linear combinations of $L(x)\omega_r$ are everywhere dense in $\mathcal{H}_r^2(G)$. However, as $\omega_r$ is Hermitian symmetric, it satisfies

$$\omega_r\left(kgk'\right) = \chi_r(k)\omega_r(g)\chi_r\left(k'\right) \tag{16.33}$$

and so $L(k^{-1})\omega_r = \chi_r(k)\omega_r$. As $A$ commutes with all $L_r(k)$, the function $\omega'$ also satisfies (33). However, for all characters $\chi$ of $K$, the subspace of $f \in \mathcal{H}_r(G)$ satisfying $\chi * f = f$ has dimension 0 or 1. Hence $A\omega_r = \lambda\omega_r$.

Finally, the representation is square integrable since its coefficient $(\omega_r|L_r(x)\omega_r) = \omega_r(x)$ is in $L^2(G)$ (Chap. XI, n° 32).

In the general case, this argument no longer holds in this precise form. But it is still possible to use operators

$$L_r\left(b^{-1}\right) f(z) = t^r f\left(t^2 z + u\right) \quad \text{for} \quad b = \begin{pmatrix} t & t^{-1}u \\ 0 & t^{-1} \end{pmatrix}, \quad t > 0 \tag{16.34}$$

and to show that any operator $A$ in $\mathcal{H}_r^2(P)$ commuting with all $L_r(b)$ is a scalar. As in the computation of the kernel function $K_r(z,w)$, by (23), we associate the function

$$K_A(z,w) = (AK_w|K_z) = AK_w(z)$$

to $A$ . It is holomorphic in $z$ and antiholomorphic in $w$. By (26) and equality $AL_r(b) = L_r(b)A$, the function $K_A$ satisfies the same functional equation (26') as $K_r$. Now, we have shown that it determines $K_r$ up to a constant factor. Hence $(AK_w|K_z) = \lambda(K_w|K_z)$, and so $A = \lambda 1$ since the functions $K_z$ generate $\mathcal{H}_r^2(P)$, qed.

*Exercise 2.* Associating the function

$$\varphi(\lambda) = \lambda^{(1-r)/2} \hat{f}(\lambda),$$

to each $f \in \mathcal{H}_r^2(P)$, show that the question reduces to $L^2(\mathbb{R}_+)$ and operators

$$\varphi(\lambda) \longmapsto \mathbf{e}(u\lambda) t^{-1} \varphi\left(t^{-2}\lambda\right). \tag{16.35}$$

Conclude that all representations $L_r$ of $B_+$ on $\mathcal{H}_r^2(P)$ are mutually equivalent.

*Exercise 3.* Associate the function

$$f_r(b) = J(b;i)^{-r} f(bi), \quad \varphi(\lambda) = \lambda^{(1-r)/2} \widehat{f}(\lambda).$$

to $f \in \mathcal{H}_r^2(P)$. Show that $f_r = J_r\varphi$ where $J_r$ is an isomorphism from $L^2(\mathbb{R}_+)$ onto a *minimal* closed invariant subspace of the left invariant representation of $B_+$, up to a constant factor. The irreducible representations of $B_+$ on these subspaces are equivalent to representation (35). (The group $B_+$ has other irreducible representations: those obtained by replacing these invariant subspaces with their conjugates, as well the one-dimensional representations

$$b \longmapsto t^{i\sigma}, \qquad \sigma \in \mathbb{R}.$$

These are the only ones).

*Exercise 4.* Fourier transforms on $\mathcal{H}_r^2(P)$ enable one to reduce operators of $B$ to formulas (35) in $L^2(\mathbb{R}_+^*)$. For integer $r$, the group $G$ acts on $\mathcal{H}_r^2(P)$, hence on the functions $\varphi(\lambda)$. Calculate the effect of the matrix $w$ on these functions. This will give an integral operator whose kernel can be written using Bessel functions; see n° 22, (iii).

Bargmann did not bother with arguments from functional analysis. By making the Lie algebra of $G$ act on an arbitrary unitary representation of $G$, he obtained symmetric operators not defined everywhere which he assumed to be essentially self-adjoint (correct but not proved) and deduced the classification of irreducible representations of $G$ using simple algebraic calculations that will be given in n° 29. He thus calculated explicitly the coefficients using differential equations whose solutions are special functions whose properties were well-known (integrable formulas, orthogonality relations, asymptotic behaviour, etc.). From this he deduced that some representations can be realized on spaces of holomorphic or antiholomorphic functions[59] and that, with respect to a privileged orthonormal basis (the functions $\zeta^n$ in the unit disc interpretation), the coefficients of the representation are square integrable and pairwise orthogonal. Some of his methods will be presented in § 8. They converted Harish-Chandra to group theory; his generalizations are extraordinary, and Bargmann's physicist arguments were soon justified. But his method remains the only one[60] and, moreover, applies to the universal covering of $G$. In n° 32 du Chap. XI on Bargmann's orthogonality methods, the steps to be followed in this case were clearly set out.

*Exercise 5.* Let $(\mathcal{H}, U)$ be a unitary representation of $G$. For any character $\chi$ of $K$, let $\mathcal{H}(\chi)$ be the subspace of $\mathbf{a} \in \mathcal{H}$ such that $U(k)\mathbf{a} = \chi(k)\mathbf{a}$, so that $\mathcal{H}$ is the Hilbert direct sum of these $\mathcal{H}(\chi)$. Assume there exists $\chi$ and $\mathbf{a} \in \mathcal{H}(\chi)$ such that $\mathcal{H}(\chi)$ is one-dimensional and that the elements $U(x)\mathbf{a}$ generate $\mathcal{H}$. Show that the representation is irreducible and that the function $\omega(x) = (U(x)\mathbf{a}|\mathbf{a})$ satisfies the functional equation

$$\omega(e) \int \omega(xky)\overline{\chi(k)}dk = \omega(x)\omega(y) .$$

(Any irreducible representation of $G$ is of the previous type: n° 30).

(vi) *Solutions of the equation $f * \omega_r = f$.* We showed at the end of section (iv) that, for integer $r$, $f = f * \omega_r$ for any function $f \in \mathcal{H}_r^2(G) = L^2(G) \cap \mathcal{H}_r(G)$. Actually, *the map $P : f \longmapsto f * \omega_r$* on the space $L^2(G)$, *is the orthogonal projection operator onto the closed subspace $\mathcal{H}_r^2(G)$.* Indeed if $f'$ is the projection of $f$ on $\mathcal{H}_r^2(G)$, then

$$f * \omega_r(x) = \int f(xy)\overline{\omega_r(y)}dy = (f|L(x)\omega_r) = (f'|L(x)\omega_r) = f'(x)$$

since $f'$ satisfies (32), whence the result. The fact that $P$ is a projection can also be deduced from the Bargmann orthogonality relations.

Hence formula (32) characterizes functions $f \in L^2(G) \cap \mathcal{H}_r(G)$, but we will need a similar result for functions about which all that we know is that they are *locally integrable* and for which, the expression $f * \omega_r$ is well-defined. For example, if $\Gamma$ is a discrete subgroup of $G$, this is the case for all $f \in L^p(\Gamma \backslash G)$ for any $p$.

**Lemma 2 (integer $r \geq 2$).** *Let $f$ be a locally integrable function such that $f * \omega_r = g$ is defined. Then*[61] *$g \in \mathcal{H}_r(G)$, and $f$ is in $\mathcal{H}_r(G)$ if and only if $f * \omega_r = f$.*

[59] A function $f(z)$ is antiholomorphic if its conjugate is holomorphic. Replacing the functions $f \in \mathcal{H}_r^2(P)$ by their conjugates and making obvious changes to the definition of $L_r(g)$ leads to another series of representations on which theorem 16 can be applied. "Antiholomorphic" representations are not equivalent to "holomorphic" ones.

[60] V. Bargmann, *Irreducible unitary representations of the Lorentz group* (Annals of Math., **48**, 1947, pp. 568–640) ; see for example Serge Lang, $SL_2(\mathbb{R})$ (Addison-Wesley, 1975 or Springer, 1985). In his *Panorama des mathématiques contemporaines* where Dieudonné proposes a list of "initiators" for each major field, in particular non-commutative harmonic analysis, I cannot find Bargmann, neither Gelfand and yet, at about the same time, together with Neumark he studied the group $SL_2(\mathbb{C})$ and influenced HC as much as Bargmann. HC told me one day: *semi-simple groups are my friends, I know them all individually.* This is not reflected in his articles, where he always from the onset only considers the most general case, but...

[61] As $f$ is not assumed to be continuous, this relation means that $g$ is equal almost everywhere to the function $g' \in \mathcal{H}_r(G)$. The same remark applies to the relation $f * \omega_r = f$.

We first recall [Chap. XI, n° 25, (iv)] that, $f * \omega_r$ is defined if and only if, for all $p \in L(G)$,

$$\iint \left|p(xy)f(x)\omega_r(y)\right|dxdy < +\infty . \tag{16.36}$$

The function

$$f * \omega_r(x) = \int f\left(xy^{-1}\right)\omega_r(y)dy = \int f(y)\omega_r\left(y^{-1}x\right)dy \tag{16.37}$$

is then defined almost everywhere and is locally integrable. Associativity of the convolution product will be used in the cases covered by lemma 3 below.

Let us first suppose that $f \in \mathcal{H}_r(G)$ and that $f * \omega_r$ exists (which is obviously not always true, for example if $f = f_r$ where $f(z) = e^z$). For every character $\chi$ of $K$, the function $\chi * f \in \mathcal{H}_r(G)$ is either trivial or corresponds to a monomial on $D$. So

$$\chi * (f * \omega_r) = (\chi * f) * \omega_r = \chi * f$$

and thus $\chi * (f * \omega_r - f) = 0$ for all $\chi$, which implies the result because 0 is the only locally integrable function $g$ such that $\chi * g = 0$ for all $\chi$ (exercise!).

Before proving the converse, let us show that *there are functions* $\varphi$ *on* $L(G)$ *such that*

$$\varphi * \omega_r = \omega_r * \widetilde{\varphi} = \omega_r . \tag{16.38}$$

To this end we consider the unitary representation $(\mathcal{H}_r^2(G), L)$ of $G$. As

$$\varphi * \omega_r(x) = \int \varphi(y)\omega_r(y^{-1}x)dy = \int \varphi(y)(L(y)\omega_r|L(x)\omega_r) = (L(\varphi)\omega_r|L(x)\omega_r) ,$$

the proof reduces to choosing $\varphi$ in such a way that $L(\varphi)\omega_r = \omega_r$. However, there exist $\varphi_n \in L(G)$ such that $L(\varphi_n)\omega_r$ converges to $\omega_r$ (take a Dirac sequence). Then

$$\omega_r = \chi_r * \omega_r = L(\chi_r)\omega_r = \lim L(\chi_r * \varphi_n)\omega_r ;$$

but the function $L(\chi_r * \varphi_n)\omega_r = \omega' \in \mathcal{H}_r^2(G)$ trivially satisfies $\chi_r * \omega' - \omega'$, and so is proportional to $\omega_r$, qed.

Let us now come back to a function $f$ for which $f * \omega_r = g$ exists. Replacing $f$ with 0 outside larger and larger compact sets leads to a sequence of functions $f_n \in L^1(G)$ converging everywhere to $f$. Then (dominated convergence)

$$f * \omega_r(x) = \lim f_n * \omega_r(x) = \lim g_n(x) \quad \text{for almost all } x . \tag{16.39}$$

Since $L^1 * L^2 \subset L^2$ [Chap. XI, n° 25, (iv), theorem 35], $g_n = f_n * \omega_r$ is in $L^2$, and so $g_n * \omega_r \in \mathcal{H}_r^2(G)$ as was seen above. However the convolution $\omega_r * \omega_r$ exists since $\omega_r \in L^2(G)$. Hence

$$g_n * \omega_r = (f_n * \omega_r) * \omega_r = f_n * (\omega_r * \omega_r) = f_n * \omega_r = g_n$$

by lemma 3 for $d\lambda(x) = f_n(x)dx$. Thus the functions $g_n$ are in $\mathcal{H}_r^2(G)$ and, being dominated by the locally integrable function

$$p(x) = \int \left|f\left(xy^{-1}\right)\right|.\left|\omega_r(y)\right|dy ,$$

converge almost everywhere to $g$. Let us then choose some $\varphi \in L(G)$ satisfying (38). By lemma 3, $g_n * \widetilde{\varphi} = f_n * \omega_r = g_n$. Since $\varphi$ is continuous with compact support, $\lim g_n * \widetilde{\varphi}(x) = g * \widetilde{\varphi}(x)$ as well (dominated convergence ), whence $g * \varphi(x) = g(x)$ almost everywhere. Furthermore,

$$|g_n * \widetilde{\varphi}(x) - g * \widetilde{\varphi}(x)| \leq \int |g_n(y) - g(y)|.|\widetilde{\varphi}(y^{-1}x)|dy\,.$$

If $M$ is the support of $\widetilde{\varphi}$, the integral is extended to the compact set $xM^{-1}$, hence to a fixed compact set $M'$ if $x$ stays in a compact subset $A$ of $G$. As the functions $|g_n(y) - g(y)|$, dominated by $2p(y)$, tend to 0, the right hand side tends uniformly to 0 on $A$. The function $g = g * \widetilde{\varphi}$ is therefore the limit of functions $g_n \in \mathcal{H}_r(G)$ with respect to compact convergence, whence $g \in \mathcal{H}_r(G)$, qed.

The associativity formula used for the convolution product remains to be justified:

**Lemma 3.** *Let $\mu$ and $\nu$ dbe two measures such that $\mu * \nu$ exists. Then*

$$\lambda * (\mu * \nu) = (\lambda * \mu) * \nu \text{ and } \mu * (\nu * \lambda) = (\mu * \nu) * \lambda \tag{16.40}$$

*for all measures $\lambda$ with compact support.*

We showed in Chap. XI, n° 25, (ii) that, generally speaking, the previous relation only involves absolute values of the measures considered and that, for positive measures, it holds provided

$$\iiint p(xyz)d\lambda(x)d\mu(y)d\nu(z) < +\infty$$

for all $p \in L_+(G)$. If $\lambda$ has compact support, then clearly $\int p(xyz)d\lambda(x) = q(yz)$ with some $q \in L_+(G)$. So if $\mu * \nu$ exists, the upper integral

$$\int p(xyz)d\lambda(x) \iint d\mu(y)d\nu(z)$$

is by definition of $\mu*\nu$ finite. Thus Lebesgue-Fubini follows by (40). The same proof holds if $\lambda * \mu$ exists and if $\nu$ has compact support.

The reader is likely to think that, as formula (28) holds for all $g_n$, it also holds for the limit $g$ making the holomorphy of $g$ obvious. To justify this argument, it is at the very least necessary to prove a general result enabling us to pass to the limit in Cauchy's formula for the functions $g_n$:

**Lemma 4.** *Let $U$ be an open subset of $\mathbb{C}$ and $(g_n)$ a sequence of holomorphic functions on $U$. Suppose that $\lim g_n(z) = g(z)$ exists almost everywhere and that there is a locally integrable function $q \geq 0$ on $U$ such that $|g_n(z)| \leq q(z)$ for all $z \in U$. Then the sequence $(g_n)$ converges uniformly in every compact set and $g$ is equal almost everywhere to a holomorphic function.*

Compact convergence being a local property, using a translation, it is possible to work in a disc $D : |z| \leq r$ contained in $U$. Choosing numbers $r'$ and $r''$ such that $r < r' < r''$ assuming the closed disc $|z| \leq r''$ to be contained in $U$, for all t $z \in D$ and all $\varrho \in [r', r'']$,

$$g_n(z) = \int g_n\left(\varrho \mathbf{e}(t)\right)\left(\varrho \mathbf{e}(t) - z\right)^{-1} \varrho \mathbf{e}(t)dt\,.$$

Using polar coordinates, we get a formula of type

(16.41)
$$g_n(z) = c \iint g_n(w)(1 - z/w)^{-1} dudv \qquad (w = u + iv)$$

where integration is over the annulus $M : r' \leq |w| \leq r''$. This being so,

$$\left| g_n(w) \left(1 - z/w\right)^{-1} \right| \leq q(w) \left|1 - z/w\right|^{-1} .$$

This is an integrable function on $M$ since, almost by definition, $q$ is integrable over all compact sets [Chap. XI, n° 5, (ii)]. It is thus possible to pass to the limit in relation (41) for the functions $g_n$, proving (41) for $g$, which is therefore holomorphic. The convergence of the functions $g_n$ on every compact set is obviously uniform: it follows from the usual upper bounds, qed.

In fact, relation (38) replaces formula (41) and can be applied in far more general situations (spherical functions) where no complex analytic structure is available.

We will also use the fact that $\mu * f \in \mathcal{H}_r(G)$ *for all* $f \in \mathcal{H}_r(G)$ *and all measures* $\mu$ *with compact support.* As $\mathcal{H}_r(G)$ is left invariant and closed with respect to the compact topology, it suffices to show that, on any compact set, for any continuous function $f$ on $G$, the function

$$\mu * f(x) = \int f\left(y^{-1}x\right) d\mu(y)$$

is the uniform limit of linear combinations of left translations of $f$. Now, if $x$ stays in a compact set $A$ and if $B$ is the support of $\mu$, then the integral is extended to the compact set $AB^{-1} = C$. As $f$ is continuous, for all $\epsilon > 0$, $B$ can be covered by finitely many open sets $U_i$ such that, for $y_1, y_2 \in U_i$,

$$\left| f\left(y_1^{-1}x\right) - f\left(y_2^{-1}x\right) \right| < \epsilon \quad \text{for tall } x \in A$$

(uniform continuity). Replacing the $U_i$ with pairwise disjoint Borel sets $E_i \subset U_i$ and choosing some $y_i \in E_i$ for all $i$, we get

$$\left| \mu * f(x) - \sum f\left(y_i^{-1}x\right) \mu\left(E_i\right) \right| \leq \sum \int_{E_i} \left| f\left(y^{-1}x\right) - f\left(y_i^{-1}x\right) \right| d|\mu|(x)$$

for all $x \in A$. The right hand side is $\leq \epsilon \sum |\mu|(E_i) \leq \epsilon |\mu|(G)$, qed.

# § 6. Modular Functions: The Classical Theory

In all of this §, $\Gamma$ will denote the modular group $SL_2(\mathbb{Z}) = G(\mathbb{Z})$, unless otherwise stated.

## 17 – Fundamental Domain, Modular Forms

(i) *Generators of the modular group.* The modular group $\Gamma$ contains the transformations

$$T : z \longmapsto z + 1\,, \qquad S : z \longmapsto -1/z$$

corresponding to matrices

$$T = \begin{pmatrix} 1 & 1 \\ 0 & 1 \end{pmatrix}, \qquad S = \begin{pmatrix} 0 & -1 \\ 1 & 0 \end{pmatrix} = w\,. \tag{17.1}$$

We start with an essential result:

**Theorem 17.** *The modular group is generated by matrices $S$ and $T$.*

This means that any $\gamma \in \Gamma$ is for the form $T^p S^q T^r \ldots$ with exponents in $\mathbb{Z}$. As $S^2 = -1$, powers of $S$ can be omitted, if need be by replacing $\gamma$ with $-\gamma$.

The proof if simple. Let $a, b, c, d$ be the entries of $\gamma$; we may assume that $c \geq 0$. We will use induction on $c$. If $c = 0$, equality $ad = 1$ requires $a = d = \pm 1$, and $\gamma$ is then a power of $T$, up to sign. When $c > 0$,

$$T^n \gamma = \begin{pmatrix} a + nc & b + nd \\ c & d \end{pmatrix}$$

and we choose $n$ in such a way that $0 \leq a + nc < c$ (Euclidian division with remainder). The entries of $\gamma' = T^n\gamma$ then satisfy $c' = c$, $0 \leq a' < c$. As

$$\begin{pmatrix} 0 & -1 \\ 1 & 0 \end{pmatrix} \begin{pmatrix} a' & b' \\ c & d' \end{pmatrix} = \begin{pmatrix} -c & -d' \\ a' & b' \end{pmatrix},$$

for $\gamma'' = S\gamma' = ST^n\gamma$, the new entry $c'' = a'$ satisfies $0 \leq c'' < c$. The induction hypothesis then shows that $\gamma''$ is a non-commutative monomial in $T$ and $S$, up to sign. Hence so is $\gamma$, qed.

**Corollary 1.** *A function $f(z)$ satisfies*

$$f(\gamma z) = J(\gamma; z)^r f(z) \quad \text{for all } \gamma \in G(\mathbb{Z})$$

*for some even integer $r$ if and only if*

$$f(z+1) = f(z)\,, \qquad f(-1/z) = z^r f(z)\,. \tag{17.2}$$

It goes without saying that for $r$ odd, $f$ is necessarily 0 since $-1 \in \Gamma$.

(ii) *Fundamental domain.* The second result is about the explicit determination of a fundamental domain for $\Gamma$ acting on the upper half-plane $P$. This is a classic, but no longer so useful exercise (seemingly resolved by Gauss in his secret papers): as such precise results are not available for arithmetic subgroups of general semisimple groups, weaker results have to make do, even for classical groups studied by Hermite, Minkowski, Siegel, etc.

The method is the same as that used for the group of the function $\theta(z)$ in § 2, n° 5. For given $z \in P$,

$$\operatorname{Im}(\gamma z) = y/|cz+d|^2 , \tag{17.3}$$

which enables us to choose $\gamma$ in such a way that $\operatorname{Im}(\gamma z)$ is maximal. As integral translations $T^n : z \longmapsto z+n$ leave $\operatorname{Im}(z)$ invariant, $\gamma$ can be required to also satisfy $|\operatorname{Re}(\gamma z)| \leq \frac{1}{2}$. But, our choice of $\gamma$ implies

$$\operatorname{Im}(\gamma z) \geq \operatorname{Im}(S\gamma z) = \operatorname{Im}(-1/\gamma z) = \operatorname{Im}(\gamma z)/|\gamma z)|^2 ,$$

so $|\gamma z| \geq 1$ follows. Finally, any $z \in P$ can be mapped by $\Gamma$ to a closed subset $F$ of $P$ defined by the inequalities

$$F : |x| \leq \frac{1}{2}, \qquad |z| \geq 1 . \tag{17.4}$$

By the way, note that *the invariant measure on $F$ is finite* as is that of any subset of $P$ contained in the strip $|x| \leq M$, $y \geq c > 0$.

To show that $F$ is really a fundamentally domain for $\Gamma$ in a strict sense, it is also necessary to check that two points $z'$, $z''$ of $F$ cannot be equivalent mod $\Gamma$ if one of them is in the *interior* of $F$. So suppose that $z'' = \gamma z'$. The entry $c$ of $\gamma$ can be assumed to be $\geq 0$ (otherwise, replace $\gamma$ by $-\gamma$), so can $y'' \geq y'$ (otherwise permute $z'$ and $z''$). By (3), the relation $y'' \geq y'$ can be written

$$|cz'+d|^2 = (cx'+d)^2 + c^2 y'^2 \leq 1$$

and implies $|cy'| \leq 1$. However it is sort of obvious that $y' \geq \sin \pi/3$ for all $z' \in F$. Hence $c^2 \leq 4/3$, and so $c = 0$ or 1.

If $c = 0$, then $z'' = z' + n$, which requires $z'$ and $z''$ to be located on the intersections of the vertical sides of $F$ with a horizontal, $\gamma$ then being $z \longmapsto z+1$ or $z \longmapsto z-1$.

If $c = 1$, then $|z'+d| \leq 1$, which requires $d \in \{-1, 0, 1\}$ as follows from the images of $F$ under integral translations. Hence there are again two possible cases.

If $d = 0$, (4) shows that $|z'| \leq 1$, hence that $|z'| = 1$, and

$$\gamma = \begin{pmatrix} a & -1 \\ 1 & 0 \end{pmatrix} = \begin{pmatrix} 1 & a \\ 0 & 1 \end{pmatrix} \begin{pmatrix} 0 & -1 \\ 1 & 0 \end{pmatrix} .$$

Thus $z'' = a - 1/z'$. For $|z'| = 1$, the point $-1/z'$ being the symmetric of $z'$ with respect to the imaginary axis, for $a - 1/z'$ to be in $F$ $a$ must be 0 unless $z'$ and $z''$ are the two vertices[62] $\exp(2\pi i/3) = j$ and $\exp(\pi i/3) = -\bar{j}$ of "triangle" $F$, in which case $a = -1$ or $+1$ may be appropriate according to the case.

If $|d| = 1$, for $|z'| = 1$, $|z'+d| \leq 1$ is possible only if either $z' = j$ and $d = 1$, or $z' = -\bar{j}$ and $d = -1$, as indicated above.

To summarize:

**Theorem 18.** *Let $F$ be the subset of the half-plane $P$ defined by inequalities*

$$|z| \geq 1 , \qquad |x| \leq \frac{1}{2} . \tag{17.5}$$

[62] In my youth, the notation $j = \exp(2\pi i/3)$ was standard in France. Freitag and Busam use the letter $\varrho$, Koblitz the letter $\omega$, etc. An international commission should be convened.

*Then* $P = \bigcup \gamma(F)$ *and two distinct points* $z', z'' \in F$ *can be mapped onto each other by a* $\gamma \in \Gamma$ *only if they are on the boundary of* $F$ *and symmetric with respect to the imaginary axis, i.e. if* $\mathrm{Im}(z') = \mathrm{Im}(z'')$. *For* $\mathrm{Im}(z) > 1$, $\mathrm{Im}(\gamma z) > 1$ *if and only if* $\gamma \in \Gamma_\infty = \Gamma \cap B$.

The arguments that have led us to case (b) show a bit more. We will say that a point $z \in P$ is a *fixed point* of $\Gamma$ if there exists $\gamma \neq \pm 1$ in $\Gamma$ such that $\gamma z = z$. The proof of the theorem shows that the only fixed points in $F$ are (a) $z = i = Sz$, (b) $z = \exp(2\pi i/3) = j = STz = T^{-1}Sz$, (c) $z = \exp(\pi i/3) = -\bar{j} = ST^{-1}z = TSz$. As the other fixed points can made mapped onto $F$ by $\Gamma$, they are the images under $\Gamma$ of the points found. It is even unnecessary to consider both $j$ and $-\bar{j}$ since $-\bar{j} = Tj$. Hence if the fixed points are divided into classes mod $\Gamma$, there are only two such classes: that of $i$ and that of $j$. Moreover, if we do not distinguish between matrices $\gamma$ and $-\gamma$ which define the same transformations on $P$, the stabilizer of $i$ in $\Gamma$ is the subgroup $\{1, S\}$ of order 2, whereas that of $j$ is the subgroup $\{1, ST, T^{-1}S\}$ of order 3.

*Exercise 1.* Let $\Gamma$ be a discrete subgroup of a *connected* locally compact group $G$. An open subset $\Omega \subset P$ will be said to be an *open fundamental* set for $\Gamma$ if (a) $P = \bigcup \gamma\Omega$, and if (b) there are finitely many matrices $\gamma$ such that $\gamma\Omega \cap \Omega \neq \emptyset$. Show that $\Gamma$ is generated by the elements $\gamma$ such that $\gamma\Omega \# \Omega$ (show that, if $\Gamma'$ is the subgroup generated by these elements of $\Gamma$, the union of $\gamma\Omega$ for $\gamma \in \Gamma'$ is both open and closed in $P$). [In generalizations of the theory to semisimple groups, only very rarely is it possible to construct genuine fundamental domains $F$. Open fundamental sets are just as useful. In the case of the modular group, one can choose any set defined by inequalities $|x| < a$, $y > T$ with sufficiently large $a$ and sufficiently small $T > 0$].

*Exercise 2.* Let $\Gamma'$ be a subgroup of finite index $n$ in the modular group $\Gamma$ and $\Gamma'\gamma_1, \ldots, \Gamma'\gamma_n$ the $n$ cosets mod $\Gamma'$ in $\Gamma$. Show that $F' = \bigcup \gamma_i F$ is a fundamental domain for $\Gamma'$. What about the group of the function $\theta(z)$? What about the subgroup[63] $\Gamma(2)$ of matrices $\equiv 1 \pmod 2$?

(iii) *The classical definition of modular forms.* The simplest version of the classical theory with respect to $\Gamma = G(\mathbb{Z})$ consists in studying meromorphic functions satisfying

$$f(\gamma z) = J(\gamma; z)^r f(z) \tag{17.6}$$

for a given even integer $r$. As will be seen, to get results, these functions must not increase too rapidly as $\mathrm{Im}(z)$ increases indefinitely.

First note that the poles of $f$ in $P$ can be divided into classes mod $\Gamma$. By the previous theorem, two poles $z, z'$ with imaginary parts $> 1$ can belong to the same class only if there exists $n \in \mathbb{Z}$ such that $z' = z + n$. The set of poles contained in the fundamental domain $F$ is a discrete subset of $F$, which does not stop them from accumulating to infinity. To get rid of this possibility, $f$ must first be required to be *holomorphic for sufficiently large* $\mathrm{Im}(z)$, so that $f$ only has finitely many poles in $F$ or, equivalently, that the poles of $f$ in $P$ are divided into finitely many classes mod $\Gamma$.

As $f(z+1) = f(z)$, there is then a Fourier series expansion

$$f(z) = \sum a_n \mathbf{e}(nz)\,, \qquad a_n = \oint f(z)\mathbf{e}(-nz)dx \tag{17.7}$$

[63] For a diagram representing a fundamental domain for $\Gamma(2)$, see Neal Koblitz, *Introduction to Elliptic Curves and Modular Forms* (Springer, 2d. ed., 1993), p. 106.

valid for sufficiently large $\text{Im}(z)$.[64] This expansion is also a Laurent series in the variable $\zeta = \mathbf{e}(z)$ since the map $z \longmapsto \mathbf{e}(z)$ transforms every half-plane $\text{Im}(z) > T > 0$ into a disc centered at 0. Note that $|\mathbf{e}(z)| = \exp(-2\pi y)$ tends to 0 as $\text{Im}(z)$ tends to infinity. This being so, the second condition that $f$ must be required to satisfy is that there be only *finitely* many terms of degree $n < 0$ in this Laurent series, in other words that

$$f(z) = O\left(e^{2\pi p y}\right), \quad y \longrightarrow +\infty \tag{17.8}$$

for some integer $p > 0$. If $k$ is the smallest integer for which $a_k \neq 0$, $f$ will be said to have a *pole of order* $-k$ *at infinity* if $k < 0$, and a *zero of order* $-k$ *at infinity* if $k > 0$. For obvious reasons, we set $k = v_\infty(f)$.

These two conditions define the *modular forms of weight* $r$ in the precise sense of the word. For $r = 0$, the expression *modular functions* is used instead. These are analogous to elliptic functions for the group $\Gamma$.

If $f$ and $g$ are modular forms of weight $p$ and $q$, the quotient $f/g$ is clearly a form of weight $p - q$.

Modular forms holomorphic at infinity are characterized by the fact that they are *bounded on* $\{\text{Im}(z) \geq T\}$ *for sufficiently large* $T$, because a Laurent series converging for $0 < |z| < R$ is a power series if and only its sum is bounded in the neighbourhood of 0. Similarly, it would suffice to have an upper bound

$$f(z) = O\left(y^N\right) \quad \text{for large } y,$$

since were this the case, then

$$|a_n| \leq \int_0^1 |f(z)\mathbf{e}(-nz)|\, dz \leq M y^N \exp(2\pi n y)$$

for large $y$, an expression tending to 0 for all $n < 0$. For such a function we denote the constant term of the Fourier series of $f$ by

$$f(\infty) = a_0 = \lim f(z). \tag{17.9}$$

Modular forms that are holomorphic everywhere, including at infinity, are said to be *entire*. They are the analogues of the theta functions of theory of elliptic functions. The Fourier series of an entire form $f$ converges and represents $f$ for all $z \in P$ and, by definition, only involves exponentials $\mathbf{e}(nz)$ with exponent $n \geq 0$. This shows that *any entire modular form is bounded on every half-plane* $\text{Im}(z) \geq T > 0$.

For any entire modular form $f$, the asymptotic behaviour of the function $f(iy)$ as $y$ tends to 0 or $+\infty$ can be easily obtained. For large $y$, $f(iy) \sim a_n \exp(-2\pi n y)$ where $n \geq 0$, which is much better than an asymptotic expansion. As $y$ tends to 0, relation $f(iy) = (-iy)^{-r} f(i/y)$ shows that $f(iy) \sim (-iy)^{-r} a_n \exp(-2\pi n/y)$. In particular,

$$f(iy) = \begin{matrix} O(1) & \text{as } y \longrightarrow +\infty, \\ O(y^{-r}) & \text{as } y \longrightarrow 0. \end{matrix} \tag{17.10}$$

This result will enable us [n° 23, (i)] to define the Mellin transform of $f(iy)$, a Dirichlet series with a simple functional equation.

[64] Chap. VII, n° 17: if $f(z) = f(z+1)$ is holomorphic on a strip $B : a < \text{Im}(z) < b$, the Fourier series of $f$ converges normally on every compact subset of $B$ and represents $f$ on all of this strip.

Entire modular forms such as Hecke *Spitzenformen* (*Spitze* = cusp) $f(\infty) = 0$, are also called *parabolic* or *cusp forms*. As $\mathrm{Im}(z)$ tends to infinity, $f(z)$ tends to 0. In fact there is an upper bound $|f(z)| \leq M(T)e^{-2\pi y}$ on the half-plane $\mathrm{Im}(z) \geq T > 0$, which replaces (10) by

$$f(iy) = \begin{matrix} O\left(e^{-2\pi y}\right) & \text{as } y \longrightarrow +\infty\,, \\ O\left(y^{-r}e^{-2\pi/y}\right) & \text{as } y \longrightarrow 0\,. \end{matrix} \tag{17.11}$$

This result makes the integral defining the Mellin transform of $f(iy)$ convergent for all $s$.

As was seen in n° 15, each solution of the functional equation $f(\gamma z) = J(\gamma; z)^r f(z)$ can be transformed into a function

$$f_r(g) = J(g; i)^{-r} f(gi) \tag{17.12}$$

on $G$, invariant under left translations $g \longmapsto \gamma g$. $f$ is holomorphic if and only if $f_r$ belongs to the space $\mathcal{H}_r(G)$ of n° 15, (ii). The set of these functions will be written $\mathcal{H}_r(\Gamma\backslash G)$. We recall the formula

$$|f_r(g)| = y^{r/2}|f(z)|\,. \tag{17.13}$$

(iv) *Eisenstein and Poincaré series.* The most obvious way to construct left $\Gamma$-invariant functions on $G$ is to use the series $\varphi^\Gamma(g) = \sum \varphi(\gamma g)$ of Chap. XI, n° 15, (vii). If $\varphi \in L^1(G)$, the series converges almost everywhere, its sum is in $L^1(\Gamma\backslash G)$ and

$$\int_{\Gamma\backslash G} \varphi^\Gamma(g)dg = \int_G \varphi(g)dg\,. \tag{17.14}$$

As here we only holomorphic functions interest us, we choose $\varphi$ to be a function in $L^1(G) \cap \mathcal{H}_r(G) = \mathcal{H}_r^1(G)$, hence of the form $f_r$ where $f \in H_r^1(P)$, the space defined in n° 16, (i). By n° 15, (ii), $\varphi^\Gamma$ is then a function of weight $r$ associated to the *Poincaré series*

$$P_{r,f}(z) = \sum_\Gamma J(\gamma; z)^{-r} f(\gamma z) = \sum L_r(\gamma)f(z)\,. \tag{17.15}$$

As these are holomorphic functions, convergence in $L^1$ implies compact convergence – details about this point will be given in a more general framework in n° 21, (i) – and the next theorem 19 will show that these series are parabolic forms. All this supposes that $\mathcal{H}_r^1(G) \neq \{0\}$. So $r > 2$ and $r$ is even, and thus $r \geq 4$. In fact, to have any chance of obtaining sums that are not identically zero, one needs to assume that $r \geq 12$, but *every parabolic form* will be shown to *be a Poincaré series* later in a more general framework.

Another type of series is obtained by grouping together the terms in sum (15) belonging to a same coset $\gamma U_\infty$, where $U_\infty = U \cap \Gamma = U(\mathbb{Z})$ is the group of translations $z \longmapsto z + n$. This gives

$$P_{r,f}(z) = \sum_{U_\infty\backslash\Gamma} J(\gamma; z)^{-r} \sum_{\mathbb{Z}} f(n + \gamma z)\,. \tag{17.16}$$

But for any $f \in \mathcal{H}_r^1(P)$ there is a Poisson summation formula

$$\sum f(z+n) = \sum_{n>0} \widehat{f}(n)\mathbf{e}(nz)$$

[n° 16, (i)]. Hence, at least formally, we find

$$P_{r,f}(z) = \sum \widehat{f}(n) E_{r,n}(z) \tag{17.17}$$

with *Poincaré-Eisenstein series*

$$E_{r,n}(z) = \sum_{U_\infty \backslash \Gamma} J(\gamma; z)^{-r} \mathbf{e}(n.\gamma z)\,. \tag{17.18}$$

They converge for $r > 2$ and $n > 0$ and again define parabolic forms. In fact, like Poincaré series, the forms of weight $r$ corresponding to them on $G$, namely

$$E_{r,n}(g) = \sum \mathbf{e}_{r,n}(\gamma g)^n\,,$$

where summation is mod $U_\infty$ and where

$$\mathbf{e}_{r,n}(g) = J(g; i)^{-r} \mathbf{e}(nz)\,,$$

are integrable mod $\Gamma$. Indeed

$$\int_{\Gamma\backslash G} |E_{r,n}(g)|\, dg \le \int_{\Gamma \backslash G} dg \sum_{U_\infty\backslash\Gamma} |\mathbf{e}_{r,n}(\gamma g)| = \int_{U_\infty \backslash G} |\mathbf{e}_{r,n}(g)| =$$

$$= \int_{U_\infty\backslash P} y^{r/2} \exp(-2\pi n y) dm(z) =$$

$$= \int_{\mathbb{R}_+^*} y^{r/2-1} \exp(-2\pi n y) d^* y < +\infty$$

since $n > 0$ and $r > 2$.

For $n = 0$, this calculation falls apart but series (18) still converge. The coset $U_\infty \gamma$ of a matrix $\gamma \in \Gamma$ is the set of all $\gamma' \in \Gamma$ whose second row $(c, d)$ is the same as that of $\gamma$, because relations $ad - bc = a'd - b'c = 1$ are equivalent to the existence of $n$ such that $a' = a + nc$, $b' = b + nd$. Hence

$$E_{r,0}(z) = \sum_{(c,d)=1} (cz+d)^{-r} = \sum_{U_\infty\backslash\Gamma} J(\gamma; z)^{-r}\,, \tag{17.19}$$

a sum extended to all ordered pairs of coprime integers. As $r$ is even, the terms in $c, d$ and $-c, -d$ are equal, and the functions

$$E_r(z) = \frac{1}{2} E_{r,0}(z)\,. \tag{17.20}$$

are called *reduced Eisenstein series*.

They are related to the first known examples, the *full Eisenstein series*

$$G_r(z) = \sum (cz+d)^{-r} = 2\zeta(r) E_r(z) \tag{17.21}$$

of the theory of elliptic functions [n° 17, (iii)], where summation is over all ordered pairs $(c, d) \neq (0, 0)$. The factor $2\zeta(r)$ is obtained by grouping together the terms in $(nc, nd)$ with $pgcd(c, d) = 1$ and $n \neq 0$. Convergence of these series for $r = 4, 6, \ldots$ is clear [n° 13, (i)], as well as the functional equation. Note that if formula (19) has a clear meaning from a group theoretic viewpoint, this is not the case of (21), a major reason for only considering series (20).

To show that functions (20) or (21) are entire modular forms, the most simple is to compute the Fourier series, a task which is anyhow essential. The usual method – it of course does not generalize – consists in starting from formula (27.10)

$$\sum_{\mathbb{Z}} (z+n)^{-s} = \frac{(-2\pi i)^s}{\Gamma(s)} \sum_{n\geq 1} n^{s-1} \mathbf{e}(nz) \tag{17.22}$$

of Chap. VII, § 6, valid for integer $s \geq 2$ and $\text{Im}(z) > 0$, and even[65] for $\text{Re}(s) > 1$ [Chap. VIII, (10.9)]. Applying it to the summation over $d$, we get

$$\frac{1}{2}G_r(z) = \zeta(r) + \frac{(-2\pi i)^r}{\Gamma(r)} \sum_{c,d\geq 1} d^{r-1}\mathbf{e}(cdz) .$$

The double series converges unconditionally, and even normally in all of the half-plane $\text{Im}(z) = y \geq T > 0$: setting $q = \exp(-2\pi T) < 1$, $|\mathbf{e}(cdz)| \leq q^{cd}$ in this half-plane, so that the proof reduces to checking the convergence of the double series $\sum d^{r-1}q^{cd}$. First summing over $c$ reduces to the series $\sum d^{r-1}q^d/(1-q^d)$ whose general term is equivalent to $d^{r-1}q^d$ for large $d$. This gives convergence for $|q| < 1$.

This being settled, it is possible to group together in the series obtained the terms for which $cd$ has a given value $n$. The coefficient of $\mathbf{e}(nz)$ then becomes equal to

$$\sigma_{r-1}(n) = \sum_{d|n} d^{r-1} ,$$

a sum extended to divisors $d \geq 1$ of $n$. Hence the final result:

$$\frac{1}{2}G_r(z) = \zeta(r) + \frac{(-2\pi i)^r}{\Gamma(r)} \sum_{n\geq 1} \sigma_{r-1}(n)\mathbf{e}(nz) . \tag{17.23}$$

We thus get a power series in $\mathbf{e}(z)$, which confirms that $G_r(z)$ is an *entire modular form*, for which $G_r(\infty) = 2\zeta(r)$.

Since $E_r(z) = G_r(z)/2\zeta(r)$, similarly

$$E_r(z) = 1 + \frac{1}{\zeta(r)} \frac{(-2\pi i)^r}{\Gamma(r)} \sum_{n\geq 1} \sigma_{r-1}(n)\mathbf{e}(nz) . \tag{17.24}$$

This expansion shows that the series $E_r$ are entire modular forms for which

$$E_r(\infty) = 1 . \tag{17.25}$$

There is a simpler form for the factor preceding series (23). Indeed, we have shown [Chap. VII, § 3, eq. (11.27)] that

$$2\zeta(2p) = (-1)^{p+1}(2\pi)^{2p}b_{2p}/(2p)\,!$$

[65] Many authors prove (22) by differentiating term by term the partial fractions expansion of the function $\text{cotg}\,\pi z$. This is not quicker than using the Poisson summation formula and this elegant method is unlikely to provide the result for complex $s$. This is also the reason why, even for an even integer $r$, I do not transform $(-2\pi i)^r$ into $(2\pi i)^r$.

where $b_{2p}$ is the Bernoulli number of index $2p$. As here $r$ is even, this can also be written

$$\zeta(r) = -\frac{(-2\pi i)^r}{\Gamma(r)}\frac{b_r}{2r},$$

whence

$$E_r(z) = 1 - \frac{2r}{b_r}\sum_{n\geq 1}\sigma_{r-1}(n)\mathbf{e}(nz). \tag{17.26}$$

The series obtained has rational coefficients. It was found in the 19th century by setting $q = \mathbf{e}(z)$, and its computation would have taken Euler half of a quarter of an hour if only he had had the idea that

$$\begin{aligned}
E_4(z) &= 1 + 240\left(q + 9q^2 + 28q^3 + \ldots\right),\\
E_6(z) &= 1 - 504\left(q + 33q^2 + 244q^3 + \ldots\right),\\
E_8(z) &= 1 + 480\left(q + 129q^2 + 2118q^3 + \ldots\right),\\
E_{10}(z) &= 1 - 264\left(q + 513q^2 + 19684q^3 + \ldots\right),\\
E_{12}(z) &= 1 + \frac{65520}{691}\left(q + 2049q^2 + 177198q^3 + \ldots\right),\\
E_{14}(z) &= 1 - 24\left(q + 8193q^2 + 1594774q^3 + \ldots\right).
\end{aligned}$$

Needless to say that the arithmetic properties of these strange coefficients have made many people cogitate and continue to make several computers run...

I will not dwell further on these series, whose study will be again taken up in the framework of Fuchsian groups. It should nonetheless be added that the theory of elliptic functions leads to a parabolic form of weight 12 defined in a very different manner, the function

$$\Delta(z) = \mathbf{e}(z)\prod_{n\geq 1}\left[1 - \mathbf{e}(nz)\right]^{24} = \eta(z)^{24}, \tag{17.27}$$

where $\eta(z) = \mathbf{e}(z/24)\prod[1 - \mathbf{e}(nz)]$ is the Dedekind function of n° 3. As

$$\eta(z+1) = \exp(\pi i/12)\eta(z), \quad \eta(-1/z) = (z/i)^{\frac{1}{2}}\eta(z),$$

and as, in Ramanujan's notation,

$$\Delta(z) = \mathbf{e}(z)\left[1 - 24\mathbf{e}(z) + \ldots\right] = \sum \tau(n)\mathbf{e}(nz) \tag{17.28}$$

is clearly a power series in $\mathbf{e}(z)$ without a constant term, $\Delta$ is obviously a parabolic form of weight 12.

It was mentioned above that Poincaré series are parabolic because they are integrable over $\Gamma\backslash G$. The next general theorem justifies this point:

**Theorem 19.** (a) *Every entire modular form of weight* $r < 0$ (resp. $r = 0$) *is trivial (resp. constant).*

*(b) For any modular form of weight* $r \geq 4$,

$$f(z) = f(\infty)E_r(z) + g(z) \tag{17.29}$$

*where g is parabolic.*

*(c) A solution of*

$$f(\gamma z) = J(\gamma; z)^r f(z)\,, \quad r \geq 2\,,$$

*holomorphic everywhere on P is a parabolic form only if (resp. if)*

$$f_r \in L^p(\Gamma\backslash G) \tag{17.30}$$

*for all (resp. for some)* $p \leq +\infty$.

*(d) The Fourier coefficients of all entire (resp. cusp) forms of weight* $r > 2$ *satisfy*

$$a_n = O\left(n^r\right) \quad (\text{resp.} \quad O(n^{r/2}))\,. \tag{17.31}$$

(a) Let $f$ be an entire modular form of weight $r$. Then, as was seen above, $f(z)$ is bounded on the fundamental domain $F$. If $r = 0$, it is bounded on $P$ since it is $\Gamma$-invariant. Removing the constant term of its Fourier series, it may be supposed to be trivial, in which case, $f(z)$ tends to 0 as $\mathrm{Im}(z)$ tends to $+\infty$. From the arguments already used in n° 5, it then follows that $|f(z)|$ reaches its maximum at a point of $F$, and as its maximum in $F$ is also its maximum in $P$, the function is constant.

If $r < 0$, like $f$ and $y^{r/2}$, the function $|y^{r/2}f(z)|$ is bounded on $F$. As it is $\Gamma$-invariant, it is bounded on $P$. Hence $f_r \in \mathcal{H}_r^\infty(G)$, a space which reduces to 0 for $r < 0$ [n° 16, (ii), theorem 14].

(b) Obvious because of (25).

(c) Formula (7) for calculating Fourier coefficients shows that

$$|a_n| \leq \exp(2\pi ny) \oint |f(x+iy)|\, dx\,.$$

For $p < +\infty$, to start with it, we deduce (Hölder for $[0,1]$) that

$$|a_n|^p \leq \exp(2\pi npy) \oint |f(x+iy)|^p\, dx\,.$$

So, setting $r/2 = k$, for all $T > 0$

$$\iint_{\substack{|x|\leq\frac{1}{2}\\ y\geq T}} \left|y^k f(z)\right|^p dm(z) \geq |a_n|^p \int_{y\geq T} \exp(-2\pi npy) y^{pk-1} d^*y\,.$$

For $n < 0$, the second integral diverges for all $p$ and $k$. Hence, if the left hand side is finite for some $p < +\infty$, then $a_n = 0$ for $n < 0$. For $n = 0$, convergence would suppose $pk < 1$ and so necessarily $k < 1$. Thus $a_0 = 0$ also holds if $r \geq 2$. Conversely if $f$ is parabolic, then $f(z) = O(\exp(-2\pi y))$ for large $y$. This condition is more than sufficient to ensure $f_r \in L^p(\Gamma\backslash G)$ for all $p$ and $k$, including $p = +\infty$.

Conversely, if $f_r(g)$ is bounded on $G$, it is in $L^p(\Gamma\backslash G)$ for all $p < +\infty$ since $\Gamma\backslash G$ has finite volume, which reduces the proof to the previous case.

(d) If $f$ is parabolic, $y^{r/2}f(z)$ is bounded on $P$ and the inequality

$$|a_n| \leq \exp(2\pi ny) \oint |f(x+iy)|\, dx$$

shows that

$$|a_n| \leq M y^{-r/2} \exp(2\pi ny)$$

for all $y$. Putting $y = 1/n$, the expected upper bound follows.[66] If $f$ is not parabolic, it can be made to be by removing $f(\infty)E_r(z)$. Now, the Fourier coefficients of Eisenstein series are proportional to the sums $\sigma_{r-1}(n)$. Such a sum having at most $n$ terms $\leq n^{r-1}$, the result is $O(n^r)$, qed.

Proposition (c) is particularly useful in the case $r = 2$. It will be seen later that there are no non-trivial entire modular forms in this case. But theorem 19 applies to far more general groups for which there are no such objections.

## 18 – Analogues of Liouville's Two Theorems

(i) *The Riemann surface of* $SL_2(\mathbb{Z})$. It was shown above that there are no non-trivial entire modular forms of weight $r < 0$ and that, for $r = 0$, they reduce to constants. This last result resembles Liouville's first theorem on elliptic functions. There is also an analogue for modular forms of Liouville's second theorem about the zeros and poles of elliptic functions. The standard proof which consists of integrating $df/f$ along the boundary of the fundamental domain $F$ can be found everywhere. Since in n° 11, we have reduced Liouville's theorems to the simpler result that is proved in the theory of Riemann surfaces (Chap. X, n° 1, theorem 1), we might as well apply the same method here, which remains valid for all generalizations of the modular group.

As a first approximation, the Riemann surface $X(\Gamma) = X$ attached to $\Gamma$ is the quotient space $\Gamma\backslash P$. We equip it with the structure of a complex manifold by requiring that for any open set $\Omega \subset \Gamma\backslash P$, the "holomorphic" functions on $\Omega$ be, by definition, the holomorphic $\Gamma$-invariant functions on the inverse image $\pi^{-1}(\Omega)$ under the canonical map $\pi : P \longrightarrow X$. To show that this gives a Riemann surface, it is necessary to exhibit local uniformizers and holomorphic chart changes (Chap. X, n° 1).

Let $\alpha = \pi(a)$ be a point of $X$ and $\Gamma_a = \Gamma \cap K_a$ the stabilizer of $a$ in $G$. The conformal representation $z \longmapsto (z-a)/(z-\overline{a})$ of $P$ on the unit disc transforms $K_a$ into the rotation group about the origin. As was seen in Chap. XI, n° 15, (vii) in a far more general case, there are arbitrarily small open neighbourhoods $W$ of $a$ having the following three properties:

(i) $W$ is stable under $K_a$,
(ii) $\gamma W \# W \Longleftrightarrow \gamma \in \Gamma_a$,
(iii) the image of $W$ under $z \longmapsto (z-a)/(z-\overline{a})$ is a disc centered at[67] 0.

For such a $W$, $\pi(W)$ is an open neighbourhood of $\alpha$ in $X$ which can be identified with the quotient space $\Gamma_a\backslash W$ by (i) and (ii). Let us consider a function $f(z)$ on $W$. It can be extended to an invariant function on $\pi^{-1}(\pi(W)) = \Gamma.W$ if and only if its restriction to $W$ is $\Gamma_a$-invariant. Indeed if this condition holds then, setting $f(z) = f(\gamma^{-1}z)$ for all $z$ belonging to some $\gamma W$, the result only depends on the choice of $\gamma$ since, by (ii), $\gamma$ is unique mod $\Gamma_a$. Besides, $f$ is clearly holomorphic on $\Gamma.W$ if and only if it is on $W$. "Holomorphic" functions on $\pi(W)$ can therefore be identified with holomorphic functions on the $\Gamma_a$-invariant set $W$.

However, the map $z \longmapsto (z-a)/(z-\overline{a})$ being a conformal representation of $W$ on a disc centered at 0, the holomorphic functions on $W$ are the power series in

[66] This argument, as well the use of Fourier series, is due to Hamburg mathematician Erich Hecke, who in the 1930s gave new life to a theory threatened to become obsolete by introducing ideas that continue to reveal their importance.

[67] So $W$ is the set of points of $\mathbb{C}$ for which the ratio of distances to points $a$ and $\overline{a}$ is less than a given number $e < 1$. The boundary of $W$ is a circle not centered at $a$.

$(z-a)/(z-\overline{a})$. As was seen after theorem 18, $\Gamma_a = \{1, -1\}$ unless $a$ is of the form $\gamma(i)$ or $\gamma(j)$. In the first case, the condition of $\Gamma_a$-invariance is empty. If $a = \gamma(i)$, $\Gamma_a$ contains two matrices 1 and $S_a$, up to the factor $\pm 1$ which acts trivially, where $S_a = \gamma S \gamma^{-1}$ satisfies $S_a^2 = 1$. As the only rotation of order 2 about the origin is $\zeta \longmapsto -\zeta$, the holomorphic functions on the $\Gamma_a$-invariant set $W$ are the power series in $[(z-a)/(z-\overline{a})]^2$. Finally if $a = \gamma(j)$, $\Gamma_a$ contains the conjugates of $\gamma$ by the three matrices 1, $ST$ and $T^{-1}S = (ST)^2$, up to the factor $\pm 1$; on the unit disc, these are the rotations of angle 0, $2\pi/3$ and $4\pi/3$, in other words the multiplications by cubic roots of unity. The $\Gamma_a$-invariant power series are then clearly the power series in $[(z-a)/(z-\overline{a})]^3$.

We then let $n(a) = 1, 2$ or 3 denote the order of $\Gamma_a/\{1, -1\}$ and set

$$\zeta_a(z) = [(z-a)/(z-\overline{a})]^{n(a)} \quad \text{for } a \in P. \tag{18.1}$$

These functions are holomorphic and invariant under the corresponding $\Gamma_a$. Furthermore, they clearly have the following property:

(iv) for $z, z' \in P$, there exists $\gamma \in \Gamma_a$ such that $z' = \gamma z$ if and only if $\zeta_a(z) = \zeta_a(z')$.

Hence there is a function $q_a$ on $\pi(W)$ for which

$$q_a \circ \pi = \zeta_a \quad \text{on } W. \tag{18.2}$$

By (iv), $q_a$ is a homeomorphism from $\pi(W)$ onto a disc centered at $O$ in $\mathbb{C}$. This gives a local topological chart $(\pi(W), q_a)$ of $X$ in the neighbourhood of $\alpha$. The neighbourhood $\pi(W)$ of $\alpha$ depends on the choice of $W$, but, for given $a$, the functions $q_a$ corresponding to the various possible choices for $W$ are clearly pairwise equal on their common domains of definition.

If "holomorphic" functions on open subsets of $X$ are defined as above, then the "holomorphic" functions on $\pi(W)$ are precisely the power series in $q_a$. A function defined on an arbitrary open subset $\Omega$ of $X$ is "holomorphic" if and only if in the neighbourhood of any $\alpha \in \Omega$, it can be holomorphically expressed in a chart $(\pi(W), q_a)$ centered at $\alpha$.

Chart changes remain to be shown to be holomorphic. This amounts to checking that, for all $a, a' \in P$ and neighbourhoods $W, W'$ of $a, a'$ satisfying conditions (i), (ii) and (iii), $\zeta_a$ is a holomorphic function of $\zeta_{a'}$ on $W \cap W'$ and conversely. This is obvious if $a = a'$, or if $\zeta_a$ and $\zeta_{a'}$ are conformal representations of $W \cap W'$ on open subsets of $\mathbb{C}$, i.e. are *injective* on $W \cap W'$ [Chap. VIII, n° 5, theorem 7]. This is always the case if $a \neq a'$.

Indeed suppose there are distinct points $a_1$ and $a_2$ in $W \cap W'$ such that $\zeta_a(a_1) = \zeta_a(a_2)$. By (iv), there exists $\gamma \in \Gamma_a$ such that $a_2 = \gamma a_1$. It follows that $\gamma W' \# W'$, hence that by (ii) applied to for $W'$, $\gamma \in \Gamma_{a'}$, so that $\gamma a' = a'$. A non-trivial elliptic matrix having a unique fixed point in $P$, $a = a'$, a contradiction.

In conclusion, the local charts introduced on $X = \Gamma \backslash P$ satisfy the conditions set *a priori*. They turn $X$ into a Riemann surface and, for any open subset $\Omega$ of $X$, a function $f$ defined on $\Omega$ is holomorphic – this term can now be used for it – if and only if $f \circ \pi$ is holomorphic on $\pi^{-1}(\Omega)$. This result says precisely that $\pi$ is a holomorphic map from $P$ onto $X$. It generalizes to meromorphic functions: in the neighbourhood of a pole $\alpha = \pi(a)$, such a function $f$ is, by definition, a Laurent series in the local uniformizer $q_a$, with finitely many terms of negative degree. Hence, in the neighbourhood of $a$, $f \circ \pi$ is a Laurent series in $[(z-a)(z-\overline{a})]^{n(a)}$ and conversely, this property is equivalent to $\Gamma_a$-invariance. .

At the same time, denoting by $v_\alpha(f)$ the order at $\alpha$ of a meromorphic function in the neighbourhood of $\alpha \in X$ and by $v_a(f)$ the order at $a$ of a meromorphic function $f(z)$ on an open subset of $P$,

$$v_\alpha(f) = v_a(f \circ \pi)/n(a)$$

for every meromorphic function in the neighbourhood of $\pi(a)$. It follows that the total order of a function $f$ defined and meromorphic on $X$ is

$$\sum_{\alpha \in X} v_\alpha(f) = \sum_{a \bmod \Gamma} v_a(f \circ \pi)/n(a) =$$

$$= v_i(f \circ \pi)/2 + v_j(f \circ \pi)/3 + \sum v_a(f \circ \pi)\,, \tag{18.3}$$

the latter $\sum$ being extended over all classes mod $\Gamma$ of points $a$ that are not fixed points of $\Gamma$.

Were $X$ compact, it would show that, like in the theory of elliptic functions, $\sum v_a(f)/n(a) = 0$ for any meromorphic modular function of weight 0. But $X$ is not compact, either because the fundamental domain $F$ is not so, either, and this is a better argument, because of a simple and general result (Poincaré) which will be proved later: a discrete subgroup $\Gamma$ of $G$ such that $\Gamma\backslash P$ is compact does not contain any parabolic matrices.

But the Riemann surface $X = \Gamma\backslash P$ can be compactified by taking into account the conditions imposed on the behaviour of modular functions at infinity. Theorem 18 shows that, for $T > 1$, the image $X(T)$ in $X$ of the half-plane $\mathfrak{S}_\infty(T) = \{\mathrm{Im}(z) > T\}$ can be identified with the quotient of $\mathfrak{S}_\infty(T)$ by the subgroup $\Gamma_\infty = \Gamma \cap B$. A function $f$ defined on $X(T)$ is, therefore, holomorphic if and only if $f \circ \pi$ is holomorphic on $\mathfrak{S}_\infty(T)$, i.e. is a Laurent series in

$$\zeta_\infty(z) = \mathbf{e}(z)\,. \tag{18.4}$$

As $\mathbf{e}(z)$ is periodic, there is a function $q_\infty$ on $X(T)$ such that

$$q_\infty \circ \pi(z) = \zeta_\infty(z)\,, \tag{18.5}$$

and $q_\infty$ is clearly a homeomorphism from $X(T)$ onto the dis $|\zeta| < \exp(-2\pi T)$ of $\mathbb{C}$ *with its centre removed.* On the other hand, all $\mathfrak{S}_\infty(T)$ clearly satisfy conditions (i), (ii), (iii) and (iv) provided $K_a$ is replaced by $U$ in (i), $\Gamma_a$ by $\Gamma_\infty = \Gamma \cap B$ in (ii) [theorem 15, (a)], $(z-a)/(z-\overline{a})$ by $\mathbf{e}(z)$ in (iii) and $\zeta_a$ by $\zeta_\infty$ in (iv). From the above, it then follows that any holomorphic function defined and periodic on $\mathfrak{S}_\infty(T)$ extends to a holomorphic $\Gamma$-invariant function on $\pi^{-1}(X(T))$.

We next adjoin a point denoted by $\infty$ to $X$, as in the definition of the Riemann sphere or of algebraic functions in Chap. X , and call a set $\Omega \subset \widehat{X} = X \cup \{\infty\}$ open if $\Omega \cap X$ is open in $X$, and in case $\infty \in \Omega$, if it contains $X(T)$ for sufficiently large $T$. Setting $\widehat{X}(T) = X(T) \cup \{\infty\}$ and defining $q_\infty(\infty) = 0$, ordered pairs $(\widehat{X}(T), q_\infty)$ are clearly charts of $\widehat{X}$ in the neighbourhood of the point $\infty$.

Let us show that these charts (which assume $T > 1$) are mutually holomorphically compatible (obvious) and with $(\pi(W), q_a)$ introduced above on $X$. The proof amounts to showing that $\zeta_\infty(z)$ and $\zeta_a(z)$ are injective on $W \cap \mathfrak{S}_\infty(T)$. Were $\zeta_a$ not so, by (iv) there would exist $\gamma \in \Gamma_a$ such that $\gamma\mathfrak{S}_\infty(T) \# \mathfrak{S}_\infty(T)$, which by (ii) applied to $\mathfrak{S}_\infty(T)$, would imply $\gamma \in \Gamma_\infty$, which is absurd. The function $\zeta_\infty$ is also injective on $W \cap \mathfrak{S}_\infty(T)$. Otherwise, by (iv) applied to $\mathfrak{S}_\infty(T)$ there would exist $\gamma \in \Gamma_\infty$ such that $\gamma W \# W$, hence such that $\gamma \in \Gamma_a$, equally absurd. Thus the charts at infinity are compatible with those already known on $X$.

As modular functions (of order 0) on $P$ have been required to be meromorphic everywhere and for $T$ sufficiently large, to have an expansion of the form $\sum_{n \geq N} a_n \mathbf{e}(z)^n$ on $\mathfrak{S}_\infty(T)$, it is now clear that the *modular functions* ($r = 0$) *are the meromorphic functions* $\widehat{X}(\Gamma)$ and that, considered as such, their order $v_\infty(f)$

at the point $\infty$ of $\widehat{X}$ is just their order as Laurent series in $\mathbf{e}(z)$. To get the total order of the corresponding meromorphic function on $\widehat{X}$, it therefore suffices to add $v_\infty(f)$ to sum (3).

To finish, we still need to show that the Riemann surface $\widehat{X}(\Gamma)$ is *compact*. To this end, let $\alpha_n$ be a sequence of points of $\widehat{X}$. If infinitely many of its terms are equal to $\infty$, extracting a convergent subsequence from it is not difficult. Hence we may suppose that $\alpha_n = \pi(a_n)$ for some $a_n \in P$, determined mod $\Gamma$ and which can be assumed to be in $F$. If $\sup \mathrm{Im}(a_n) < +\infty$, the points $a_n$ belong to a compact subset of $P$. Thus Bolzano-Weierstrass holds. If $\sup \mathrm{Im}(a_n) = +\infty$, extracting a subsequence, we may assume that $\lim \mathrm{Im}(a_n) = +\infty$, hence that, for all $T > 0$, $a_n \in \mathfrak{S}_\infty(T)$ for large $n$. This says precisely that $\lim \alpha_n = \infty$, qed.

(ii) *Zeros and poles.* We are now ready to prove the classic result:

**Theorem 20.** *For any meromorphic modular form $f$ of weight $r$,*

$$\nu(f) = \sum_{a \bmod \Gamma} v_a(f)/n(a) =$$

$$= v_i(f)/2 + v_j(f)/3 + v_\infty(f) + \sum_{a \neq i,j,\infty} v_a(f) = r/12 . \tag{18.6}$$

If $r = 0$, as shown above, $\nu(f)$ is the sum of the orders of zeros and poles of the meromorphic function which corresponds to $f$ on $\widehat{X}(\Gamma)$. Hence $\nu(f) = 0$ in this case (Chap. X, n° 1, theorem 1).

In the general case, if $f$ and $g$ are modular forms of arbitrary weights, then it is clear that $\nu(fg) = \nu(f) + \nu(g)$. If $f$ and $g$ have equal weight, $f/g$ has weight 0, whence $\nu(f) = \nu(g)$. As a result, $\nu(f)$ is an additive function of $r \in \mathbb{Z}$, and so $\nu(f) = cr$ with a constant $c$ which needs to be determined. To do this, we choose the form $\Delta(z) = \eta(z)^{24}$ of weight 12 already introduced in n° 16. It is holomorphic everywhere, including infinity where $v_\infty(\Delta) = 1$, and has no zeros in $P$. Hence $\nu(\Delta) = 1$. As a result, $12c = 1$, $c = 1/12$ and the proof of (6) follows using the Dedekind function.

A less miraculous argument consists in considering a meromorphic differential form $\omega$ on $\widehat{X}$. The map $\pi : P \longmapsto X$ being holomorphic, $\omega$ has a meromorphic inverse image $\omega \circ \pi = f(z)dz$. As $\omega \circ \pi$ is $\Gamma$-invariant, the meromorphic function $f$ satisfies the functional equation $f(\gamma z) = J(\gamma; z)^{-2} f(z)$ for modular forms of weight 2. Let $\alpha = \pi(a)$ be a point of $X$ and $(\pi(W), q_a)$ a local chart of $X$ at $\alpha$. Then $\omega = h_a.dq_a$ on $\pi(W)$ where $h_a$, meromorphic on $\pi(W)$, is a Laurent series in $q_a$. If it starts with a term $q_a^p$, by definition, the number $p$ is the order $v_\alpha(\omega)$ of $\omega$ at $\alpha$ (Chap. X, n° 1). Then by (2), $\omega \circ \pi = (h_a \circ \pi) d(q_a \circ \pi) = (h_a \circ \pi) d\zeta_a$ on $W$. By (1), $d\zeta_a = (z-a)^{n(a)-1} c_a(z) dz$ where $c_a(a) \neq 0$. On the other hand, $h_a \circ \pi$ is a Laurent series in $\zeta_a$ starting with a term of degree $v_\alpha(\omega)$ in $\zeta_a$, hence of degree $n(a)v_\alpha(\omega)$ in $z - a$. Thus

$$v_a(f) = n(a)v_\alpha(\omega) + n(a) - 1 . \tag{18.7}$$

Let us now work in the neighbourhood of the point $\infty$ of $\widehat{X}$. The same arguments and the relation

$$f(z)dz = \omega \circ \pi = (h_\infty \circ \pi)\, d\zeta_\infty = 2\pi i\, (h_\infty \circ \pi)\, \mathbf{e}(z) dz$$

show that, for sufficient large $T$, the Fourier series of $F$ on $\mathfrak{S}_\infty(T)$ starts with a term of degree $v_\infty(\omega) + 1$, whence

$$(18.8) \qquad v_\infty(f) = v_\infty(\omega) + 1\,.$$

This also shows that at infinity $f$ satisfies the definition of modular forms of weight 2.

As a result, adding relations (7) and (8) and taking into account the fact that $n(a) = 1$ unless $n(i) = 2$ and $n(j) = 3$, we see that

$$\nu(f) = v_\infty(f) + \sum_{a \in P} v_a(f)/n(a) =$$

$$(18.9) \qquad = \sum_{\alpha \in \widehat{X}} v_\alpha(\omega) + 1 + (1 - 1/2) + (1 - 1/3)\,.$$

By definition of the genus $g$ of $\widehat{X}$, (Chap. X, n° 1), $\sum v_\alpha(\omega) = 2g - 2$. So the right hand side of (9) equals

$$2g - 2 + 1 + 1/2 + 2/3 = 2g + 1/6\,.$$

But since $f$ is a form of weight 2, the left hand side equals $2c$, where $c$ if the constant such that $\nu(f) = cr$ for all forms of weight $r$. Hence $c = g + 1/12$. In conclusion, theorem 20 says that *the compact Riemann surface* $\widehat{X}(\Gamma)$ *has genus* 0.

Proving directly that $g = 0$ is a different matter. If a compact Riemann surface is known to be of genus 0 if and only if it is *homeomorphic* to the Riemann sphere – which was not proved in Chap. X –, then we only need to check that $\widehat{X}$ falls in this case. However, $\widehat{X}$ is obtained from the fundamental domain $F$ by identifying two points of $F$ that are mapped to each other by $\Gamma$. This amounts to identifying points on the boundary of $F$ having the same $y$ coordinate and to adjoining the point at infinity to the result. Very gifted for selling their wares, topologists will explain that, to understand what is going on, you need to cut the paper of infinite length on which you have drawn $F$ along the boundary of $F$, then to glue together its vertical sides as well as the two circular arcs $|z| = 1$ located on both sides of the $Oy$ axis. You thus get a tube of infinite length,[68] and you then contract to a unique point its vaguely circular "boundary" with which it "ends" at infinity. The result can then "clearly" be deformed into a sphere. Elementary, my dear Watson.

A less fanciful proof consists in showing that there is a meromorphic function $J$ on $\widehat{X}$ with a simple pole at infinity and holomorphic elsewhere. For such a function, the map $J$ from $\widehat{X}$ to the Riemann sphere $\widehat{\mathbb{C}}$ is bijective because, for any meromorphic function $f$ on a compact Riemann surface and any $z \in \widehat{\mathbb{C}}$, the number of solution of the equation $f(\zeta) = z$ does not depend on $z$ (Chap. X, n° 1). As $J : \widehat{X} \longrightarrow \widehat{\mathbb{C}}$ is continuous, $J$ is a homeomorphism – and even an *isomorphism* of analytic manifolds since a holomorphic and injective map is a conformal representation. Whence $g = 0$. But $J$ needs to be constructed. We will do this later, again unfortunately, using the miraculous Dedekind function: $J = E_4^3/\Delta$.

*Exercise 1.* If $\omega$ is a meromorphic differential form on $\widehat{X}$, we showed that $\omega \circ \pi = f(z)dz$ where $f$ is a meromorphic modular form of weight 2. Prove the converse.

*Exercise 2.* Keeping the notation of the previous exercise, express the fact that the sum of residues of $\omega$ is zero (Chap. X, theorem 1) using $f$ and show that there is an analogous formula for $f$ of arbitrary weight $r$.

---

[68] Transforming $P$ into the unit disc by $z \longmapsto (z-i)/(z+i)$, $F$ becomes a curvilinear triangle bounded by three circular arcs orthogonal to $|z| = 1$ and with one vertex on the latter. This vertex plays the role of the point at infinity of $F$, which simplifies the topological argument.

(iii) *Construction of modular forms using $\Delta(z)$ and Eisenstein series.* Theorem 20 has immediate consequences.

If $f$ is an entire modular form of weight $r \geq 2$, the integers $v_i(f)$, etc. of theorem 20 must all be $\geq 0$. For $r = 2$, this is clearly impossible, and so there are no non-trivial *entire* modular forms[69] of weight 2. Besides, such a form would correspond to a differential form holomorphic everywhere on $\widehat{X}$, i.e. on the Riemann sphere. For small values of $r$, it is easy to convince oneself that the only possibilities are given by the following table, where $v_\infty(f)$ is included in the sum $\sum v_a(f)$:

(18.10)

| $r$ | $v_i(f)$ | $v_j(f)$ | $\sum v_a(f)$ |
|---|---|---|---|
| 4 | 0 | 1 | 0 |
| 6 | 1 | 0 | 0 |
| 8 | 0 | 2 | 0 |
| 10 | 1 | 1 | 0 |
| 14 | 1 | 2 | 0 |

However, if $f$ and $g$ are modular forms of equal weight such that $v_a(f) = v_a(g)$ for all $a$, including $a = \infty$, the quotient $f/g$ is a $\Gamma$-invariant function without any zeros or poles, hence a constant. So for $r = 4, 6, 8, 10$ or 14, there is at most one entire modular form of weight $r$, up to a constant factor. The existence of such a form is obvious: for all $r \geq 4$, the reduced Eisenstein series $E_r(z) = \sum 1/(cz+d)^r$ satisfies the functional equation, is holomorphic everywhere on $P$ and tends to 1 as $z$ tends to the vertex $\infty$ of the fundamental domain $F$.

New strange identities follow from this. If $f$ and $g$ are entire modular forms of weight $r$ and $s$, $f(z)g(z)$ is clearly entire of weight $r+s$. Hence

$$E_4(z)^2 = E_8(z)\,, \quad E_4(z)E_6(z) = E_{10}(z)\,, \quad E_6(z)E_8(z) = E_{14}(z) \tag{18.11}$$

since, in each case, the left hand sides take value 1 at infinity.

Let us now suppose $r = 12$. First we have the case where $v_a(f) = 0$ for all $a$ except $v_\infty(f) = 1$. The series $E_{12}(z)$ falls in this case and is the only one in it, up to a constant factor. As it takes value 1 at infinity, $f - f(\infty)E_{12}$ is parabolic for every entire modular form of weight 12. However, $\Delta(z) = \eta(z)^{24}$ is a parabolic form of weight 12 which is zero only at infinity. Hence, if $f$ is another parabolic form of weight 12, the function $f/\Delta$ is holomorphic everywhere on $P$, $\Gamma$-invariant and holomorphic at infinity since it is of the form $(a_1q+\ldots)/q(1+\ldots)$ where $q = \mathbf{e}(z)$, hence regular. As a result, $\Delta$ is the only parabolic form of weight 12, up to a constant factor. So the only entire modular forms of weight 12 are the linear combinations of $E_{12}$ and $\Delta$.

For $r = 12$, $E_4(z)^3$, $E_4(z)E_8(z)$ and $E_6(z)^2$ are necessarily equal to $E_{12}(z) + c\Delta(z)$, with a constant $c$ depending on the case. It can be calculated using the first few terms of their Fourier series which were computed in the previous n°. For example,

$$\begin{aligned} E_6(z)^2 &= [1 - 504\mathbf{e}(z) + \ldots]^2 = 1 - 1008\mathbf{e}(z) + \ldots\,, \\ E_{12}(z) &= 1 + 65520\mathbf{e}(z)/691 + \ldots\,. \end{aligned}$$

As $\Delta(z) = \mathbf{e}(z) + \ldots$,

$$E_{12}(z) - E_6(z)^2 = c\Delta(z) \quad \text{with } c = 2^6 3^5 7^2/691\,. \tag{18.12}$$

The same type of argument shows that $E_4(z)^3 - E_6(z)^2$ is proportional to $\Delta(z)$. As $E_4(z) = 1 + 240\mathbf{e}(z) + \ldots$ and $E_6(z) = 1 - 504\mathbf{e}(z) + \ldots$,

[69] Naturally, there are meromorphic forms of weight 2, for example $E_{14}/\Delta$.

$$E_4(z)^3 - E_6(z)^2 = (3.240 + 2.504)\mathbf{e}(z) + \ldots = 1728\mathbf{e}(z) + \ldots,$$

whence

$$E_4(z)^3 - E_6(z)^2 = 1728\Delta(z). \tag{18.13}$$

For a form $f$ of weight $r > 14$, we use $\Delta$ to reduce to forms of weight $< 14$. To start with, $f = f(\infty)E_r + h$ where $h$ is parabolic; So the quotient $h/\Delta$ is an entire modular form of weight $r - 12$. Thus $f = f(\infty)E_r + g\Delta$ where $g$ is of weight $r - 12$. If $r - 12 \leq 14$, $g$ is proportional to $E_{r-12}$ if $r \neq 24$ and, if $r = 24$, a linear combination of $E_{12}$ and $\Delta$. Redoing the operation if $r > 26$ leads to the next result:

**Theorem 21.** *Every entire modular form of weight $r$ is a linear combination of functions $E_r$, $E_{r-12}\Delta$, $E_{r-24}\Delta^2$,...*

The sequence ends when it gives nonexistent forms of weight $\leq 2$. These functions are linearly independent because their power series in $q = \mathbf{e}(z)$ start respectively with the term $1, q, q^2$, etc. So they form a basis for the finite dimensional vector space of entire modular forms of weight $r$.

*Exercise 3.* Every entire modular form is a polynomial in $E_4$ and $E_6$.

(iv) *Application to elliptic functions.* Let $L$ be a period lattice in $\mathbb{C}$ and let us consider the Weierstrass function $y = \wp_L(u)$. It satisfies the differential equation

$$y'^2 = 4y^3 - py - q \tag{18.13}$$

where

$$p = g_2(L) = 60\sum 1/\omega^4, \quad q = g_3(L) = 140\sum \omega^6. \tag{18.14}$$

We are going to show that, for any set of given coefficients, there is a corresponding lattice $L$, except for one restriction.

To this end we choose a basis $\omega_1$, $\omega_2$ of $L$ such that

$$\omega_2/\omega_1 = z \in P.$$

So

$$p = c_2 E_4(z) \quad \text{with } c_2 = 120\zeta(4)/\omega_1^4 = 4\pi^4/3\omega_1^4, \tag{18.15'}$$

$$q = c_3 E_6(z) \quad \text{with } c_3 = 280\zeta(6)/\omega_1^6 = 8\pi^6/27\omega_1^6. \tag{18.15''}$$

Incidentally, notice the useful detail that $c_2^3 = c_3^2$. We saw above that

$$\begin{aligned}1728\Delta(z) &= E_4(z)^3 - E_6(z)^2 = (\pi/\omega_1)^{-12}\left[(3/4)^3p^3 - (27/8)^2q^2\right] = \\ &= (\pi/\omega_1)^{-12}(p^3 - 27q^2)3^3/2^6,\end{aligned}$$

whence

$$p^3 - 27q^2 = (2\pi/\omega_1)^{12}\Delta(z). \tag{18.16}$$

As $\Delta(z) \neq 0$ for all $z \in P$, we need to assume

$$p^3 - 27q^2 \neq 0.$$

This relation says that the equation $4x^3 - px - q = 0$ does not have double roots. This is an exceptional case where the Weierstrass differential equation can be solved using elementary algebraic functions.

So let us assume this condition holds. We need to show the existence of $z \in P$ and $\omega_1$ satisfying (15') and (15"). However, the ratio

$$J(z) = 1728p^3/(p^3 - 27q^2) = E_4(z)^3/\Delta(z) \tag{18.17}$$

does not depend on $\omega_1$. If we can find some $z$ for which it takes the value required by those fixed for $p$ and $q$, then we can find some $\omega_1$ satisfying (16), which determines it up to a $12th$ root of unity. As

$$\begin{aligned} p^3 &= E_4(z)^3(p^3 - 27q^2)/1728\Delta(z) = (\pi/\omega_1)^{12}2^6 3^{-3} E_4(z)^3 = \\ &= \left[4\pi^4 E_4(z)/3\omega_1^4\right]^3 , \end{aligned}$$

relation (15') then holds modulo a cubic root of unit, which can be removed by multiplying it by $\omega_1$. As (15') holds, the relation

$$E_4(z)^3[p^3 - 27q^2) = 1728p^3\Delta(z) = p^3[E_4(z)^3 - E_6(z)^2]$$

shows that

$$27q^2 E_4(z)^3 = p^2 E_6(z)^2 = c_2^3 E_4(z)^3 E_6(z)^2 .$$

If $E_4(z) \neq 0$, it follows that $q^2 = c_3^2 E_6(z)^2$ giving (15") up to sign, an ambiguity which can once again be removed by multiplying $\omega_1$ by a $12th$ root of unity, which solves the problem in this case. If $E_4(z) = 0$ and so $p = 0$, it is easier.

Finally, the proof reduces to showing that there always exists $z \in P$ where ratio (17) has a given value. Now, this ratio, the *modular function* $J(z) = E_4(z)^3/\Delta(z)$ that Charles Hermite hoped to encounter on arrival in Heaven, is strictly $\Gamma$-invariant and holomorphic everywhere on $P$. Its Fourier series at infinity is

$$[1 + \ldots]/[\mathbf{e}(z) + \ldots] = \mathbf{e}(z)^{-1} + \ldots ,$$

where as usual the degrees of the omitted terms are greater than those of the term explicitly written. In fact

$$J(z) = \mathbf{e}(-z) + 744 + 196884\mathbf{e}(z) + 21493760\mathbf{e}(2z) + \ldots .$$

As a result, like $E_4(z)$, $J(z)$ is a modular function ($r = 0$) for which $v_\infty(J) = -1$, and $v_a(J) \geq 0$ elsewhere. So is the function $f(z) = J(z) - c$ for all $c \in \mathbb{C}$. For $r = 0$, by theorem 20, this implies the existence of some $a \in P$ such that $v_a(f) > 0$, hence of a root of $J(z) = c$, qed.

The number $z$ is obviously not unique since $\gamma z$ is equally suitable for all $\gamma \in \Gamma$, but thanks to theorem 18, it is the only solution when $c \neq J(i), J(j)$.

When $c = J(i)$, $v_i(f) = 2$ because only then does

$$v_i(f)/2 + v_j(f)/3 + \sum v_a(f) = -v_\infty(f) = 1$$

have integer solutions. Hence in this case, $z$ is again the unique solution $\bmod \Gamma$. As $E_6(i) = 0$ and $E_4(i) \neq 0$, it follows that $p \neq 0$ and $q = 0$ in this case.

When $c = J(j)$, the results are the same: $v_j(f) = 3$, $p = 0$ and $q \neq 0$, and there is a unique solution $\bmod \Gamma$.

Coming back to differential equation (13) with $p^3 - 27q^2 \neq 0$, the previous results can be applied. Choose a solution $z$ of (15) then $\omega \in \mathbb{C}$ satisfying (14). The squares and cubes of formulas (3) then hold. If there is an added factor $-1$ in (3"), it can be made to vanish by multiplying $\omega$ by an arbitrary $6th$ root of $-1$, which is then chosen to make sure (15') holds. It remains to observe that when $z$ is replaced by $(az + b)/(cz + d)$, this replaces $z = \omega'/\omega$ by $(a\omega + b\omega')/(c\omega + d\omega')$, which is equivalent to taking the basis $(a\omega + b\omega, c\omega + d\omega')$ of $L$ instead of $(\omega, \omega')$ and does not change $L$. Therefore, the lattice $L$ only depends on the class of $z \bmod \Gamma$. To summarize:

**Theorem 22.** *Let $p$ and $q$ be two complex numbers such that $p^3 - 27q^2 \neq 0$. There is a unique lattice $L$ such that $p = g_2(L)$, $q = g_3(L)$.*

To conclude, let us also state the next result:

**Theorem 23.** *Every modular function is a rational function of $J(z)$. The function $J$ is an isomorphism from the Riemann surface $\widehat{X}(\Gamma)$ onto the Riemann sphere $\widehat{\mathbb{C}}$.*

To see this, we again consider the relation $\sum \nu_a(f) = 0$ which holds for every modular form of weight 0, where $\nu_a(f)$ is an *integer* obtained by dividing the usual order $v_a(f)$ by 2 if $a = \gamma(i)$, by 3 if $a = \gamma(j)$, and by 1 otherwise. The function

$$\prod [J(z) - J(a)]^{\nu_a(f)}$$

where the product is extended to all $a$ (including $\infty$) where $\nu_a(f) \neq 0$, has exactly the same zeros and poles as $f$, including at infinity, and with the same multiplicities. So it is proportional to $f$. The second proposition is obvious since, taking the quotient and setting $J(\infty) = \infty$ gives a bijective analytic map, hence a conformal representation, qed.

All this is only the beginning of the *abc* of a theory that, at the end of the 19th century, gave rise to a great body of work related to the algebra of number fields, Galois theory, the geometry of algebraic curves of genus 1, the "division" of elliptic functions, etc. After almost a half-century of hibernation, it again expanded after the war under the impulse of Shimura, Taniyama, Weil, etc. who naturally had methods and ideas at their disposal that were unknown or not well understood by Felix Klein's contemporaries.[70]

[70] For the former period, see for example Felix Klein and Robert Fricke, *Die Elliptischen Funktionen und ihre Anwendungen* (Teubner, 1916); for the relatively recent period, Goro Shimura, *Introduction to the Arithmetic Theory of Automorphic Functions* (Princeton UP, 1971), Serge Lang, *Introduction to Modular Forms* (Springer, 1976), Anthony W. Knapp, *Elliptic Curves* (Princeton UP, 1992).

# § 7. Fuchsian Groups

## 19 – Generalities on Automorphic Forms

(i) *Two lemmas on discrete subgroups.* If $\Gamma$ is a discrete subgroup of $G$, $z \in P$ is said to be a *fixed elliptic point* of $\Gamma$ if the stabilizer $\Gamma_z$ of $z$ in $\Gamma$ is not contained in $\{1, -1\}$. As $\Gamma_z$ is the intersection of $\Gamma$ with the compact subgroup $K_z$, it is a finite group. The set of these fixed point is a *discrete* subset of $P$ (but not of $\mathbb{C}$: they may accumulate on $\mathbb{R}$). Indeed as $\Gamma$ acts properly on $P$, any $z \in P$ has a neighbourhood $W$ such that the relation $\gamma W \# W$ implies $\gamma z = z$ [Chap. XI, n° 15, (vii)]. Hence $\Gamma_{z'} \subset \Gamma_z$ for all $z' \in W$, and as matrices 1 and $-1$ are the only $k \in K_z$ leaving points other than $z$ fixed, it follows that $z'$ is not a fixed point of $\Gamma$ if $z' \neq z$, whence the result.

The *parabolic fixed points* of $\Gamma$ are equally important. To define them, we first observe that the elements $b \in B$ are characterized by the fact that

$$\lim_{z=\infty} bz = \infty \,.$$

On the other hand, for a matrix $g \notin B$, $\lim gz = a/c \in \mathbb{R}$. We are thus led to make $G$ act on a set vaguely analogous to the Riemann sphere, namely

$$\hat{P} = P \cup \mathbb{R} \cup \{\infty\} = P \cup P_1(\mathbb{R})$$

where $P_1(\mathbb{R}) = \mathbb{R} \cup \{\infty\}$ is the projective line.[71] By definition, we set $g\infty = a/c$ if $c \neq 0$ and $g\infty = \infty$ if $c = 0$. As, for all $\xi \in P_1(\mathbb{R})$, there exists $x \in G$ such that $\xi = x(\infty)$, the relation $g(\xi) = \xi$ is equivalent to $g^{-1}xg(\infty) = \infty$. The stabilizers $G_\xi$ of the points of $P_1(\mathbb{R})$ are therefore the Borel subgroups of $G$.

$\xi \in P_1(\mathbb{R})$ is then said to be a *parabolic fixed point* of $\Gamma$ if $\Gamma_\xi = \Gamma \cap G_\xi$ contains parabolic matrices $\neq \pm 1$. For example, parabolic fixed points of $SL_2(\mathbb{Z})$, or of a congruence group, are those of $\mathbb{Q} \cup \{\infty\}$, because the only fixed point of a parabolic matrix is $(a - d)/2c$.

The reader will probably expect there to be also hyperbolic fixed points of $\Gamma$. They certainly exist but will not occur in this chapter and present very complex problems when, for some extraordinary reason, they are considered.

We will need some simple results for groups having parabolic fixed points. They are almost trivial in the case of $G(\mathbb{Z})$.

**Lemma 1.** *Let $\Gamma$ be a discrete subgroup of $G$ such that $\Gamma \cap U \neq \{e\}$.*

a) *For all compact sets $M \subset P$, there exists $T < +\infty$ such that*

$$\Gamma.M \subset \{\mathrm{Im}(z) \leq T\} \,.$$

b) *For sufficiently large $T$,*

$$\mathrm{Im}(z) > T \quad \& \quad \mathrm{Im}(\gamma z) > T$$

*if and only if $\gamma \in \Gamma \cap B = \Gamma_\infty$.*

Let us first show that

[71] $G$ and more generally $GL_2(\mathbb{R})$ are made to act on $P_1(\mathbb{R})$ by considering $G$ as a group of linear transformations on $\mathbb{R}^2$. This group acts on the set of 1-dimensional vector subspaces, which by definition is $P_1(\mathbb{R})$, and can be identified with $\mathbb{R} \cup \{\infty\}$ by associating its slope $t$, in the elementary sense of the term, to each line $y = tx$ with initial point the origin [Chap. IX, § 4, n° 11, (iii)].

$$(19.1) \qquad \inf_{\gamma \notin \Gamma \cap B} |c(\gamma)| > 0\,.$$

Any matrix $\gamma$ such that $c \neq 0$ can indeed be written as

$$\gamma = x\left(u'\right)\begin{pmatrix} 1 & 0 \\ c & 1 \end{pmatrix} x\left(u''\right)$$

by (15.22). Since $U/\Gamma \cap U$ is compact, leaving $c$ invariant, $u'$ and $u''$ may be assumed to stay in a fixed compact subset of $\mathbb{R}$. If there exist $\gamma_n$ such that $c_n \neq 0$ tend to 0, the corresponding $u'_n$ and $u''_n$ can be assumed to converge to limits $u'$ and $u''$. Then $(\gamma_n)$ converges to $x(u')x(u'')$, and so $\gamma_n \in U$ for large $n$ since $\Gamma$ is *discrete*, a contradiction which proves (1).

This being settled, the relation $|cz + d|^2 = (cx + d)^2 + c^2y^2 = y/\operatorname{Im}(\gamma z)$ shows that, for $c \neq 0$, $\operatorname{Im}(z)\operatorname{Im}(\gamma z) \leq 1/c^2 \leq m$ where $m > 0$ is a constant. Thus proposition b) of the lemma follows. If $z$ stays in a compact set $M$, so that $\inf \operatorname{Im}(z) > 0$, then $\sup \operatorname{Im}(\gamma z) < +\infty$, proving proposition a), qed.

These results shows that *if $\Gamma \backslash P$ is compact, $\Gamma$ only contains elliptic of hyperbolic matrices.* Replacing $\Gamma$ with a conjugate, it suffices to show that $\Gamma \cap U = \{e\}$. This follows from proposition a) of the lemma since, by assumption, there exists a compact set $M \subset P$ such that $\Gamma.M = P$. (Alternate proof: show that, for all $\gamma \in \Gamma$, the set of conjugates $g\gamma g^{-1}$ is *closed* in $G$ and deduce the result).

The next lemma gives more details about the structure of the subgroup of $\gamma \in \Gamma$ such that $c(\gamma) = 0$: for these matrices, $d(\gamma) = \pm 1$.

**Lemma 2.** *Let $\Gamma$ be a discrete subgroup of $G$ having $\infty$ as a parabolic fixed point. There exists a unique $\omega > 0$ such that $U_\infty = U(\omega\mathbb{Z})$. The subgroup $\Gamma_\infty = \Gamma \cap B$ is then one of the three following groups:*

$$(19.2) \qquad \pm\begin{pmatrix} 1 & n\omega \\ 0 & 1 \end{pmatrix}^n \qquad \text{(case where } -1 \in \Gamma\text{)}\,,$$

$$(19.2') \qquad \begin{pmatrix} 1 & n\omega \\ 0 & 1 \end{pmatrix}^n \qquad \text{(case where } -1 \notin \Gamma\text{)}\,,$$

$$(19.2'') \qquad \begin{pmatrix} -1 & \omega/2 \\ 0 & -1 \end{pmatrix}^n \qquad \text{(case where } -1 \notin \Gamma\text{)}$$

Let $x(\omega), \omega > 0$ be the generator of $\Gamma \cap U = U_\infty$. If the first row of $\gamma \in \Gamma_\infty$ is $(a\ b)$, then $\gamma^n x(\omega)\gamma^{-n} = x(a^{2n}\omega)$, whence $a^2 = 1$.

If $-1 \in \Gamma$, for example in the case of $G(\mathbb{Z})$, then $\pm\gamma \in U_\infty$ and we are in case (2).

If $-1 \notin \Gamma$, then either $\Gamma_\infty = U_\infty$, or else $\Gamma_\infty \neq U_\infty$. The first case is (2'). In the second case, $\Gamma$ contains matrices of the form

$$(19.3) \qquad \gamma = \begin{pmatrix} -1 & b \\ 0 & -1 \end{pmatrix}$$

with $b \neq 0$. As $\gamma^2 = x(-2b) \in \Gamma \cap U$, $2b = k\omega$ for some integer $k$. But $\Gamma_\infty$ also contains matrices

$$(19.4) \qquad \gamma\begin{pmatrix} 1 & -p\omega \\ 0 & 1 \end{pmatrix} = \begin{pmatrix} -1 & (p-k/2)\,\omega \\ 0 & -1 \end{pmatrix}.$$

If $k$ is even, then one may choose $p = k/2$, whence $-1 \in \Gamma$, but this case is excluded. So $b = k\omega/2$ with $k$ odd for all matrices (3) in $\Gamma$. Relation (4) then shows that $\Gamma_\infty$ contains the matrix

$$\gamma_\infty = \begin{pmatrix} -1 & -\omega/2 \\ 0 & -1 \end{pmatrix}. \tag{19.5}$$

Its square is the matrix $x(\omega)$, and so $\gamma = \gamma_\infty^{2p}$ for all $\gamma \in U_\infty$. For $\gamma \notin \Gamma \cap U$, hence of form (3), $b = (2p+1)\omega/2$; thus $\gamma = \gamma_\infty^{2p+1}$. As a result, $\gamma_\infty$ generates $\Gamma_\infty$, which finishes the proof.

This lemma shows that, for solutions of

$$f(\gamma z) = J(\gamma; z)^r f(z),$$

there are three possible cases for the action of $\Gamma_\infty$.

If $-1 \in \Gamma$, then $f(z+\omega) = f(z)$ and $r$ must be even.

If $-1 \notin \Gamma$, then no restrictions apply to $r$, and $f(z+\omega) = f(z)$ again; but if $\infty$ is *irregular* [case (2")], the function $f$ must also satisfy $f(z+\omega/2) = (-1)^r f(z)$.

*Exercise 1.* Let $\Gamma$ be the subgroup of $\gamma \in G(\mathbb{Z})$ satisfying

$$b \equiv c, \quad a \equiv d \equiv (-1)^b \bmod 4.$$

Show that $\infty$ is an irregular parabolic fixed point of $\Gamma$.

Lemmas 1 and 2 concern the point at infinity. There are obviously similar result for any other parabolic fixed point $\xi = x(\infty)$: apply the lemmas to the group $x^{-1}\Gamma x = \Gamma'$. Setting $z = xz'$ and

$$\mathfrak{S}_\infty(T) = \{\mathrm{Im}(z) > T\} \tag{19.6}$$

for all $T > 0$, lemma 1 applied to $\Gamma'$ shows that, for all compact sets $M \subset P$, $x^{-1}\Gamma x.M \cap \mathfrak{S}_\infty(T) = \varnothing$ for large $T$. Hence, choosing $M' = x^{-1}M$ and setting

$$\mathfrak{S}_\xi(T) = x\mathfrak{S}_\infty(T), \tag{19.7}$$

for any compact set $M \subset G$,

$$\Gamma.M \cap \mathfrak{S}_\xi(T) = \varnothing \quad \text{for large } T. \tag{19.8}$$

Sets like in (7) are *horocycles centered at*[72] $\xi$. Finding its shape is easy. If $x \in B$, then $\xi = \infty$ and $x$ transforms (6) into another horocycle centered at $\infty$. If $x \notin B$, we set $x = uwb$. Since $b$ transforms $\mathfrak{S}_\infty(T)$ into a similar set, it may be assumed that $x = uw$ where $u$ maps 0 onto $\xi$ since $w(\infty) = 0$. As $\mathrm{Im}(-1/z) = y/|z|^2$, the set $w\mathfrak{S}_\infty(T)$ is defined by a relation $y/|z|^2 > T$. It is thus a tangent disc to the real axis at the origin. Making it undergo the translation $z \longmapsto z + \xi$ gives a horocycle centered at $\xi$ represented by the figure below.

To work in $G$, we replace horocycles by their inverse images in $G$. These are *Siegel domains* in $G$. Those corresponding to horocycles centered at $\infty$ are the sets

$$\mathfrak{S}(T) : \alpha(g) > T > 0 \tag{19.9}$$

where we have set $\alpha(g) = \mathrm{Im}(gi)$ as in n° 15. Obviously

$$U\mathfrak{S}(T) = \mathfrak{S}(T)$$

for all $T$. Siegel domains corresponding to horocycles centered at $\xi = x(\infty)$ are translations $x\mathfrak{S}(T)$ of the previous sets.

[72] The notation $\mathfrak{S}(\xi)$ will also be used to denote an unspecified horocycle centered at $\xi$.

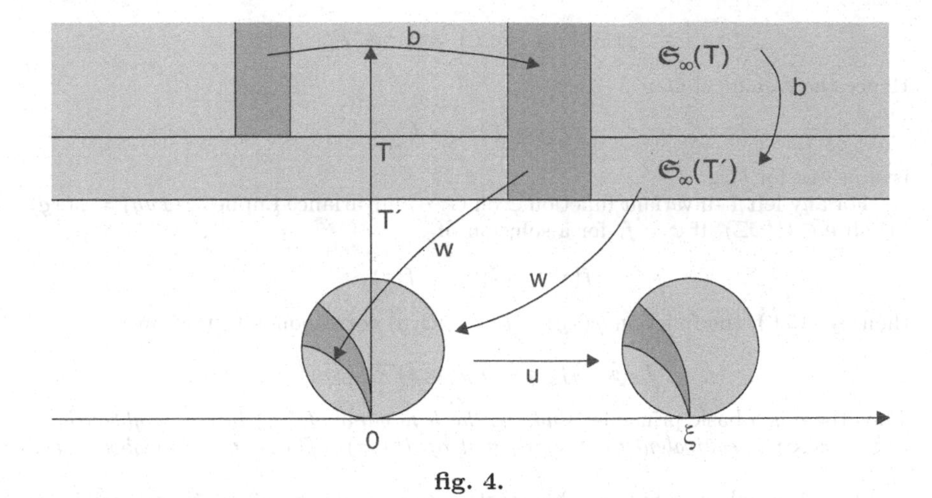


**fig. 4.**

(ii) *Generalities on automorphic forms.* We saw that in the case of the group $G(\mathbb{Z})$ to obtain theorems about holomorphic or even simply meromorphic solutions of the functional equation

$$f(\gamma z) = J(\gamma; z)^r f(z)\,, \tag{19.10}$$

restrictions must be imposed on their behaviour in the neighbourhood of the point $\infty$, the only parabolic fixed point of $G(\mathbb{Z})$, up to equivalence. A general discrete group $\Gamma$ has parabolic fixed points if and only if it contains unipotent matrices. It may have none, for example if the quotient $\Gamma\backslash P$ is compact, for there is then a compact set $M \subset P$ such that $\Gamma M = P$, which contradicts (8). There is then no point in distinguishing "entire" forms from "cusp" ones.

The question arises in the other case, but as already seen in n° 5, the point $\infty$ is not the only one that matters for the group of the function $\theta(z)$. For every parabolic fixed point $\zeta = x(\infty)$, it is also necessary to consider its stabilizer subgroup $\Gamma_\xi = xBx^{-1} \cap \Gamma$ in $\Gamma$, and $U_\xi = xUx^{-1} \cap \Gamma$ which plays the same role for $\xi$ as $U_\infty$ does for $\infty$. If $\xi$ is such a point, so are $\eta = \sigma\xi$ for all $\sigma \in \Gamma$ and

$$U_\eta = \sigma U_\xi \sigma^{-1}\,, \qquad \Gamma_\eta = \sigma \Gamma_\xi \sigma^{-1}\,.$$

Two such fixed points are said to be *equivalent.*

The structure of the subgroups $\Gamma_\xi$ is easy to work out. For the group

$$\Gamma' = x^{-1}\Gamma x\,,$$

$\infty$ is a parabolic fixed point. Besides, it is clear that

$$\Gamma_\xi = x\Gamma'_\infty x^{-1}\,, \qquad U_\xi = xU'_\infty x^{-1}$$

with the obvious notation. Lemma 2 applies to $\Gamma'$ and gives the result. In all cases,

$$U_\xi = xU(\omega\mathbb{Z})x^{-1}$$

for some $\omega > 0$, which is obvious, the three cases of lemma 2 being possible for $\Gamma_\xi$.

Lemma 1 applied to $\Gamma'$ shows that if $\mathfrak{S}(T)$ is a sufficiently small Siegel domain in $G$, then

$$\gamma x\mathfrak{S}(T) \# x\mathfrak{S}(T) \Longleftrightarrow \gamma \in \Gamma_\xi .$$

Hence the canonical map

$$\Gamma_\xi \backslash x\mathfrak{S}(T) \longrightarrow \Gamma \backslash G$$

is *injective* for large $T$.

For any left $\Gamma$-invariant function $\varphi$ on $G$, $U_\xi$-invariance implies $\varphi(xug) = \varphi(xg)$ for all $u \in U(\omega\mathbb{Z})$. If $\varphi = f_r$ for a solution of

$$f(\gamma z) = J(\gamma; z)^r f(z) ,$$

then by (15.6), the function $\varphi(xg) = L(x^{-1})\varphi(g)$ corresponds to the function

$$L_r(x^{-1})f : z \longmapsto J(x; z)^{-r} f(xz) .$$

Thus there is a basic principle: *studying the behaviour of $f(z)$ in the neighbourhood of $\xi = x(\infty)$ is equivalent to studying that of $J(x; z)^{-r} f(xz)$ in the neighbourhood of $\infty$.*

If $f(z)$ is holomorphic on a horocycle centered at $\xi = x(\infty)$, then there is an expansion

$$\text{(19.11)} \qquad J(x; z)^{-r} f(xz) = \sum a_n \mathbf{e}(z/\omega)^n$$

involving all $n \in Z$ and valid for large $\mathrm{Im}(z)$, and even for all $z$ if $f$ is holomorphic on $P$. $f$ will be said to be *holomorphic (resp. parabolic) at* $\xi$ if $a_n = 0$ for all $n < 0$ (resp. $n \leq 0$).

Expansion (11) only takes $U_\xi$-invariance into account. $\Gamma_\xi$-invariance does not impose any further conditions except if the fixed point $\xi$ is irregular. Indeed, the translation $z \longmapsto z + \omega/2$ multiplies function (11) by $(-1)^r$. Hence $a_n \neq 0$ only if $n \equiv r \bmod 2$. In particular, *if $r$ is odd and $\xi$ irregular, any holomorphic form in $\xi$ is parabolic in $\xi$.*

Denoting by

$$\mathbf{e}_r(g, t) = J(g; i)^{-r} e(tz) , \quad z = gi$$

the function of weight $r$ which corresponds on $G$ to the exponential $\mathbf{e}(tz)$ for all $t \in \mathbb{R}$, expansion (12) becomes

$$\text{(19.11')} \qquad f_r(xg) = \sum a_n \mathbf{e}_r(g, n/\omega) \quad \text{for } g \in x\mathfrak{S}(T)$$

and sufficiently large $T$. The $J$ factors have disappeared.

This expansion only depends on the class of the point $\xi$, because if $\xi$ is replaced by $\eta = \sigma\xi$ with $\sigma \in \Gamma$, the function $f_r(xg)$, replaced by $f_r(\sigma xg)$, remains invariant and the expansions with respect to $\xi$ and $\eta$ are mapped to each other by the translation $g \longmapsto \sigma g$.

We also need to know the effect of a change of matrix $x$ on (11) or (11') at $\xi = x(\infty)$. The only possibility is replacing $x$ with $x' = xb$ where $b \in B$, which replaces $U(\omega\mathbb{Z})$ by $b^{-1}U(\omega\mathbb{Z})b$. Simple calculations then show that $\omega$ is multiplied by a factor $> 0$ and the coefficients $a_n$ by non-trivial factors only depending on $b$. It follows that a relation such as $a_n \neq 0$ is independent of the choice of $x$. Saying that the function $f$ is holomorphic or parabolic at $\xi$ has an absolute meaning.

Besides, the matrix $x$ such that $\xi = x(\infty)$ being only defined modulo $B$, one may assume that $x = uw$ if $\xi \neq \infty$, which removes all arbitrariness. Then

$$x^{-1} = \begin{pmatrix} 0 & -1 \\ 1 & -\xi \end{pmatrix}$$

and $x^{-1}z = -1/(z-\xi)$, $J(x^{-1}; x^{-1}z) = (z-\xi)^{-1}$ for all $z$. So, replacing $z$ by $x^{-1}z$ in (11) gives

$$(z-\xi)^r f(z) = \sum a_n \mathbf{e}\left[-1/\omega(z-\xi)\right]^n .$$

If $f$ is holomorphic at $\xi$, it is convenient to set

$$f(\xi) = a_0 .$$

Hence, if $a_0 \neq 0$,

$$f(z) \approx f(\xi)(z-\xi)^{-r} \tag{19.13}$$

as $z$ tends to $\xi$ *in such a way that* $\mathbf{e}[-1/\omega(z-\xi)]$ *tends to 0.* For this the variable $-1/(z-\xi) = x^{-1}z$ only needs to tend to infinity without leaving a strip

$$a \leq u \leq b\,, \qquad v \geq T > 0 \tag{19.14}$$

of finite width. A set of this type is said to be a parabolic cusp at infinity in $P$, and its image under the matrix $x$ a *a parabolic cusp at the vertex* $\xi$. Similarly, we define parabolic cusps with vertex $\xi$ in $G$: these are the inverse images of the parabolic cusps with vertex $\xi$ in $P$ under $g \longmapsto gi$ . They are deduced from the parabolic cusps at $\infty$ by applying the translation $g \longmapsto xg$. Hence a parabolic cusp in $G$, with vertex $\xi = x\infty$, is defined by inequalities of the form

$$|\mathrm{Re}\,\alpha(xg)| \leq c < +\infty\,, \quad \mathrm{Im}\,\alpha(xg) \geq T > 0\,. \tag{19.14'}$$

It is obvious that, in $P$ or in $G$, *the invariant measure on any parabolic cusp is finite.*

As the images under $x$ of the verticals $a$ and $b$ are circles orthogonal to the real axis at $\xi$, while $x$ transforms the horocycle $y > T$ into a horocycle centered at $\xi$, a parabolic cusp at $\xi$ is a subset of a horocycle centered at $\xi$ contained between two circular arcs orthogonal to the real axis at $\xi$. See the figure at the end of section (i). Relation (13) is exact on every parabolic cusp at $\xi$, but any other type of convergence to $\xi$ gives rise to uncontrollable variations of $f(z)$.

For $r > 0$, automorphic forms holomorphic (meromorphic) or parabolic at a given fixed point $\xi$ can be characterized as in theorem 19, (c). More generally, let us consider a solution $f$ of the usual functional equation defined on an invariant open subset of $P$, and suppose it is defined and holomorphic on a horocycle centered at $\xi = x(\infty)$. This gives expansions (11) and (11'). As $\alpha(g) = \mathrm{Im}(gi)$ tends to $+\infty$, the following possibilities may arise, where $\mathfrak{S}(T)$ denotes a sufficiently small Siegel domain:

(1) $f_r(xg) = O[\alpha(g)^N]$ for some $N \geq 0$;
(2) $f(z)$ is holomorphic at $\xi$;
(3) $f(z)$ is parabolic at $\xi$;
(4) $f_r(xg)$ is bounded on $\mathfrak{S}(T)$;
(5) $|f_r(g)|^p$ is integrable mod $\Gamma_\xi$ over $x\mathfrak{S}(T)$ for all $p < +\infty$;
(6) there exists $p < +\infty$ such that $|f_r(xg)|^p$ is integrable mod $\Gamma_\xi$ over $\mathfrak{S}(T)$.

To find the logical connections between these statements, we may assume that $x = e$, hence work in the half-plane $\mathrm{Im}(z) > T$ using the relation

$$|f_r(g)| = y^{r/2}\,|f(z)|$$

and the formula

$$a_n = \oint f(x+iy)\mathbf{e}(-nz/\omega)dx = \exp(2\pi ny/\omega)\oint f(x+iy)\mathbf{e}(-nx/\omega)dx\,,$$

where integration is over $\mathbb{R}/\omega\mathbb{Z}$.

To start with,

$$(1) \Longleftrightarrow (2) \quad \text{for all } r \in \mathbb{Z}\,,$$

because then

$$|a_n| \leq My^{N-r/2}\exp(2\pi ny/\omega)\,.$$

This expression tends to 0 at infinity for all $n < 0$.

On the other hand,

$$(3) \Longleftrightarrow (4) \Longrightarrow (5) \quad \text{if } r > 0\,. \tag{19.15}$$

If $f$ is parabolic, then $f(z) = O[\exp(-2\pi y/\omega)]$, whence (4). Conversely, (4) implies $|a_0| \leq My^{-r/2}$, whence (3). (4) implies (5) because a bounded function belongs to all $L^p$ when integration is over a set of finite measure.

Finally,

$$(3) \Longleftrightarrow (4) \Longleftrightarrow (5) \Longleftrightarrow (6) \quad \text{if } r \geq 2\,. \tag{19.15'}$$

As (5) $\Longrightarrow$ (6), by (15), it suffices to show that (6) $\Longrightarrow$ (3). The proof of theorem 19, (c), based on the inequality

$$\iint_{\substack{|x|\leq\omega/2\\ y\geq T}} |y^{r/2}f(z)|^p dm(z) \geq |a_n|^p \int_{y\geq T} \exp(-2\pi npy/\omega)y^{\frac{1}{2}pr-1}d^*y\,, \tag{19.16}$$

applies without any changes since the integral on the right hand side always diverges if $n < 0$. For $n = 0$, convergence of this integral assumes that $r < 2/p \leq 2$. So the result follows if $r \geq 2$.

The $p = 2$ case will be needed later. The integral with respect to $a_0$ then converges if and only if $r < 1$. Hence,

$$(3) \Longleftrightarrow (4) \Longleftrightarrow (6) \quad \text{if } r \geq 1\,, \quad p = 2\,. \tag{19.15''}$$

*Exercise 2.* Suppose that $\Gamma = U(\mathbb{Z})$ and set

$$\mathcal{H}_r^p(\Gamma\backslash G) = \mathcal{H}_r(G) \cap L^p(\Gamma\backslash G)\,.$$

Under what condition is this space non-trivial? Find properties of the spaces $\mathcal{H}_r^2(\Gamma\backslash P)$ similar to those of $\mathcal{H}_r^2(P)$ of n° 16.

These calculations enable us to generalize all the definitions made about the modular group to $\Gamma$. In the precise sense of the word, an *automorphic form of weight* $r$ for $\Gamma$, must satisfying the following conditions:

(FA 1): $f$ is meromorphic on $P$;
(FA 2): $f(\gamma z) = J(\gamma; z)^r f(z)$ for all $\gamma \in \Gamma$;
(FA 3): the zeros and poles of $f$ in $P$ are divided into finitely many classes mod $\Gamma$;
(FA 4): in the neighbourhood of a parabolic fixed point of $\Gamma$, the Fourier series of $f$ only has finitely many terms of degree $< 0$.

$f$ is said to be *entire* (resp. *parabolic*) if it is holomorphic on $P$ and holomorphic (resp. parabolic) at parabolic fixed points of $\Gamma$. Generalizing n° 18, (i) about the modular group, we are now going to show that automorphic functions ($r = 0$) on $\Gamma$ can be identified with meromorphic functions on a Riemann surface associated to $\Gamma$.

(iii) *The topology of horocycles. The Riemann surface of $\Gamma$.* As already mentioned, the group $G$ acts on $\widehat{P} = P \cup P_1(\mathbb{R})$. A strange topology but one fitted to to the theory of automorphic functions can be defined on $\widehat{P}$[73] because, in this topology, automorphic forms on a group $\Gamma$ holomorphic at some parabolic fixed point $\xi$ become continuous at $\xi$. To define it, we declare a set $E \subset \widehat{P}$ to be open if $E \cap P$ is open in $P$ in the usual sense and if, for all $\xi \in E \cap P_1(\mathbb{R})$, $E$ contains a horocycle centered at $\xi$. The neighbourhoods of $\xi \in P^1(\mathbb{R})$ are then the sets containing a horocycle centered at $\xi$. The elements of the group act on $\widehat{P}$ as homeomorphisms as they map horocycles to horocycles.

A sequence of $z_n \in P$ converges in this topology to another point $\xi = x(\infty)$ if and only if

$$\lim \operatorname{Im}\left[x^{-1}(z_n)\right] = +\infty\,.$$

Indeed, for all $T$, we need to state that $z_n \in x\mathfrak{S}_\infty(T)$, i.e. $\operatorname{Im}(x^{-1}z_n) > T$, for large $n$. For this, it is enough that $z_n$ converges to $\xi$ in the usual sense *staying all the while in a parabolic cusp with vertex* $\xi$, a notion which like above can be defined without referring to any group $\Gamma$.

Obviously, $\widehat{P}$, equipped with this topology, satisfies the Hausdorff axiom. $P_1(\mathbb{R})$ is *discrete* in this topology because the only point of $P_1(\mathbb{R})$ contained in a a horocycle centered at $\xi$ is $\xi$ itself. But $\widehat{P}$ *is not locally compact*: the sequence $z_n = i + n$ does not contain any subsequence converging to a point of $P$, nor any subsequence converging to $\infty$ in the topology of $\widehat{P}$ since this would imply that $\lim \operatorname{Im}(z_n) = +\infty$. Hence no horocycle satisfies BW.

On the other hand, every parabolic cusp

$$E : -\infty < a \leq x \leq b < +\infty\,, \quad y \geq T > 0$$

at $\infty$ is compact in $\widehat{P}$. Indeed for a sequence $z_n \in E \cap P$, either $\sup y_n < +\infty$ and it then contains a subsequence converging in $P$ to a point $E$ since $x_n$ stays in a compact subset of $\mathbb{R}$, or $\sup y_n = +\infty$ and it then contains a subsequence converging to the point $\infty \in E$ in the topology of horocycles. Thus $E$ is compact. As the elements of $G$ act on $\widehat{P}$ as homeomorphisms, we conclude that *every parabolic cusp is compact in* $\widehat{P}$.

To obtain the Riemann surface associated to a discrete group $\Gamma$, we replace $\widehat{P}$ by the union $\widehat{P}(\Gamma)$ of $P$ and the *countable* set of parabolic fixed points of $\Gamma$. Hence for $\Gamma = G(\mathbb{Z})$, $\widehat{P}(\Gamma) = P \cup P_1(\mathbb{Q})$. The group $\Gamma$ acts on $\widehat{P}(\Gamma)$ by homeomorphisms, which enables us to define quotient spaces

[73] I am unable to say who had the idea first, but its inventor must have clearly been familiar with baroque topologies, which, before the 1960s, excludes modular function specialists dedicated to classical analysis. G. Shimura uses it in his excellent *Introduction to the Arithmetic Theory of Automorphic Functions* (Princeton UP, 1971), I presented it in my lectures in 1971–72, but Henri Cartan was already familiar with it in the 1950s.

$$X(\Gamma) = \Gamma\backslash P(\Gamma), \quad \widehat{X}(\Gamma) = \Gamma\backslash\widehat{P}(\Gamma).$$

**Theorem 24.** *Let $\Gamma$ be a discrete subgroup of G and $a, a'$ two points of $\widehat{P}(\Gamma)$.*

(a) *If W and W' are two sufficiently small neighbourhoods of a and a' in $\widehat{P}(\Gamma)$, then*

$$\gamma W \# W' \Longleftrightarrow \gamma a = a'. \tag{19.17}$$

(b) *The quotient space $\widehat{X}(\Gamma) = \Gamma\backslash\widehat{P}(\Gamma)$ is locally compact.*

To prove the first proposition, we need to distinguish three cases.

(1) If $a, a' \in P$, there is a more general result available [Chap. XI, n° 15, (vi), theorem 24].

(2) If $a \in P$ and if $a' = x(\infty)$ is a parabolic fixed point of $\Gamma$, replacing $\Gamma$ by $x\Gamma x^{-1}$ and $a$ by $x^{-1}a$ reduces the proof to the case where $a' = \infty$ and $W' = \mathfrak{S}_\infty(T)$. It then suffices to show that, for any compact set $W \subset P$ and sufficiently large $T > 0$, $\Gamma W \cap \mathfrak{S}_\infty(T) = \varnothing$. This is lemma 1 of section (i). By the way, observe that the proof of this lemma also shows that

$$\inf_{\gamma\notin\Gamma_\infty} |c(\gamma)| > 0, \tag{19.18}$$

where $c(\gamma)$ denotes the coefficient $c$ of $\gamma$, so that $c(\gamma) = 0$ is equivalent to $\gamma \in \Gamma_\infty$.

(3) Finally suppose that $a = \xi$ and $a' = \xi'$ are parabolic fixed points of $\Gamma$. This immediately reduces the proof to the case $\xi' = \infty$. Setting $\xi = x(\infty)$ for some $x \in G$, we need to show that if $\mathfrak{S} = \{\mathrm{Im}(z) > T\}$ and $\mathfrak{S}' = \{\mathrm{Im}(z) \geq T'\}$ are closed horocycles centered at $\infty$ and if $T$ and $T'$ are sufficiently large, the relation $\gamma x\mathfrak{S} \# \mathfrak{S}'$ requires $\gamma\xi = \infty$, i.e. $\gamma x \in B$.

If $\xi = \sigma(\infty)$ for some $\sigma \in \Gamma$, then $x = \sigma b$ for some $b \in B$, and as $b$ transforms $\mathfrak{S}$ into another horocycle centered at $\infty$, $\gamma x\mathfrak{S} \# \mathfrak{S}$ if and only if $\gamma\sigma\mathfrak{S} \# \mathfrak{S}$ for some other $\mathfrak{S}$. Hence the proof reduces to showing that

$$\gamma\mathfrak{S} \# \mathfrak{S} \Longrightarrow \gamma \in \Gamma_\infty$$

if $\mathfrak{S}$ is sufficiently small. This is point (b) of lemma 1 of section (i).

If $\xi$ does not belong to the same class of fixed points as $\infty$, i.e. if $x \notin B$, the horocycles $\gamma x\mathfrak{S} = \gamma\mathfrak{S}(\xi)$ are tangent discs to the real axis. If such a disc has non-trivial intersection with $\mathfrak{S}'$, then this is also the case of its intersection with the limit horizontal $\mathrm{Im}(z) = T'$. Multiplying $\gamma$ on the left by an element of $\Gamma \cap U$, $\gamma\mathfrak{S}(\xi)$ can even be assumed to have non-trivial intersection with some fixed compact subset $C'$ of this horizontal. But then the intersection of $\gamma^{-1}C'$ and $\mathfrak{S}(\xi)$ is non-trivial, which, by lemma 1, is impossible if $\mathfrak{S}(\xi)$ is sufficiently small.

Let us now prove the second proposition of the theorem. As $\Gamma$ is a group of homeomorphisms of $\widehat{P}(\Gamma)$, the canonical map

$$\pi : \widehat{P}(\Gamma) \longrightarrow \widehat{X}(\Gamma)$$

is open. Proposition (a) of the theorem then shows that $\widehat{X}(\Gamma)$ is separated. On the other hand, the image in $\widehat{X}(\Gamma)$ of any $z \in P$ clearly has a compact neighbourhood, for example the image of the closed disc centered at $z$. If $\xi$ is a parabolic fixed point, the image of any horocycle[74] $\mathfrak{S}(\xi)$ is a neighbourhood of $\pi(\xi)$. But $U_\xi$ enables us to put all $z \in \mathfrak{S}(\xi)$ in the same parabolic cusp at $\xi$. As any parabolic cusp is compact in the topology of $\widehat{P}$, so is the image of $\mathfrak{S}(\xi)$, qed.

[74] We set $\xi \in \mathfrak{S}(\xi)$.

Applied to a parabolic fixed point $\xi = x(\infty)$, proposition (a) of the theorem says that if $\mathfrak{S}(\xi)$ is a sufficiently small horocycle centered at $\xi$, then

$$\gamma\mathfrak{S}(\xi) \# \mathfrak{S}(\xi) \Longleftrightarrow \gamma \in \Gamma_\xi .$$

A function $f$ defined on $\mathfrak{S}(\xi)$ can be extended into a solution of

$$f(\gamma z) = J(\gamma; z)^r f(z)$$

defined on $\Gamma\mathfrak{S}(\xi)$ if and only if, as in n° 18, this relation is satisfied by all $z \in \mathfrak{S}(\xi)$ and all $\gamma \in \Gamma_\xi$. $U_\xi$-invariance means there is (if $f$ is holomorphic) an expansion

$$J(x; z)^{-r} f(xz) = \sum a_n \mathbf{e}(nz/\omega) .$$

But this condition is sufficient for ensuring $\Gamma_\xi$-invariance only if $\xi$ is a regular parabolic fixed point. If $\xi$ is irregular, $\Gamma_\xi$ contains the conjugate under $x$ of matrix (5) of lemma 2. This gives the additional condition $f[x(z + \omega/2)] = (-1)^r f(xz)$, which, as was seen above is equivalent to $a_n = (-1)^{r-n} a_n$. For $r$ even (resp. odd), we, therefore, find an expansion depending on the *even* (resp. *odd*) powers of $\mathbf{e}(z/\omega)$.

The different cases can be unified by setting

$$\zeta_\xi(z) = \mathbf{e}\left[x^{-1}(z)/\omega_\xi\right] , \tag{19.19}$$

where

$$\omega_\xi = \begin{cases} \omega & \text{if } \xi \text{ is regular}, \\ \omega/2 & \text{if } \xi \text{ is irregular}. \end{cases} \tag{19.19'}$$

For all $r$ if $\xi$ is regular, for all even $r$ if $\xi$ is irregular, any holomorphic form of weight $r$ on a horocycle $\mathfrak{S}(\xi)$ centered at $\xi$ is then a Laurent series in variable (19) in this horocycle. Conversely, any Laurent series in $\zeta_x$ defined in $\mathfrak{S}(\xi)$ can be extended into a solution of the functional equation on $\Gamma.\mathfrak{S}(\xi)$. Using variable (19), we define the *order* $v_\xi(f)$ *at the point* $\xi$ of a function of weight $r$ (even or odd) for $\Gamma$ meromorphic at $\xi$: it is an integer if $r$ is even and of the form $\frac{1}{2} + n$ if $r$ is odd and $\xi$ irregular.

To equip $\widehat{X}(\Gamma)$ with the structure of a complex manifold, we proceed as we did in n° 18, (i) for the modular group, so that we need not spell everything out in detail.

To start with, this structure has to be defined on $X(\Gamma) = \Gamma\backslash P$. Hence we need local uniformizers at each point $\alpha = \pi(a)$. For this we use open neighbourhoods $W$ of $a$ in $P$ satisfying conditions (i), (ii) and (iii) of n° 18, (i). Since the conformal representation

$$z \longmapsto (z - a)/(z - \overline{a})$$

transforms the actions of the stabilizer $\Gamma_a$ into a group of rotations of finite order $n(a)$, holomorphic $\Gamma_a$-invariant functions on $W$ are, as in n° 18, power series in the variable

$$\zeta_a(z) = [(z - a)/(z - \overline{a})]^{n(a)}$$

and property (iv) of n° 18 continues to hold here. Denoting by $q_\alpha$ the function which, in $\pi(W)$, corresponds to $\zeta_a$, as in n° 18, $(\pi(W), q_\alpha)$ is a local chart of $X(\Gamma)$ at $\alpha$ and, properties (i) to (iv) of n° 18 tell us that these local charts are holomorphically compatible.

We then need to extend this structure to $\widehat{X}(\Gamma) = \Gamma\backslash\widehat{P}(\Gamma)$ and so define local charts in the neighbourhood of $\pi(\xi)$ for $\xi = x(\infty)$. It was shown above that a holomorphic function $f(z)$ defined in a neighbourhood $W = \mathfrak{S}(\xi)$ of $\xi$ is $\Gamma_\xi$-invariant if and only if it is a power series in the $\Gamma_\xi$-invariant function $\zeta_\xi(z)$ which only depends on $\xi$ up to a constant factor. Since it is clear that

$$\zeta_\xi(z) = \zeta_\xi(z') \Longleftrightarrow \pi(z) = \pi(z')\,,$$

the local chart $(\pi(W), q_\xi)$ of $\widehat{X}(\Gamma)$ at $\pi(\xi)$ is then obtained by transforming $\zeta_\xi$ into a function $q_\xi$ on $\pi(W)$.

As in n° 18, (i), it follows that the charts of $\widehat{X}(\Gamma)$ thereby obtained are mutually holomorphically compatible. More precisely, if $a$ and $a'$ are two distinct points of $\widehat{P}(\Gamma)$ and $W$ and $W'$ are neighbourhoods of $a$ and $a'$ satisfying the required conditions, then $\zeta_a$ and $\zeta_{a'}$ are injective on $W \cap W'$ (hence holomorphic functions of each other). Indeed, if $\zeta_a(z_1) = \zeta_a(z_2)$ for distinct points in $W \cap W'$, then there would be some $\gamma \in \Gamma_a$ such that $z_2 = \gamma z_1$, whence $\gamma W' \# W'$ and so $\gamma \in \Gamma_{a'}$. This is impossible for $a \neq a'$ since a parabolic matrix has a unique fixed point in $\widehat{P}$.

(iv) *Fuchsian groups*. The Riemann surface of the modular group is compact because it is covered by the image of any sufficiently large horocycle. Because of this trivial remark, the following definition of Poincaré's *Fuchsian groups* becomes natural: these are discrete groups $\Gamma$ satisfying conditions

(GF 1) the parabolic fixed points of $\Gamma$ are divided into finitely many classes $\Gamma\xi_i$,

(GF 2) there exist a compact subset $M \subset P$ and finitely many horocycles $\mathfrak{S}(\xi_i)$ such that

$$P = \bigcup \gamma F \quad \text{where } F = M \cup \bigcup \mathfrak{S}(\xi_i)\,.$$

The reader will probably observe that the first condition could have been omitted since it clearly follows from the second one. The compact set $M$ is superfluous if $\Gamma$ has parabolic fixed points: it suffices to choose sufficiently large horocycles. Besides, condition (GF 2) can be immediately restated for $G$: there exist a compact set $M \subset G$ and a Siegel domain $\mathfrak{S}$ such that $G = \Gamma F$ where $F$ is the union of $M$ and of finitely many translations $x_i\mathfrak{S}$ of $\mathfrak{S}$.

Clearly, $G(\mathbb{Z})$ satisfies these conditions, with $n = 1$ in this case. In the trivial case from this point of view of a group with a compact quotient, we choose $F = M$. Any subgroup of finite index in a Fuchsian group $\Gamma$ is again a Fuchsian group (obvious). In particular, all congruence subgroups of the modular group are Fuchsian groups.

The set $F$ of condition (GF 2) is obviously not a proper fundamental domain for $\Gamma$. To start with, each horocycle $\mathfrak{S}(\xi_i)$ is invariant under the stabilizer of the point $\xi_i$ in $\Gamma$. If $\xi_i = \infty$, this is a discrete group of horizontal translations, enabling us to put every $z \in \mathfrak{S}(\infty)$ in a strip of finite width, i.e. in a parabolic cusp at $\infty$. The same holds for the other $\xi_i$. In other words, *condition (GF 2) can be satisfied by replacing each* $\mathfrak{S}(\xi_i)$ *by a parabolic cusp at* $\xi_i$. Since a closed parabolic cusp is compact in the topology of horocycles, as expected, the Riemann surface $\widehat{X}(\Gamma)$ is compact.

Every parabolic cusp in $P$ having a finite and invariant measure, so does $F$. However, by assumption, the map from $P$ onto $\Gamma\backslash P$ is surjective when it is restricted to $F$. So theorem 24 can be completed as follows:

**Theorem 25**. *The Riemann surface $\widehat{X}(\Gamma)$ of every Fuchsian group $\Gamma$ is compact. The total mass of the invariant measure on $\Gamma\backslash G$ is finite.*

The converse holds and is mostly interesting because of its proof. The easiest is to show that, if $\widehat{X}(\Gamma)$ is compact, then $\Gamma$ is a Fuchsian group. Indeed, we saw that the set of parabolic fixed points of $\Gamma$ is discrete in $\widehat{P}(\Gamma)$. It is also closed. Its image in $\widehat{X}(\Gamma)$ having the same properties, it is finite, giving (GF 1). Then for each $\xi$ let us choose an open horocycle $\mathfrak{S}(\xi)$ in such a way that $\mathfrak{S}(\gamma\xi) = \gamma\mathfrak{S}(\xi)$. The image of $P - \bigcup \mathfrak{S}(\xi)$ in $\widehat{X}(\Gamma)$ is clearly a set contained in $X(\Gamma)$, closed in $\widehat{X}(\Gamma)$, hence compact, and so the image of a compact subset[75] $M$ of $P$. Then $P - \Gamma\mathfrak{S}(\xi) = \Gamma M$, whence (GF 2).

Showing that the relation $m(\Gamma\backslash G) < +\infty$ implies (GF 1) and (GF 2) is far from being as easy. The proof is mainly interesting because it requires constructing genuine fundamental domains bounded by finitely many circular arcs or rays orthogonal to the real axis. This amounts to non-Euclidean geometry open to considerable generalizations. But all this would take up too much space.[76]

For a Fuchsian group, these constructions show that, for $r = 0$, automorphic forms on $\Gamma$ can be identified with meromorphic functions on the compact Riemann surface $\widehat{X}(\Gamma)$. Poincaré had already announced in his early works that, conversely, any compact Riemann surface $X$ of genus $g \geq 2$ is isomorphic to $X(\Gamma)$ for some group *with compact quotient acting without fixed points,* hence not containing any hyperbolic matrices. The proof consists in considering the universal covering of $X$ and in proving that any simply connected Riemann surface is isomorphic to the Riemann sphere or, if it is not compact, to some open subset of $\mathbb{C}$, hence, by Riemann's theorem on simply connected domains, either to $\mathbb{C}$, or to $P$. The fundamental group of the compact surface $X$ considered then becomes a discrete group $\Gamma$ of analytic automorphisms of $\mathbb{C}$ or of $P$, so that either $X = \mathbb{C}/L$, or $X = \Gamma\backslash P$ where, in the former case, $L$ is the period lattice in $\mathbb{C}$ (elliptic functions, $g = 1$) and, in the latter, a Fuchsian group with compact quotient acting without fixed points (automorphic functions, $g \geq 2$). In the latter case, the rational function field of $X$ is just the field of automorphic functions on $\Gamma(r = 0)$, which, for any Fuchsian group, is indeed an algebraic function field of one variable as in the theory of elliptic functions. Applying the result to the Riemann surface associated in Chap. X to an irreducible algebraic equation $P(X, Y) = 0$, we conclude that there are two automorphic functions $f$ and $g$ on $\Gamma$ such that all solutions of $P(z, w) = 0$ are given by $z = f(u), w = g(u)$ for some $u \in P$ unique mod $\Gamma$. This is the generalization of theorem 12 of n° 14, (ii) about elliptic functions, a result which could be called Poincaré's youthful dream by analogy to Kronecker's in another domain.

Relations between the French and Germans, even in the case of mathematicians, was somewhat ambiguous between 1870 and 1914. It has needed at least a quarter century of cooperation-competition between Poincaré and some Germans (H.-A. Schwarz and especially Paul Koebe) to prove this difficult result using methods from potential theory (existence of harmonic functions whose only singularity is a simple pole) at a time when this domain of analysis was still sufficiently primitive for Poincaré himself to make mistakes. In the early 1950s, another Franco-German undertaking made all this far more comprehensible by using algebraic topology and "coherent sheaves" of analytic functions.

At a much more accessible level, using general theorems on compact Riemann surfaces, presented in Chap. X, would enable us to obtain immediately proposi-

[75] Every point $\pi(a)$ of $X(\Gamma)$ has neighbourhoods of the form $\pi(W)$ where $W$ is a compact neighbourhood of $a$ in $P$. Hence the existence of $M$ by BL.

[76] The construction is in Poincaré, but the first fully correct proof is probably the one found in C. L. Siegel, *Topics in Complex Function Theory*, vol. 2 (Wiley, 1971).

tions (a) and (e) of the following result, which generalizes theorem 19 of n° 17. But it is possible to proceed differently. Proposition (e) and its proof can be generalized to situations where neither Riemann surfaces nor holomorphic functions are encountered.

**Theorem 26**. *Let $\Gamma$ be a Fuchsian group.*

(a) *Every entire automorphic form of weight $r$ is trivial (resp. constant) if $r < 0$ (resp. $r = 0$).*

(c) *An everywhere holomorphic solution $f$ on $P$ of*

$$f(\gamma z) = J(\gamma; z)^r f(z)$$

*is a parabolic form if and only if $f_r \in L^\infty(\Gamma\backslash G)$, when $r > 0$. When $r \geq 2$, only if (resp. if)*

$$f_r \in L^p(\Gamma\backslash G) \tag{19.22}$$

*for all (resp. for some) $p \leq +\infty$.*

(d) *The set of parabolic forms of weight $r \geq 1$ is identical to the space*

$$\mathcal{H}^2_r(\Gamma\backslash G) = L^2(\Gamma\backslash G) \cap \mathcal{H}_r(G)\,.$$

(e) *If $f$ is parabolic, its Fourier coefficients at every parabolic fixed point of $\Gamma_\infty$ satisfy $a_n = O(n^{r/2})$.*

(f) *The space of entire forms of weight $r$ is finite-dimensional.*

(a) If $r = 0$, the function $f(z)$ is holomorphic everywhere in $\widehat{X}(\Gamma)$, hence constant. If $r < 0$ and if $\infty$ is a parabolic fixed point, $f(z)$ is bounded for $y > T$, hence also $y^{r/2}|f(z)| = |f_r(g)|$ since $r < 0$. For an arbitrary parabolic fixed point $\xi = x(\infty)$, the function $f_r(xg)$ is bounded for $y > T$. So $f_r(g)$ is bounded on $x\mathfrak{S}$ for any Siegel domain (9) in $G$. By (GF 2), $\Gamma\backslash G$ is the union of finitely many images of sets of this type. As a result, $f_r$ is bounded on $G$, especially if $\Gamma\backslash G$ is compact, and so is in $\mathcal{H}^\infty_r(G)$. Now, this space reduces to 0 if $r < 0$ (n° 16, theorem 14 for $p = +\infty$).

(c), (d) If $f$ is parabolic and if $r > 0$, the invariant function $y^{r/2}|f(z)|$ is bounded on every horocycle of $\widehat{P}(\Gamma)$, hence on $P$ by (GF 2). So parabolic forms belong to $L^\infty(\Gamma\backslash G)$, and thus necessarily to $L^p$ since $m(\Gamma\backslash G) < +\infty$. Conversely, if $f_r$ is bounded, $f_r$ is parabolic by relation (15) of section (ii).

To make sure that $f_r \in L^p(\Gamma\backslash G)$ for given $p < +\infty$, it suffices to check convergence in every parabolic cusp since $\Gamma\backslash G$ is the union of a compact set and images of *finitely* many parabolic cusps. Propositions (c) and (d) then follow from equivalences (15') and (15").

(e) Same proof for the modular group.

(f) An entire form $f$ is parabolic if and only if $f(\xi) = 0$ for every parabolic fixed point of $\Gamma$, which forces a finite number of linear conditions on $f$. Hence it is enough to prove (e) for the subspace $\mathcal{S}_r(\Gamma\backslash G)$ of parabolic forms, the latter being trivial if $r \leq 0$. If $r \geq 1$, by (15) and (15")

$$\mathcal{S}_r(\Gamma\backslash G) = \mathcal{H}_r(\Gamma\backslash G) \cap L^2(\Gamma\backslash G) = \mathcal{H}_r(\Gamma\backslash G) \cap L^\infty(\Gamma\backslash G)\,.$$

Moreover, $\mathcal{S}_r(\Gamma\backslash G)$ is closed in $L^2$ and $L^\infty$ since, in all $L^p(\Gamma\backslash G)$, convergence in mean for holomorphic functions implies compact convergence, and so preserves holomorphy. Now, $m(\Gamma\backslash G) < +\infty$. So all we need is to prove a general lemma:

**Lemma 3**. *Let $X$ be a locally compact space, $\mu$ a* bounded *positive measure on $X$ and $\mathcal{H}$ a subspace of $L^2(X;\mu) \cap L^\infty(X;\mu)$ closed in $L^2(X;\mu)$ and $L^\infty(X;\mu)$. Assume that all classes $f \in \mathcal{H}$ contain a continuous function. Then* $\dim(\mathcal{H}) < +\infty$.

We denote by $\mathcal{H}^2$ (resp. $\mathcal{H}^\infty$) the space $\mathcal{H}$ equipped with the $L^2$ (resp. $L^\infty$) norm. By the Fischer-Riesz theorem, the set of pairs $(f, f)$ where $f \in \mathcal{H}$ is closed in the Banach space $\mathcal{H}^2 \times \mathcal{H}^\infty$. Thus the two topologies introduced on $\mathcal{H}$ are the same (closed graph theorem), and so there is an upper bound

$$\|f\|_\infty \leq M\|f\|_2$$

valid for all $f \in \mathcal{H}$. If $S$ is the support of $\mu$ (complement of the union of open null subsets) and if every class $f \in \mathcal{H}$ is identified with the unique continuous function on $S$ which it contains, the left hand side is the *uniform* norm of $f$ on $S$. So for all $x \in S$, $|f(x)| \leq M\|f\|_2$. This gives an $\omega_x \in \mathcal{H}$ such that

$$f(x) = (f|\omega_x)$$

for all $f \in \mathcal{H}$. Moreover, by the previous inequality, $\|\omega_x\|_2 \leq M$. Then let $(e_i)$ be an orthonormal basis for $\mathcal{H}$. For any finite subset $F$ of the set of indices $i$,

$$\sum_F |e_i(x)|^2 = \sum_F |(e_i|\omega_x)|^2 \leq \|\omega_x\|_2^2 \leq M^2 .$$

Thus, integrating, $\sum_F \|e_i\|_2^2 \leq M^2\mu(X)$. The left hand side equals Card$(F)$, qed.

Using the Riemann-Roch theorem on Riemann surfaces, the dimension of the space of entire or cusp forms of weight $r$ can be explicitly calculated.[77] In particular, for $r = 2$, parabolic forms correspond precisely to everywhere holomorphic differential forms on $\widehat{X}(\Gamma)$. So the space of parabolic forms of weight 2 has dimension $g$. The space of entire forms of weight 2 can also be shown to have dimension $g + N - 1$, where $N$ is the number of classes of parabolic fixed points. This shows that ( as will be seen, contrary to what happens for $r > 2$), the values $f(\xi)$ of an entire form of weight 2 cannot be chosen arbitrarily at parabolic fixed points. This is no great surprise. If $f(z) = \sum a_n q^n$ where $q = \mathbf{e}(z/\omega)$ or $\mathbf{e}(z/\omega)^2$ is the local uniformizer at infinity and if $\omega_f$ is the differential form corresponding to $f(z)dz$ on $\widehat{X}(\Gamma)$, then

$$\omega_f = \sum a_n q^n dq/q$$

up to a factor $2\pi i$ or $4\pi i$, so that $\omega_f$ has a simple pole at $\pi(\infty)$, where its residue is equal to $f(\infty)$. Hence $\sum f(\xi) = 0$, which gives a non-trivial linear relation between the $f(\xi)$. Saying that the dimension is exactly $g+N-1$, therefore, means that there are meromorphic differential forms on $\widehat{X}(\Gamma)$ whose only singularities are simple poles at given points with given residues, provided their sum is zero. This is a general result about Riemann surfaces. In the case at hand, standard methods (Poincaré or Eisenstein series) do not lead to it because they always assume $r > 2$.

Theorem 20 on zeros and poles of meromorphic modular forms can also be generalized to Fuchsian groups. We will see later (theorem 27) that they exist for all $r$, even if $-1 \in \Gamma$. First of all, it is clear that automorphic forms of weight 0 are precisely the meromorphic functions on $\widehat{X}(\Gamma)$. So, for any form $f$ of weight 0, $\sum v_a(f)/n(a) = 0$, the sum being extended to $\Gamma\backslash\widehat{P}(\Gamma)$, with $n(a) = 1$ except if $a$ is an elliptic fixed point of $\Gamma$. The arguments used to obtain theorem 20 show that, for $f$ of weight $r$,

$$\sum v_a(f)/n(a) = cr$$

[77] See for example T. Miyake, *Modular Forms* (Springer, 1989), Chap. 2.

where the constant $c$ is independent of $r$. We calculate it for $r = 2$, in which case $f(z)dz = \omega_f \circ \pi$ for some meromorphic differential form $\omega_f$ on $\widehat{X}(\Gamma)$. The problem amounts to comparing the usual order $v_a(f)$ of $f$ at some point $a \in \widehat{P}(\Gamma)$ and the order $v_\alpha(\omega_f)$ of $\omega_f$ at $\alpha = \pi(a)$. The arguments and the result are the same as in n° 18:

$$v_a(f) = n(a)v_\alpha(\omega_f) + n(a) - 1\,, \quad \text{or } v_\alpha(f) = v_a(\omega_f) + 1\,,$$

according to whether $a$ is in $P$ or a parabolic fixed point. Hence

$$\sum v_a(f)/n(a) = \sum v_\alpha(\omega_f) + \sum [1 - 1/n(a)] + N$$

where $N$ is the number of classes of parabolic fixed points and where the latter $\sum$ is extended to all classes of elliptic fixed points. Thus, if $\widehat{X}(\Gamma)$ has genus $g$, then setting $n(a) = +\infty$ if $a$ is a parabolic fixed point,

$$2c = 2g - 2 + \sum [1 - 1/n(a)]\,.$$

So the problem reduces to the calculation of the genus of $X(\Gamma)$. I will not discuss these questions at any greater length as they as they have little to do with group theory.

*Exercise 3.* (a) Show that if $f(z)$ is an automorphic form of weight $r$ for the modular group, the function $f[(z+1)/2]$ is an automorphic form of weight $r$ for the group $\Gamma(\theta)$ of the Jacobi function (see exercise 3 of n° 3).

(b) Using the arguments of n° 5, show that

$$\pi^4\theta(z)^8 = 48G_4(z) - 3G_4\left[(z+1)/2\right]\,.$$

Deduce a formula giving the number of representations of an integer as a sum of eight squares.

## 20 – Parabolic Forms and Representations of $G$

Theorem 26 shows in particular that, for a Fuchsian group,

$$\varphi \in L^2(\Gamma\backslash G) \Longleftrightarrow \varphi \text{ parabolic if } r \geq 1\,.$$

However, for any discrete subgroup $\Gamma$, the group $G$ acts on $L^2(\Gamma\backslash G)$ by *right* translations $V(x)$ given by

$$V(x)\varphi(g) = \varphi(gx)\,.$$

This gives a unitary representation of $G$. The closed subspaces

$$\mathcal{H}_r^2(\Gamma\backslash G) = \mathcal{H}_r(G) \cap L^2(\Gamma\backslash G)$$

are not invariant, in particular because the translations $V(x)$ destroy the relation

$$\varphi(gk) = \varphi(g)\chi_r(k) \quad \text{or} \quad V(k)\varphi = \chi_r(k)\varphi \tag{20.1}$$

satisfied by all $\varphi \in \mathcal{H}_r^2(\Gamma\backslash G)$, not to mention the holomorphy condition. But for all $\varphi \in L^2(\Gamma\backslash G)$, in particular holomorphic, the elements $V(x)\varphi$ generated a closed invariant subspace $\mathcal{H}(\varphi)$ in $L^2(\Gamma\backslash G)$. Similarly, As $x \in G$ and $\varphi \in \mathcal{H}_r^2(\Gamma\backslash G)$ vary, the elements $V(x)\varphi$ generate a closed invariant subspace $\mathcal{D}_r$. It is the closure of the sum of $\mathcal{H}(\varphi)$ for $\varphi \in \mathcal{H}^2_r(\Gamma\backslash G)$. One of the results of the new theory is that representations of $G$ obtained on $\mathcal{H}(\varphi)$ are all isomorphic to the conjugate[78] of the representation by left translations on the space $\mathcal{H}^2_r(G)$ defined in n° 16. This at the very least supposes $r > 1$ for otherwise $\mathcal{H}^2_r(G)$ would be trivial.

Henceforth, as Harish-Chandra would have done, we will denote by $\pi_r$ the representation of $G$ on $\mathcal{H}^2_r(G)$. So $\pi_r(x)$ is the restriction to $\mathcal{H}^2_r(G)$ of the left translation operator $L(x)$ on $L^2(G)$. For all $r > 1$, $\omega_r$ will denote the kernel function of $\mathcal{H}^2_r(G)$, defined in n° 16, (iv). It is integrable for $r > 2$. Because of this property calculations are well-defined for $r = 2$, which they otherwise would not be. $\Gamma$ will only be assumed to be a Fuchsian group simply because this assumption is perfectly useless: the theorem applies to the group $\Gamma = U(\mathbb{Z})$ for example and to $\Gamma = \{e\}$, a case not without significance. We will only need to know that $\mathcal{H}^2_r(\Gamma\backslash G)$ is a closed subspace of $L^2(\Gamma\backslash G)$. This is clear since, as it concerns holomorphic functions, convergence in mean implies compact convergence. We assume $r$ to be an integer because we are working in $SL_2(\mathbb{R})$. Similar results can be derived for a real $r$ by using its universal covering $\widehat{G}$ provided the centre of $\widehat{G}$ is taken account of as in Chap. XI by proving the Bargmann orthogonality relations.

**Theorem 27.** *Let $\Gamma$ be a discrete subgroup of $G$ and $x \longmapsto V(x)$ the representation of $G$ on $L^2(\Gamma\backslash G)$ by right translations. For any $\varphi \in \mathcal{H}^2_r(\Gamma\backslash G)$, let $\mathcal{H}(\varphi)$ be the smallest closed invariant subspace of $L^2(\Gamma\backslash G)$ containing $\varphi$.*

(i) ($r \in \mathbb{Z}$) *The representation $(\mathcal{H}(\varphi), V)$ of $G$ is irreducible.*

(ii) ($r > 1$) *It is equivalent to the conjugate of $\pi_r$ and is square integrable and*

$$(20.2)\qquad (V(x)\varphi|\varphi) = \int_{\Gamma\backslash G} \varphi(yx)\overline{\varphi(y)}dy = \frac{(\varphi|\varphi)}{\omega_r(e)}\omega_r(x) \quad \text{if } r \geq 2\,.$$

(iii) ($r \geq 2$) *The semilinear map $J$ given by*

$$(20.3)\qquad Jf(x) = \omega_r(e)^{\frac{1}{2}}\,(V(x)\varphi|f) = \omega_r(e)^{\frac{1}{2}} \int_{\Gamma\backslash G} \varphi(yx)\overline{f(y)}dy$$

*is an isometric isomorphism from $\mathcal{H}(\varphi)$ onto $\mathcal{H}^2_r(G)$ compatible with the representations of $G$ on these spaces.*

(iv) ($r > 2$) *The orthogonal projection $L^2(\Gamma\backslash G) \longrightarrow \mathcal{H}^2_r(\Gamma\backslash G)$ is the map $f \longmapsto f * \omega_r$.*

(v) ($r > 1$) *For any closed invariant subspace $\mathcal{H}$ of $L^2(\Gamma\backslash G)$ such that the representation $(\mathcal{H}, V)$ is isomorphic to the conjugate of $\pi_r$, there exists $\varphi \in \mathcal{H}^2_r(G)$, unique up to a constant factor, such that $\mathcal{H} = \mathcal{H}(\varphi)$.*

(vi) ($r > 1$) *If $(f_i)$ is an orthogonal basis of $\mathcal{H}^2_r(\Gamma\backslash G)$, the subspaces $\mathcal{H}(f_i)$ are pairwise orthogonal and their direct sum is the closed invariant subspace $\mathcal{D}_r$ of $L^2(\Gamma\backslash G)$ generated by $\mathcal{H}^2_r(\Gamma\backslash G)$.*

The proof found everywhere uses differential operators associated to the Lie algebra of $G$. It will be presented in n° 30. The one given here uses the simplest notions about unitary representations – *abstract nonsense* – and, instead of Lie algebras, integral methods of n° 16. As the spaces $\mathcal{H}^2_r(G)$ can be generalized to all semisimple groups $G$ for which $G/K$ has a complex analytic structure as shown by Harish-Chandra in the 1950s, this is surely also the case of theorem 27. There are

[78] The conjugate of a unitary representation $(\mathcal{H}, U)$ is obtained with replacing the map $(\lambda, f) \longmapsto \lambda f$ by $(\lambda, f) \longmapsto \lambda^* f$, the inner product $(f|g)$ with $(g|f)$, and by preserving the operators $U(x)$. By setting the functions $(U(x)f|g)$ to be the coefficients of $(\mathcal{H}, U)$, those of the conjugate representation are the functions $(g|U(x)f)$.

also reasons to suppose that these results, once suitably reformulated, will apply to all locally compact groups by replacing the representation $\pi_r$ with a square integrable irreducible unitary representation, as in the case, of course trivial, of compact groups.

(a) Let us consider the representation of $G$ by right translations $V(x)$ on $L^2(\Gamma\backslash G)$ and show that, for any $\varphi \in \mathcal{H}_r^2(\Gamma\backslash G)$ and all $f \in L^2(\Gamma\backslash G)$, the coefficient

$$\Phi_f(x) = (V(x)\varphi|f) = \int_{\Gamma\backslash G} \varphi(yx)\overline{f(y)}dy \tag{20.4}$$

is in $\mathcal{H}_r(G)$. If $f$ is continuous with compact support mod $\Gamma$, there is [Chap. XI, n° 15, (iv)] a function $p \in L(G)$ such that

$$f(x) = \sum p(\gamma x).$$

For given $x$, the function $y \longmapsto \varphi(yx)p(y)$ is locally $L^2$ like $\varphi$, with compact support like $p$, and so integrable over $G$. As a result (generalized LF),

$$\int_{\Gamma\backslash G} \varphi(yx)\overline{f(y)}dy = \int_{\Gamma\backslash G} dy \sum_{\Gamma} \varphi(\gamma yx)\overline{p(\gamma y)}dy =$$

$$= \int_G \varphi(yx)\overline{p(y)}dy = \widetilde{p} * \varphi(x)\,; \tag{20.5}$$

So, by the final remark of n° 16, the left hand side is in $\mathcal{H}_r(G)$ like $\varphi$. If now $f \in L^2(\Gamma\backslash G)$, there exist $f_n \in L(\Gamma\backslash G)$ such that $\lim \|f - f_n\|_2 = 0$. As

$$|\Phi_f(x)| \leq \|\varphi\|_2.\|f\|_2$$

for all $x$ and $f$, the functions $(V(x)\varphi|f_n)$ converge uniformly to $(V(x)\varphi|f)$, whence the result.

(b) As $V(k)\varphi = \chi_r(k)\varphi$, the function $\Phi(x) = (V(x)\varphi\backslash\varphi)$ satisfies

$$\Phi(kxk') = \chi_r(k)\Phi(x)\chi_r(k')\,. \tag{20.6}$$

However, in $\mathcal{H}_r(G)$, (6) determines $\Phi$ up to a constant factor: $\Phi$ corresponds to a constant function on the unit disc $D$ [n° 16, (ii)]. We then consider the operator $A$ in $\mathcal{H}(\varphi)$ commuting with all $V(x)$ and set $A\varphi = f$. The function $(V(x)\varphi|A\varphi)$ is in $\mathcal{H}_r(G)$ and a trivial calculation shows that it satisfies (6), and so is proportional to $(V(x)\varphi|\varphi)$. Since the $V(x)\varphi$ generate $\mathcal{H}(\varphi)$ topologically, $A\varphi = \lambda\varphi$. Hence $AV(x)\varphi = \lambda V(x)\varphi$ and $A = \lambda 1$. So for all $r$, the representation is irreducible, even when $\mathcal{H}_r^2(\Gamma\backslash G) = \{0\}$...

(6) also shows that $\Phi \in \mathcal{H}_r^2(G)$ whenever $\mathcal{H}_r^2(G)$ is non-trivial, i.e. for $r > 1$. The representation $(\mathcal{H}(\varphi), V)$ is then square integrable since one of its coefficients is in $L^2(G)$, from which it follows (Bargmann) that the function $(V(x)f|f')$ is square integrable over $G$ for all $f, f' \in \mathcal{H}(\varphi)$.

(c) If $r > 1$, simply by definition, the kernel function $\omega_r$ of the space $\mathcal{H}_r^2(G)$ satisfies $\omega_r * \widetilde{\omega_r} = \omega_r = \widetilde{\omega_r}$, i.e.

$$\omega_r(x) = \int_G \omega_r(xy)\overline{\omega_r(y)}dy = (\omega_r|L(x)\omega_r)\,,$$

the inner product being on $L^2(G)$. As $\omega_r$ is in $\mathcal{H}_r(G)$ and satisfies (6), the functions $\Phi$ and $\omega_r$ are proportional, and the coefficient of proportionality is necessarily indicated in (2): put $x = e$. But if two irreducible unitary representations have a common diagonal coefficient, they are equivalent as was seen in Chap. XI, n° 31, (iii) following the proof of lemma 1. The representation $(\mathcal{H}(\varphi), V)$ is therefore equivalent to the conjugate of $\omega_r$. Propositions (i), (ii) and (iii) of the theorem are now proved.

(d) Proposition (iv) is, here too, a matter of trivial facts about representations admitting a generating vector. Since $(V(x)\varphi|\varphi) = \lambda(\omega_r|L(x)\omega_r)$, there is a semilinear isomorphism $J$ from $\mathcal{H}(\varphi)$ onto $\mathcal{H}_r^2(G)$ commuting with the representations and, omitting a factor $\lambda^{\frac{1}{2}}$ without any significance, mapping $\varphi$ onto $\omega_r$. By assumption,

$$JV(x) = L(x)J\,, \quad (Jf|Jf') = (f'|f)$$

for $f, f' \in \mathcal{H}(\varphi)$. For $f \in \mathcal{H}(\varphi)$,

$$Jf(x) = (Jf|L(x)\omega_r)\,,$$

the main property of the kernel function [n° 16, (iv)]. Thus

$$Jf(x) = (Jf|L(x)J\varphi) = (Jf|JV(x)\varphi) = (V(x)\varphi|f)\,,$$

which is formula (3).

*Exercise 1* ($r > 2$). Show that on the everywhere dense subspace $\mathcal{H}_r^1(G)$, the map $J^{-1}$ is given by $J^{-1}f = \varphi * \widetilde{f}$ up to a constant factor. (On the face of it, it is not obvious that this map extends continuously to $\mathcal{H}_r^2(G)$. This follows from the existence of $J$).

*Exercise 2* ($r > 1$). Using the Bargmann orthogonality relations, show directly that there is a constant $c$ such that

$$(\Phi_f|\Phi_{f'}) = \overline{c\,(f|f')}$$

for all $f, f' \in \mathcal{H}(\varphi)$. The inner products are calculated on $L^2(G)$ and $L^2(\Gamma\backslash G)$ respectively.

(e) Before proving proposition (iv), we recall that it is possible to associate to any bounded measure $\mu$ on $G$ an operator $V(\mu) = \int V(y)d\mu(y)$ in $L^2(\Gamma\backslash G)$ [Chap. XI, n° 25, (i)]. We have $\|V(\mu)\| \leq \|\mu\|$,

$$V(\mu * \nu) = V(\mu)V(\nu)\,, \quad V(\widetilde{\mu}) = V(\mu)^*\,.$$

All closed invariant subspace of $L^2(\Gamma\backslash G)$, in particular the subspaces $\mathcal{H}(\varphi)$, are stable under the operators $V(\mu)$. For $f, f' \in L^2(\Gamma\backslash G)$,

$$\begin{aligned}(V(\overline{\mu})f|f') &= \int_G (V(x)f|f')\,\overline{d\mu(x)} = \int \overline{d\mu(x)} \int_{\Gamma\backslash G} f(yx)\overline{f'(y)}dy = \\ &= \int_{\Gamma\backslash G} \overline{f'(y)}dy \int f(yx)\overline{d\mu(x)} = (f * \widetilde{\mu}|f')\,,\end{aligned} \tag{20.7}$$

but we need to justify the permutation of integrals needs to be justified and show that $f * \widetilde{\mu} \in L^2(\Gamma\backslash G)$ in order to prove that this formal calculation is legitimate. Now, $f$ and $f'$ being in $L^2(\Gamma\backslash G)$, the function

$$x \longmapsto \int |f(yx)f'(y)|\,dy$$

is bounded above by $\|f\|_2\|f'\|_2$. As $\mu$ is bounded,

$$\int d|\mu|(x) \int |f(yx)f'(y)|\,dy \leq \|\mu\|.\|f\|_2\|f'\|_2 < +\infty\,. \tag{20.8}$$

So the integrals can be permuted by LF.

Moreover as the latter integral in (7) is well-defined for all $f' \in L(\Gamma\backslash G)$, the function

$$\int f(yx)\overline{d\mu(x)} = f * \widetilde{\mu}(y)\,,$$

is defined for almost all $y$ by LF, and is locally integrable by definition [Chap. XI, n° 12, (ii)]. By (7) and (8),

$$\int_{\Gamma\backslash G} \left|f'(y).f * \widetilde{\mu}(y)\right| dy \leq \|\mu\|.\|f\|_2\|f'\|_2$$

for all $f' \in L^2(\Gamma\backslash G)$. As a result, $f * \widetilde{\mu}$ belongs to $L^2(\Gamma\backslash G)$ [Chap. XI, n° 18, corollary of theorem 30] and

$$(20.9) \qquad f \in L^2(\Gamma\backslash G) \Longrightarrow f * \widetilde{\mu}(x) = V\left(\overline{\mu}\right) f(x) \text{ ae.}$$

(f) This result shows that $\varphi * \widetilde{\mu} \in \mathcal{H}(\varphi)$. If $r > 2$, one may choose $d\mu(x) = \omega_r(x)dx$ since $\omega_r \in L^1(G)$. As

$$\widetilde{\omega_r} = \omega_r = \omega_r * \omega_r\,,$$

the operator

$$(20.10) \qquad f \longmapsto f * \omega_r = V\left(\overline{\omega_r}\right) f$$

is an orthogonal projection. However, if $f \in L^2(\Gamma\backslash G)$, lemma 2 of n° 16, (vi) shows that $f * \omega_r = f$ if and only if $f \in \mathcal{H}^2_r(\Gamma\backslash G)$. Hence for $f \in L^2(\Gamma\backslash G)$,

$$(20.11) \qquad f \in \mathcal{H}^2_r(\Gamma\backslash G) \Longleftrightarrow V\left(\overline{\omega_r}\right) f = f \quad (r > 2)\,,$$

proving proposition (iv).

(g) Let $J^{-1} : \mathcal{H}^2_r(G) \longrightarrow L^2(\Gamma\backslash G)$ be a semilinear isomorphism compatible with the representations. We set $\varphi = J^{-1}\omega_r$ and let $\mathcal{H}(\varphi)$ be the image of $\mathcal{H}^2_r(G)$ under $J^{-1}$, i.e. the closed invariant subspace of $L^2(\Gamma\backslash G)$ generated by the $V(x)\varphi$. Proposition (v) of the theorem reduces to proving that $\varphi$ belongs to $\mathcal{H}^2_r(\Gamma\backslash G)$ and is the only function $f \in \mathcal{H}(\varphi)$ with this property, up to a constant factor.

Let $J : \mathcal{H}(\varphi) \longrightarrow \mathcal{H}^2_r(G)$ be the inverse map of $J^{-1}$, whence $\omega_r = J\varphi$. For all $f \in \mathcal{H}(\varphi)$,

$$(V(x)\varphi|f) = \left(Jf|JV(x)J^{-1}\omega_r\right) = (Jf|L(x)\omega_r) = Jf(x)$$

by definition of the kernel function of $\mathcal{H}^2_r(G)$. The function $\Phi_f(x) = (V(x)\varphi|f)$ is, therefore, in $\mathcal{H}_r(G)$ for all $f \in \mathcal{H}(\varphi)$, and even for all $f \in L^2(\Gamma\backslash G)$: replace $f$ with its projection on $\mathcal{H}(\varphi)$. Like in point (a) of the proof, one may choose $f(y) = \sum p(\gamma y)$ where $p \in L(G)$. By (5),

$$\Phi_f = \varphi * \widetilde{p} \in L^2(\Gamma\backslash G) \cap \mathcal{H}_r(G) = \mathcal{H}^2_r(\Gamma\backslash G)\,.$$

Choosing the $p_n$ so that the measures $p_n(x)dx$ form a Dirac sequence at $e$, like in any unitary representation of a locally compact group, [Chap. XI, n° 25, (v)],

$$\varphi = \text{l.i.m.}^2\, V\left(\overline{p_n}\right) \varphi = \text{l.i.m.}^2\, \varphi * \tilde{p}_n\,,$$

where this is the limit in $L^2(\Gamma\backslash G)$. However all $\varphi * \tilde{p}_n$ belong to the closed subspace $\mathcal{H}^2_r(\Gamma\backslash G)$ of $L^2(\Gamma\backslash G)$ for all $\Gamma$ (and finite-dimensional if $\Gamma$ is a Fuchsian group). So the same holds for $\varphi$.

All that remains to be shown is that every $\varphi' \in \mathcal{H}(\varphi) \cap \mathcal{H}^2_r(\Gamma\backslash G)$ is proportional to $\varphi$. As the representation $V$ on $\mathcal{H}(\varphi)$, equivalent to the conjugate of $\pi_r$, is irreducible, $\mathcal{H}(\varphi') = \mathcal{H}(\varphi)$. Hence, by proposition (ii) of the theorem, there is an isomorphism $J' : \mathcal{H}(\varphi') \longrightarrow \mathcal{H}^2_r(G)$ mapping $\varphi'$ onto $\omega_r$. $J$ and $J'$ differ by an automorphism of the representation on $\mathcal{H}^2_r(G)$, so by a scalar since it is irreducible. This gives proposition (v) of the theorem.

Alternative version: $V(k)\varphi' = \chi_r(k)\varphi'$ since $\varphi' \in \mathcal{H}^2_r(\Gamma\backslash G)$. But the subspace of $f \in \mathcal{H}^2_r(G)$ such that $L(k)f = \chi_r(k)f$ has dimension one (constant functions on the unit disc) and contains $\omega_r$, qed. It is even possible to refer to the general result that will be proved in n° 30: in *every* irreducible unitary representation $(\mathcal{H}, U)$ of $G$, the subspace of solutions of $U(k)\mathbf{a} = \chi(k)\mathbf{a}$ has dimension $\leq 1$ for all characters $\chi$ of $K$.

*Exercise 3.* For $r > 2$, show directly that $J(f * \omega_r) = Jf$ for all $f \in \mathcal{H}(\varphi)$.

(h) To finish the proof, consider two functions $\varphi, \psi \in \mathcal{H}^2_r(\Gamma\backslash G)$ and the projection $P$ on $\mathcal{H}(\varphi)$ in $L^2(\Gamma\backslash G)$. It commutes with all $V(x)$, and so maps $\mathcal{H}(\psi)$ onto a subspace in which the linear combinations of $PV(x)\psi = V(x)P\psi$ are everywhere dense. The subspaces $\mathcal{H}(\varphi)$ and $\mathcal{H}(\psi)$ are, therefore, orthogonal if and only if $P\psi = 0$. This condition remains to be shown to be equivalent to $(\varphi|\psi) = 0$ and, to this end, that $P\psi$ is proportional to $\varphi$. But as $P$ commutes with all $V(k)$,

$$V(k)P\psi = PV(k)\psi = \chi_r(k)P\psi$$

and we remarked above that the only elements of $\mathcal{H}(\varphi)$ satisfying this relation are the multiples of $\varphi$.

Finally, let $(\varphi_i)$ be an orthonormal basis of the space $\mathcal{H}^2_r(\Gamma\backslash G)$. The spaces $\mathcal{H}(\varphi_i)$ are pairwise orthogonal and contained in the closed invariant subspace $\mathcal{D}_r$ generated by $\mathcal{H}^2_r(\Gamma\backslash G)$. Now, the finite sums of functions $V(x)\varphi$, where $\varphi \in \mathcal{H}^2_r(\Gamma\backslash G)$, are everywhere dense in $\mathcal{D}_r$. As every $\varphi \in \mathcal{H}^2_r(\Gamma\backslash G)$ is the limit of finite linear combinations of the functions $\varphi_i$, these $V(x)\varphi$ are in the Hilbert sum of the spaces $\mathcal{H}(\varphi_i)$. Thus $\mathcal{D}_r = \widehat{\bigoplus}\mathcal{H}(\varphi_i)$. This is a Hilbert direct sum for all $\Gamma$, and a finite direct sum if $\Gamma$ is Fuchsian.

*Exercise 4.* Interpret the theorem in the case $\Gamma = \{e\}$.

## 21 – Poincaré, Eisenstein and Maaß-Selberg Series

The reader will probably have noticed that, contrary to the statement of theorem 19, that of theorem 26 does not include any point (b). It could be, but is not, the author's "malfunction", as is politely said nowadays. In this n°, we will present methods that, together with very simple series, enable us, like in the case of the modular group, to effectively construct automorphic forms and in particular analogues for Fuchsian groups for Eisenstein series, following which the omitted proposition (b) will become as obvious as for the modular group.

(i) *Poincaré series*[79]. In 1880, wishing to extend the theory to discrete groups to which conventional calculations cannot be generalized, Poincaré found a less

[79] Most of the arguments and calculations of this n° can be generalized to the much more general case of the Siegel modular group; see my talks in the Cartan Seminar on automorphic functions of several variables (1957–1958). They have been reproduced for $SL_2$ in Miyake, *Modular Forms,* chap. 6, probably because through Ichiro Satake's participation in the Cartan Seminar a copy of it reached Japan! See also Walter L. Baily, Jr., *Introductory Lectures on Automorphic Forms* (Princeton UP, 1973), interesting in more than one respect.

miraculous way than arithmetic for constructing entire (and even, as will be seen, parabolic) automorphic forms. An arbitrary discrete subgroup $\Gamma$ of $G$ being given, starting from a holomorphic function $f(z)$ on $P$ or even only on an open $\Gamma$-invariant subset $\Omega \subset P$, it consists in considering the series

$$P_{r,f}(z) = \sum_{\Gamma} J(\gamma; z)^{-r} f(\gamma z) = \sum L_r(\gamma) f(z) \tag{21.1}$$

hoping they converge. If that is the case, (15.10') shows that the function of weight $r$ associated to $P_{r,f}$ on the inverse image of $\Omega$ under $g \longmapsto z = gi$ is

$$P_{r,f}(g) = \sum f_r(\gamma g)\,, \tag{21.1'}$$

which, hundred years later, makes Poincaré's idea trivial. If the function $f_r$ is integrable over $\pi^{-1}(\Omega)$, the series (1) or(1') converges in mean – which, if $f$ is holomorphic, most probably implies compact convergence – and its sum is integrable mod $\Gamma$. If $\Gamma$ is a Fuchsian group and if $\Omega = P$, it will therefore be a parabolic form by theorem 26, (c). These heuristic considerations remain to be justified. We do this by comparing series (1) to an integral. This method appears somewhat implicitly (see below) in Poincaré and was systematically exploited much later. Both Cartan Seminars of 1953/54 and 1957/58 on automorphic functions use it extensively, as well as Siegel.[80]

We again start from the fact that, for any compact set $A \subset \Omega$ and any compact neighbourhood $A' \subset \Omega$ of $A$, there is a constant $M_1$ such that the uniform norm on $A$ of any holomorphic function $f$ on $\Omega$ has an upper bound

$$\|f\|_A \leq M_1 \iint_{A'} |f(z)|\, dxdy\,.$$

To formulate this result in terms of $f_r$, we use (15.30)

$$\int |f_r(g)|\, dg = \int \left| y^{r/2} f(z) \right| dm(z)\,.$$

As the variable $y$ stays away from 0 and $+\infty$ in $A'$, in the previous inequality, the Euclidean measure $dxdy$ can be replaced by the invariant measure $dm(z)$, so that it can equally well be written

$$\|f_r\|_C \leq M_2 \int_{C'} |f_r(g)|\, dg \tag{21.2}$$

where $C$ and $C'$ are the inverse images of $A$ and $A'$ under $g \longmapsto gi$. For given $A$ and $A'$, the constant $M_2$ being the same for all holomorphic functions on $P$, (2) may be applied to the left translations $L(\gamma)f_r$ of $f_r$. Thus

$$\sum_{\Gamma} \|L(\gamma)f_r\|_C \leq M_2 \sum_{\Gamma} \int_{C'} |f_r(\gamma g) dg = M_2 \sum_{\Gamma} \int_{\gamma C'} |f_r(g)| dg\,.$$

However, $\gamma C' \# \gamma' C'$ implies $\gamma^{-1}\gamma' \in C' C'^{-1} \cap \Gamma$. The latter, being a compact discrete set, is finite with $N$ elements. Hence each $\gamma C'$ has non-trivial intersection

[80] Carl Ludwig Siegel, *Einführung in die Theorie der Modulfunktionen n-ten Grades* (i.e. related to the group $Sp(n, \mathbb{R})$), Math. Annalen, **116**, 1939, pp. 617–657. There are probably older references by the same author, in whose complete works can be found considerable ideas and beautiful calculations despite the hostility Siegel showed in his old days for new methods (representations and adeles).

with at most $N$ sets $\gamma' C'$, so that the sum of the integrals extended to all $\gamma C'$ is equal to at most $N$ times the integral extended to the union $\Gamma.C'$ of these $\gamma C'$. This is an easy exercise of measure theory. Series (1) therefore converges normally in every compact subset provided

$$\int_{\Gamma A} y^{r/2} |f(z)| \, dm(z) < +\infty \tag{21.3}$$

for all compact sets $A \subset \Omega$. This condition is also necessary since

$$\int_{\gamma C} |f_r(g)| \, dg \leq m(C) \sup_{x \in C} |f(\gamma x)| \, .$$

The problem is then to show that there are functions $f$ satisfying condition (3). To evaluate the obstacle, let us suppose that $\Omega = P$ and that the quotient $\Gamma \backslash P$ is compact. $A$ can then be chosen so that $\Gamma A = P$, in which case the condition means that $f \in \mathcal{H}_r^1(P)$. The latter is a non-trivial space if and only if $r > 2$. In the case of an arbitrary group $\Gamma$, condition $f \in \mathcal{H}_r^1(P)$ is obviously sufficient. Moreover, as this assumption means that the function $f_r$ is integrable over $G$, the sum of series (1') is integrable over $\Gamma \backslash G$. As a result:

**Theorem 28.** *For all discrete subgroups $\Gamma \subset G$, all open $\Gamma$-invariant subsets $\Omega \subset P$, all $r > 2$ and all holomorphic functions such that*

$$\int_{\Omega} y^{r/2} |f(z)| \, dxdy < +\infty \, ,$$

*the Poincaré series $P_{r,f}(z) = \sum J(\gamma; z)^{-r} f(\gamma z)$ converges normally in every compact subset of $\Omega$ and its sum belongs to $H_r^1(\Gamma \backslash \Omega)$.*

In fact, initially Poincaré only considered series $P_{r,f}(z)$ for functions $f$ which, on the unit disc $D$, correspond to *polynomials* in $z$. His arguments are a bit different. The conformal representation

$$z \longmapsto \zeta = (z - i)/(z + i) = s^{-1} z$$

used in n° 15, (vi) transforms $\Gamma$ into a subgroup $\Gamma' = s^{-1} \Gamma s$ of the group $G'$ of automorphisms (15.34) of $D$. Associating the function

$$\varphi(\zeta) = J(s; \zeta)^{-r} f(s\zeta)$$

to $f$, series (1) becomes

$$\sum J(\gamma'; \zeta)^{-r} \varphi(\gamma' \zeta) \, , \tag{21.4}$$

where summation is over $\Gamma'$. This does not appear to simplify the problem. But instead of comparing the terms of the new series to integrals with respect to the *invariant* measure of $D$, Poincaré uses the *Euclidean* measure $d\xi d\eta = d\mu(\zeta)$ on the complex plane. For any sufficiently small disc $A$ centered at $\zeta$, the area $\mu(\gamma' A)$ is approximately equal to the product of $\mu(A)$ and the Jacobian of $\gamma'$ calculated somewhere in $A$, namely[81]

$$|d(\gamma'\zeta)/d\zeta|^2 = |J(\gamma'; \zeta)|^{-4} \, .$$

[81] The Jacobian of a holomorphic map $z \longmapsto f(z)$ is $|f'(z)|^2$ because of Cauchy's holomorphy condition: Chap. III, formula (24.9).

As $\mu(D) < +\infty$, without strictly speaking using integrals, he concludes that $\sum \mu(\gamma' A) < +\infty$ and as a result that

$$\sum |J(\gamma';\zeta)|^{-4} < +\infty .$$

So it suffices to assume that $\varphi$ is *bounded* on $D$, for example is polynomial, in order to make series (4) convergent for $r = 4$. And if it converges for $r = 4$, it necessarily converges for $r > 4$.

This argument, based on

$$\sum |d(\gamma'\zeta)/d\zeta|^2 < +\infty ,$$

is simpler that the previous one since it only uses spaces $\mathcal{H}_1^r$; but it does not include the case $r = 3$, which occurs for groups that do not contain $-1$ (as well as non-integral $r \in ]2, 4[$), and the assumption that $\varphi$ is bounded is too restrictive. In fact it is enough for $\varphi$ to belong to the space $H_r^1(D)$. It can be immediately generalized to groups acting on a *bounded* domain of a space $\mathbb{C}^n$.

Characterizing polynomials to which Poincaré confined himself is easy. On $D$, these are precisely the holomorphic functions whose images under rotations centered at 0 generate a finite-dimensional vector space. Returning to the half-plane and to $G$, we get functions $f_r$ whose *left* translations $g \longmapsto f_r(kg)$ under $k \in K$ generate a finite-dimensional space. They will be called *$K$-finite functions*[82] on $G$. Formula (16.15), with $n = r + 2p$, $p \geq 0$, provides a basis for it for all $r$.

*Exercise 1.* What is relation (5) in $P$?

Let us consider the simplest case of the function $f(z) = (z - w)^{-p}$ where $p > 0$ and choose $\Omega = P - \Gamma w$. Any compact subset of $\Omega$ is contained in $P - \Gamma V$ where $V$ is a compact neighbourhood of $w$ in $P$. As $\Omega \subset P - V$, to show that the corresponding Poincaré series converges normally in every compact subset of $\Omega$, it suffices to check that $y^{r/2}|f(z)|$ is integrable in $P - \Gamma V$, or even more so in $P - V$. In this open set, the ratio $|(z - w)/(z - \overline{w})|$ stays in a compact subset of $\mathbb{R}_+$ since

$$z \longmapsto (z - w)/(z - \overline{w})$$

transforms $P - V$ into $D - V'$, where $V'$ is a neighbourhood of 0. Hence it is enough to check that the function $y^{r/2}(z - \overline{w})^{-p}$ is integrable in $P - V$. Now, we know [Chap. VIII, n° 12, (iv)] that for $p > r/2 > 1$, it is in the space $\mathcal{H}_r^1(P)$, which is better. The corresponding Poincaré series $P_{r,f}$ therefore converges normally in every compact subset of $P - \Gamma w$ if $p > r/2 > 1$.

If $w$ is not a fixed point of $\Gamma$, the only terms of the series having a pole at $w$ correspond to $\gamma = \pm 1$. If $-1 \notin \Gamma$, $w$ and its images are poles of order $p$ of $P_{r,f}$; if $-1 \in \Gamma$, the result is the same for $r$ even. To deduce that $P_{r,f}$ is a (non-entire) automorphic form of weight $r$, its behaviour at parabolic fixed points of $\Gamma$ needs to be analyzed. Since $f$ is integrable in $P - \Gamma V$, the series $P_{r,f}$ is integrable mod $\Gamma$ in this open set. However, thanks to lemma 1 of the previous n°, for every parabolic fixed point $\xi$ of $\Gamma$, $P - \Gamma V$ contains a horocycle centered at $\xi$. So the function $P_{r,f}$ is not only holomorphic, but also parabolic at each of these points.

[82] Their importance is not only due to the fact that they are known explicitly. $K$-finite functions can be defined for all groups $G$ and all compact subgroups $K$ of $G$. Their role is essential in Harish-Chandra's theory ($G$ semisimple, $K$ maximal compact) and in automorphic function theory *à la* Maaß-Selberg-Langlands, because the Lie algebra of $G$ acts almost algebraically on $K$-finite vectors of an irreducible representation. See § 8 of this chapter.

**Theorem 29**. *Let $\Gamma$ be a Fuchsian group. For all $r \in \mathbb{Z}$, even if $-1 \in \Gamma$, there exist non-trivial automorphic forms of weight $r$ for $\Gamma$.*

the previous argument proves the theorem for all $r > 2$: choose $f(z) = (z-w)^{-p}$ with $p > r$. If $w' \notin \Gamma w$ is not an elliptic fixed point, the function $f'(z) = (z-w')^{-p'}$ is also appropriate for all $r' > 2$ and all $p' > r'$. The quotient $P_{r,f}(z)/P_{r',f'}(z)$ is then a non-trivial automorphic form of weight $r - r'$ having poles of order $\geq p$ at the points $\gamma w$ and zeros of order $\geq p'$ at the points $\gamma w'$, qed.

These results can be refined; see for example Miyake, § 2.6. I will not go any further in this direction and henceforth will keep to *entire* automorphic forms.

For $f \in \mathcal{H}_r^1(P)$, though it is already known, using results of n° 16, (i) the series $P_{r,f}$ can be checked to indeed be parabolic forms for all Fuchsian subgroups $\Gamma$.

More generally, let us consider a group $\Gamma$ admitting $\infty$ as a parabolic fixed point and set

$$U_\infty = \Gamma \cap U$$

to be the subgroup of matrices $x(n\omega)$ for some $\omega > 0$. Then any holomorphic solution of $f(\gamma z) = J(\gamma; z)^r f(z)$ has an expansion

$$f(z) = \sum a_n \mathbf{e}(z/\omega)^n \,, \qquad a_n = \omega^{-1} \oint f(z)\mathbf{e}(z/\omega)^{-n} dx \tag{21.5}$$

where integration is modulo $\omega$. We need to show directly that if $f$ is a Poincaré series, then $a_n = 0$ for all $n \leq 0$.

The easiest is to argue in $G$. We have to expand the function $u \longmapsto P_{r,f}(ug) = \sum f_r(\gamma ug)$, as a $U_\infty$-invariant Fourier series. Identifying the matrix $x(u) \in U$ with the number $u \in \mathbb{R}$, the expected Fourier series can be written

$$P_{r,f}(ug) = \sum \mathbf{e}(u/\omega)^n a_n(g) = \sum a_n(ug)$$

where

$$a_n(g) = \omega^{-1} \int_{U_\infty \backslash U} \mathbf{e}(u/\omega)^{-n} du \sum_{\Gamma} f_r(\gamma ug) \,.$$

Grouping together the terms $\gamma\gamma'$ where $\gamma' \in U_\infty$ and observing that $\mathbf{e}(u/\omega)$ is $U_\infty$-invariant, we get

$$\begin{aligned} \omega a_n(g) &= \int_{U_\infty \backslash U} du \sum_{\gamma . U_\infty} \sum_{\gamma' \in U_\infty} \mathbf{e}\left[\gamma'(u)/\omega]\right)^{-n} f_r\left(\gamma\gamma' ug\right) = \\ &= \sum_{\gamma . U_\infty} \int_{U_\infty \backslash U} du \sum_{\gamma'} \mathbf{e}\left[\gamma'(u)/\omega\right]^{-n} f_r\left(\gamma\gamma' ug\right) \,, \end{aligned}$$

whence

$$\omega a_n(g) = \sum_{\Gamma \backslash U_\infty} \int_U \mathbf{e}(u/\omega)^{-n} f_r(\gamma ug) du \,.$$

This formal calculation, which does not assume $f$ to be holomorphic, is justified because exponential factors have absolute value 1. Like the Poincaré series, for all $g$, the series being integrate thus converges normally over any compact subset of $Ug$. For $\gamma = e$,

$$\int_U f_r(ug)\mathbf{e}(u/\omega)^{-n}du = \int_{\mathbb{R}} J(g;i)^{-r} f(z+u)\mathbf{e}(u/\omega)^{-n}du =$$
$$= J(g;i)^{-r}\mathbf{e}(z/\omega)^n \widehat{f}(n/\omega)$$

as was seen in n° 16, (i). There are similar formulas for the translations $g \longmapsto f_r(\gamma g)$ of $f_r$ under $\gamma \in \Gamma$ as they are also in $\mathcal{H}_r^1(G)$. Hence, since $\widehat{f}(t) = 0$ for $t \leq 0$ for all $f \in \mathcal{H}_r^1(P)$, $a_n(g) = 0$ for $n \leq 0$, qed.

This computation method does not provide any explicit formulas for the coefficients $a_n$, since the effect of left translations by $\gamma \in \Gamma$ on $\widehat{f}$ is not simple [n° 16, (v), exercise 3]. To get the coefficients $a_n$, summing over the cosets $U_\infty \gamma$ is necessary as will be seen later.

We can at last address the question whether, for a Fuchsian group $\Gamma$, a parabolic form of weight $r > 2$ is a Poincaré series. Let $f \in \mathcal{H}_r^1(G)$ and $\varphi \in \mathcal{H}_r^2(\Gamma\backslash G)$ be a parabolic form. Let us calculate the inner product

$$(\varphi|P_{r,f}) = \int_{\Gamma\backslash G} \varphi(g)\overline{P_{r,f}(g)}dg = \int_{\Gamma\backslash G} dg \sum_\Gamma \varphi(\gamma g)\overline{f_r(\gamma g)} =$$
$$= \int_G \varphi(g)\overline{f_r(g)}dg\,.$$

This formal calculation supposes $\varphi(g)f_r(g)$ to be integrable over $G$, which is the case since $\varphi$ is bounded on $G$ (theorem 26) and $f_r$ is integrable. Hence if $\varphi$ is orthogonal to all Poincaré series, then it is also orthogonal to $\mathcal{H}_r^1(P)$. However, every $\varphi \in \mathcal{H}_r(G)$ orthogonal to $\mathcal{H}_r^1(G)$ is trivial because the corresponding holomorphic function $\varphi_D$ on the unit disc is orthogonal to monomials $\zeta^n$. By the way, this shows that it is enough to consider the series $P_{r,f}$ of $K$-finite functions.

Hence, the subspace orthogonal to Poincaré series in $\mathcal{H}_r^2(\Gamma\backslash G)$ is trivial. If $\Gamma$ is a Fuchsian group, $\mathcal{H}_r^2(\Gamma\backslash G)$ is finite-dimensional, whence $f = 0$. *Every $f \in \mathcal{H}_r^2(\Gamma\backslash G)$ is therefore a Poincaré series.* In the next section, we will be able to give more details about this result due to $H$. Petersson.

The previous calculations lead to the construction of a kernel function for the space $\mathcal{H}_r^2(\Gamma\backslash G)$ when $\Gamma$ is a Fuchsian group.

*Exercise 1.* Suppose $r > 2$. Let $\omega_r$ be the kernel function of $\mathcal{H}_r^2(G)$. It is integrable. (a) Set

$$\omega_r^\Gamma(x,y) = \sum \omega_r\left(x^{-1}\gamma y\right)\,.$$

Show that, for all $x$, the function $y \longmapsto \omega_r^\Gamma(x,y)$ is in $H_r^1(\Gamma\backslash G)$. (b) Show that

$$(*) \qquad \varphi(x) = \int_{\Gamma\backslash G} \overline{\omega_r^\Gamma(x,y)}\varphi(y)dy$$

for all $\varphi \in \mathcal{H}_r^2(\Gamma\backslash G)$. In particular,

$$\int_{\Gamma\backslash G} \omega_r^\Gamma(x,z)\overline{\omega_r^\Gamma}(y,z)dz = \omega_r^\Gamma(x,y)\,.$$

(c) Using the fact that, for all $\varphi \in \mathcal{H}_r^2(\Gamma\backslash G) = \mathcal{H}_r^\infty(\Gamma\backslash G)$, the second integral (*) is a bounded function of $x$, show that

$$\sup_{x\in\Gamma} \int_{\Gamma\backslash G} \left|\omega_r^\Gamma(x,y)\right|^2 dy < +\infty$$

and that $\omega_r^\Gamma(x,y)$ is bounded on $G \times G$. (d) Set

$$E_r\varphi(x) = \int_{\Gamma\backslash G} \omega_r^\Gamma(x,z)\varphi(z)dz$$

for all $\varphi \in L^2(\Gamma\backslash G)$. Show that $E_r$ is the orthogonal projection of $L^2(\Gamma\backslash G)$ on $\mathcal{H}_r^2(\Gamma\backslash G)$. Deduce that

$$\dim \mathcal{H}_r^2(\Gamma\backslash G) = \int_{\Gamma\backslash G} dx \sum_{\Gamma} \omega_r\left(x^{-1}\gamma x\right) .$$

The rest of the calculation consists in grouping together the terms of the series according to the conjugacy classes in $\Gamma$, in calculating the centralizers of the $\gamma$ in $G$ (if $\gamma \neq \pm 1$, these are well determined conjugates of the subgroups $K$, $A$ or $U$ according to whether $\gamma$ is elliptic, hyperbolic or parabolic), in transforming the integral over $\Gamma\backslash G$ of the subseries extended to a class into an integral modulo the centralizer of $\gamma$, in checking that everything converges, etc., in other words of techniques developed by Selberg for his "trace formula".

(ii) *Poincaré-Eisenstein series.* Supposing that $\Gamma$ admits $\infty$ as a parabolic fixed point, which implies $U_\infty = U(\omega\mathbb{Z})$ for some $\omega > 0$, we are going to show that the Poincaré series can be written in terms of standard series involving no "arbitrary" function $f$. For $\Gamma = U(\mathbb{Z})$, this result will reducc to formula $\sum f(z+n) = \sum \widehat{f}(n)\mathbf{e}(nz)$ of n° 16, (i) and for $SL_2(\mathbb{Z})$, to calculations of n° 17, (iii).

Grouping the terms of the series into cosets $\gamma U_\infty$,

$$P_{r,f}(z) = \sum_{U_\infty\backslash\Gamma} J(\gamma;z)^{-r} \sum_{\mathbb{Z}} f\left[\gamma(z) + n\omega\right] =$$

$$= \omega^{-1} \sum J(\gamma;z)^{-r} \sum \widehat{f}(n/\omega)\mathbf{e}\left[\gamma(z)/\omega\right]^n . \tag{21.6}$$

Thus, calculating formally,

$$\omega P_{r,f}(z) = \sum_{n>0} \widehat{f}(n/\omega) E_{r,n}(z) \tag{21.7}$$

with *Poincaré-Eisenstein series*

$$E_{r,n}(z) = \sum_{U_\infty\backslash\Gamma} J(\gamma;z)^{-r}\mathbf{e}(\gamma z)^n \quad (r > 2, n > 0) \tag{21.8}$$

which should be written $E_{r,n,\infty}$ to indicate that they are with respect to the fixed point $\infty$. To get functions that are not trivially zero, we need to assume that $r$ is even if $-1 \in \Gamma$. If $\infty$ is irregular, $\Gamma$ contains the matrix

$$\gamma_0 = \begin{pmatrix} -1 & \omega/2 \\ 0 & -1 \end{pmatrix}$$

and as $\Gamma_\infty\gamma = U_\infty\gamma \cup U_\infty\gamma_0\gamma$, the terms of series (8) in $\gamma$ and $\gamma_0\gamma$ can be grouped together and it can then be summed mod $\Gamma_\infty$. As $J(\gamma_0\gamma;z) = -J(\gamma;z)$ and

$$\mathbf{e}\left[\gamma_0\gamma(z)/\omega\right] = \mathbf{e}\big[\gamma(z)/\omega + \frac{1}{2}\big] = -\mathbf{e}\left[\gamma(z)/\omega\right] ,$$

this gives a factor $1 + (-1)^{n+r}$ in each term of the sum mod $\Gamma_\infty$. As a result,

(21.9) $$E_{r,n}(z) = 0 \quad \text{if } \infty \text{ is irregular and } n + r \text{ odd}.$$

To justify the permutation of the $\sum$ in (6) and, at the same time, to ensure compact convergence of series (8), it suffices to prove that series (6) converges unconditionally. For $z$ in a fixed compact set, there is (lemma 1) an upper bound $\mathrm{Im}(\gamma z) \leq T$ for all $\gamma$ and so

$$|\mathbf{e}[\gamma(z)/\omega]| \leq \exp(-2\pi T).$$

As a result,

$$\sum_n \left|\widehat{f}(n/\omega)\mathbf{e}[\gamma(z)/\omega]^n\right| \leq \sum \left|\widehat{f}(n/\omega)\exp(-2\pi n T)\right| = M < +\infty.$$

Like $\sum \widehat{f}(n)\mathbf{e}(nz)$, the second series converges for all $f \in \mathcal{H}_r^1(P)$ [n° 16, (i)]. Then

$$\sum_{U_\infty\backslash\Gamma} |J(\gamma;z)|^{-r} \sum_n \left|\widehat{f}(n/\omega)\mathbf{e}[\gamma(z)/\omega]^n\right| \leq M \sum_{U_\infty\backslash\Gamma} |J(\gamma;z)|^{-r},$$

so that to justify going from (6) to (7) and compact convergence of series (8) for $n > 0$, it suffices to show that

(21.10) $$\sum_{U_\infty\backslash\Gamma} |J(\gamma;z)|^{-r} < +\infty,$$

which brings us back to the convergence of the *Eisenstein series*

(21.11) $$E_{r,\infty}(z) = \sum_{U_\infty\backslash\Gamma} J(\gamma;z)^{-r} \quad (r > 2)$$

associated to the parabolic fixed point $\infty$.

If this point, which will be proved in section (iii) in the general case, is admitted for now, relation (7) and theorem 26, (f) show that the series $P_{r,f}$ are limits of linear combinations of series (8). But in finite dimension, every vector subspace is closed. As shown at the end of section (i), every $f \in \mathcal{H}_r^2(\Gamma\backslash G)$ is a Poincaré series. Thus the results of these two sections can be summarized as follows:

**Theorem 30 (H. Petersson).** *Let $\Gamma$ be a Fuchsian group.*

(i) *For all $r > 2$, the space $\mathcal{H}_r^2(\Gamma\backslash G)$ of parabolic forms of weight $r$ for $\Gamma$ is the set of Poincaré series $P_{r,f}$ associated to $K$-finite functions $f \in \mathcal{H}_r^1(P)$.*

(ii) *Every parabolic form of weight $r > 2$ is a linear combination of Poincaré-Eisenstein series.*

If $\Gamma$ contains the matrix $-1$, $r$ is even and the general term of (11) only depends on the coset of $\gamma$ modulo the subgroup

$$\Gamma_\infty = \Gamma \cap B = U_\infty \cup -U_\infty$$

of matrices for which $c = 0$, $d = \pm 1$. So the sum is twice that obtained by summing $\bmod \Gamma_\infty$. To avoid this factor 2, in the case of $SL_2(\mathbb{Z})$, a factor $\frac{1}{2}$ was introduced in (17.20) in definition (11). We will not do so here in order to get formulas for all $\Gamma$. If the fixed point $\infty$ is irregular, the formal argument leading to (9) applies equally for $n = 0$. Thus the series $E_{r,\infty}$ are trivial for $r$ odd.

*Exercise 2.* (a) Show that

$$\sum \omega_r(z+n) = \sum_{n>0} n^{r-1}\mathbf{e}[n(z+i)]$$

up to a constant factor. (b) Show that

$$\sum \omega_r(\gamma g) = \sum_{n>0} \exp(-2\pi/\omega)^n n^{r-1} E_{r,n}(g)$$

up to a constant factor. (c) Using exercise 1, recover theorem 30.

Another reason for the importance of the series $E_{r,n}$ studied by Petersson but also found in Poincaré[83] for the modular group, follows from the next calculation. Let $f$ be a holomorphic solution of the functional equation $f(\gamma z) = J(\gamma; z)^r f(z)$, without any other restrictions for the moment, and let us compute the integral

$$(f|E_{r,n})_r = \int_{\Gamma\backslash P} y^r f(z)\overline{E_{r,n}(z)}dm(z) = \int_{\Gamma\backslash G} f_r(g)\overline{E_{r,n}(g)}dg\,, \tag{21.12}$$

where

$$E_{r,n}(g) = J(g;i)^{-r} E_{r,n}(z) = \sum \mathbf{e}_r(\gamma g; n/\omega)$$

and

$$\mathbf{e}_r(g;t) = J(g;i)^{-r}\mathbf{e}(tz)$$

are the functions of weight $r$ associated to $E_{r,n}(z)$ and $\mathbf{e}(tz)$ on $G$. Let us suppose that the function under the $\int$ sign in (12) is integrable. As $f_r(\gamma g) = f_r(g)$,

$$(f|E_{r,n})_r = \int_{\Gamma\backslash G} dg \sum_{U_\infty\backslash\Gamma} f_r(\gamma g)\overline{\mathbf{e}_r(\gamma g; n/\omega)}\,.$$

Thus

$$\begin{aligned}(f|E_{r,n})_r &= \int_{U_\infty\backslash G} f_r(g)\overline{\mathbf{e}_r(g;n/\omega)}dg = \\ &= \int_{U_\infty\backslash P} y^{r-1} f(z)\overline{\mathbf{e}(nz/\omega)}dxd^*y\,.\end{aligned} \tag{21.13}$$

Using the Fourier series $f(z) = \sum a_p \mathbf{e}(pz/\omega)$ and integrating term by term, we get

$$(f|E_{r,n})_r = \sum_p a_p \int_{U_\infty\backslash P} y^r \mathbf{e}(pz/\omega)\overline{\mathbf{e}(nz/\omega)}dm(z)\,. \tag{21.14}$$

Replacing $z$ by $\omega z$, leaves the measure invariant, replaces $U_\infty$ by $U(\mathbb{Z})$ and multiplies $y^r$ by $\omega^r$. Integration is then via an integral over $x \bmod \mathbb{Z}$, which is zero for $n \neq p$. For $n = p$, the product of the two exponentials equals $\exp(-4\pi ny)$, thus the final result follows readily:

$$(f|E_{r,n})_r = (4\pi n)^{1-r}\,\Gamma(r-1)\omega^r a_n\,. \tag{21.15}$$

[83] See *Fonctions modulaires et fonctions fuchsiennes* (Œuvres complètes, end of volume II), a 1912 article sent to Ann. of the Fac. of Sc. of Toulouse on the eve of his admission to a clinic for a prostrate operation which went wrong... Formula (12) is already in there for particular functions $f$ (those corresponding to monomials $\zeta^p$ on the unit disc), and so are the Fourier series of the functions $E_{r,n}$ $(n \geq 0)$ on the modular group and the arithmetic sums occurring in their coefficients.

These calculations still remain to be justified assuming $f$ to be parabolic. To get (13), it suffices to show that the integral obtained extended to the strip $|\operatorname{Re}(z)| \leq \omega/2$ of $P$ is convergent. Now, the function $y^{r/2}f(z)$ is bounded on $P$ if $\Gamma$ is a Fuchsian group, the exponential is rapidly decreasing at infinity since $n > 0$ and the function $y^{r-1}$ is integrable with respect to $d^*y$ in the neighbourhood of 0 since $r > 1$. So it clearly converges.

To justify going from (14) to (15), it suffices to show that

$$\sum_p |a_p| \int_{U_\infty \backslash P} y^r \, |\mathbf{e}(pz/\omega)\mathbf{e}(nz/\omega)| \, dm(z) < +\infty .$$

The integral is proportional to $(n+p)^{1-r} \approx p^{1-r}$, whence the condition

$$\sum |a_p|/p^{r-1} < +\infty .$$

As we know that $a_p = O(p^{r/2})$ for any parabolic form, assumption $r > 4$ justifies the calculation, but leaves open the cases $r = 3$ and $r = 4$. The question is academic for the modular group. If $\Gamma$ is a congruence group, Deligne's upper bound

$$a_p = O\left(p^{\frac{1}{2}(r-1)+\epsilon}\right)$$

can be used. This result, which does not come very cheap, takes care of the question for $r = 4$ but not for $r = 3$. To make the series convergent in this case, we need an estimation of the genus in $O(p^{1-\epsilon})$, but these gentlemen say that it is not possible to go beyond $O(p^{1+\epsilon})$.

In conclusion, *The Fourier coefficient with index $n > 0$ of any parabolic form $f$ of weight $r$ is the inner product of $f$ and the Poincaré-Eisenstein series $E_{r,n}$, up to a simple factor*, at least for $r > 4$.

(iii) *Eisenstein series.* They do not generalize the series $G_r(z)$ of the modular group, which are not well-defined for non-arithmetic groups $\Gamma$, but the reduced series

$$E_{r,\infty}(z) = \sum_{U_\infty \backslash \Gamma} J(\gamma; z)^{-r} \tag{21.16}$$

introduced in (17.20) with a factor $\frac{1}{2}$. They are defined for any discrete group $\Gamma$ having $\infty$ as a parabolic fixed point. For $r$ odd, as was seen above, the series is trivial if $-1 \in \Gamma$ or if the fixed point $\infty$ is irregular.

The convergence of these series, obvious if $\Gamma \subset SL_2(\mathbb{Z})$, requires a more difficult proof for other groups, but its principle can be applied to multiple Eisenstein series associated to arithmetic subgroups of general semisimple groups. In fact we will study the *Maaß-Selberg series*[84]

[84] Hans Maaß, *Über eine neue Art von nichtanalytischen automorphen Funktionen und die Bestimmung Dirichletscher Reihen durch Funktionalgleichungen* (Math. Annalen, **121**, 1949, pp. 141–183), Atle Selberg, *Harmonic analysis and discontinuous groups in weakly symmetric Riemannian spaces* (J. Indian Math. Soc., **20**, 1956). There is also, for all $r \in \mathbb{Z}$, a series

$$M_{r,\infty}(z,s) = \sum J(\gamma; z)^{-r} \operatorname{Im}(\gamma z)^{s-r/2} .$$

These series, which satisfy differential equations which we shall prove in § 8, enable us to write down explicitly the " continuous spectrum " of the representation of $G$ on $L^2(\Gamma \backslash G)$, hence their importance.

(21.17) $$M_\infty(z;s) = \sum_{\Gamma_\infty\backslash\Gamma} \operatorname{Im}(\gamma z)^s = \sum_{\Gamma_\infty\backslash\Gamma} \alpha(\gamma g)^s$$

where $s$ is a complex exponent. As $|J(\gamma;z)|^{-2} = \operatorname{Im}(\gamma z)/\operatorname{Im}(z)$, convergence will be obtained for (16) when $r > 2$ by showing that series (17) converges for $\operatorname{Re}(s) > 1$. Summation mod $\Gamma_\infty$ aims to ensure that the "constant" term, i.e. independent of $x$, of the Fourier series

$$M_\infty(z;s) = \sum a_n(y)\mathbf{e}(nx/\omega)$$

of the function is, as will be seen, of the form $a_0(y) = y^s + a(s)y^{1-s}$ without any superfluous factor 2 if $U_\infty \neq \Gamma_\infty$. However, summing mod $\Gamma_\infty$ is problematic for the series

(21.17') $$M_{r,\infty}(z;s) = \sum_{\Gamma_\infty\backslash\Gamma} J(\gamma;z)^{-r}\operatorname{Im}(\gamma z)^{s-r/2}$$

when $\Gamma_\infty \neq U_\infty$. One obviously needs to assume $r$ is even if $-1 \in \Gamma$ or else if the fixed point $\infty$ is irregular. In the latter case, summing mod $U_\infty$ is certainly possible for $r$ odd, but the result would be zero.

We will set

$$\alpha(g) = |J(g;i)|^{-2} = \operatorname{Im}(gi)$$

as in n° 15, whose notation we will keep. For $k \in K$, $u \in U$, $h = h(t) \in A$ and $z = gi$,

$$uhgi = uhz = t^2 z + u = \alpha(h)z + u\,,$$

whence

$$\alpha(uhgk) = \alpha(h)\alpha(g)\,.$$

The Haar measure of $G$ is given by

$$\begin{aligned}\varphi(g)dg &= \iint \varphi(bk)dbdk = \iiint \varphi(huk)dhdudk = \\ &= \iiint \varphi(uhk)\alpha(h)^{-1}dudhdk\,.\end{aligned}$$

$g \longmapsto gi$ transforms the horocycles centered at $\infty$ into Siegel domains

$$\mathfrak{S}(T) : g = uhk \quad \text{with } \alpha(h) \geq T$$

of (19.9) and the horocycles centered at $\xi = x\infty$ into their translations $x\mathfrak{S}(T)$.

**Theorem 31**. *For every discrete subgroup $\Gamma$ of $G$ having a parabolic fixed point at infinity, the series*

(21.18) $$M_\infty(g;s) = \sum_{\Gamma_\infty\backslash\Gamma} \alpha(\gamma g)^s = \sum \operatorname{Im}(\gamma z)^s$$

*converges absolutely for* $\operatorname{Re}(s) > 1$. *For any parabolic fixed point $\xi = x\infty$ of $\Gamma$, the series converges normally in every horocycle $\mathfrak{S}(\xi)$, unless $\xi = \sigma\infty$ ($\sigma \in \Gamma$) belongs to the same class as the point $\infty$. In this case the series with the term $\alpha(\sigma^{-1}g)^s$ removed converges normally in $\mathfrak{S}(\xi)$.*

Again, the proof amounts to comparing the series with an integral.

**Lemma 1.** *Normal convergence in $\mathfrak{S}(\xi)$ is equivalent to convergence at $x \in G$.*

To prove this it suffices to find an upper bound

$$\alpha(\gamma x g) \le M\alpha(\gamma x)$$

valid for $\gamma \notin \Gamma_\infty$ and $g \in \mathfrak{S}(T)$.

Let us first show that there is a constant $M$ such that

$$\alpha(yg) \le M\alpha(y) \tag{21.19}$$

for all $y \in G - B$ and all $g \in \mathfrak{S}(T)$. The ratio between both sides being invariant under $y \longmapsto uhy$, one may assume that $y \in K$, a compact set in which $\alpha(y)$ stays away from 0 and $+\infty$. So it suffices to prove that $\alpha(yg)$ is bounded for $y \in K - K \cap B$ and $g \in \mathfrak{S}(T)$. Then (Bruhat) $y = huwu'$ where, $h$, $u$ and $u'$, like $y$, stay in compact sets. In $\alpha(yg) = \alpha(h)\alpha(wu'g)$, $u'g$ stays in $U\mathfrak{S}(T) = \mathfrak{S}(T)$, which gets rid of $u'$. On the other hand, the factor $\alpha(h)$ is bounded since $y \in K$. Hence the function

$$g \longmapsto \alpha(wg) = \mathrm{Im}(wgi) = \mathrm{Im}(-1/z) \quad \text{where } z = gi$$

remains to be shown to be bounded on $\mathfrak{S}(T)$, but this is obvious.

This being settled, let us return to series (18) in $x\mathfrak{S}(T)$. By (19),

$$\alpha(\gamma x g) \le M\alpha(\gamma x) \quad \text{for } \gamma x \notin B\,, \quad g \in \mathfrak{S}(T)\,. \tag{21.20}$$

If the fixed point $\xi$ belongs to the same class as $\infty$, i.e. if $x = \sigma b$ for some $\sigma \in \Gamma$ and some $b \in B$, clearly,

$$\gamma x \notin B \Longleftrightarrow \gamma\sigma \notin \Gamma_\infty\,.$$

Thus (20) applies to all terms of the series except $\alpha(\sigma^{-1}g)$. Therefore, normal convergence in $\mathfrak{S}(T)$ of the series with this term removed follows from its convergence at $x$ (or any other point). If, on the other hand, $\xi$ is not equivalent to the fixed point $\infty$, since $\gamma x \in B$, $\gamma\xi = \infty$, which is impossible. In this case, (20) applies to all $\gamma$, qed.

**Lemma 2.** *Let $\varphi$ be a continuous function with values $> 0$ on $G$ such that*

$$\varphi(uhg) = \alpha(h)^s\varphi(g) \quad \textit{for some } s \in \mathbb{R}\,. \tag{21.21}$$

*For all compact sets $C \subset G$ with measure $> 0$, there exist constants $m, M > 0$ such that*

$$m\varphi(x) \le \int_C \varphi(xg)dg \le M\varphi(x)$$

*for all $x \in G$.*

Let $\varphi'(x)$ denote the integral. The function $\varphi'$ is continuous, satisfies (21) like $\varphi$ and is everywhere $> 0$ since $m(C) > 0$. The ratio $\varphi(x)/\varphi'(x) > 0$ is invariant under $x \longrightarrow uhx$, and so takes the same values in $G$ as in $K$. Since $K$ is compact, in $G$, it stays away from 0 and $+\infty$, qed.

As the Maaß series is of the form $\sum \varphi(\gamma g)$ where $\varphi$ satisfies (21), we might as well consider the general case. For all $x \in G$ and all $C \subset G$,

$$\varphi(\gamma x) \asymp \int_C \varphi(\gamma x g)dg = \int_{\gamma x C} \varphi(g)dg = \int_{\gamma C'} \varphi(g)dg$$

where $C' = xC$, which reduces to the proof to the convergence of the series

$$\sum_{\Gamma_\infty \backslash \Gamma} \int_{\gamma C'} \varphi(g)dg\,.$$

As $C$ is only required to be compact and have measure $> 0$ and as $\Gamma$ is discrete, we may suppose that $C$ is sufficiently small so that, for given $x$, the $\gamma C'$ are pairwise disjoint. Convergence of the previous series then amounts to saying that $\varphi(g)$ is integrable over $\Gamma_\infty \backslash \Gamma C'$.

But the map $g \longrightarrow gi = z$ from $G$ to $P$ transforms $C'$ into a compact subset $W$ of $P$ and $\Gamma C'$ into a set $\Gamma W$ contained in the strip $\mathrm{Im}(z) < T$ for some $T < +\infty$ by lemma 1 of n° 19, (i). Therefore, the set $\Gamma C'$ is contained in the *complement* of a Siegel domain. So it suffices to show that $\varphi(g)$ is integrable over the set of $g = uhk$ where $u$ stays in a compact set (since integration is mod $\Gamma_\infty$) and where $\alpha(h) < T$. As was seen above,

$$dg = \alpha(h)^{-1} du dh dk \quad \text{for } g = uhk\,.$$

Hence convergence reduces to that of the integral

$$\int_a^b du \int_K \varphi(k)dk \int_{\alpha(h)<T} \alpha(h)^{s-1} dh$$

where $[a, b]$ is a compact interval of $\mathbb{R}$. For $h = \mathrm{diag}(t, t^{-1})$, $\alpha(h)^{s-1} = t^{2(s-1)}$ and $dh = d^*t$. Thus the condition $\mathrm{Re}(s) > 1$ follows, ending the proof of the theorem.

The latter applies to series with respect to the point at $\infty$, but a similar series can be associated to any other parabolic cusp $\xi = x\infty$ of $\Gamma$. It suffices to apply the formula to the group $\Gamma' = x^{-1}\Gamma x$ for which $\infty$ is a parabolic fixed point. This gives the series

$$\sum_{\Gamma'_\infty \backslash \Gamma'} \alpha(\gamma' g)^s = \sum_{\Gamma_\xi \backslash \Gamma} \alpha\left(x^{-1}\gamma x g\right)^s\,.$$

To get a $\Gamma$-invariant function, we replace $g$ by $x^{-1}g$, whence the series

$$\begin{aligned} \sum \alpha\left(x^{-1}\gamma g\right)^s &= \sum \left|J(x^{-1}; \gamma z)\right|^{-2s} \mathrm{Im}(\gamma z)^s = \\ &= \sum \mathrm{Im}\left(x^{-1}\gamma z\right)^s \end{aligned} \tag{21.22}$$

where summation is mod $\Gamma_\xi$. By the previous theorem, the series associated to $\Gamma'$, with possibly one term removed, converges normally in every horocycle of $\widehat{P}(\Gamma')$. However, the parabolic fixed points of $\Gamma'$ are the $\xi'$ such that $x^{-1}\gamma x\xi' = \xi'$ for some $\gamma \neq \pm 1$. Hence these are the points $x\eta$ where $\eta \in \widehat{P}(\Gamma)$, which, by the way, shows that $\widehat{P}(\Gamma') = x\widehat{P}(\Gamma)$. Since $g \longmapsto x^{-1}g$ transforms the horocycles of $\Gamma'$ into those of $\Gamma$, it follows that, for any parabolic fixed point $\eta$ of $\Gamma$, series (22) converges normally in $\mathfrak{S}(\eta)$ unless obviously $\eta = \sigma\xi$ is equivalent to $\xi$, in which case the result continues to hold provided the term $\alpha(x^{-1}\sigma g)^s$ which increases indefinitely as $z$ tends to $\eta$ in the topology of horocycles is removed from the series.

Series (22) depends on the choice of $x$ such that $\xi = x\infty$. But since $x \longmapsto xb$ multiplies $\alpha(x^{-1}g)$ by $\alpha(b^{-1})$ for all $g$, changing $x$ multiplies (34) by a constant. A result independent of $x$ is obtained by dividing all the terms of the series by $\alpha(x^{-1})^s$, which leads to the series

$$\alpha\left(x^{-1}\right)^{-s}\sum_{\Gamma_\xi\backslash\Gamma}\alpha\left(x^{-1}\gamma g\right)^{s}=\left(c^2+d^2\right)^{s}\sum\mathrm{Im}\left(x^{-1}\gamma z\right)^{s}$$

where $(c\ d)$ is the second row of the matrix $x^{-1}$. For $(c\ d)=(1\ -\xi)$, (22) is the function

$$M_\xi(z;s)=\sum_{\Gamma_\xi\backslash\Gamma}|\gamma z-\xi|^{-2s}\,\mathrm{Im}(\gamma z)^s=\sum\mathrm{Im}\left[-1/(\gamma z-\xi)\right]^s . \tag{21.23}$$

This obviously supposes $\xi\neq\infty$. As the transformation $z\longmapsto -1/(z-\xi)$ of $G$ maps the point $\xi$ to infinity, series (23) with respect to $\Gamma$ and to $\xi$ is clearly obtained by applying $x$ to the analogous series with respect to $\Gamma'=x^{-1}\Gamma x$ and the point at infinity.

We can now return to classical Eisenstein series

$$E_{r,\infty}(z)=\sum J(\gamma;z)^{-r}$$

where summation is mod $U_\infty$ or, up to a possible factor 2, mod $\Gamma_\infty$. As

$$y^{r/2}\,|J(\gamma;z)|^{-r}=\mathrm{Im}(\gamma z)^{r/2},$$

comparison with the Maaß series is immediate: the series converges for $r>2$. The Maaß series converges normally in all parabolic cusps with vertex $\xi$ *not equivalent to* $\infty$, so that

$$\left|y^{r/2}E_{r,\infty}(z)\right|\leq M_{r/2,\infty}(z)$$

is bounded and as a result parabolic at these points. On the contrary, in a cusp at a vertex $\xi$ *equivalent to* $\infty$, we have to remove the $\nu\leq 2$ equal terms if the series is not trivially zero, for which $U_\infty\gamma\subset\Gamma_\infty$. So, for $\xi=\infty$,

$$y^{r/2}E_{r,\infty}(z)=\nu y^{r/2}+O(1)\quad\text{for large } y,$$

whence a Fourier series expansion of the form

$$E_{r,\infty}(z)=\nu+\sum_{n\geq 1}a_n\mathbf{e}(nz/\omega)=\nu+O\left[y^{-r/2}\exp(-2\pi y/\omega)\right].$$

The series is, therefore, holomorphic but not parabolic at $\infty$. On this matter, recall [n° 19, (ii)] that a holomorphic form $f$ at a fixed parabolic point $\xi$ is parabolic at $\xi$ if and only if the function $y^{r/2}|f(z)|$, and not $f(z)$, is bounded in a horocycle centered at $\xi$.

Denoting by

$$\alpha_r(g)=J(g;i)^{-r}\alpha(g)^{r/2}$$

the function of weight $r$ on $G$ associated to $y^{r/2}$, the function associated to $E_{r,\infty}(z)$ clearly becomes

$$E_{r,\infty}(g)=\sum\alpha_r(\gamma g),$$

where summation is mod $U_\infty$. To associate an Eisenstein series to a parabolic fixed point $\xi=x(\infty)$, we replace $\Gamma$ by $\Gamma'=x^{-1}\Gamma x$. This gives a series

$$E'_{r,\infty}(g) = \sum \alpha_r(\gamma' g)$$

with respect to $\Gamma'$. In accordance with the basic principle of n° 19, (ii), the expected series $E_{r,\xi}(g)$ is given by

$$E_{r,\xi}(g) = E'_{r,\infty}\left(x^{-1}g\right) .$$

Thus it easily follows that

$$E_{r,\xi}(z) = \sum_{U_\xi \backslash \Gamma} J\left(x^{-1}\gamma; z\right)^{-r} .$$

Conclusions are obviously the same as earlier: if $\xi$ is irregular and $r$ odd, the series is identically zero; if $\xi$ is regular or if $r$ is even, then

$$E_{r,\xi}(\eta) \begin{array}{ll} \neq 0 & \text{if } \eta \text{ is equivalent to } \xi , \\ = 0 & \text{otherwise} . \end{array}$$

If $f$ is an entire automorphic form of weight $r$, $f$ can therefore be made parabolic at $\xi$ by subtracting a scalar multiple of $E_{r,\xi}$ from it. This operation does not alter $f(\eta)$ at parabolic fixed points $\eta$ not equivalent to $\xi$. Hence the next result which generalizes what we already know about the modular group:

**Theorem 32.** *Let $\Gamma$ be a Fuchsian group. Every entire automorphic form of weight $r > 2$ is, in a unique way, the sum of a parabolic form and of a linear combination of Eisenstein series associated to parabolic fixed points of $\Gamma$.*

For $r$ even, the dimension of the vector space generated by the Eisenstein series is the number of classes of parabolic fixed points. For $r$ odd, it is zero if $-1 \in \Gamma$ and equal to the number of *regular* fixed points otherwise.

## 22 – Fourier Series Expansions

(i) *General method.* The computation of the Fourier series of the functions $E_r(z)$ in n° 17, (iii) uses the series $G_r(z)$ and, on this account, cannot be applied to general discrete groups. Computing the Fourier coefficients of an automorphic function is a problem encountered in many other situations, either when the assumption of holomorphy is abandoned, or when more general groups $\Gamma$ than the modular group are considered. To unify them, we consider $G$ and a series of the form

$$E_\varphi(g) = \sum \varphi(\gamma g) . \tag{22.1}$$

Summation is over $\Gamma$ in the case of Poincaré series and of a function $\varphi$ integrable over $G$. In the case of a series associated to a parabolic fixed point which will be assumed to be $\infty$, the function $\varphi$ is left $\Gamma_\infty$-invariant and summation is over $\Gamma_\infty \backslash \Gamma$. We will consider the second case since replacing $\varphi(g)$ by $\sum \varphi(\gamma g)$, where summation is over $\Gamma_\infty$, reduces the first one to it, as was seen for Poincaré series.

Expanding $E_\varphi$ in a a Fourier series in the neighbourhood of a parabolic fixed point $\xi = s(\infty)$ of $\Gamma$ reduces this case to expanding the function

$$E_\varphi(sg) = \sum_{\Gamma_\infty \backslash \Gamma} \varphi(\gamma s g)$$

at infinity. To this end, we group together the terms belonging to a same double coset $\Gamma_\infty c U_\xi$, hence of the form $\Gamma_\infty \gamma\gamma'$ with $\gamma' \in U_\xi$. For $\gamma', \gamma'' \in U_\xi$,

$$\Gamma_\infty \gamma\gamma' = \Gamma_\infty \gamma\gamma'' \iff \gamma'\gamma''^{-1} \in U_\xi \cap \gamma^{-1}\Gamma_\infty\gamma = U_\xi(\gamma)\,.$$

This relation is equivalent to $U_\xi(\gamma)\gamma' = U_\xi(\gamma)\gamma''$. So one has to sum over the classes $U_\xi(\gamma)\gamma'$, and so

$$E_\varphi(sg) = \sum_{\Gamma_\infty\gamma U_\xi} \sum_{U_\xi(\gamma)\backslash U_\xi} \varphi\left(\gamma\gamma' sg\right)\,.$$

However, the $\gamma' \in U_\xi(\gamma)$ satisfy $\gamma'\xi = \xi$ and $\gamma\gamma'\gamma^{-1}\infty = \infty$, hence leave invariant $\xi$ and $\gamma^{-1}\infty$. Hence, since a parabolic matrix other than $\pm 1$ has a unique fixed point,

$$U_\xi(\gamma) = \{e\} \quad \text{if } \gamma\xi \neq \infty \tag{22.2}$$

and summation over $\gamma'$ is then extended to $U_\xi$. Therefore,

$$E_\varphi(sg) = \sum_{\Gamma_\infty\gamma U_\xi} \sum_{U_\xi} \varphi\left(\gamma\gamma' sg\right) \quad \text{if } \xi \text{ not equivalent to } \infty\,. \tag{22.3}$$

On the contrary,

$$U_\xi(\gamma) = U_\xi \quad \text{if } \gamma\xi = \infty\,. \tag{22.4}$$

In this case, one may assume $s = \gamma^{-1}$. Thus $E_\varphi(sg) = E_\varphi(g)$. Then one might as well keep to the case $\xi = \infty$. The relation $\gamma\infty = \infty$ is equivalent to $\gamma \in \Gamma_\infty$, which gives a term equal to $\varphi(g)$ in the series $E_\varphi(g)$. If $\gamma \notin \Gamma_\infty$, then again $U_\infty(\gamma) = \{e\}$ and summing over all $\gamma' \in U_\infty$ is necessary. So

$$E_\varphi(g) = \varphi(g) + \sum_{\substack{\Gamma_\infty\gamma U_\infty \\ \gamma\notin\Gamma_\infty}} \sum_{U_\infty} \varphi\left(\gamma\gamma' g\right)\,. \tag{22.3'}$$

The formula is the same as in the previous case except for a constant term. These formal calculations are no problem if the series $E_\varphi(g)$ converges unconditionally.

In both cases, $U_\xi = sU(\omega\mathbb{Z})s^{-1}$ for some $\omega > 0$, whence

$$\sum_{U_\xi} \varphi\left(\gamma\gamma' sg\right) = \sum_{\mathbb{Z}} \varphi\left[\gamma s x(n\omega) g\right]\,.$$

Introducing the Fourier transform

$$a_t\left(g', g\right) = \int \varphi\left[g' x(u) g\right] \mathbf{e}(-tu)du \tag{22.5}$$

defined independently of all discrete groups for $g' \in G - B$,

$$a_t\left[g', x(v)g\right] = a_t\left(g', g\right)\mathbf{e}(tv)\,. \tag{22.6}$$

At least formally, the Poisson summation formula shows that

$$\sum \varphi\left[\gamma s x(n\omega) g\right] = \omega^{-1} \sum a_{n/\omega}(\gamma s, g)\,. \tag{22.7}$$

It then follows that

$$E_\varphi(sg) = \delta(\xi)\varphi(sg) + \sum_{\Gamma_\infty\gamma U_\xi} \sum_{\mathbb{Z}} \omega^{-1} a_{n/\omega}(\gamma s, g)\,, \tag{22.8}$$

where $\delta(\xi) = 1$ or $0$ according to whether $\xi$ is or is not equivalent to $\infty$, and where summation is over all cosets except the coset $\Gamma_\infty$ if $\xi$ is equivalent to $\infty$. Therefore, by (6), the expected Fourier series expansion is given by

$$E_\varphi\left[sx(u)g\right] = \delta(\xi)\varphi(sg) + \sum a_n(g)\mathbf{e}(nu/\omega)\,, \tag{22.9}$$

$$a_n(g) = \sum_{\Gamma_\infty\gamma U_\xi} \omega^{-1} a_{n/\omega}(\gamma s, g)\,, \tag{22.9'}$$

with the same restriction on the summation as above.

These formal calculations remain to be justified. Supposing the series $E_\varphi(g)$ converges normally in every compact set, the series

$$\sum_n \varphi\left[\gamma sx(\omega u + \omega n)g\right]\,,$$

obtained by grouping the terms of the coset $\Gamma_\infty\gamma U_x$ considered and by replacing $g$ by $x(\omega u)g$, converges normally in all compact subsets of $U$, so that formula (5) is well-defined. To go from here to (7), it suffices to assume that the series on the right hand side of (7) converges absolutely.[85] As going from (8) to (9) presumes a permutation of summations, the easiest is to assume that

$$\sum_{\Gamma_\infty\gamma U_\xi}\sum_{\mathbb{Z}} \left|a_{n/\omega}(\gamma s, g)\right| < +\infty\,. \tag{22.10}$$

If $\varphi = f_r$ for a $U_\infty$-invariant function $f(z)$, then

$$E_\varphi(g) = J(g;i)^{-r}E_{r,f}(z) \quad \text{where } E_{r,f}(z) = \sum_{\Gamma_\infty\backslash\Gamma} J(\gamma;z)^{-r}f(\gamma z)\,. \tag{22.11}$$

In this case,

$$\varphi\left[g'x(u)g\right] = J\left(g';z+u\right)^{-r}J(g;i)^{-r}f\left[g'(z+u)\right]\,.$$

Thus,

$$\begin{aligned} a_t\left(g',g\right) &= J(g;i)^{-r}\int J\left(g';z+u\right)^{-r}f\left[g'(z+u)\right]\mathbf{e}(-tu)du = \\ &= J(g;i)^{-r}a_t\left(g',z\right)\,, \end{aligned}$$

where we have set

$$a_t\left(g',z\right) = \int J\left(g';z+u\right)^{-r}f\left[g'(z+u)\right]\mathbf{e}(-tu)du$$

or, for $g \in G - B$ with second row $(c\, d)$ and $z \in P$,

$$\mathbf{e}(-tz)a_t(g,z) = \int_{\mathrm{Im}(\zeta)=y}(c\zeta+d)^{-r}f(g\zeta)\mathbf{e}(-t\zeta)d\zeta\,. \tag{22.12}$$

[85] For a continuous function $f$ on $\mathbb{R}$ such that the series $\sum f(x+n)$ converges normally on every compact set, $\sum f(x+n) = \sum \widehat{f}(n)\mathbf{e}(nx)$ if $\sum|\widehat{f}(n)| < +\infty$ because every periodic continuous function whose Fourier series converges absolutely is equal to its sum (Chap. VII, n° 6, theorem 2 or n° 27, theorem 24).

The integral does not depend on $x$, but on $y$ if $f$ is not holomorphic. Using ($g$) and ($g'$), and setting

(22.13') $$a_n(z) = \omega^{-1} \sum_{\Gamma_\infty \gamma U_\xi} a_{n/\omega}(\gamma s, z) = a_n(iy)\mathbf{e}(nx/\omega)\,,$$

finally,

(22.13) $$J(s;z)^{-r} E_{r,f}(sz) = \delta(\xi) f(sz) + \sum a_n(z)\,.$$

The general term on the right hand side is indeed invariant under $z \longmapsto z+\omega$, and so is the term $f(sz)$ since $sU(\omega\mathbb{Z}) = U_\infty s$ when $\xi$ is equivalent to $\infty$. If the function $f$ is holomorphic and bounded on $P$, we will see that integral (12) is independent of $y$ and zero for $t \leq 0$, the function $a_{n/\omega}(\gamma s, z)$ is proportional to $\mathbf{e}(nz/\omega)$ and (13") reduces directly to a series of the form $\sum_{n\geq 0} a_n \mathbf{e}(nz/\omega)$.

(ii) *The case of Poincaré-Eisenstein series.* Let us consider the series

$$E_{r,p}(z) = \sum_{\Gamma_\infty \backslash \Gamma} J(\gamma;z)^{-r} \mathbf{e}(\gamma z)^p \qquad (r > 2\,, \quad p \geq 0)\,.$$

This is series $E_{r,f}$ where $f(z) = \mathbf{e}(pz)$ is holomorphic and bounded on $P$. For $g \in G - B$,

(22.15) $$\mathbf{e}(-tz)a_t(g,z) = \int (c\zeta + d)^{-r} \mathbf{e}\left[p.g(\zeta) - t\zeta\right] d\zeta\,.$$

Integration is over the horizontal $\mathrm{Im}(\zeta) = y$ of $P$, but the result does not depend on $y$. $|\mathbf{e}(-t\zeta)| = \exp(2\pi t\eta)$ is bounded on all of the closed horizontal strip $B \subset P$ of finite width, and so is $f[g(\zeta)]$. Thus the function being integrated is $O(\zeta^{-r})$ at infinity on $B$ with $r > 2$. The result then follows from Cauchy.

This being so, the Bruhat decomposition

$$gz = (az+b)/(cz+d) = -1/c^2(z + d/c) + a/c$$

shows that

(22.15') $$\mathbf{e}(-tz)a_t(g,z) = c^{-r}\mathbf{e}\left[(pa+td)/c\right] \int \zeta^{-r} \mathbf{e}\left(-p/c^2\zeta - t\zeta\right) d\zeta\,.$$

As $p \geq 0$ and $-1/c^2\zeta \in P$, the function $\mathbf{e}(-p/c^2\zeta)$ is bounded on $P$. So is $\mathbf{e}(-t\zeta)$ if $t \leq 0$. The function being integrated is then holomorphic and $O(\zeta^{-r})$ on $\mathrm{Im}(z) \geq T > 0$, whence

(22.16) $$a_t(g,z) = 0 \quad \text{for } t \leq 0$$

again thanks to Cauchy.

For $t > 0$, two cases need to be distinguished.

(a) *The case of Eisenstein series $E_r(z)$.* As $f(z) = 1$,

$$\mathbf{e}(-tz)a_t(g,z) = c^{-r}\mathbf{e}(td/c) \int \zeta^{-r}\mathbf{e}(-t\zeta)d\zeta\,.$$

The function being integrated is $O(\zeta^{-r})$ at infinity on all of the half-plane $\mathrm{Im}(\zeta) < T$ and holomorphic for $\zeta \neq 0$. Hence integration over the horizontal can be replaced by integration counterclockwise around a circle centered at 0. The result is the product of $-2\pi i$ by the residue at $\zeta = 0$, namely

$$\mathbf{e}(-tz)a_t(g,z) = \frac{(-2\pi i)^r}{\Gamma(r)} c^{-r}\mathbf{e}(td/c)t^{r-1}. \tag{22.17}$$

Taking into account the factors $\omega$, expansion (13) in the neighbourhood of the cusp $\xi = s\infty$ is therefore

$$J(s;z)^{-r}E_r(sz) = \delta(\xi) + \sum_{n>0} a_n \mathbf{e}(nz/\omega), \tag{22.18}$$

where

$$a_n = \frac{(-2\pi i)^r}{\Gamma(r)}\omega^{-1} \sum_{\Gamma_\infty\gamma U_\xi} \mathbf{e}(nd/c\omega)c^{-r} \tag{22.18'}$$

and where $(c\ d)$ is the second row of the matrix $\gamma s$, but not of $\gamma$. As

$$\Gamma_\infty \gamma U_\xi s = \Gamma_\infty \gamma s U(\omega\mathbb{Z})$$

by definition of $\omega$, the coefficient $c$ of $\gamma s$ only depends on the coset $\Gamma_\infty\gamma U_\xi$. The same holds for the exponential since multiplying $\gamma$ by an element of $U_\xi = sU(\omega\mathbb{Z})s^{-1}$ adds a multiple of $c\omega$ to the coefficient $d$ of $\gamma s$. Finally, we sum over all cosets $\Gamma_\infty\gamma U_\xi$ if $\xi$ is not equivalent to $\infty$ and all cosets other than $\Gamma_\infty$ if $\xi = \infty$.

Supposing $\xi = \infty$, let us for example consider the Eisenstein series of the modular group. Summing over $\gamma$ for which $c > 0$ is sufficient. The arithmetic nature of $\Gamma$ plays a role here:

**Lemma 1** $[\Gamma = SL_2(\mathbb{Z})]$. *Let* $\gamma$ and $\gamma'$ *be two matrices of* $\Gamma$ *such that* $c > 0$, $c' > 0$. $\Gamma_\infty\gamma U_\infty = \Gamma_\infty\gamma' U_\infty$ *if and only if*

$$c' = c, \qquad d' \equiv d \bmod c. \tag{22.19}$$

Obviously, $\Gamma_\infty\gamma U_\infty = \Gamma_\infty\gamma' U_\infty$ if and only if $U_\infty\gamma U_\infty = U_\infty\gamma' U_\infty$ since $c > 0$, $c' > 0$. On the other hand,

$$x(m)\gamma x(m') = \begin{pmatrix} 1 & m \\ 0 & 1 \end{pmatrix}\begin{pmatrix} a & b \\ c & d \end{pmatrix}\begin{pmatrix} 1 & m' \\ 0 & 1 \end{pmatrix} = \begin{pmatrix} a' & b' \\ c' & d' \end{pmatrix} = \gamma'$$

is equivalent to $c' = c, a' = a + mc$, $d' = d + m'c$, hence proving the necessity of (19). Conversely, (19) implies $ad \equiv a'd' \equiv 1 \bmod c$. Applying arguments to the group $(\mathbb{Z}/c\mathbb{Z})^*$, we see that $a' \equiv a \bmod c$ is equivalent to $d' \equiv d \bmod c$. Conditions (19) are therefore sufficient, qed.

Hence

$$\begin{aligned} a_n &= (-2\pi i)^r n^{r-1}/\Gamma(r) \sum_{c>0} c^{-r} \sum_{\substack{d \bmod c \\ (c,d)=1}} \mathbf{e}(nd/c) = \\ &= (-2\pi i)^r n^{r-1}/\Gamma(r) \sum_{c>0} \Gamma_c(n) c^{-r}, \end{aligned}$$

where these are Gauss sums. Since it was shown in (9.34) that

$$\sum_{c>0} \Gamma_c(n)c^{-s} = n^{1-s}\sigma_{s-1}(n)/\zeta(s)$$

for $\mathrm{Re}(s) > 1$, the sum of the previous series is equal to

$$\frac{1}{\zeta(r)}\frac{(-2\pi i)^r}{\Gamma(r)}\sigma_{r-1}(n)$$

and we recover (17.25).

(b) *The case of Poincaré-Eisenstein series $E_{r,p}(z), p > 0$.* Now $f(z) = \mathbf{e}(pz)$ and (15') includes an integral of the form

$$\int \zeta^{-r}\mathbf{e}\left(-u\zeta - v/\zeta\right)d\zeta \tag{22.20}$$

taken along a horizontal $\mathrm{Im}(\zeta) = T$ of $P$, with $u > 0$ and $v > 0$. To reduce it to a standard form, we analyze the function under the $\int$ sign on the set

$$\mathrm{Im}(\zeta) \leq T\,, \quad |\zeta| \geq \varrho > 0\,.$$

$|\mathbf{e}(-u\zeta)| = \exp(2\pi u\eta) \leq \exp(2\pi uT)$, and as $|1/\zeta| \leq 1/\varrho$, the function being integrated is $O(\zeta^{-r})$. The integration horizontal can thus be replaced by a circle centered at 0 followed counterclockwise. Choosing the circle $\zeta = (v/u)^{\frac{1}{2}}\exp(i\varphi)$, $d\zeta = i\zeta d\varphi$ and

$$-u\zeta - v/\zeta = -2(uv)^{\frac{1}{2}}\cos\varphi\,.$$

As a result,

$$\begin{aligned} &\int \zeta^{-r}\mathbf{e}\left(-u\zeta - v/\zeta\right)d\zeta = \\ &= -i(v/u)^{(1-r)/2}\int \exp\left[-4\pi i\lambda\cos\varphi + (1-r)\varphi\right]d\varphi \end{aligned} \tag{22.21}$$

where integration is from 0 to $2\pi$ and where $\lambda = (uv)^{\frac{1}{2}}$. The simplest Bessel functions

$$\begin{aligned} J_k(w) &= \int e^{iw\sin\varphi - ki\varphi}d\varphi = i^{-k}\int e^{-iw\cos\varphi - ki\varphi}d\varphi = \\ &= (w/2)^k\sum_{n\geq 0}\left(-w^2\right)^n/2^{2n}n!(n+k)! \quad (k \geq 0) \end{aligned} \tag{22.22}$$

occur and we find

$$\int \zeta^{-r}\mathbf{e}\left(-u\zeta - v/\zeta\right)d\zeta = -i^r\,(v/u)^{(1-r)/2}\,J_{r-1}(4\pi\lambda)\,. \tag{22.23}$$

This being settled, since

$$\mathbf{e}(-tz)a_t(g,z) = c^{-r}\mathbf{e}\left[(pa+td)/c\right]\int \zeta^{-r}\mathbf{e}\left(-p/c^2\zeta - t\zeta\right)d\zeta\,,$$

$$\begin{aligned} &\mathbf{e}(-tz)a_t(g,z) = \\ &= -i^r\,(t/p)^{(r-1)/2}c^{-r}|c|^{r-1}\mathbf{e}\left[(pa+td)/c\right]J_{r-1}\left(4\pi\sqrt{tp/c^2}\right)\,. \end{aligned} \tag{22.24}$$

Then the general formula

$$J(s;z)^{-r}E_{r,f}(sz) = \delta(\xi)f(sz) + \omega^{-1}\sum a_{n/\omega}(\gamma s, z)$$

involves the exponentials

$$\mathbf{e}\left[(pa\omega + nd)/c\omega\right],$$

where $a$, $c$ and $d$ are the coefficients of $\gamma s$. Multiplying $\gamma$ on the left by a matrix $x(m) \in \Gamma_\infty$ and on the right by an element of $U_\xi$, hence of the form $sx(m'\omega)s^{-1}$ with $m, m' \in \mathbb{Z}$, the matrix $\gamma s$ is multiplied on the left by $x(m)$ and on the right by $x(m'\omega)$. As

$$\begin{pmatrix} 1 & m \\ 0 & 1 \end{pmatrix}\begin{pmatrix} a & b \\ c & d \end{pmatrix}\begin{pmatrix} 1 & m'\omega \\ 0 & 1 \end{pmatrix} = \begin{pmatrix} a+mc & ? \\ c & d+m'c\omega \end{pmatrix},$$

the value of the exponential does not change, which forebodes well for the exactness of calculations. Moreover, the coset $U_\infty\gamma U_\xi$ is characterized by the value of $c$ and the values of $a \bmod c\mathbb{Z}$ and $d \bmod c\omega\mathbb{Z}$. Then the cosets $\Gamma_\infty\gamma U_\xi$ can be grouped together in (13') according to the value of the coefficient $c$ of $\gamma s$. The sum

$$\sum_{\substack{a \bmod c\mathbb{Z} \\ d \bmod c\omega\mathbb{Z}}} \mathbf{e}\left(\frac{pa\omega + nd}{c\omega}\right) = K_{p,n}\left(U_\infty\gamma U_\xi\right) \tag{22.25}$$

occurs as a factor. Thus

$$J(s;z)^{-r}E_{r,p}(sz) = \delta(\xi)\mathbf{e}(p.sz) + \omega^{-1}\sum a_n\mathbf{e}(nz/\omega), \tag{22.26}$$

$$\begin{aligned} a_n = &-i^r(n/p\omega)^{(r-1)/2}\times \\ &\times \sum_{\Gamma_\infty\gamma U_\xi} \frac{\mathrm{sgn}(c)^r}{|c|} K_{p,n}\left(U_\infty\gamma U_\xi\right) J_{r-1}\left[4\pi\sqrt{p/nc^2}\right]. \end{aligned} \tag{22.26'}$$

It is not very easy.

For $\Gamma = SL_2(\mathbb{Z})$ and $\xi = \infty$, $s = e$, the coefficients $a$, $c$ and $d$ are integers and lemma 1 shows that sums (25) reduce to *Kloosterman sums*

$$K_n(u,v) = \sum_{\substack{x,y \bmod n \\ xy\equiv 1 \bmod n}} \mathbf{e}\left[(ux+vy)/n\right]$$

that occur elsewhere in number theory and that, for $v = 0$, reduce to Gauss sums $\Gamma$. All this can already be found in Poincaré's work of 1912. For given $n \neq 0$, the expression $\mathbf{e}(x/n)$ only depends on the class of $x \bmod n$ and as a result defines a character $x \longmapsto \mathbf{e}(x)$ of the additive group of the ring $A = \mathbb{Z}/n\mathbb{Z}$. Kloosterman sums only depend on classes of $u$ and $v$ and are then written as

$$K_A(u,v) = \int_{A^*} \mathbf{e}\left(ux + vx^{-1}\right) d^*x,$$

where integration is over the multiplicative group $A^*$ with respect to the invariant measure (weight 1 at each point) on this group. It may seem ridiculously scholarly, but shows that these sums are arithmetic analogues of classical Bessel integrals. Similarly, Gauss sums are arithmetic analogues of the $\Gamma$ function. Adelic methods would enormously simplify these calculations and enable us to deal with all congruence subgroups of $SL_2(\mathbb{Z})$ simultaneously. The same remark applies to the following.

(iii) *The case of Maaß-Selberg forms; analytic extensions.* For the functions

$$M_\infty(z;s) = \sum_{\Gamma_\infty\backslash\Gamma} \mathrm{Im}(\gamma z)^s = y^s \sum |J(\gamma;z)|^{-2s}\,,$$

$r = 0$ and $f(z) = \mathrm{Im}(z)^s$. So now, setting

$$W_s(t) = \int_{\mathbb{R}} \left(1+u^2\right)^{-s} \mathbf{e}(-tu)du = W_s(|t|) \tag{22.27}$$

for $\mathrm{Re}(s) > \frac{1}{2}$, $t \in \mathbb{R}$, an unorthodox but convenient notation, we get

$$\begin{aligned} a_t(g,z) &= \int \mathrm{Im}\left[g(z+u)\right]^s \mathbf{e}(-tu)du = \\ &= y^s \int \left[(cx+cu+d)^2 + c^2y^2\right]^{-s} \mathbf{e}(-tu)du = \\ &= y^{1-s}|c|^{-2s}\mathbf{e}(td/c)W_s(ty)\mathbf{e}(tx)\,. \end{aligned}$$

*Exercise 2.* Show that

$$\sum y^s/|z+n|^{2s} = y^{1-s}\sum W_s\left(|n|y\right)\mathbf{e}(nx)$$

pour $\mathrm{Im}(z) > 0$, $\mathrm{Re}(s) > 1$.

*Exercise 3.* As usual set $\alpha(g) = y$ for $z = gi$. Show that

$$\int \alpha\left[wx(u)g\right]^s \mathbf{e}(-tu)du = y^s W_s(ty)\mathbf{e}(tx)\,.$$

The coefficients of the Fourier series

$$M_\infty(xz;s) = \sum_{\mathbb{Z}} a_n(y;s)\mathbf{e}(nx/\omega) \tag{22.28}$$

at a parabolic cusp[86] $\xi = x\infty$, are of the form

$$a_0(y;s) = \delta(\xi)y^s + a_0(s)W_s(0)y^{1-s}\,, \tag{22.29}$$

$$a_n(y;s) = a_n(s)W_s\left(|n|y/\omega\right)(n \neq 0)\,. \tag{22.29'}$$

The first term $y^s$ on the right hand side of (29) is the contribution of the coset $\Gamma_\infty U_\xi$ in the case where $\xi$ is equivalent to $\infty$. The numerical coefficients $a_n(s)$ are the series

$$a_n(s) = \sum_{\Gamma_\infty\gamma U_\xi} |c|^{-2s}\mathbf{e}\left(nd/c\omega\right)\,, \tag{22.30}$$

where $(c\ d)$ is the second row of $\gamma x$ and where the sum is subject to the usual restrictions.

In the *case of the modular group*, lemma 1 shows that

---

[86] Change $s$ to $x$ so as to avoid confusion with the variable $s$. $x$ was changed to $s$ above in order not to have to write expressions such as $xx(t)$.

$$a_n(s) = \sum_{c>0} c^{-2s} \sum_{\substack{d \bmod c \\ (c,d)=1}} \mathbf{e}(nd/c) = \sigma_{1-2s}(|n|)/\zeta(2s)$$

if $n \neq 0$. For $n = 0$, $\Gamma_c(0) = \text{Card}[(\mathbb{Z}/c\mathbb{Z})^*] = \varphi(c)$, i.e. the Euler indicator (number of classes). Thus

$$a_0(s) = \sum \varphi(c)c^{-2s} = \zeta(2s-1)/\zeta(2s)$$

by (9.33). Hence, the "constant" term, i.e. independent of $x$, of the Fourier series of

$$M(z,s) = \sum \text{Im}(\gamma z)^s = M_\infty(z;s),$$

is

$$\zeta(2s)a_0(y;s) = \zeta(2s)y^s + \zeta(2s-1)W_s(0)y^{1-s} \tag{22.31}$$

with (see further down)

$$\begin{aligned} W_s(0) &= \pi^{\frac{1}{2}}\Gamma(s-1/2)/\Gamma(s) = \\ &= \pi^{-(s-\frac{1}{2})}\Gamma(s-1/2)/\pi^{-s}\Gamma(s). \end{aligned} \tag{22.32}$$

Introducing the function $\xi(s) = \pi^{-s/2}\Gamma(s/2)\zeta(s)$, it follows that

$$\begin{aligned} \xi(2s)M(z;s) &= \xi(2s)y^s + \xi(2-2s)y^{1-s} + \\ &+ \pi^{-s}\Gamma(s)y^{1-s}\sum_{n\neq 0}\sigma_{1-2s}(n)W_s(|n|y)\,\mathbf{e}(nx) \end{aligned} \tag{22.33}$$

or

$$\xi(2s)M(z;s) = \alpha_0(y;s) + \sum \alpha_n(y;s)\mathbf{e}(nx) \tag{22.34}$$

with

$$\alpha_0(y;s) = \xi(2s)y^s + \xi(2-2s)y^{1-s}, \tag{22.35}$$

$$\alpha_n(y;s) = \sigma_{1-2s}(n)\pi^{-s}\Gamma(s)y^{1-s}W_s(|n|y) \quad (n \neq 0). \tag{22.35'}$$

The term $\alpha_0(y;s)$ being clearly invariant under $s \longmapsto 1-s$, conjecturing that the same holds for $\alpha_n(y;s)$ makes sense. This requires a simple relation between $W_s$ and $W_{1-s}$. Integrals (27) have been the focus of extensive studies, but it is best to follow Jacquet and Langlands.[87] We start by writing that

$$\pi^{-s}\Gamma(s)\left(u^2+1\right)^{-s} = 2\int_0^{+\infty} \exp\left[-\pi t^2\left(u^2+1\right)\right]t^{2s}d^*t \quad \text{for } \text{Re}(s) > 0.$$

So

$$\begin{aligned} \frac{1}{2}\pi^{-s}\Gamma(s)W_s(y) &= \int \mathbf{e}(uy)du \int \exp\left[-\pi t^2\left(u^2+1\right)\right]t^{2s}d^*t = \\ &= \int \exp\left[-\pi\left(y^2/t^2+t^2\right)\right]t^{2s-1}d^*t \end{aligned}$$

[87] *Automorphic Forms on GL*(2) (Springer, Lecture Notes 114, 1970).

for all $y \in \mathbb{R}$. This leads to (32) for $y = 0$ and, for $y > 0$,

$$\frac{1}{2}\pi^{-s}\Gamma(s)W_s(y) = y^{s-\frac{1}{2}}\int \exp\left[-\pi y\left(t^2+t^{-2}\right)\right]t^{2s-1}d^*t\,. \tag{22.36}$$

Introducing Bessel functions

$$K_\nu(y) = \frac{1}{2}\int_0^{+\infty} \exp\left[-y\left(t+t^{-1}\right)/2\right]t^\nu d^*t\,,$$

defined for all $\nu \in \mathbb{C}$ and $y > 0$, (35') can be written as the Maaß form

$$\alpha_n(y;s) = 2|n|^{s-\frac{1}{2}}\sigma_{1-2s}(n)y^{\frac{1}{2}}K_{s-\frac{1}{2}}\left(2\pi|n|y\right) \quad (n \neq 0)\,, \tag{22.35''}$$

but it is best to preserve the $W_s$, especially so as not to fall under the temptation of becoming absorbed in the contemplation of the 12.345 formulas obtained by specialists of Bessel functions who have never come across $SL_2(\mathbb{R})$ and omit the factor $2\pi$ in the exponentials defining the Fourier transforms.

This calculation is justified provided

$$\int t^{2\,\mathrm{Re}(s)}\exp\left(-\pi t^2\right)d^*t\int \exp\left(-\pi t^2u^2\right)du < +\infty\,,$$

hence for $\mathrm{Re}(s) > 1/2$. But integral (36) converges for all $s$ since, at infinity or 0, the exponential under the $\int$ sign tends to 0 more rapidly than any power of $t$. As a result,

$$\pi^{-s}\Gamma(s)W_s(y)y^{\frac{1}{2}-s}$$

*is an entire function of $s$ invariant under $s \longmapsto 1-s$.* As $1/\Gamma(s)$ is an entire function, so is $W_s(y)$ for all $y > 0$. Observing that

$$|n|^{s-1}\sigma_{1-2s}(n) = \sum_{d_1d_2=|n|} d_1^s d_2^{1-s}$$

is invariant under $s \longmapsto 1-s$, we readily see that, for $n \neq 0$, *$\alpha_n(y;s)$ is an entire function of $s$ invariant under $s \longmapsto 1-s$.*

To deduce a similar result for the function $\xi(2s)M(z;s)$, it suffices to show that, for all $y > 0$, the series $\sum \alpha_n(y;s)\mathbf{e}(nx)$ converges normally in all compact subsets of the plane of the variable $s$. Now,

$$\frac{1}{2}\pi^{-s}\Gamma(s)W_s(y) = \exp(-2\pi y)y^{s-\frac{1}{2}}\int \exp\left[-\pi y\left(t-t^{-1}\right)^2\right]t^{2s-1}d^*t\,.$$

The exponential is a decreasing function of $y$, while, on the strip $1-\sigma \leq \mathrm{Re}(s) \leq \sigma$, the function $|t^{2s-1}|$ remains between its values at $\sigma$ and $1-\sigma$. So in a set such as

$$y \geq T > 0\,, \quad 1-\sigma \leq \mathrm{Re}(s) \leq \sigma\,, \tag{22.37}$$

there is an upper bound of the form

$$\left|\frac{1}{2}\pi^{-s}\Gamma(s)W_s(y)\right| \leq c(T;\sigma)\exp(-2\pi y)y^{\sigma-\frac{1}{2}}\,, \tag{22.38}$$

where, using (35') and a trivial upper bound of $\sigma_{1-2s}(n)$,

$$|\alpha_n(y;s)| = y^{\frac{1}{2}} O\left(n^N \exp(-2\pi|n|y)\right).$$

The precise value of the exponent $N$ does not matter much since the result obtained suffices to prove normal convergence in (37) of the Fourier series of $\xi(2s)M(z;s)$ with its "constant" term

$$\xi(2s)y^s + \xi(2-2s)y^{1-s} = \xi(2s)y^s + \xi(2s-1)y^{1-s} \tag{22.39}$$

removed. Its singularities are [n° 2, theorem 1] those of the function

$$y^s\left(\frac{1}{2s-1} - \frac{1}{2s}\right) + y^{1-s}\left(\frac{1}{2s-2} - \frac{1}{2s-1}\right). \tag{22.40}$$

The two simple poles at $s = \frac{1}{2}$ cancel out and the residues at 1 and 0 are $\frac{1}{2}$ and $-\frac{1}{2}$. In conclusion:

**Theorem 33** [$\Gamma = SL_2(\mathbb{Z})$, **Maaß**]. *The function $\xi(2s)M(z;s)$ extends analytically to a meromorphic function of $s$ invariant under $s \longmapsto 1-s$ and whose only singularities are simple poles at $s=1$ and $s=0$, where the residues are equal to $\frac{1}{2}$ and $-\frac{1}{2}$.*

Since $1/\xi(2s)$ is zero at $s=0$, this point is not a pole of $M(z;s)$. All the zeros of $\xi(2s)$ being located in $\mathrm{Re}(s) < \frac{1}{2}$, the only pole of $M(z;s)$ in $\mathrm{Re}(s) \geq \frac{1}{2}$ is $s=1$. As

$$M(z;s) = y^s + \xi(2s-1)y^{1-s}/\xi(2s) + \varphi(z;s)/\xi(2s) \tag{22.41}$$

and as

$$\varphi(z;s) = \sum_{n\neq 0} \alpha_n(y;s)\mathbf{e}(nx)$$

is an entire function of $s$, the residue of $M(z;s)$ at $s=1$ can be calculated by writing that

$$\begin{aligned}\xi(2s-1)/\xi(2s) &= [1/2(s-1) + a + \ldots] / [\xi(2) + 2\xi'(2)(s-1) + \ldots] = \\ &= 1/2\xi(2)(s-1) + b + \ldots\end{aligned}$$

where $b = a\xi(2) + \xi'(2)$, whence

$$\lim_{s=1}(s-1)M(z;s) = 1/2\xi(2) = 3/\pi.$$

Moreover,

$$\begin{aligned}&\lim M(z;s) - 3/\pi(s-1) = y + \varphi(z;1)/\xi(2) + \\ &+ \lim\left[\xi(2s-1)y^{1-s}/\xi(2s) - 1/2\xi(2)(s-1)\right].\end{aligned}$$

However,

$$\begin{aligned}\xi(2s-1)y^{1-s}/\xi(2s) &= [1/2\xi(2)(s-1) + b + \ldots][1 - (s-1)\log y + \ldots] = \\ &= 1/2\xi(2)(s-1) + b - \log y/2\xi(2) + \ldots.\end{aligned}$$

As a consequence,

$$\lim_{s=1}\left[M(z;s) - \frac{3}{\pi(s-1)}\right] = y + b - \frac{3}{\pi}\log y + \frac{1}{\xi(2)}\sum a_n(y;1)\mathbf{e}(nx),$$

a $\Gamma$-invariant result.

Similarly, in the neighbourhood of $s = \frac{1}{2}$,

$$\xi(2s-1)/\xi(2s) = \frac{-1/(2s-1) - a + \ldots}{1/(2s-1) + a + \ldots} = -1 + a^2(2s-1)^2 + \ldots,$$

$$y^{1-s} = y^{\frac{1}{2}}\left[1 + (\frac{1}{2} - s)\log y + \ldots\right],$$

$$y^s = y^{\frac{1}{2}}\left[1 + (s - \frac{1}{2})\log y + \ldots\right].$$

So

$$y^s + \xi(2s-1)y^{1-s}/\xi(2s) = (2s-1)y^{\frac{1}{2}}\log y + \ldots$$

and

$$M(z;s) = (2s-1)y^{\frac{1}{2}}\log y + \ldots + [(2s-1) + \ldots]\left[\varphi(z;\frac{1}{2}) + \ldots\right].$$

Hence $\lim M(z;s) = 0$ and

$$\lim_{s=\frac{1}{2}} M(z;s)/(2s-1) = y^{\frac{1}{2}}\log y + \sum_{n\neq 0} \alpha_n(y;\frac{1}{2})\mathbf{e}(nx).$$

Like the Maaß series, this function is $\Gamma$-invariant.

*Exercise 4.* Set $a(s) = W_s(0)\zeta(2s-1)/\zeta(2s)$. Show that

$$M(z;s) - \left[y^s + a(s)y^{1-s}\right] = O\left[y^{\frac{1}{2}}\exp(-2\pi y)\right]$$

for large $y$.

*Exercise 5.* Let

$$f(z) = \sum a_n(y)\mathbf{e}(nx)$$

be an invariant function under the modular group and $O(y^{-N})$ at infinite for all $N$. Calculating the inner product (on $\Gamma\backslash G$) of $f$ and $M(z;s)$, show that the Mellin transform of the function $a_0(y)$ can be analytically extended to a functional equation in $s \longmapsto 1-s$.

A far simpler method for obtaining a more general result than the previous theorem consists in associating the series

$$\theta_F(t;g) = \sum_{\xi\in\mathbb{Z}^2} F\left[t.g^{-1}(\xi)\right] = F(0) + \sum_{U_\infty\backslash\Gamma}\sum_{n>0} F\left[nt.g^{-1}\gamma^{-1}(e_1)\right]$$

to every function $F$ on the Schwartz space $\mathcal{S}(\mathbb{R}^2)$, where $e_1$ is the first basis vector of $\mathbb{R}^2$. $G$ acts linearly, $\Gamma$ transforms $e_1$ into primitive vectors of the lattice $\mathbb{Z}^2$ and $U_\infty$ is the stabilizer of $e_1$. Convergence is obvious for $t \neq 0$. Defining the Fourier transform by

$$\widehat{F}(x,y) = \int F(u,v)\mathbf{e}(yu - xv)\,du\,dv \tag{22.42}$$

and using the Poisson summation formula in $\mathbb{R}^2$, we get

$$\theta_F\left(t^{-1};g\right) = t^2\theta_{\widehat{F}}(t;g)\,.$$

Thus the series behaves asymptotically in the neighbourhood of $t = 0$ since this is obvious for large $t$. Setting

$$\theta_F^* = \theta_F - F(0)\,,$$

$$\Gamma_F(g;s) = \int F\left(t.g^{-1}e_1\right)t^s d^*t \quad \text{for } \operatorname{Re}(s) > 0\,,$$

the method used for the Riemann function then leads immediately to the analytic extension of the Mellin transform

$$M_F(g;s) = \int \theta_F^*(t;g)t^{2s}d^*t = \zeta(2s)\sum_{U_\infty\backslash\Gamma}\Gamma_F(\gamma g;2s)\,. \tag{22.43}$$

It has at most simple poles at 1 and 0 and

$$M_{\widehat{F}}(g;1-s) = M_F(g;s)\,. \tag{22.44}$$

Supposing that $F[k(x)] = \overline{\chi_r(k)}F(x)$ for some character $K$, $\Gamma_F(g;2s)$ is [n° 15, (i), theorem 13] the function of weight $r$ associated to

$$f(z) = J(g;i)^r\Gamma_F(g;2s)\,.$$

As

$$\Gamma_F\left[\begin{pmatrix} y^{\frac{1}{2}} & y^{-\frac{1}{2}}x \\ 0 & y^{-\frac{1}{2}}\end{pmatrix};2s\right] = \int F\left(ty^{-\frac{1}{2}}e_1\right)t^{2s}d^*t = \Gamma_F(e;2s)y^s$$

implies that

$$f(z) = \Gamma_F(e;2s)y^{s-r/2}\,,$$

$M_F(g;s)$ is then, up to the factor $\Gamma_F(e;s)\zeta(2s)$, the function $M_r(g;s)$ of weight $r$ associated to the series

$$M_r(z;s) = \sum(cz+d)^{-r}\operatorname{Im}(\gamma z)^{s-r/2}\,, \tag{22.45}$$

which is obviously zero for $r$ odd. Choosing

$$F(u,v) = \begin{matrix} (u+iv)^r\exp\left[-\pi\left(u^2+v^2\right)\right] & \text{if } r \geq 0\,, \\ (u-iv)^{-r}\exp\left[-\pi\left(u^2+v^2\right)\right] & \text{if } r \leq 0\,, \end{matrix} \tag{22.46}$$

we find

$$\Gamma_F(e;2s) = \int t^{|r|}\exp\left(-\pi t^2\right)t^{2s}d^*t = \frac{1}{2}\pi^{-s+|r|/2}\Gamma\left(s-|r|/2\right)\,.$$

On the other hand, the general relation

$$(d/du \pm id/dv)\int F(x,y)\mathbf{e}(yu-xv)dxdy =$$
$$= 2\pi\int(x\pm iy)F(x,y)\mathbf{e}(yu-xv)dxdy$$

shows that Fourier transforms (42) of functions (46) are given by $\widehat{F} = (-1)^r F$. In conclusion, for $r$ even, *the function*

$$\pi^{-s+|r|/2}\Gamma(s - |r|/2)\,\zeta(2s) M_r(z; s)$$

*extends analytically and is invariant under* $s \longmapsto 1 - s$.

We leave it to the reader to fill in the details and to generalize exercise 5 to this case by assuming $f$ to be of weight $r$. In particular, *series (45) are orthogonal to parabolic forms of weight* $r$, as well as to their conjugates. This method can also be used to compute the Fourier series of $M_r(z; s)$ and more generally of $M_F(g; s)$ for all $F \in \mathcal{S}(\mathbb{R}^2)$.

Here too, adelic calculations, inaugurated by John Tate in his thesis on zeta functions are far simpler and enable us not only to generalize results to congruence groups, but also to replace $\mathbb{Q}$ by an arbitrary algebraic number field.[88] The far harder analytic extension of Maaß-Selberg series for an arbitrary Fuchsian group was announced by A. Selberg in 1956 in a famous article whose proofs were put off till the recent publication of his complete works. But other authors have dealt with the subject.[89]

[88] R. Godement, *Analyse spectrale des fonctions modulaires* (Séminaire Bourbaki, 1964/65), I.M. Gelfand, M. Graev and I. Piateski-Shapiro, *Representation Theory and Automorphic Functions* (trad. Saunders, 1968, or Academic Press, 1990), S. Gelbart, *Automorphic Forms on Adele Groups* (Princeton UP, 1975).

[89] S. Lang, $SL_2(\mathbb{R})$, who presents Fadeev's the method, Peter Lax & Ralph S. Philips, *Scattering Theory for Automorphic Functions* (Annals of Math. Studies, 1976) T. Kubota, *Elementary Theory of Eisenstein Series* (Tokyo, Kodansha, 1973), A. Venkov, *Spectral Theory of Automorphic Functions* (Kluwer, 1990), A. Borel, *Automorphic Forms on* $SL(2, \mathbb{R})$ (Cambridge UP, 1998). And all this was generalized to general semisimple groups by R. P. Langlands at the start of the 1960s (Springer Lecture Notes, **544**, 1976).

# § 8. Hecke Theory

## 23 – Modular Forms and Dirichlet Series

(i) *Hecke series.* If

$$f(z) = \sum_{\mathbb{N}} a_n \mathbf{e}(nz) \tag{23.1}$$

is an entire modular form of weight $r$, the function

$$g(y) = f(iy) - a_0 = \sum_{n \geq 1} a_n \exp(-2\pi n y)$$

decreases rapidly as $y$ tends to $+\infty$. Since $f(-1/z) = z^r f(z)$,

$$g\left(y^{-1}\right) = (iy)^r g(y) + a_0 (iy)^r - a_0 \,,$$

and so $g(y) = a_0(iy)^{-r} - a_0 + O(y^N)$ in the neighbourhood of $y = 0$ for all $N$. Therefore, the Mellin transform

$$\Lambda_f(s) = \int_0^{+\infty} [f(iy) - a_0]\, y^s d^* y \tag{23.2}$$

falls within the compass of the general theorems of Chap. VIII, n° 12: integral (2) converges for $\mathrm{Re}(s) > r$, extends analytically to a meromorphic function on $\mathbb{C}$ and its only singularities are simple poles at $s = r$ and $s = 0$, with

$$\mathrm{Res}_r\,(\Lambda_f) = a_0 i^r \,, \quad \mathrm{Res}_0\,(\Lambda_f) = -a_0 \,. \tag{23.3}$$

Besides, the arguments used for the zeta function apply here in too obvious a manner for there to be any need to give details: introducing the entire function

$$\Lambda_f^+(s) = \int_1^{+\infty} g(y) y^s d^* y \,, \tag{23.4}$$

$$\Lambda_f(s) = \Lambda_f^+(s) + i^r \Lambda_f^+(r - s) - a_0\left[1/s + i^r/(r-s)\right] \tag{23.5}$$

for $\mathrm{Re}(s) > r$, for all $s$. Hence the functional equation

$$\Lambda_f(r - s) = i^r \Lambda_f(s)$$

and the poles and residues of $\Lambda_f$.

To compute $\Lambda_f(s)$ in terms of the coefficients $a_n$, we first write formally that

$$\Lambda_f(s) = \sum a_n \int y^s \exp(-2\pi n y) d^* y = (2\pi)^{-s}\Gamma(s) \sum a_n/n^s \,, \tag{23.6}$$

where summation is over $n \geq 1$. To justify this computation it suffices to check that

$$\sum_{n \geq 1} \int |a_n y^s \exp(-2\pi n y)|\, d^* y < +\infty \,,$$

which reduces to checking that the series $\sum a_n/n^s$ converges absolutely. For this, we need information about the order of magnitude of the $a_n$. Now, as was seen (n° 17, theorem 16) earlier, $a_n = O(n^{r/2})$ if $f$ is *parabolic*. In this case, the calculation is justified for $\mathrm{Re}(s) > 1 + r/2$, and even for $\mathrm{Re}(s) > \frac{1}{2} + r/2$ using Deligne's results. If $f$ is not parabolic, we subtract $f(\infty)E_r(z)$ from it to make it parabolic. Since, for $E_r(z)$, $a_n$ is proportional to $\sigma_{r-1}(n)$, it suffices to analyze the corresponding series, namely

$$\sum_{n\geq 1}\sum_{d|n} d^{r-1}/n^s$$

up to a constant factor. Setting $n = md$, reduces the question to the double series

$$\sum d^{r-1}/(md)^s = \sum 1/d^{s-r+1}m^s = \zeta(s-r+1)\zeta(s)\,,$$

which converges for $\mathrm{Re}(s) > r$. The associativity theorem for unconditional convergence (Chap. II, § 2, n° 18, theorem 13) then justifies (6). In conclusion:

**Theorem 34.** *Let $f(z) = \sum a_n\mathbf{e}(nz)$ be an entire (resp. parabolic) modular form of weight $r$. Then the Dirichlet series*

$$L_f(s) = \sum_{n\geq 1} a_n/n^s \tag{23.7}$$

*converges for* $\mathrm{Re}(s) > r$ *(resp.* $\mathrm{Re}(s) > (r+1)/2$*). The function*

$$\Lambda_f(s) = (2\pi)^{-s}\Gamma(s)L_f(s) \tag{23.8}$$

*extends analytically to a meromorphic (resp. entire) function on $\mathbb{C}$ whose only singularities are simple poles at 0 and $r$, where its residues equal $i^r a_0$ and $-a_0$. It satisfies the functional equation*

$$\Lambda_f(r-s) = i^r\Lambda_f(s)\,. \tag{23.9}$$

For example, if one considers the parabolic form

$$\Delta(z) = \sum \tau(n)\mathbf{e}(nz)\,,$$

one notices that the series $\sum \tau(n)/n^s$ converges for $\mathrm{Re}(s) > 13/2$, extends to an entire function and satisfies a functional equation in $s \longmapsto 12 - s$. These results were conjectured by Ramanujan in 1916 and proved by Mordell in 1917, except obviously for the abscissa of convergence which is based on rather difficult Deligne theory, and for Ramanujan's conjecture according to which non-trivial zeros of the Dirichlet series are located on the "critical line" $\mathrm{Re}(s) = 6$. It will be seen later that, besides, the series $\sum \tau(n)/n^s$ has an infinite product expansion.

(ii) *Weil series.* All of the above is in Hecke (Math. Annalen, **114**, 1937) ; André Weil (Math. Annalen, **168**, 1967) discovered that it can be generalized to series

$$L_f(s;\chi) = \sum a_n\chi(n)/n^s \tag{23.10}$$

where $\chi$ is a character modulo an arbitrary integer $m$ (§ 3, n° 9). Calculations resemble, and for good reason, those developed in n° 10 to obtain the functional equation of the Dirichlet series $L(s;\chi)$.

In fact, Weil's calculations are to a large extent valid for all solutions, not just holomorphic ones, of the functional equation $f(\gamma z) = J(\gamma;z)^r f(z)$. We set

(23.11) $$f(z;\chi) = \sum_{\substack{a \bmod m \\ (a,m)=1}} \chi(a) f\left[(z+a)/m\right] = f(z+m;\chi)\,.$$

Recall that $\chi(a) = 0$ if $a$ is not coprime to $m$. As every so often, it will be useful to only use left invariance of the function

$$f_r(g) = J(g,i)^{-r} f(z) \quad (z = gi)\,.$$

Using notations (15.16) and (15.18), the translations $z \longmapsto z + u$ are defined by matrices $x(u) \in U$ and the homothety $z \longmapsto z/m$ by the diagonal matrix[90] $h\left(m^{-\frac{1}{2}}\right)$, and $J[h(t)x(u), z] = t^{-1}$ for all $u$, $t$ and $z$. As a result,

$$\begin{aligned} f\left[(z+a)/m\right] &= f\left[h\left(m^{-\frac{1}{2}}\right) x(a) z\right] = \\ &= J\left[h\left(m^{-\frac{1}{2}}\right) x(a) g, i\right]^r f_r\left[h\left(m^{-\frac{1}{2}}\right) x(a) g\right] = \\ &= J\left[h\left(m^{-\frac{1}{2}}\right) x(a), z\right]^r J(g,i)^r f_r\left[h\left(m^{-\frac{1}{2}}\right) x(a) g\right] = \\ &= J(g,i)^r m^{r/2} f_r\left[h\left(m^{-\frac{1}{2}}\right) x(a) g\right]. \end{aligned}$$

Relation (11), multiplied by $J(g,i)^{-r}$, becomes

(23.12) $$J(g,i)^{-r} f(z;\chi) = m^{r/2} \sum_{\substack{a \bmod m \\ (a,m)=1}} \chi(a) f_r\left[h\left(m^{-\frac{1}{2}}\right) x(a) g\right].$$

The factor $m^{-r/2}$ being well-defined only for functions of weight $r$ on $G$, we are led to more generally set

(23.12') $$\varphi(g;\chi) = \sum_{\substack{a \bmod m \\ (a,m)=1}} \chi(a) \varphi\left[h\left(m^{-\frac{1}{2}}\right) x(a) g\right]$$

for all left $\Gamma$-invariant functions $\varphi$ on $G$, for the moment without any other assumption. Using the Fourier series expansion

(23.13) $$\varphi(ug) = \sum_{\mathbb{Z}} a_n(g) \mathbf{e}(u)^n\,, \quad u \in U\,,$$

whose coefficients

(23.14) $$a_n(g) = \int_{U_\infty \backslash U} \varphi(ug) \overline{\mathbf{e}(u)}^n$$

satisfy

(23.15) $$a_n(ug) = \mathbf{e}(u)^n a_n(g)\,,$$

we get

---

[90] Abhorring useless square roots, conservative arithmeticians prefer to consider the group $GL_2(\mathbb{R})$ of $2 \times 2$ matrices with determinant $> 0$, i.e. the direct product of $G$ and of $\mathbb{R}^*_+$ ; see the beginning of the next n°.

$$\begin{aligned}\varphi(g;\chi) &= \sum_{a \bmod m} \chi(a) \sum_n a_n \left[h\left(m^{-\frac{1}{2}}\right) x(a) g\right] = \\ &= \sum \chi(a) \sum a_n \left[x(a/m) h\left(m^{-\frac{1}{2}}\right) g\right] = \\ &= \sum \chi(a) \sum a_n \left[h\left(m^{-\frac{1}{2}}\right) g\right] \mathbf{e}(na/m) = \\ &= \sum a_n \left[h\left(m^{-\frac{1}{2}}\right) g\right] \sum \chi(a) \mathbf{e}(na/m) .\end{aligned}$$

Hence, introducing the Gauss sums

$$\Gamma_m(n,\chi) = \sum_{\substack{a \bmod m \\ (a,m)=1}} \chi(a) \mathbf{e}(na/m) \tag{23.16}$$

of n° 9, (v),

$$\varphi(g;\chi) = \sum_{n \neq 0} \Gamma_m(n,\chi) a_n \left[h\left(m^{-\frac{1}{2}}\right) g\right] ,$$

expression from which $a_0$ has disappeared because $\sum \chi(a) = 0$ if we assume $\chi$ is not the unit character and, a fortiori, as we will do to simplify the presentation, that $\chi$ is a *proper* character mod $m$. Theorem 5 of n° 9 then shows that

$$\varphi(g;\chi) = \Gamma_m(\chi) \sum a_n \left[h\left(m^{-\frac{1}{2}}\right) g\right] \overline{\chi(n)} \tag{23.17}$$

where $\Gamma_m(\chi) = \Gamma_m(1,\chi) = \sum \chi(a) \mathbf{e}(a/m)$ satisfies $|\Gamma_m(\chi)| = m^{\frac{1}{2}}$ by (10.6').

When $\varphi(g)$ arises from an entire modular form $f(z) = \sum a_n \mathbf{e}(nz)$ of weight $r$,

$$a_n(g) = J(g,i)^{-r} a_n \mathbf{e}(nz) , \quad z = gi ,$$

and so

$$\begin{aligned}a_n \left[h\left(m^{-\frac{1}{2}}\right) g\right] &= J\left[h\left(m^{-\frac{1}{2}}\right), z\right]^{-r} J(g,i)^{-r} m^r a_n \mathbf{e}(nz/m) = \\ &= J(g,i)^{-r} m^{r/2} a_n \mathbf{e}(nz/m) .\end{aligned}$$

Since $\varphi(g;\chi) = J(g,i)^{-r} m^{r/2} f(z;\chi)$,

$$f(z;\chi) = \Gamma_m(\chi) \sum_{n \geq 1} a_n \overline{\chi(n)} \mathbf{e}(nz/m) , \tag{23.17'}$$

a result that could have been obtained by a direct calculation starting from (11). As $a_0$ disappears from the result, $f(z;\chi)$ is an entire series in $\mathbf{e}(z/m)$ without a constant term, hence decreasing exponentially as $\operatorname{Im}(z)$ tends to $+\infty$.

Let us now show that, for any $\Gamma$-invariant function $\varphi(g)$,

$$\varphi(wg;\chi) = \chi(-1) \varphi(g;\overline{\chi}) . \tag{23.18}$$

The left hand side of (18) is obtained by replacing $g$ with $wg$ in (12). Denoting by $a'$ a general solution of $aa' \equiv 1 \bmod m$, the right hand side of (18) is obtained by replacing $\chi(a)$ with $\chi(-a')$ in definition (12') of $\varphi(g;\chi)$ or, what is equivalent since $a \longmapsto -a'$ permutes the summation, by replacing $a$ with $-a'$ in the term $x(a/m)$. Hence to prove (18), it suffices to show that the $\varphi$ takes the same value at points

$h\left(m^{-\frac{1}{2}}\right)x(a)wg$ and $h\left(m^{-\frac{1}{2}}\right)x(-a')g$ of $G$. The only way to do this is to check that, for all $a$ and $g$,

$$h\left(m^{-\frac{1}{2}}\right)x(a)wg \in \Gamma.h\left(m^{-\frac{1}{2}}\right)x\left(-a'\right)g\,.$$

However, this means that the entries of the matrix

$$h\left(m^{-\frac{1}{2}}\right)x(a)wx\left(a'\right)h\left(m^{-\frac{1}{2}}\right) = \begin{pmatrix} a & \left(aa'-1\right)/m \\ m & a' \end{pmatrix}$$

are integers, which is clear.

When $\varphi = f_r$ is associated to a generalized modular form $f(z)$ of weight $r$, going from $g$ to $wg$ transforms $z$ into $-1/z$. As

$$J(wg,i) = J(w,z)J(g,i) = J(g,i)z\,,$$

(18) becomes

$$f\left(-1/z;\chi\right) = \chi(-1)z^r f\left(z;\overline{\chi}\right)\,. \tag{23.18'}$$

If the right hand side decreases exponentially to 0 as $y$ tends to $+\infty$, this relation shows that so does the left hand side and hence the right hand side as $y$ tends to 0. This is the case of entire modular forms and of Maaß series since going from $f(z)$ to $f(z;\chi)$ removes the constant term of polynomial growth from their Fourier series.

This being done, we associate the Mellin transform

$$\Lambda_f(g;s,\chi) = \int \varphi(hg;\chi)\alpha(h)^s dh = \Lambda_f(-g;s,\chi) \tag{23.19}$$

to $\varphi$, where $\alpha(h) = \mathrm{Im}(hi) = t^2$ for $h = h(t)$. Convergence for all $s$ is obvious if $\varphi = f_r$ for some entire modular form of weight $r$ or a Maaß series (21.17) or if, trivially, it is continuous and with compact support mod $\Gamma$ for then a continuous function of $h$ with compact support is being integrated. In fact, the correct assumption to use is $\varphi(g) = O[\alpha(g)^N]$. So using (18), we find

$$\Lambda_\varphi\left(g;s,\overline{\chi}\right) = \int \varphi\left(hg;\overline{\chi}\right)\alpha(h)^s dh = \chi(-1)\int \varphi(whg;\chi)\alpha(h)^s dh =$$
$$= \chi(-1)\int \varphi(hwg;\chi)\alpha(h)^{-s}dh\,,$$

i.e.

$$\Lambda_\varphi\left(g;s,\overline{\chi}\right) = \chi(-1)\Lambda_\varphi(wg;-s,\chi)\,. \tag{23.20}$$

To obtain Weil's result, we apply (20) for $g = e$ to a function

$$\varphi(g) = J(g,i)^{-r}f(z) = J(g,i)^{-r}\sum a_n \mathbf{e}(nz)$$

of weight $r$. For $h = h(t)$,

$$\varphi(h;\chi) = J(h,i)^{-r}m^{r/2}f(iy;\chi) = m^{r/2}t^r f\left(t^2 i;\overline{\chi}\right)$$

by (12) and (12'), and setting by definition,

$$\Lambda_f(s;\chi) = \int_0^{+\infty} f(iy;\chi)y^s d^*y$$

for every generalized modular form $f(z)$, for example entire or Maaß, for which the integral converges, we get

$$\begin{aligned}\Lambda_\varphi\left(e;s,\overline{\chi}\right) &= m^{r/2}\int t^r f\left(t^2 i;\overline{\chi}\right)t^{2s}d^*t = \\ &= \frac{1}{2}m^{r/2}\int f\left(iy;\overline{\chi}\right)y^{s+r/2}d^*y = \frac{1}{2}m^{r/2}\Lambda_f\left(s+r/2;\overline{\chi}\right).\end{aligned}$$

On the right hand side of (20), we need to compute

$$\Lambda_f(w;-s,\chi) = \int \varphi(hw;\chi)\alpha(h)^{-s}dh\,,$$

where

$$m^{-r/2}\varphi(hw;\chi) = m^{r/2}J(hw,i)^{-r}f(hwi;\chi) = i^{-r}t^r f(hi;\chi) = i^{-r}y^{r/2}f(iy;\chi)\,.$$

Hence

$$\begin{aligned}m^{-r/2}\Lambda_f(w;-s,\chi) &= i^{-r}\int y^{r/2}f(iy;\chi)y^{-s}d^*y = \\ &= i^{-r}\int f(iy;\chi)y^{r/2-s}d^*y = i^{-r}\Lambda_f(r/2-s,\chi)\,.\end{aligned}$$

Relation (20) then becomes

$$\Lambda_f\left(s+r/2,\overline{\chi}\right) = \chi(-1)\Lambda_f(r/2-s,\chi)$$

or, in Weil's form,

$$\Lambda_f\left(r-s,\overline{\chi}\right) = \chi(-1)\Lambda_f(s,\chi)\,. \tag{23.20'}$$

Here too, (20') can be easily deduced from (18').

As

$$\begin{aligned}\int f(iy;\chi)y^s d^*y &= \Gamma_m(\chi)\sum a_n\int \overline{\chi(n)}\mathbf{e}\left(-2\pi ny/m\right)y^s d^*y = \\ &= \Gamma_m(\chi)\left(2\pi/m\right)^{-s}\Gamma(s)\sum a_n\overline{\chi(n)}n^{-s}\,,\end{aligned}$$

where summation is over $n > 0$, the usual arguments together with (14) lead to the final result:

**Theorem 35 (A. Weil).** *Let $f(z) = \sum_{n\geq 0} a_n\mathbf{e}(nz)$ be an entire modular form of weight $r$ and $\chi$ a proper character modulo $m > 1$. The function*

$$f(z;\chi) = \sum_{\substack{a \bmod m \\ (a,m)=1}} \chi(a)f\left[(z+a)/m\right]$$

*satisfies functional equations*

$$f(z+m;\chi) = f(z;\chi)\,, \quad f\left(-1/z;\overline{\chi}\right) = \chi(-1)z^r f(z;\chi)\,.$$

*The integral*

$$\begin{aligned}\Lambda_f(s;\chi) &= \int f(iy;\chi)y^s d^*y = \\ &= \Gamma_m\left(\overline{\chi}\right)\left(2\pi/m\right)^{-s}\Gamma(s)\sum_{n\geq 1} a_n\chi(n)/n^s \end{aligned}\tag{23.21}$$

*converges for all $s \in \mathbb{C}$, is an entire function and*

$$\Lambda_f(r-s;\chi) = i^r\chi(-1)\Lambda_f(s;\overline{\chi}) \,. \tag{23.22}$$

*Exercise.* Write down (21) explicitly when $f$ is an Eisenstein series.

Weil's far more complete results are mostly about congruence groups and admit a converse which says that if the series $\sum a_n\chi(n)/n^s$, with given $a_n = O(n^k)$, satisfy (22) for "sufficiently many" characters $\chi$, then $f(z) = \sum a_n \mathbf{e}(nz)$ is a modular form[91] for a congruence group. Weil's aim was to show that zeta functions of elliptic curves over $\mathbb{Q}$ are obtained in this manner using modular forms with respect to congruence subgroups. Once you get on this slippery slope, you run the risk of going, like Deligne, Serre and others, as far as the now proved Taniyama, Shimura and Weil conjectures, Fermat's theorem, etc. Mercy!

(iii) *Generalization to non-holomorphic forms.* Apart from functional equation (22), we would like to put $\Lambda_\varphi(g;s;\chi)$ in the form of a Dirichlet series for an "arbitrary" function $\varphi$ on $\Gamma\backslash G$ as we did when $\varphi$ arises from an entire modular form. This is possible using an assumption known to hold in the case of Maaß-Selberg series.

Writing the Fourier series expansion

$$\varphi(ug) = \sum a_n(g)\mathbf{e}(nu) \tag{23.23}$$

of $\varphi$ gives

$$\begin{aligned}\varphi(ug;\chi) &= \Gamma_m(\chi)\sum a_n\left[h\left(m^{-\frac{1}{2}}\right)x(u)g\right]\overline{\chi(n)} = \\ &= \Gamma_m(\chi)\sum a_n\left[x(u/m)h\left(m^{-\frac{1}{2}}\right)g\right]\overline{\chi(n)} = \\ &= \Gamma_m(\chi)\sum a_n\left[h\left(m^{-\frac{1}{2}}\right)g\right]\overline{\chi(n)}\mathbf{e}(nu/m)\,.\end{aligned}$$

Thus

$$\begin{aligned}\Lambda_\varphi(g;s,\chi) &= \int \varphi(hg;\chi)\alpha(h)^s dh = \\ &= \Gamma_m(\chi)\int\sum a_n\left[h\left(m^{-\frac{1}{2}}t\right)g\right]\overline{\chi(n)}t^{2s}d^*t = \\ &= m^s\Gamma_m(\chi)\sum\overline{\chi(n)}\int a_n\left[h(t)g\right]t^{2s}d^*t\,.\end{aligned}$$

When $\varphi(g) = J(g,i)^{-r}f(z) = J(g,i)^{-r}\sum_n \mathbf{e}(nz)$ arises from an entire modular form $f$ of weight $r$,

$$a_n(g) = J(g,i)^{-r}a_n\exp(-2\pi ny)\mathbf{e}(nx)\,. \tag{23.24}$$

So

[91] A. Weil, *Über die Bestimmung Dirichletscher Reihen durch Funktionalgleichungen* (Math. Annalen, **168**, 1967, pp. 255–261). See also Andrew Ogg, *Modular Forms and Dirichlet Series* (Benjamin, 1969) and Serge Lang, *Modular Forms* (Springer, 1976). All this has been fully revolutionized in an adelic framework by Hervé Jacquet and R. P. Langlands, *Automorphic Forms on GL*(2) (Springer, 1970, Lecture Notes n° **114**, 548 pp.) and Hervé Jacquet, *Automorphic Forms on GL*(2), *Part II* (Lecture Notes n° **278**, 1972).

$$a_n\left[h(t)g\right] = J(g,i)^{-r}a_n t^r \mathbf{e}\left(nt^2z\right)$$

and

$$\int a_n\left[h(t)g\right]t^{2s}d^*t = J(g,i)^{-r}\int a_n \mathbf{e}\left(nt^2z\right)t^{2s+r}d^*t\,.$$

As

$$\int_0^{+\infty}\exp(-wt)t^s d^*t = \Gamma(s)w^{-s} \quad \text{for } \operatorname{Re}(w) > 0\,,$$

and equality showed in Chap. VIII, n° 10, it follows that

$$\int a_n\left[h(t)g\right]t^{2s}d^*t = \frac{1}{2}J(g,i)^{-r}a_n\Gamma\left(s+r/2\right)\left(-2\pi inz\right)^{-r/2-s}\,.$$

In particular, for $g = e$ and $z = i$,

$$\begin{aligned}\int a_n\left[h(t)\right]t^{2s}d^*t &= \frac{1}{2}i^{-r}a_n\Gamma\left(s+r/2\right)\left(2\pi n\right)^{-r/2-s} = \\ &= \frac{1}{2}i^{-r}\left(2\pi\right)^{-s-r/2}\Gamma\left(s+r/2\right)a_n/n^{s+r/2}\,,\end{aligned}$$

whence a Dirichlet series in accordance with the previous theorem.

In the case of a function $\varphi(g) = J(g,i)^{-r}f(z) = \sum a_n(g)$ associated to a not necessarily holomorphic function $f$, the previous calculation again leads to a Dirichlet series if, instead of (24), we have a relation of the form

$$a_n(g) = J(g,i)^{-r}a_n W\left(2\pi|n|y\right)\exp(2\pi inx)$$

i.e.

$$f(z) = \sum_{\mathbb{Z}} a_n W\left(2\pi|n|y\right)\mathbf{e}(nx)\,, \tag{23.25}$$

with constants $a_n$ and a function $W(2\pi|n|y)\mathbf{e}(nx)$ replacing the function $\mathbf{e}(nz) = \exp(-2\pi ny)\mathbf{e}(nx)$ of (24) for $n > 0$, as in the case of the Maaß series. Indeed relation (25) shows that

$$\begin{aligned}\int a_n\left[h(t)g\right]t^{2s}d^*t &= \int J(g,i)^{-r}t^r a_n W\left(2\pi|n|t^2y\right)\mathbf{e}\left(nt^2x\right)t^{2s}d^*t = \\ &= \frac{1}{2}J(g,i)^{-r}a_n|n|^{-s-r/2}(2\pi)^{-s-r/2} \\ &\quad\times\int W(ty)\mathbf{e}(tx)t^{s+r/2}d^*t\end{aligned}$$

and in particular that for $g = e$,

$$\int a_n\left[h(t)\right]t^{2s}d^*t = \frac{1}{2}(2\pi)^{-s-r/2}\Gamma_W\left(s+r/2\right)a_n|n|^{-s-r/2} \tag{23.26}$$

where

$$\Gamma_W(s) = \int W(t)t^s d^*t \tag{23.27}$$

is the Mellin transform of $W$.

In the case of the Maaß series $M_\infty(z;u)$, for $u > \frac{1}{2}$, the function

$$W(t) = W_u(t) = \int \left(1+x^2\right)^{-u} \mathbf{e}(-tx)dx$$

is the Fourier transform of $(1+x^2)^{-u}$. By the change of variable $1 + x = 1/y$, its Mellin transform is

$$\begin{aligned}\int_0^{+\infty} \left(1+x^2\right)^{-u} x^s d^*x &= \frac{1}{2}\int_0^{+\infty} (1+x)^{-u} x^{s/2} d^*x = \\ &= \frac{1}{2}\int_0^1 y^{u-s/2-1}(1-y)^{s/2-1}dy = \\ &= \frac{1}{2}\Gamma\left(u - s/2\right)\Gamma\left(s/2\right)/\Gamma(u)\,.\end{aligned}$$

The general relation

$$\pi^{-s/2}\Gamma(s/2)\Gamma_f(1-s) = \pi^{-(1-s)/2}\Gamma\left[(1-s)/2\right]\Gamma_{\hat{f}}(s)$$

of n° 1 of this chapter, applied to $f(x) = (1+x^2)^{-u}$, then shows that[92]

$$\begin{aligned}&\pi^{-(1-s)/2}\Gamma\left[(1-s)/2\right]\Gamma_W(s) = \\ &= \tfrac{1}{2}\pi^{-s/2}\Gamma(s/2)\Gamma\left[u-(1-s)/2\right]\Gamma\left[(1-s)/2\right]/\Gamma(u)\,.\end{aligned}$$

Thus

$$\Gamma_W(s) = \pi^{\frac{1}{2}-s}\Gamma\left(u - \frac{1-s}{2}\right)\Gamma\left(\frac{s}{2}\right)\Big/\,\Gamma(u) \quad \text{if } W = W_u\,.$$

This relation makes it possible to compute the series $L_f(s;\chi)$ corresponding to Maaß functions $f(z) = M(z;u)$. The result is not of much interest since, as was seen in n° 22, eq. (35'), in this case

$$\xi(2u)a_n(y;u) = \sigma_{1-2u}(n)\pi^{-u}\Gamma(u)y^{1-u}W_u\left(|n|y\right)\,.$$

From what precedes, this is a function whose Mellin transform at $y$ is

$$\sigma_{1-2u}(n)\pi^{-u}\Gamma(u)|n|^{u-s-1}\Gamma_W(s+1-u)\,.$$

We, therefore, obtain, up to factors $\Gamma$, the series $\sum \sigma_{1-2u}(n)/|n|^{s+1-u}$, whose functional equation is obvious. But as an exercise, the reader can give the details of all the computations in this case.

The existence of a Fourier series like (25) means that the integral

$$\int_{x \bmod \mathbb{Z}} f(x+iy)\overline{\mathbf{e}(nx)}dx$$

only depends on the product $|n|y$. If $W_f(2\pi y)$ denotes its value for $n = 1$, then

$$\int_0^1 f(x+iy)\mathbf{e}(-nx)dx = W_f\left(2\pi|n|y\right)\,.$$

This property is related to Jacquet-Langlands "Whittaker models".

[92] Restrictions obviously need to be imposed on $s$ and $u$ to ensure that the integrals converge.

## 24 – Hecke Operators

(i) *Operators $T(x)$ for an abstract group.* If a function $\varphi(g)$ on a group $G$ is invariant under a subgroup $\Gamma$ of $G$, it is no longer in general the case for its left translations under $L(x)\varphi : g \longmapsto \varphi(x^{-1}g)$. Indeed

$$L(\gamma)L(x)\varphi = L(x)L\left(x^{-1}\gamma x\right)\varphi\,,$$

so that invariance of $L(x)\varphi$ under $\gamma$ presupposes $x^{-1}\gamma x \in \Gamma$ and so

$$\gamma \in x\Gamma x^{-1} \cap \Gamma = \Gamma(x)\,.$$

Hence only

$$\gamma \in \Gamma(x) \Longrightarrow L(\gamma x)\varphi = L(x)\varphi \tag{24.1}$$

can be guaranteed. Article 1 of group theory rules having been recalled, let us recall article 2 (Poisson-Eisenstein law): to make $L(x)\varphi$ $\Gamma$-invariant, replacing it by the sum

$$T(x)\varphi = \sum L(\gamma x)\varphi \tag{24.2}$$

extended to $\gamma \in \Gamma$. But nothing ensures the convergence of the series, everything even shows that in general it diverges. Formula (2) is nonetheless well-defined if we only consider matrices $x$ for which series (2) can be replaced by a finite sum. Since, by (1), $L(\gamma x)\varphi$ only depends on the coset $\gamma\Gamma(x)$, it therefore suffices to assume that there finitely many such cosets in $\Gamma$. In fact, as will be seen, it is better to assume that

$$\Gamma(x) \quad \text{and} \quad \Gamma(x^{-1}) \quad \text{have finite index in } \Gamma \tag{24.3}$$

if we want results. We then also have

$$\left[x\Gamma x^{-1} : \Gamma(x)\right] < +\infty\,, \quad \left[x^{-1}\Gamma x : \Gamma\left(x^{-1}\right)\right] < +\infty \tag{24.3'}$$

because the automorphism $g \longmapsto xgx^{-1}$ maps $\Gamma$ onto $x\Gamma x^{-1}$ and $\Gamma(x^{-1})$ onto

$$x\Gamma\left(x^{-1}\right)x^{-1} = x\left(x^{-1}\Gamma x \cap \Gamma\right)x^{-1} = \Gamma \cap x\Gamma x^{-1} = \Gamma(x)\,.$$

From now on

$$C(x) = \Gamma x^{-1}\Gamma\,. \tag{24.4}$$

The group $\Gamma$ acts on the right and on the left on $C(x)$, so there is a quotient set $\Gamma\backslash C(x)$, which is also the quotient of $x^{-1}\Gamma$ by the equivalence relation $\Gamma x^{-1}\gamma = \Gamma x^{-1}\gamma'$. However this is equivalent to $x^{-1}\gamma'\gamma^{-1}x \in \Gamma$, i.e. to $\gamma'\gamma^{-1} \in x\Gamma x^{-1} \cap \Gamma = \Gamma(x)$. Hence a bijection

$$\Gamma(x)\backslash\Gamma \longrightarrow \Gamma\backslash C(x) \tag{24.5}$$

transforming the coset $\Gamma(x)\gamma$ into the coset $\Gamma x^{-1}\gamma$.

We will denote by $\Gamma'$ the set of $x$ satisfying (3). These are the elements of $G$ for which operators

$$T(x)\varphi = \sum_{\gamma\Gamma(x)} L(\gamma x)\varphi \tag{24.6}$$

can be defined or, more explicitly, using map (5),

$$T(x)\varphi(g) = \sum_{\gamma\Gamma(x)} \varphi\left(x^{-1}\gamma^{-1}g\right) = \sum_{\Gamma(x)\gamma} \varphi\left(x^{-1}\gamma g\right) = \sum_{\Gamma\backslash C(x)} \varphi(ug)\,. \tag{24.6'}$$

A first trivial property of operators $T(x)$ is that *they commute with right translations*, which act on left $\Gamma$-invariant functions. Another is that $T(x)$ *only depends on the double coset* $C(x) = \Gamma x^{-1}\Gamma$ as shown by the latter sum (6'). Others will be useful for the modular group, but can be proved in all generality, which avoids questions unrelated to the problem or restrictions to operators associated to elements $x \in \Gamma'$ that are too specific to make calculations comprehensible.[93]

**Lemma 1.** *$\Gamma'$ is a subgroup of $G$.*

Clearly, $x \in \Gamma'$ implies $x^{-1} \in \Gamma'$. On the other hand,

$$\Gamma(xy) = xy\Gamma y^{-1}x^{-1} \cap \Gamma = x\left(y\Gamma y^{-1} \cap x^{-1}\Gamma x\right)x^{-1}.$$

But if $\Gamma(x)$ has finite index in $\Gamma$, $y\Gamma(x)y^{-1}$ has finite index in $y\Gamma y^{-1}$, so $\Gamma \cap y\Gamma(x)y$ is of finite index in $\Gamma \cap y\Gamma y^{-1} = \Gamma(y)$. However,

$$\begin{aligned}\Gamma \cap y\Gamma(x)y^{-1} &= \Gamma \cap y\left(\Gamma \cap x\Gamma x^{-1}\right)y^{-1} = \\ &= \Gamma \cap y\Gamma y^{-1} \cap yx\Gamma x^{-1}y^{-1} = \Gamma(y) \cap yx\Gamma x^{-1}y^{-1}.\end{aligned}$$

As this group is contained in $\Gamma$, it is equal to

$$\Gamma(y) \cap \Gamma \cap yx\Gamma x^{-1}y^{-1} = \Gamma(y) \cap \Gamma(yx)\,,$$

which is of finite index in $\Gamma(y)$, hence in $\Gamma$. As it contains $\Gamma(yx)$, it follows that $\Gamma(yx)$ is nenecessarily of finite index in $\Gamma$. So $yx$ satisfies the first condition (3), and so does $xy$, thus also $xy^{-1}$, qed.

**Lemma 2.** *For all $x, y \in \Gamma'$, the operator $T(x)T(y)$ is a linear combination with integer coefficients $\geq 0$ of $T(z), z \in \Gamma'$.*

By (5),

$$T(x)T(y)\varphi(g) = \sum_{\Gamma\backslash C(x)} T(y)\varphi(ug) = \sum_{\Gamma\backslash C(x)} \sum_{\Gamma\backslash C(x)} \varphi(uvg)\,.$$

The result is of the form $\sum c(x,y;w)\varphi(wg)$ with integers $c(x,y;w)$ not depending on $\varphi$ for any $w \in \Gamma'$. To calculate them, we count the ordered pairs $(u,v) \in C(x)\times C(y)$ such that $uv = w$ taking into account the fact that only the *cosets* $\Gamma u$ and $\Gamma v$ matter. Since replacing $u$ by $\gamma u$ replaces $w$ by $\gamma w$, the product $uv = w$ only occurs via the coset $\Gamma w$, which means that

$$c(x,y;\gamma w) = c(x,y;w)\,. \tag{24.7}$$

The substance of the proof consists in showing that we also have

$$c(x,y;w\gamma) = c(x,y;w)\,. \tag{24.7'}$$

[93] In 2000 or thereabout, some authors still for example, replace summation over $\Gamma(x)\backslash\Gamma$ by summation over a "system of representatives" of cosets $\Gamma(x)\gamma$ in $\Gamma$, i.e. over $\gamma_i$ such that each coset $\Gamma(x)\gamma$ contains a unique $\gamma_i$. Quotient sets were invented more than sixty years ago to get rid of this type of calculations.

As $C(x)^{-1} = C(x^{-1})$, $uv = w$ implies that

$$v = u^{-1}w \in C\left(x^{-1}\right) w \cap C(y)\,,$$

where this is a left $\Gamma$-invariant set dependent only on $\Gamma w$. Conversely, every element $v$ of this set is in $C(y)$ and satisfies $uv = w$ for some $u \in C(x)$. However, knowing $v$ determines $\Gamma u = \Gamma wv^{-1}$. Replacing $v$ by $\gamma v$ replaces $u = wv^{-1}$ by $u\gamma^{-1} = wv^{-1}\gamma^{-1} \in wv^{-1}\Gamma$, and conversely. So, for given $\Gamma v$, the cosets $\Gamma u$ involved in the calculation of $c(x,y;w)$ are those having non-trivial intersection with $wv^{-1}\Gamma$, i.e. those corresponding to elements of $\Gamma wv^{-1}\Gamma$. There are

$$\text{Card}\left(\Gamma\backslash\Gamma wv^{-1}\Gamma\right) = \text{Card}\left[\Gamma\backslash\Gamma w(\Gamma v)^{-1}\right]$$

of them. Hence

$$c(x,y;w) = \sum_{\Gamma v \subset C(x)w\cap C(y)} \text{Card}\left[\Gamma\backslash\Gamma w(\Gamma v)^{-1}\right]\,, \tag{24.8}$$

where summation is over cosets $\Gamma v$.

To deduce (7'), we write

$$\begin{aligned} c(x,y;w\gamma) &= \sum_{\Gamma v\subset C(x)w\gamma\cap C(y)} \text{Card}\left[\Gamma\backslash\Gamma w\gamma(\Gamma v)^{-1}\right] = \\ &= \sum_{\Gamma v\subset [C(x)w\cap C(y)]\gamma} \text{Card}\left[\Gamma\backslash\Gamma w\left(\Gamma v\gamma^{-1}\right)^{-1}\right] = \\ &= \sum_{\Gamma v\gamma^{-1}\subset C(x)w\cap C(y)} \text{Card}\left[\Gamma\backslash\Gamma w\left(\Gamma v\gamma^{-1}\right)^{-1}\right]. \end{aligned}$$

The sum obtained can be deduced from (8) by the permutation $\Gamma v \longmapsto \Gamma v\gamma^{-1}$, whence (7').

This being settled, let us calculate the contribution of a double coset

$$\Gamma w\Gamma = \bigcup_{\Gamma(w^{-1})\backslash\Gamma} \Gamma w\gamma$$

to the sum

$$T(x)T(y)\varphi(g) = \sum c(x,y;w)\varphi(wg)$$

extended to all cosets $\Gamma w \in \Gamma'$, but with finitely many non-trivial coefficients. It is clearly

$$c(x,y;w) \sum_{\Gamma(w^{-1})\backslash\Gamma} \varphi(w\gamma g) = c(x,y;w)T\left(w^{-1}\right)\varphi(g)\,.$$

Summing over all $\Gamma w\Gamma$ now gives the final formula

$$T(x)T(y) = \sum_{\Gamma\backslash\Gamma'/\Gamma} c(x,y;w)T\left(w^{-1}\right). \tag{24.9}$$

This finishes an "elementary" proof in which, believe the author, it is easy to get lost in dead-ends.

(ii) *Operators $T(x)$ for a locally compact group.* We now assume $G$ to be locally compact unimodular and $\Gamma$ to be discrete.

**Lemma 3.** *For all $x \in \Gamma'$, the operator $T(x)$ maps $L(\Gamma\backslash G)$ to $L(\Gamma\backslash G)$.*

For $\varphi \in L(\Gamma\backslash G)$, let $M$ be a compact set such that $\varphi$ is zero outside $\Gamma M$. $T(x)\varphi(g) \neq 0$ only if $x^{-1}\Gamma g \# \Gamma M$, hence only if $g \in \Gamma x \Gamma M = \Gamma x \Gamma x^{-1}.xM$. By (3'), $x\Gamma x^{-1}$ is a finite union of cosets $\Gamma(x)h_i$. Thus $\Gamma x \Gamma x^{-1}.xM = \bigcup \Gamma h_i xM = \Gamma M'$, where $M' = \bigcup h_i xM$ is compact, qed.

**Theorem 36.** *For all $x \in \Gamma'$ and all $p \geq 1$, the operator $T(x) : L(\Gamma\backslash G) \longrightarrow L(\Gamma\backslash G)$ has a continuous extension to $L^p(\Gamma\backslash G)$ and*

$$(T(x)\varphi|\psi) = \left(\varphi|T\left(x^{-1}\right)\psi\right) \tag{24.10}$$

*for all $\varphi \in L^p(\Gamma\backslash G)$ and $\psi \in L^q(\Gamma\backslash G)$. In particular,*

$$T(x)^* = T\left(x^{-1}\right) \tag{24.10'}$$

*in $L^2(\Gamma\backslash G)$.*

Let us start by remarking that the automorphism $g \longmapsto x^{-1}gx$ preserves the measure $dg$ and that its inverse transforms $\Gamma$ into $x\Gamma x^{-1}$. Hence for all $p \geq 1$ we get an isomorphism from $L^p(\Gamma\backslash G)$ onto $L^p(x\Gamma x^{-1}\backslash G)$ by transforming every function $\varphi(g)$ of the former space to a function $\varphi(x^{-1}gx)$ of the latter one.[94] It follows that

$$\begin{aligned}\int_{\Gamma\backslash G} |\varphi(g)|^p\, dg &= \int_{x\Gamma x^{-1}\backslash G} \left|\varphi\left(x^{-1}gx\right)\right|^p dg = \\ &= \int_{x\Gamma x^{-1}\backslash G} \left|\varphi\left(x^{-1}g\right)\right|^p dg\end{aligned} \tag{24.11}$$

since quotient measures are right invariant. The same argument implies that

$$\int_{\Gamma\backslash G} \varphi(g)dg = \int_{x^{-1}\Gamma x\backslash G} \varphi(xg)dg \tag{24.11'}$$

for all $\varphi \in L^1(\Gamma\backslash G)$.

---

[94] The attractive formula

$$\int_{\Gamma\backslash G} \varphi\left(x^{-1}g\right)\psi(dg) = \int_{\Gamma\backslash G} \varphi(g)\psi(xg)dg\,,$$

where $\varphi$ and $\psi$ are left $\Gamma$-invariant and where $x \notin \Gamma$, is not well-defined since the function $g \longmapsto \varphi(x^{-1}g)$ is not $\Gamma$-invariant. But formula (11') is correct. More generally: let $G$ be a lcg, $H$ a closed subgroup of $G$ and $s$ an automorphism of $G$ preserving the measure $dg$. Then, for any left $H$-invariant function $\varphi$,

$$\int_{H\backslash G} \varphi(g)dg = \int_{s^{-1}(H)\backslash G} \varphi(sg)dg\,.$$

This relation is connected to integration formulas for images of measures (Chap. XI, n° 13, theorem 23) and, if $H$ is discrete, reduces to them by replacing integration over $H\backslash G$ by integration over a "fundamental domain" of $H$ in $G$. See the change of variable formula for the simplest Riemann integrals over $\mathbb{R}$.

Then let $\varphi$ and $\psi$ be two continuous functions with compact support mod $\Gamma$. (11') applied to $T(x)\varphi(g).\psi(g)$ gives

$$\int_{\Gamma\backslash G} T(x)\varphi(g).\psi(g)dg = \int_{x^{-1}\Gamma x\backslash G} T(x)\varphi(xg).\psi(xg)dg =$$

$$= \int_{x^{-1}\Gamma x\backslash G} dg \sum_{\Gamma(x)\backslash\Gamma} \varphi\left(x^{-1}\gamma xg\right)\psi(\gamma xg)$$

$$\text{car } \psi(\gamma g) = \psi(g)$$

$$= \int_{x^{-1}\Gamma x\backslash G} dg \sum_{\Gamma(x^{-1})\backslash x^{-1}\Gamma x} \varphi(\gamma g)\psi(x\gamma g)\,,$$

which is seen by transforming the sum by $\gamma \longmapsto x^{-1}\gamma x$. Thus

$$\int_{\Gamma\backslash G} T(x)\varphi(g).\psi(g)dg = \int_{\Gamma(x^{-1})\backslash G} \varphi(g)\psi(xg)dg =$$

$$= \int_{\Gamma\backslash G} \varphi(g)dg \sum_{\Gamma(x^{-1})\backslash\Gamma} \psi(x\gamma g) =$$

$$= \int_{\Gamma\backslash G} \varphi(g).T\left(x^{-1}\right)\psi(g)dg\,.$$

Hence, replacing $\psi$ with its conjugate,

$$(T(x)\varphi|\psi) = \left(\varphi|T\left(x^{-1}\right)\psi\right)\,, \tag{24.12}$$

where this is the inner product on $L^2(\Gamma\backslash G)$ or, more generally, the duality between $L^p$ and $L^q$. Therefore, to finish the proof, the continuity of $T(x)$ on $L^p(\Gamma\backslash G)$ remains to be shown.

To do this, we start from the relation (same proof)

$$\int_{\Gamma(x)\backslash G} L(x)\varphi(g).\psi(g)dg = \int_{\Gamma\backslash G} T(x)\varphi(g).\psi(g)dg\,,$$

where $\varphi$ and $\psi$ are continuous with compact support mod $\Gamma$. It shows that $|(T(x)\varphi|\psi)|$ is bounded above by the product of the norms of $L(x)\varphi$ on $L^p(\Gamma(x)\backslash G)$ and of $\psi$ on $L^q(\Gamma(x)\backslash G)$. But by (11),

$$\int_{\Gamma(x)\backslash G} |\psi(g)|^q\, dg = [\Gamma : \Gamma(x)]\,.\|\psi\|_q^q\,,$$

where this is the norm on $L^q(\Gamma\backslash G)$. Hence (Hölder)

$$|(T(x)\varphi|\psi)| \leq [\Gamma : \Gamma(x)]^{1/q}\,.\|\psi\|_q \left(\int_{\Gamma(x)\backslash G} \left|\varphi\left(x^{-1}g\right)\right|^p dg\right)^{1/p}\,,$$

where the norm of $\psi$ is calculated on $L^q(\Gamma\backslash G)$. Applying $g \longmapsto xgx^{-1}$ to the latter integral transforms it to

$$\int_{\Gamma(x^{-1})\backslash G} \left|\varphi\left(gx^{-1}\right)\right|^p dg = \int_{\Gamma(x^{-1})\backslash G} |\varphi(g)|^p\, dg =$$

$$= \left[\Gamma : \Gamma\left(x^{-1}\right)\right] \int_{\Gamma\backslash G} |\varphi(g)|^p\, dg\,.$$

So finally,

$$|(T(x)\varphi|\psi)| \leq [\Gamma : \Gamma(x)]^{1/q} \left[\Gamma : \Gamma\left(x^{-1}\right)\right]^{1/p} \|\varphi\|_p \|\psi\|_q .$$

As this inequality holds for all $\varphi$, $\psi \in L(\Gamma\backslash G)$, it follows that the operator $T(x)$ extends to $L^p(\Gamma\backslash G)$ and satisfies

$$\|T(x)\| \leq [\Gamma : \Gamma(x)]^{1/q} \left[\Gamma : \Gamma\left(x^{-1}\right)\right]^{1/p} \tag{24.13}$$

on $L^p(\Gamma\backslash G)$, which gives the theorem proved for holomorphic modular forms in 1939 by its inventor, Hans Petersson by repeatedly using fundamental domains.[95]

(iii) *Operators $T(x)$ for the modular group.* If $G = SL_2(\mathbb{R})$ and if $\Gamma$ is the modular group, as will be seen, $x$ can be chosen to be any matrix proportional to a rational matrix or equivalently, an integer matrix with determinant $> 0$. Such a matrix can be written as

$$x = n^{-\frac{1}{2}} \begin{pmatrix} a & b \\ c & d \end{pmatrix}, \quad a, b, c, d \in \mathbb{Z}, \quad ad - bc = n > 0 \tag{24.14}$$

and only these matrices will be used in what follows. Replacing $G = SL_2(\mathbb{R})$ with $GL_2(\mathbb{R})$ would avoid square roots, disliked by arithmeticians; it is the point of view taken by Jacquet and Langlands, but here I will keep to $SL_2(\mathbb{R})$.

As the operators $T(x)$ only depend on cosets $\Gamma x \Gamma$, we start by determining them. The answer is provided by the elementary divisor theorem:[96] if $L$ is a lattice in $\mathbb{Z}^n$, there is a basis $(a_i)$ of $\mathbb{Z}^n$ and integers $d_i > 0$ satisfying

$$d_1 | d_2 | \ldots | d_n , \tag{24.15}$$

fully determined by $L$ and such that the vectors $b_i = d_i a_i$ form a basis for the lattice $L$. This being so, if $g \in M_n(\mathbb{Z})$ is a matrix with determinant $> 0$ and if this result is applied to $L = g(\mathbb{Z}^n)$, one can deduce that there are matrices $\gamma, \gamma' \in SL_n(\mathbb{Z})$ such that

$$\gamma g \gamma'^{-1} = \text{diag}\,(d_1, \ldots, d_n) .$$

Let us come back to $SL_2$. Matrix (14) can, therefore, be written as

$$x = \gamma \begin{pmatrix} d_1 & 0 \\ 0 & d_2 \end{pmatrix} \gamma' \quad \text{with } 0 < d_1 | d_2 ,$$

up to the factor $n^{-\frac{1}{2}}$, where $\gamma, \gamma' \in \Gamma = SL_2(\mathbb{Z})$. So it is possible to only consider diagonal $x$. The matrix $x^{-1}$ is then proportional to $wxw^{-1}$. Thus

$$\Gamma x \Gamma = \Gamma x^{-1} \Gamma, \quad T\left(x^{-1}\right) = T(x), \quad \Gamma\left(x^{-1}\right) = \Gamma(x) \tag{24.16}$$

for all $x \in \Gamma'$. (13) then shows that $\|T(x)\| \leq [\Gamma : \Gamma(x)]$.

This can also be proved without knowing anything. To this end, let us consider the involution $\sigma$ of $G$ given by

[95] The same could be done here provided $G$ is known to contain a *measurable* set $F$ meeting each coset $\Gamma g$ at precisely one point. True, but you will not be able to prove this in the general case without using the methods of Chap. XI, n° 12.

[96] See for example exercises in § 31 of my *Cours d'algèbre*, or, of course, Serge Lang, *Algebra*.

(24.17) $$\sigma\begin{pmatrix} a & b \\ c & d\end{pmatrix} = \begin{pmatrix} d & b \\ c & a\end{pmatrix} = \omega g^{-1}\omega^{-1} \quad \text{where } \omega = \begin{pmatrix} 1 & 0 \\ 0 & -1\end{pmatrix}.$$

We have $\sigma(\Gamma) = \Gamma$, whence $\Gamma x\Gamma = \sigma(\Gamma\sigma(x)\Gamma)$, i.e.

$$\Gamma x\Gamma = \omega^{-1}\Gamma\omega x^{-1}\omega^{-1}\Gamma\omega = \omega^{-1}\Gamma\omega.x^{-1}.\omega^{-1}\Gamma\omega = \Gamma x^{-1}\Gamma.$$

**Theorem 37 (Hecke-Petersson).** *The operators $T(x)$ commute and are Hermitian on $L^2(\Gamma\backslash G)$.*

The latter point is obvious: $T(x)^* = T(x^{-1})$ (theorem 36) for all $x \in \Gamma'$, and $T(x^{-1}) = T(x)$.

To prove the former, we start from (9)

$$T(x)T(y) = \sum c(x,y;z)T\left(z^{-1}\right) = \sum c(x,y;z)T(z).$$

The coefficients being real, it follows that

$$T(y)^*T(x)^* = \sum c(x,y;z)T(z)^*$$

on $L^2(\Gamma\backslash G)$, and so

$$T(y)T(x) = \sum c(x,y;z)T(z),$$

qed.[97]

(iv) *Operators $T(p)$: the case of functions on $\Gamma\backslash G$.* As the substance of the theory is obtained by supposing $n$ to be *prime* in (14), for the moment this restriction will apply. Still without knowing anything, let us first show that $\Gamma x\Gamma = \Gamma h_p\Gamma$, where generally speaking, we set

(24.18) $$h_m = \begin{pmatrix} |m|^{\frac{1}{2}} & 0 \\ 0 & |m|^{-\frac{1}{2}}\end{pmatrix}.$$

If $x$ is as in (14) with $n = p$, then $ad - bc = p$ and the *gcd* of $c$ and $d$ is either 1 or $p$. If $(c,d) = 1$, $a'd - b'c = 1$ can be solved; so there is some $u \in \mathbb{Z}$ such that $a = pa' + cu$ and $b = pb' + du$; then

$$x = p^{-\frac{1}{2}}\begin{pmatrix} a & b \\ c & d\end{pmatrix} = p^{-\frac{1}{2}}\begin{pmatrix} 1 & u \\ 0 & 1\end{pmatrix}\begin{pmatrix} p & 0 \\ 0 & 1\end{pmatrix}\begin{pmatrix} a' & b' \\ c & d\end{pmatrix} = x(u)h_p\gamma'$$

where $\gamma' \in \Gamma$, and so $\Gamma x\Gamma = \Gamma h_p\Gamma$ since $x(u) \in \Gamma$. If $(c,d) = p$, then $(a,b) = 1$, the matrix $wx$ falls in the previous case and as $w \in \Gamma$, once again $\Gamma x\Gamma = \Gamma wx\Gamma = \Gamma h_p\Gamma$, qed.

We will set $T(p) = T(h_p) = T(h_p^{-1})$, whence

(24.19) $$T(p)\varphi(g) = \sum_{\Gamma(h_p)\gamma} \varphi\left(h_p^{-1}\gamma g\right),$$

[97] Some readers will probably regret that this proof of a purely algebraic result uses calculus from theorem 36. But a substitute for this theorem could be use by considering $G$ as an "abstract" or discrete group and by replacing $L(\Gamma\backslash G)$ with the space of $\Gamma$-invariant functions having finite support mod $\Gamma$. Integrals are then replaced by finite sums.

where summation is over $\Gamma(h_p)\backslash\Gamma$ and where $\Gamma(h_p)$ is the set of $\gamma \in \Gamma$ such that $h_p^{-1}\gamma h_p \in \Gamma$.

$$\begin{pmatrix} p^{-\frac{1}{2}} & 0 \\ 0 & p^{\frac{1}{2}} \end{pmatrix} \begin{pmatrix} a & b \\ c & d \end{pmatrix} \begin{pmatrix} p^{\frac{1}{2}} & 0 \\ 0 & p^{-\frac{1}{2}} \end{pmatrix} = \begin{pmatrix} a & b/p \\ pc & d \end{pmatrix},$$

shows that $\Gamma(h_p)$ is the congruence subgroup $b \equiv 0 \bmod p$ of $\Gamma$, which corresponds to a Borel subgroup in $SL_2(\mathbb{Z}/p\mathbb{Z})$. The Bruhat decomposition being valid over any field, cosets $\Gamma(h_p)\gamma$ are represented by matrices 1 and $wy(m) = x(m)w$ where $m$ varies $\bmod p$, or by matrices $w$ and $x(m)$ since $\gamma \longmapsto \gamma w$ permutes these cosets. There are $p+1$ of them and

$$T(p)\varphi(g) = \varphi\left(h_p^{-1}wg\right) + \sum \varphi\left[h_p^{-1}x(m)g\right].$$

As $\varphi$ is $w$-invariant, we finally find

$$T(p)\varphi(g) = \varphi\left(h_p g\right) + \sum_{m \bmod p} \varphi\left[h_p^{-1}x(m)g\right]. \tag{24.20}$$

Let us now analyze the effect of the operators $T(p)$ on the Fourier series

$$\varphi\left[x(u)g\right] = \sum a_n(g)\mathbf{e}(nu) \tag{24.21}$$

of $\varphi$, whence

$$a_n\left[x(u)g\right] = a_n(g)\mathbf{e}(nu).$$

By (20),

$$\begin{aligned} T(p)\varphi\left[x(u)g\right] &= \varphi\left[h_p x(u)g\right] + \sum \varphi\left[h_p^{-1}x(u+m)g\right] = \\ &= \varphi\left[x(pu)h_p g\right] + \sum \varphi\left[x\left(\frac{u+m}{p}\right)h_p^{-1}g\right] = \\ &= \sum_n a_n(h_p g)\,\mathbf{e}(pnu) + \sum_n a_n\left(h_p^{-1}g\right)\mathbf{e}(nu/p) \sum_{m \bmod p} \mathbf{e}\left(m/p\right)^n. \end{aligned}$$

On the right hand side, the sum over $m$ equals 0 if $p$ does not divide $n$ and $p$ if it does. Hence

$$T(p)\varphi\left[x(u)g\right] = \sum a_n\left(h_p g\right)\mathbf{e}(pnu) + \sum p a_{pn}\left(h_p^{-1}g\right)\mathbf{e}(nu).$$

The coefficients of the expected expansion

$$T(p)\varphi\left[x(u)g\right] = \sum a_{p;n}(g)\mathbf{e}(nu) \tag{24.22}$$

are therefore functions

$$a_{p;n}(g) = \begin{cases} p a_{pn}\left(h_p^{-1}g\right) + a_{n/p}\left(h_p g\right) & \text{if } p|n, \\ p a_{pn}\left(h_p^{-1}g\right) & \text{if } (p,n) = 1. \end{cases} \tag{24.23}$$

In particular,

$$a_{p;0}(g) = p a_0\left(h_p^{-1}g\right) + a_0\left(h_p g\right).$$

So equality $a_0(g) = 0$ is preserved.

Since $h_m h_n = h_{mn}$, relations (23) can also be written

$$a_{p;n}\left(h_n^{-1}g\right) = \begin{cases} pa_{pn}\left(h_{pn}^{-1}g\right) + a_{n/p}\left(h_{n/p}^{-1}g\right) & \text{if } p|n\,, \\ pa_{pn}\left(h_{pn}^{-1}g\right) & \text{if } (p,n)=1\,. \end{cases} \tag{24.23'}$$

(v) *Eigenfunctions of Hecke operators.* Eigenfunctions of operators $T(p)$ have curious properties. Let us suppose that a function

$$\varphi(g) = a_0(g) + \sum a_n(g)$$

satisfies

$$T(p)\varphi(g) = \lambda(p)\varphi(g)$$

for all $p$ and set

$$W_n(g) = na_n\left(h_n^{-1}g\right) \quad \text{for } n \neq 0\,, \tag{24.24}$$

whence $W_n(gu) = W_n(g)\mathbf{e}(nu)$ and

$$\varphi(g) = a_0(g) + \sum_{n\neq 0} n^{-1}W_n\left(h_n g\right)\,.$$

Relations (23') multiplied by $n$ become

$$\lambda(p)W_n(g) = \begin{cases} W_{pn}(g) + pW_{n/p}(g) \\ W_{pn}(g) \end{cases} \tag{24.25}$$

according to whether $p$ divides $n$ or not. In particular,

$$W_{\pm p}(g) = \lambda(p)W_{\pm 1}(g) \tag{24.25'}$$

for $p$ prime, as well as

$$\lambda(p)a_0(g) = pa_0\left(h_p^{-1}g\right) + a_0\left(h_p g\right)\,. \tag{24.25''}$$

**Lemma 4.** *Every function $\lambda(p)$ defined on the set of prime numbers extends to a unique function $\lambda(n)$ defined for $n \geq 1$ and such that $\lambda(1) = 1$,*

$$\lambda(p)\lambda(n) = \lambda(pn) \qquad \text{if } (p,n) = 1\,, \tag{24.26}$$

$$\lambda(p)\lambda(n) = \lambda(pn) + p\lambda(n/p) \qquad \text{if } p|n\,. \tag{24.26'}$$

*Moreover, $\lambda(mn) = \lambda(m)\lambda(n)$ if $(m,n) = 1$.*

The proof is obvious. Since $\lambda(1) = 1$, (26) holds for $n = 1$ for all $p$. Because of

$$\lambda\left(p^{k+1}\right) = \lambda(p)\lambda\left(p^k\right) - p\lambda\left(p^{k-1}\right)\,,$$

it is possible to calculate the numbers $\lambda(p^k)$, hence the numbers $\lambda(n) = \prod\lambda(p^k)$ starting from the prime factor decomposition of $n$. The function obtained obviously satisfies the conditions of the lemma.

Similar arguments and relations (25) and (25') show that, for all $g$, $W_1(g) = 0$ implies $W_n(g) = 0$ for all $n > 0$. If, on the other hand, $W_1(g) \neq 0$, the lemma applies to expressions $W_n(g)/W_1(g)$. Hence

$$W_{\pm n}(g) = \lambda(|n|)W_{\pm 1}(g) \quad \text{for all } n > 0 \tag{24.27}$$

or, coming back to the $a_n(g)$ and to (24),

$$na_n(g) = \begin{array}{ll} \lambda(n)a_1(h_n g) & \text{for } n > 0\,, \\ \lambda(|n|)a_{-1}(h_n g) & \text{for } n < 0\,, \end{array} \tag{24.28}$$

as well as

$$\lambda(p)a_0(g) = a_0(h_p g) + pa_0\left(h_p^{-1}g\right). \tag{24.28'}$$

The Fourier series of $\varphi$ is, therefore, of the form

$$\varphi(g) = \sum_{n<0} \lambda(|n|)n^{-1}W_{-1}(h_n g) + a_0(g) + \sum_{n>0} \lambda(n)n^{-1}W_1(h_n g)\,. \tag{24.29}$$

**Lemma 5.** *Relations* (26) *and* (26') *are equivalent to the identity*

$$\sum_{n\geq 0} \lambda(n)X^n = \prod \left[1 - \lambda(p)X + pX^2\right]^{-1} \tag{24.30}$$

*between formal series.*

Setting

$$F_p(X) = \sum \lambda\left(p^k\right)X^k\,,$$

it is clear that the series $F(X) = \sum \lambda(n)X^n$ satisfies

$$F(X) = \prod F_p(X)\,.$$

But (26') for $n = p^{k+1}$ implies that

$$\begin{aligned} F_p(X) &= 1 + \lambda(p)X + \sum \lambda\left(p^{k+2}\right)X^{k+2} = \\ &= 1 + \lambda(p)X + \sum \left[\lambda(p)\lambda\left(p^{k+1}\right) - p\lambda\left(p^k\right)\right]X^{k+2} = \\ &= 1 + \lambda(p)X + \lambda(p)X\left[F_p(X) - 1\right] - pX^2F_p(X) = \\ &= 1 + \lambda(p)XF_p(X) - pX^2F_p(X) \end{aligned}$$

i.e.

$$\left[1 - \lambda(p)X + pX^2\right]F_p(X) = 1\,, \tag{24.31}$$

whence (30). The converse is obvious.

Identity (30) between formal series can be transformed into an identity between convergent series if $(\lambda(n))$ does not increase too rapidly. It suffices to replace $X^n$ with any other function $\alpha(n)$ satisfying $\alpha(1) = 1$, $\alpha(mn) = \alpha(m)\alpha(n)$ if $(m, n) = 1$ and $\alpha(p^k) = \alpha(p)^k$. $F_p(X)$ is then replaced with

$$\sum \lambda\left(p^k\right)\alpha(p)^k = \left[1 - \lambda(p)\alpha(p) + p\alpha(p)^2\right]^{-1}.$$

Computing the product gives

$$\sum_{n>0} \lambda(n)\alpha(n) = \prod \left[1 - \lambda(p)\alpha(p) + p\alpha(p)^2\right]^{-1}. \tag{24.32}$$

For example, we may choose $\alpha(n) = \chi(n)n^{-s}$, where $\chi$ is a character modulo an arbitrary integer $m > 0$. Then

$$\textstyle\sum_{n>0} \lambda(n)\chi(n)/n^s = \prod \left[1 - \lambda(p)\chi(p)p^{-s} + \chi\left(p^2\right)p^{1-2s}\right]^{-1}. \tag{24.33}$$

This supposes $\lambda(n) = O(n^\sigma)$ for sufficiently large $\sigma$.

Multiplying (33) by $W_{\pm1}(g)$, we get two relations

$$\sum_{n>0} W_{\pm n}(g)\chi(n)n^{-s} =$$

$$= W_{\pm1}(g) \prod \left[1 - \lambda(p)\chi(p)p^{-s} + \chi\left(p^2\right)p^{1-2s}\right]^{-1}. \tag{24.33'}$$

(vi) *Applications to modular forms.* All of the above only assumes that $\varphi(g)$ is an eigenfunction of the operators $T(p)$. Calculations now need to be made more specific in order to get results about modular forms.

First of all, the operators $T(x)$ commute with right translations and preserve the subspace $\mathcal{F}_r(\Gamma\backslash G)$ of functions of weight $r$. Hence, if

$$\varphi(g) = J(g;i)^{-r} f(gi) = f_r(g)$$

is a solution of $f(\gamma z) = (cz+d)^r f(z)$,

$$T(p)f_r(g) = J(g;i)^{-r} T_r(p)f(z) \tag{24.34}$$

where the function $T_r(p)f$ has the same property. It is given by

$$\begin{aligned} T_r(p)f(z) &= J(g;i)^r f_r\left(h_p g\right) + \sum J(g;i)^r f_r\left[h_p^{-1}x(m)g\right] = \\ &= J(g;i)^r J\left(h_p g;i\right)^{-r} f\left(h_p gi\right) + \\ &\quad + J(g;i)^r \sum J\left[h_p^{-1}x(m)g;i\right]^{-r} f\left[h_p^{-1}x(m)gi\right] = \\ &= J\left(h_p;z\right)^{-r} f(pz) + \sum J\left(h_p^{-1};z+m\right)^{-r} f\left[(z+m)/p\right], \end{aligned}$$

whence

$$T_r(p)f(z) = p^{r/2} f(pz) + p^{-r/2} \sum_{m \bmod p} f\left[(z+m)/p\right]. \tag{24.35}$$

The standard *Hecke operator*

$$T_p f(z) = p^{r-1} f(pz) + p^{-1} \sum_{m \bmod p} f\left[(z+m)/p\right], \tag{24.35'}$$

which it would be more appropriate to denote by $T_{r;p}$, is the product of $p^{r/2-1}$ and the previous operator. It applies to all solutions of the functional equation

$$(*) \qquad f(\gamma z) = (cz+d)^r f(z),$$

whether holomorphic or not. To avoid confusion, for functions on $G$, I will keep to definition (20), which does not assume the weight of $\varphi$ to be $r$, and for functions on $P$ to (35), even if this means reformulating results in Hecke's notation to obtained somewhat simpler formulas, for example by using the fact that

$$(24.36) \qquad T_r(p)f = \lambda(p)f \Longleftrightarrow T_p f = \lambda_p f \quad \text{where } \lambda_p = p^{r/2-1}\lambda(p).$$

Because of (35) the operators $T_r(p)$ clearly preserve the set of entire (resp. parabolic) modular forms of weight $r$. So they act on the space $\mathcal{H}_r^2(\Gamma\backslash G) \subset L^2(\Gamma\backslash G)$ of parabolic forms of weight $r$. The operators $T(p)$ are Hermitian and commute pairwise (theorem 37) on this finite dimensional space. As a result, *the eigenfunctions of the operators* $T(p)$ *generate the vector space* $\mathcal{H}_r^2(\Gamma\backslash G)$, a result due to Petersson.

If $f(z) = \sum a_n(y)\mathbf{e}(nx)$ is the Fourier series of a solution of (*), comparing with (21) shows that, for $z = gi$,

$$(24.37) \qquad a_n(y)\mathbf{e}(nx) = J(g;i)^r a_n(g) = J(g;i)^r n^{-1} W_n(h_n g).$$

Replacing $g$ with $h_n^{-1}g$ replaces $J(g;i)$ with $J(h_n^{-1}g;i) = |n|^{\frac{1}{2}} J(g;i)$ and $z$ with $h_n^{-1}z = z/|n|$. Hence

$$(24.38) \qquad W_n(g) = n|n|^{-r/2} J(g;i)^{-r} a_n(y/|n|)\,\mathbf{e}(\pm x),$$

$$(24.39) \qquad W_n(ug) = W_n(g)\mathbf{e}(\pm u)$$

for all $u \in U$, the sign that should be chosen being that of $n$.

In the case of an *entire modular form* $f(z) = a_n\mathbf{e}(nz)$,

$$a_n(y) = a_n \exp(-2\pi n y)$$

with non-zero $a_n$ for $n < 0$. Thus

$$(24.40) \qquad W_n(g) = n^{1-r/2} a_n J(g;i)^{-r} \exp(-2\pi y)\,\mathbf{e}(x)$$

for $n > 0$, a result proportional to $n^{1-r/2}a_n = a'_n$. Setting

$$(24.41) \qquad T_r(p)f(z) = \sum a_{p;n}\mathbf{e}(nz),$$

according to the case, relations (25) transform into

$$(24.42) \qquad \lambda(p)a'_n = a'_{pn} + pa'_{n/p} \quad \text{or } a'_{pn}$$

if $T_r(p)f = \lambda(p)f$ for all $p$. Lemma 4 then shows that

$$(24.43) \qquad a'_n = \lambda(n)a'_1, \quad \text{i.e. } a_n = \lambda_n a_1 \quad \text{for } n > 0$$

by (36). Incidentally, observe that *if the form* $f(z)$ *is parabolic, it is determined by* $\lambda_p$*, up to a constant factor.*

For $n = 0$, (42) shows that

$$(24.44) \qquad a_0 \neq 0 \Longrightarrow \lambda_p = 1 + p^{r-1} = \sigma_{r-1}(p).$$

This expression is the Fourier coefficient $a_p$ of the Eisenstein series $E_r(z)$ of (17.25), up to the factor $-2r/b^r$. To deduce

$$T_p E_r(z) = \lambda_p E_r(z)\,, \tag{24.45}$$

it suffices to check (42) for all $a'_n$ corresponding to $a_n = \sigma_{r-1}(n)$, which is easy. As Maaß series also satisfy (46), with other eigenvalues, this proof of (46) is obviously not correct...

To conclude this n°, let us summarize the main results obtained:

**Theorem 38.** *For every even integer $r > 2$, the space of parabolic forms of weight $r$ for $SL_2(\mathbb{Z})$ has a basis whose elements satisfy the following properties:*

*(i) $T_p f = \lambda_p f$ for all prime $p$;*
*(ii) $f(z) = \sum \lambda_n \mathbf{e}(nz)$;*
*(iii) for all $m \geq 1$ and all characters $\chi \bmod m$,*

$$\sum \lambda_n \chi(n) n^{-s} = \prod \left[1 - \lambda_p \chi(p) p^{-s} + \chi(p)^2 p^{r-1-2s}\right]^{-1}\,; \tag{24.46}$$

*(iv) the function $f$ is fully determined by the eigenvalues $\lambda_p$.*

These results complete Weil's theorem [n° 23, (ii)]. As $\|T(p)\| = \|T(h_p)\| \leq [\Gamma : \Gamma(h_p)] = p + 1$, $|\lambda_p| \leq p^{\frac{r}{2}-1}(p+1)$, a "trivial" result proving the convergence of both sides of (46) for $\mathrm{Re}(s) > r/2 + 1$.

For $r = 12$, the only parabolic form is the function

$$\Delta(z) = \sum \tau(n)\mathbf{e}(nz)\,.$$

This is obviously an eigenfunction of the operators $T_{12}(p)$, whence the relation

$$\sum \tau(n)\chi(n)n^{-s} = \prod \left[1 - \tau(p)\chi(p)p^{-s} + \chi(p)^2 p^{11-2s}\right]^{-1}\,,$$

conjectured by Ramanujan in 1916 and proved by Mordell in 1917, both of whom only considered the trivial character.

There are also Hecke operators associated to congruence groups. The Jacquet-Langlands method reduces them to convolution products on the adelic group $SL_2$ or $GL_2$, but it does not avoid explicit computations to get precise results about the associated $L$-series. See Koblitz's book, already referred to, which keeps to the most classical methods.

# § 9. $SL_2(\mathbb{R})$ as a Lie Group

## 25 – Lie Groups

(i) *Definition and examples.* A Lie group $G$ is both a $C^\infty$ manifold and a group, the map $(x, y) \longmapsto xy^{-1}$ from $G \times G$ to $G$ having to be $C^\infty$. Then so are $(x, y) \longmapsto xy$, $x \longmapsto x^{-1}$, etc. Maps $x \longmapsto ax$ or $xa$ or $x^{-1}$ are diffeomorphisms from $G$ onto itself. If $G$ and $H$ are to Lie groups, a homomorphism from $G$ to $H$ is a $C^\infty$ map $f$ such that $f(xy) = f(x)f(y)$.

Additive groups $\mathbb{R}^n$, multiplicative groups $\mathbb{R}^*$ and $\mathbb{C}^*$, $n \times n$ triangular matrix groups with non-trivial diagonal entries, or else $> 0$, or else equal to 1, groups $GL_n(\mathbb{R})$, open in $M_n(\mathbb{R})$, groups $GL_n(\mathbb{C})$, open in $M_n(\mathbb{C})$, are Lie groups.

A subgroup $H$ of a Lie group $G$ that is a submanifold of $G$ is a Lie group; $H$ will be said to be *a Lie subgroup* of $G$, Indeed the map $(x, y) \longmapsto xy^{-1}$ from $G \times G$ to $G$ maps the submanifold $H \times H$ to $H$. It is therefore $C^\infty$ as a map from $H \times H$ to $H$ (Chap. IX, n° 13, exercise 3). *A Lie subgroup is closed* (and conversely, but it is far less obvious). Its closure $H'$ in $G$ is indeed a subgroup in which $H$ is open. This is a general property of submanifolds [Chap. IX, n° 13, end of section (i)]. However, $H'$ is the union of cosets $xH$, which, like $H$, are open in $H'$ and pairwise disjoint, hence closed in $H'$. In particular, $H$ is closed in $H'$, hence in $G$, qed.

The next general result leads to many examples of Lie groups. Let us suppose that a Lie group $G$ acts on a manifold $X$ in such a way that the map $(g, x) \longmapsto gx$ is $C^\infty$. Then the map $j : g \longmapsto gx$ has constant rank for all $x \in X$, i.e. is a *subimmersion* [Chap. IX, n° 12, (iii), end]. Indeed for all $a \in G$, $j(ag) = aj(g)$. The tangent linear map to $j$ at $ag$ is therefore the composition of the tangent map to $j$ at $g$ and of the tangent map to $x \longmapsto ax$ at $j(g)$. Now, the latter is an isomorphism like $x \longmapsto ax$, qed.

This being settled, it follows that, for all $x \in X$, the subgroup $G_x$ of $g$ such that $gx = x$ is a submanifold of $G$ [Chap. IX, n° 13, (ii)], hence a Lie group.

For example, the subgroup of $G$ defined by $gs = sg$ is a Lie group for all $s \in G$: make $G$ act on $G$ by $(g, x) \longmapsto gxg^{-1}$. If $G = GL_n(\mathbb{R})$, $G$ can also be made to act on the right by $(g, s) \longmapsto g'sg$, where $g'$ is the transpose of $g$. The group of automorphisms of a non-degenerate bilinear form, for example *orthogonal groups* in the symmetric case or *symplectic groups* in the alternate one, is therefore a Lie group. Making $G$ act on $\mathbb{R}^*$ by $(g, x) \longmapsto \det(g)x$ shows that $SL_n(\mathbb{R})$ is a Lie subgroup of $G$.

Every $C^\infty$ homomorphism $f : H \longmapsto G$ from a Lie group to another is a subimmersion: make $H$ act on $G$ by f $(h, g) \longmapsto f(h)g$. The *kernel* of $f$ is therefore a Lie subgroup of $H$. On the other hand, the image of $f$ is not necessarily a Lie subgroup of $G$.

Indeed, apart form subgroups that are genuine closed submanifolds, a Lie group $G$ can contain subgroups $H$ which, though not submanifolds, are equipped with the structure of a Lie group such that the identity map $H \longrightarrow G$ is an *immersion.* Lie group theory cannot be constructed without giving as much importance to these "bad" subgroups as to the "good" ones.

In particular, *one-parameter subgroups* whose elements are homomorphisms $t \longmapsto \gamma(t)$ from $\mathbb{R}$ to $G$, play a fundamental role. Such a homomorphism only need to be continuous to be $C^\infty$ – obvious if $G = GL_n(\mathbb{R})$. It is then an immersion, unless $\gamma(t) = e$ for all $t$, since the map has rank 1 everywhere or rank 0 everywhere. In the general case, the kernel of $\gamma$ is either $\{0\}$, or a discrete subgroup $\omega\mathbb{Z}$ of $\mathbb{R}$. In the latter case, the image $\gamma(\mathbb{R})$ is a compact subgroup, and $\gamma(\mathbb{R})$ is a 1-dimensional Lie subgroup isomorphic to $\mathbb{R}/\omega\mathbb{R}$. But in the former case, $\gamma(\mathbb{R})$ may be everywhere dense [Chap. IX, n° 13, (iii)].

We show that a continuous homomorphism from a Lie group to another is $C^\infty$. This is particularly easy to prove for *finite*-dimensional continuous linear representations $(\mathcal{H}, \pi)$ of a Lie group $G$, which are homomorphisms to some $GL_n$. For any function $\varphi \in \mathcal{D}(G)$, i.e. $C^\infty$ and with compact support, the function

$$\pi(x)\pi(\varphi) = \pi(x) \int \pi(y)\varphi(y)dy = \int \pi(xy)\varphi(y)dy = \int \pi(y)\varphi\left(x^{-1}y\right) dy$$

is indeed $C^\infty$. This follows from the simplest theorems on differentiation under the $\int$ sign. On the other hand, there exist $\varphi$ whose support is an arbitrarily small neighbourhood of $e$, hence Dirac sequences composed of functions $\varphi_n \in \mathcal{D}(G)$. So the unit operator is a limit of operators $\pi(\varphi)$. But in finite dimension, an everywhere dense vector subspace is the whole of the space. Hence there exist $\varphi$ such that $\pi(\varphi) = 1$, qed.

A final remark: every Lie group can be equipped with an *analytic* structure compatible with its $C^\infty$ structure, i.e. for which change of charts functions and the composition law $(x, y) \longmapsto xy^{-1}$ have real power series expansions. Such an analytic structure is unique.

(ii) *Operations on tangent vectors.* Let $G$ be a Lie group. Since, for all $a \in G$, the map $g \longmapsto ag$ is a diffeomorphism, the tangent map $G'(g) \longrightarrow G'(ag)$ is bijective. It is then natural to denote by $ah$ the image of $h \in G'(g)$ under this application. $ha$ can be similarly defined. We check that $a(hb) = (ah)b$ by differentiating the identity $a(xb) = (ax)b$. Hence we can talk of $ahb$ without ambiguity. If $h$ is the tangent vector at $t = 0$ to some path $\gamma(t)$ on $G$, then

$$df(agb; ahb) = \frac{d}{dt} f\left[a\gamma(t)b\right] \quad \text{for } t = 0\,. \tag{25.1}$$

Let us now set $m(x, y) = xy$ and calculate

$$m'(a, b) : G'(a) \times G'(b) \longrightarrow G'(ab)\,.$$

To do this, we need to differentiate $(x, y) \longmapsto f(xy)$, hence[98] add the differentials of $x \longmapsto f(xb)$ and $y \longmapsto f(ay)$. As a result,

$$m'(a, b) : (h, k) \longmapsto hb + ak\,. \tag{25.2}$$

In Leibniz style: $d(xy) = dx.y + x.dy$. In particular, $m'(e, e)$ is addition $(h, k) \longmapsto h + k$ in $G'(e)$ even if $G$ is not commutative. If $h$ and $k$ are given by $h = \gamma'(0)$, $k = \delta'(0)$ where $\gamma$ and $\delta$ are curves over $G$, vector (2) is tangent to $t \longmapsto \gamma(t)\delta(t)$ at $t = 0$. Generally speaking, setting

[98] We recall the calculus rule of Chap. IX, n° 12, (iv): let $f(x, y)$ be a map from a product manifold $X \times Y$ to a manifold $Z$. Denote by $d_X f(a, b)$ the differential of $x \longmapsto f(x, b)$ at $a$ and by $d_Y f(a, b)$ the differential of $y \longmapsto f(a, y)$ at $b$. Then the "total" differential of $f$ is given by

$$df\left((a, b); (h, k)\right) = d_X f\left((a, b); h\right) + d_Y f\left((a, b); k\right)\,.$$

For obvious reasons, I will call this result the *differentiation formula of a product.* Together with the chain rule, it shows that given maps $\gamma : T \longrightarrow X$ and $\delta : T \longrightarrow Y$ from a manifold $T$ to $X$ and $Y$ and setting $F(t) = f[\gamma(t), \delta(t)]$, the tangent linear map $F'(t_0)$ at a point $t_0$ is the sum of the tangent maps to $t \longmapsto F[\gamma(t), \delta(t_0)]$ and $t \longmapsto F[\gamma(t_0), \delta(t)]$.

$$D_0 f(t) = f'(0) = \lim [f(t) - f(0)] / t$$

for any differentiable function at $t = 0$, which contradicts taboos related to free variable, we therefore get

$$df(ab; hb + ak) = D_0 f [\gamma(t)\delta(t)] \tag{25.3}$$

for all $f \in C^1(G)$. Setting $F(x, y) = f(xy)$, formula (3) becomes

$$D_0 F [\gamma(t), \delta(t)] = D_0 F [\gamma(t), b] + D_0 F [a, \delta(t)] \tag{25.4}$$

as a consequence of the differentiation rule of a product. More generally,

$$\frac{d}{dt}\gamma(t)\delta(t) = \gamma'(t)\delta(t) + \gamma(t)\delta'(t) \tag{25.5}$$

for all $t$, both sides being in the tangent space at $\gamma(t)\delta(t)$. In particular, if $\delta(t) = \gamma(t)^{-1}$, then

$$\gamma'(t)\delta(t) + \gamma(t)\delta'(t) = 0,$$

whence

$$\frac{d}{dt}\gamma(t)^{-1} = -\gamma(t)^{-1}\gamma'(t)\gamma(t)^{-1}. \tag{25.6}$$

Even more particularly,

$$D_0 \left[\gamma(t)^{-1}\right] = -D_0 [\gamma(t)] \quad \text{if } \gamma(0) = e. \tag{25.6'}$$

This result says that the tangent map to $x \longmapsto x^{-1}$ at $e$ is $X \longmapsto -X$.

Since $g \longmapsto ga$ induces an isomorphism from $G'(e)$ onto $G'(a)$, the manifold $T(G)$ of tangent vectors to $G$ can be identified to the product $G'(e) \times G$ by associating the vector $Xa \in G'(a)$ to the ordered pair $(X, a)$. The law of composition $m'$ of $T(G)$ being associative,

$$XaYb = X.aYa^{-1}.ab = X.ad(a)Y.ab$$

where the image vector

$$ad(a)Y = aYa^{-1}, \tag{25.7}$$

under the tangent map to $x \longmapsto axa^{-1}$ at $e$, is still in $G'(e)$. The map $g \longmapsto ad(g)X$ is a linear representation of $G$ on $G'(e)$, the *adjoint representation* of $G$. If $G$ is a linear group and if the elements $X \in G'(e)$ are identified to matrices or endomorphisms, then obviously

$$ad(g)X = gXg^{-1}.$$

(iii) *Differentiation and invariant vector fields.* Translations on $G$ enable us to associate two $C^\infty$ vector fields, namely

$$L_X : g \longmapsto -Xg, \qquad R_X : g \longmapsto gX, \tag{25.8}$$

to every $X \in G'(e)$. They satisfy relations

$$L(ga) = L(g)a, \qquad R(ag) = aR(g)$$

and these clearly characterize them (put $g = e$). If $(X_i)$ is a basis for $G'(e)$, the vectors $gX_i$ (or $X_i g$) form a basis for $G'(g)$ for all $g$, so that any vector field on $G$ can be written as a linear combination with variable coefficients of $gX_i$ or $X_i g$.

However, we saw in Chap. IX, n° 14 that to every vector field $L$ on a manifold, one can associate a differential operator given by

$$Lf(x) = df(x; L(x))$$

for all $x \in G$ and all function $f$ differentiable everywhere with values in $\mathbb{C}$ (or even in a Banach space). We will denote by $L(X)$ and $R(X)$ the operators associated to the vector fields $L_X$ and $R_X$. Hence

$$R(X)f(g) = df(g; gX) = D_0 f\,[g\gamma(t)]\,, \tag{25.9}$$

$$L(X)f(g) = df(g; -Xg) = D_0 f\,\left[\gamma(t)^{-1}g\right] \tag{25.9'}$$

if $X = \gamma'(0)$. Clearly $R(X)$ (resp. $L(X)$) commutes will left (resp. right) translations. Operators $R(X)$ commute with operators $L(Y)$, for if $X = \gamma'(0)$ and $Y = \delta'(0)$, we get $R(X)L(Y)f(g)$ by differentiating $L(Y)f[g\gamma(s)]$ at $s = 0$, hence by differentiating $f[\delta(t)^{-1}g\gamma(s)]$ with respect to $s$ and $t$ at $s = t = 0$. The order of differentiation is irrelevant. Denoting by $R(x)$ and $L(x)$ the translations operators given by[99]

$$R(x)f(g) = f(gx)\,, \quad L(x)f(g) = f\left(x^{-1}g\right) \quad \text{for } x \in G\,,$$

it is at least formally possible to write that

$$R(X) = D_0 R\,[\gamma(t)]\,, \quad L(Y) = D_0 L\,[\delta(t)]\,. \tag{25.10}$$

These vector fields are related to one-parameter subgroups of $G$. Indeed, since $R_X(ag) = aR_X(g)$, left translations permute the maximal integral curves of the vector field $R_X$ (Chap. IX, n° 15). If $\gamma(t)$ is the maximal integral such that $\gamma(0) = e$, $\gamma'(0) = X$, then by definition $\gamma'(t) = \gamma(t)X$. However, if $I$ is the definition interval of $\gamma(t)$, for all $s \in I$ and all vector fields, the curve $t \longmapsto \gamma(s+t)$ is the maximal integral with initial point $\gamma(s)$. The curve $t \longmapsto \gamma(s)^{-1}\gamma(s+t) = \delta(t)$ is therefore also a maximal integral of $R_X$, such that $\delta(0) = e$, $\delta'(0) = \gamma(s)^{-1}\gamma'(s) = X$ and defined in $I - s$. As a result, $I = I - s$. Thus $I = \mathbb{R}$ and $\gamma(s+t) = \gamma(s)\gamma(t)$ for all $s$, $t$. Hence the integral curve $\gamma(t)$ is a one-parameter subgroup, which will be denoted by $\gamma_X(t)$.

Conversely, for any one-parameter subgroup, $\gamma(t)$ is obtained in this manner: by differentiating the equality $\gamma(s+t) = \gamma(s)\gamma(t)$ at $s = 0$, we find $\gamma'(t) = X\gamma(t) = \gamma(t)X = R_X[\gamma(t)]$ where $X = \gamma'(0)$.

If $G$ is a Lie subgroup of $GL_n(\mathbb{R})$ and $X$ is identified to a matrix, then

$$\gamma_X(t) = \exp(tX) = \sum (tX)^{[n]}$$

[Chap. IX, n° 15, (v)], which, in the general case, explains the conventional notation

$$\gamma_X(t) = \exp_G(tX)\,. \tag{25.11}$$

As a homomorphism $\pi : G \longrightarrow H$ of Lie groups transforms one-parameter subgroups of $G$ into their analogues in $H$, clearly, $\pi \circ \exp_G = \exp_H \circ \pi'(e)$. For $G = H$ and $\pi(x) = gxg^{-1}$,

[99] Attention should be paid to the definition of operators $R(x)$ adopted here: it is different from that of Chap. XI, n° 31.

$$g.\exp(X)g^{-1} = \exp[ad(g)X] \tag{25.12}$$

follows. So $g$ such that $ad(g) = e$ are the elements of $G$ commuting with $\exp(X)$ for all $X$, hence with the subgroup they generate. We will see later that it is in $G$ if $G$ is *connected.* In this case, the kernel of the adjoint representation is, therefore, the centre of $G$.

If $(\mathcal{H}, \pi)$ is a *finite* dimensional continuous linear representation of $G$, hence $C^\infty$, then, setting

$$\pi(X) = \pi'(e)X = D_0\pi[\gamma(t)] \tag{25.13}$$

for all $X \in G'(e)$,

$$\pi[\exp_G(X)] = \exp[\pi(X)]$$

where the right hand side is the usual exponential of a matrix. One readily deduces that, *if $G$ is connected, then an operator $T : \mathcal{H} \longrightarrow \mathcal{H}$ commutes with all $\pi(g)$ if and only if it commutes with all $\pi(X)$.* Likewise, subspaces of $\mathcal{H}$ invariant under the $\pi(g)$ or the $\pi(X)$ can be shown to be the same.

In fact, invoking the exponential map to get these results is unnecessary. Indeed if $G$ is made to act on the manifold of endomorphisms of $\mathcal{H}$ by $(g, T) \longmapsto \pi(g)T\pi(g)^{-1}$, by section (i) above, for all $T$, we get a map of constant rank. To show that it is constant in the neighbourhood of $e$ (hence in $G$ if $G$ is connected), it is sufficient to show that its rank at the origin is zero. But if $\gamma'(0) = X$, then

$$D_0\left[\pi(\gamma(t))\,T\pi(\gamma(t))^{-1}\right] = \pi(X)T - T\pi(X) \tag{25.14}$$

by (6') and the differential rule of a product, whence the result.

(iv) *Canonical coordinates.* As the vector field $R_X$ depends in a $C^\infty$ manner on the parameter $X$, the map $(X, t) \longmapsto \exp_G(tX)$ is $C^\infty$ (Chap. IX, n° 15), hence so is $X \longmapsto \exp_G(X)$. On a neighbourhood of 0, but not globally in general, it is a diffeomorphism because the tangent map to $\exp_G$ at 0 is $X \longmapsto D_0 \exp_G(tX) = X$. The subgroup generated by the image of $\exp_G$, i.e. the set of products

$$\exp(X_1)\ldots\exp(X_n)\,,$$

is therefore open in $G$; it is $G$ if $G$ is connected.

For any sufficiently small neighbourhood $V$ of $e$ in $G$, there is a unique diffeomorphism $\log_G$ from $V$ onto a neighbourhood $W$ of 0 in $G'(e)$ such that $\exp_G \circ \log_G$ is the identity on $V$, whence a chart $(V, \log_G)$, called *canonical,*[100] from $G$ to $V$. For all $g \in G$, the map $Y \longmapsto g.\exp(Y)$ is, therefore, a chart of $G$ in the open set $gV$. It follows that a function $f$ is of class $C^p$ on $G$ if and only if, for all $g$, the map $Y \longmapsto f[g.\exp(Y)]$ is of class $C^p$ on $W$, i.e. has continuous partial derivatives of order $\leq p$. For example, the first order derivatives are the functions

$$Y \longmapsto D_0 f[g.\exp(Y + tX)] \tag{25.15}$$

where $X$ varies in a basis of $G'(e)$. For $Y = 0$, we recover the functions

$$R(X)f(g) = D_0 f[g.\exp(tX)]$$

defined above by supposing that $f$ is differentiable, but due to the previous formula, they can be defined whenever they are well-defined. If $f$ is of class $C^p$, these functions exist and are of class $C^{p-1}$.

[100] The notation exp and log will be used when there is no possibility of confusion.

The converse is obvious if $f$ is *differentiable* at every point of $G$, since then $R(X)f(g) = df(g; gX)$. However, in every local chart, differentiation operators with respect to these coordinates are linear combinations with $C^\infty$ coefficients of the operators $R(X)$ associated to the vectors $X$ in a basis of $G'(e)$. But it is not a priori obvious that the existence and the continuity of functions $R(X)f$ imply the differentiability of $f$, because the derivatives $R(X)f(g)$ are computed in a chart depending on $g$ contrary to derivatives (15). Hence a proof is needed.

**Theorem 39**. *Let $G$ be a Lie group. A function $f$ defined on $G$ is of class $C^p$ if and only if, for all $X \in G'(e)$, the derivative*

$$R(X)f(g) = D_0 f\left[g.\exp(tX)\right] \tag{25.16}$$

*exists and is a function of class $C^{p-1}$ on $G$.*

Let us first show that the map $X \longmapsto R(X)f$ is linear. Indeed, suppose that, for given $X$, the derivative $R(X)f$ exists and is a continuous function of $g$. Replacing $g$ by $g.\exp(tX)$, the function $t \longmapsto f[g.\exp(tX)]$ is seen to be differentiable and its derivative

$$Df\left[g.\exp(tX)\right] = R(X)f\left[g.\exp(tX)\right], \quad \text{où } D = d/dt,$$

to be a continuous function of $(g, t)$. Hence, if $M$ is a compact subset of $G$, the right hand side is bounded for $g \in M$ and $|t| \leq 1$. Integrating with respect to $t$ then gives an inequality of the form

$$|f\left[g.\exp(tX)\right] - f(g)| \leq c(M)|t|$$

which holds for all $g \in M$ and $|t| \leq 1$. Then for any $\varphi \in \mathcal{D}(G)$, (dominated convergence)

$$\lim_{t=0} \int \frac{f\left[g.\exp(tX)\right] - f(g)}{t}\varphi(g)d_r g = \int R(X)f(g).\varphi(g)d_r g\,.$$

But the left hand side also equals

$$\lim_{t=0} \int f(g)\frac{\varphi\left[g.\exp(-tX)\right] - \varphi(g)}{t}d_r g\,.$$

As $\varphi \in \mathcal{D}(G)$, the quotient converges uniformly to the function $-R(X)\varphi$ staying zero outside a fixed compact set. It is therefore possible to pass to the limit under the $\int$ sign. Hence

$$\int R(X)f(g).\varphi(g)d_r g + \int f(g).R(X)\varphi(g)d_r g = 0 \tag{25.17}$$

for all $\varphi \in \mathcal{D}(G)$. The right hand side being a linear function of $X$, the linearity of $R(X)f$ follows from the assumption of the theorem.

To show that $f \in C^p(G)$, it is convenient to use canonical charts "of the second kind". They are defined using a basis $(X_i)_{1 \leq i \leq n}$ of $G'(e)$ and a map

$$X = t_1 X_1 + \ldots + t_n X_n \longmapsto \exp(t_1 X_1) \ldots \exp(t_n X_n)\,. \tag{25.18}$$

As the tangent vector to $t \longmapsto \exp(tX)$ at $t$ is, by definition, $\exp(tX)X$, the differentiation formula of a product shows that the tangent map to (18) at $X = 0$ is the identity. Map (18) is thus a diffeomorphism in the neighbourhood of 0. A function $f$ is then of class $C^p$ if and only if so is the function

$$F(t_1, \dots, t_n) = f\left[g.\exp(t_1 X_1) \dots \exp(t_n X_n)\right]$$

on $\mathbb{R}^n$ for all $g$.

To compute the derivative $D_i F$ with respect to $t_i$, we use by definition the expression

$$f\left[g \exp(t_1 X_1) \dots \exp(t_i X_i + h X_i) \dots \exp(t_n X_n)\right].$$

Setting

$$x = \exp(t_1 X_1) \dots \exp(t_n X_n), \quad x_i = \exp(t_{i+1} X_{i+1}) \dots \exp(t_n X_n),$$

it equals

$$f\left[gx.x_i^{-1} \exp(hX_i)\, x_i\right] = f\left[gx.\exp(hX_i')\right] \quad \text{où } X_i' = ad(x_i)^{-1} X_i.$$

Thus

$$D_i F(t) = D_0 f\left[gx.\exp(tX_i')\right] = R(X_i') f(gx),$$

the existence of the left hand side being equivalent to that of the right hand one. Hence these derivatives exist if $R(X)f$ is well-defined for all $X$. However, there are relations of the form

$$ad(g)^{-1} X_i = \sum a_{ij}(g) X_j \tag{25.19}$$

with coefficients $a_{ij} \in C^\infty(G)$, whence

$$X_i' = \sum a_{ij}(x_i) X_j.$$

As $R(X)f(g)$ is a linear function of $X$, it follows that

$$D_i F(t) = \sum a_{ij}(x_i) R(X_j) f(gx)$$

where $a_{ij}(x_i)$ are $C^\infty$ functions of the coordinates $t_k$. For given $g$ and under the assumption of the theorem, the right hand side is, therefore, a function of class $C^{p-1}$ of the $t_k$, qed.

We will return to this "obvious" theorem in the next n° in the context of $C^\infty$ vectors in infinite dimensional representations.

(v) *The Lie algebra of a group.* There are several infinitesimal methods to emphasize the non-commutativity of multiplication in a Lie group $G$. The simplest consists in considering vectors $X, Y \in G'(e)$ defined by curves $\gamma(t)$ and $\delta(t)$. Applying formula (9') twice shows that

$$R(X)R(Y)f(g) = \frac{d^2}{dsdt} f\left[g\gamma(s)\delta(t)\right] \quad \text{for } s = t = 0 \tag{25.20}$$

$$R(Y)R(X)f(g) = \frac{d^2}{dsdt} f\left[g\delta(t)\gamma(s)\right] \quad \text{for } s = t = 0. \tag{25.20'}$$

The Jacobi bracket $R = [R(X), R(Y)] = R(X)R(Y) - R(Y)R(X)$ of the two operators considered is, therefore, given by

$$Rf(g) = \frac{d^2}{dsdt} \left\{ f\left[g\gamma(s)\delta(t)\right] - f\left[g\delta(t)\gamma(s)\right] \right\} \tag{25.21}$$

at $s = t = 0$. Now, like $R(X)$ and $R(Y)$, $R$ is defined by a vector field (Chap. IX, n° 14). Like $R_X$ and $R_Y$, it is right invariant. Hence there exists $Z \in G'(e)$ such that $R = R(Z)$. We set $Z = [X, Y]$. So by definition,

$$R([X, Y]) = [R(X), R(Y)] \; . \tag{25.22}$$

For $g = e$, the previous formula shows that, for all $f \in C^2(G)$,

$$df\,(e; [X, Y]) = \frac{d^2}{dsdt} \{f[\gamma(s)\delta(t)] - f[\delta(t)\gamma(s)]\} \; . \tag{25.23}$$

This expression needs to be calculated at $s = t = 0$. A similar calculation would also show that

$$L([X, Y]) = [L(X), L(Y)] \; . \tag{25.23'}$$

If $\pi$ is a homomorphism from $G$ to a Lie group $H$, then denoting by $\pi'(e)$ the tangent map to $\pi$ at the origin, once again

$$\pi'(e)([X, Y]) = [\pi'(e)X, \pi'(e)Y] \tag{25.24}$$

because $\pi$ transforms the product $\gamma(s)\delta(t)$ of two paths defining $X$ and $Y$ into the product of paths $\pi \circ \gamma(s)$ and $\pi \circ \delta(t)$, which are tangent to the vectors $\pi'(e)X$ and $\pi'(e)Y$ at the origin. If $\pi$ us a finite-dimensional representation of $G$, for $f = \pi$, the function to be differentiated in (23) equals

$$\pi[\gamma(s)]\,\pi[\delta(t)] - \pi[\delta(t)]\,\pi[\gamma(s)] \; .$$

The derivative of $s \longmapsto \pi[\gamma(s)]$ at the origin being by definition $\pi(X)$,

$$\pi([X, Y]) = [\pi(X), \pi(Y)]$$

follows, where these are now linear operators.

*Exercise 1.* Let $\gamma(t)$ and $\delta(t)$ be two paths with initial point $e$ tangent to $X$ and $Y$. Differentiating the expression $f[\gamma(t)\delta(s)\gamma(t)^{-1}]$ with respect to $s$ and $t$, show that

$$[X, Y] = D_0 ad\,[\gamma(t)Y] \; . \tag{25.25}$$

Hence, setting $ad(X)Y = [X, Y]$,

$$ad(X) = D_0 ad\,[\gamma(t)] \tag{25.25'}$$

for every path defining $X$, for example $\gamma_X(t) = \exp_G(tX)$. We have

$$ad([X, Y]) = [ad(X), ad(Y)] \; .$$

*Exercise 2.* If $x, y$ are elements of a group $G$, their *commutator* is the element $[x, y] = xyx^{-1}y^{-1}$ which measures the deviation of $yx$ from $xy$. Suppose that $G \subset GL_n(\mathbb{R})$ and consider paths $\gamma$ and $\delta$ with initial point $e$ defining vectors $X, Y$ in $G'(e) \subset M_n(\mathbb{R})$. Using limited expansions of order 2 of $\gamma(t)$ and $\delta(t)$, show that

$$[\gamma(t), \delta(t)] = 1 + t^2(XY - YX) + o(t^2)$$

and that $XY - YX$ is the tangent vector to $t \longmapsto [\gamma(t^{\frac{1}{2}}), \delta(t^{\frac{1}{2}})]$.

The vector space $G'(e)$, equipped with the Jacobi bracket $[X, Y]$, is the *Lie algebra* of $G$ and is usually written $\mathfrak{g}$. In the rest of this §, it is better to keep this notation for the corresponding *complex* vector space. The elements of $\mathfrak{g}$ can be identified with differentiation operators having left and right invariant complex coefficients.

The bracket $[X, Y]$ is bilinear in $X, Y$, alternate and, like the corresponding differential operators, satisfy the Jacobi identity

$$[X, [Y, Z]] + [Y, [Z, X]] + [Z, [X, Y]] = 0 .$$

Setting

$$ad(X)Y = [X, Y] ,$$

the previous relation means that

$$[ad(X), ad(Y)] = ad ([X, Y]) .$$

A vector space equipped with a bilinear composition law $(X, Y) \longmapsto [X, Y]$ satisfying the preceding formulas is a *Lie algebra.* A linear map $\pi$ from the latter to the space of operators of a vector space $\mathcal{H}$ is a *representation* if $\pi([X, Y]) = [\pi(X), \pi(Y)]$.

(vi) *Lie algebras of classical groups.* Let us show how to determine the Lie algebras of the groups defined in section (i) using subimmersions. If a Lie group acts on a manifold $X$, the maps $g \longmapsto ga$ have been seen to be subimmersions. If $H$ is the stabilizer of $a$ in $G$, the Lie algebra of $H$ is thus the kernel of the tangent map to $g \longmapsto ga$ at $e$.

For the map $x \longmapsto xsx^{-1} = y$ from $G$ to $G$, we differentiate equality $xs = yx$. This gives $dx.s = dy.x + y.dx$ and, for $x = e$,

$$dy = dx.s - y.dx = dx.s - sdx .$$

The kernel of $dx \longmapsto dy$ is thus defined by $dx.s = s.dx$, whence $H'(e) = \text{Ker}[ad(s) - 1]$.

For $G = GL_n(\mathbb{R})$ acting on itself by $x \longmapsto x'sx = y$ (where $x'$ is the transpose of $x$), $dy = dx'.s + s.dx$ at $x = e$. So $H'(e)$ is the set of $X \in G'(e) = M_n(\mathbb{R})$ such that $X's + sX = 0$. If $s = s'$, this means that the matrix $sX$ is antisymmetric, and if $s' = -s$, that it is symmetric. This gives the Lie algebras of the orthogonal and symplectic groups.

The group $SL_n(\mathbb{R})$ of real matrices with determinant 1 is obtained by making $GL_n(\mathbb{R})$ act on $\mathbb{R}^*$ by$(g, x) \longmapsto \det(g)x$. All that is needed is to compute the tangent map to the determinant at $g = e$. If $(e_i)$ is the canonical basis of $\mathbb{R}^n$, then

$$\det(g) = D [g (e_1) , \ldots , g (e_n)] ,$$

where $D$ is the $n$-linear alternate form equal to 1 at $(e_1, \ldots, e_n)$. The differentials of the functions $g \longmapsto g(e_i)$ at the origin being $X \longmapsto X(e_i)$, the differential of the determinant at $e$ is the linear functional

$$X \longmapsto D [X (e_1) , e_2, \ldots , e_n] + \ldots + D [e_1, \ldots , e_{n-1}, X (e_n)] = Tr(X)$$

on $G'(e)$. The Lie algebra of the group $G = SL_n(\mathbb{R})$ is therefore the set of matrices with zero trace.

In all these examples of subgroups of $GL_n$, the Jacobi bracket $[X, Y]$ is obviously the matrix $XY - YX$.

(vii) *Invariant distributions and differential operators.* For every manifold $G$ of class $C^\infty$, it is possible to define a Schwartz space $\mathcal{D}(G) \subset C^\infty(G)$ of $C^\infty$ functions with compact support and, for every compact set $M \subset G$, the subspace $\mathcal{D}(G;M)$ of functions vanishing outside $M$. These subspaces are stable under all differential operators[101] defined in Chap. IX, n° 14. This enables us to equip each $\mathcal{D}(G;M)$ with a locally convex topology using seminorms

$$(25.26) \qquad p_L(\varphi) = \sup |L\varphi(g)| = \|L\varphi\|_\infty \,,$$

where $L$ is such a differential operator. As $L$ is a linear combination of products of operators $R(X)$ or $L(X)$, whichever, we may only consider seminorms

$$\|R(X_1)\dots R(X_p)\,\varphi(g)\|_\infty \,,$$

where the $X_i$ are in $G'(e)$, or even belong to a basis of $G'(e)$. Likewise, the seminorms $\sup_{g\in M} |L\varphi(g)|$, where $M$ is an arbitrary compact subset, enable us to define a topology on $C^\infty(G)$, that of compact convergence of functions and all their derivatives.

Based on this, the notion of a *distribution* on $G$ becomes obvious: it is a linear functional[102]

$$\mu : \varphi \longmapsto \mu(\varphi) = \int \varphi(x)d\mu(x) = \langle \varphi, \mu\rangle$$

on $\mathcal{D}(G)$ whose restriction to each $\mathcal{D}(G;M)$ is continuous. This means that, for any compact set $M$, there are finitely many differential operators $L_i$ such that

$$(25.27) \qquad |\mu(\varphi)| \le \sum \|L_i\varphi\|_\infty \quad \text{for all } \varphi \in \mathcal{D}(G;M)\,.$$

Here we will mostly use distributions *with compact support*, i.e. for which there is a compact set $M$ such that $\mu(\varphi) = 0$ if $\varphi = 0$ in a *neighbourhood*[103] of $M$. The smallest possible $M$ is the support of $\mu$. $\mu(\varphi)$ can then be defined for all $\varphi \in C^\infty(G)$ by setting $\mu(\varphi) = \mu(\varphi')$ for all $\varphi' \in \mathcal{D}(G)$ equal to $\varphi$ in a neighbourhood of $M$, the result being a continuous linear functional in $C^\infty(G)$: inequality (27), where the right hand side is computed in a fixed compact neighbourhood of $\mu$, then holds for all $\varphi \in C^\infty(G)$. Such a distribution is of the form[104]

---

[101] Linear, of arbitrary order and with $C^\infty$ coefficient will always be implied.

[102] The integral notion is far more convenient than that of Schwartz, especially in the context of Lie groups. The shortest presentation of distribution theory is probably that given by Lars Hormänder in the beginning of his *Linear Differential Operators*.

[103] If $h \in G'(a)$, $\varphi(a) = 0$ does not imply $d\varphi(a;h) = 0\dots$

[104] Let $M$ be a neighbourhood of of the support of $\mu$ and suppose there is an upper bound $|\mu(\varphi)| \le \sum_{1\le i\le n} \|L_i\varphi\|_M$ in $\mathcal{D}(G;M)$, where this the norm of uniform convergence on $M$. Consider the product space $\mathcal{H} = L(M)^n$ equipped with the topology of uniform convergence, where $L(M)$ is, as usual, the set of continuous functions on $M$. $\mu$ can be considered a continuous linear functional on the subspace $\mathcal{M} \subset \mathcal{H}$ of $(L_i\varphi)_{1\le i\le n}$ where $\varphi \in \mathcal{D}(G,M)$. The Hahn-Banach theorem then proves the existence of a continuous linear functional on $\mathcal{H}$ extending $\mu$. Such a form is obviously given by

$$(\varphi_1, \dots, \varphi_n) \longmapsto \sum \mu_i(\varphi_i)$$

where the $\mu_i$ are measures on $M$. Thus $\mu(f) = \sum \mu_i(L_i f)$ for all $f \in \mathcal{D}(G;M)$. Generalization to all of $\mathcal{D}(G)$ is immediate.

$$\mu(\varphi) = \sum \mu_i\,(L_i\varphi) \tag{25.28}$$

where the $\mu_i$ are finitely many*measures* with compact support and the $L_i$ are finitely many differential operators, which may be assumed to be invariant if $G$ is a Lie group. The converse in obvious.

Any tangent vector at $t = 0$ to a curve $\gamma(t)$ with initial point $a \in G$ is such a distribution, namely $\varphi \longmapsto d\varphi(a; h)$. This can be written as

$$d\varphi(a;h) = \int \varphi(x)dh(x) = D_0\varphi[\gamma(t)]\,. \tag{25.29}$$

However strange this *notation* may seen – it is only a notation –, it makes it possible to unify integral and infinitesimal methods in representation theory.

Like in measure theory, the product of two distributions $\lambda$ and $\mu$ is defined by the formula

$$\varphi \longmapsto \iint \varphi(x,y)d\lambda(x)d\mu(y) = \int d\mu(y) \int \varphi(x,y)d\lambda(x)\,.$$

It is always well-defined if one of the two distributions has compact support, because if $\varphi$ and $\lambda$, for example, have compact support, so does the function $y \longmapsto \int \varphi(x,y)d\lambda(x)$ and it is also $C^\infty$. This can be seen using a generalization to distributions of theorems on differentiation under the $\int$ sign. The essential step consists in showing that, if $\gamma(t)$ is a $C^\infty$ curve, if $\varphi \in \mathcal{D}(G,M)$ and if $M'$ is a compact neighbourhood of $M$, then the function

$$\frac{\varphi\left[x,\gamma(t)\right] - \varphi\left[x,\gamma(0)\right]}{t}$$

converges to its limit in the topology of $\mathcal{D}(G,M')$ as $t$ tends to 0.

The Lebesgue-Fubini formula applies for all $\varphi \in C^\infty(G)$ if $\lambda$ and $\mu$ have compact supports; and (28) reduces the question to the case of measures.

Let us next suppose that $G$ is a Lie group. Like in Chap. XI, n° 25, it is then possible to associate to $\lambda$ and $\mu$ their *convolution product*, namely the distribution

$$\lambda * \mu : \varphi \longmapsto \iint \varphi(xy)d\lambda(x)d\mu(y)\,, \tag{25.30}$$

at least if $\lambda$ or $\mu$ has compact support. The product of three distributions is associative *if* two of them have compact support. It is even possible to associated convolution operators in $C^\infty(G)$ to every distribution $\mu$ with compact support, namely

$$L(\mu)f(x) = \int f\left(y^{-1}x\right) d\mu(y) = \mu * f(x)\,, \tag{25.31}$$

$$R(\mu)f(x) = \int f(xy)d\mu(y) = f * \mu'(x)\,, \tag{25.31'}$$

where $d\mu'(x) = d\mu(x^{-1})$. Since any locally integrable function $f$ defines a distribution $\varphi \longmapsto \int \varphi(x)f(x)dx$, $\mu * f$ and $f * \mu$ can be defined for any distribution $\mu$ with compact support. But the result may be a genuine distribution, not a function.

Operators $L(\mu)$ (resp. $R(\mu)$) clearly commute with right (resp. left) translations. If $\mu$ is the distribution with support $\{e\}$ defined by some $X = \gamma'(0) \in G'(e)$, then

$$X * \varphi(g) = \int \varphi\left(x^{-1}g\right) dX(x) = D_0\varphi\left[\gamma(t)^{-1}g\right]\,,$$

whence

$$X * \varphi(g) = -d\varphi(g; Xg) = L(X)\varphi(g)\,. \tag{25.32}$$

Similarly,

$$R(X)\varphi(g) = -\varphi * X(g) = d\varphi(g; gX)\,. \tag{25.32'}$$

This supposes $\varphi \in C^\infty(G)$.

Since $R(X)\varphi = -\varphi * X$,

$$R(X)R(Y)\varphi - R(Y)R(X)\varphi = (\varphi * Y) * X - (\varphi * X) * Y\,.$$

Hence, the left hand side being equal to $R([X,Y])\varphi = -\varphi * [X,Y]$, in terms of distributions,

$$[X,Y] = X * Y - Y * X\,. \tag{25.33}$$

Formulas (32) and (32') suggest a generalization to distributions: for any distribution $\mu$ on $G$, we will set

$$L(X)\mu = X * \mu\,, \quad R(X)\mu = -\mu * X\,. \tag{25.34}$$

By (30), $X * \mu$ is the distribution

$$\varphi \longmapsto \iint \varphi(xy)dX(x)d\mu(y) = \int d\mu(y) \int f(xy)dX(x) =$$

$$= \int D_0\varphi\,[\gamma(t)y]\,d\mu(y) = -\int L(X)\varphi(y)d\mu(y)\,.$$

Hence, using Schwartz's notation $\langle\varphi,\mu\rangle = \mu(\varphi)$, we get equivalent formulas

$$\langle\varphi, L(X)\mu\rangle + \langle L(X)\varphi,\mu\rangle = 0\,, \quad \langle\varphi, X * \mu\rangle + \langle X * \varphi, \mu\rangle = 0 \tag{25.35}$$

similar (and for good reason...) to the definition of the derivative of a distribution of a variable [Chap. V, formula (35.3)]. It can similarly be shown that

$$\langle\varphi, R(X)\mu\rangle + \langle R(X)\varphi,\mu\rangle = 0\,, \quad \langle\varphi, \mu * X\rangle + \langle\varphi * X, \mu\rangle = 0\,. \tag{25.35'}$$

This relation can in particular be applied if $d\mu(g) = f(g)d_rg$ where $f$ is a locally integrable function, i.e. if $\langle\varphi,\mu\rangle = \int\varphi(x)f(x)d_rx$ for all $\varphi \in \mathcal{D}(G)$. Formula (17) obtained by proving theorem 39 therefore shows that, if the *function* $R(X)f(g) = D_0f[g.\exp(tX)]$ exists and is continuous, then

$$\int R(X)f(g).\varphi(g)d_rg + \int f(g).R(X)\varphi(g)d_rg = 0 \tag{25.36}$$

for all $\varphi \in \mathcal{D}(G)$. This means that *existence and continuity of the function* $R(X)f(g)$ *imply*

$$R(X)f = -f * X \qquad \text{in the sense of distributions}\,. \tag{25.36'}$$

The two possible interpretations of $R(X)f$ are therefore compatible in this case. It can similarly be shown that by identifying $f$ with the distribution $f(g)d_lg$, existence and la continuity of $L(X)f(g) = D_0f[g.\exp(-tX)]$ imply

$$L(X)f = X * f \qquad \text{in the sense of distributions}\,. \tag{25.36''}$$

Incidentally, observe an important consequence of (36). Using the inner product on the space $L^2(G)$ of the measure $d_rg$, (36) shows that $(R(X)\varphi|\psi) = (\varphi|R(X)\psi)$ for all $\varphi, \psi \in \mathcal{D}(G)$. Considered an operator in $\mathcal{D}(G)$, $iR(X)$ is therefore symmetric. Hence, if $(X_i)$ is a basis of $G'(e)$, the operator $\sum R(X_i)^2$ is symmetric $> 0$. In the language of PDE theory, this means that it is an *elliptic* differential order of the second order, analogous to the usual Laplacian.

Linear combinations of distributions $X_1 * \ldots * X_p$, with $X_i$ in $G'(e)$, are precisely the distributions with support $\{e\}$ or, equivalently in any manifold, those for which $\mu(f)$ is a linear combination of partial derivatives of $f$ at $e$, in every chart at $e$. Definition (30) of a convolution product shows that if $X_i = \gamma_i'(0)$, then $X_1 * \ldots * X_p$ is a linear functional

$$\varphi \longmapsto D_1 \ldots D_p \varphi \left[\gamma_1(t_1) \ldots \gamma_p(t_p)\right],$$

the derivatives $D_i$ with respect to $t_i$ being calculated at 0.

The set $\mathcal{U}(\mathfrak{g})$ of these distributions is an infinite dimensional associative algebra generated by $X \in G'(e)$. It is the *universal enveloping algebra* of the complex Lie algebra[105] $\mathfrak{g}$, systematically used from the start by Harish-Chandra, who was probably inspired by Chevalley. Associating the operator $L(\mu)$ (resp. $R(\mu)$) defined in (34) to each $\mu \in \mathcal{U}(\mathfrak{g})$ gives *all* the differential operators on $G$ commuting with the translations, because such an operator $L$ defines a distribution $\varphi \longmapsto L\varphi(e)$ with support $\{e\}$ which it is fully determined by due to its invariance. So $\mathcal{U}(\mathfrak{g})$ can be identified with *the algebra of left (resp. right) invariant differential operators*, which I will instead denote by $\mathcal{U}(G)$. For all $p$, the set $\mathcal{U}_p(G)$ of operators of order $\leq p$ is finite-dimensional.

The definition of convolution products readily shows that, for all $X \in G'(e)$ and all $a \in G$,

$$aX = \epsilon_a * X, \quad Xa = X * \epsilon_a$$

provided tangent vectors $aX$ and $Xa$ to $G$ at $a$ are identified with distributions having support $\{a\}$. This gives the formula

$$ad(a)X = \epsilon_a * X * \epsilon_{a^{-1}}.$$

A similar formula allows $ad(a)\mu$ to be defined for all distributions on $G$. It is the image of $\mu$ under the automorphism $x \longmapsto axa^{-1}$ of $G$. If $\mu$ is of the form $f(x)dx$, $ad(a)\mu$ is defined by the function $f(a^{-1}xa)$ which we could denote by $ad(a)f(x)$.

*Central* distributions, i.e. invariant under operators $ad(a)$, are easily characterized: they must satisfy

$$X * \mu = \mu * X$$

for all $X \in G'(e)$, and this condition is sufficient if $G$ is *connected.* To see this, we compute the differential of the function

$$g \longmapsto \langle \varphi, ad(g)\mu \rangle = \int \varphi\left(gxg^{-1}\right) d\mu(x)$$

[105] To tell the truth, this how an "abstract" algebra associated to any Lie algebra is called. But for Lie groups, it is isomorphic to the one in the text. On this topic, see J. Dixmier's book *Algèbres enveloppantes* (Gauthier-Villars, 1974), or N. Bourbaki's first volume, *Groupes et algèbres de Lie* (the chapter on Lie groups, dedicated to infinite-dimensional analytic groups over more general fields than $\mathbb{R}$ or $\mathbb{C}$, is practically unreadable. The founding members, who had left the group when it was written, would never have allowed such a monster...).

for $\varphi \in \mathcal{D}(G)$. It is possible to differentiate under the $\int$ sign and to use the differentiation formula of a product. So the result is

$$\begin{aligned} D_0\varphi\left[\gamma(t)gxg^{-1}\gamma(t)^{-1}\right] &= D_0\varphi\left[\gamma(t)gxg^{-1}\right] + D_0\varphi\left[gxg^{-1}\gamma(t)^{-1}\right] \\ &= -X * \varphi\left(gxg^{-1}\right) + \varphi * X\left(gxg^{-1}\right) \end{aligned}$$

if $X = \gamma'(0)$. Hence, if $ad(g)\mu = \mu$,

$$\langle X * \varphi, \mu\rangle = \langle \varphi * X, \mu\rangle .$$

So by (35) and (36), the condition is necessary. The converse follows from similar computations.

The operators $ad(g)$ are homomorphisms with respect to the convolution product. On the other hand, they map $\mathcal{U}_p(G)$ to $\mathcal{U}_p(G)$ for all $p$, thus defining linear representations of $G$ on the spaces $\mathcal{U}_p(G)$. Denoting them by $ad_p(g)$, there are operators

$$ad_p(X) = ad_p'(e)X$$

corresponding to them. Previous calculations show that

$$ad_p(X)\mu = X * \mu - \mu * X$$

like in $G'(e)$. Calculating $ad_p(\nu)\mu$ for an arbitrary distribution $\nu \in \mathcal{U}(G)$ is a different problem altogether, as the reader will convince himself easily.

If $\pi$ is a linear representation of $G$ on a finite-dimensional vector space $\mathcal{H}$, as in the case of the measure in Chap. XI, the operator

$$\pi(\mu) = \int \pi(g)d\mu(g) \tag{25.37}$$

can be associated to every distribution $\mu$ with compact support. The formula is well-defined since $\pi(g)$ is $C^\infty$. The definition of convolution products shows that

$$\pi(\lambda * \mu) = \pi(\lambda)\pi(\mu) . \tag{25.38}$$

If $\mu$ is a vector $X \in G'(e)$, we recover $\pi(X) = \pi'(e)X$.

## 26 – Lie Algebras in Infinite-Dimension

(i) *The subspace* $\mathcal{H}^\infty$. It was shown above that every finite-dimensional representation $(\mathcal{H}, \pi)$ of $G$ has an associated Lie algebra representation $\mathfrak{g}$ obtained by setting

$$\pi(X)\mathbf{a} = D_0\pi\left[\gamma(t)\right]\mathbf{a} = \lim \frac{\pi\left[\gamma(t)\right]\mathbf{a} - \mathbf{a}}{t} \tag{26.1}$$

for every tangent curve to $X$ at $t = 0$. In infinite dimension, though the situation is not intractable, it is considerably more complicated. It has given rise to some difficult theorems[106] and entails many pitfalls.

Let us consider a continuous representation $\pi$ of $G$ on a Banach space $\mathcal{H}$ and, for $\mathbf{a} \in \mathcal{H}$, the function $p(g) = \pi(g)\mathbf{a}$. Even if derivative (1) exists for all tangent

[106] This section follows closely the more complete though older presentation of G. Warner's in *Harmonic Analysis on Semisimple Lie Groups* (Springer, 1972), vol. I, pp. 252–304.

curves $\gamma(t)$ to a given $X$, it is not a priori obvious that the result only depends on $X$. To avoid this ambiguity, we set

$$(26.2) \qquad \pi(X)\mathbf{a} = D_0\pi\left[\gamma_X(t)\right]\mathbf{a}$$

each time this derivative exists. As

$$\pi\left[g\gamma(t)\right]\mathbf{a} = \pi(g)\pi\left[\gamma(t)\right]\mathbf{a},$$

the operator $R(X)$ defined by (25.16) can be applied to the function $\pi(g)\mathbf{a}$. The result, namely

$$(26.3) \qquad R(X)\pi(g)\mathbf{a} = \pi(g)\pi(X)\mathbf{a},$$

proves that $g \longmapsto R(X)\pi(g)\mathbf{a}$ is continuous. Theorem 39 then shows that *the existence of derivative (2) for all $X$ is necessary and sufficient*[107] *for the function $\pi(g)\mathbf{a}$ to be of class $C^1$.*

The set $\mathcal{H}^\infty$ of $\mathbf{a} \in \mathcal{H}$ such that the function $\pi(g)\mathbf{a}$ is $C^\infty$ is, therefore, the set of vectors for which the expression

$$(26.4) \qquad \pi(X_1)\dots\pi(X_p)\mathbf{a}$$

is well-defined for all $p$ and $X_i$. We will denote by $\pi^\infty(X)$ the restriction of $\pi(X)$ to $\mathcal{H}^\infty$ for all $X$. Generalization to distributions $\mu \in \mathcal{U}(G)$ is obvious, with

$$\pi^\infty(\mu * \nu) = \pi^\infty(\mu)\pi^\infty(\nu).$$

The subspace $\mathcal{H}^\infty$ is $\pi(g)$-invariant since

[107] This point is not always properly justified. See for example A. Knapp, *Representation theory of semisimple groups...* (Princeton UP, 1986), proof of lemma 3.13, p. 55: *For each* **a** *the map $g \longmapsto \pi(g)\pi(X)\mathbf{a}$ is continuous, and thus $\pi(g)\mathbf{a}$ has continuous first partials everywhere and must be of class $C^1$* (I have changed Knapp's notation) ; N. Wallach, *Real Reductive Groups* (Academic Press, 1988), vol I, pp. 33, lemma 1.6.4, uses the same argument to show that $\mathcal{H}^\infty$ is complete ; Warner, *Harmonic Analysis...*, vol I, p. 252, Example, also uses it to show that, in the regular representation of $G$ on $L^2(G)$, the $f \in \mathcal{D}(G)$ are $C^\infty$ vectors (true, but for other reasons). D. Bump's arguments, *Automorphic Forms and Representations*, lemma 2.4.2, p. 188 are hardly more convincing.

Knapp's argument would be correct if the functions $R(X)f(g) = D_0 f[g\exp(tX)]$ were, as he says, *partials*, i.e. partial derivatives of $f$ at $g$ with respect to a local chart *independent of $g$*. However, as remarked above, they are calculated in a chart which depends on it. The whole problem is showing that if derivatives

$$D_0 f\left[\exp(Y)\exp(tX)\right]$$

exist and are continuous functions of $Y \in W$, then the same is true for genuine *partials*

$$D_0 f\left[\exp(Y + tX)\right]$$

which Knapp does not write down.

By the way, note that F. Bruhat, *Sur les représentations induites des groupes de Lie* (Bull. SMF, 84, 1956, pp. 97–205) is the first author to have systematically used $C^\infty$ vectors, distribution theory and that of locally convex spaces to prove theorems (irreducibility of "almost all" principal series representations).

$$D_0\pi\left[\gamma(t)\right]\pi(g)\mathbf{a} = \pi(g)D_0\pi\left[g^{-1}\gamma(t)g\right]\mathbf{a} = \pi(g)D_0\pi\left[\delta(t)\right]\mathbf{a},$$

where $\delta'(0) = ad(g)^{-1}\gamma'(0)$, whence the existence of

$$\pi(X)\pi(g)\mathbf{a} = \pi(g)\pi\left[ad(g)^{-1}X\right]\mathbf{a}.$$

the result and the formula

$$\pi^{\infty}\left[ad(g)X\right] = \pi(g)\pi^{\infty}(X)\pi(g)^{-1} \tag{26.5}$$

follow by iteration. We will see in the next section (ii) that operators $\pi^{\infty}(\mu)$ can also be defined for every distribution with compact support.

Then, like in finite dimension,

$$\pi^{\infty}\left([X,Y]\right) = \left[\pi^{\infty}(X), \pi^{\infty}(Y)\right] \quad \text{in } \mathcal{H}^{\infty}, \tag{26.6}$$

because (3) shows first that $R(Y)\pi(g)\mathbf{a} = \pi(g)\pi^{\infty}(Y)\mathbf{a}$ then, applied to $\pi^{\infty}(Y)\mathbf{a}$, that

$$R(X)R(Y)\pi(g)\mathbf{a} = R(X)\pi(g)\pi^{\infty}(Y)\mathbf{a} = \pi(g)\pi^{\infty}(X)\pi^{\infty}(Y)\mathbf{a}.$$

This give (6) since $R([X,Y]) = [R(X), R(Y)]$ even when these differential operators are applied to functions with values in a Banach space.

(ii) *Weak differentiability and strong differentiability.* In finite dimension, a function with values in $\mathcal{H}$ is checked to be $C^p$ by verifying that so are its components with respect to some basis. It is then a trivial result applicable to all $C^{\infty}$ manifolds. This result can be generalized here, but less trivially.

Let $\mathcal{H}'$ denote the topological dual of the Banach space $\mathcal{H}$. It is the set of continuous linear functionals on $\mathcal{H}$ equipped with the obvious norm. For $\mathbf{x} \in \mathcal{H}$ and $\mathbf{y} \in \mathcal{H}'$, we set $\langle \mathbf{x}|\mathbf{y}\rangle = \mathbf{y}(\mathbf{x})$. If $p$ is a $C^{\infty}$ map from $G$ to $\mathcal{H}$, for any $\mathbf{x} \in \mathcal{H}'$, the function

$$p_{\mathbf{x}}(g) = \langle p(g)|\mathbf{x}\rangle \tag{26.7}$$

is clearly in $C^{\infty}(G)$ since it is the composition of $p$ and a continuous linear map from $\mathcal{H}$ to $\mathbb{C}$.

To prove the converse, we first note that every $\mathbf{a} \in \mathcal{H}$ defines a continuous linear functional on $\mathcal{H}'$, namely $\mathbf{x} \longmapsto \langle \mathbf{a}|\mathbf{x}\rangle$, which gives a canonical map from $\mathcal{H}$ to $\mathcal{H}'' = (\mathcal{H}')'$. It is easily shown (Hahn-Banach) to be isometric. So $\mathcal{H}$ is a *closed* subspace of $\mathcal{H}''$. Hence to show that a map with values in $\mathcal{H}$ is $C^{\infty}$, it suffices to show that it is as a map in the dual of $\mathcal{H}'$. Replacing $\mathcal{H}$ by $\mathcal{H}'$ reduces the proof to the next result whose author is unknown to me,:

**Theorem 40.** *Let $\mathcal{H}$ be a Banach space, $G$ a $C^{\infty}$ manifold and $p$ a map from $G$ to the dual $\mathcal{H}'$ of $\mathcal{H}$. $p$ is $C^{\infty}$ if and only if the function $\langle \mathbf{x}|p(g)\rangle$ is $C^{\infty}$ for all $\mathbf{x} \in \mathcal{H}$.*

The question being local, $G$ may be assumed to be open in $\mathbb{R}^n$. We first suppose that $n = 1$ and set $D = d/dt$. By its very definition, the derivative $D\langle \mathbf{x}|p(t)\rangle$ is the limit of a sequence of continuous linear functionals in $\mathbf{x}$. Hence it is continuous.[108] Thus there is a vector $p'(t) \in \mathcal{H}'$ such that

[108] The proof given in Chap. XI, n° 19, (v), lemma 6 for Hilbert spaces readily generalizes. The Banach-Steinhaus theorem [Chap. XI, n° 15, (ii), note 52] can also be used. It shows that if $\lim\langle \mathbf{x}|\mathbf{a}_n\rangle = f(\mathbf{x})$ exists for all $\mathbf{x} \in \mathcal{H}$, then $\sup \|\mathbf{a}_n\| < +\infty$ and as a consequence $|f(\mathbf{x})| \leq M\|\mathbf{x}\|$. More generally if $M$ is a subset of $\mathcal{H}'$ on which the function $\mathbf{y} \longmapsto \langle \mathbf{x}|\mathbf{y}\rangle$ is bounded for all $\mathbf{x}$, then so is the function $\|\mathbf{y}\|$ ("any weakly bounded subset is strongly bounded"). These arguments extend to more general spaces, for example Fréchet spaces. N. Bourbaki's book, *Espaces vectoriels topologiques*, is the unavoidable reference for maximum generality and, almost always, for clearness and simplicity of proofs .

$$(26.8) \qquad D\langle \mathbf{x}|p(t)\rangle = \big\langle \mathbf{x}|p'(t)\big\rangle \quad \text{for all } \mathbf{x}.$$

The left hand side being $C^\infty(G)$, this result applies to $p'$. Thus there is a vector $p''(t) \in \mathcal{H}$ such that $D^2\langle \mathbf{x}|p(t)\rangle = \langle \mathbf{x}|p''(t)\rangle$, etc.

The left hand side of (8) being bounded on every compact subset of $G$ for all $\mathbf{x}$, so is the right hand side, hence so is $\|p'(t)\|$ (Banach-Steinhaus). In

$$\langle \mathbf{x}|p(t+h) - p(t)\rangle = \int_0^h \big\langle \mathbf{x}|p'(u)\big\rangle\, du\,,$$

the right hand side is thus bounded above by $M|h|.\|\mathbf{x}\|$ for sufficiently small $|h|$. Therefore,

$$\|p(t+h) - p(t)\| = O\left(|h|\right)$$

for sufficiently small $|h|$. Hence the functions $p, p', p'', \ldots$ are continuous.

$$\big\langle \mathbf{x}|p(u+h) - p(u) - p'(u)h\big\rangle = \int_0^h \big\langle \mathbf{x}|p'(u+t) - p'(u)\big\rangle\, dt$$

then likewise shows that

$$\big\|p(u+h) - p(u) - hp'(u)\big\| = o(h)\,.$$

As a result, $p'$ is the derivative of $p$ in the strong topology, the function $p$ is of class $C^1$, hence so are $p', p'', \ldots$, proving the theorem for $n = 1$.

For arbitrary $n$, the previous result, applied to the function $p(u + tX)$ for given $u \in G$ and $X \in \mathbb{R}^n$, first of all shows the existence of the derivative $D_0 p(u + tX)$ with respect to the strong topology. It depends linearly on $X$ since, for all $\mathbf{x}$, the differential of $\langle \mathbf{x}|p(u)\rangle$ is linear in $X$. Setting $D_0 p(u + tX) = p'(u)X$ defines a linear map $p'(u)$ de $\mathbb{R}^n$ on $\mathcal{H}'$. The function $\|p'(u)\|$ is bounded on every compact subset of $G$ since so is the function $\langle \mathbf{x}|p'(u)X\rangle = D_0\langle \mathbf{x}|p(u + tX)\rangle$ for all $X$ and $\mathbf{x}$ (Banach-Steinhaus). For sufficiently small $\|h\|$,

$$\langle \mathbf{x}|p(u+h) - p(u)\rangle - \int_0^1 \langle \mathbf{x}|p'(u+th)h\rangle\, dt\,,$$

then shows that $\|p(u+h) - p(u)\| = O(\|h\|)$. Thus $p$ is continuous with respect to the strong topology of $\mathcal{H}'$. However, for all $X \in \mathbb{R}^n$ and all $\mathbf{x} \in \mathcal{H}$, the function

$$\big\langle \mathbf{x}|p'(u)X\big\rangle = D_0\,\langle \mathbf{x}|p(u+tX)\rangle$$

is in $C^\infty(G)$ since $\langle \mathbf{x}|p(u)\rangle$. Hence, like $p$, $p'$ is a continuous function with respect to the strong topology and $\|p'(u)\|$ is bounded on every compact set. Writing

$$\big\langle \mathbf{x}|p(u+h) - p(u) - p'(u)h\big\rangle = \int_0^1 \big\langle \mathbf{x}|p'(u+th)h - p'(u)h\big\rangle\, dt\,,$$

we deduce that

$$(26.9) \qquad \big\|p(u+h) - p(u) - p'(u)h\big\| = o\left(\|h\|\right)$$

This shows that $p$ is *differentiable* at $u$ and that $p'(u)$ is the tangent linear map to $p$ at $u$. Since $p'$ is continuous, $p$ is of class $C^1$. Applying these arguments to $p', p''$, etc., the existence and continuity of all partial derivatives of $p$ follow, qed.

**Corollary 1.** *Let $p$ be a map from a manifold $G$ to a Banach space $\mathcal{H}$. $p$ is $C^\infty$ if and only if $\langle p(g)|\mathbf{x}\rangle$ is $C^\infty$ for all $\mathbf{x} \in \mathcal{H}'$.*

The proof follows by regarding $p$ as a map with values in the dual[109] of $\mathcal{H}'$.

**Corollary 2.** *Let $(\mathcal{H}, \pi)$ be a continuous linear representation of a Lie group $G$ on a Banach space $\mathcal{H}$. Then $\mathcal{H}^\infty$ is the set of $\mathbf{a} \in \mathcal{H}$ such that $\langle \pi(g)\mathbf{a}|\mathbf{x}\rangle$ is $C^\infty$ for all $\mathbf{x} \in \mathcal{H}'$.*

Apply corollary 1 to $p(g) = \pi(g)\mathbf{a}$.

(iii) *Convolution operators on* $\mathcal{H}^\infty$. Let us first show how for any manifold, $C^\infty$ functions with values in a Banach space $\mathcal{H}$ can be "integrated" with respect to a distribution $\mu$ with compact support: it suffices to define $\mu(p)$ by the condition

$$\langle \mu(p)|\mathbf{x}\rangle = \int \langle p(g)|\mathbf{x}\rangle \, d\mu(g) \tag{26.10}$$

for all $\mathbf{x} \in \mathcal{H}'$. To prove the existence of a vector $\mu(p)$ satisfying this condition, the easiest is to write $\mu$ as (25.28). The right hand side of (10) then equals

$$\sum \int \langle L_i p(g)|\mathbf{x}\rangle \, d\mu_i(g) \, .$$

As the function $L_i p(g)$ is continuous, the proof reduces to integrating a continuous function with values in $\mathcal{H}$ with respect to a genuine measure with compact support: Chap. XI, n° 4, (i). Hence

$$\int p(g) d\mu(g) = \sum \int L_i p(g).d\mu_i(g) \, .$$

$G$ denoting again a Lie group and $(\mathcal{H}, \pi)$ a continuous representation on a Banach space, one can associate an operator $\pi^\infty(\mu)$ defined by

$$\pi^\infty(\mu)\mathbf{a} = \int \pi(g)\mathbf{a}.d\mu(g) \quad \text{for all } \mathbf{a} \in \mathcal{H}^\infty \, ,$$

i.e. by

[109] At first glance, the theorem appears to follow from its corollary. It is not so: applied to $\mathcal{H}'$, the corollary would require that $\langle \mathbf{x}|p(g)\rangle$ is $C^\infty$ for all $\mathbf{x} \in \mathcal{H}''$, which is impossible to check in general if $\mathcal{H}'' \neq \mathcal{H}$. Incidentally, the following mistake should be avoided. If $\pi$ is a (continuous) representation from a group $G$ to $\mathcal{H}$ and if, for any continuous operator $T : \mathcal{H} \longmapsto \mathcal{H}$, $T' : \mathcal{H}' \longmapsto \mathcal{H}'$ denotes the transpose operator, the "contragredient" map $g \longmapsto \pi(g^{-1})'$ is a representation of $G$ on $\mathcal{H}'$. But in general, it *is not* continuous with respect to the strong topology. Counterexample: $G = \mathbb{R}$, $\mathcal{H} = L^1(\mathbb{R})$, $\pi$ being the representation by translations. Denoting by $f_t$ the function $x \longmapsto f(x-t)$, strong continuity of $t \longmapsto f_t$ for some $f \in L^\infty(G)$ means that $\|f_t - f\|_\infty$ tends to 0 as $t$ tends to 0. If $f$ is continuous, this requires $f$ to be uniformly continuous on $\mathbb{R}$.

N. Bourbaki's booklet, *Variétés différentiables et analytiques*, 2.6.2, mentions in two lines the corollary for every *almost complete* locally convex space $\mathcal{H}$, i.e. in which every closed and bounded set is complete (a set $B$ is *bounded* if all continuous seminorms on $B$ are bounded). This general result can easily be deduced from the corollary by observing that every continuous seminorm on $\mathcal{H}$ defines a continuous map from $\mathcal{H}$ to a Banach space.

(26.11) $$\langle\pi^{\infty}(\mu)\mathbf{a}|\mathbf{x}\rangle = \int \langle\pi(g)\mathbf{a}|\mathbf{x}\rangle\, d\mu(g) \quad \text{for all } \mathbf{x} \in \mathcal{H}'$$

to every distribution with compact support $\mu$. For $g \in G$ and $\mathbf{x} \in \mathcal{H}'$,

$$\begin{aligned}\langle\pi(g)\pi^{\infty}(\mu)\mathbf{a}|\mathbf{x}\rangle &= \langle\pi^{\infty}(\mu)\mathbf{a}|\pi(g)'\mathbf{x}\rangle = \\ &= \int \langle\pi(y)\mathbf{a}|\pi(g)'\mathbf{x}\rangle\, d\mu(y) = \int \langle\pi(gy)\mathbf{a}|\mathbf{x}\rangle\, d\mu(y)\,.\end{aligned}$$

This is the convolution of a $C^{\infty}$ function and a distribution with compact support. The result is $C^{\infty}$. So $\pi^{\infty}(\mu)$ maps $\mathcal{H}^{\infty}$ to $\mathcal{H}^{\infty}$. For all $\mathbf{a} \in \mathcal{H}^{\infty}$, $\pi^{\infty}(\mu)\mathbf{a}$ belongs to the closed subspace $\mathcal{H}(\mathbf{a})$ generated by the $\pi(g)\mathbf{a}$, because every $\mathbf{x} \in \mathcal{H}'$ orthogonal to these is orthogonal to $\pi^{\infty}(\mu)\mathbf{a}$ by (11). If $\mathcal{H}$ is infinite-dimensional, the vectors $\pi^{\infty}(\mu)\mathbf{a}$ associated to $\mu \in \mathcal{U}(G)$ are not necessarily everywhere dense in $\mathcal{H}(\mathbf{a})$. Counterexample: take $G = \mathbb{R}$, $\mathcal{H} = L^2(G)$ and a function $\mathbf{a} \in \mathcal{D}(G)$.

Calculations used for measures [Chap. XI, n° 25, (i)] also easily show that $\pi^{\infty}(\lambda)\pi^{\infty}(\mu) = \pi^{\infty}(\lambda * \mu)$. Hence in $\mathcal{H}^{\infty}$ we get all the noncommutative polynomials in $\pi^{\infty}(X)$ and $\pi(\mu)$ associated to measures with compact support. Generalizing definition (25.31') of $R(\mu)$ in an obvious manner to all functions with values in $\mathcal{H}$, relations (25.34) and (25.36) imply

(26.12) $$R(\mu)\pi(g)\mathbf{a} = \pi(g)\pi^{\infty}(\mu)\mathbf{a}\,,$$

(26.13) $$\pi^{\infty}\left[ad(g)\mu\right] = \pi(g)\pi^{\infty}(\mu)\pi(g)^{-1}\,.$$

$\mathcal{H}^{\infty}$ can be equipped with a topology using seminorms

$$p_{\mu}(\mathbf{a}) = \|\pi^{\infty}(\mu)\mathbf{a}\|$$

where $\mu \in \mathcal{U}(G)$, distributions with support $\{e\}$. Until further notice, we will only consider these. We could even confine ourselves to measures $\mu$ such as

$$\mu = X_1 * \ldots * X_p$$

where all $X_i$ belong to a given basis of $G'(e)$. As the function $\|\pi(g)\|$ is bounded on every compact set, relation $\lim \mathbf{a}_n = \mathbf{a}$ means that the functions $R(\mu)\pi(g)\mathbf{a}_n$, or the successive derivatives of $\pi(g)\mathbf{a}_n$ in any chart, converge uniformly on every compact set. The topology of $\mathcal{H}^{\infty}$ is thus identical to that of Schwartz on the space $C^{\infty}(G;\mathcal{H})$ of $C^{\infty}$ maps with values in $\mathcal{H}$. Applying definitions, it is not very difficult to show that (i) $\mathcal{H}^{\infty}$ is*complete* (hence is a Fréchet space, because its topology is defined by a countable family of seminorms; see Dieudonné, XII), (ii) for every distribution $\mu$ with compact support, the operator $\pi^{\infty}(\mu)$ is *continuous* on $\mathcal{H}^{\infty}$, (iii) for all $\mathbf{a} \in \mathcal{H}^{\infty}$, the map $g \longmapsto \pi(g)\mathbf{a}$ from $G$ to $\mathcal{H}^{\infty}$ is $C^{\infty}$ with respect to the topology of $\mathcal{H}^{\infty}$.

(iv) *The Dixmier-Malliavin theorem.* Let us first show that $\mathcal{H}^{\infty}$ is everywhere dense in $\mathcal{H}$, otherwise previous considerations would be devoid of interest. The proof is the same, slightly less trivial, than the one given above in finite dimension. Indeed every $\varphi \in \mathcal{D}(G)$ defines a continuous operator

$$\pi(\varphi) = \int \pi(x)\varphi(x)dx$$

on $\mathcal{H}$, and

$$\pi(g)\pi(\varphi)\mathbf{a} = \int \pi(gx)\mathbf{a}.\varphi(x)dx = \int \pi(x)\mathbf{a}.\varphi\left(g^{-1}x\right)dx\,.$$

This convolution product is $C^\infty$ since it is the integral of a continuous function with compact support which depends in a $C^\infty$ manner on the parameter $g$. The fact that its values are in a Banach space is not a problem. Hence $\pi(\varphi)\mathbf{a} \in \mathcal{H}^\infty$, and as $\varphi \in \mathcal{D}(G)$ enables us to construct Dirac sequences $(\varphi_n)$, every $\mathbf{a} \in \mathcal{H}$ is the *limit* of vectors belonging to $\mathcal{H}^\infty$, namely the vectors $\pi(\varphi_n)\mathbf{a}$ [Chap. XI, n° 25, (v)], qed. Besides, clearly

$$\pi^\infty(\mu)\pi(\varphi) = \pi(\mu * \varphi)$$

for every distribution $\mu$ with compact support.

This result, spectacular for its time but easy,[110] has been greatly improved on:

**Theorem 41.** *For any representation $(\pi, \mathcal{H})$ of a Lie group $G$ on a Banach space, the Gårding subspace is identical to the subspace $\mathcal{H}^\infty$: every $\mathbf{a} \in \mathcal{H}^\infty$ is of the form*

$$\mathbf{a} = \pi(\varphi_1)\,\mathbf{a}_1 + \ldots + \pi(\varphi_p)\,\mathbf{a}_p$$

*with $\mathbf{a}_i \in \mathcal{H}^\infty$ and $\varphi_i \in \mathcal{D}(G)$ for all $i$.*

For all $i$, the supports of $\varphi_i$ can also be required to be in a given arbitrary neighbourhood of $e$. Inspired by methods from PDE theory, Dixmier's and Malliavin's very ingenious ten page proof requires little knowledge and even covers the case of representations on Fréchet spaces. The first step consists in proving the next result, starting with $G = \mathbb{R}$:

**Theorem 42.** *If $G$ is a Lie group, any function $p \in \mathcal{D}(G)$ is a finite sum of convolutions $\varphi * \psi$ where $\varphi, \psi \in \mathcal{D}(G)$.*

$G$ being assumed to be unimodular, we consider a discrete subgroup $\Gamma$ of $G$ (possibly reduced to $\{e\}$) and the representation $R$ of $G$ on $\mathcal{H} = L^p(\Gamma\backslash G)$ with $1 \le p < +\infty$, given by right translations.[111] $C^\infty$ vectors are functions $f$ for which the inner product

$$(R(x)f|g) = \int_{\Gamma\backslash G} f(yx)\overline{g(y)}dy \tag{26.14}$$

is $C^\infty$ for all $g \in L^q(\Gamma\backslash G)$. Hence $\mathcal{D}(\Gamma\backslash G) \subset \mathcal{H}^\infty$. For $\varphi \in \mathcal{D}(G)$, $R(\varphi)f = f * \varphi'$ where $\varphi'(x) = \varphi(x^{-1})$. By the Dixmier-Malliavin theorem, $\mathcal{H}^\infty$ is therefore the set of sums of products $f * \varphi$ with $\varphi \in \mathcal{D}(G)$ and $f \in L^p(\Gamma\backslash G)$. Thus

$$\mathcal{D}(\Gamma\backslash G) \subset \mathcal{H}^\infty \subset C^\infty(\Gamma\backslash G)\,.$$

If $\mu$ is a distribution with compact support, by (11), the operator $R^\infty(\mu)$ associated to representation $R$ is given by

$$(R^\infty(\mu)f|g) = \int_G (R(x)f|g)\,d\mu(x) = \int_G d\mu(x)\int_{\Gamma\backslash G} f(yx)\overline{g(y)}dy$$

[110] In Nancy, in 1947, Schwartz and I were rather surprised, not to say upset, when Lars Gårding published it. We could have proved it " in less than fifteen minutes " if only we had thought of it. Gårding's reputation is based on far harder successes in PDE theory. For the next two theorems see J. Dixmier and P. Malliavin, *Factorisations de fonctions et de vecteurs indéfiniment différentiables* (Bull. Sc. Math., **102**, 1978, pp. 305–330).

[111] The case $p = +\infty$ needs to be excluded because the representation of $G$ on $L^\infty$ is not continuous with respect to the strong topology.

for all $f \in \mathcal{H}^\infty$ and all $g \in L^q(G)$. If $g \in L(\Gamma\backslash G)$, differentiation under the $\int$ sign of the integral over $\Gamma\backslash G$ is possible since $f$ is $C^\infty$. (25.28) then shows that $(R^\infty(\mu)f|g) = (f * \mu'|g)$ where $d\mu'(g) = d\mu(g^{-1})$. As a result,

$$R^\infty(\mu)f = f * \mu' \quad \text{for all } f \in \mathcal{H}^\infty . \tag{26.15}$$

These vectors belong to the closed subspace $\mathcal{H}(f)$ generated by all $R(x)f$.

Let us give an outline of the proof that conversely all $f \in C^\infty(\Gamma\backslash G)$ such that $f * \mu \in L^p(\Gamma\backslash G)$ for all $\mu \in \mathcal{U}(G)$ are in $\mathcal{H}^\infty$. The proof resembles that of theorem 40. For given $X \in G'(e)$, we set

$$R(t) = R\left[\gamma_X(t)\right]$$

to be the right translation operator by $\gamma_X(t)$. For all $\varphi \in \mathcal{D}(\Gamma\backslash G)$, clearly $DR(t)\varphi = -R(t)(\varphi * X)$ in the topology of $\mathcal{H}$. Integrating, we deduce that

$$\|R(u)\varphi - \varphi\|_q = \|\varphi * X\|_q .O(u)$$

However,

$$\left(\frac{R(t)f - f}{t}\bigg|\varphi\right) = \left(f\bigg|\frac{R\left(t^{-1}\right)\varphi - \varphi}{t}\right).$$

As $t \longmapsto R(t)\varphi$ is $C^\infty$, the right hand side tends to $-(f|\varphi * X)$, which implies that $D_0(R(t)f|\varphi) = -(f|\varphi * X)$. But formula (25.35) for the distribution $d\mu(x) = f(x)dx$ shows that $(f|\varphi * X) + (f * X|\varphi) = 0$. Hence $D_0(R(t)f|\varphi) = (f * X|\varphi)$ for all $\varphi \in \mathcal{D}(\Gamma\backslash G)$ and more generally $D(R(t)f|\varphi) = (R(t)f * X|\varphi)$. Since $f * X \in L^p$, iterating is possible. Thus $D^2(R(t)f|\varphi) = (R(t)f * X * X|\varphi)$ is a continuous function of $t$. As a result,

$$(R(t)f - f + t.f * X|\varphi) = \int_0^t du \int_0^u (R(v)f * X * X|\varphi)\, dv$$

for all $\varphi \in \mathcal{D}(\Gamma\backslash G)$. The function being integrated being bounded above by $\|f * X * X\|_p \|\varphi\|_q$ for all $v$, it follows that

$$\|R(t)f - f + t.f * X\|_p = \|f * X * X\|_p O\left(t^2\right),$$

proving the existence of

$$D_0 R(t)f = -f * X$$

in the topology of $L^p$. The operator $\pi(X)$ of the general theory, defined by (2) when the relation is well-defined, is thus defined at $f$. But as $f * \mu \in L^p(\Gamma\backslash G)$ is assumed to hold for all $\mu \in \mathcal{U}(G)$, all functions $\pi(X_1)\ldots\pi(X_n)f$ are clearly defined, whence $f \in \mathcal{H}^\infty$. In conclusion:

**Theorem 43.** *Let $\Gamma$ be a discrete subgroup*[112] *of a Lie group $G$ and $R$ the representation of $G$ on $\mathcal{H} = L^p(\Gamma\backslash G)$, $1 \le p < +\infty$. Then $\mathcal{H}^\infty$ is the set of functions such that*

$$f * \mu \in \mathcal{H} \quad \text{for all } \mu \in \mathcal{U}(G)$$

*and then $R^\infty(\mu)f = f * \mu'$.*

---

[112] This assumption is obviously far too restrictive.

(v) *Analytic vectors.* Thanks to the exponential map, every Lie group $G$ can be equipped with an *analytic structure*. In a sufficiently small neighbourhood $V$ of 0, there is indeed a $C^\infty$ function $H(X,Y) \in G'(e)$ such that

$$\exp(X)\exp(Y) = \exp\left(H(X,Y)\right) .$$

The Campbell-Hausdorff formula

$$\begin{aligned} H(X,Y) = X + Y + [X,Y]/2 + \\ +[X,[X,Y]]/12 + [Y,[Y,X]]/12 - \\ -[X,[Y,[X,Y]]]/24 + \dots \end{aligned}$$

enables us to write $H(X,Y)$ as a series $\sum H_n(X,Y)$ having homogeneous functions of degree $n$ in $X$ and $Y$ as its terms and obtained by applying products of $n$ operators of the form $ad(X)$ or $ad(Y)$ to $X$ or $Y$. The formula is the same for *all* Lie groups and converges for sufficiently small $X$, $Y$. The analyticity of $H(X,Y)$ is a consequence of this,[113] and all $C^\infty$ (or even continuous) homomorphisms of Lie groups are analytic.

This raises the question of whether in every representation $(\mathcal{H},\pi)$ of $G$ there are vectors $\mathbf{a} \in \mathcal{H}$ such that $g \longmapsto \pi(g)\mathbf{a}$ is analytic.[114] The answer is that the set $\mathcal{H}^\omega$ of these vectors is everywhere dense in $\mathcal{H}$. This difficult result obtained by Harish-Chandra for semisimple groups was one of his first successes. P. Cartier and J. Dixmier generalized it to almost all groups, but like HC using structure theorems. The general case, due to E. Nelson, only uses techniques from partial differential equations (heat equation) and has given rise to surprising theorems.[115]

The operator $\pi^\infty(\mu)$ preserves $\mathcal{H}^\omega$ if $\mu \in \mathcal{U}(G)$ (but not if $\mu$ is a distribution with an arbitrary compact support). Then let $\mathcal{M}$ be a closed invariant subspace of $\mathcal{H}$. Considering the restrictions of $\pi(g)$ to $\mathcal{M}$, we get a representation of $G$ on $\mathcal{M}$ for which $\mathcal{M}^\omega = \mathcal{M} \cap \mathcal{H}^\omega$. These subspace is stable under operators $\pi(X)$.

Conversely, *if $\mathcal{E}$ is a stable subspace of $\mathcal{H}^\omega$ under the operators $\pi(X)$, the closure $\mathcal{M}$ of $\mathcal{E}$ in $\mathcal{H}$ is $\pi(g)$-invariant if $G$ is connected.* It suffices (Hahn-Banach) to show that every $\mathbf{x} \in \mathcal{H}'$ orthogonal to $\pi^\infty(\mu)\mathbf{a}$ for all $\mu \in \mathcal{U}(G)$ is also orthogonal to all $\pi(g)\mathbf{a}$. It is, however, obvious that this condition is equivalent to all derivatives of the function $\langle \pi(g)\mathbf{a}, \mathbf{x}\rangle$ being zero at the origin. As the latter is analytic, it is zero on the connected component of $e$, whence the result.[116] In particular, for all $\mathbf{a} \in \mathcal{H}^\omega$, the closed subspace generated by all $\pi(g)\mathbf{a}$ is the closure of the subspace of $\pi(\mu)\mathbf{a}$, $\mu \in \mathcal{U}(G)$.

When $G$ is semisimple with finite centre, there are always compact subgroups $K$ in $G$ (the pairwise conjugate, maximal compact subgroups) with the following property:[117] if $(\mathcal{H},\pi)$ is a reasonable (for example unitary) irreducible representation of $G$, and if the restriction of $\pi$ to $K$ is decomposed [Chap. XI, n° 29, (iii)] into subspaces $\mathcal{H}(\chi)$ corresponding to the characters of the irreducible representations of $K$, then $\dim \mathcal{H}(\chi) < +\infty$.

[113] Detailed presentation in my *Introduction à la théorie des groupes de Lie* (Publ. Math. Université Paris VII, 1982 or Springer, 2003), § 6.

[114] Theorem 40 remains valid for analytic maps on a *Banach* space. See Browder, Amer. J. Math., **84**, 1962, beginning of pp. 666–710, or Serge Lang, $SL_2(\mathbb{R})$, appendix 5, § 2.

[115] P. Cartier and J. Dixmier, Amer. J. Math., **80**, 1958, pp. 131–145; R. Nelson, Annals of Math., **70**, 1959, pp. 572–614. See also Lars Gårding, Bull. SMF, **88**, 1960, pp. 73–93. Warner's book presents most of these results.

[116] There is no analogous result for $\mathcal{H}^\infty$.

[117] For a fairly simple proof of this result of Harish-Chandra, see R. Godement, *A Theory of Spherical Functions* (Trans. AMS, **73**, 1952, pp. 496–556).

**Theorem 44.** *Let $G$ be a Lie group, $K$ a compact subgroup of $G$ and $(\mathcal{H}, \pi)$ a continuous representation of $G$ on a Banach space $\mathcal{H}$. Suppose that, for some character $\chi$ of $K$, the subspace $\mathcal{H}(\chi)$ is finite-dimensional. Then all $\mathbf{a} \in \mathcal{H}(\chi)$ are analytic.*

(a) Let us first show that $\mathbf{a} \in \mathcal{H}(\chi)$ is $C^\infty$. To this end, we consider the subspace of $\pi(\varphi)\mathbf{a}$ where $\varphi \in \mathcal{D}(G)$. The operator

$$E(\chi) = \pi\left(\overline{\chi}\right) = \int \pi(k)\overline{\chi(k)}dk$$

projects $\mathcal{H}$ onto $\mathcal{H}(\chi)$ (at least if $\chi$ is normalized in such a way that $\chi * \chi = \chi$). Since $E(\chi)\mathbf{a} = \mathbf{a}$, $\mathbf{a}$ is the limit of vectors of type $E(\chi)\pi(\varphi)\mathbf{a} = \pi(\overline{\chi})\pi(\varphi)\mathbf{a} = \pi(\overline{\chi} * \varphi)\mathbf{a}$. The set of these vectors is a vector subspace of $\mathcal{H}(\chi)$. Hence, since $\dim \mathcal{H}(\chi) < +\infty$, there exists $\varphi \in \mathcal{D}(G)$ such that $\mathbf{a} = \pi(\overline{\chi} * \varphi)\mathbf{a}$. Thus $\mathbf{a} \in \mathcal{H}^\infty$.

(b) Let us show that there is a basis $(X_i)$ of $G'(e)$ such that the element $\Delta = \sum X_i^2$ of $\mathcal{U}(G)$ is $ad_2(k)$-invariant for all $k \in K$. For this note that, $K$ being compact, there is an $ad(k)$-invariant positive definite quadratic form on $G'(e)$. Every orthonormal basis with respect to this quadratic form is suitable. Indeed there are equalities $ad(k)X_i = \sum a_i^p(k)X_p$ where the matrix of $a_i^p$ is orthogonal, whence $ad_2(k)\Delta = \sum_i a_i^p(k)a_i^q(k)X_pX_q = \Delta$.

(c) As $\pi^\infty(\Delta)$ commutes with all $\pi(k)$, $\pi^\infty(\Delta)\mathcal{H}(\chi) \subset \mathcal{H}(\chi)$. Since $\mathcal{H}(\chi)$ is finite-dimensional, there is a basis for $\mathcal{H}(\chi)$ all of whose elements satisfy a relation of the form $[\pi^\infty(\Delta) - \lambda]^p\mathbf{a} = 0$ for some $\lambda \in \mathbb{C}$ and some integer $p \geq 1$. For all $\mathbf{x} \in \mathcal{H}'$, the function $f(g) = \langle \pi(g)\mathbf{a}, \mathbf{x}\rangle$ then satisfies the differential equation $[R(\Delta) - \lambda]^p f = 0$. However, in the previous n°, using (25.35'), the differential operator $R(\Delta)$ was found to be elliptic (and obviously with analytic coefficients). PDE specialists will explain that then the same is true for $[R(\Delta) - \lambda]^p$ and that every solution (function or distribution) of an elliptic PDE with analytic coefficients is an analytic function. The function $\langle \pi(g)\mathbf{a}, \mathbf{x}\rangle$ is therefore analytic for all $\mathbf{x} \in \mathcal{H}$, hence so are $\mathbf{a}$ and as a result so are all vectors in $\mathcal{H}(\chi)$, qed.

(vi) *The case of unitary representations.* All of the above applies to what is the most important case, that of unitary representations. As $\mathcal{H}' = \mathcal{H}$, corollary 2 of theorem 40 immediately gives the characterization of $C^\infty$ vectors in $\mathcal{H}$.

For $\mathbf{a}, \mathbf{a}' \in \mathcal{H}^\infty$ and any distribution $\mu$ with compact support,

$$\left(\pi^\infty(\mu)\mathbf{a}|\mathbf{a}'\right) = \left(\mathbf{a}|\pi^\infty(\widetilde{\mu})\mathbf{a}'\right) \tag{26.16}$$

like in the case of measures and by the same calculations, and in particular

$$\left(\pi^\infty(X)\mathbf{a}|\mathbf{a}'\right) + \left(\mathbf{a}|\pi^\infty(X)\mathbf{a}'\right) = 0 \tag{26.16'}$$

for all $X \in G'(e)$, generalizing (25.36). So the operator $i\pi^\infty(X)$ is *symmetric* on $\mathcal{H}^\infty$. The next result can then be deduced:

**Theorem 45.** *Let $(\mathcal{H}, \pi)$ be an irreducible unitary representation of $G$ and $\mu$ a distribution with compact support. If $\mu$ is central, the operator $\pi^\infty(\mu)$ is a scalar.*

For all scalars $\lambda$, the operator

$$S = \left[\pi^\infty(\widetilde{\mu}) + \overline{\lambda}1\right]\left[\pi^\infty(\mu) + \lambda 1\right] : \mathcal{H}^\infty \longrightarrow \mathcal{H}^\infty$$

is positive symmetric by (16) and commutes with all $\pi(g)$, hence is a scalar [Chap. XI, n° 23, (ii), Schur's lemma II: use the self-adjoint canonical extension of $S$]. As this is the case for $\lambda = 0$, the operator $\lambda\pi^\infty(\widetilde{\mu}) + \overline{\lambda}\pi^\infty(\mu)$ is a scalar for all $\lambda$, for example for $\lambda = \pm 1$, qed.

Let us show that, in every unitary representation, the operator $i\pi(X)$, defined by (2) and extending $i\pi^\infty(X)$, is *self-adjoint.* As $\pi(X)\mathbf{a}$ is the derivative of the function $\pi[\gamma_X(t)]\mathbf{a}$ at $t = 0$ when it exists, it suffices to prove it for $G = \mathbb{R}$. Proposition (ii) of the next theorem will be sufficient.

**Theorem 46 (M. H. Stone).** *Let $t \longmapsto U(t)$ be a unitary representation of $\mathbb{R}$ on a Hilbert space $\mathcal{H}$.*

(i) *There is a spectral measure $M$ on $\mathbb{R}$ such that*

$$U(t) = \int \exp(i\lambda t) dM(\lambda) \quad \text{for all } t. \tag{26.17}$$

(ii) *$D_0 U(t)\mathbf{a}$ exists if and only if $\mathbf{a}$ belongs to the domain of definition of the self-adjoint operator*

$$H = \int \lambda dM(\lambda)\,; \tag{26.18}$$

*then*

$$D_0\left[U(t)\mathbf{a}\right] = iH\mathbf{a}\,. \tag{26.19}$$

(iii) *The function $U(t)\mathbf{a}$ is $C^\infty$ if and only if*

$$\mathbf{a} \in \bigcap \operatorname{Def}(H^n)\,.$$

Proposition (i) holds for every commutative lcg. It can be obtained directly from the spectral theory of u Chap. XI, n° 22 applied to the GN algebra generated by operators $U(f) = \int f(t)dt$ where $f \in L^1$. Like F. Riesz for $G = \mathbb{R}$, one can also deduce Bochner's theorem [Chap. XI, n° 30, (iii), exercise 4].

To prove point (ii), use spectral measures $d\mu_{a,b}(\lambda)$ associated to $M$. Hence

$$(U(t)\mathbf{a}|\mathbf{b}) = \int \exp(i\lambda t) d\mu_{a,b}(\lambda)\,.$$

As was seen in n° 23 of Chap. XI, the subspace $\operatorname{Def}(H)$ is the set of $a \in \mathcal{H}$ such that

$$\int \lambda^2 d\mu_{a,a}(\lambda) < +\infty\,. \tag{26.20}$$

Then the value of the left hand side is $\|H\mathbf{a}\|^2$ and

$$(H\mathbf{a}|\mathbf{x}) = \int \lambda d\mu_{a,x}(\lambda)$$

for all $\mathbf{x} \in \mathcal{H}$, the integral being convergent.

Calculation rules of Chap. XI, n° 22, (ii) then show that, for all $\mathbf{a} \in \operatorname{Def}(H)$,

$$\left\| \frac{U(t) - 1}{t}\mathbf{a} - iH\mathbf{a} \right\|^2 = \int \left| \frac{\exp(i\lambda t) - 1}{t} - i\lambda \right|^2 d\mu_{a,a}(\lambda)\,. \tag{26.21}$$

But equality

$$\exp(it) - 1 - it = i\int_0^t \left[\exp(iu) - 1\right] du$$

shows that the function integrated in (21) is dominated by the integrable function $4\lambda^2$. Thus the left hand side tends to 0 as t tends to 0 if (20) holds. Therefore, the derivative $D_0U(t)\mathbf{a} = i\mathbf{a}'$ exists for all $\mathbf{a} \in \mathrm{Def}(H)$, with $\mathbf{a}' = H\mathbf{a}$.

Conversely let us suppose it exists. For all $\mathbf{x} \in \mathcal{H}$,

$$(\mathbf{x}|i\mathbf{a}') = \lim\,(\mathbf{x}|\,(U(t)\mathbf{a} - \mathbf{a})\,/t) = \lim\,((U(-t)\mathbf{x} - \mathbf{x})\,/t|\mathbf{a})$$

since $U(t)$ is unitary. If $\mathbf{x} \in \mathrm{Def}(H)$, the last term tends to $(-iH\mathbf{x}|\mathbf{a})$ by what has been proved above. Hence

$$(\mathbf{x}|i\mathbf{a}') = -\,(iH\mathbf{x}|\mathbf{a}) \quad \text{for all } \mathbf{x} \in \mathrm{Def}(H)\,.$$

As a result, the adjoint $H^*$ of $H$ is defined at $\mathbf{a}$ and $H^*\mathbf{a} = i\mathbf{a}'$. But $iH$ is self-adjoint. So $\mathbf{a} \in \mathrm{Def}(H)$ and $i\mathbf{a}' = H\mathbf{a}$.

Finally, let us suppose that $U(t)\mathbf{a}$ soit $C^\infty$. Setting $D = d/dt$,

$$iU(t)H\mathbf{a} = U(t)D_0U(t)\mathbf{a} = DU(t)\mathbf{a}\,.$$

This is a $C^\infty$ function of $t$. As a consequence, $H\mathbf{a} \in \mathrm{Def}(H)$ and $\mathbf{a} \in \mathrm{Def}(H^2)$. Iterating the calculation gives $\mathbf{a} \in \mathrm{Def}(H^n)$ for all $n$. Conversely, we suppose $\mathbf{a} \in \mathrm{Def}(H^n)$ for all $n$. As $\mathbf{a} \in \mathrm{Def}(H)$, the function $U(t)\mathbf{a}$ is differentiable and has derivative $iU(t)H\mathbf{a}$. As $H\mathbf{a} \in \mathrm{Def}(H)$, the result can be applied to $H\mathbf{a}$. So

$$D^2U(t)\mathbf{a} = -U(t)H^2\mathbf{a}$$

exists, etc, proving point (iii) of Stone's theorem.

Observe that conversely, any self-adjoint operator

$$H = \int \lambda dM(\lambda)$$

arises from a unitary representation

$$U(t) = \int \exp(it\lambda)dM(\lambda)$$

of $\mathbb{R}$. We set $U(t) = \exp(itH)$, not to be confused with the series, which is not well-defined.

*Exercise 5.* Suppose $G = \mathbb{R}$. Show that, for the regular representation of $G$, $\mathcal{H}^\infty$ is the space of functions $f$ whose Fourier transforms satisfy $\int|\widehat{f}(\lambda)|^2\lambda^{2n}d\lambda < +\infty$ for all $n$. These functions are $C^\infty$. All their derivatives are in $L^2$ and have integrable Fourier transforms. Interpret the Dixmier-Malliavin theorem.

*Exercise 6.* Let $U(t) = \int \mathbf{e}(\lambda t)dM(\lambda)$ be a unitary representation of $\mathbb{R}$. Show that, for any compact set $\omega \subset \mathbb{R}$, the elements of the spectral manifold $\mathcal{H}(\omega)$ are analytic and the infinitesimal generator $H = \int \lambda dM(\lambda)$ is defined and continuous on $\mathcal{H}(\omega)$. Show that the operator $A = \int \exp(-\pi\lambda^2)dM(\lambda)$ maps $\mathcal{H}$ to the subspace of analytic vectors of $\mathcal{H}$.

*Exercise 7.* Consider the group, named after Heisenberg, of real matrices

$$g = \begin{pmatrix} 1 & x & z \\ 0 & 1 & y \\ 0 & 0 & 1 \end{pmatrix}.$$

(i) Show that there is a basis $(X, Y, Z)$ for $G'(e)$ such that

$$[X, Z] = [Y, Z] = 0\,, \quad [X, Y] = Z\,.$$

(ii) In $\mathcal{H} = L^2(\mathbb{R})$, consider operators

$$\pi(g)f(t) = \mathbf{e}\left[a(z+ty)\right] f(t+x),$$

where $a \in \mathbb{R}$ is given and non-zero. Show that $\pi$ is an irreducible unitary representation of $G$ and that, for all $\varphi \in L(G)$, $\pi(\varphi)$ is a Hilbert-Schmidt operator.[118]
(iii) Show that $\mathcal{H}^\infty = \mathcal{S}(\mathbb{R})$, the Schwartz space (Chap. VII, n° 31) of rapidly decreasing $C^\infty$ functions and of their successive derivatives.

*Exercise 8.* Let $(\mathcal{H}, \pi)$ be a unitary representation of a connected Lie group and $A : \mathcal{H}^\infty \longrightarrow \mathcal{H}^\infty$ a continuous operator with respect to the topology of $\mathcal{H}^\infty$ defined at the end of section (iii) ($A$ is not assumed to be defined on all of $\mathcal{H}$). Show that, like in finite dimension,

$$A\pi(g) = \pi(g)A \Longleftrightarrow A\pi^\infty(X) = \pi^\infty(X)A. \tag{26.22}$$

## 27 – Differential Operators on $SL_2(\mathbb{R})$

From now on, nr 15 and 16 of this chapter will be frequently used. Almost all of this end of chapter is taken from a course I gave in Paris in 1971–1972.

(i) *The Lie algebra of* $SL_2(\mathbb{R})$. As the space of real matrices with zero trace has dimension 3, for example the matrices

$$X = \begin{pmatrix} 0 & 1 \\ 0 & 0 \end{pmatrix}, \quad H = \begin{pmatrix} 1 & 0 \\ 0 & -1 \end{pmatrix}, \quad Y = \begin{pmatrix} 0 & 0 \\ 1 & 0 \end{pmatrix} \tag{27.1}$$

form a basis for the Lie algebra $G'(e)$ of $G = SL_2(\mathbb{R})$. $X$, $H$ and $Y$ are the tangent vectors at $t = 0$ to the one-parameter subgroups

$$x(t) = \begin{pmatrix} 1 & t \\ 0 & 1 \end{pmatrix}, \quad h(t) = \begin{pmatrix} e^t & 0 \\ 0 & e^{-t} \end{pmatrix}, \quad y(t) = \begin{pmatrix} 1 & 0 \\ t & 1 \end{pmatrix} \tag{27.2}$$

already extensively used with a different parametrization for $h(t)$. Hence

$$x(t) = \exp(tX), \quad h(t) = \exp(tH), \quad y(t) = \exp(tY). \tag{27.3}$$

These subgroups act on the upper half-plane $P$ by

$$x(t)z = z + t, \quad h(t)z = e^{2t}z, \quad y(t)z = 1/(tz+1). \tag{27.4}$$

The commutation formulas

$$[H, X] = 2X, \quad [H, Y] = -2Y, \quad [X, Y] = H \tag{27.5}$$

occur in every semisimple groups.

The tangent vector to the one-parameter subgroup of orthogonal matrices at the origin will also be needed. Differentiating the matrix

$$k(t) = \begin{pmatrix} \cos 2\pi t & \sin 2\pi t \\ -\sin 2\pi t & \cos 2\pi t \end{pmatrix}, \tag{27.6}$$

we get the matrix

---

[118] Plancherel's formula for this group will be found in my *mémoire du Journal de Liouville* (XXX, 1951, pp. 92–101). All this was generalized to unipotent groups by J. Dixmier, A. Kirillov, etc.

$$(27.7) \quad A = 2\pi \begin{pmatrix} 0 & 1 \\ -1 & 0 \end{pmatrix} = 2\pi i W, \quad \text{where } W = \begin{pmatrix} 0 & -i \\ i & 0 \end{pmatrix} = i(Y - X).$$

$W$ is in the complex Lie algebra $\mathfrak{g}$ of $G$. Hence

$$k(t) = \exp(tA) = \exp(2\pi i t W).$$

The operator $R(W)$ corresponding to $W$ is defined by

$$(27.8) \quad 2\pi i R(W) f(g) = D_0 f\,[gk(t)]\,.$$

For every function $f \in \mathcal{F}_r^\infty = \mathcal{F}_r \cap C^\infty(G)$, i.e. every $C^\infty$ solution of the functional equation

$$f\,[gk(t)] = f(g)\chi_r(t) = f(g)\exp(2\pi i r t),$$

$$(27.9) \quad R(W)f = rf$$

and this property characterizes $f \in \mathcal{F}_r^\infty$.

Together with $W$, the complex matrices

$$Z = \tfrac{1}{2}\begin{pmatrix} 1 & i \\ i & -1 \end{pmatrix} = \tfrac{1}{2}(H + 2iX + W),$$

$$(27.10) \quad \overline{Z} = \tfrac{1}{2}\begin{pmatrix} 1 & -i \\ -i & -1 \end{pmatrix} = \tfrac{1}{2}(H - 2iX - W),$$

form a basis for $\mathfrak{g}$ and once again

$$(27.11) \quad [W, Z] = 2Z, \quad [W, \overline{Z}] = -2\overline{Z}, \quad [Z, \overline{Z}] = W.$$

These formulas being valid for all corresponding operators, if $f \in \mathcal{F}_r^\infty$, then

$$(27.12') \quad R(W)R(Z)f = 2R(Z)f + R(Z)R(W)f = (r + 2)R(Z)f,$$

$$(27.12'') \quad R(W)R\left(\overline{Z}\right)f = -2R\left(\overline{Z}\right)f + R\left(\overline{Z}\right)R(W)f = (r - 2)R\left(\overline{Z}\right)f.$$

Setting as we will generally do to simplify notations

$$(27.13) \quad Mf = R(M)f = -f * M, \quad MNf = R(M)R(N)f, \quad \text{etc.}$$

for all $M, N \in \mathfrak{g}$ and all function $f$ on $G$, these results can be presented in a diagram

$$(27.14) \quad \begin{array}{ccc} & & \mathcal{F}_{r+2}^\infty \\ & \nearrow^{Z} & \\ \mathcal{F}_r^\infty & \xrightarrow{W} & \mathcal{F}_r^\infty \\ & \searrow_{\overline{Z}} & \\ & & \mathcal{F}_{r-2}^\infty \end{array}$$

which is going to play an important role.

Using convention (13) and commutation formulas, we get

$$[W^2, Z] = 2(WZ + ZW); \quad [W^2, \overline{Z}] = -2\left(W\overline{Z} + \overline{Z}W\right),$$
$$[\overline{Z}Z, W] = [Z\overline{Z}, W] = 0,$$
$$[\overline{Z}Z, Z] = -WZ, \quad [\overline{Z}Z, \overline{Z}] = \overline{Z}W,$$
$$[Z\overline{Z}, Z] = -ZW, \quad [Z\overline{Z}, \overline{Z}] = W\overline{Z}.$$

These are calculations in $\mathcal{U}(G)$, and not of matrix products. Hence

$$[\overline{Z}Z + Z\overline{Z}, Z] = -\frac{1}{2}[W^2, Z], \quad [\overline{Z}Z + Z\overline{Z}, \overline{Z}] = -\frac{1}{2}[W^2, \overline{Z}].$$

Thus the element

$$\Omega = \frac{1}{2}\left(\overline{Z}Z + Z\overline{Z}\right) + W^2/4 = Z\overline{Z} + W^2/4 - W/2 \tag{27.15}$$

of $\mathcal{U}(G)$ commutes with $W$, $Z$ and $\overline{Z}$. It is the *Casimir operator*[119] of $G$. As it belongs to the centre of $\mathcal{U}(G)$, the differential operator $R(\Omega)$ commutes with left and right translations.

(ii) *Differential operators on the half-plane.* In n° 15, (ii), a function

$$f_r(g) = (ci + d)^{-r} f(z), \quad z = gi$$

belonging to the space $\mathcal{F}_r^\infty$ of functions of weight $r$ on $G$ was associated to every $C^\infty$ function $f(z)$ on the half plane $P$ and to every $r \in \mathbb{Z}$. Letting operators $Z$ and $\overline{Z}$ act on $f_r$, we get functions in $\mathcal{F}_{r+2}^\infty$ and $\mathcal{F}_{r-2}^\infty$ to which functions on $P$ are therefore inversely associated. They will be denoted by $Z_r f$ and $\overline{Z}_r f$:

$$g = Z_r f \Longleftrightarrow g_{r+2} = Zf_r, \quad g = \overline{Z}_r f \Longleftrightarrow g_{r-2} = \overline{Z} f_r. \tag{27.16}$$

These operators arise from right translations on $G$, hence commute with left translations. However, these are the operators $L_r(g)$ of n° 15, (ii) acting on the half-plane $P$. Hence

$$Z_r L_r(g) = L_{r+2}(g) Z_{r+2}, \quad \overline{Z}_r L_r(g) = L_{r-2}(g) \overline{Z}_{r-2}.$$

An immediate consequence of these relations is that, if $\Gamma$ is a discrete subgroup of $G$, the operator $Z_r$ (resp. $\overline{Z}_r$) transforms every generalized automorphic form[120] of weight $r$ for $\Gamma$ into a generalized automorphic form of weight $r+2$ (resp. $r-2$) for $\Gamma$.

To express these operators explicitly, we start from the relation

---

[119] If $\mathfrak{g}$ is a Lie algebra, the bilinear form $K(X, Y) = Tr[ad(X)ad(Y)]$ is invariant under the adjoint representation. By definition, $\mathfrak{g}$ (or $G$) is *semisimple* if $K$ is non-degenerate. If $(X_i)$ and $(Y^i)$ are bases for $\mathfrak{g}$ such that $K(X_i, Y^j) = \delta_i^j$, the element $\Omega = X_i Y^i$ is in the centre of the enveloping algebra and does not depend on the chosen basis. It the Casimir operator of $\mathfrak{g}$, invented in the 1930s by a physicist who later became the technical director of Philips. H.B. Casimir and B.L. van der Waerden used it at the time to prove purely algebraically that all finite-dimensional representations of $\mathfrak{g}$ are direct sums of irreducible representations.

[120] i.e., recall that these are the $C^\infty$ solutions of $f(\gamma z) = (cz + d)^r f(z)$ and are not necessarily holomorphic. But operators $Z_r$ do not preserve holomorphy.

(27.17) $$f(x+iy) = y^{-r/2} f_r(g) \quad \text{if } g = \begin{pmatrix} y^{\frac{1}{2}} & y^{-\frac{1}{2}}x \\ 0 & y^{-\frac{1}{2}} \end{pmatrix} \in B$$

of n° 15, (ii). Hence calculations for this value of $g$ are sufficient. As the action is on functions $f(z)$ that actually are functions $f(x,y)$, since they are not necessarily holomorphic, in what follows we will set

$$D_1 = d/dx\,, \quad D_2 = d/dy\,,$$

(27.18) $$D = D_2 + iD_1 = 2i\partial/\partial z\,, \quad \overline{D} = D_2 - iD_1 = -2i\partial/\partial\overline{z}\,.$$

So holomorphic functions are solutions of $\overline{D}f = 0$.

This being so,

$$2Z = H + W + 2iX\,, \quad 2\overline{Z} = H - W - 2iX$$

and $R(W) = r$ in $F_r$. Computing the real operators $R(X)$ and $R(H)$ using the general definition (25.9) is therefore sufficient.

As $X = x'(0)$, differentiating the function $t \longmapsto f_r[gx(t)] = y^{r/2} f(x + ty + iy)$ at $t = 0$, $R(X)f_r(g) = Xf_r(g)$. Thus

(27.19) $$Xf_r(g) = y^{(r+2)/2} D_1 f(z)\,.$$

For $H = h'(0)$, $f_r[gh(t)] = (e^{2t}y)^{r/2} f(x + ie^{2t}y)$, whence

(27.20) $$Hf_r(g) = ry^{r/2} f(z) + 2y^{(r+2)/2} D_2 f(z)\,.$$

Finally, $Wf_r(g) = rf_r(g) = ry^{r/2} f(z)$. So

(27.21) $$2Zf_r(g) = 2ry^{r/2} f(z) + 2y^{(r+2)/2} Df(z)\,,$$

(27.21') $$2\overline{Z}f_r(g) = 2y^{(r+2)/2} \overline{D}f(z)\,,$$

and using (17) for $r$ and $r+2$, we finally get

(27.22) $$Z_r f(z) = \left(ry^{-1} + D\right) f(z)\,,$$

(27.22') $$\overline{Z}_r f(z) = y^2 \overline{D} f(z)\,.$$

The second relation proves that, for every function $\varphi \in C^\infty(G)$,

(27.23) $$\varphi \in \mathcal{H}_r \Longleftrightarrow W\varphi = r\varphi \;\&\; \overline{Z}\varphi = 0\,.$$

Operators (22) and (22') were introduced by Maaß in 1953 in the context of modular functions without any reference to Lie groups, but they already appeared in Bargmann (1947) for the unit disc, which comes to the same.

Setting

(27.24) $$\Delta = y^2 D\overline{D} = y^2 \left(d^2/dx^2 + d^2/dy^2\right)\,,$$

easy calculations show that

(27.25) $$Z_{r-2}\overline{Z}_r = \Delta + ry\overline{D}\,.$$

Similarly, using (24) and $[Z, \overline{Z}] = W$, we get

(27.24')
$$\overline{Z}_{r+2}Z_r = \Delta + ry\overline{D} - r\,.$$

A differential operator of the second order $\Omega$ commuting with $W$, $Z$ and $\overline{Z}$ was introduced in (15). Since it in particular commutes with $W$, it acts on each space $\mathcal{F}_r^\infty(G)$, hence on functions $f(z)$ on the upper half-plane. Since it commutes with left translations, it also acts on $\mathcal{F}_r^\infty(\Gamma\backslash G)$. For given integer $r$, associating the corresponding function $f_r(g)$ to a function $f(z)$, $\Omega$ transforms it into a function which, by the same formula, will define a function on the half–plane which we will write $\Omega_r f(z)$. Since $\Omega$ commutes with left translations, $\Omega_r$ *commutes with operators* $L_r(g)$ of n° 15.

To compute $\Omega_r$ explicitly, we write $\Omega$ as

(27.26)
$$\Omega = Z\overline{Z} + W^2/4 - W/2\,.$$

Then, since $W = r$ on $F_r^\infty$,

(27.27)
$$\Omega_r f(z) = \frac{r}{2}\left(\frac{r}{2} - 1\right) f(z) + Z_{r-2}\overline{Z}_r f(z)\,.$$

Taking (24) into account, it follows that

(27.28)
$$\Omega_r = \frac{r}{2}\left(\frac{r}{2} - 1\right) + ry\overline{D} + \Delta\,.$$

In particular, $\Omega_0 = \Delta$, the *invariant Laplacian* of the half-plane, which commutes with operators $L_0(g)f(z) = f(g^{-1}z)$. In conclusion:

**Theorem 47**. *Let $\Gamma$ be a discrete subgroup of $G$ and $f$ a generalized automorphic form of weight $r$ for $\Gamma$. Then the functions*

$$Z_r f(z) = rf(z)y^{-1} + Df(z)\,,$$
$$\overline{Z}_r f(z) = y^2\overline{D}f(z)\,,$$
$$\Omega_r f(z) = \Delta f(z) + ry\overline{D}f(z) + \frac{r}{2}\left(\frac{r}{2} - 1\right) f(z)$$

*are generalized automorphic form of weight $r+2$, $r-2$ and $r$ for $\Gamma$.*

Like Maaß in 1953, these results could be obtained by direct calculations. Half a century of experience shows that "elementary" and "explicit" calculations generally become inextricable[121] beyond groups of dimension 3 or 4 and that by disregarding the impressive technique of semisimple groups, we become cut off from fundamental ideas. Besides, even in the present case, checking "without knowing anything" that operators $Z_r$ and $\overline{Z}_r$ act on automorphic forms is an exercise in which it is easy to get lost and which is omitted by all authors I am acquainted with. So let us indicate how to proceed for $\overline{Z}_r$, the other case being left to the reader.

Starting from the equality $f(\gamma z) = (cz+d)^r f(z)$, we get

$$\overline{D}\left[f(\gamma z)\right] = (cz+d)^r\overline{D}f(z)$$

since $(cz+d)^r$ is holomorphic. As $\overline{D}$ is a differentiation operator, the chain rule shows that[122]

[121] This can be realized by reading H. Maaß, *Lectures on Siegel's modular functions* (Tata Institute, 1954–1955), § 15.

[122] Recall that a notation such as $Df(\gamma z)$ denotes the value of the function $Df(z)$ at $\gamma z$, whereas the notation $D[f(\gamma z)]$ denotes the effect of $D$ on the function $z \longmapsto f(\gamma z)$.

$$\overline{D}\left[f(\gamma z)\right] = D_1 f(\gamma z).\overline{D}\left[\mathrm{Re}(\gamma z)\right] + D_2 f(\gamma z).\overline{D}\left[\mathrm{Im}(\gamma z)\right].$$

However, $\overline{D}(\gamma z) = 0$ and so $\overline{D}[\mathrm{Re}(\gamma z)] = -i\overline{D}[\mathrm{Im}(\gamma z)]$, whence

$$\begin{aligned}\overline{D}\left[f(\gamma z)\right] &= \overline{D}f(\gamma z).\overline{D}\left[\mathrm{Im}(\gamma z)\right] = \overline{D}f(\gamma z).\overline{D}\left(\gamma z - \gamma\overline{z}\right)/2i = \\ &= -\overline{D}f(\gamma z).\overline{D}\left(\gamma\overline{z}\right)/2i = \overline{D}f(\gamma z).\left(c\overline{z}+d\right)^{-2}.\end{aligned}$$

Thus

$$\overline{D}f(\gamma z) = \left(c\overline{z}+d\right)^2 (cz+d)^r \overline{D}f(z).$$

It follows that

$$\begin{aligned}\overline{Z}_r f(\gamma z) &= \mathrm{Im}(\gamma z)^2 \overline{D}f(\gamma z) = y^2 |cz+d|^{-4} \left(c\overline{z}+d\right)^2 (cz+d)^r \overline{D}f(z) = \\ &= (cz+d)^{r-2} \overline{Z}_r f(z)\end{aligned}$$

as predicted.

The simplest eigenfunction of $\Delta$ is $f(z) = y^s$, which satisfies

$$\Delta\left(y^s\right) = s(s-1)y^s. \tag{27.29}$$

Hence, for the corresponding function

$$f_0(g) = \alpha(g)^s \tag{27.30}$$

of weight 0,

$$\Omega f_0 = s(s-1)f_0. \tag{27.29'}$$

As the Casimir operator commutes with left and right translations, the Maaß series

$$M(z;s) = \sum \mathrm{Im}(\gamma z)^s \tag{27.31}$$

also satisfies (29) when it converges and even, by analytic extension, when it does not.

To get eigenfunctions of $\Omega_r$ for any even $r$, it suffices to repeatedly apply operators $Z$ and $\overline{Z}$ on $f_0$. Due to diagram (14), this amounts to applying operators $Z_{2k-2}\dots Z_2 Z_0$ and $\overline{Z}_{2k-2}\dots\overline{Z}_2\overline{Z}_0$ on $y^s$. We find

$$Z_0 f(z) = sy^{s-1}, \quad Z_2 Z_0 f(z) = s(s+1)y^{s-2}, \dots \tag{27.32}$$

and

$$\overline{Z}_0 f(z) = sy^{s+1}, \quad \overline{Z}_2\overline{Z}_0 f(z) = s(s+1)y^{s+2}, \dots. \tag{27.32'}$$

As we get the functions $y^{s-k}$, up to a constant factor (sometimes zero...), the operators used transform the first series (31) of weight 0 into the series of weight $2k$ associated to $y^{s-k}$, namely

$$M_{2k}(z;s) = \sum J(\gamma;z)^{-2k}\,\mathrm{Im}(\gamma z)^{s-k} \quad (k \geq 0), \tag{27.33}$$

and the second one into the analogous series of weight $-2k$. The corresponding functions on $G$ are therefore all eigenfunctions of $\Omega$. We will return to these calculations in n° 29.

*Exercise 1.* How should the previous calculations be modified so as to obtain eigenfunctions of odd weight?

## 28 – The Representation of $\mathfrak{g}$ Associated to a Holomorphic Function

(i) *The* $\mathfrak{g}$*-module* $HC_r(f) = HC(f_r)$. For a given integer $r$, let us start from a non-trivial holomorphic function on the half-plane and associate the function $f_r \in \mathcal{H}_r$ on $G$ to it. Hence $\overline{Z} f_r = 0$, $W f_r = r f_r$. We show that

$$\overline{Z} Z^n f_r = \lambda_n Z^{n-1} f_r \tag{28.1}$$

with scalars $\lambda_n \in \mathbb{C}$. For $n = 0$, this is obvious, with $\lambda_0 = 0$. For $n > 0$, we write

$$\overline{Z} Z^n = \overline{Z} Z Z^{n-1} = \left[\overline{Z}, Z\right] Z^{n-1} + Z \overline{Z} Z^{n-1} = -W Z^{n-1} + Z \overline{Z} Z^{n-1} .$$

Hence, supposing that (1) holds for $n-1$, the induction relation $\lambda_n = \lambda_{n-1} - (r + 2n - 2)$ proves it for $n$. As $\lambda_0 = 0$,

$$\lambda_n = -n(n + r - 1) .$$

Therefore, setting

$$g_{r+2n} = Z^n f_r \quad \text{for } n \geq 0 , \tag{28.2}$$

we get the following formulas:

$$\begin{aligned} Z g_{r+2n} &= g_{r+2n+2} \\ W g_{r+2n} &= (r + 2n) g_{r+2n} \\ \overline{Z} g_{r+2n} &= -n(r + n - 1) g_{r+2n-2} . \end{aligned} \tag{28.3}$$

$HC(f_r)$ or $HC_r(f)$ will denote the set of finite linear combinations of the functions $g_{r+2n}$. It is stable under operators of $\mathfrak{g}$ and so under $R(\mu)$, $\mu \in \mathcal{U}(\mathfrak{g})$.

Since

$$Z \overline{Z} g_{r+2n} = -n(r + n - 1) g_{r+2n} \tag{28.4'}$$

$$\overline{Z} Z g_{r+2n} = -(n + 1)(r + n) g_{r+2n} \tag{28.4''}$$

follow from (3),

$$\Omega g_{r+2n} = s(s - 1) g_{r+2n} \quad \text{with } s = r/2 . \tag{28.5}$$

So the Casimir operator reduces to a scalar in $HC_r(f)$.

For $r > 0$, these formulas can be simplified by setting

$$e_r = f_r , \quad e_{r+2n} = g_{r+2n} / r(r + 1) \ldots (r + n - 1) \quad \text{for } n \geq 1 . \tag{28.6}$$

Then

$$\begin{aligned} Z e_m &= (s + m/2) e_{m+2} , \\ W e_m &= m e_m , \\ \overline{Z} e_m &= (s - m/2) e_{m-2} \end{aligned} \tag{28.7}$$

for $m = r, r + 2, \ldots$ The vectors $e_m$ are all $\neq 0$, since the coefficients $s - m/2 = \frac{1}{2}(r - m)$ being $\neq 0$ for $m > r$, the relation $e_m = 0$ would imply $e_{m-2} = \ldots = e_r = 0$. For $r > 0$, the space $HC_r(f)$ is therefore infinite-dimensional, but ,as we will see, not necessarily for $r \leq 0$.

Furthermore, for $r > 0$, $HC_r(f)$ *does not contain any non-trivial subspaces stable under* $\mathfrak{g}$. Being stable under $W$, such a subspace is indeed generated by the vectors $e_m$ it contains. But formulas (7) with all coefficients $\neq 0$, show that it then contains all $e_m$. The Lie algebra $\mathfrak{g}$ of $G$ therefore acts in an *algebraically irreducible* manner on $HC_r(f)$. So $HC_r(f)$ is a simple $\mathcal{U}(\mathfrak{g})$-module, as one says in Algebra. Since only $r$ occurs in formulas (7), the structure of $HC_r(f)$ as a $\mathfrak{g}$-module is independent of the choice of $f$. We thereby get the following result:

**Theorem 48.** *Let $\Gamma$ be a discrete subgroup of $G$, $f(z)$ a* holomorphic *automorphic form of weight $r > 0$ for $\Gamma$ and*

$$\varphi(g) = f_r(g) = (ci + d)^{-r} f(gi)$$

*the associated function on $G$. Let $HC(\varphi)$ be the vector space generated by the functions $R(Z)^n\varphi$. Then the spaces $\psi \in HC(\varphi)$ are left $\Gamma$-invariant, $HC(\varphi)$ is stable under $\mathfrak{g}$ and a simple $\mathfrak{g}$-module. For given $r$, all the simple $\mathfrak{g}$-modules thus obtained are isomorphic.*

For parabolic forms, i.e. such that $\varphi \in L^2(\Gamma\backslash G)$, this result is the infinitesimal analogue of theorem 27 of n° 20 which says that the representation of $G$ on the closed invariant subspace $\mathcal{H}(\varphi)$ of $L^2(\Gamma\backslash G)$ generated by $\varphi$ is irreducible. By point (b) in the proof of theorem 27, $\varphi$ is the only element of weight $r$ of $\mathcal{H}(\varphi)$, up to a constant factor. So by theorem 43, it is an *analytic* element of the representation $R$ of $G$ on $\mathcal{H}(\varphi)$ or $L^2(\Gamma\backslash G)$. For all $\mu \in \mathcal{U}(G)$, $R^\infty(\mu)\varphi = \varphi * \mu' \in \mathcal{H}(\varphi)$ by (26.15), whence $HC(\varphi) \subset \mathcal{H}(\varphi)$. Moreover, $HC(\varphi)$ is everywhere dense in $\mathcal{H}(\varphi)$, because if there is some $g \in \mathcal{H}(\varphi)$ orthogonal to all $\varphi * \mu$, all derivatives of the analytic function $(R(x)\varphi|g)$ are zero at the origin. So $g$ is orthogonal to all $R(x)\varphi$, hence zero.

For $\Gamma = e$ and $r \geq 2$, theorem 47 applies to $\varphi \in \mathcal{H}^2_r(G)$ and in particular to the kernel function $\omega_r$ [n° 16, (iv)]. Transforming the latter by left (resp. right) translations generates the closed subspaces $\mathcal{H}^2_r(G)$ [resp. $\mathcal{H}(\omega_r)$] of n° 16 and 20 in $L^2(G)$. As in both cases $\omega_r$ is an analytic vector (theorem 41), the former (resp. latter) space is the closure in $L^2(G)$ of the set of functions $\mu * \omega_r$ (resp. $\omega_r * \mu$) where $\mu \in \mathcal{U}(G)$. Since $(\omega_r * \mu) = \tilde{\mu} * \omega_r$, the map $\varphi \longmapsto \tilde{\varphi}$ is a semilinear isomorphism from $\mathcal{H}(\omega_r)$ onto $\mathcal{H}^2_r(G)$ transforming the representation $R$ of $G$ on the former space into the representation $L$ on the latter one. The same conclusion follows concerning representations of $\mathfrak{g}$ on these spaces, defined by $L(\mu)f = \mu * f$ in the former case, by $R(\mu)f = f * \mu'$ in the latter. Relation $R(W)\omega_r = r\omega_r$ satisfied by all $f \in \mathcal{H}_r(G)$ becomes $L(W)\omega_r = -r\omega_r$, so that $\mathcal{H}^2_r(G)$ has a basis consisting of eigenfunctions of $L(W)$, the eigenvalues being the integers $-r - 2n$ $(n \geq 0)$.

(ii) *The case* $r = -p \leq 0$. Formulas (3) become

$$Zg_{2n-p} = g_{2(n+1)-p}$$

$$\overline{Z}g_{2n-p} = n(p + 1 - n)g_{2(n-1)-p}\,. \tag{28.8}$$

There are then two possible cases.

If $g_{2n-p} \neq 0$ for all $n > 0$, the space $HC(f_r)$ admits as basis the functions $g_{-p}, g_{-p+2}, \ldots, g_p, g_{p+2}, \ldots$ and is infinite-dimensional. But it is no longer a simple $\mathfrak{g}$-module. Indeed $\overline{Z}g_{p+2} = 0$ by (8). Thus the subspace $\mathcal{F}' = HC(g_{p+2})$ generated by $g_{p+2}, g_{p+4}, \ldots$ is stable and corresponds to case (i) for $r = p + 2$. As the Lie algebra acts both on $HC_r(f)$ and $\mathcal{F}'$, it also acts on the quotient space $HC_r(f)/\mathcal{F}'$ which, as a $\mathfrak{g}$-module, is simple and of dimension $p + 1$.

If on the other hand $g_{2n-p} = 0$ for some $n$, then, applying $Z$, $g_{2n+2-p} = 0$ as well, so $g_{2m-p} = 0$ for all $m > n$. Hence if $n$ denotes the smallest integer such that $g_{2n+2-p} = 0$, whence $g_{2n-p} \neq 0$, the second relation (8) applied to $n + 1$ shows that $n = p$. So, this case occurs if and only if $g_{p+2} = Z^{p+1}f_r = 0$. The

$p+1$ vectors $g_{-p}, g_{-p+2}, \ldots, g_p$ then form a basis for the space $HC(f_r)$, which is of dimension $p+1$ and is again a simple $\mathfrak{g}$-module (same arguments as above). In this case too, formulas (6) can be used since denominators are $\neq 0$ for $0 < n < p$. As $s = r/2 = -p/2$, we recover (7) in the form

$$\begin{aligned} Ze_m &= -\frac{1}{2}(p-m)e_{m+2}\,, \\ We_m &= me_m\,, \quad (m = -p, -p+2, \ldots, p) \\ \overline{Z}e_m &= -\frac{1}{2}(p+m)e_{m-2}\,. \end{aligned} \tag{28.9}$$

(iii) *Finite-dimensional simple* $\mathfrak{g}$*-modules.* Every simple $\mathfrak{g}$-module $\mathcal{E}$ of finite dimension $p+1$ is easily expressible by formulas (9). For this, observe that $W$ has at least one eigenvector $a$ in $\mathcal{E}$ and that the relation $Wa = \lambda a$ implies $W\overline{Z}a = (\lambda-2)\overline{Z}a$ because of commutation formulas. As $W$ only has finitely many eigenvalues, there exists $a \neq 0$ in $\mathcal{E}$ and $r \in \mathbb{C}$ such that $Wa = ra$, $\overline{Z}a = 0$. As above, the subspace generated by the $Z^n a$ is stable under $\mathfrak{g}$, hence equal to $\mathcal{E}$. Thus the non-zero $Z^n a = a_{r+2n}$ form a basis since they belong to eigenvalues $r+2n$ of $W$. Formulas (3)and (4) continue to hold since they are only based on commutation relations. Since $\mathcal{E}$ has dimension $p+1$, as above $a_r, a_{r+2}, \ldots, a_{r+2p}$ are non-zero and form a basis for $\mathcal{E}$. Hence $\overline{Z}Za_{r+2p} = 0$, and so $r = -p$, a negative integer etc.

The existence of a basis satisfying (9) is uniquely based on commutation relations satisfied by $W$, $Z$ and $\overline{Z}$. Hence the arguments could also be applied to the operators associated to the matrices $H$, $X$ and $Y$ defined in (27.6). Hence, if $\mathcal{E}$ is a simple $\mathfrak{g}$-module of dimension $p+1$, there is also a basis $u_{-p}, u_{-p+2}, \ldots, u_p$ in $\mathcal{E}$ for which, instead of (9), we have

$$\begin{aligned} Xu_m &= -\frac{1}{2}(p-m)u_{m+2}\,, \\ Hu_m &= mu_m\,, \\ Yu_m &= -\frac{1}{2}(p+m)u_{m-2}\,. \end{aligned} \tag{28.10}$$

Conversely, these formulas obviously always define a simple $\mathfrak{g}$-module of dimension $p+1$.

(iv) *Condition for* $\dim HC_r(f) < +\infty$. If we knew when the $\mathfrak{g}$-module $HC_r(f)$ of a holomorphic function $f$ has finite dimension $p+1$, then these calculations would be complete. This supposes $r = -p$ and is equivalent to $Z^{p+1} f_{-p} = 0$, hence to

$$Z_p Z_{p-2} \ldots Z_{-p} f(z) = 0\,. \tag{28.11}$$

Hence this product of differential operators has to be computed taking into account definition $Z_n f = ny^{-1} f + Df$, where $D = d/dy + id/dx$. We are going to do this for all $C^\infty$ functions $f$.

First it is clear that $Z_0 f = Df$. Now let us calculate

$$\begin{aligned} Z_1 Z_{-1} f &= \left(y^{-1} + D\right)\left(-y^{-1} f + Df\right) = \\ &= -y^{-2} f + y^{-1} Df - y^{-1} Df + y^{-2} f + D^2 f = D^2 f\,. \end{aligned}$$

Hence

$$Z_p Z_{p-2} \ldots Z_{-p} = D^{p+1} \tag{28.12}$$

for $p = 0$ or 1. We show by induction that this formula holds for all $p$. Since

$$Z_p Z_{p-2} \ldots Z_{-p} = Z_p \left(Z_{p-2} \ldots Z_{-p+2}\right) Z_{-p} = Z_p D^{p-1} Z_{-p}$$

we only need to prove that

$$Z_p D^{p-1} Z_{-p} = D^{p+1}\,, \tag{28.13}$$

in other words that

$$\left(py^{-1} + D\right) D^{p-1} \left(-py^{-1} f + Df\right) = D^{p+1} f\,.$$

The left hand side being equal to

$$\begin{aligned}\left(py^{-1} + D\right)\left[D^p f - pD^{p-1}\left(y^{-1} f\right)\right] = D^{p+1} f - pD^p\left(y^{-1} f\right) + \\ + py^{-1} D^p f - p^2 y^{-1} D^{p-1}\left(y^{-1} f\right)\,,\end{aligned}$$

the proof reduces to showing that

$$D^p\left(y^{-1} f\right) = y^{-1} D^p f - py^{-1} D^{p-1}\left(y^{-1} f\right)\,,$$

or, replacing $f$ by $yf$, that $D^p(yf) = yD^p f + pD^{p-1} f$, qed.

In conclusion, the space $HC_r(f)$ of a *holomorphic* function $f(z)$ is a simple $p+1$-dimensional $\mathfrak{g}$-module if and only if $r = -p$ and

$$D^{p+1} f(z) = 0\,, \tag{28.14}$$

in other words if and only if $f$ is a *polynomial of degree* $p$ (and not $< p$ since $Z^p f_r \neq 0$). The images $L_{-p}(g) f(z) = (cz + d)^p f(gz)$ are then polynomials of degree $\leq p$.

(v) *A theorem of Maaß.* Applied to a holomorphic or even to a meromorphic form, formula (12) leads to the function

$$D^{p+1} f(z) = Z_p Z_{p-2} \ldots Z_{-p} f(z)\,,$$

proportional to $f^{(p+1)}(z)$. On the other hand, we know that, for every discrete subgroup $\Gamma$ of $G$, the operator $Z$ transforms generalized automorphic forms of weight $r$ into generalized automorphic forms of weight $r+2$. So, for $r = -p > 0$, the operator $D^{p+1}$ transforms forms of weight $-p$ into forms of weight $p+2$. As a result:

**Theorem 49 (Maaß).** *If $f$ is a generalized (resp. meromorphic) automorphic form of weight $-r < 0$ for a discrete subgroup $\Gamma$ of $G$, then $\partial^{r+1} f / \partial z^{r+1}$ is a generalized (resp. meromorphic) automorphic form of weight $r+2$ for $\Gamma$.*

The reader will check that with the strict definition of meromorphic forms (behaviour at parabolic fixed points) the theorem holds for Fuschian groups.

Theorem 49 leads to other strange identities. For example let us take $\Gamma = G(\mathbb{Z})$ and

$$f(z) = 1/\Delta(z) = \mathbf{e}(-z) + b_0 + b_1 \mathbf{e}(z) + \ldots\,;$$

$f$ is of weight $-12$, holomorphic on $P$ and has a simple pole at infinity. The function

$$F(z) = f^{(13)}(z) = (-2\pi i)^{13} \mathbf{e}(-z) + (2\pi i)^{13} b_1 \mathbf{e}(z) + \ldots$$

is, therefore, a modular form of weight 14, holomorphic everywhere on $P$ and having a simple pole at infinity. As a result, $\Delta(z)F(z) = (-2\pi i)^{13} + \ldots$ is an entire modular form of weight $14 + 12 = 26$. Thus (n° 18, theorem 21)

$$\Delta(z)\frac{d^{13}}{dz^{13}}\Delta(z)^{-1} = (-2\pi i)^{13}E_{26}(z) + c\Delta(z)E_{14}(z)$$

with a constant $c$ whose calculation we leave to the reader.

The case of inverses of Eisenstein series can also be dealt with in principle. Let us for example choose $f(z) = E_4(z)^{-1} = 1 - 240\mathbf{e}(z) + \ldots$ [n° 17, (iii)], whence $F(z) = f^{(5)}(z) = -240(2\pi i)^5\mathbf{e}(z) + \ldots$. This is a modular form of weight 6 whose only singularity, $\bmod \Gamma$, is at $j$, where $v_j(F) = -6$ since $v_j(1/E_4) = -1$. The function $F(z)E_4(z)^6$ is, therefore, an entire modular form of weight 30. As a result,

$$E_4(z)^6\frac{d^5}{dz^5}E_4(z)^{-1} = aE_{30}(z) + b\Delta(z)E_{18}(z) + c\Delta(z)^2E_6(z)$$

with constants $a$, $b$, $c$ that can be calculated explicitly by investigating the first few terms of the Fourier series of both sides.

## 29 – Irreducible Representations of $\mathfrak{g}$

(i) *Classification.* For all $s \in \mathbb{C}$, formulas

$$\begin{aligned} Ze_m &= (s + m/2)e_{m+2}\,, \\ We_m &= me_m\,, \\ \overline{Z}e_m &= (s - m/2)e_{m-2} \end{aligned} \tag{29.1}$$

obtained above in (28.7) can be used to construct other $\mathfrak{g}$-modules different from the modules $HC_r(f)$ associated to holomorphic functions to $P$. To this end, we consider a vector space $\mathcal{E}$ admitting a basis $(e_m)$, $m \in \mathbb{Z}$, and define $W$, $Z$ and $\overline{Z}$ on $\mathcal{E}$ by these formulas. Trivial calculations show that indeed we get a $\mathfrak{g}$-module on which the Casimir operator is given by

$$\Omega = s(s-1)\,.$$

This $\mathfrak{g}$-module $\mathcal{E}_s$ is not simple: the subspace $\mathcal{E}_s^+$ (resp. $\mathcal{E}_s^-$) generated by vectors of even (resp. odd) weight is invariant.

Since $W$ is diagonalizable, every submodule of $\mathcal{E}_s^+$ is generated by the vectors $e_m$ it contains. If $s \notin \mathbb{Z}$, the coefficients $s + m/2$ of (1) are non-trivial, so that applying the monomials $Z^p\overline{Z}^q$ to some $e_m$ for $m$ even, we get all the $e_n$ with $n$ even. A similar conclusion holds for the vectors $e_m$, with $m$ odd, if $s - \frac{1}{2} \notin \mathbb{Z}$. As a result, *the* $\mathfrak{g}$*-module* $\mathcal{E}_s^+$ (resp. $\mathcal{E}_s^-$) *is simple if* $s$ (resp. $s - \frac{1}{2}$) *is not an integer.*

If $s = r/2$ for some even integer $r$, then $\overline{Z}e_r = 0$ and $Ze_{-r} = 0$. The vectors $e_{r+2m}$, $m \geq 0$, generate a submodule $\mathcal{F}'$ in $\mathcal{E}_s^+$ isomorphic to the modules $HC(f_r)$ of the preceding n°. It is irreducible if and only $r > 0$. Similarly, the vectors $e_{-r-2m}$, $m \geq 0$ generate a submodule $\mathcal{F}''$, which is irreducible if and only if $r > 0$. In this case, the quotient module $\mathcal{E}_s^+/(\mathcal{F}' + \mathcal{F}'')$ is irreducible and finite-dimensional. On the other hand for $r \leq 0$, the vectors $e_r, \ldots, e_{-r}$ generate an irreducible submodule, namely $\mathcal{F}' \cap \mathcal{F}''$. The same conclusions follow for $\mathcal{E}_s^-$ if $s = r/2$ for an odd integer.

As will be seen without going into detailed easy calculations, all simple $\mathfrak{g}$-modules simples are obtained in this way provided they are required to be *admissible*: this means that $W$ is diagonalization, with integer eigenvalues[123] and that, for all $r$, the subspace $\mathcal{E}(r)$ defined by $Wf = rf$ is finite-dimensional.

Indeed let $\mathcal{E}$ be such a $\mathfrak{g}$-module and let us suppose it is simple. A first result is that the Casimir operator is a scalar. Indeed let $r$ be an eigenvalue of $W$ and $\mathcal{E}(r)$ the space of solutions of $Wa = ra$. Since $\Omega$ commutes with $W$, it acts on $\mathcal{E}(r)$, hence has at least one eigenvector. Since $\Omega$ commutes with $W$, $Z$ and $\overline{Z}$, the set of vectors of $\mathcal{E}$ associated to a given eigenvalue of $\Omega$ is $\mathfrak{g}$-invariant, whence the result.

This being settled, the formula

$$\Omega = Z\overline{Z} + W^2/4 - W/2 = \overline{Z}Z + W^2/4 + W/2$$

shows that, for any eigenvector $a$ of $W$, $Z\overline{Z}a$ and $\overline{Z}Za$ are proportional to $a$. The subspace generated by vectors $Z^p a$ and $\overline{Z}^q a$ is thus stable under $Z$ and $\overline{Z}$, hence invariant. So, choosing a non-trivial $a_r$ such that $Wa_r = ra_r$ and setting

$$a_{r+2p} = Z^p a_r\,, \quad a_{r-2q} = \overline{Z}^q a_r\,,$$

the vectors $a_m \neq 0$ form a basis for $\mathcal{E}$. Setting $\Omega = s(s-1)$ and using commutation formulas,

$$Za_m = (s + m/2)(s - m/2 - 1)a_{m+2} \quad \text{if } m < r\,, \tag{29.2}$$

$$\overline{Z}a_m = (s - m/2)(s + m/2 - 1)a_{m+2} \quad \text{if } m > r \tag{29.2'}$$

easily follow. If all the coefficients occurring in these formulas are non-trivial, which supposes that $2s$ is not an integer with the same parity as $r$, then the same is true for all vectors $a_{r\pm 2n}$ and the module is simple. The only possibilities are easily seen to be as follows:

(a) $m = \ldots, -4, -2, 0, 2, 4, \ldots$ and non-integer $s$;
(b) $m = \ldots, -3, -1, 1, 3, \ldots$ and non-integer $s - \frac{1}{2}$;
(c) $m = r, r+2, \ldots$ and $s = r/2$, $r \geq 1$;
(d) $m = -r, -r-2, \ldots$ and $s = -r/2$, $r \geq 1$;
(e) $m = -p, -p+2, \ldots, p-2, p$ et $s = -p/2$ or $1 + p/2$.

In cases (a) and (b), the vectors $e_m$ of (1) are obtained by setting $e_r = a_r$ and

$$e_{r+2n} = (s + r/2)^{-1} \ldots (s + r/2 + n - 1)^{-1} a_{r+2n}\,, \tag{29.3}$$

$$e_{r-2n} = (s - r/2)^{-1} \ldots (s - r/2 + n - 1)^{-1} a_{r+2n} \tag{29.3'}$$

for $n \geq 1$. In case (c), the second formula is not well-defined and the expected basis of $\mathcal{E}$ is provided by the first one, the second one being, however, suitable for case (d). Note that the value of $\Omega$ is not sufficient for characterizing the representation and that, in cases (a) and (b), nothing is changed by replacing $s$ by $1 - s$. This is more obvious for relations (2) and (2') than for formulas (1).

---

[123] This condition follows from the fact that, in any representation of the *group* $G$, characters of the compact subgroup $K$ which occur are functions

$$\chi_r\,[\exp(2\pi itW)] = \exp(2\pi irt)$$

with $r \in \mathbb{Z}$. In representations of the universal covering of $G$, where $K$ is replaced by a group isomorphic to $\mathbb{R}$, the eigenvalues of $W$ associated to $C^\infty$ vectors can be real numbers. The following calculations continue to hold with obvious changes.

Representations of type (a) (resp. (b)) constitute the even (resp. odd) *principal series*. Types (c) (resp. (d)) cover the holomorphic (resp. antiholomorphic) *discrete series*. Finally case (e) corresponds to finite-dimensional representations.

(ii) *Function space models for representations of* $\mathfrak{g}$. Whether a group or a Lie algebra, one of the basic problems of representation theory consists in finding concrete realizations. One always tries to construct function space models, where the space $\mathcal{H}$ of the representation is composed of functions (possibly with values in Banach spaces) or even of distributions on $G$ on which $G$ acts by right (or left) translations in the case of $G$, by differential operators $R(X)$ [or $L(X)$] in the case of $\mathfrak{g}$.

The case of $SL_2(\mathbb{R})$ is particularly simple. Observe first that any function space model (on the right) of an irreducible representation of $\mathfrak{g}$ has a base consisting of distributions $\mu$ such that

$$R(W)\mu = -\mu * W = r\mu, \quad \mu * \Omega = s(s-1)\mu .$$

Operators $iR(M)$ are symmetric on the subspace $\mathcal{D}(G)$ of $L^2(G)$, for all $M \in G'(e)$. Hence $R(M)^* = -R(\overline{M})$ for all $M \in \mathfrak{g}$. As a result, $2\Omega = -(ZZ^*+Z^*Z)+W^*W/2$. However, $ZZ^* + Z^*Z + W^*W/2$ is positive-definite symmetric on $\mathcal{D}(G)$ and thus corresponds to a an *elliptic* differential operator on $G$ with analytic coefficients. Since any eigenfunction or distribution $\mu$ of $W$ *and* $\Omega$ is clearly an eigenfunction of this operator, the theory of PDEs enables us to conclude that $\mu$ is in fact an analytic function on $G$. With a slightly less simple argument, this result can be generalized to function space models of any admissible representation of $\mathfrak{g}$ (or of the Lie algebra of a general semisimple group).

The spaces $HC(f_r)$ associated to a holomorphic function on $P$ are obviously function space models of representations of the holomorphic discrete series (c).

To obtain series (d), it suffices to transform a model for series (c) by the map $f \longmapsto \overline{f}$. As $\overline{W} = -W$, relation $f * W = -rf$ implies $\overline{f} * W = +r\overline{f}$.

Another method consists in transforming a right model for series (c) by $f \longmapsto \widetilde{f}$. This gives a left model for series (d). Indeed $\widetilde{W} = -\overline{W} = W$, so that

$$R(W)f = -f * W = rf \Longrightarrow L(W)\widetilde{f} = W * \widetilde{f} = (f * W) = -r\widetilde{f} .$$

For example let us choose the kernel function $f = \omega_r$ of the space $\mathcal{H}^2_r(G)$ [n° 16, (iv), defined by (16.31)]. The $\mathfrak{g}$-module $HC(\omega_r)$, i.e. the set of $\omega_r * \mu$ where $\mu \in \mathcal{U}(\mathfrak{g})$, belongs to series (c). As $\widetilde{\omega}_r = \omega_r$, the functions $\mu * \omega_r$, therefore, form a model for series (d).

For finite-dimensional representations, we make $G$ act on $\mathbb{R}^2$ by $(x,y) \longmapsto (ax+by, cx+dy)$ and use the obvious representation of $\mathfrak{g}$ (or of $G$) on the space of homogeneous polynomials of a given degree.

The construction of the function spaces $\mathcal{H}$ which are transformed in accordance with the principal series by operators $R(M)$, $M \in \mathfrak{g}$, is then immediate. In case (a), $\mathcal{H}$ has a vector $f_0$ satisfying

$$Wf_0 = 0, \quad \Omega f_0 = s(s-1)f_0 .$$

In a function space model, $f_0$ is, therefore, the function on $P$ of weight 0 corresponding to a function $f(z)$ such that

$$y^2 \Delta f = s(s-1)f .$$

To satisfy these conditions, for $f(z)$ it suffices to choose a solution of this PDE, for example $y^s$ or $y^{1-s}$, and to denote by $f_0$ the weight 0 function corresponding to $f$, for example

$$f_0(g) = \mathrm{Im}(gi)^s = \alpha(g)^s .$$

We then obtain a basis for $HC(f_0)$ by applying the powers of the differential operators $Z$ and $\overline{Z}$ on $f_0$, i.e. by applying operators $Z_{2p} \ldots Z_2 Z_0$ and $\overline{Z}_{2q} \ldots \overline{Z}_2 \overline{Z}_0$ on $f(z)$. Formulas

$$Z_r f(z) = \left(r y^{-1} + D\right) f(z) , \quad \overline{Z}_r f(z) = y^2 \overline{D} f(z)$$

show that, for $f(z) = y^s$,

$$Z_0 f(z) = s y^{s-1} , \quad Z_2 Z_0 f(z) = s(s+1) y^{s-2} , \ldots \tag{29.4}$$

and that

$$\overline{Z}_0 f(z) = s y^{s+1} , \quad \overline{Z}_2 \overline{Z}_0 f(z) = s(s+1) y^{s+2} , \ldots \tag{29.4'}$$

On $G$, the left hand sides of (3) and (3') correspond to functions

$$Z^n f_0(g) = s(s+1) \ldots (s+n-1)(ci+d)^{-2n} |ci+d|^{-2s+2n} \quad (n \geq 0) ,$$
$$\overline{Z}^n f_0(g) = s(s+1) \ldots (s-n+1)(ci+d)^{2n} |ci+d|^{-2s-2n} \quad (n \geq 0) .$$

of weight $2n$ and $-2n$. Formulas (4) and (4′) for $r = 0$ then show that the functions

$$e_{2n}(g) = (ci+d)^{-2n} |ci+d|^{-2s+2n} \quad (n \in \mathbb{Z}) \tag{29.5}$$

satisfy relations (1) for non-integer $s \in \mathbb{C}$.

For $f_0(g) = \mathrm{Im}(gi)^s$, the functions $\varphi \in HC(f_0)$ are obviously right $K$-finite and are transformed by the subgroup $UH$ according to formula

$$\varphi(uhg) = |t|^{2s} \varphi(g) = \alpha(h)^s \varphi(g) \tag{29.6}$$

for $u \in U$, $h = \mathrm{diag}(t, t^{-1})$, $t \neq 0$. Conversely, any $K$-finite solution of this functional equation satisfies $\varphi(-g) = \varphi(g)$, hence is a linear combination of functions (5). The space of these functions is, therefore, a function space model for the even principal series of representations of $\mathfrak{g}$.

The odd principal series is obtained likewise by replacing $f_0$ with the weight 1 function corresponding to $y^{s-\frac{1}{2}}$, namely

$$f_1(g) = (ci+d)^{-1} \mathrm{Im}(z)^{s-\frac{1}{2}} = (ci+d)^{-1} |ci+d|^{1-2s} .$$

Then

$$\varphi(uhg) = \mathrm{sgn}(t) |t|^{2s} \varphi(g) \tag{29.7}$$

for all $\varphi \in \mathcal{H}$.

There are also models composed of automorphic functions for a discrete group $\Gamma$ having a parabolic cusp at infinity. In case (c), any (holomorphic) automorphic form of weight $r$ is suitable. To realize case (a), it suffices to choose the Maaß series $M(z; s)$ associated to an arbitrary parabolic cusp of $\Gamma$. As it is obtained by making a left translation of the above function $\alpha(g)^s$ left $\Gamma$-invariant and as the Lie algebra acts on the right, the formulas are necessarily the same as before.[124] This supposes $\mathrm{Re}(s) > 1$, but the analytic extension enables us to remove this restriction. From the point of view of representation theory, these properties of Maaß series justify a posteriori their introduction and allowed Langlands to generalize the theory to all semisimple groups.

Lastly, let us note that, in Jacquet-Langlands theory,[125] *Whittaker models* composed of functions are used for the groups $GL_2$ associated to algebraic number fields over $\mathbb{Q}$. Instead of (6), these functions satisfy the equality $W(ugk) = \mathbf{e}(u)W(g)\chi_r(k)$. They must be eigenfunctions of $\Omega_r$ (or of $\Omega$ for the corresponding functions on $G$) and satisfy "tempered increase" conditions at infinity. The Fourier series expansion of Maaß series reveal their origin. Presenting the theory would take us too far.

## 30 – Irreducible Representations of $G$

To obtain representations of $G$ corresponding to a simple $\mathfrak{g}$-module using right translations, we must renounce $K$-finite functions and choose Banach spaces of functions in which a model of the given $\mathfrak{g}$-module is everywhere dense.

(i) *The multiplicity one theorem.* Let us first state an important result that was often announced:

**Theorem 50.** *For any irreducible unitary representation $(\mathcal{H}, \pi)$ of $G$ and any character $\chi$ of $K$, the subspace $\mathcal{H}(\chi)$ of solutions of*

$$\pi(k)\mathbf{a} = \chi(k)\mathbf{a}$$

*has dimension $\leq 1$.*

Let us suppose that $\mathcal{H}(\chi)$ is non-trivial and let $\mathbf{a} \neq 0$ and $\mathbf{b}$ be two elements of this space. The representation being irreducible, $\mathbf{b}$ is the limit of vectors $\pi(f)\mathbf{a}$ where $f \in L(G)$. If $E(\chi) = \int \pi(k)\chi(k)^{-1}dk$ is the orthogonal projection operator on $\mathcal{H}(\chi)$, $\pi(f)\mathbf{a}$ can be replaced by

$$E(\chi)\pi(f)E(\chi)\mathbf{a} = \pi\left(\overline{\chi} * f * \overline{\chi}\right)\mathbf{a}\,.$$

If $L(G;\chi)$ is the set of $f \in L(G)$ such that

$$\overline{\chi} * f * \overline{\chi} = f\,, \quad \text{i.e. } f\left(k'xk''\right) = \chi\left(k'k''\right)f(x)\,,$$

and if, $\pi(f)$ is also made to abusively denote the operator defined on $\mathcal{H}(\chi)$ by$\pi(f)$, we get a homomorphism from the algebra $L(G;\chi)$ equipped with the convolution product to the algebra of continuous operators on $\mathcal{H}(\chi)$. The previous argument

---

[124] In the case where $\Gamma\backslash G$ is compact, the representation of $G$ on $L^2(\Gamma\backslash G)$ is a *discrete* direct sum of irreducible representations since convolutions by $f \in L(G)$ are Hilbert-Schmidt operators. They correspond either to holomorphic or antiholomorphic forms, or to invariant solutions of $\Omega f = s(1-s)f$, which only exist for a countably many values of $s$ and cannot be obtained using Maaß series. Note that, even when $\Gamma$ has parabolic fixed points, $\Omega f = s(1-s)f$ may well have solutions in $L^2(\Gamma\backslash G)$ that cannot be obtained from Maaß series either.

[125] *Automorphic Forms on $GL(2)$* (Springer, 1972, Lecture Notes **278**). See also R. Godement, *Notes on Jacquet-Langlands theory* (Institute for Advanced Study, Princeton, 1970) and Daniel Bump, *Automorphic Forms and Representations* (Cambridge UP, 1996), 2.8, much less systematic on this point.

shows that the only closed subspace of $\mathcal{H}(\chi)$ invariant under the algebra $A$ of $\pi(f)$ is the trivial one.

But as $\pi$ is assumed to be unitary, $A$ is self-adjoint. Von Neumann's density theorem [Chap. XI, n° 19, (vii)] then shows that $A$ is everywhere dense in the algebra of continuous operators on $\mathcal{H}(\chi)$. Since the latter is commutative only if $\dim \mathcal{H}(\chi) \leq 1$, it will, therefore, suffice to show that $L(G;\chi)$ is *commutative* with respect to the convolution product and, more generally (?), that this is the case of the algebra of continuous functions with compact support such that

$$f(xk) = f(kx)\,. \tag{30.1}$$

To prove the commutativity of Hecke operators,[126] the map $\sigma : G \longmapsto G$ given by

$$\sigma\begin{pmatrix} a & b \\ c & d \end{pmatrix} = \begin{pmatrix} d & b \\ c & a \end{pmatrix} = \omega g^{-1}\omega^{-1} \quad \text{where } \omega = \begin{pmatrix} 1 & 0 \\ 0 & -1 \end{pmatrix} \tag{30.2}$$

has already been used in n° 24, (iii).

The operation transforming all functions $f(x)$ into $f[\sigma(x)]$ reverses the order of every convolution product. Hence it is sufficient to show that $f[\sigma(x)] = f(x)$ for every solution of (1). To this end, let us write $x = k'hk''$ with $k', k'' \in K$ and $h$ diagonal. Clearly, $\sigma(k) = k$ and $\sigma(h) = h^{-1}$ for all $k$ and $h$. Since $K$ is commutative and contains $w$, by (1),

$$\begin{aligned} f\left[\sigma(x)\right] &= f\left(k''h^{-1}k'\right) = f\left(k''whw^{-1}k'\right) = \\ &= f\left(w^{-1}k'k''wh\right) = f\left(k''k'h\right) = f\left(k'hk''\right)\,, \end{aligned}$$

qed.[127]

The reader will probably think that there is a result similar to theorem 49 for non-unitary representations. This is correct if all continuous operators $T$ on $\mathcal{H}$ are strong limits of operators $\pi(f), f \in L(G)$, i.e. if, for all finitely many $n$ elements $\mathbf{a}_i \in \mathcal{H}$ and $r > 0$, there exists $f$ such that $\|T\mathbf{a}_i - \pi(f)\mathbf{a}_i\| \leq r$ for all $i$. For $n = 1$, this amounts to the non-existence of closed invariant subspaces. In finite dimension, this property suffices to show that any operator $T$ is a $\pi(f)$ using a rather old result

[126] On functions invariant under the modular group, Hecke operators, which are sums of left translations, clearly commute with operators $L(f) = \int L(g)f(g)dg$. So the operators $T(p)$ and $L(f)$ for $f(kx) = f(xk)$ generate a self-adjoint commutative algebra of continuous operators in $L^2(\Gamma\backslash G)$. Its discrete spectrum is of multiplicity 1. Jacquet-Langlands adelic theory explains this remarkable phenomenon far better than can be done here.

[127] A less miraculous proof of commutativity consists in checking the theorem for every finite-dimensional irreducible representation $\pi$ [obvious: case (e) of the classification], so that, for all functions (1), operators $\pi(f)$ commute pairwise. The result generalizes to finite-dimensional representations as they are direct sums of irreducible representations (true but not obvious). However, the coefficients of these representations form a function algebra which enables us to "separate" the points of $G$ (obvious: it is the algebra of polynomial functions on $G$). Hence if, for some solution of (1), $\pi(f) = 0$ for all finite-dimensional representations, then $f = 0$ (Stone-Weierstrass). As $\pi(g * g) = \pi(g * f)$, it follows that functions (1) commute, qed. This method generalizes to all semisimple linear groups; see R. Godement, *A theory of spherical functions* (Trans. AMS, **73**, 1952, pp. 496–556).

of Burnside. But, as far I know, there are no simple theorems covering the case of general Banach spaces.

Following Harish-Chandra let us say that a representation $(\mathcal{H}, \pi)$ of $G$ on a Banach space is *admissible* if the subspaces $\mathcal{H}(\chi)$ are finite-dimensional. Let $\mathcal{H}_K$ be their algebraic direct sum, everywhere dense in $\mathcal{H}$. If $(\mathcal{H}, \pi)$, not necessarily unitary, is topologically irreducible, the operators $\pi(f)$ associated to functions (1) act on each $\mathcal{H}(\chi)$ in a topologically irreducible manner, hence algebraically irreducible since $\dim \mathcal{H}(\chi) < +\infty$. As these operators commute, they have a common eigenvalue in $\mathcal{H}(\chi)$, whence $\dim \mathcal{H}(\chi) \leq 1$. If $(\mathcal{H}, \pi)$ is only admissible, all $\mathbf{a} \in \mathcal{H}_K$ are analytic [n° 26, (v), theorem 43]. Thus there is a representation of $\mathfrak{g}$ on this subspace. For all $\mathbf{a} \in \mathcal{H}_K$, the $G$-invariant closed subspace generated by $\mathbf{a}$ is, therefore, the closure of the $\mathfrak{g}$-invariant subspace generated by $\mathbf{a}$. As every $G$-invariant closed subspace $\mathcal{E}$ is the closure of its intersection with $\mathcal{H}_K$, which is $\mathfrak{g}$-invariant, the correspondence between $G$-invariant closed subspaces $\mathcal{E}$ and $\mathfrak{g}$-invariant subspaces of $\mathcal{H}_K$ must be bijective. Consequently, the representation of $G$ on $\mathcal{H}$ is *topologically* irreducible (no closed invariant subspaces) if and only if the representation of $\mathfrak{g}$ on $\mathcal{H}_K$ is *algebraically* irreducible. This result was obtained very early on for all semisimple groups by Harish-Chandra. As $\mathcal{H}_K$ is obviously generated by eigenvectors of $\pi(k)$ and hence of $\pi(W)$, the representation of $\mathfrak{g}$ on $\mathcal{H}_K$ belongs to one of the types obtained above.

(ii) *Function space models for G: the discrete series.* Consider the holomorphic discrete series. Supposing $r \geq 2$, theorem 27 of n° 20 shows that, for all $\varphi \in \mathcal{H}^2_r(G)$, the representation of $G$ on the right invariant closed subspace $\mathcal{H}(\varphi)$ generated by $\varphi$ in $L^2(G)$ is unitary and irreducible. Choosing $\varphi = \omega_r$, $\omega_r(kg) = \omega_r(gk) = \chi_r(k)\omega_r(g)$ and, in conformity with the previous theorem, $\omega_r$ is the only $f \in \mathcal{H}(\varphi)$ with this property, up to a constant factor. $HC(\varphi)$ is everywhere dense in $\mathcal{H}(\varphi)$ and it is the subspace $\mathcal{H}(\varphi)_K$.

If, instead of considering the closed subspace generated by all $R(g)\omega_r$, we consider the subspace generated by all $L(g)\omega_r$, we get $\mathcal{H}^2_r(G)$ and the irreducible unitary representation of $G$ on this space described in n° 16. $\mathcal{H}^2_r(G)$ is therefore the closure in $L^2(G)$ of the subspace of $L(\mu)\omega_r = \mu * \omega_r$, $\mu \in \mathcal{U}(\mathfrak{g})$, which plays the role of $HC(\omega_r)$ for left translations. It is, however, the image of $HC(\omega_r)$ under $f \longmapsto \widetilde{f}$. In consequence, the subspaces $\mathcal{H}(\omega_r)$ and $\mathcal{H}^2_r(G)$ are mapped to each other by $f \longmapsto \widetilde{f}$ or, equivalently, the representation of $\mathfrak{g}$ on the $K$-finite vectors of the representation $(\mathcal{H}^2_r(G), L)$ of $G$ belongs to the discrete series (d).

Condition $r > 1$ is compulsory because of theorem 14 of n° 16, (ii). If $r = 1$, the space $\mathcal{H} = \mathcal{H}^p_1(G)$ is non-trivial for $2 < p \leq +\infty$, and as the spaces $\mathcal{H}(\chi)$ obviously have dimension 1, it is easy to see that all topologically irreducible representations of $G$ corresponding to the representation $r = 1$ of the holomorphic discrete series are obtained in this way. For $p = +\infty$, in the half-plane, as already seen at the start of n° 16, the space obtained corresponds to the set of holomorphic functions such that

$$\sup \left| y^{\frac{1}{2}} f(z) \right| < +\infty .$$

But it is not a Hilbert space. To obtain one, we build on theorem 15 of n° 16: for $r \geq 2$, the complex Fourier transform shows that $\mathcal{H}^2_r(P)$ is isomorphic to the $L^2$ subspace of $\mathbb{R}^*_+$ with measure $t^{1-r}dt$. Now, the complex Fourier transform applies equally for $r = 1$ since, for $\mathrm{Im}(z) > 0$, the function $\mathbf{e}(tz)$ decreases rapidly to $+\infty$. For any function $\varphi(t), t \geq 0$, satisfying $\int |\varphi(t)|^2 dt < +\infty$, the function

$$f(z) = \int \varphi(t)\mathbf{e}(tz)dt = \int \varphi(t) \exp(-2\pi ty)\mathbf{e}(tx)dt \tag{30.3}$$

satisfies

$$\int |f(x+iy)|^2\,dx = \int |\varphi(t)|^2 \exp(-4\pi t y)dt \leq \int |\varphi(t)|^2\,dt\,.$$

So the functions obtained satisfy

$$\sup_{y>0} \int |f(x+iy)|^2\,dx = \|\varphi\|_2^2 < +\infty\,. \tag{30.4}$$

The converse is "the" classical Paley-Wiener theorem proved in n° 16. Hence it is possible to conjecture that, for $r = 1$, the discrete series (c) can be realized on this space of holomorphic functions using operators

$$L_1\left(g^{-1}\right) : f(z) \longmapsto (cz+d)^{-1} f(gz) \tag{30.5}$$

of n° 15, (ii). Condition (4) is obviously preserved if $g \in B$. To justify the construction, it is also necessary to check the theorem if $g = w$. I leave this "detail" to the reader's meditations; see exercise 4 of n° 16, (v).

Note also that the left hand side of (4) is in fact the limit of the integral as $y$ tends to 0, i.e. (Plancherel) the norm on $L^2(\mathbb{R})$ of the function

$$f(x) = \text{l.i.m.}^2 f(x+iy) = \widehat{\varphi}(-x)\,,$$

which is well-defined. This indeed gives a unitary representation $U$ of $G$ on $L^2(\mathbb{R})$ by setting, as in (5),

$$U\left(g^{-1}\right) f(x) = (cx+d)^{-1} f\left[(ax+b)/(cx+d)\right]\,.$$

That these operators preserve the subspace of functions whose Fourier transforms are zero for $t \leq 0$ remains to be checked. All this is related to another function space model for the odd principal series for $\text{Re}(s) = \frac{1}{2}$, obtained using operators

$$U_s\left(g^{-1}\right) f(x) = (cx+d)^{-1} |cx+d|^{s-\frac{1}{2}} f\left[(ax+b)/(cx+d)\right]\,. \tag{30.5'}$$

(iii) *Function space models for $G$: the even principal series.* Let us consider the set $\mathcal{V}^+(s)$ of solutions of

$$f(uhg) = \alpha(h)^s f(g)$$

which, instead of being $K$-finite, are continuous and let us make $G$ act by right translations $R(g)$. Setting as above $f_0(g) = \alpha(g)^s$, the function space model $HC(f_0)$ for the representation of $\mathfrak{g}$ is obviously everywhere dense in $\mathcal{V}^+(s)$ (expand as a Fourier series on $K$), and is the set of $K$-finite vectors of the representation $R$ of $G$ on $\mathcal{V}^+(s)$. It is therefore topologically irreducible if $HC(f_0)$ is algebraically irreducible, i.e. if $s \notin \mathbb{Z}$.

As replacing $s$ by $1-s$ is known to leave the representation invariant, there must be an isomorphism from $\mathcal{V}^+(s)$ onto $\mathcal{V}^+(1-s)$ commuting with all $R(g)$. For $\text{Re}(s) > \frac{1}{2}$, this is the *intertwining operator*[128] $M(s)$ which, associates the function

[128] See some calculations in Daniel Bump, *Automorphic Forms and Representations* (Cambridge UP, 1998), pp. 227–232. All this theory has been generalized, to start with by A. Knapp and E. M. Stein, *Intertwining operators for semi-simple groups* (Ann. Math., **93**, 1971, pp. 489–578).

(30.6) $$M(s)\varphi(g) = \int \varphi(wug)du$$

to every $\varphi \in \mathcal{V}^+(s)$. Formally it is in $\mathcal{V}^+(1-s)$. To check that the integral converges, note that the weight 0 function $\alpha(g)^s = \mathrm{Im}(z)^s$ is in $\mathcal{V}^+(s)$. As $\varphi(g)/\alpha(g)^s$ is continuous and $B$-invariant, there is a uniform upper bound $|\varphi(g)| \leq M|\alpha(g)^s| = M|\mathrm{Im}(z)^s|$. Then a simple calculation enables us to deduce that

(30.7) $$|\varphi(wug)| \leq M\,|\alpha(wug)^s| = O\left(|u|^{-2s}\right) \quad \text{for large } u\,.$$

As the sign $O$ is even an asymptotic equivalence if $\varphi$ is never zero, we must have $\mathrm{Re}(s) > \frac{1}{2}$. Finally, (7) shows that integral (6) converges normally. So, like $\varphi$, it is a continuous function of $g$, whence $M(s)\varphi \in \mathcal{V}^+(1-s)$ for $\mathrm{Re}(s) > \frac{1}{2}$.

If $\varphi$ is $K$-finite, for example if

(30.8) $$\varphi(uhk) = \alpha(h)^s \chi_r(k) = \varphi_{r,s}(g)\,,$$

then $M(s)\varphi_{r,s}(g) = c_r(s)\varphi_{r,1-s}(g)$, where the function $c_r(s)$ is meromorphic on $\mathbb{C}$ and can be computed using the Euler $\Gamma$ function. The intertwining operator can then be defined by analytic extension and it can be checked to be bijective at least on $K$-finite vectors if $s \notin \mathbb{Z}$. Without any calculation, $M(1-s)M(s)$ can even be presumed to reduce to a scalar ... a computable one. We could make exercises out of this, for example:

*Exercise 1.* Set

$$\varphi(g) = \int \Phi\left[g\left(e_1\right)t\right]|t|^{2s}d^*t\,,$$

where $\Phi \in \mathcal{S}(\mathbb{R}^2)$ is the function

$$\Phi(x,y) = (x \pm iy)^r \exp\left[-\pi\left(x^2+y^2\right)\right]\,.$$

Calculate $\varphi$, $M(s)\varphi$ and $M(1-s)M(s)$.

Let us now indicate how to determine those, among these models for the principal series, that could in some way be unitary. $K$-finite vectors of $\mathcal{V}^+(s)$ form a model for the series (a) of irreducible representations of $\mathfrak{g}$. Hence, on this space, there must exist a positive Hermitian form $(\ |\ )$ with respect to which functions of weight $r$ are pairwise orthogonal and operators $i\pi(X)$ symmetric for all $X \in G'(e)$, which amounts to saying that $\pi(\overline{Z})$ is the adjoint of $\pi(Z)$. Setting $(e_r|e_r) = c_r$ in the notation of (29.1),

$$(s + r/2 - 1)\,c_r = -\,(\overline{s} - r/2)\,c_{r-2}$$

and as it is necessary to be able to choose $c_r > 0$, we must have

$$(s + r/2 - 1)\,(s - r/2) < 0$$

for all $r$ occurring in the representation. For the even principal series and $s \notin \mathbb{Z}$, this is equivalent to

$$\mathrm{Re}(s) = \frac{1}{2} \quad \text{or} \quad s \in\, ]0,1[\,.$$

For the odd principal series and $s - \frac{1}{2} \notin \mathbb{Z}$, the condition is the same. In all cases, the expected Hermitian form is unique up to a constant factor. But the essential point is calculating the scalar products on the corresponding spaces $\mathcal{V}^+(s)$. Let us do it for the even principal series.

First observe that, for $\varphi \in \mathcal{V}^+(s)$ and $\psi \in \mathcal{V}^+(1-s)$, $\varphi\psi \in \mathcal{V}^+(1)$, the space of solutions of

$$\varphi(bg) = \alpha(b)\varphi(g) \quad \text{where } \alpha(huk) = \alpha(h)\,.$$

On the other hand, every $\varphi \in \mathcal{V}^+(s)$ is clearly of the form

$$\varphi(g) = \int f(bg)\alpha(b)^{-s}d_rb\,,$$

where $f \in L(G)$. Let us now show that[129] *there exists a right invariant positive linear functional $\mu$ on $\mathcal{V}^+(1)$*. The $B$-invariant measures are given by

$$b = uh \Longrightarrow d_rb = dudh\,, \quad b = hu \Longrightarrow d_lb = dhdu = \alpha(h)^{-1}d_rb\,,$$

whence $d_rb = \alpha(b)d_lb$. Then let us associate the functions

$$f^B(g) = \int f(bg)d_rb\,,$$

$$f_B(g) = \int f(bg)d_lb = \int f(bg)\alpha(b)^{-1}d_rb \tag{30.9}$$

to every $f \in L(G)$. The former one is in $\mathcal{V}^+(0)$ and the latter in $\mathcal{V}^+(1)$. However, an immediate calculation shows that

$$\int p^B(g)f(g)dg = \int p(g)f_B(g)dg$$

for all $f, p \in L(G)$. As $p$ can be chosen so that $p^B = 1$, $f_B = 0$ implies $\int f(g)dg = 0$. The expected linear functional $\mu$ is then given by

$$\mu(f_B) = \int f(g)dg \quad \text{for all } f \in L(G) \tag{30.10}$$

in conformity with Chap. XI. Note that $\mu(\varphi)$ is well-defined if $\varphi \in \mathcal{V}^+(1)$, but not if $\varphi \in \mathcal{V}^+(0)$.

To calculate $\mu$ more explicitly, we write that $G = BK$, whence

$$\mu(f_B) = \int f(g)dg = \iint f(bk)d_lbdk = \int f_B(k)dk\,. \tag{30.11}$$

One can also write $G = BwU$ up to a null set. As $U$ is unimodular, there is a left invariant measure on $G/U$, which can only be $d_lb$. Hence

$$\mu(f_B) = \iint f(bwu)d_lbdu = \int f_B(wu)du = M(1)f_B(e) \tag{30.11'}$$

up to a constant factor. This is an integral of type (6) with $s = 1$, hence converges if $f$ is continuous. Besides, $f_B(g) = O[\mathrm{Im}(z)]$, and so $f_B(wu) = O(u^{-2})$ at infinity.

Then for $\varphi \in \mathcal{V}^+(s)$ and $\psi \in \mathcal{V}^+(1-s)$, let us set

$$(\varphi, \psi) = \mu(\varphi\psi) = \int \varphi(k)\psi(k)dk = \int \varphi(wu)\psi(wu)du\,. \tag{30.12}$$

We thereby get an invariant bilinear form under right translations. Similarly, setting

[129] Here, I resume Chap. XI, n° 15, (iv).

$$(\varphi|\psi) = \mu\left(\varphi\overline{\psi}\right) \tag{30.13}$$

for $\varphi \in \mathcal{V}^+(s)$ and $\psi \in \mathcal{V}^+(1-\overline{s})$, we get an invariant sesquilinear form. For $\mathrm{Re}(s) = \frac{1}{2}, 1-\overline{s} = s$, whence an invariant positive-definite Hermitian form on $\mathcal{V}^+(s)$. As expected, the representations of the even principal series are therefore unitary for $\mathrm{Re}(s) = \frac{1}{2}$. To obtain a Hilbert model, we replace the continuity condition imposed for convenience on $f \in \mathcal{V}^+(s)$ by the condition that $\varphi$ be square integrable with respect to $\mu$, i.e. that $\int |\varphi(k)|^2 dk < +\infty$ or $\int |\varphi(wu)|^2 du < +\infty$.

The case $s \in \left]0,1\right[$ is hard to tackle but more instructive. As we are in the previous case if $s = \frac{1}{2}$, we may suppose that $s > \frac{1}{2}$ because of the symmetry $s \longleftrightarrow 1-s$. The intertwining operator $M(s)$ maps $\mathcal{V}^+(s)$ to $\mathcal{V}^+(1-s)$, which, for $\varphi, \psi \in \mathcal{V}^+(s)$, makes the expression

$$\langle\varphi,\psi\rangle = \mu\left[\varphi.M(s)\psi\right] = \iint \varphi(k).M(s)\psi(k)dk =$$

$$= \int \varphi(wu).M(s)\psi(wu)du \tag{30.14}$$

well-defined. It is an invariant bilinear form on $\mathcal{V}^+(s)$, whence an invariant sesquilinear form

$$(\varphi|\psi) = \langle\varphi|\overline{\psi}\rangle\,,$$

which is Hermitian if $s$ is real. Let us show that it is *positive-definite* if $\frac{1}{2} < s < 1$.

Since $\psi$ is even,

$$M(s)\psi(wu) = \int \psi\left[w\begin{pmatrix}1 & v\\ 0 & 1\end{pmatrix}w^{-1}\begin{pmatrix}1 & u\\ 0 & 1\end{pmatrix}\right]dv = \int \psi\left[\begin{pmatrix}1 & 0\\ v & 1\end{pmatrix}\begin{pmatrix}1 & u\\ 0 & 1\end{pmatrix}\right]dv\,.$$

On the other hand, obviously

$$\begin{pmatrix}1 & 0\\ v & 1\end{pmatrix} = \begin{pmatrix}v^{-1} & 1\\ 0 & v\end{pmatrix}w\begin{pmatrix}1 & v^{-1}\\ 0 & 1\end{pmatrix} = bw\begin{pmatrix}1 & v^{-1}\\ 0 & 1\end{pmatrix} \quad \text{with } \alpha(b) = v^{-2}\,.$$

Hence, identifying $v \in \mathbb{R}$ with the matrix with parameter $v$ in $U$,

$$M(s)\psi(wu) = \int \psi\left[w\begin{pmatrix}1 & u+v^{-1}\\ 0 & 1\end{pmatrix}\right]|v|^{-2s}dv =$$

$$= \int \psi\left[w\begin{pmatrix}1 & u+v\\ 0 & 1\end{pmatrix}\right]|v|^{2s-2}dv$$

and so

$$\langle\varphi,\psi\rangle = \iint \varphi(wu)\psi(wvu)|v|^{2s-2}dudv$$

for $\mathrm{Re}(s) > \frac{1}{2}$. The double integral converges since Lebesgue-Fubini reduces it to the latter integral in (14). Hence, if $\frac{1}{2} < s < 1$,

$$(\varphi|\varphi) = \iint \varphi(wu)\overline{\varphi(wuv)}|v|^{2s-2}dudv = \iint \Phi(u)\overline{\Phi(u+v)}|v|^{2s-2}dudv =$$

$$= \int \Phi * \widetilde{\Phi}(v)|v|^{2s-2}dv\,.$$

The function $\Phi(u) = \varphi(wu)$ is continuous on $\mathbb{R}$, bounded, and because of (7), integrable. The function $|v|^{2s-2}$ is locally integrable. Finally, as mentioned above, the latter integral converges. To prove the positivity of this expression, it is then sufficient to show that the measure $|v|^{2s-2}dv$ is *of positive type* on $\mathbb{R}$. This is indeed the definition for $\Phi \in L(\mathbb{R})$ and the expected result will be obtained by approximating $\Phi$ by functions $\Phi_n \in L(\mathbb{R})$ bounded above by $|\Phi|$ (dominated convergence). Showing that $|v|^{2s-2}dv = |v|^{2s-1}d^*v$ is the Fourier transform of a positive measure is therefore sufficient [Chap. XI, n° 30].

We have therefore returned to the entrance of the Garden of modular delights, which will allow the author to come out of it. If the Mellin transform

$$\Gamma_f^0(s) = \int f(x)|x|^s d^*x = \int f(x)d\mu_s(x), \quad \text{Re}(s) > 0$$

is associated to every reasonable function $f$, then, by (1.8),

$$\pi \int \hat{f}(x)d\mu_s(x) = (2\pi)^{1-s}\cos(\pi s/2)\Gamma(s)\int f(x)d\mu_{1-s}(x)$$

for $0 < \text{Re}(s) < 1$. This means precisely that the Fourier transform of $d\mu_s(x)$ is $\pi^{-1}(2\pi)^{1-s}\cos(\pi s/2)\Gamma(s)d\mu_{1-s}(x)$. For the measure $|v|^{2s-2}dv$ that concerns us here, the parameter is $2s-1$. To conclude, I leave it to the reader to prove that

$$\frac{1}{2} < s < 1 \Longrightarrow 0 < 2s-1 < 1 \quad \& \quad \cos\left[\pi(2s-1)/2\right] > 0\,.$$

End

It only remains for me to again express my gratitude to Springer-Verlag – they don't make editors like that anymore at a time when financiers govern everything – and my admiration for the forbearance shown me by Catriona Byrne: at the beginning, she could not imagine what venture she was getting into. I could not imagine it either. All mathematicians know what they owe her.

Without knowing any French, Uwe Matrisch created the typography in Leipzig from a NotaBene version encoded in my own way in order to somewhat simplify the transcription into TEX. M. Matrisch dealt with countless corrected proofs and delivery delays of various parts of the manuscript without ever complaining. May he too be thanked.

# Index

Algebra
- bicommutator, 132
- commutator algebra, 132
- Gelfand-Naimark (GN), 141
- Hilbert, 237
- normed, 134
- self-adjoint, 123
- spectral measures, 145
- von Neumann, 134
Almost everywhere, 14
Analytic (vector), 484
Automorphic form
- generalized, 350
- of weight $r$, 406
- of weight $r$, 400, 416, 427
- parabolic, 400

Bargmann (orthogonality relations), 254
Bitrace, 235

Canonical coordinates, 467
Character
- of a lcg, 247
- of a commutative lcg, 185
- of a compact group, 213
- conductor of, 304
- of type (I), 249
- modulo an integer, 303
- primitive, 304
Continuous sum
- of Hilbert spaces, 167
- of measures, 73
Convergence
- dominated, 26, 29
- in mean of order $p$, 17
Convolution, 72, 175, 178, 181, 373, 473
$C^\infty$ (vector), 477

Decomposition
- Bruhat, 354
- Iwasawa, 354
- Lebesgue, 112
Density (theorem), 132
Invariant differentiation, 465
Dirac (sequence), 183
Distribution, 471
Domain
- fundamental, 100, 376
- Siegel, 396
Dominated (sequence of functions), 26
Dual
- of a Banach space, 116
- of a commutative lcg, 187
- of a locally convex space, 140
Duplication (formula), 264

Equimeasurable, 49
Euler indicator, 303

Fixed points, 394
- equivalent, 397
- irregular parabolics, 396
Formula
- differentiation of a product, 464
- duplication of $\Gamma$, 264
- Fourier inversion, 35, 201
- Möbius inversion, 310
- Mellin inversion, 275
Fourier transform
- of a measure of positive type, 227
- on $L^1$, 34, 188
- on $L^2$, 34, 200
Function
- central, 209
- elementary of type $\gg 0$, 230
- elliptic, 316
- integrable, 20
- $K$-finite, 416
- locally integrable, 33
- modular $J(z)$, 392
- null, 14
- Bergman kernel, 368, 418
- Weierstrass $\wp$, 329, 332

- of weight $r$, 350
- $p^{th}$ power integrable, 23
- of positive type ($\gg 0$), 227
- lower semi-continuous (lsc), 8, 67
- spherical, 223
- Jacobi theta, 282, 324
- Jacobi theta , 270
- general theta, 321
- $\Delta(z)$, 383, 391, 442, 462
- $Dedekind\eta(z)$, 299, 325
- Dedekind $\eta(z)$, 273, 278
- $G_2(z)$, 335
- $\zeta$, 266
- Möbius, 309

Gauss sum, 291
- reciprocity law for, 291

Gauss sums, 306
Gelfand transform, 139
Generators of the modular group, 376
Group
- acting properly, 91
- Borel, 353
- central, 207, 220
- congruence, 352
- Fuchsian, 404
- $G(m)$, 302
- Lie, 463
- locally compact (lcg), 72
- modular, 284, 376
- of the function $\theta(z)$, 283
- properly discontinuous, 98
- unimodular, 85, 93

Hölder (inequality), 16, 32
Hilbert algebra, 237
- traces on a h. a., 242

Horocycle, 396
- topology of, 401

Ideal of a ring, 137
- maximal, 137

Image of a measure, 72
Infinite product
- $of\theta(z)$, 288
- Hecke series, 462
- of $\theta(z)$, 289
- of $L(s,\chi)$, 315

Integrable
- Function, 20
- Set, 37

Integral of a continuous function with values in a Banach space, 21

Legendre symbol, 296
Lie algebra
- enveloping algebra of a, 475
- of $SL_2(\mathbb{R})$, 488
- of a Lie group, 471
- representation, 471
- semisimple, 490

Map
- Borel, 44
- measurable, 42, 49
- proper, 71
- vaguely continuous, 76
- vaguely measurable, 76

Matrix
- elliptic, 355
- hyperbolic, 355
- parabolic, 355

Measurable
- Lusin, 49
- map, 42
- set, 39

Measure, 5
- absolutely continuous, 71, 107
- bounded, 39
- central, 233
- with base $\mu$, 71, 103
- Haar, 72, 84, 356
- of positive type, 224
- defined by a density, 70
- of an integrable set, 37
- outer, 11
- Image of a measure, 72, 80
- induced, 41
- product, 53
- quotient, 72, 91
- spectral, 145
- absolute value, 110

Mellin (transform), 261
Mellin Transform, 261
Modular form, 378
- entire, 379
- parabolic, 380

One-parameter subgroup, 463
Operator
- adjoint, 122
- self-adjoint, 156
- Casimir, 490
- intertwining, 505
- invariant differential, 475, 490
- Hilbert-Schmidt, 129
- Hermitian, 123
- positive Hermitian, 123

- normal, 143
- spectrum, 125
- symmetric, 156
- Cayley transform, 125
- $L(x)$, $R(x)$, 86
- $L_r(x)$, 350
- $L(X)$, $R(X)$, 466
- Hecke $T(x)$, 450, 455
- $U(f)$, $U(\mu)$, 174
- $U(\mu)$, 174
Orthogonality (relations), 208, 254, 304
Orthonormal bases, 121

Parabolic cusp, 401
Philtre, 6
Positive type
- function of, 227
- measure of, 224

quadratic reciprocity law, 298
Quotient
- of a space by a group, 89
- of an invariant measure, 91

Representation
- adjoint, 465
- admissible, 504
- square integrable, 253
- defined by a function $\gg 0$, 228
- defined by a measure $\gg 0$, 226
- defined by a central measure of positive type, 224
- irreductible, 88
- continuous linear, 86
- function model of a, 500
- function space model for a, 504, 505
- regular, 86, 195, 224
- discrete series, 370, 500
- principal series, 506
- prinicpal series, 500
- unitary, 87, 151, 204, 506

Schur's Lemma, 151, 161, 206
Self-adjoint extension, 159
Seminorm, 126
Semisimple (group or Lie algebra), 490
Serie
- Maaß-Selberg, 422
Series
- Dirichlet $L(s, \chi)$, 312
- Eisenstein, 333, 381, 422
- Hecke of a modular form, 442
- Jacobi, 324
- Jacobi theta, 270
- Maaß-Selberg, 437
- Poincaré, 380, 413
- Poincaré-Eisenstein, 381, 419
- Riemann, 266
- Weil of a modular form, 442
Series
- Jacobi theta, 323
Set
- analytic, 69
- Borel, 40
- integrable, 37
- measurable, 39
- measure-zero, null, 13
- reasonable, 14, 38
Space
- Hilbert, 118
- of countable type, 42
- countable at infinity, 15
- homogeneous, 96
- Polish, 61
- separable, 41
- totally discontinuous, 69
- $\mathcal{H}_r^p(P)$, $\mathcal{H}_r^p(G)$, 359, 366
- $\mathcal{H}_r(G)$, 352, 370
- $L(X)$, 5
- $L^p(X;\mu)$, 23
Spectral decomposition
- of the regular representation, 216
- of the regular representation, 195
- of a normal operator, 151
- of a GN algebra, 150
Spectral manifold, 149
Spectrum
- of an element of a normed algebra, 137
- operator, 125
Stabilizer, 349
Support
- of a distribution, 472
- of a function, 5

Theorem
- Abel, 318
- Baire, 87
- Banach-Steinhaus, 87
- Bochner, 231
- Dini, 7
- Dixmier-Malliavin, 481
- Egorov, 45
- Gelfand-Mazur, 137
- generalized LF , 76
- Hahn-Banach, 22
- Krein-Milman, 232
- Krull, 138

– Lebesgue, 29
– Lebesgue-Fubini (LF), 54
– Lebesgue-Nikodym (LN), 107
– Lusin, 46
– Plancherel, 200, 219, 222
– Riesz-Fischer, 24
– Stone, 486
– Urysohn, 112
– von Neumann density, 132
– Zorn, 138
Topology
– strong, 127, 131
– ultrastrong, 131
– uniform, 131
– vague, 76
– weak, 126, 131, 140, 188
Tribe of sets, 40

Upper envelope, 7
Upper integral, 9, 10

Vector
– $C^\infty$ in a representation, 477
– analytic, 484

THEOREM of Chap. XI
– Theorem 40, 198
THEOREMS of Chap. XI
– Theorem 27, 107
THEOREMS of Chap. XI
– Theorem 1, 9
– Theorem 2, 12
– Theorem 3, 13
– Theorem 4, 14
– Theorem 5, 17
– Theorem 6, 24
– Theorem 7, 24
– Theorem 8, 28
– Theorem 9, 29
– Theorem 10, 30
– Theorem 11, 32
– Theorem 12, 43
– Theorem 13, 43
– Theorem 14, 43
– Theorem 15, 44
– Theorem 16, 45
– Theorem 17, 46
– Theorem 18, 50
– Theorem 19, 54
– Theorem 20, 65
– Theorem 20 bis, 65
– Theorem 21, 67
– Theorem 22, 76
– Theorem 23, 80
– Theorem 24, 96
– Theorem 25, 103
– Theorem 26, 104
– Theorem 28, 109
– Theorem 29, 109
– Theorem 30, 116
– Theorem 31, 119
– Theorem 32, 138
– Theorem 33, 142
– Theorem 34, 154
– Theorem 35, 157
– Theorem 36, 179
– Theorem 37, 182
– Theorem 38, 191
– Theorem 39, 193
– Theorem 41, 201
– Theorem 42, 202
– Theorem 43, 208
– Theorem 44, 226
– Theorem 45, 231
– Theorem 46, 235
– Theorem 47, 238
– Theorem 48, 243
– Theorem 49, 246
– Theorem 50, 250
– Theorem 51, 254

THEOREMS du Chap. XII
– Theorem 5, 307
THEOREMS of Chap. XII
– Theorem 1, 269
– Theorem 2, 273
– Theorem 3, 287
– Theorem 4, 289
– Theorem 6, 314
– Theorem 7, 317
– Theorem 8, 318
– Theorem 9, 329
– Theorem 10, 332
– Theorem 11, 341
– Theorem 12, 343
– Theorem 13, 352
– Theorem 14, 365
– Theorem 15, 366
– Theorem 16, 370
– Theorem 17, 376
– Theorem 18, 377
– Theorem 19, 383
– Theorem 20, 388
– Theorem 21, 391
– Theorem 22, 393
– Theorem 23, 393
– Theorem 24, 402
– Theorem 25, 404

– Theorem 26, 406
– Theorem 27, 409
– Theorem 28, 415
– Theorem 29, 417
– Theorem 30, 420
– Theorem 31, 423
– Theorem 32, 427
– Theorem 33, 437
– Theorem 34, 442
– Theorem 35, 446
– Theorem 36, 453
– Theorem 37, 456
– Theorem 38, 462
– Theorem 39, 468
– Theorem 40, 478
– Theorem 41, 482
– Theorem 42, 482
– Theorem 43, 483
– Theorem 44, 485
– Theorem 45, 485
– Theorem 46, 486
– Theorem 47, 492
– Theorem 48, 495
– Theorem 49, 497
– Theorem 50, 502

# Table of Contents of Volume I

**Preface** ........ V

**I – Sets and Functions** ........ 1
§1. Set Theory ........ 7
1 – Membership, equality, empty set ........ 7
2 – The set defined by a relation. Intersections and unions ... 10
3 – Whole numbers. Infinite sets ........ 13
4 – Ordered pairs, Cartesian products, sets of subsets ........ 17
5 – Functions, maps, correspondences ........ 19
6 – Injections, surjections, bijections ........ 23
7 – Equipotent sets. Countable sets ........ 25
8 – The different types of infinity ........ 28
9 – Ordinals and cardinals ........ 31
§2. The logic of logicians ........ 39

**II – Convergence: Discrete variables** ........ 45
§1. Convergent sequences and series ........ 45
0 – Introduction: what is a real number? ........ 45
1 – Algebraic operations and the order relation: axioms of $\mathbb{R}$ . 53
2 – Inequalities and intervals ........ 56
3 – Local or asymptotic properties ........ 59
4 – The concept of limit. Continuity and differentiability ........ 63
5 – Convergent sequences: definition and examples ........ 67
6 – The language of series ........ 76
7 – The marvels of the harmonic series ........ 81
8 – Algebraic operations on limits ........ 95
§2. Absolutely convergent series ........ 98
9 – Increasing sequences. Upper bound of a set of real numbers 98
10 – The function $\log x$. Roots of a positive number ........ 103
11 – What is an integral? ........ 110
12 – Series with positive terms ........ 114
13 – Alternating series ........ 119
14 – Classical absolutely convergent series ........ 123
15 – Unconditional convergence: general case ........ 127

16 – Comparison relations. Criteria of Cauchy and d'Alembert 132
17 – Infinite limits ........ 138
18 – Unconditional convergence: associativity ........ 139
§3. First concepts of analytic functions ........ 148
19 – The Taylor series ........ 148
20 – The principle of analytic continuation ........ 158
21 – The function $\cot x$ and the series $\sum 1/n^{2k}$ ........ 162
22 – Multiplication of series. Composition of analytic functions. Formal series ........ 167
23 – The elliptic functions of Weierstrass ........ 178

**III – Convergence: Continuous variables** ........ 187
§1. The intermediate value theorem ........ 187
1 – Limit values of a function. Open and closed sets ........ 187
2 – Continuous functions ........ 192
3 – Right and left limits of a monotone function ........ 197
4 – The intermediate value theorem ........ 200
§2. Uniform convergence ........ 205
5 – Limits of continuous functions ........ 205
6 – A slip up of Cauchy's ........ 211
7 – The uniform metric ........ 216
8 – Series of continuous functions. Normal convergence ........ 220
§3. Bolzano-Weierstrass and Cauchy's criterion ........ 225
9 – Nested intervals, Bolzano-Weierstrass, compact sets ........ 225
10 – Cauchy's general convergence criterion ........ 228
11 – Cauchy's criterion for series: examples ........ 234
12 – Limits of limits ........ 239
13 – Passing to the limit in a series of functions ........ 241
§4. Differentiable functions ........ 244
14 – Derivatives of a function ........ 244
15 – Rules for calculating derivatives ........ 252
16 – The mean value theorem ........ 260
17 – Sequences and series of differentiable functions ........ 265
18 – Extensions to unconditional convergence ........ 270
§5. Differentiable functions of several variables ........ 273
19 – Partial derivatives and differentials ........ 273
20 – Differentiability of functions of class $C^1$ ........ 276
21 – Differentiation of composite functions ........ 279
22 - Limits of differentiable functions ........ 284
23 – Interchanging the order of differentiation ........ 287
24 – Implicit functions ........ 290

**Appendix to Chapter III** .......... 303
1 – Cartesian spaces and general metric spaces .......... 303
2 – Open and closed sets .......... 306
3 – Limits and Cauchy's criterion in a metric space; complete spaces .......... 308
4 – Continuous functions .......... 311
5 – Absolutely convergent series in a Banach space .......... 313
6 – Continuous linear maps .......... 317
7 – Compact spaces .......... 320
8 – Topological spaces .......... 322

**IV – Powers, Exponentials, Logarithms, Trigonometric Functions** .......... 325
§1. Direct construction .......... 325
1 – Rational exponents .......... 325
2 – Definition of real powers .......... 327
3 – The calculus of real exponents .......... 330
4 – Logarithms to base $a$. Power functions .......... 332
5 – Asymptotic behaviour .......... 333
6 – Characterisations of the exponential, power and logarithmic functions .......... 336
7 – Derivatives of the exponential functions: direct method .. 339
8 – Derivatives of exponential functions, powers and logarithms 342
§2. Series expansions .......... 345
9 – The number $e$. Napierian logarithms .......... 345
10 – Exponential and logarithmic series: direct method .......... 346
11 – Newton's binomial series .......... 351
12 – The power series for the logarithm .......... 359
13 – The exponential function as a limit .......... 368
14 – Imaginary exponentials and trigonometric functions .......... 372
15 – Euler's relation *chez* Euler .......... 383
16 – Hyperbolic functions .......... 388
§3. Infinite products .......... 394
17 – Absolutely convergent infinite products .......... 394
18 – The infinite product for the sine function .......... 397
19 – Expansion of an infinite product in series .......... 403
20 – Strange identities .......... 407
§4. The topology of the functions $\mathcal{A}\mathrm{rg}(z)$ and $\mathcal{L}\mathrm{og}\, z$ .......... 414

**Index** .......... 425

# Table of Contents of Volume II

**V – Differential and Integral Calculus** ........ 1
§ 1. The Riemann Integral ........ 1
1 – Upper and lower integrals of a bounded function ........ 1
2 – Elementary properties of integrals ........ 5
3 – Riemann sums. The integral notation ........ 14
4 – Uniform limits of integrable functions ........ 16
5 – Application to Fourier series and to power series ........ 21
§ 2. Integrability Conditions ........ 26
6 – The Borel-Lebesgue Theorem ........ 26
7 – Integrability of regulated or continuous functions ........ 29
8 – Uniform continuity and its consequences ........ 31
9 – Differentiation and integration under the $\int$ sign ........ 36
10 – Semicontinuous functions ........ 41
11 – Integration of semicontinuous functions ........ 48
§ 3. The "Fundamental Theorem" (FT) ........ 52
12 – The fundamental theorem of the differential and integral calculus ........ 52
13 – Extension of the fundamental theorem to regulated functions ........ 59
14 – Convex functions; Hölder and Minkowski inequalities ... 65
§ 4. Integration by parts ........ 74
15 – Integration by parts ........ 74
16 – The square wave Fourier series ........ 77
17 – Wallis' formula ........ 80
§ 5. Taylor's Formula ........ 82
18 – Taylor's Formula ........ 82
§ 6. The change of variable formula ........ 91
19 – Change of variable in an integral ........ 91
20 – Integration of rational fractions ........ 95
§ 7. Generalised Riemann integrals ........ 102
21 – Convergent integrals: examples and definitions ........ 102
22 – Absolutely convergent integrals ........ 104
23 – Passage to the limit under the $\int$ sign ........ 109
24 – Series and integrals ........ 115

25 – Differentiation under the $\int$ sign ........ 118
26 – Integration under the $\int$ sign ........ 124
§ 8. Approximation Theorems ........ 129
27 – How to make $C^\infty$ a function which is not ........ 129
28 – Approximation by polynomials ........ 135
29 – Functions having given derivatives at a point ........ 138
§ 9. Radon measures in $\mathbb{R}$ or $\mathbb{C}$ ........ 141
30 – Radon measures on a compact set ........ 141
31 – Measures on a locally compact set ........ 150
32 – The Stieltjes construction ........ 157
33 – Application to double integrals ........ 164
§ 10. Schwartz distributions ........ 168
34 – Definition and examples ........ 168
35 – Derivatives of a distribution ........ 173

**Appendix to Chapter V – Introduction to the Lebesgue Theory** 179

**VI – Asymptotic Analysis** ........ 195
§ 1. Truncated expansions ........ 195
1 – Comparison relations ........ 195
2 – Rules of calculation ........ 197
3 – Truncated expansions ........ 198
4 – Truncated expansion of a quotient ........ 200
5 – Gauss' convergence criterion ........ 202
6 – The hypergeometric series ........ 204
7 – Asymptotic study of the equation $xe^x = t$ ........ 206
8 – Asymptotics of the roots of $\sin x . \log x = 1$ ........ 208
9 – Kepler's equation ........ 210
10 – Asymptotics of the Bessel functions ........ 213
§ 2. Summation formulae ........ 224
11 – Cavalieri and the sums $1^k + 2^k + \ldots + n^k$ ........ 224
12 – Jakob Bernoulli ........ 226
13 – The power series for $\cot z$ ........ 231
14 – Euler and the power series for $\arctan x$ ........ 234
15 – Euler, Maclaurin and their summation formula ........ 238
16 – The Euler-Maclaurin formula with remainder ........ 239
17 – Calculating an integral by the trapezoidal rule ........ 241
18 – The sum $1+1/2+\ldots+1/n$, the infinite product for the $\Gamma$ function, and Stirling's formula ........ 242
19 – Analytic continuation of the zeta function ........ 247

**VII – Harmonic Analysis and Holomorphic Functions** ......... 251
1 – Cauchy's integral formula for a circle .................. 251
§ 1. Analysis on the unit circle ............................ 255
2 – Functions and measures on the unit circle ............. 255
3 – Fourier coefficients ..................................... 261
4 – Convolution product on $\mathbb{T}$ ........................... 266
5 – Dirac sequences in $\mathbb{T}$ ................................ 270
§ 2. Elementary theorems on Fourier series .................. 274
6 – Absolutely convergent Fourier series .................. 274
7 – Hilbertian calculations ................................. 275
8 – The Parseval-Bessel equality ........................... 277
9 – Fourier series of differentiable functions ............... 283
10 – Distributions on $\mathbb{T}$ ................................. 287
§ 3. Dirichlet's method ......................................... 295
11 – Dirichlet's theorem ..................................... 295
12 – Fejér's theorem ........................................ 301
13 – Uniformly convergent Fourier series .................. 303
§ 4. Analytic and holomorphic functions .................... 307
14 – Analyticity of the holomorphic functions ............. 308
15 – The maximum principle ................................. 310
16 – Functions analytic in an annulus. Singular points. Meromorphic functions ................................. 313
17 – Periodic holomorphic functions ....................... 319
18 – The theorems of Liouville and of d'Alembert-Gauss ..... 320
19 – Limits of holomorphic functions ....................... 330
20 – Infinite products of holomorphic functions ............ 332
§ 5. Harmonic functions and Fourier series .................. 340
21 – Analytic functions defined by a Cauchy integral ....... 340
22 – Poisson's function ..................................... 342
23 – Applications to Fourier series ......................... 344
24 – Harmonic functions ..................................... 347
25 – Limits of harmonic functions .......................... 351
26 – The Dirichlet problem for a disc ...................... 354
§ 6. From Fourier series to integrals ........................ 357
27 – The Poisson summation formula ....................... 357
28 – Jacobi's theta function ................................. 361
29 – Fundamental formulae for the Fourier transform ....... 365
30 – Extensions of the inversion formula ................... 369
31 – The Fourier transform and differentiation ............ 374
32 – Tempered distributions ................................ 378

**Postface. Science, technology, arms** ......................... 387

**Index** ............................................................ 436

**Table of Contents of Volume I** ................................ 441

# Table of Contents of Volume III

**VIII – Cauchy Theory** .......... 1
§ 1. Integrals of Holomorphic Functions .......... 3
1 – Preliminary Results .......... 3
2 – The Problem of Primitives .......... 8
3 – Homotopy Invariance of Integrals .......... 18
§ 2. Cauchy's Integral Formulas .......... 31
4 – Integral Formula for a Circle .......... 31
5 – The Residue Formula .......... 42
6 – Dixon's Theorem .......... 56
7 – Integrals Depending Holomorphically on a Parameter .......... 59
§ 3. Some Applications of Cauchy's Method .......... 63
8 – Fourier Transform of a Rational Fraction .......... 65
9 – Summation Formulas .......... 75
10 –The Gamma Function, The Fourier Transform of $e^{-x}x_+^{s-1}$ and the Hankel Integral .......... 78
11 – The Dirichlet problem for the half-plane .......... 86
12 – The Complex Fourier Transform .......... 94
13 – The Mellin Transform .......... 103
14 – Stirling's Formula for the Gamma Function .......... 118
15 – The Fourier Transform of $1/\cosh \pi x$ .......... 125

**IX – Multivariate Differential and Integral Calculus** .......... 133
§ 1. Classical Differential Calculus .......... 133
1 – Linear Algebra and Tensors .......... 133
(i) *Finite-dimensional vector spaces* .......... 133
(ii) *Tensor notation* .......... 135
2 – Differential Calculus of $n$ Variables .......... 146
(i) *Differential functions* .......... 146
(ii) *Multivariate chain rule* .......... 149
(iii) *Partial differentials* .......... 151
(iv) *Diffeomorphisms* .......... 154
(v) *Immersions, submersions, subimmersions* .......... 155
3 – Calculations in Local Coordinates .......... 155

(i) *Diffeomorphisms and local charts* ... 155
(ii) *Moving frames and tensor fields* ... 157
(iii) *Covariant derivatives on a Cartesian space* ... 161
§ 2. Differential Forms of Degree 1 ... 166
4 – Differential Forms of Degree 1 ... 166
5 – Local Primitives ... 168
(i) *Existence: calculations in terms of coordinates* ... 168
(ii) *Existence of local primitives: intrinsic formulas* ... 170
6 – Integration Along a Path. Inverse Images ... 172
(i) *Integrals of a differential form* ... 172
(ii) *Inverse image of a differential form* ... 173
7 – Effect of a Homotopy on an Integral ... 175
(i) *Differentiation with respect to a path* ... 175
(ii) *Effect of a homotopy on an integral* ... 177
(iii) *The Banach space* $C^{1/2}(I;E)$ ... 179
§ 3. Integration of Differential Forms ... 182
8 – Exterior Derivative of a Form of Degree 1 ... 182
(i) *Physicists' Vector Analysis* ... 182
(ii) *Differential forms of degree 2* ... 183
(iii) *Forms of degree p* ... 186
9 – Extended Integrals over a 2-Dimensional Path ... 191
(i) *The exterior derivative as an infinitesimal integral* ... 192
(ii) *Stokes' formula for a 2-dimensional path* ... 195
(iii) *Integral of an inverse image* ... 197
(iv) *A planar example* ... 199
(v) *Classical version* ... 200
10 – Change of Variables in a Multiple Integral ... 203
(i) *Case where* $\varphi$ *is linear* ... 204
(ii) *Approximation Lemmas* ... 208
(iii) *Change of variable formula* ... 213
(iv) *Stokes' formula for a p-dimensional path* ... 215
§ 4. Differential Manifolds ... 218
11 – What is a Manifold? ... 218
(i) *The sphere in* $\mathbb{R}^3$ ... 218
(ii) *The notion of a manifold of class* $C^r$ *and dimension* $d$ ... 219
(iii) *Some Examples* ... 221
(iv) *Differentiable maps* ... 224
12 – Tangent vectors and Differentials ... 225
(i) *Vectors and tangent vector spaces* ... 225
(ii) *Tangent vector to a curve* ... 228
(iii) *Differential of a map* ... 230
(iv) *Partial differentials* ... 233
(v) *The manifold of tangent vectors* ... 233
13 – Submanifolds and Subimmersions ... 235
(i) *Submanifolds* ... 236

(ii) *Submanifolds defined by a subimmersion* ... 239
(iii) *One-Parameter Subgroups of a Torus* ... 242
(iv) *Submanifolds of a Cartesian space: tangent vectors* ... 246
(v) *Riemann spaces* ... 248
14 – Vector Fields and Differential Operators ... 250
15 – Vector Fields and Differential Equations ... 252
(i) *Reduction to an integral equation* ... 253
(ii) *Existence of solutions* ... 254
(iii) *Uniqueness of the solution* ... 255
(iv) *Dependence on initial conditions* ... 255
(v) *Matrix exponential* ... 258
16 – Differential Forms on a Manifold ... 261
17 – Integration of Differentiable Forms ... 262
(i) *Orientable manifolds* ... 263
(ii) *Integration of differential forms* ... 266
18 – Stokes' Formula ... 269

**X – The Riemann Surface of an Algebraic Function** ... 275
1 – Riemann Surfaces ... 275
2 – Algebraic Functions ... 280
3 – Coverings of a Topological Space ... 285
(i) *Definition of a covering* ... 285
(ii) *Sections of a covering space* ... 287
(iii) *Path-lifting* ... 288
(iv) *Coverings of a simply connected space* ... 292
(v) *Coverings of a pointed disc* ... 295
4 – The Riemann Surface of an Algebraic Function ... 297
(i) *Global uniform branches* ... 297
(ii) *Definition of a Riemann surface* $\hat{X}$ ... 298
(iii) *The algebraic function* $\mathcal{F}(z)$ *as a meromorphic function on* $\hat{X}$ ... 300
(iv) *Connectedness of* $\hat{X}$ ... 303
(v) *Meromorphic functions on* $\hat{X}$ ... 305
(vi) *The purely algebraic point of view* ... 306

**Index** ... 311

**Table of Contents of Volume I** ... 315

**Table of Contents of Volume II** ... 319

Printed in the United States
By Bookmasters